Science in the Twentieth Century

Science in the Twentieth Century

edited by
John Krige & Dominique Pestre
CRHST, Paris, France

Harwood Academic Publishers
Australia • Canada • China • France • Germany • India• Japan • Luxembourg • Malaysia
The Netherlands • Russia• Singapore • Switzerland • Thailand • United Kingdom

Published in The Netherlands by Harwood Academic Publishers.

Printed in Singapore

Amsteldijk 166
1st Floor
1079 LH Amsterdam
The Netherlands

British Library Cataloguing in Publication data

Science in the twentieth century
1. Science – History – 20th century
I. Krige, John II. Pestre, Dominique
509'.04

ISBN 90-5702-172-2

Jacket Illustrations:

Projection into orbit of first two cluster spacecraft of Ariane 5. Courtesy of ESA.

German soldier wearing gas mask. Courtesy of Imperial War Museum.

Advertisement for Mulford's Diphtheria Antitoxin. Courtesy of Merck & Co., Inc.

Fly heads. Courtesy of Cornelia Hesse-Honegger.

Earth Science in the Laboratory: High Pressure Research at the Carnegie Institution of Washington, 1906. Courtesy of Carnegie Institution of Washington, Geophysical Laboratory.

Contents

List of Figures

List of Tables

List of Contributors

Pnina G. Abir-Am
The Dibner Institute for the History of Science and Technology, Massachusetts, USA

James R. Bartholomew
Department of History, The Ohio State University, Ohio, USA

Bernadette Bensaude Vincent
Laboratoire de Genetique Moleculaire, Ecole Normale Superieure, Paris, France

Frans Berkhout
Science Policy Research Unit, University of Sussex, Brighton, UK

Christophe Bonneuil
REHSEIS, Paris, France

Paolo Brenni
Istituto e Musco di Storia della Scienza, Florence, Italy

W. Bernard Carlson
School of Engineering, University of Virginia, Virginia, USA

Yves Cohen
Ecole des Hautes Etudes en Sciences Sociales, Paris, France

John Peter Collett
Historisk Institutt, University of Oslo, Oslo, Norway

Susan E. Cozzens
Department of Science and Technology Studies, Rensselaer Polytechnic Institute, New York USA

Amy D. Dalmedico
CNRS, Paris, France

Michael Aaron Dennis
Science and Technology Studies, Cornell University, New York, USA

Ronald E. Doel
Department of History, University of Alaska, Alaska, USA

David Edgerton
Centre for the History of Science, Technology and Medicine, Imperial College of Science, Technology and Medicine, London, UK

Wendy Faulkner
Department of Sociology, Edinburgh University, Edinburgh, UK

Deborah Fitzgerald
Department of Science and Technology, Massachusetts Institute of Technology, Massachusetts, USA

Yasu Furukawa
Department of Humanities and Social Sciences, Tokyo Denki University, Tokyo, Japan

Louis Galambos
Department of History, John Hopkins University, Maryland, USA

Peter Galison
Department of History, University of Harvard, Massachusetts, USA

Jean-Paul Gaudilliere
Department of History, Princeton University, New Jersey, USA

Roger L. Geiger
Higher Education Program, The Pennsylvania State University, Pennsylvania, USA

Lillian Hoddeson
Department of History, University of Illinois, Illinois, USA

Karl Hufbauer
Department of History, University of California at Irvine, California, USA

Ashok Jain
National Institute of Science, Technology and Development Studies, New Delhi, India

Harmke Kamminga
Wellcome Unit for History of Medicine, Cambridge, UK

Martin Kemp
Department of Art History, University of Oxford, Oxford, UK

Elizabeth A. Kerr
Department of Sociology, Edinburgh University, Edinburgh, UK

Daniel J. Kevles
Division of the Humanities and Social Sciences, California Institute of Technology, California, USA

Sharon Kingsland
Department of History of Science, Johns Hopkins University, Maryland, USA

Nikolai Krementsov
Institute for the History of Science, St. Petersburg, Russia,

John Krige
CHRST, Paris, France

Christopher Lawrence
Wellcome Institute for the History of Medicine, London, UK

Ilana Löwy
Hospital des Enfants Malades, Paris, France

Michael S. Mahoney
Princeton University, New Jersey, USA

Everett Mendelsohn
Department of the History of Science, Harvard University, Massachusetts, USA

Peter J.T. Morris
The Science Museum, London, UK

Anne Marie Moulin
Hospital des Enfants Malades, Paris, France

Mary Jo Nye
History Department, Oregon State University, Oregon, USA

Larry Owens
Department of History, University of Massachusetts, Massachusetts, USA

Theodore M. Porter
Department of History, University of California at Los Angeles, California, USA

Keith Pavitt
Science Policy Research Unit, University of Sussex, Brighton, UK

Dominique Pestre
CHRST, Cite des Sciences et de L'Industrie, Paris, France

Dhruv Raina
National Institute of Science, Technology and Development Studies, New Delhi, India

Simon Schaffer
Department of History and Philosophy of Science, Cambridge University, Cambridge, UK

Sam S. Schweber
Martin Fisher School of Physics, Brandeis University, USA

Karlheinz Steinmüller
Secretariat for Futures Studies, Gelsenkirchen, Germany

Jeffrey L. Sturchio
Department of History, John Hopkins University, Maryland, USA

ANTHONY S. TRAVIS
The Sidney M. Edelstein Center for the History and Philosophy of Science, Technology and Medicine, Hebrew University of Jerusalem, Israel.

HEBE M.C. VESSURI
Departamento Estudio de la Ciencia, Instituto Venezolano de Investigaciones Cientificas, Caracas Venezuela

MARK WALKER
History Department, Union College, New York, USA

Introduction

John Krige and Dominique Pestre

'What is Science?' is surely the first question anyone has in mind when opening a book entitled *Science in the Twentieth Century*. And on perusing this volume readers will be struck by the infinite variety of the answers which are proposed. This is hardly surprising. It is indicative of the fact that 'science' is not a 'thing' which can be grasped by one description or one experience; not an object belonging to only one realm of human activity. We want to begin therefore by reflecting on how different social actors talk about the sciences.

The current definition which scientists give is of course one answer, and an answer to which we cannot but give particular weight since they are situated at the core of the process of knowledge production. Epistemologists, historians and sociologists have, however, also put forward analyses and approaches which are not without interest — as have also, through their actions and their discourses, engineers, professional popularizers, politicians, managers responsible for industrial innovation or 'the man in the street.' It would certainly be too quick, and most unwise to reject these out of hand, to ignore them as having nothing of interest to say about this complex and heterogeneous object that is 'science.' Such a presumption would imply that we are the only ones who see matters clearly. In refusing to take seriously the experience of others we would privilege for no good reason our situation and our (inevitably restricted) field of vision.

It is also worth remembering that the characteristics which we isolate as typical of science are heavily context-laden. Working at the laboratory bench the scientist can easily speak of his presuppositions and will accept that he has difficulty stabilizing his results; speaking to his peers at the Academy he cannot but present a more edifying picture. Our point of departure is thus to refuse to consider science as an object whose essence has simply to be unveiled. We prefer to see it as an ensemble of almost infinite relations between people, objects and statements. Science is to be situated in a field in which truth is distributed among the mass of actors who are the subjects of science one day, its objects the next. Behind the word, and depending on where we are, it is we who confirm the boundaries, who specify the definitions and the principles which we take to be crucial. None of us is at the center of the world, none of us grasps the essence of things (perhaps because there is no essence to be grasped), and all we can hope for is to establish connections between those activities which we deem to be relevant. As Paul Veyne has said of historical work, it is we who construct our problem, it is we who create certain associations and who ignore others (if only because our time is limited), it is we who produce certain pictures and who are blind to others. It is we too who build an object which we call science, who define its contour and who say things about it which are not without merit, but which could not conceivably exhaust what others would want to say about that which is defined and lived, collectively and contradictorily, as 'science.'[1]

We could readily accept that the epistemological question is a valid one (i.e., that it is capital for scientists, philosophers, historians and intellectuals interested in how knowledge is produced), but it is surely justifiable to insist that quite different questions are just as legitimate. The sciences also appear, after all, as systems enabling us to act on the material and social world. They are systems of values enabling us to fix ideals and norms for the conduct of our lives. They function as systems of representations underlying discourses and postures of authority. In short we suggest that science be thought of as something we build through our life experiences, that we construct through word and deed. Without presuming to be exhaustive, nor pretending to define the major historical periods in which any particular conception of science has reigned supreme, we would quickly like to review some of the ways in which science has been described by different members of the body politic.

Images of Science

Science is, for example, a rational practice guided by the search for Truth and a critical approach. It aims to build logical interconnected systems of propositions (as in many of the physical sciences, for example), but can also express itself through clusters or arguments (as in cosmology or natural history). Science is also a personal, an aesthetic, an exhilarating experience for those who practice it — think of the feelings expressed by a mathematician faced by a particularly beautiful proof, or the sentiments of a Richard Feynman who tells us that 'physics is fun.'[2] Rather like poetry, science can be a way for the savant to be-in-the-world.

For many scientists in the twentieth century science is also an ethic, a source of moral values which inform social and political engagement. We see this in the behavior of those French savants who insisted, at the turn of the century, that the Dreyfus trial be re-opened. Initially at least, their intervention was based less on the conviction that Dreyfus was innocent, than on the feeling that justice had not been done, that Dreyfus had been treated unfairly. For them being scientific meant to be open to all possibilities, to collect with care and attention all facts, and then to draw a coherent and plausible conclusion. Science was the locus where one learnt what objectivity and justice meant, it was an education in virtue. Being the unrestricted and disinterested research into truth, science was a major resource for democracy and for the Republic in its battle against malicious doctrines which exploited the preconceptions and the ignorance of the populace. It was a weapon in the fight against social injustice, it was a recognition that it was possible to better understand the world: and the vistas it opened had to be defended and transmitted to others. Since their conception of politics was not that of the simple balance of interests but rested rather on the putting into practice of the common good which science could help define they were educators and they were involved, for example, in the 'universitiés populaires.'[3]

The sciences are also the most influential knowledge-system in our societies, the system to which social and political authorities can appeal. Scientists are thus transformed into experts, experts from whom the state demands definition of health or safety standards. Experts to whom the military turn for advice on the proposals for inventions which they have received, or whom they place at the core of the development of their new weapons systems (since the Second World War). They are experts called to give an opinion in court and whose judgments control the life or death of others (for example experts in handwriting who enabled the prosecutor to use Bertillon to 'confuse' Dreyfus). They are experts in anthropometry who offer their knowledge to the elites of the day; and also the experts tied to the nuclear who repeatedly reassure us that science can manage the future. Certain in their beliefs, they are no longer educators but lawmakers, indeed the prince's councillors.[4]

Symmetrically, science can be, for an anti-nuclear militant or an ecologist, a sign of man's hubris, the incarnation of an excess which he can label as Promethean or Faustian. Extending (or anticipating) these attitudes, some have suggested that science (or the image of rationality which it embodies) is at the heart of modernity, including the 'rational' practices of totalitarian regimes in this century. For the savant, science is the way of escaping from doxa and the contingent, a way of becoming (almost) equal to God. For others, by contrast, it can symbolize a deplorable insensitivity, a way of working and of being which is at once imperious and dominating, a practice which, its denials notwithstanding, does not know that there are many things that it does not know. Because it lacks humility, science is a menace and scientists are the new sorcerer's apprentices.[5]

For those who are dominated by it science can be a way of domesticating them, of making them behave in certain ways, of forcing them, in the name of a higher good, into patterns of behavior and attitudes which they cannot but internalize. This happens when science takes it upon itself to tell people what counts as the model of rational behavior. It happens in the laboratory and in the factory — for example, through the practices of scientific management so dear to the Taylorians and post-Taylorians.

Science is many other things besides. It is a system of institutions; it is what we see in museums and 'science centers'; it is what is displayed in laboratories during 'open days.' Here science is an ongoing success story, embracing always more of the world around us, revealing that which ordinary mortals could not see: a non-debatable system of knowledge from which the profane are excluded. Presented thus, science is at once the universal criterion of knowledge, the mode of knowledge par excellence, but also a quasi-magical practice since it remains at once incomprehensible and effective.

Then again science is an ensemble of buildings, an ensemble of representations majestically inscribed in the decor of the city, a collection of paintings and sculptures which are bearers of social values. We are thinking here of the fresco by Dufy at the City of Paris' Museum of Modern Art or the classical statue of 'Nature unveiling herself before Science,' the former usually represented as a woman as beautiful as she is dominated and without power. Science is a collection of memorials which we visit, producing emotion (like Pasteur's house), solidifying national pride (like the dome of the Academy), provoking hope or fear in progress (like the Atomium at the Brussels exhibition in 1958). Science can also give one a sense of time, of the human condition and of tradition, as does the statue in front of Stockholm's Royal Institute of Technology which presents the past as disconnected from the present and the future, the latter being fused together and showing the direction in which to go.

Science is an ensemble of discourses, a repertoire of models, a rhetorical resource. It can underpin a conception of Human Nature (is this not what we have in primatology which, while putatively describing the life of baboons is speaking above all about the relationship between men and women) and it can underwrite a set of social norms (as when it speaks of the biological basis of homosexuality). Science can also be a range of practices which can be opposed on moral grounds. Is it justifiable to apply double blind protocols, 'scientifically' more reliable, when patients are facing almost certain death, as in the case of AIDS? It can be a set of procedures which shock the conscience and which could be called to account in the name of other values; the physicist Rasetti who abandoned his field after Hiroshima, or the condemnation of animal experiments because they are brutal and could lead to human experimentation.[6]

Science in also a means for climbing the social ladder (the '*boursiers de la Republique*,' those young people in France who obtain financial support and often choose to enter science faculties), a way of selecting an elite and a way of defining what counts as scholarly excellence. For many academics (and for us as well) it is a way of life, a career, a practice which we learnt at school because we were good at it, a practice which became second nature to us.

Finally science could be said to be a collection of instruments and machines which have penetrated into every corner of our lives (allied, in a usually non-conflictual relationship with technologies which have sprung from science and which are in turn ever present in laboratory life), a practice which is itself regulated by these instruments and technical objects. 'Western science' during the last few centuries can even be legitimately characterized by the exponential production of 'artifacts,' by the ever more organic integration of science, technology and industry. This has enabled our societies to dominate the world. In this sense (Western) science is above all a way of controlling the material world through machines (and weapons), a system for managing production which has enabled an economic system, called liberal, to become ever more efficient. We are surely far from morality and epistemology now — but who would dare say that science has nothing to do with this too?

The Organization of the Book

The structure of this book reflects our refusal to adopt a single definition of science, the recognition that it would not be very interesting to proceed in this way. There are four main sections each looking at science from a different point of view. The first main section, "Science and the Social Fabric," treats the sciences as systems, which constantly redefine the social, as systems of knowledges and discourses, as systems of know-hows and norms, of actions and interventions which are mobilized in different worlds — political, administrative, managerial, industrial, technical and academic. The second main section, entitled "Research Dynamics," shifts the gaze to knowledge producing institutions themselves or, more precisely, to the frameworks and the systems of thought articulated by scientists during the century. The sciences are here described from studies which focus on the evolution and the reconfigurations of disciplinary fields. The third section, "Science and its Practices," stays with the scientists but is less concerned with their conceptual schemes and interpretative systems than with their ways of doing research, their material and instrumental practices in the laboratory and in the field. The final section, "Regional and National Contexts," again treats the sciences as institutions embedded in specific socio-political contexts. Now it is the national (or regional) dimension which is emphasized. The dynamics of the fabrication of knowledge can thus be reconsidered in all their contradictory aspects: between center and periphery, between international developments and the protection of national interests, between interests of states and economic and cognitive interests.

Turning now to specifics, three introductory chapters by Dennis, Schaffer, and Faulkner and Kerr, which are grouped under the title "Images of Science," provide different insights into how the sciences have been described during this century. Dennis does this by looking at our discipline itself, by reflecting on our practices: those of historians of the sciences of the last two or three generations. Concentrating on key moments in the (mainly American) historiography of science he argues that successive definitions of nature and shifts in the boundaries of science were 'coproduced' with changing political situations between the Great Depression and the end of the Cold War. Schaffer next considers the most common images produced at different times in different social spaces since the beginning of the century. His sources are encyclopedias and influential essays, treatises and books written by scientists, engineers and intellectuals and, above all, exhibitions and museums. Finally in this section Faulkner and Kerr analyze the research of contemporary feminists on the relationships between gender and science. Among other things they discuss the possible masculinity of scientists' practices and knowledge constructions. Conceived as a general introduction to the book, these chapters freely highlight political, social and gender questions in relation to the way science has been constructed and promoted in this century.

Section "Science and the Social Fabric"

The section "Science and the Social Fabric" opens with articles by Pestre, Bonneuil, Porter and Cohen. These essays have one theme in common, that of a scientificity claimed by many actors to be the privileged way to understand the modern world. In the twentieth century, politicians and colonial administrators, state officials and industrial managers have all wanted to be increasingly scientific. To do so they have come to treat their objects with the same tools, notably quantification, with the same determination to objectify and the same desire to control by associating scientists with their undertakings. In the colonies, for example, science was placed at the core of plunder. It legitimated the imperial dream of the rational exploitation of the planet, using botany, agronomy and anthropology. It supplied the practical means, with the support of the military to be sure, to render the natural and social worlds at once more transparent and more productive. In the major powers strategies for managing people and economies (originating in economic practices, industrial shopfloors as well as state bureaucracies) have helped make the claim to be scientific a widely respected value. Through quantifying practices, seen as the quintessential form of neutral and objective information, the state and Taylorian mangers built a faith in science, optimized productivity, and invented new ways of disciplining the social: means which were subsequently put to use to control the practice of scientists themselves.

The quantifying approach to scientific activities is explored by Cozzens and Pavitt. Derived from wartime needs (it was scientific rationality which underpinned the organization of production for victory by treating huge quantities of data), the counting of the available resources of scientists and engineers became increasingly common in the 1950s and 1960s. It led, as one might expect, to the orientation of these resources towards the needs of economic development and national security. Pavitt's article also introduces the idea, explored in detail in the five following chapters, that it makes no sense to speak of scientific knowledge if one does not differentiate between the sites where it is produced. Despite persistent difference between industrial sectors, and notwithstanding some spectacular examples of close links between basic research and some technologies (in quantum electronics and biotechnologies, for example), Pavitt insists that the worlds of academic research and technical development are above all regulated by their own, quite different, logics of reproduction.

The articles by Geiger, Mendelsohn, Carlson, Galambos & Sturchio and Collett each analyze a specific site of knowledge production. The first considers science as an investigative activity located in universities seen as educational institutions (even if they have also been 'objects' of philanthropical patronage and have been modeled by the 'consumers' of the research they produce). Mendelsohn looks at the military as a decisive locus for the production of science, notably after World War II, while the next three articles concentrate on industrial settings: that of Carlson mainly on the first half of the century, that of Galambos and Sturchio on the post-World War II pharmaceutical industry, that of Collett on electronics across the century. Clearly the military and industry have been major sites of knowledge production since 1900. While knowledge has been produced at many different sites for centuries, and is not simply to be found in academies and universities, the phenomenon has become so widespread since 1914 that a qualitative change has taken place.

The section ends with four chapters. That by Berkhout treats the way in which the problem of the management of high-level radioactive waste has been dealt with. It shows how science has been used strategically in public disputes. Berkhout shows how the definition of an acceptable risk has required extensive architectures of reasoning which could not but be contradictorily assessed. From the moment science serves as a tool to put in place systems from which no one can escape (this is the case with the nuclear, but also with agrobusiness and biotechnologies), scientific and technological claims are open to public scrutiny. It is essential to the democratic process that this should be so. Kevles, for his part, explores the movement from eugenic practices to genetic manipulation. He shows how the use of quantification, in conjunction with strong genetic convictions, could give a scientific veneer to the most banal types of racism; how it could be used to justify forced sterilization (or even the holocaust). The essay also raises a number of questions about the current possibility of a 'homemade genetics' based on biotechnology.

Finally Bensuade-Vincent studies a group of phenomena collectively labeled the "popularization of science," while Steinmuller considers the science fiction literature, "the mythology of the modern, scientific age." The former stresses the radical disymmetry between the profane and those who speak "in the name of science." She reminds us of the missionary dimension that accompanies the act of popularization; the fact that it mostly serves to persuade the public of the social value of science. The study of science fiction helps identify some of the public images of the scientist which have permeated the twentieth century, this literature being one kind of vehicle for expressing some of the hopes and fears surrounding technology. The chapter is followed by a survey by Kemp of the many modes of visualization used in the sciences.

Section "Research Dynamics"

The section entitled "Research Dynamics" focuses on the major results, the most important achievements, the grand visions of scientists in this century, concentrating on the mathematical, physico-chemical, biological and medical sciences. Since classifications were always open to discussion (in 1910 the field of radioactivity as defined by Ernest Rutherford and by Marie Curie was very different, for example), and since they have undergone important changes through the century (radioactivity only became the 'precursor' of nuclear physics in the discourses of physicists in the 1930s), there was no single way to cut the sciences which would have been both systematic and which did not do violence to the historical record. We have accordingly chosen to take the categories and questions current today as our point of departure, and have invited our authors to adopt that historical perspective which seemed pertinent to them. Fourteen fields have been covered, from clinical research to polymer chemistry and computing, from natural history to mathematics and immunology, from astronomy to atomic physics and the earth sciences. Some of these are treated from more than one point of view to bring out the richness of their historical connections. Macromolecules, a central object in twentieth-century science, are a perfect case in point. They are described by Nye from a physico-chemical perspective, in the tradition of industrial organic chemistry by Furukawa, from the point of view of biochemistry by Kamminga, and finally within the framework of the molecularization of biology by Abir Am.

These chapters use the actors' categories and follow scientists as they redefine and modify these categories. They show how knowledge advances, how knowledge builds on, reconfigures or rejects what came before. Since science aims to be systematic and critical, because it is driven by the desire to say something always new, science moves forward. That said, the texts in this section insist on the non-linearity of this process, on the fact that these advances are made in a space of many dimensions, and along a multitude of crisscrossing pathways. They show how some questions are suddenly dropped and declared to be insoluble or uninteresting, how some knowledge is lost or forgotten as new ideas emerge. Two other points also come out clearly. Firstly, that the feeling of progress which the actors have is not easily recaptured when one looks at developments from without. Löwy demonstrates this in her study of cancer research, where the overall result, in terms of effective cures, is disappointing. Secondly, the way in which science defines its own object of analysis is determined from the start by social and institutional attitudes. Lawrence shows that clinical medicine has defined disease as disorders which can be dealt with by surgery or drugs, and that it has systematically neglected chronically debilitating and diffuse ailments or disturbances where the clinician's intervention would have called for more global and social approaches.

The section begins with a text by Doel dealing with the earth sciences and geophysics, a disciplinary ensemble which has only recently emerged. This field, like those established long before and discussed by Kingsland ("NeoDarwinism and Natural History") and by Hufbauer ("Astronomy and Astrophysics") is largely organized around big historical pictures and, in particular, accounts of origins. Even if, in all three cases, the tools and the techniques of research have been dramatically modified throughout the century (they are indeed changed unrecognizably), the narrative dimension and the construction of 'scenarios' gives these disciplines an extraordinary uniformity of tone. They remind us that science is also the descendant of philosophical and theological enquiries, that it is inspired by a 'saintly curiosity' (as Langevin, the great French physicist, was to put it at the beginning of the century). These sciences can be embedded in big political projects like the Apollo programme, however, and they remain very popular. Indeed the appeal to the testimony of 'ordinary people' is always a success. One example among many is the 6000 replies received by Danjon, the director of the Paris observatory, to the appeal which he launched through the press in 1952 for literary or pictorial descriptions of the passage of a comet which had just burnt out over France.

This fascination with 'ultimate' questions can be found in many passages in this section. We see it in the chapter by Schweber describing the physicist's quest for the ultimate constituents of matter, in that by Nye on the atomic and molecular sciences, or in that by Moulin on immunology. In many others what emerges is that scientific activity can also be described as the resolution of gigantic practico-theoretical puzzles with elements drawn from many different sources. The plausibility of these elements is constantly reinforced (or undermined) by proofs built with very different kinds of tools, tools which one needs to combine in a coherent way. The identification of complex molecular structures (the hemoglobin discussed by Abir Am and Kamminga), or of solids (Hoddeson) are perfect examples of this. Making use of a variety of techniques (X-ray or neutron diffraction, electron microscopy, or ultracentrifuges, as well as theoretical systems like quantum mechanics) scientists in these fields ended up with more and more 'artificial' worlds.

The practical mastery of nature is indeed a key dimension for most of the sciences in the twentieth century. More precisely, the aim often became to create an 'unnatural nature' better fitted to our purposes and interests. Biotechnologies, the electrons of transistors or quantum optics (lasers and masers) are typical here: molecular biology *is* biotechnology, from the start solid state electronics *is* industrial science; and the new materials which were constantly manufactured opened new research avenues which, in turn, led to new products. To conclude this section Mahoney and Dahan discuss mathematics and its uses. The former studies one particular case, computer science, an "amalgam of mathematical theory, engineering practice and craft skill." Dahan, for her part, offers a panorama from the "supremacy of the Hilbert School" to the "return to the concrete" in the 1970s. Drawing comparisons between different cultural contexts, she insists on the growing interaction between mathematics and other human activities, their ever more intimate insertion in the increasing scientification/mathematization of the natural and social worlds which has occurred in this century.

Section "Science and its Practices"

"The question of how [scientists] *work* has become an important topic" writes Gaudilliere. Indeed historians are currently taking a more systematic look than heretofore at the mundane and everyday ways of doing things in the scientific laboratory or the field. They have become interested in the mechanisms whereby knowledge is appropriated and transferred from one site to another (for example, how a theoretical or instrumental system is deployed in specific ways by new sets of actors). They have taken a close look at how research materials are collected in different places (slaughter houses or army barracks in the case of hormones) before being purified and used for medical purposes.[7] They have studied the organization of the physical space in the laboratory, of the agricultural plot, and of the 'white' solid state electronics laboratory for solid state research, showing how results are shaped by this spatial dimension. They have followed the circulation of material objects and living organisms between different research sites, the increasingly important role of instrumentation in scientific work, and so on. These are the kinds of issues treated in this section, it being understood that the distinction between these practices and the major developments discussed in the previous section is only analytical and provisional, and that some of the chapters found there would have been at home here to.

Starting from the burgeoning field of anthropological linguistics, Galison considers what happens at the boundaries between cultures, be they theoretical, experimental or instrumental. Introducing the notion of a trading zone, he suggests how we might think of the ways in which scientists (mainly physicists) go about coordinating their theories with experimental results. Refusing simple formulae like epistemic breaks or paradigm shifts, he suggests that we take seriously more complex concepts like the pidginization or creolization of scientific languages to take into consideration the myriad of productive and heterogeneous confrontations which define the practice of physics.

Gaudilliere deals with the life sciences. He draws attention to the crucial role played by the standardization of materials (test procedures, for example, which are often of industrial origin), and how they enable very different communities to converge on 'scientific facts.' He also shows the importance of the mass production of calibrated living organisms (*drosophila* or laboratory mice) for large sectors of the biomedical sciences in the twentieth century. Some of these became universal models for humans, and they served as such in the screening campaigns for drugs produced by pharmaceutical companies.

Gaudilliere's chapter ends by drawing our attention to the penetration of physics-based instruments into the biology laboratory, as do the next two papers in this section. Morris and Travis bring out forcefully the transformation which the chemistry laboratory underwent after the war thanks to the massive introduction of new instrumentation developed from wartime work. They speak of infrared and ultraviolet spectroscopy, of nuclear magnetic resonance, and the major changes introduced by the use of computers to routinize the previously tedious task of the determination of molecular structures. The article by Brenni is dedicated to the world of instruments and those who conceived them, and identifies three key moments: that of the laboratory at the end of the last century organized around its instruments made of brass and glass; that of the inter-war period, noteworthy for the introduction of new materials and equipment of a greater anonymity; and that of the post-war period. Now we find instrument systems (and not just instruments): systems which are often blackboxed for researchers and which are built around industrial electronics and computer programs.

To conclude, Fitzgerald looks at the major transformations undergone in agricultural practices and agrobusiness. She recounts, *inter alia,* the field work of seed men and plant breeders before World War II, and the transformation of stock raising through breeding, mechanization and pharmaceuticals. Fitzgerald stresses that despite all the progress that has been made, agricultural science still relies upon and resides within the fields and feedlots: "cattle still wander through meadows and fruit still grows on trees." Secondly the most lasting invention of the agricultural sciences in this century might eventually be the *system* of agricultural research and production itself.

SECTION "REGIONAL AND NATIONAL SITUATIONS"

The level of analysis in the final section of the book shifts back from the micro, the laboratory and the field, to the macro, science as it is embedded in national and regional structures. Europe is examined by looking at two liberal societies, both of which have declined in relative importance as world and scientific powers through the century (Britain and Germany), one a winner in war and the other a loser. Socialism is represented by the Soviet Union/Russia. The second major bloc deals with the Americas: first the United States, which began to assert its scientific and technological leadership in the 1930s and which, as the century closes, is the dominant economic, political and military force on the planet. Alongside the USA there is Latin America, where science and technology remain marginal to national political objectives. Asia is 'represented' by Japan and India, the former an expanding scientific and technological nation, the latter a society in which science and technology are assuming increasing importance. The final chapter deals with collaborative European scientific research, the attempt to organize science and technology in a political framework which transcends the boundaries of the nation state.

A number of common themes emerge from these chapters. Firstly as the century has progressed science and the state have become steadily and massively entangled with each other. That entanglement has been particularly intimate in countries which have embarked on major military programs. Correlatively where science and national security have not been synonymous, the links between science and state have been far more vulnerable to changing political and ideological currents.

The state, for its part, first had to learn to deal with an elite which had become indispensable to it. Stalinist and Nazi purges of scientists in the 1930s, and the less brutal 'Lysenko' affair and McCarthyist loyalty trials in the first decade after World War II, all attest to a political class both anxious to promote science but nervous about the autonomy of a community which it felt it could not yet trust ideologically. Scientists learnt to weather these storms. Essentially opportunistic, they adapted and adjusted their strategies and tactics to the prevailing political and ideological climate, be that capitalist or communist, fascist or liberal. Pragmatism, not principle, has infused the science-state alliance.

The first three chapters are by Edgerton, Walker and Krementsov. Edgerton reminds us forcibly that the link between science and nation state is not natural or inevitable, but is itself the result of a historical development which has accelerated enormously in this century. The 'nationalization' of science has been a response to the demands of war and particular economic policies. Walker traces the changing institutional structures put in place in Germany, stressing the ability of the scientific community to adapt to the violent fluctuations in political regime which have occurred. Krementsov describes the symbiotic relationship built up between science and the state in the Soviet Union, showing how the scientific community became co-opted into it. Contrary to what has often been said in the West, the asymmetry between political power and science was never total.

Owens associates the rise of American science with the demands of political and economic elites to control and to rationalize an increasingly complex world. This dynamic was expressed above all through a fascination with the 'big' (telescopes and accelerators, but also public works like the TVA and weapons systems), all of which demanded complex hierarchical organizational structures. Vessuri points out that although political elites in Latin America have enthusiastically adopted science as the bearer of progress and order, major investments in science have seldom been forthcoming. Instead of the steady consolidation and expansion of a science base we tend to find a number of important, but isolated developments generally pursued thanks to the determination and skill of one or two individuals or, more recently, to the demands of state enterprises.

Finally we come to chapters by Bartholomew, Raina and Jain, and Krige. Both world wars played a key role in the development of Japanese science and its research base. Defeated in 1945, and seeking self-reliance, the post-war period saw a de-emphasis of basic research and the university and the rise of support for applied research in corporate laboratories, a trend now being reversed again. In India the universities have had a chequered existence. Poorly developed in the colonial period, they were subsequently swamped by large scale mission-oriented programs in the nuclear and in space, with their related institutions, which have left the universities on the periphery of the knowledge producing systems. Big science and technology is also the main concern in the last chapter. It shows how scientists, administrators and politicians in post-war Europe combined scientific need (to remain competitive with the USA) and foreign policy concerns (to build a European political and economic space) to develop new forms of scientific collaboration upstream from the market and from possible military applications.

The Twentieth Century and Science

To conclude this introduction we should just like to mention again — though who does not know it? — how important World War II was for science, and for the relationship between science and the state in this century. This war shared with the Great War both the systematic application of science and technology to the improvement of the war fighting capabilities of the major protagonists, and the rational organization of production. In these respects it was no different to its predecessor. What sets it apart, however, is that, whereas in the former case, most scientists 'returned to their laboratories' once the conflict was over, now the intimate link between science, the state and the management of civil society remained in place. A new social contract centered on science and its potential was put in place. The armed services in all the major powers continued to sponsor research through organizations of their own, like the Office of Naval Research in the USA, and expand their presence in universities, industry, national laboratories (notably in Europe) and various defense research establishments (notably in Britain and France). Seen in terms of the relationship between science and the state, and the military in particular, World War II did not punctuate two periods of (relative) peace. On the contrary, it inaugurated 50 years in which the two world systems and their main allies, those countries in which science and technology were concentrated, were either fighting a war or in a state of readiness for war. In fact the war in Europe did not end in 1945: it was pursued, at least in its technico-scientific dimension (and though local wars in surrounding 'trouble spots') by the Cold War.

This state of permanent war fundamentally changed the relationships between scientists, the state and industry. To acquire the material resources which only the state could provide, scientists skillfully exploited the new identification between science and technology and state power and prestige. Moving rapidly into the corridors of power they built networks through which they acquired the authority to impose their priorities, while always maintaining the 'freedom of research.' A regime was built in which economic progress, political development and military readiness came to be quantified in terms of the percentage of the Gross National Product spent on research and development. Science was the symbol of modernity.

The nature and direction of scientific work changed as a result, even if, subjectively, the desire to produce knowledge, to know more about 'nature' still remained the main motive of the practitioners. Entire fields like oceanography or geophysics were profoundly reconfigured in the 1950s and '60s, not to say created, thanks to the military's direct interest in the seabed (for the submarine war) and the upper atmosphere and beyond (for missiles). Other fields, like solid state physics, which had already started bringing together, in the 1930s, a variety of previously scattered practices, benefited during the war and the Cold War from the extensive exchange of artifacts and knowhow between military and commercial research programs. Since national security on both sides of the Iron Curtain was premised on the triad missiles/nuclear warheads/advanced guidance and detection, solid-state electronic components were central to the reliability of the systems put in place. In other areas like the early development of particle accelerators or lasers, the military kept a watchful eye (and often financed) the research only intervening to shape its direction when persuaded by entrepreneurial scientists of the strategic value of the work. In short the physical sciences and their associated technologies since World War II have been practiced in a space whose contours have been defined by a state and a military apparatus convinced that competitive advantage over the enemy depended on scientific excellence and technological prowess.

The biomedical sciences similarly underwent major changes in orientation, work style, techniques and relations to society and political power. The benefits of scientific medicine were given an enormous boost by the massive, and propaganda-laden war time use of penicillin and by the discovery of antibiotics. The development of new drugs demanded not only extensive laboratory research and a huge increase in the number of practitioners, but also large-scale randomized clinical trials, which by the mid 1950s were being funded mainly by the state. The war on cancer required not only the production of hundreds of thousands of various standardized mice but also the careful control of patients by task forces which brought together academic and industrial scientists, clinical investigators, statisticians and the members of the state apparatus. Goal-oriented, large scale, multidisciplinary programs such as those put in place during World War II became increasingly widespread and accepted as the most appropriate way to solve politically and medically urgent problems of public health.

Essential during World War I, taken up again and institutionalized in a more systematic fashion during the Great Depression of the 1930s, the role of the state in the management of society thus became of central importance after World War II. The belief in the need for state intervention in economic and social planning, intrinsic to the socialist system, and practiced as a solution in capitalist society to the poverty and turmoil of the thirties, intensified and became normal after 1945 notably in Northern Europe, Britain and France. It led to an increasing quantification and regulation of social life. Science, with its objectifying, calculating tools seeped into the pores of the body politic, estimating input and output, profit and loss, cost and benefit, growth and decline. The invisible hand of the market was partially replaced by the conscious, if scarcely more visible hand of the expert.

REFERENCES

This project benefited from financial support from the Research Council of the European University Institute, Florence, and DGXII of the European Commission in Brussels.

1. Veyne, P., *Comment on écrit l'histoire,* (Paris: Gallimard: 1971).
2. Feynman, R., *"Surely You're Joking, Mr. Feynman!": Adventures of a Curious Character,* (W.W. Norton. 1985).
3. Duclert, V., *L'Usage des savoirs: les engagements des savants et l'affaire Dreyfus (1894–1994)*: thesis (forthcoming) and Charles, C., *Naissance des 'intellectuels' 1880–1900,* (Paris: Minuit. 1990).
4. Pestre, D., 'Les physiciens dans les societes occidentales de l'après-guerre. une mutation des pratiques techniques et des comportements sociaux et culturels', *Revue d'Histoire Moderne et Contemporaine,* **39** (1), 56–72 (1992).
5. Lecourt. L., *Prométhée, Faust, Frankenstein. Fondements imaginaires de l'éthique,* (Paris: Les empôcheurs de penser en rond, Synthelabo: 1996).
6. Haraway. D., *Primate Visions. Gender, Race and Nature in the World of Modern Science,* (New York: Routledge: 1989) and Epstein, S.G. *Impure Science: AIDS, Activism and the Politics of Knowledge* (PhD thesis, University of California at Berkeley: 1993); Lederer S.E., *Subjected to Science. Human Experimentation in America Before the Second World War,* (Johns Hopkins: 1995).
7. Oudshoorn. N., *Beyond the Natural Body: The Making of Sex Hormone,* (London: Routledge: 1994) shows how the material at one's disposal define what one studies and what one ignores.

CHAPTER 1

Historiography of Science

An American Perspective

MICHAEL AARON DENNIS

Why is the historiography of science important? In part, claims about the proper method for writing the history of science are simultaneously claims about the relations between the producers and the consumers of scientific knowledge as well as the relationship between historians of science and their object of study. Different historiographic perspectives yield divergent conceptions of 'science' as well as distinctive forms of historical inquiry. In the historian's workshop, the scholar produces both a historical work and a map of knowledge delineating the boundaries separating the topic at hand, the investigator, and the work of other researchers. However, the historian of science, like the scientist, also engages in another demarcation: the identification of the scientific and the natural. Establishing and maintaining such a demarcation is of more than academic interest. As a major resource for the legitimization of existing social orders, science differs from other forms of knowledge because of its public professions of truthfulness and universality. Take the case of the electron. Measuring an electron's charge in a Berkeley laboratory or a laboratory in the Balkans should produce identical results. How historians of science account for that identity is one of the great historiographic changes of this century. If an earlier generation of scholarship assumed the electron's charge a 'natural' consequence of the particle's existence as well as the genius of the discoverer, a more recent generation has seen the discovery, identification, and persistence of the charge as the upshot of a complex array of institutions and practices ranging from the organization of work in Cambridge's Cavendish Laboratory to the design and development of imperial telegraphy and contemporary electric power systems. In other words, for the historian of science the electron's existence is a necessary, yet insufficient cause for the discovery of its charge. No longer is the history of science the story of individual researchers confronting a 'nature' that must be persuaded to reveal its secrets. Instead, the new historiography has made the identification of the natural and its boundaries one of the prime areas of inquiry. What has taken place is nothing less than a redrawing of the boundaries separating the natural and the artificial.

Intellectual change has been accompanied by an equally significant social transformation. History of science, once the domain of retired researchers, practicing scientists, and the rare specialist, has developed into an academic discipline, replete with graduate schools, research institutes, public and private patronage, peer-reviewed journals, and professional societies. In the United States, George Sarton, the founding editor of *Isis: Revue consacrée à l'histoire et à l'organisation de la science,* was the discipline's heroic founding figure complete with a compelling personal story. After burying his precious notes, Sarton and his family fled Belgium as the Germans occupied his home during World War I. Following a brief stay in England, Sarton arrived in America, penniless and unemployed. Lecturing part-time at several academic institutions and, beginning in 1916, at Harvard University, Sarton managed to remain solvent. By 1918 he was once again in desperate straits; his appointment at Harvard was scheduled to expire with little chance of renewal, and he confronted a bleak choice: "stop that work or starve."[1] Sarton made that impassioned declaration in a final appeal to Robert S. Woodward, president of the Carnegie Institution of Washington (CIW), one of the first private research institutions in the United States and one of the few patrons that might support Sarton's work in the history of science. Upon arriving in the United States in 1914, Sarton had approached Woodward, only to find the CIW interested in supporting a version of "Whewell's 'History of the Inductive Sciences' for the present," but incapable of funding the work. By Easter 1918, Sarton needed a savior. Woodward easily played the role, offering the unemployed scholar a two year position and, in 1920, a permanent tenured appointment as a Research Associate in the Institution's Department of History. In exchange for space in Harvard's Widener library, Sarton continued to teach a single undergraduate lecture course each year.

The drama of Sarton's travails left little impact on his field's intellectual development. In one sense, Sarton is more like an Aby Warburg than a Ranke of the history of science. Like Warburg, Sarton provided his chosen field with the instruments through which scholarship might continue. Although Warburg's writings remain important to the student of art's historiography, his significance for most students of art history lies in his magnificent library, now located in the Warburg and Courtauld Institutes at the University of London, an instrument fashioned for the express purpose of understanding Warburg's wondrous obsession, "*das Nachleben der Antike,*" literally, "the after-life of classical antiquity." Contemporary scholars need not share Warburg's interpretive practices to make use of his unique repository; similarly, historians of science need not accept Sarton's vision of the history of science as the only genuine example of human progress to continue using the instruments he bequeathed to the field: the journal, *Isis*; the annual volume, *Osiris;* plus an annual analytical bibliography that Sarton believed to be among his most important scholarly innovations.

The fundamental engine of historiographic change during the twentieth century has been, and remains in one or another guise, a contest between internalists and externalists. Although never adequately defined nor existing in 'pure' forms, internal and external approaches to the history of science distinguished themselves through the array of resources available to the historian to explain scientific change. For the internalist, the immanent logic of science itself generated the forces producing change, yet the problem of demarcating science from non-science was far from obvious. George Sarton, the founding editor of *Isis*, made clear the ambiguous character of the boundaries when he explained that

> the progress of science is absolutely dependent upon its emancipation from non-scientific issues, whatever they be, and in particular, upon its laicization.[2]

Writing nearly forty years later in 1968, Thomas S. Kuhn, perhaps the best known historian of science in the US, could assert that the "best interpretive scholarship in the history of science" over the previous thirty years displayed what he described as the maxims of the "new internal historiography." For Kuhn, the core of this new approach was its goal of understanding what natural philosophers and scientists had understood themselves to be doing, rather than what textbook writers and contemporary researchers projected onto the past in their search for either simple pedagogical narratives or equally inventive predecessors.[3] In Kuhn's historiography, internalism's success stemmed from science's own dynamics, for when

> compared with other professional and creative pursuits, the practitioners of a mature science are effectively insulated from the cultural milieu in which they live their extra-professional lives.

Such insulation also effectively circumscribed the set of potential historians of science and their research areas. Those possessing a twentieth-century undergraduate education might write insightful studies of the development of early modern natural philosophy, but they would find themselves incapable of writing the history of more recent science, especially the giant of the twentieth century, theoretical physics.

While the "internal approach" elucidated science as knowledge, the "external approach" focused "the activities of scientists as a social group within a wider culture." Externalism's most disturbing feature lay in the attempts to demonstrate the effect of society upon not only the organization but the actual practice of science and the content of scientific knowledge. For Kuhn, Robert K. Merton's pioneering work on science and society in the seventeenth century exemplified the problems of the external approach. Always an insightful reader, Kuhn recognized that Merton argued two distinct yet related theses: first, that economic and military influences greatly affected the selection of research problems during the great efflorescence of science in seventeenth century England. Second, following Max Weber's suggestion, Merton argued that there were significant links between English scientific developments and Puritan versions of English Protestantism. Far from being enemies, Puritan theology provided especially fertile ground for the development of the new science. A sensitive reading of Merton paved the way for Kuhn's belief that while there was something to Merton's theses, they failed to take into account the major substantive findings of the new internal historiography: the fact that such external matters had little to do with the so-called "scientific revolution." New instruments, experiments and observation were irrelevant to the fundamental changes of the period. If explaining the scientific revolution required the use of external or cultural novelties, then those were primarily intellectual resources, for example Renaissance neoplatonism or the rediscovery of Archimedes. Ideas remained far more important than actions; a point exemplified by Alexandre Koyré's famous claim that Galileo's experiments were thought experiments. At best, an external approach might supplement the internal historiography, but nothing could replace "immersion in the literature of science." In 1968, Kuhn claimed that the fusion of the two approaches was the greatest challenge facing the young discipline; in 1971, he would observe that the growing enthusiasm for the external approach was a consequence of "the increasingly virulent antiscientific climate of these times."

Today, the social study of science – including the history and sociology of science – is increasingly the focus of attack by practicing scientists and engineers. Friends and allies have become enemies and opponents; accounts of science that diverge from the accounts offered by contemporary scientists are subject to censure, public ridicule and the powerful charge of being 'anti–science.' Historians of science could once ask, with a twinkle in their eye, if their work should be 'rated X' and declared unsuitable for aspiring scientists; today students of science and technology are told by researchers, pundits, and other cultural critics that our works are simply wrong and that we must shoulder some responsibility for the growing 'anti–science' attitude present in the American polity. What is going on here? How and why has the historiography of science become important? In what follows I outline a history of the public and scholarly discourse on science that locates this speech and writing in several specific socio-political contexts, especially in the United States. In other words, I make the historiography of science itself the topic of research in the sociology of knowledge.

To write historiography historically is to acknowledge another novel element within the social studies of science and technology: the concept of coproduction. There is no simple, nor any single definition of coproduction, but with this term scholars signify the inseparability of the political and the intellectual, or the technical and the social. In this chapter, coproduction refers to the process through which a new political economy of science in post-war America emerges at the same time as a new way to conceptualize the history of science. Invoking coproduction as an explanation does not mean that history is reduced to politics or that science is only politics by other means. Coproduction attempts to make clear what historical actors take for granted: the need to create new forms of understanding as they create new political forms. Political and conceptual innovation are related and integral aspects of historical and technical change.

THE PAPER HEARD AROUND THE WORLD

As "it is widely acknowledged that the 1930s were the seminal decade for our present understanding of science," what was its legacy?[4] As a period of intense ferment it remains unmatched. Koyré, Merton, Mumford, Hessen, Bernal, Borkenau, Fleck and Zilsel all made their first appearances, and at Harvard George Sarton began training the first American doctoral students in the history of science. However, most of this has passed from disciplinary memory. The thirties are remembered as a time of 'crude' attempts to link science and society. These are exemplified by Robert Merton's work, and to a lesser extent, the infamous visit of the Soviet delegation, led by N.I. Buhkarin, to the Second International Congress of the History of Science in London in 1931. The Soviet presentations and papers, especially Boris Hessen's seminal work on Newton, have been labeled "vulgar Marxism" and dismissed by practitioners as largely propagandistic efforts to legitimize the new Soviet state. What made these papers and other work in the thirties important was that they were the first sustained challenge to the idea of the history of science as the history of individual genius in action, a view dominant since the publication of Whewell's *History of the Inductive Sciences* in 1837. Unfortunately, contemporary attitudes ignore the deep interest taken in Hessen's work, not only by British researchers with a socialist bent, but by such scholars as Robert Merton, G.N. Clark, and Henry Guerlac.[5] What did they, and others see in Hessen?

It is common to argue that Hessen reduced Newton's *Principia* to nothing more than the solution of technical problems of the bourgeoisie. Although it is now possible to parody the text in this fashion, and to recognize its profound limitations, we cannot underestimate its impact. As J.G. Crowther commented, no one had bothered to think about Newton's *Principia* as anything but the product of genius. Or as Hessen observed, quoting Pope: "Nature and nature's laws lay hid in night/God said 'Let Newton be!' and all was light."[6] There is no denying the heavy handed treatment that Hessen gives to Newton. After all, a large part of the essay is devoted to an explanation of why the absence of a steam driven industrial society prevented Newton from articulating the laws related to the conservation of energy. Most readers of Hessen's text were not prepared to accept a rigid connection between economy and knowledge, yet they were quite willing to accept another connection which Hessen made as part of his argument: that of a close connection between the growth of knowledge and the art of war. The transformations of defensive technologies in the sixteenth and seventeenth centuries were well known to historians. New styles of fortress design demanded new means of attack, especially the development of new forms of artillery. In Hessen's universe, ballistics became a central part of physics and of Newton's new world.

Whether the audience was reading H.G. Wells' *When the Sleeper Wakes*, watching his film *Things to Come*, or following the debates over the usefulness of strategic bombing, science and technology were clearly linked with the instruments of destruction, the machinery of war. In the US, this relationship was made explicit in a variety of media ranging from the declarations of Technocracy Inc., one of the many groups which emerged in the US claiming to have a cure for the nation's economic paralysis, to other far less radical critics of the industrial system, like Lewis Mumford, and popularizers of science, such as Waldemar Kaempffert, the influential science writer of the *New York Times*.

Henry Guerlac, a Harvard graduate student in history, wrote his dissertation on "Science and War in the Old Regime," in which he argued that the anglocentric bias of the history of science prevented historians from understanding that

> modern science developed in part at least by borrowing from practical and humble craft traditions; that these craft traditions became increasingly significant with the development of the modern state; and that among these mechanical arts the art of war played a significant role, especially through military engineering.[7]

Even Hessen's academic critics, like the Cambridge historian G.N. Clark, accepted that warfare might affect the growth and development of science. What Clark and others found objectionable was the reduction of all scientific knowledge to economics. Clark wished to disaggregate Hessen's category of economics into at least six separate domains, including art, war, religion, medicine, trade and the disinterested search for truth.

Robert Merton's famous work on science, technology and society in seventeenth-century England might also be read, in part, as both a reaction to and emendation of Hessen's argument. First, Merton sought to introduce an additional category, Puritanism, into the explanation for the spectacular growth of science in this period. Second, rather than concentrate on a single figure and text, as in Hessen's case, Merton deployed a form of quantitative analysis, seeking to show that his arguments did not rely on theoretical constructions about class, but on empirical materials available to all who would do the laborious work of counting. Third, like Clark, Merton chose a form of disaggregation, splitting economics into transportation and mining, as well as military technique. Furthermore, Merton attempted to demonstrate quantitatively how various factors may have affected the development of science. Yet, even with all these qualifications Merton was keen to note how "heavily indebted" he was to Hessen.[8] Where Merton and Hessen parted interpretive company was in the move to link the content of knowledge with its social context. For Merton, the "general types of problems" which interest researchers may be suggested by extra-scientific factors, but the researcher's interests were driven by "the internal history of the science in question." Context might suggest the area of work, but the content of the sciences themselves would determine the solutions individual scientists developed. It was a logic demarcating disciplinary boundaries: context was the domain of the sociologist of science, content, the historian of science.

ONE WAY STREET?

Vannevar Bush's famous report, *Science — The Endless Frontier*, written in 1945 as a response to a request from President Roosevelt asking how the nation might apply the lessons of war during the coming peace, is often regarded as the fountain of American science policy and the original constitution of the National Science Foundation.[9] That nothing Bush recommended came to pass, including his National Research Foundation, is much less important than the way in which the language of the report became part and parcel of post-war US policy.

Central to Bush's report was a new taxonomy of knowledge in which "basic scientific research" appeared as the most valuable form of technical knowledge, essential for national security, economic growth, and the survival of democratic American values. Pursued "without any thought of practical ends," basic research was the "pacemaker of technological progress." However, this definition revealed a fundamental paradox. Basic research was only identifiable through its applications. Industrial power, national security, democracy, and national wealth were all secondary indicators of the state of basic research as well as the result of its application. Furthermore, the war had redrawn the geography of science. Post-war America could no longer rely upon war "ravaged Europe as a source of fundamental knowledge." Only the federal government could provide the "new impetus" for research.

Bush's report was as much a history of science as a blueprint for government support of the disciplines. In particular, it presented a history in which the US did little basic research prior to World War II, relying instead on imported basic research. Bush commissioned, through the Office of Scientific Research and Development which he directed during the war, histories of its various divisions and activities: histories which would lend support to his claims about the need for publicly financing university-based research. The report and the histories were important to the project of erecting boundaries between 'science' and its various possible contexts; hence, the incorporation within Bush's research foundation of divisions devoted to national defense, medicine, and the natural sciences as well as supply-side issues such as education and the dissemination of technical information.[10] Here we see a set of origins for contemporary beliefs about the relationship between knowledge and its context; it was a relationship embodied in the name *National Research Foundation*. While the word "foundation" captured the interwar sensibility in which the private philanthropies supported the sciences, Bush's nomenclature secured research as the foundation of the post-war American state's power. Implicit in such an arrangement was the temporal logic of basic research, which always preceded worthwhile applications. Bush's stress on the "flow of scientific knowledge" acted to make such knowledge analogous to scientific theories or concepts; the world of technology vanished. Differentiating science from technology was not an academic problem, but a genuine problem in politics; as the authors of one Bush report appendix explained, "applied research invariably drives out pure." In the immediate context in which the report was produced, the dominant fear was that military needs might drive the research enterprise independent of the needs or desires of researchers, but it was a suspicion that could easily extend to other branches of government and private industry. Invoking Gresham's law inadvertently highlighted a far more profound transformation that Bush certainly feared: the removal of the sciences from the marketplace and their inadvertent transformation into wards of the state.

Concomitant with the need to differentiate between science and technology was a need to distinguish between scientists and engineers or technicians. Citizens and politicians could interpret the wartime research and development effort, especially the Manhattan Project, in at least two distinct ways. First, one might accept the language of Bush's report and view Allied victory as the result of successfully applying basic research derived from largely European sources. Or one might read the wartime lesson in far more base terms: a willingness to spend immense sums of money could buy new, more powerful weapons. If Americans accepted the latter view, then researchers were no more than technicians guided to research topics of interest by their patrons; directed research guaranteed victory. Independence and autonomy, virtues which Bush sought to embed in the foundation to support basic research, became irrelevant. Within this particular political context a new use appeared for a previously esoteric field: the history of science might serve as as apologetic for a new organization of knowledge in a polity where only an elite actually understood the practices explored and defended in Bush's text.

History — Servant of Science?

Post World War II history of science, at least from the end of the war until approximately 1970, articulated a version of the past that fit comfortably with the framework Bush established. Superficially, the discipline's overwhelming emphasis on the 'Scientific Revolution' of the seventeenth century reinforced the claim that America had imported its knowledge from the past as well as Europe. More importantly for our purposes was the rapid destruction of anything resembling the pre-war view that science and war might possess a fundamental connection. For Koyré and his American disciples, context was a constellation of ideas rather than a world in which individuals worked and lived. Richard Shryock's seminal 1948 essay on "American Indifference to Basic Research" lent further legitimacy to the claims of the Bush report, adding evidence to the recasting of Gresham's law while arguing that basic science was the casualty of an American addiction to utilitarianism. And, in a model of ironic understatement, he wrote that "[i]t seems clear, therefore, that economic interests and technology will not of themselves lead automatically into basic investigations."[11] Finally, A.R. Hall's 1952 monograph *Ballistics in the Seventeenth Century* declared that "the practice of artillery contributed nothing to seventeenth-century science" despite the claims of those writing a "sociological history of science."[12] Far removed from the corridors of Washington power and writing under Herbert Butterfield's direction, Hall was most likely unaware of the Bush report or the logic that made it possible for science to influence war, but impossible for the war to affect science; however, he was quite clear that Hessen's theses were unworthy of respect. That the theory of projectiles figured in both Aristotle's and Newton's work resulted from the "evolution" of science rather than the "strong hand of economic necessity."

Morphological similarities do not imply evolutionary nor substantive relation. At best, such similarities might serve as the first indicators of a deeper connection, but such a linkage must be established. How might we move from examples drawn from the canon of post-war history of science to the world of politics and patronage that was so central to the era, especially in the US? One possible path may lie in asking about history of science that was not directed towards other members of a young and developing discipline, but aimed at a different audience. What was the perceived value of the history of science as both a discipline and a field of inquiry? Why did universities create positions for historians of science to teach undergraduates, and in turn, train doctoral students? We can only sketch an answer here, but we must follow two separate, yet related historical trails: the process through which the US National Academy of Sciences sought to produce a history publicizing the relation of "fundamental research" to the so-called comforts of everyday life and the ways in which Harvard president and wartime science administrator James Bryant Conant, sought to harness the history of science to the problem of educating citizens for the atomic polity. Both trails meet in Madison, Wisconsin, in summer 1957 at the famous conference, "Critical Problems in the History of Science."

In April 1941, the National Academy of Sciences established a National Science Fund to secure monies from private sources for the support of fundamental scientific research. The Fund sought to supplement the private philanthropies, then dominating the patronage of the sciences, by convincing individuals to leave the Fund monies in their estates and private firms to provide annual gifts. Publicizing the Fund fell upon a small committee which planned a public relations campaign centered around a historical study demonstrating "the fundamental research behind modern necessities and conveniences" and a small leaflet, entitled "Science — Our Modern Frontier," that would be circulated among potential donors. Both George Sarton and Richard Shryock recommended I. Bernard Cohen, a student of Sarton's at Harvard. Cohen had already begun writing a brief book, *American Science and War*

> to inform the public concerning the organization of American scientists during the present emergency, it also presents materials concerning the organization of our scientists in the past — especially in war, but also in peace-time. In a sense, then, it may be considered a history of the relations between scientists and our government, and the history of our national scientific organizations and institutions.

Cohen's manuscript floundered on the reef of military secrecy, but with the Fund's support and two assistants Cohen began to gather materials on pure and applied research. Nor was initial writing a waste. Cohen believed that much of it might make its way into the Fund's volume, especially since

> the service rendered to the nation in time of crisis or emergency belongs to the story of the practical use to which pure science is put. Especially is this the case in the present war, when many of our new and important devices are merely simple applications of knowledge gathered entirely for the sake of knowing — which is the criterion of pure science.[13]

By mid-1944, the volume existed as a sixty page outline, but demands on Cohen's time, especially in teaching physics, made it difficult to complete the Fund's publicity materials. Exasperation at the Fund grew as Cohen was conscripted to assist one of the committees set up to answer the questions posed to Vannevar Bush by President Roosevelt. For Cohen, "the material uncovered in the investigation" might figure in the book then tentatively entitled, *The Practical Applications of Fundamental Research.* All that remained was research, writing and the search for a more exciting title.

Case Study Fever

When, in 1948, Little, Brown, and Company of Boston, the publishers of the official wartime histories of the OSRD, published *Science, Servant of Man: A Layman's Primer for the Age of Science,* the book's very form and content embodied not only the original message of the value of fundamental research, but an approach to the study of science for the layman developed at Harvard by university president James Bryant Conant. In Spring 1943, Conant had appointed a University Committee on the Objectives of a General Education in a Free Society. The committee's final report, *General Education in Free Society,* issued in June 1945, remains among the most significant texts in the history of American higher education. General education was the perceived means to foster the "necessary bonds and common ground" among citizens in a "technological age" which accelerated democracy's natural tendency to produce difference and division. Such education was not to be restricted to those attending college, but available to the majority of citizens who attended secondary school. Nor was the use of "free society" insignificant. While we might read the report as an early adumbration of cold war concerns, we should recall that it was written while the US was still fighting against fascism. We should read the report as an attempt to articulate a general education to ensure against any form of authoritarianism taking root in American soil.

Attempts to create a common intellectual culture were not new to Conant or to Harvard. Both Columbia University and the University of Chicago had attempted to define such a common intellectual culture. And George Sarton had been arguing for over a decade that the history of science was the only available instrument for bridging the chasm separating the scientist from the rest of contemporary culture. It was the "two cultures" problem long before C.P. Snow uttered his famous phrase. Sarton's "new humanism" was quite similar to the views articulated by the committee and later by Conant. It was ironic that while Sarton saw the new humanism as the vehicle through which the history of science might rise in the university's disciplinary economy, the authors of the Harvard report argued that "the claim of general education is that the history of science is part of science."[14] What was new was that the Harvard report sought to make the history of science the vehicle through which an appreciation and understanding of science might be brought to the citizens, especially those whose education ended with a high school diploma.

Conant's own views on the education of the layman and the relation of science to society appear as early as November 1943, when he told the American Philosophical Society that the "lay public" would determine the future of scholarship.[15] Only a free society in the US could guarantee the appropriate attitudes; victory against the Axis was essential, but so was a victory in the "second battle of freedom," a struggle to make capitalism work despite the "complexities of modern industrial life" and the demobilization of the war's vast "military undertaking." Success in each struggle would guarantee the bedrock of scholarship in any field: free inquiry. Conant did not equate free inquiry with a social vacuum; instead he chose to introduce a "controversy" over the "variety and kinds of social forces which have conditioned the relation of science and society." Speaking of Hessen's work as well as that of G.N. Clark, Conant argued that "the scholar may imagine that he is as free as a pioneer in a virgin forest, yet those who trace his wandering from a distance can discern the effect of many varied social forces." For Conant, Merton's connection of the founding of the Royal Society and English Puritanism was a plausible example. An excursion into the history and sociology of science was at one with Conant's belief that without "certain types of strong social forces" free inquiry would become "an aimless, leisurely ramble amidst delightful scenery" undertaken by only the "old and intellectually infirm." Relevance to the "future of our civilization" was the source of intellectual vitality in Conant's mind. The point was that relevance would make science attractive and worthy of support, just as Puritanism made natural philosophy theologically attractive in the seventeenth century. A populace sharing similar ideas about the nature of science was a populace that would recognize how the "flame of war" made America the home for future scholarly activities and the University the institution in which future science and scholarship would flourish. With these social and intellectual predicates, the nation would eagerly support the American university as a venue for education and as a community of scholars. What Conant failed to realize was that wartime technical successes might serve as an even more powerful source of relevance.

The dropping of the atomic bomb dramatically changed the context in which Cohen's and Conant's work would appear. Wartime technical successes heightened the public's interest in science and technology, as did public discussion of Bush's report and the growing debate over the organization of science. Cohen's text began, "Science today is everybody's business," and Conant in *On Understanding Science* observed that

> [t]o write a book about science in the year 1946 without some consideration of the atomic bomb may seem the academic equivalent of fiddling while Rome burns.[16]

The atomic bomb was the price paid for "health and comfort in this scientific age." Unless the polity as a whole possessed an understanding of science, there was little chance of walking "boldly along the tightrope of the atomic age." Both Conant and Cohen sought to use particular "case studies" to make readers and students aware of what Conant called the "tactics and strategy of science." It was a phrase with special resonance given the just concluded war, but it was also a clever inversion of the traditional phrasing [strategy and tactics] and an attempt to convey the workings of this new approach to science, one that would work up from the specifics of technical life to the generalities of science. At one with this phrase was Conant's disdain for a conception of "scientific method" and anything resembling a philosophical analysis. In a locution anticipating Thomas Kuhn's famous work on the structure of scientific revolutions, Conant asked why if the method of science were so clear to philosophers "did it take so long a period of fumbling before scientists were clear on some very familiar matters?" Conant argued that Renaissance humanism with its goal of understanding antiquity was the source of our belief in the need for exact and impartial understanding. Contemporary science represented the migration of the "accumulated fruits" of humanistic inquiries into fields "ripe for cultivation" which once self-propagating were relatively easy to maintain. In a colorful example, Conant explained that

> in the natural sciences today the given social environment has made it very easy for even an emotionally unstable person to be exact and impartial in his laboratory.

Here was the key to Conant's use of the history of science. If the "tactics and strategy of science" was itself a clever metaphor for induction; then so was the goal of the education itself a form of social induction. "Understanding science," as David Hollinger has observed, would allow the layman to "behave scientifically in social environments very different from the one in which science actually proceeds."[17] Science was not necessarily more important than other forms of knowledge, but it was different. As Conant argued

> a dictator wishing to mold the thoughts and actions of a literate people could afford to leave the scientists and scholars alone, but he must win over to his side or destroy the philosophers, the writers and the artists.

A society educated through the case studies Conant envisioned would possess an immunity to totalitarianism as well as a desire to support scientific research.

Cohen and Conant's common invocation of the primacy of case studies was far from coincidental; Conant thanked Cohen for the historical materials in his text, and Cohen explicitly linked his book to Conant's vision of teaching science to non-scientists through case studies. Harvard's president returned to the classroom in 1949 with Natural Science 4 (NatSci 4), a course embodying the beliefs of *On Understanding Science.* NatSci 4 was the birthplace of modern history of science in the United States. Those who assisted Conant in teaching as well as preparing the course materials included Cohen, Kuhn, Duane Roller, Leonard Nash, and Gerald Holton. Indeed, it is not too much to claim that the Harvard Case Studies, as well as Cohen's work, continue to shape areas of inquiry within the history of science. Boyle's pneumatics, the overthrow of the phologiston theory, origins of the atomic and molecular framework for understanding matter, Pasteur's work on fermentation and spontaneous generation and the development of the idea of electric charge remain vibrant areas of research in contemporary history of science. Detailed case studies were difficult to develop and even harder to teach, despite the fact that the first audiences for NatSci 4 were among the most mature generation ever to grace any American campus, including Harvard's. The initial recipients of the course were none other than returning veterans replenishing academic treasuries with their government subsidized tuition payments. Cases chosen for NatSci 4 reflected Conant's belief that students with little interest in science would find cases from the early development of science much more accessible than cases about more recent science which would require more background knowledge. The availability of primary materials in English was also quite important. Furthermore, the case studies themselves were insufficient; a fact that Conant made clear in the final lectures on the Hessen thesis and the contemporary organization of the sciences.

NatSci 4 informed a new generation of citizens that science was largely a foreign enterprise and worthy of support in the US. Throughout *On Understanding Science,* Conant made it clear that the course he proposed was <u>not</u> a course in the history of science or even European intellectual history. Certainly, the history of science played a necessary role in the development of the case studies, but furthering the history of science as an academic enterprise was never Conant's goal. Instead, the case studies would illuminate the "tactics and strategy of science" which Conant grouped under three main headings:

- New concepts evolve from experiments or observations and are fruitful of new experiments or observations.
- Significant observations are the result of 'controlled experiments' or observations; the difficulties of experimentation must not be overlooked.
- New techniques arise as a result of experimentation and influence further experimentation.

We need not examine each corollary to these fundamental theses. Contemporary readers will recognize the role of contingency, the idea that "a well-established concept may prove a barrier to the acceptance of a new one," and the fundamental assertion that

> advances in the practical arts are not the same as advances in science. No new concepts or conceptual schemes are evolved; likewise the amassing of data does not constitute advance in science.[18]

If nothing else, NatSci 4 was the crucible in which Thomas Kuhn forged the ideas that would become the basis for his best selling work, *The Structure of Scientific Revolutions.* Important here for our purposes is Conant's definition of science as "the development of conceptual schemes." At pains to avoid the use of the word 'theory' or any language that might appear philosophical, Conant defined science as a dynamic interplay between conceptual schemes and experiment in which fruitful conceptual schemes generated new experiments which yielded progressively better conceptual schemes. Such a perspective located the sources of intellectual change within science, lending further credence to the claim that scientists were not simply technicians whose work was easily managed and planned.

Why choose historical case studies as the vehicle through which one would educate the layman about science? Writing in 1957, in an introduction to a collection of the case studies used in NatSci 4, Conant offered a partial explanation:

> a direct study of the methods of modern science presents great difficulties. A visitor to a laboratory, unless he is himself a scientist, will find it impossible to understand the work in progress; he will comprehend neither the objectives nor the implications of the measurements and observations that the investigator is making.

If the complexities of modern science made it unlikely that one might learn "by looking over the shoulder of the scientist at work," then what if one "could transport a visitor to a laboratory where significant results were being obtained at the early stages of a particular science." At some origin point in every science there would be a time and place where the "scientist's knowledge would not be much greater than that of his inquiring guest." Scientist and layman would not confront each other as immigrant and native, but as members of a family catching up on the news of seldom seen relatives.

Conant was among the key proponents of the official report on the Manhattan Project, Henry DeWolf Smyth's *Atomic Energy for Military Purposes.* Neither Smyth, the pre-war chair of the Princeton University physics department and associate director of the Manhattan Project's Metallurgical Laboratory, nor Conant could recall whose idea it was to prepare a public report or history documenting the work of the super secret project. But Conant's support for the work was essential for its completion, as was the support of the project's military leader, Leslie R. Groves. Although Groves and Conant agreed on the need for such a history, if only to explain how much money the project had spent, it was never clear how much of Smyth's report would ultimately find its way into the public domain. Ultimately, President Truman decided to release a version of the report consistent with the needs of military secrecy and censorship. Smyth's report was among the strangest press releases in military history, but most remarkable was its last page and section, "The Questions Before the People."

Military security had heretofore made it impossible for a public discussion of the issues surrounding the bomb, "a new tool for mankind" as well as a tool of "unimaginable destructive power." Although the scientists involved in the project had debated the weapon's military and diplomatic effects these questions were political and social rather than technical. While the scientists and administrators had addressed these questions as citizens of the United States it was imperative that

> in a free country like ours, such questions should be debated by the people and decisions must be made by the people through their representatives. This is one reason for the release of this report. It is a semi-technical report which it is hoped men of science in this country can use to help their fellow citizens in reaching wise decisions. The people of the country must be informed if they are to discharge their responsibilities wisely.[19]

As General Groves made clear, Smyth's report was an "administrative history of the Atomic Bomb Project and the basic scientific knowledge on which the several developments were based." Like a case study in NatSci 4, the Smyth report edited the past to make points clearer to the lay reader. Groves' censors, like Conant's historians of science, edited the past to highlight the truly important and significant. Here lay the great irony of the case studies, both in the Smyth report and in NatSci 4. Certainly the Smyth report helped the citizen to act and to understand the actions of those in the Project, but the citizen was still dependent upon the scientists and the government officials who knew not only the contents of the report, but all that remained shrouded in secrecy. In this sense, so did the students in NatSci 4 still have to depend upon the instructor to interpolate among the vast spaces in the case studies. Undergirding both studies was a faith in both the interpreter's veracity and the reader's ability to comprehend the texts in question. Understanding science required not so much an understanding of scientific practice, but the acquisition of a faith in the scientist to act correctly, even when the audience did not know all the details.

Although originating in the National Science Fund's attempt to secure private patronage for the sciences, I.B. Cohen's *Science, Servant of Man* became enmeshed in the post-war debate over the organization and funding of the sciences. The prospect of government support for scientific research transformed each taxpayer into a prospective "lay administrator of science." Following Conant, Cohen chose a case study approach that would educate the layman about the workings of science as well as the impossibility of planning fundamental research. There is much to discuss in Cohen's work, but let us focus on two related components — the justification for the use of the history of science as an aid in understanding contemporary science and the claims made about the relationship between science and war. Almost in passing, Cohen made a revealing claim:

> no one would question that the paths of scientific discovery are the same now as they were in the preceding centuries, an example from the eighteenth century may serve to illustrate a particular point as well as, if not better than, one from the twentieth century; with the obvious advantage that the materials are easier to grasp because little previous factual knowledge is required.[20]

Here was a key to the post-war historiography of science: a belief in an unchanging set of practices that might fall under the label of science. Equally important, here was a source of the discipline's great affliction: Whig history, the writing of history as if the present were the inevitable consequence of the past. During the seventies and eighties, the discipline would work to eradicate such practices as part of its own attempt to forge an identity separate from that of the sciences. Yet this was bound to prove difficult, if not impossible, since the very idea of legitimizing a present-centered account was embedded in the groundwork of the discipline's earliest writers. Progress, for both Conant and Cohen as well as Sarton, defined science as separate and separable from other parts of culture. Here was an argument for the autonomy of science that also homogenized the past so that any historical figure considered a scientist in post-war America might illuminate the workings of current science. Such an approach had the interesting consequence of making historical actors not all that different from contemporary practitioners and the latter became far less threatening.

Cohen took care not to use military examples, save for the development of radar and its apparent dependence upon inter-war ionospheric research. Acknowledging the contribution science had made to warfare, as well as the "socioeconomic fact" that war had been a "great stimulus for particular lines of scientific development and for accelerating the application of known findings of science," Cohen changed the question of the reciprocal relations between science and war into how might science bear some special responsibility for modern warfare? For Cohen, only if scientists in all countries abandoned research that might lead to new and more powerful weapons could one separate scientists and war; but such a move required a heretofore unprecedented level of international cooperation and rested upon a fundamental misunderstanding:

> Science does not, nay cannot, recognize of itself any distinction between right and wrong, and hence it cannot concern itself with moral issues. Science is never an *immoral* activity, but rather *amoral* — completely removed from the moral sphere.[21] [italics in original]

As a tool "no blame or praise should be imputed to science itself," he argued for "science acts simply as a most efficient servant." Nor was it possible to argue that science and scientists bore some special guilt in modern warfare for war was not the "responsibility of scientists any more than of bakers and candlestick makers." Science produced knowledge, but society decided how to use the product. It was a one-way street in which the autonomous world of science communicated to the world at large through people, papers, and instruments. It was also a world where the pre-war beliefs in a profound connection between science and war could no longer exist. The new political taxonomy of knowledge embedded in both the Bush report, and NatSci 4 separated basic or fundamental research from applied research or technology. The latter might possess a profound connection to warfare, but the existence of the new domains effectively circumscribed the range of contexts which might influence or affect the researcher. Paradoxically, solving the problem of providing for the support of the material foundations of science — salaries, laboratories, instruments — effectively eviscerated the possibility of anything even remotely resembling the materialist historiography of science that had developed between the wars.

For Cohen, and other supporters of the Bush proposals, the history of science demonstrated the impossibility of planning research. In the book's final chapter, "How to Get the Most Out of Science," Cohen strenuously protested against those who saw the need to continue the wartime organization and direction of the sciences in peacetime. Responding to those who argued for greater planning of research, Cohen suggested that his reader reflect on the "glorious pages of the history of science." Reflection and recollection of Cohen's text, like Conant's case studies, had an ambiguous effect. Scientists were like other citizens and their work worthy of support, but after providing the funds the lay administrator had best leave the researchers to decide among themselves how to use the once public money. What kind of servant was science? Ironically, in both Cohen and Conant, understanding science was similar to dining at a magnificent *prix fixé* restaurant where the chef chose both the food and wine without consulting the customer. One could only imagine such an asymmetrical relation between science and society by ignoring the very reason for a discussion of the planning of science — the massive government support for university-based research. Citizens, like our mystified diners, would have to pay for whatever was served. In Bush's ideal universe, as well as the world described by Cohen and Conant, science was not the servant of man; instead, researchers were accountable only to each other rather than to their public patron. Reviewers recognized both Conant's and Cohen's works as apologetics for the new political economy of science. Cohen's primer, like NatSci 4, was not for an age of science, but for an age of scientists.

Boundaries, Boundaries Everywhere!

The growth and development of the history of science in the early post-war era rested upon the demand for the knowledge historians of science contributed with respect to the larger problem of teaching non-scientists about science, rather than the intrinsic skills of the nascent discipline's practitioners. At one with such a demand was a new political economy of knowledge in post-war America, as the US military and the federal government replaced the private foundations as the dominant patrons of the sciences. Constituting the history of science as both an apologetic and as part of the larger project of general education had both institutional and intellectual consequences. How were the "heirs of Sarton" to build an academic discipline complete with graduate students and research agendas while fulfilling their perceived roles as ambassadors of science, domesticating scientific knowledge for the lay student? Where would the resources come from that would allow for the training of graduate students and their subsequent employment as professional scholars? Finally, what would be the appropriate intellectual foci for the new discipline and how would students investigate those areas? As Steven Shapin has observed, the development of the famous external-internal language (e/i) was an integral element of this early stage of discipline building.[22] Certainly such a language worked well with the desire to demarcate science from both non-science and technology. Nor were such distinctions simply academic issues. The boundary problem cut across various sectors of society and politics. In Britain, Michael Polanyi's famous conception of "tacit knowledge" was clearly at one with his attack on government planning of scientific research. Tacit or personal knowledge was by its very nature knowledge that the state could not possess. Therefore, attempts to plan science were bound to fail since non-scientist administrators would never possess all the requisite knowledge. States might fund successful science, but they could neither control nor direct its development.[23]

In the US, Kenneth Arrow and Richard Nelson of the RAND Corporation developed the idea of science as a "public good," lending further credibility to the government's reconstruction of the post-war political economy. In the language of neo-classical economics, national defense is the classic example of a public good. Each citizen, regardless of his or her beliefs, is protected by the state. No single citizen can claim all the benefits of defense expenditures for his or herself. One's neighbors will no doubt pay for their defense, so how is one to convince any given individual to pay his or her share? Or as Arrow put it:

> Thus basic research, the output of which is only used as an information input into other inventive activities, is especially unlikely to be rewarded. In fact, it is likely to be of commercial value to the firm undertaking it only if other firms are prevented from using the information obtained. But such restriction on the transmittal of information will reduce the efficiency of inventive activity in general and will therefore reduce its quantity also.[24]

Given private underinvestment in basic research, only government funding could supply the appropriate level of resources for basic research. Arrow and Nelson's work rested upon the post-war assumption that basic research was both identifiable and separate from other activities. It was known by its funding. However it remains unclear why no one pointed out the great problem with this piece of boundary work: for better or worse, basic research had become identical with national defense. Perhaps this was the result of Arrow's and Nelson's location in RAND, a US Air Force think tank.

American researchers could believe themselves separate from society because the public monies lavished upon science effectively insulated researchers from the world at large. Freed from the problems of pre-war researchers, the post-war generations simply accepted the massive public expenditure as a given rather than the remarkable accomplishment it represented. It was a generational difference manifesting itself in metaphor. Bush and Oppenheimer both used architectural or construction metaphors to describe the development of science. In his classic essay, "The Builders," Bush compared the growth of knowledge to mining a quarry to build an edifice. It was never entirely clear what the structure was, or who might live within it; and it appeared that the world consisted almost entirely of the quarry and edifice, save for Bush's acknowledgment that "the edifice is not built by quarrymen and the masons alone." Where those who provided food and water lived remained unclear, but Bush recognized that the scientist did not labor in solitude. At a 1959 symposium on basic research, J. Robert Oppenheimer compared the growth of science to a "great house, and that those who work in it lay the bricks."[25] The latter's house was built on a featureless plain, without streets, sewers, or even the rudiments of zoning. So effective were the boundaries separating science and society that the researcher need no longer acknowledge society's existence.

Remembering the Future?

Understanding science was not the same as understanding history. Arriving at Cornell as a full professor after the war, Guerlac had developed a year-long required course in the history of science for students in the Engineering School that was also open to all other undergraduates as an elective. Speaking at a Princeton conference in 1947, Guerlac acknowledged the value of using historical examples to teach the lay person about science, but he emphasized that the history of science was no substitute for a science course. More important for the young historian was the recognition that in a course such as Conant's NatSci 4, it may prove "impossible to convey the historical, as opposed to the scientific, significance of the examples chosen."[26] Yet it was the problem of educating science students about history which most exercised Guerlac and which he saw as the discipline's greatest possible contribution to the present. History of science could offer the science student nothing of technical importance; instead, the value of such a course lay in imparting a recognition

> of the spirit, the tradition and the intangible assets of the scientific enterprise. He must, however, proceed further and show how these intangible assets have always been associated with, and dependent upon, certain general attitudes and values in society. Where we do not find in certain periods of history the critical spirit, the spirit of tolerance, and

> the uncurtailed right to investigate and to differ, then we are likely to find the scientific spirit on the defensive if not wholly absent.

Making the science student aware of these intangible assets was the key to success for Guerlac since it was only through historical study that one might recognize these assets as accomplishments and learn not to take them for granted. For Conant, the lay person's failure to understand science would diminish their ability to act responsibly as citizens and patrons of the sciences; for Guerlac, the problem was that the scientist failed to understand the political culture which made his work possible. In other words, the very issues of autonomy and trust which Conant and Cohen sought to embed in their pedagogical project were the converse of the values and views Guerlac wished to embed in his students. Clearly, much of the difference lay in the different audiences for each course. Guerlac could take much of the basic technical knowledge for granted: the very basic knowledge which Conant and Cohen sought to impart. Yet Guerlac believed there was something more to the history of science. It would interest those so immersed in their technical studies that they seldom left the laboratory, and expose them to the historical point of view in which the past forever constrained the present's possibilities, often in unseen and unsuspected ways. No matter how complex social problems might appear, historians had developed tools developed to minimize these difficulties. Despite their reliance on the "treacherous instruments" of "ordinary speech," students might at least understand that the methods of "historical criticism and historical synthesis" commanded respect in the world outside the laboratory. And this, the young historian explained, "should contribute in no small measure towards bridging the great educational gulf" separating the sciences from the humanities. Here, history was no longer the servant of science, but its equal partner in understanding our most fundamental problems.

Guerlac cleverly inverted the project embedded within the Harvard agenda, but he did not oppose it. On the contrary, Guerlac and Cohen were the major academic entrepreneurs in the development of the history of science in the post-war era. Yet e/i talk performed a very different function in this period, perhaps most clearly illustrated in the famous 1955 conference on the history, philosophy, and sociology of science sponsored by the American Philosophical Society and the five-year old US National Science Foundation. Others have explained the importance of this meeting for the development of federal patronage for the history of science, support which was essential to the growth and development of the discipline and its expansion from its origins at Harvard, Cornell, and the University of Wisconsin at Madison. I.B. Cohen appropriated the distinction of pure and applied research for his still struggling discipline. Echoing his earlier work, Cohen now argued that just as the funding of science should not be dictated by the possible uses of the knowledge, so should research in the history of science not be restricted to those problems and areas that non-professional historians thought important. What was good for science was equally good for the history of science. Cohen's major claim on federal resources lay in the production of more historians of science. As interesting as Cohen's formal paper are Guerlac's notes on Cohen's presentation. Cohen began by stating that "the history of science is a specialized form of history," to which Guerlac observed there were at least two sorts:

> internal history of a special science, or portion of a science; and external history: social forces exerting an influence on science; applications of scientific ideas to other intellectual fields; applications of science to the useful arts (technology).[27]

However, what was important about this distinction was not that one was better than another, but that each variety required different skills. Internal history was done by scientists or historians with special training; external history was done by these individuals as well as scholars trained primarily in history, philosophy or another of the humanities. Here is another forgotten key to so much of the post-war historiography — e/i talk was an enlistment technique for a small and poorly defined field of study. Such talk legitimized the presence of scientists and historians in the new field while broadening the base of potential support for such work. External history was not anti-science; it was the site where the historian might offer the scientist knowledge and insights he might not otherwise acquire. For Guerlac, it was external history, what he also called "the social history of science" which held out the promise of resolving the complex questions of the relations of science and society. As he explained only a few years before Cohen's paper: "it is my contention that we who are working in the history of science have hesitated at the threshold of our greatest and most important assignment."[28] Attempting to develop rationales for federal funding of the history of science, Guerlac lauded the goals of the Harvard project, but he also maintained that the real value would lie in alleviating some of the educational problems related to what Guerlac labeled the "mass education of scientists" at both the undergraduate and graduate level. History of science could combat the "technological scholasticism" of undergraduate science education; aiding those students who would work in technical jobs yet never do research. At the graduate level, Guerlac watched as education shifted away from a traditional apprenticeship where the student learned from the professor's actions at the laboratory bench. The culture and "tradition" of science could be transmitted at such close quarters by first-rate minds, but the supply of such minds was limited. The few great minds were swamped with other duties, including university administration, industrial contracts, and "secret consulting work for the government." Concluding his comments on Cohen's paper, Guerlac wrote that

> history and philosophy of science can give some vicarious contact with first-class scientific brains at work. No substitute for the real thing, but better a second hand contact with a first-rate scientific mind in action, than a first hand contact with an overworked lesser man. Firmly believe that careful study of achievements of Galileo or Newton or Lavoisier or Faraday can contribute vastly to deepen and enrich the future research student. Give him perspective on his work and problems. Civilize him.

Those last two words mean so much. For what Guerlac clearly envisioned was a form of culture we no longer think of when we think of science. Contemporary culture does not classify scientists as intellectuals, but Guerlac saw in the history of science the possibility of transforming the mass-produced researcher into a bearer of culture.

Among the many consequences of the 1955 meeting was NSF support for the conference on the "Critical Problems in the History of Science," held in Madison in August 1957. The sixty-odd individuals in attendance included nearly everyone active in the field. Important for our purposes are the essays and commentaries on the role of the history of science in the liberal arts and the education of engineers. Guerlac addressed the latter and Dean Dorothy Stimson of Goucher College addressed the former. Their presence at this pivotal meeting underscored the pedagogical origins of the history and historiography of science in the post-war era, but it is revealing that thirty-four years later, in 1991, when a second Critical Problems conference was held in Madison there were no papers about teaching or the place of the history of science in the curriculum.[29]

Conclusion

What has been accomplished by locating the origins of the history and historiography of science in the political economy of post-war America and various pedagogical projects? And, more importantly, why the great change between the 1957 and 1991 Madison meetings on Critical Problems? Certainly, one simple answer lies in using a concept like 'professionalization.' That is, we might imagine that during this particular thirty-odd year period, the history of science became an independent discipline free of its pedagogical obligations and historical justifications. NSF patronage did not pay for the development of undergraduate studies, but it funded graduate fellowships, and massive research projects, including the Archives for the History of Quantum Physics (AHQP), the discipline's first example of 'Big History.' Interested readers may also believe that any account of the change depicted here must take into account Thomas Kuhn's famous 1962 text, *The Structure of Scientific Revolutions.* Indeed, that work has made several appearances, but attentive readers must already realize that I view it as part of Conant's legacy of boundary work, with an important caveat. Even the most cursory reading of *Structure* reveals Kuhn's concern for demarcating science from non-science as well as one paradigm from another. Too much has already been written about Kuhn and more is on its way, but recall that Conant criticized Kuhn's excessive reliance on the word "paradigm" to avoid discussing the multidimensional character of any given scientific community. For Conant, Kuhn had obliterated the very boundaries worth studying and preserving. To add insult to injury, Kuhn had forgotten that the history of science was the means to an end, not an end in and of itself. However, by 1962 historians of science could believe themselves practicing in an independent discipline; that was one significance of *Structure.*

Like US scientists, historians of science became accustomed to federal funding. If government patronage encouraged the growth of Big Science, it also encouraged the growth of bigger history. Once again, Kuhn matters, since he led what has become a model for large-scale history projects — the AHQP. Interested in their history, but lacking the time, the AHQP represented a fundamentally new relationship between historians and scientists. Researchers no longer need to write memoirs or publish their correspondence. All that was over, replaced by the historian with his tape-recorder, microfilm camera and copy machine. Kuhn and Guerlac could lament the passing of the scientist-historian, but they could only mourn its passing as the loss of another source, rather than the loss of a professional equal.[30] At the same time, the emergence of such projects provided support for the growing number of graduate students in the field. Of the three remaining principals of the AHQP — John Heilbron, Paul Forman, and Lini Allen — Heilbron and Forman went on to distinguished careers in the field. Big history could lay the foundations for even larger reputations.

Despite the vast sums spent on intelligence and espionage, no one in the West forecast the dissolution of the Soviet bloc in 1989. Those events, perhaps more than any other, have created the new, hostile climate for the history and social study of science. Trained to believe that our research is everything, we have forgotten the value of undergraduate pedagogy in the creation and maintenance of an academic field in the American university system. Brought up in the cocoon of government support, we are no different from the scientist, save that since our funding levels were never quite so large we have a smaller distance to fall. Mediocre works like Gross and Levitt's *Higher Superstition* are only symptoms, not the source of the critique of our historiographic practice. More upsetting is Gerald Holton's *Science and Anti-Science,* a text which conveniently erases its author's participation in Conant's pedagogical project. After all, part of the problem is that both Conant's and Guerlac's projects failed miserably. The lay public knows neither more nor less about science than they did in 1950; and the practicing scientist still knows little about the history of science. Here, however, coproduction reappears. For just as the conditions which made it possible for the history of science to flourish are now gone, should we not expect a new way to write and talk about science? That new way is only now developing in science and technology studies.

Assume for a moment that coproduction is the case. Assume again that the rise of the modern nation state in the early modern period is at one with the control of religious confessional practices and the rise of the new philosophy, what we have called modern science. Today, the nation state is still standing, but international organizations — the European community, NAFTA, GATT, and the ubiquitous multinational corporation — slowly transform its power. If we do accept coproduction, should we not expect a change in the historiography of science on the order of the change which Conant ushered in fifty years ago? Is our past the future? Or should we once again follow the scientists? Claiming that public understanding of science is once again the problem, leading scientists and the NSF are designing a dramatic television show to interest Americans in science.[31] Armed with a $100,000 grant, physicists like Leon Lederman propose a show with plots rooted in real-life research, but as exciting as any science fiction. As a further demonstration of the present climate's effects, the drama will take place in a private research institute, GRALE [General Research at the Leading Edge]. GRALE is also the name of the prospective series and a curious recognition that the new holy grail of science is privatization. Should historians of science assist in the development of the show, or should we propose our own series about our own work — Science Studies: The Series? The historiographic possibilities are mind-boggling.

References

My thanks to Susan Vasquez of the CIW for assistance in using these materials. Special thanks to the editors as well as John Carson, Peter Dear, Sheila Jasanoff, Henrika Kuklick, Anne C. Rose, and Robert Westman for their assistance on early drafts of this chapter.

1. George Sarton Folder, Carnegie Institution of Washington Archives, Washington, DC.
2. George Sarton, *Introduction to the History of Science*, 4 volumes. (Baltimore: Carnegie Institution of Washington, 1928), **1**: 28.
3. Thomas S. Kuhn, "The History of Science," in *International Encyclopedia of the Social Sciences*, vol. 14. (New York: Crowell Collier and Macmillan, 1968) 71–68; reprinted in *idem, The Essential Tension Selected Studies in Scientific Tradition and Change.* (Chicago: University of Chicago Press, 1977), 105–126, 110–111.
4. Arnold Thackray, "Scientific Knowledge and its Historical Problems," *Minerva* (1972), **10**: 491–495.
5. J.G. Crowther, *Fifty Years with Science.* (London: Barrie and Jenkins, 1970), 76–82; on the scientific socialists, the standard work remains Gary Werskey. *The Visible College: The Collective Biography of British Scientific Socialists of the 1930s.* (New York: Holt, Rinehart and Winston, 1978).
6. Pope quoted in Boris Hessen, "The Social and Economic Roots of Newton's 'Principia'," in *Science at the Cross Roads*, N.I. Bukharin (Ed.). (London: Frank Cass and Company, Ltd., 1931; 1971), **151**: 151–212.
7. Henry Guerlac, "Science and War in the Old Regime: The Development of Science in an Armed Society". (Ph.D. Dissertation, Harvard University, 1941), **364**.
8. See, Robert K. Merton, *Science, Technology and Society in Seventeenth Century England.* (New York: Howard Fertig, 1970), **206**: 142–3.
9. Vannevar Bush, *Science — The Endless Frontier.* (Washington, DC: GPO, 1945).
10. Daniel J. Kevles, "The National Science Foundation and the Debate over Post-war Research Policy, 1942–1945." *Isis* (1977), **68**: 5–26.
11. Richard Shryock, "American Indifference to Basic Science during the Nineteenth Century," *Archives Internationales d'Histoire des Sciences* (1948) reprinted in Shryock, *Medicine in America Historical Essays.* (Baltimore: The Johns Hopkins University Press, 1966) **87**: 71–89.
12. A. Rupert Hall, *Ballistics in the Seventeenth Century.* (Cambridge: Cambridge University Press, 1952) 161–2.
13. See 20 November 1942, Cohen to Robbins, Jewett Files, 50.1341, National Science Fund, Volume 1, "Cohen, I.B." NAS-NRC.
14. Quote from Harvard University, *General Education in a free Society: Report of the Committee.* (Cambridge, MA: Harvard University Press, 1945), 222.
15. James B. Conant, "The Advancement of Learning in the United States in the Post-War World," *Proceedings of the American Philosophical Society* (1994), **87**: 291–298.

16. James B. Conant, *On Understanding Science: An Historical Approach.* (New Haven: Yale University Press, 1947), xi.
17. See David Hollinger, "Science as a Weapon in the *Kulturämpfen* in the United States During and After World War II," *Isis* (1995), **86**: 440–455, 446.
18. Conant, *On Understanding Science*, 101–106.
19. Henry DeWolf Smyth, *Atomic Energy for Military Purposes.* (Washington; Stanford: GPO: Stanford University Press, 1945; 1989), 226.
20. I.B. Cohen, *Science, Servant of Man: A Layman's Primer for the Age of Science.* (Boston: Little, Brown and Company, 1948), 14.
21. Cohen, *Science, Servant of Man,* 287.
22. Steven Shapin, "Discipline and Bounding: The History and Sociology of Science as Seen Through the Externalism-Internalism Debate," *History of Science* (1992), **30**: 333–369.
23. See Michael Polanyi, *Science, Faith and Society.* (Chicago: University of Chicago Press, 1946); and *idem, Personal Knowledge: Towards a Post-Critical Philosophy.* (Chicago: University of Chicago Press, 1958).
24. Kenneth J. Arrow, "Economic Welfare and the Allocation of Resources for Invention," in *The Rate and Direction of Inventive Activity: Economic and Social Factors,* National Bureau of Economic Research (Ed.). (Princeton: Princeton University Press, 1962), 609–625, 618; and Richard Nelson, "The Simple Economics of Basic Research," *Journal of Political Economy* (June, 1959), 297–306.
25. Originally published in 1945, "The Builders" was reprinted in Vannevar Bush, *Science is not Enough.* (New York: William Morrow and Company, 1967), 11–13. J. Robert Oppenheimer, "The Need for New Knowledge," in *Symposium on Basic Research,* Dael Wolfle (Ed.). (Washington, DC: AAAS, 1959), **5**: 1–15.
26. Henry Guerlac, "History and the Sciences," December 1947, Henry Guerlac Papers (HGP), Cornell University, Box 26/10–11.
27. See I.B. Cohen, "Present Status and Needs of the History of Science," *Proceedings of the American Philosophical Society* (1955), **99**: no. 5, 343–347; Guerlac's commentary is found in HGP, Box 4–43, page 1.
28. This quote is taken from an undated document that I believe is from the early fifties. See Henry Guerlac, "Development and Present Prospects of the History of Science," HGP Box 26–17, 40.
29. Compare Marshall Clagett (Ed.), *Critical Problems in the History of Science.* (Madison: University of Wisconsin Press, 1969) with "Conference on Critical Problems and Research Frontiers in History of Science and History of Technology" (Madison, 1991; distributed at the meeting).
30. Thomas S. Kuhn *et al., Sources for the History of Quantum Physics.* (Philadelphia: American Philosophical Society, 1967); and Kuhn, The Turn to Recent Science." *Isis* (1967), **58**: 409–418.
31. Barbara Rosewicz. "Cops have 'NYPD Blue'; Doctors, 'ER.' Now Scientists Want a Show," *The Wall Street Journal* (20 March, 1996), B2.

Further Reading

Terrence Ball, James Farr, and Russell L. Hanson (Eds.), *Political Innovation and Conceptual Change.* (New York: Cambridge University Press, 1989).

Peter S. Buck and Barbara G. Rosenkrantz, "The worm in the core: science and general education," in *Transformation and Tradition in the Sciences: Essays in Honor of I. Bernard Cohen,* Everett Mendelsohn (Ed.). (New York: Cambridge University Press, 1984), 371–394.

I. Bernard Cohen (Ed.), *Puritanism and the Rise of Modern Science.* (New Brunswick: Rutgers University Press, 1990).

H. Floris Cohen, *The Scientific Revolution: A Historiographical Inquiry.* (Chicago: University of Chicago Press, 1994).

David Gooding, Trevor Pinch, and Simon Schaffer (Eds.), *The Uses of Experiment: Studies in the Natural Sciences.* (Cambridge: Cambridge University Press, 1989).

E.H. Gombrich, *Aby Warburg An Intellectual Biography.* (Chicago: University of Chicago Press, 1986).

Sheila Jasanoff, Gerald E. Markle, James C. Petersen, and Trevor Pinch (Eds.), *Handbook of Science and Technology Studies.* (London: Sage Publications, 1995).

Alexandre Koyré, *Metaphysics and Measurement: Essays in Scientific Revolution.* (Cambridge, Mass.: Harvard University Press, 1968.

Thomas S. Kuhn, *The Structure of Scientific Revolutions.* 2 ed. (Chicago: University of Chicago Press, 1970 [1962]).

Bruno Latour, *Science in action: how to follow scientists and engineers through society.* (Cambridge, MA: Harvard University Press, 1987).

David C. Lindberg and Robert S. Westman (Eds.), *Reappraisals of the Scientific Revolution.* (Cambridge: Cambridge University Press, 1990).

Lewis Mumford, *Technics and Civilization.* (New York: Harcourt, Brace and Co., 1934).

Peter Novick, *That Noble Dream: The "Objectivity Question" and the American Historical Profession.* (New York: Cambridge University Press, 1988).

George Sarton, *The History of Science and the New Humanism.* (Cambridge, MA: Harvard University Press, 1937).

Steven Shapin, *The Scientific Revolution.* (Chicago: University of Chicago Press, 1996).

Richard Yeo, *Defining Science: William Whewell, Natural Knowledge and Public Debate in Early Victorian Britain.* (New York: Cambridge University Press, 1994).

CHAPTER 2

What is Science?

SIMON SCHAFFER

Hero or villain, maniac or rationalist, savior or fool, the figure of the scientist much preoccupied the imaginative landscape of the twentieth century. As part of this preoccupation, efforts were made to forge an account of the sciences' essence. In the oddly desiccated version of American English which displaced Latin or German as the shared tongue of international scientific communication, the term 'science' was often used in the singular for the organized methodical investigation of nature's capacities. There seemed to be but one common form of such inquiry and a common language for its communication. It was also supposed that the same form of science could and should be pursued worldwide. Yet such images of scientific unity and universality were difficult to sustain. Most scientists were in the position of lay spectators with respect to other specialists' work. Their labors were entangled with major technological and social enterprizes. The globalization of the sciences by no means effaced local differences of style and content. New groups began to demand their part in the making and distribution of scientific knowledge. So one puzzle was to locate the sciences and define their practitioners' role. This chapter sketches some ways in which such location and definition worked.

Throughout the century it seemed important to maintain certain boundaries between pure and applied knowledge, between the sciences and society, between experts and the laity. But at the same time the permeability of these boundaries was also stressed: pure sciences were to be applied to human purposes, hinge on the conditions of social order, and thus help integrate lay and expert understanding. Defining and crossing such boundaries was a public affair, never the monopoly of cloistered specialists. Though the secluded laboratory became a dominant institution of scientific work, museums, gardens, zoos or field stations all remained vital sites for the sciences as they had been in previous epochs. While the label 'science' was rarely allowed for systems of ethics or morals, the social and human sciences were widely cultivated, the new expertize in fields such as 'sports science' or 'food science' was established. The task of situating scientists and the sciences was carried on in lectures, exhibitions and other media of exposition. They helped define the sciences' scope and meaning, their relation with everyday life, the claims of capital and the range of human purposes. Even apparently rarified issues such as the distinction between science and technology or the link between science and practical reason were closely bound up with this world of public display. Several different images of the sciences were developed during the century, and these differences can be understood through the function of these images and the places where they were made. Equally important, these were places where the principles scientists described were put to work and their efficacy demonstrated.

In schematic terms, the sciences could be seen as unified or diverse, as mundane aspects of shared human capacity or as rare, distinctive activities, as impersonal forces of modernization or as skilful forms of human labor and sociability. Among these, one dominant version claimed that the sciences shared the common-sense practices of everyday life. There was nothing special about the scientific attitude; the problems science addressed were those which presented themselves to everyone. Making their way around their world, it was argued, scientists simply observed, calculated and theorized in a manner rather similar to, if occasionally more carefully than their fellows. Many twentieth century displays reinforced this view by presenting the sciences as activities in which all could equally engage; then showing that such activities would generate a wondrous common future. These were the themes of the French physicist Jean Perrin's plans for a Palace of Discovery built in Paris during the 1937 International Exhibition. With the support of the socialist premier Léon Blum, Perrin and his colleagues helped start a modern form of state science funding and exposition "to bring to everyone's attention the progress of science and technology; to develop the scientific spirit and hence the qualities of honest criticism, of free judgment." Similar motives marked the host of science installations at the Festival of Britain staged in London in 1951 to celebrate post-war recovery and a safe nuclear future of science and technology. One contemporary socialist apostle of the common-sense of science was the eminent British embryologist C.H. Waddington: "Science is largely common-sense . . . When people discover what science is, they find it is something they have known all along." This was an old theme given new salience by twentieth-century technologies and cultures. Speaking to large audiences in wartime Britain, Waddington lectured that "the way to get results is science" and that in a scientific society based on socialism, Darwinism and engineering, social progress would be unlimited.[1] With very different political implications, these were common arguments of innumerable post-war public exhibitions in such countries as Japan (which during the 1980s, partly through its huge private overseas investment in Europe and America, became a major patron of world-wide scientific research) or Brazil (which by that decade of economic and ecological crisis had more than 250 scientists per million of population). An analyst of the new Japanese "technopolis" explained in 1986 that a "romanticized notion of the rugged individual, the lone scientist-inventor" must give way "to the more pragmatic reality of large, corporate research and development teams conducting joint research." Such accounts were promoted by state and industry to underline the worth of widely-diffused science and technology in the name of economic development.[2]

The image of a universally distributed scientific spirit became increasingly plausible as the works of the sciences permeated more aspects of society, and the everyday world began to resemble that which scientists claimed to master and describe. In an influential contribution to a utopian *Encyclopedia of Unified Science* (1938), a project which most coherently expressed the thesis of a shared language for the sciences, the American philosopher and educator John Dewey supposed that "science is an elaboration of everyday operations" and that "the home, the school, the shop, the bedside and hospital present [scientific] problems as truly as does the laboratory."[3] For several powerful scientific programs, the household as much as the hospital would become a site of science, stocked with sophisticated and robust technologies. The status of citizens as producers and consumers within national economies hinged on making these worlds into places of science. In 1930 the Austrian novelist Robert Musil, a quizzical observer of visions of unification and technocracy, asked "in an unprejudiced way how science came to have its present day aspect. This is itself important, since after all science dominates us, not even illiterates being safe from it, because they learn to live with countless things that are born of science ... One might think, and not quite without justification, that we now find ourselves in the midst of the miracle of the Antichrist."[4]

After the catastrophes of the mid-century, national investments in the sciences became an increasingly salient feature of public life, especially between 1945 and 1975, when global manufacturing output quadrupled and manufacturing trade increased tenfold. In 1959 two Soviet science writers, Mikhail Vassiliev and Sergei Guschev, published a widely-read account of life in the twenty-first century based on interviews with eminent Soviet scientists. "The present day level of science and the dynamism of its constant development allow us to make with confidence the most optimistic forecasts for the future," they were told. This optimism mixed admiration for science with a certain dislike of nature's currently feeble capacities: "the construction of the Earth can be criticized from various points of view ... We shall make a strict list of the constructional faults of our world just as technical commissions can catalogue the defects of a certain machine."[5] Soviet scientists had long been concerned with the scientific management of the "biosphere," a term invented by the mineralogist Vladimir Vernadsky early in the century. But in the midst of the so-called Green Revolution, the Arms Race and the Space Race, the sciences were then seen as proffering limitless power to the technocratic state. No doubt such hubris was characteristic of a very important account of the sciences' capacities throughout the century and carried with it obvious political connotations. By the last decade, there were about 5 million scientists and engineers in the world. That number had doubled in the previous twenty years. Scientists were no longer overwhelmingly European, as they had been at the start of the century, nor almost exclusively men, though in such nations as the United States of the 1980s barely a quarter were women. "The appeal of utopianism makes it difficult to criticize scientific research; after all, who wants to discourage our hope for a more perfect future?" remarked the American feminist and biochemist Ruth Doell in 1991. While the tone of voice typified by Vassiliev and Guschev became distinctly less assured, an age of science for all mediated by universally accessible information networks was still widely prophesied.[6]

The globalization of the sciences became especially marked in the twentieth century. At its start, it was still possible to pretend that scientific activity was properly European, both in personnel and institutions. Indeed, many Europeans justified their own sense of cultural (and racial) superiority by appeal to the apparent success of their own sciences and technologies. Some modern commentators continued to describe a uniquely scientific method which, despite resistance and incomprehension, had spread worldwide because it was the superior manner of establishing reliable knowledge. The imperial entrenchment of European institutions and economies elsewhere in the world promoted the spread of sciences and of their claims to worldwide efficacy. Globalization had paradoxical effects. As the numbers of scientists rapidly increased and the age of European science ended, so appeals to a specific ethnocentric factor on whose force the sciences depended lost plausibility. Such developments therefore helped the search for a distinct account of the sciences, which, rather than seeing them as a form of widely-distributed common sense, instead made them the finest achievements of humanity, uniquely deserving of society's support and of generating solutions to society's needs.

According to many twentieth-century advocates, the term 'scientific' was to be applied to the knowledge and practices of specialist experts. Other forms of knowing and acting were labeled vulgar, folk or primitive. Scientific method picked out statements which were rational, testable and legitimate from those which lacked meaning or reason. Some philosophers marked the extraordinary (according to some, fatal) distinction between the shared life-world of ordinary experience and the abstracted and rational universe of the natural sciences. The scientists who plied their trade in such a universe were thus rather *unlike* their fellows not merely or even mainly because of their training, though that mattered a great deal. During the twentieth century, the professionalization of the sciences was completed. Systematic training, typically culminating with a doctoral degree awarded by a recognized establishment of further education, was an almost universal prerequisite for full-time participation in the world scientific community. But amateur groups watching birds, recording the weather, hunting comets or collecting fossils all grew in numbers and activity. In the years after World War II, for example, Japanese geologists mobilized huge numbers of amateur participants in geological mapping and cognate tasks (a program known by its acronym *Chidanken*). It has been argued that these egalitarian initiatives provided precedents for similarly large-scale citizens' movements monitoring pollution in Japan in the 1970s.[7] The mass media encouraged such enterprizes in natural history, domestic technology or various forms of medical self-help (though it was remarked that they also propagated practices dismissed by scientists as superstitious hokum).

Sciences' protagonists referred to a general public whose endorsement they required and whose apparent ignorance they sought to correct or challenge. The British writer H.G. Wells, one of the century's more influential promoters of the sciences' public image, complained in 1908 that "science stands, a too-competent servant behind her wrangling under-bred masters, holding out resources, devices and remedies they are too stupid to use."[8] At least as much was made of the specific moral features of the scientific community, its intolerance of fraud, bias or vice, its unique methods for securing reliable and effective knowledge, and its capacity to settle matters of dispute and interest. Many emphasized the sciences' difference from any other means humans had devised for finding out about their world. Thus the French sociologist Emile Durkheim conceded in 1912 that "the essential ideas of scientific logic are of . . . social origin," yet he reckoned that "science purges them of all accidental elements. It brings a spirit of criticism into all its doings, which religion ignores, it surrounds itself with precautions to escape precipitation and bias and to hold aside all the passions, prejudices and all subjective influences."[9] Such views implied that the institutions of the sciences had, or at least should have, the function of preserving scientific activity from such prejudices and influences. Under the aegis of the German sociologist Max Weber and his heirs, value-freedom became a critical theme in twentieth-century assessment of the sciences.

Just as it was common in this alternative version of the sciences to argue that their methods and ontologies were unique rather than common, much was made of the difference between science, a search for knowledge, and technology, the search for control. The German physicist Emil Warburg, powerful boss in the early years of the century of the Imperial Physical-Technical Institute, held that "scientific work is occasioned and conducted by excluding every practical secondary aim. Natural science is motivated solely by the desire implanted in the human mind spurring us to find connections between natural phenomena." Scientific purity was insulated from, though it might generate, technical utility. This difference was rather hard to sustain. The century saw the emergence of unprecedented institutions without which the sciences could scarcely function and which marked them out as distinct yet hitched them to technological ends. Warburg's Physical-Technical Institute was a good example of such an organization.[10] The theoreticians, who now began to appear as distinct groups in their own right, also found it important to build their own institutions. In European physics, for example, specialist sites for theory were rare at the century's start, and when the Danish physicist Niels Bohr set up an Institute for Theoretical Physics in 1917, he readily acknowledged that "we could perhaps much better have called in an institute for atomic physics."[11] Atomic physics was an important field in which the boundaries of theory and practice were hard to define.

In the Manhattan project of the early 1940s, in order to produce a weapon of mass destruction, the United States government employed more than a quarter of a million workers in a new complex of scientists, engineers, managers and military men. "There disappeared forever," according to the Soviet physicist Peter Kapitza, "the happy days of free scientific work which gave us such delight in our youth. Science has lost her freedom. Science has become a productive force. She has become rich but she had become enslaved." The gendered language was significant — the bombs themselves were nicknamed "babies" by some of their makers.[12] Though the scale and investment were rarely matched afterwards, the form of institution rapidly became typical of most large scientific and technical enterprises, and, in turn, began to dominate the world of scientific work. By the century's end, the model was used to organize and distribute research in genetics (as, for example, the Human Genome Project), in work in space, the oceans, in large-scale collaborative projects in nuclear physics, epidemiology and surveys of world biodiversity. The term 'engineering' became pervasive in scientific research. Molecular biologists worked in 'genetic engineering,' and, for example, in the 1960s the distinguished theoretician at the European Center for Nuclear Research, John Bell, described himself as a "quantum engineer."[13]

Both accounts of the sciences — whether commonplace or exceptional — were explicitly designed to account for the sciences' remarkable efficacy. The very scale of the promises made by twentieth-century sciences, and the effects of their pursuit, occasionally generated a rather apocalyptic tone. Throughout the century, several eminent authorities proclaimed the imminent attainment of a grand unified theory, an end of the sciences, especially in physics. The best-selling *Brief History of Time* by the British mathematical physicist Stephen Hawking, soon turned into a movie and accompanied by mass marketing campaigns, ended with exactly such a vision. Hawking reckoned "there are grounds for cautious optimism that we may now be near the end of the search for the ultimate laws of nature" and such a theory "would bring to an end a long and glorious chapter in the history of humanity's intellectual struggle to understand the universe. But it would also revolutionize the ordinary person's understanding of the laws that govern the universe."[14] Such frontier talk became commonplace at the end of the century. Academic philosophers tracking the sciences reckoned their chief task was to find a criterion which demarcated the natural sciences from other forms of knowledge and belief and, especially during the Cold War and its aftermath, a way of showing what forces were properly internal and which external to the sciences' culture and change.

Often the costs — social, environmental, financial, moral — of transgressing such frontiers determined the directions and limits of scientific life. This was apparent in large-scale scientific projects such as programs to develop larger particle accelerators, weapons systems, agrochemical technologies or manned space flight. The price of such ventures sometimes looked too high and backers of Big Science cast around for better arguments to promote their projects and for the enemies who were allegedly undermining confidence in scientific enterprize. Following a notorious controversy in the late 1950s between the public scientist C.P. Snow and the literary critic F.R. Leavis, a simple-minded distinction between "two cultures", scientific and humanist, was used to make sense of some of these concerns. Similar issues were raised in the modish debates on political ecology and the 'end of history' which imagined the dire consequences of the disenchantment of the world and the growing determinism of purely technical and scientific reason. When he reflected on the twentieth century mobilization of the sciences by the politics of nuclear war, environmental decay and the military-industrial complex, the French philosopher Michel Serres judged that "science has left history behind . . . It is utterly in the grip of the death instinct."[15] The collapse, in the 1980s, of a Soviet regime which had long based its legitimacy on the science of history, and the strengthened hegemony of a globally capitalist system which based its legitimacy on the science of intelligent productive machines, seemed perversely to inspire this bleak mood. Protagonists of the sciences therefore often held that the value of the sciences was under attack — whether from those who complained that the sciences had been absorbed by warlike and economic interests, or from those who sought to make this absorption even more effective.

For many reasons, therefore, the *definition* of the sciences and the *mapping* of their frontiers were important components of scientific activity. This work of definition and mapping took place in public forums rather than within places of scientific work. Answers to the question 'what is science?' were more often provided by powerful institutions devoted to the analysis, planning, definition and public display of the sciences' scope. The contributors to this book belong to such institutions. Throughout the century, exhibitions and other public displays were crucial sites where public interests worked out their account of the sciences. For example, the major industrial, commercial, imperial and scientific exhibitions of the later nineteenth and early twentieth centuries helped define the term 'scientific instrument' as the apparatus required for experimental work and instruction, and the concept of 'human evolution' as an account of global history. Sometimes such manifestations helped define the scope of the sciences; sometimes they were sites where conflicts over that scope and authority were conducted. The Social and Economic Museum of Vienna, directed in the 1920s by the logical positivist Otto Neurath, founder of the *Encyclopedia of Unified Science*, had as its goal the construction and dissemination of an image of unified science in an avowedly socialist struggle against metaphysics and for popular unity. Neurath's enemy was defined as modish philosophical idealism and irrational mystification. The establishment of acausal quantum mechanics in the German-speaking lands during the

1920s may, in part, have been a response to these forms of idealism and mysticism. Public fights about the right image of science had their effects on the pathways groups of scientists chose for their research.[16]

These fights also took place in the natural history museums of the period where the scientific investigation of subject peoples and of other species fitted into a wider picture of imperial and national development. Henry Fairfield Osborne, president of the American Museum of Natural History, explained in 1924 that the function of his stage-managed shows of human evolution was to "restore the vision and inspiration of Nature as well as the compelling force of the struggle for existence in education." Osborne challenged the rights of laboratory scientists to speak about evolutionary processes. His imagery turned scientists into warlike frontiersmen, uniquely privileged interpreters of past and current states of nature. His enemies included the migrant alien poor, radicals who denied the principles of human evolution and scientific racism, and religious fundamentalists alike. In these places the very meaning and place of sciences such as paleontology, economics, evolutionary biology or physics were at stake. In Neurath's Vienna an apparently unified common-sensical science was a socializing resource for making utopia in the workers' city in place of immiseration and urban struggle. In Osborne's New York a self-styled science of specialist experts was a naturalizing resource for escape from the dystopian alien city to a patriarchal frontier wilderness.[17]

Such models of sciences were typically based on some kind of ideal social order such as the market, when the virtues of scientific competition and reward were to be stressed; the agora, when scientists were to be seen as engaged in free and democratic self-expression; or the workshop, when the practical skills of scientists were foregrounded. Exhibitions then organized and classified these social ideals for persuasive purposes. Just as freak shows had helped sustain the public view of nature's inexhaustibly strange capacities, so the display of science-based commodities proffered utopian visions of a mechanized and transformed scientific future. The more than fifty million visitors to the Paris exhibition of 1900 were inundated with the visible marvels of contemporary sciences and technologies, machines performing as much as possible in public. Its international culmination was exemplified in the modern movement of the architect-engineer Le Corbusier, who in the 1920s foresaw scientifically-planned city streets "as well equipped as a factory," "a machine for traffic" in which "cafés and the places of recreation will no longer be the fungus that eats up the streets of Paris."[18] This vision of the scientific world survived to the century's end. In Paris, Perrin's Palace of Discovery was eventually succeeded by a vast City of Science and Industry on the site of a disused abattoir, with huge cinema screens, computer installations and high-tech showmanship. Yet there was also another quite different mode of public science, which instead stressed the intimate, mundane and domestic aspects of scientists' lives — the sciences as part of café society. During the 1990s a team of science curators designed a "historical walk" around a district of central Paris through a sequence of institutions of teaching and research, including the Curie Institute, laboratories of agronomy, physics, chemistry and oceanography, evoking the patronage, instruction and sociability on which, it was argued, the sciences depended. "Hospital workers, researchers, engineers and technicians have all taken part in the scientific life," whether that life involved ethnography, cancer research, geography or mathematics. The tour showed the artisan quality of scientific labour and retold anecdotes of the intimacy of scientists, literati and artists. Instead of displaying the sciences as asocial, inhuman forces of production, the sciences resembled a network of intricate workshops and studios, bohemian rather than technocratic, staffed by "strong personalities" and their support staffs, closely integrated with "our modern society."[19]

Integration of modern society with the worlds of science allowed the establishment of a number of public versions of the sciences; it also helped sustain the efficacy of the large-scale claims which scientists made. Scientists and their allies made settings where the principles they described were at work. Public exhibitions were important examples of these settings. For example, the early twentiet-century public shows were a crucial site for the marketing of electric lighting systems. Incandescent and fluorescent light became part of the new natural order, along with what its advocates called "a new Science of Seeing" which "presented a solid front to the public."[20] Twentieth-century physics, for example, referred to phenomena apparent principally in the machine-complexes put on show in the displays of the sciences. One performer at the Paris Exhibition of 1900 was the mathematician Henri Poincaré, who started his career as an engineer from the School of Mines managing the railway network of northern France, but then became a preeminent Sorbonne professor and member of the Parisian elite. In 1900 he delivered a celebrated lecture at the physics conference which accompanied the International Exposition. "Does our ether actually exist?" he asked. In experiments involving spinning toothed wheels to detect the relative displacement of the Earth in the ether, Poincaré reckoned "we believe we can touch the ether with our fingers." He kept on using the public platforms of turn-of-the-century expositions to hammer home these messages about the reality of the constructs of physics in the workings of machines. Poincaré lectured at the St Louis International Congress of 1904 attacking the possibility of an absolute etherial reference frame against which velocities could be measured. Between the St Louis exhibition and the following summer of 1905, he began to argue that "there is no absolute time; to say that two durations are equal in an assertion which has by itself no meaning and which can acquire one only by convention." The local use of machines such as tuning forks and regulated clocks were the sole basis of time measurement, and founded what Poincaré reckoned would be "a new mechanics, of which we can only catch a glimpse, in which the velocity of light would become an unpassable limit."[21] Poincaré's lectures helped define the aims and structure of the exact sciences and their temporal technologies. Between 1900 and 1905 Albert Einstein closely studied Poincaré's lectures, got a job in 1902 as patent office technical expert in Bern, and produced a series of eight papers on the energy changes in molecular physics and light emission and absorption. The ninth of these papers, submitted in June 1905, was entitled "On the electrodynamics of moving bodies." Many historians have pointed out that the mechanisms of these papers rely on familiar machinery of physics and technology in 1900. As the technology historian Thomas Hughes has put it, "for Einstein a hard-and-fast line between technology and science did not exist."[22] Einstein's and Poincaré's technology was the complex of 1900: trains, gyroscopes, measuring rods, and electric clocks.

Just as twentieth-century physics referred to the behavior of the machinery with which it stocked its world, so other major twentieth-century sciences helped change the world to make their principles effective. This was the aim of another performer at the 1900 exhibition, the American engineer Frederick Winslow Taylor. What he put on show in Paris was a method for cutting metal, the best possible way of making the kind of toothed gears like those used in the new physics experiments described by Poincaré. The point of this show, a sensation in both America and France, was to put on stage the secrets of what Taylor called "scientific management," a rational method for maximising productivity through the disciplinary management of the workforce. "Our method is scientific," Taylor's followers explained, "because it determines exactly — scientifically — the length of time in which a man can do a piece of work." The system involved the decomposition of a task into unit jobs, the measurement of the minimum time required for this unit through chronometric vigilance (Taylor set up a firm to manufacture stopwatches for distribution to all factories), and a calculation of wage rates fitted to these time measures. Taylorism embodied scientific control over work-rate in the new factory: planned routing, systematic inspection, printed instructions, cost-accounting, inventory control and functional foremanship, in which the stopwatch was the key device for turning the whole works into one unified timed machine. As the historian Siegfried Giedion explained, "Taylor was a specialist of the 1900 type: he conceives the object of his research — the factory — as a closed organism, as a goal in itself."[23]

These specialists helped produced controlled systems with strategies which helped simultaneously define the meaning of the sciences and the fields in which they would be applied. The sciences were understood as indispensable conditions of economic, military and moral success. The exhibition complex which dominated the public presentation of twentieth-century sciences both traded on, and reinforced, this modernist sensibility. The huge Century of Progress Exposition at Chicago in 1933, with a Hall of Science (containing exhibits on mathematics, physics, biology and chemistry) at its center, chose as its official slogan "Science Finds — Industry Applies — Man Conforms." In the significantly named "Hall of Social Sciences," it was claimed that "you see the reasons for the prices of things, the cost of making and the profit. Moving lights show you the governments to which your money goes and the estimated percentage of it actually returned to you." Contemporary Chicago experts were at work making these quantitative estimates into the basis of a new science of society.[24] This was the decade which saw the word 'scientism' in use (in French first) as a label for the overextensive application of the sciences and an excessive regard for their power. By then, the most sustained example of science presentation was that of the Deutsches Museum in Munich, planned from 1911 but not completed until 1925. The Munich museum directly inspired a comparable institution in Chicago, and in many other cities worldwide. Under the direction of the electrical engineer Oskar von Mueller, the Deutsches Museum helped define science as a national and productive force, entwined with the aims of industry and generating high-tech solutions for a range of economic, social and political problems. Support was won from such major national-industrial concerns as Zeiss and Siemens. Both the apparent purity of the sciences' search for universal natural laws and the intimate linkage with military and commercial advance were thus on show and became part of the meaning of the sciences themselves in the catastrophic decades around World War II.

The mid-century mobilization of the scientists made it necessary to complement the image of the '1900 specialist,' who would collaborate in the production of insulated systems where scientific and technical work could be confidently applied and assessed, with new accounts of the social significance and life of the sciences. In 1942 the American sociologist Robert Merton significantly argued that this mobilization meant that scientists could no longer "consider science as a self-validating enterprise which was in society but not of it." So he formulated an account of the scientific ethos to show that the irrationalism and centralism of "modern totalitarian society" made it incompatible with scientific activity.[25] Merton's work inspired much functionalist analysis of the social institutions of the sciences. Definitions of scientific life acquired specific political charges during the Cold War and its tortuous ideological career. Scientists became important lobbyists and villain-heroes of industrial culture. Disciplines such as science policy and science history burgeoned, directed at the analysis and dissemination of the scientific principles of this culture. One increasingly fashionable model of the sciences linked scientific work with ingenious craft skill, thus effectively insisting on (or wishing for) the sciences' comparative distance from ideological struggle, large technological systems and political planning. Several anti-Communist polemicists agreed that the sciences must be unplanned and depended on the intrinsic spontaneity of individual skill. This was also the image of physics which appeared in the highly influential lectures delivered by the Caltech physicist Richard Feynman in 1962. The search for natural laws, the genial Feynman argued, was like the individual interpretation of moves in a chess game. In his exactly contemporary contribution to a late volume of *Encyclopedia of Unified Science*, the American physics historian T.S. Kuhn characterized normal science as a complex of local puzzle-solving and practical techniques. Like many other analysts of professional sciences in the 1950s and 1960s, Kuhn stressed the training of scientists in specialist subcultures which generated their own standards of behavior, capacity to handle the right apparatus, and accounts of urgent questions and the appropriate means (and persons) to solve them. This was an influential picture of a self-validating scientific life which successfully mastered the world it helped produce.[26] The appearance in 1968 of James Watson's autobiographical account of the discovery of the structure of DNA at Cambridge in the early 1950s, a parable designed to show the ferociously competitive and risky course of scientific work, did much to reinforce such a vision of everyday science as playful, puzzling and personal. At the start of this widely-read story of the origin of one of the century's most important new industrial and medical technologies, Watson explained that "science seldom proceeds in the straightforward logical manner imagined by outsiders. Its steps forward (and sometimes backward) are often very human events in which personalities and cultural traditions play major roles."[27]

In promoting such an account of the sciences as ingenious play, the workshop and the studio were used to counter the factory and the office-block. Crucial work to achieve this end was pursued by the Communist, nuclear physicist and science educator Frank Oppenheimer, veteran of the Manhattan project and inspirational founder of the San Francisco Exploratorium in 1969. This new-style science center set out to break with the standard model of science shows and thus to rework the definition of science. It found its home in the Palace of Fine Arts building left after the Panama Pacific International Exposition of 1915. Hands-on science centers, inspired by the San Francisco model, soon burgeoned worldwide. They gained considerable support from governments keen to cultivate higher levels of scientific literacy among their working populations and from industrial sponsors who saw science centers as useful shop windows for high-tech installations linking the sciences with profitability. For Oppenheimer, however, the sciences were to be understood rather as systems of informal skill, appealing most to images of free play and exploration. He remained strongly opposed to the installation of sequences of computer screens which, he reckoned, concealed rather than revealed the genuine processes of labor on which the sciences depended. He argued that science shows should break the image of scientific technologies as incomprehensible pieces of wizardry whose inner workings were opaque and which reduced their users to the status of automata. Initial installations included arrays of cybernetic devices and focused most on visual illusion and the puzzles of perception. In 1981, the Exploratorium began planning electricity displays designed to render familiar to their visitors the science and technology of energy systems. "Everyone agreed," according to the Exploratorium's historian, "that unless people are made to feel at ease with electrical phenomena rather than frightened by them, they will not discover the astonishing unity of nature that the study of electricity can reveal." The aim was to move from reassurance through play to an account of how scientific understanding mobilized industrial success.[28]

Characteristically, in the Exploratorium the task of defining the sciences for their public thus concentrated on seeking metaphors which would lucidly make esoteric techniques more evident and comforting. At the century's end, this was by no means the only way in which such a scientific and technological complex as the electrical power and light system could be put on show. Differences in display indicated differences in the definition of the sciences, their value and public meaning. Visitor centers were set up in many scientific institutions; the figure of the public scientist reoccupied the stages of political and cultural debate. In 1989 the Museum of Civilization in Québec City staged a major show on Electricity, sponsored by the state-owned conglomerate Hydro Québec. Ingenious displays of the science's past heroes, Gilbert, Franklin and Coulomb, of twentieth-century lasers and electronics, were juxtaposed with washing machines and refrigerators. The Museum and its sponsor also addressed a pre-eminent site of public scientific and technological interest at that time, the huge new hydroelectric station and dam at James Bay in the far north of the province. In 1971 the Québec government had announced its plans for this installation, at once facing an unprecedented campaign for ecological surveillance of the effects of the industrialisation of the Bay. James Bay was to be designated "a vast natural laboratory." So the following year the James Bay Energy Company set up an environmental service employing almost one hundred scientists on permanent or temporary secondment in fields such as aquatic ecology, zoology and climatology. Historians have identified this moment as decisive for the emergence of ecological science as a field of research. During the same years the Québec government and its hydroelectric firms also encountered active resistance and protest from the Cree, the indigenous population. In 1973 a court ordered the temporary cessation of all work in James Bay while negotiations were conducted with the Cree. These protests were mentioned at the Electricity exhibition fifteen years later, but then quickly subsumed within more optimistic assessments of the environmental, political and economic impact of the hydro schemes in James Bay and elsewhere. Scientific techniques were used to estimate the future energy needs of Québec, to plan the restocking and reafforestation of the James Bay region, and to assess the impact of the dam on the area's inhabitants. Questions of the ownership of scientific knowledge and the legitimacy of rival forms of expertise and experience were essential to these programs. In the context of widespread campaigns for Québec secession from Canada, resisted by the Cree and asserted by the Québecois, such stories about energy supply and the scientific and technical prowess of francophone culture carried a major political message.[29] The extent of the properly scientific elements in such projects was highly controversial. In such shows as that at Québec the scope of such sciences as electricity, economics, ecology and anthropology were all redefined then reinforced for the public and for scientists in national and commercially sponsored displays. Other voices and other politics contested all these definitions and made themselves increasingly audible and significant.

If nineteenth-century sciences were typified by the institution of the Museum, those of the twentieth century were similarly dominated by the Laboratory. In the manifold transitions between museums and laboratories, the twentieth century produced at least two distinctive complexes. In one such system, the important theme was the simultaneous definition of the figure of the scientist-genius, disembodied and objective, and of the immense intellectual and economic consequences of the exploitation of these heroes' works. So in such exemplary shows as Paris 1900 and in the Deutsches Museum and its cognate institutions the public saw the great dynasties of intellectual science and the machines whose workings hinged on the scientific laws described by these intellectuals. The theories of knowledge which accompanied such displays stressed the unity of the sciences, their capacity to express nature's and culture's behavior in elegant and precise principles and thus the legitimacy of extending natural scientific models to all realms of action. The juxtaposition of scientific management and postclassical physics seemed highly significant. The physical sciences acquired their provisionally high status through this intriguing mix of intellectual purity and economic and military prowess.

A second system of display made scientific work look like other forms of craft and showed the practical ingenuity and artisanal elegance of scientists' labors. The public was introduced to the world of the sciences by participating in seductive games and provocative simulations. This system's officially sanctioned theories made the sciences look like a part of everyday life, drawing attention to the permeation of that life by sciences' products. While in San Francisco, for example, the museum became a laboratory, at Québec it was envisaged that entire tracts of indigenous culture and nature would be best understood as such a laboratory. In such systems choices between scientific and political strategies were represented as choices between consumer lifestyles and, in the century's most important eschatology, styles of extinction. There, for example, the techniques of biomedicine and of ecology found much of their legitimacy. In public discourse the sciences could look like the source of the crises to which they also offered solutions. One of the few utopias to serve the twentieth century was that of the sciences. As the century ended it was thus more important than ever to understand where the vision of the sciences was made and how it was to be directed.

References

1. Jean Perrin in V. Danilov. *Science and Technology Centres.* (Cambridge, MA.: MIT Press, 1982), 30; C.H. Waddington, *The Scientific Attitude.* (Harmondsworth: Penguin, 1941), 98.
2. Sheridan Tatsumo. *The Technopolis Strategy: Japan, High Technology and the Control of the Twenty-first Century.* (New York: Brady, 1986), 42.
3. John Dewey. 'Unity of Science as a Social Problem', in *International Encyclopedia of Unified Science,* Otto Neurath, Rudolf Carnap and Charles Morris (eds.), volume 1, numbers 1–5. (Chicago: University of Chicago Press, 1955), 31.
4. Robert Musil. *The Man Without Qualities,* 3 vols. (London: Picador, 1979), volume 1, 359–60 (translation altered).
5. M. Vassiliev and S. Guschev. *Life in the Twenty-first Century.* (London: Penguin, 1961), 12, p. 185.

6. Ruth Doell. "Whose research is this? Values and Biology", in Joan Hartman and Ellen Messer-Davidow (eds.), *(En)Gendering Knowledge.* (Knoxville: University of Tennessee, 1991), 121–39, p. 123.
7. Shigeru Nakayama. *Science, Technology and Society in Postwar Japan.* (London: Kegan Paul, 1991), 23–26.
8. H.G. Wells. *A Modern Utopia.* (London: Unwin, 1926), 92.
9. Emile Durkheim. *The Elementary Forms of Religious Life* (London: Unwin, 1915), 429.
10. David Cahan. *An Institute for an Empire: the Physikalisch-Technische Reichsanstalt 1871–1918.* (Cambridge: Cambridge University Press, 1989), 181.
11. Christa Jungnickel and Russell McCormmach. *Intellectual Mastery of Nature: Theoretical Physics from Ohm to Einstein.* (Chicago: Chicago University Press, 1986), 2 volumes, 2: 159; Abraham Pais, *Niels Bohr's Times in Physics, Philosophy and Polity.* (Oxford: Oxford University Press, 1991), 168.
12. Mike Hales. *Science or Society? The Politics of the Work of Scientists.* (London: Pan Books, 1982), 70; Brian Easlea, *Fathering the Unthinkable.* (London: Pluto, 1983), 95–6.
13. Jeremy Berstein. *Quantum Profiles.* (Princeton: Princeton University Press, 1991), 12.
14. Steven Hawking. *A Brief History of Time.* (London: Bantam, 1988), 156, 167.
15. Lutz Nietzhammer. *Posthistoire: Has History come to an End?.* (London: Verso, 1992), 48.
16. Jordi Cat, Nancy Cartwright and Hasok Chang. 'Otto Neurath: Politics and the Unity of Science', in Peter Galison and David Stump (eds.), *The Disunity of Science.* (Stanford: Stanford University Press, 1996), 347–69, p. 367; Paul Forman. "Weimar culture, causality and quantum theory 1918–1927", *Historical studies in the physical sciences* (1971), **3**: 1–115.
17. Donna Haraway. *Primate Visions: Gender, Race and Nature in the World of Modern Science.* (New York: Routledge, 1989), 26–58; Timothy Lenoir and Cheryl Lynn Ross. 'The Naturalized History Museum,' in Galison and Stump, *Disunity of Science*, 370–97, p. 394.
18. Marshall Berman. *All that is solid melts into air: the experience of modernity.* (London: Verso, 1983), 167.
19. Jacqueline Eidelman *et al. Parcours culturels à travers la recherche scientifique sur la Montagne Sainte Geneviève.* (Paris: Musée Curie, 1995).
20. Wiebe Bijker. "The Social Construction of Fluorescent Lighting," in Wiebe Bijker and John Law (eds.), *Shaping Technology/Building Society.* (Cambridge, MA.: MIT Press, 1992), 75–1002, p. 88.
21. Henri Poincaré. *Science and Hypothesis.* (London: Scott, 1905), 169–82; Abraham Pais. *Subtle is the Lord.* (Oxford: Oxford University Press, 1982), 126–30.
22. Thomas Hughes. 'Einstein, Inventors and Invention', *Science in Context* (1993), **6**: 25–42.
23. Hugh Aitken. *Taylorism at Watertown Arsenal.* (Cambridge, MA.: Harvard, 1960), 19–29; Siegfried Giedion. *Mechanization Takes Command.* (New York: Norton, 1969), 96–101; Yves Cohen. 'Le XXe siècle commence en 1900' *Alliage* (1994/5), **21**: 88–104.
24. Folke Kihlstedt. 'Utopia Realized: the World's Fairs of the 1930s', in Joseph Conn (ed.), *Imagining Tomorrow.* (Cambridge, MA.: MIT Press, 1986), 97–118; Burton Benedict. *The Anthropology of World's Fairs.* (Berkeley: Scolar, 1983), 34–5.
25. Robert K. Merton. *The Sociology of Science.* (Chicago: Chicago University Press, 1973), 267–78.
26. Richard P. Feynman, Robert B. Leighton and Matthew Sands. *The Feynman Lectures on Physics.* (Reading: Addison Wesley, 1963), chapter 2: Thomas B. Kuhn. *The Structure of Scientific Revolutions.* (Chicago: Chicago University Press, 1962).
27. James Watson. *The Double Helix.* (New York: Signet, 1968), ix; Edward Yoxen, 'Speaking out about competition: an essay on *The Double Helix* as popularization,' in T. Shinn and R.D. Whitley (eds.), *Expository Science.* (Dordrecht: Reidel, 1985), 162–81.
28. Hilde Hein. *The Exploratorium: the Museum as Laboratory.* (Washington: Smithsonian, 1990), 115.
29. Danielle Ouellet, 'L' hydro-électricité au Québec', in *Electrique.* (Quebec: Musée de la Civilisation, 1988), 40–75; Luc Chartrand, Raymond Duchesne and Yves Gingras, *Histoire des sciences au Québec.* (Montréal: Boréal, 1987), 336–8.

Further Reading

Barry Barnes. *About Science.* (Oxford: Blackwell, 1985)
David Harvey. *The Condition of Postmodernity.* (Oxford: Blackwell, 1989)
Roslynn Haynes. *From Faust to Strangelove.* (Baltimore: Johns Hopkins, 1994)
Armand Mattelart. *L'Invention de la Communication.* (Paris: Découverte, 1994)
Jerry Ravetz. *Scientific Knowledge and its Social Problems.* (Oxford: Oxford University Press, 1971).

CHAPTER 3

On Seeing Brockenspectres

Sex and Gender in Twentieth-Century Science

W. FAULKNER AND E.A. KERR

From the outset, the founding fathers of modern science in Europe excluded women from their powerful 'new' philosophy and, even after all legal barriers have been removed, women constitute only a small minority in all but a few of the life sciences. Those that have won sufficient prominence to warrant mention in the chapters of this book, for example, are notable exceptions. Feminists have naturally been keen to increase access for women into science — from the 'first wave' feminists who battled to gain for women the right to university education and the right to vote in the late nineteenth and early twentieth centuries, to the 'second wave' feminists of the women's liberation movement which gained ground from the late 1960s onwards.

Significantly, this latter generation of feminists have not only been concerned with issues of access to science; they have also mounted powerful critiques of science itself. As philosopher of science Sandra Harding notes, feminists of diverse persuasions have shifted their attention from the 'woman question' in science to the 'science question' in feminism.[1] From the 1970s onwards, feminist scientists and activists pointed to numerous ways in which scientific evidence and theories, especially in the biological sciences, are 'biased' in favor of men — thus challenging conventional assumptions about science being objective. In the early 1980s, ecofeminists and others characterized the very notion of objectivity, and the scientific project more broadly, as somehow essentially masculine — raising the possibility that women may have good reason to opt out of science. Later in the 1980s, feminist philosophers and scientists, rather than rejecting science and objectivity altogether, have advocated alternative 'feminist epistemologies' of science as part of a program to transform the very practice of science.

Our main focus in this chapter is on the diverse feminist critiques of science which have been developed in the latter decades of the twentieth century. Accordingly, we are concerned not so much with the position of women in science, as with the implications of science for women — or, more accurately, for gender relations — and with the implications of the feminist challenge for science. As we will show, a central theme in the feminist critiques and epistemologies is the relationship between sex and gender, in particular between the male domination of science on the one hand and its apparent masculinity on the other. For us this raises four crucial questions: In what sense is science a masculine endeavor? How much is the masculinity of science due to the fact that it is dominated by men? Or, is science male dominated because of its masculine image and culture, in a self-reinforcing way? Finally, if there were more women in science, would this in any way alter either the way science is practiced or the knowledge produced?

The body of the chapter opens, in section 2, with examples of 'sexist science' where scientific knowledge both reflects and promotes the interests of men to the detriment of women. Feminists have debated whether these examples can be dismissed as cases of 'bad science,' to be remedied by 'better' methodologies and the like, or whether the bias evident in them is an inevitable function of 'science as usual.' This debate is reflected in the 'radical science' critique and the 'new' sociology of science which emerged in the 1960s and 1970s. In section 3, we draw two important conclusions from these critical perspectives: first, that all knowledge claims (in science as elsewhere) are 'relative' to their claimants' social context and thus subjective; and second, that claims to objectivity may be rhetorically important myths, even if they bear little resemblance to the practice of science. The dilemma which we and many feminists face is the desire to maintain this critical stance on objectivity and science, whilst at the same time finding criteria with which to make choices between competing knowledge claims. We therefore consider alternative practices in science which attempt to combine elements of objectivity with an insistence that scientists be reflexive about their 'situatedness.'

Other feminists have taken a rather different tack to this integrating approach, suggesting that science — and especially objectivity — are in some crucial and dangerous way masculine, or that a 'feminine' approach to science is both possible and desirable. In section 4, we argue that the tendency to 'essentialize' notions of masculinity and femininity which these diverse positions share to varying degrees is not supportable given the diversity of men's and women's lives and experiences. On the other hand, it seems entirely possible that women may in specific arenas share perspectives which influence their practice as scientists; this appears to be supported by the case of primatology. As we show in section 5, however, the intriguing suggestion that women 'do science differently' from men is not generally supported by the evidence — at least not on the level of methodology. Our conclusion draws together these rather disparate threads by exploring the vexed relationship between the male dominance of science and the masculinity (or otherwise) of its practice and products.

Before we start, some terminological clarification is called for. By convention in feminist and social science scholarship, the terms male and female are taken to refer to *biological sex*, whilst the term *gender* refers to the culturally or socially shaped categories of masculinity and femininity. This is not an absolute distinction: as feminist biologists have argued, there is significant interaction or mutual shaping between 'the biological' and 'the social'; the very definition of biological sex is itself a social act. Nevertheless, the notion of gender as a substantially social, and thus malleable, category is crucial. It means that what we take as masculine and feminine traits and behavior may vary from one cultural setting to another and from one individual to another; there is no monolithic or universal 'masculinity.' Indeed gender may be constructed in quite contradictory ways even in the same cultural settings: witness the association of men with technology, epitomized by images of oily car engines for some men and pristine high technology environments for others. One final complication is that the terms masculinity and femininity can be used in two ways: either to refer to those socialized traits held respectively by men and women; or to refer to stereotypical traits which might be held variously by men *or* women. We try to distinguish our meaning by referring explicitly to stereotypical gender (masculine or feminine) when intending the latter.

Sexist Science

Feminists have long had cause to be wary of scientific evidence and theories which attempt to explain iniquitous patterns of social behavior in biological terms; the effect of such biological determinism is to lend support to ideologies which see sexual (and other) inequalities as 'natural,' and so preclude social and political change to remove them.

To take a case in point, Aristotelian assumptions about women's intellectual inferiority to men were amplified in the nineteenth century by evidence from the field of phrenology that women's brains were smaller than men's. Middle class women in Britain and the US — including the first wave feminists — were frequently warned by doctors that education and other intellectual activities would drain essential energy from their wombs or cause their ovaries to shrivel! At the same time, the suppression of female sexuality in this class was reflected in theories linking all forms of female malady and 'hysteria' — including, potentially, that brought on by frustrated ambitions — to over active sex organs for which cliterodectomy or other forms of sexual mutilation was performed as 'treatment.' A full century later, research into sex differences renewed the credibility of theories about psychological differences in intellectual ability: in particular, the theory that the low numbers of women in science and engineering could be explained by marginal differences in visual spatial ability in teenagers.

The resurgence of sex differences research and the birth of the new field of sociobiology in the 1970s are both widely believed to have been prompted by the emergence of second wave feminism. Sociobiologists attempt to explain animal social behavior in terms of evolutionary theory. (Although virtually all of the research is on animals, extrapolation is frequently made to humans.) Members of the field have thus sought to 'explain' such behavior as male polyandry and male violence in terms of the 'strategies' of genes to optimize their continuity. Understandably, given the wider gender significance of such 'naturalizing' theories, the primary concern of early feminist critiques was to counter the biological determinism, largely by drawing on evidence from anthropology and other social sciences. During the 1980s, however, this approach gave way to a deeper analysis of sexist 'bias' in science and what this means for the production of scientific knowledge more generally.

Two fields in which sexist bias has been most thoroughly exposed are research into sex differences and the overlapping fields of primatology and human evolution. Whereas in sex differences research feminists seem to be suggesting that bias can be eliminated by 'better' research methods and theoretical assumptions, in primatology and human evolution the feminist challenge is often seen as providing a 'counter bias' to sexist science.[2]

As an example of the former case, take the research findings that fetally androgenized females tend to become 'tomboys' as teenagers, which is taken to suggest that masculine and female behavior are caused by hormonal influences during fetal growth in the womb. This is a classic illustration of how the determinism of biologists often involves a reductionist step in which explanations of complex phenomena are sought in terms of micro-level biological phenomena, usually molecular or genetic, without appropriately locating these in the wider organism, social or environmental context. Thus, detailed examination reveals that there is a long distance between evidence and hypothesis in this work. Quite apart from methodological weaknesses concerning sample size and issues of measurement, the linking hypotheses are all heavily shaped by value judgements — most notably, that 'tomboyism' is a universal version of masculinity, and that such aspects of teenage behavior are hormonally caused rather than a product of parental attitudes, sibling play, etc.

In contrast, the very limited availability of data on human evolution means that disputes in this field occur mostly at the level of theorizing and are often impossible to resolve empirically. For example, debates about the role of tool use in the development of bipedalism can be cast in terms of 'man the hunter' or 'woman the gatherer' — the influence of scientists' gender perspectives are pretty transparent here! During the 1960s and 1970s, women began to enter the field of primatology in noticeable numbers; they are credited (and credit themselves) with 'canceling out' the bias of their male counterparts by shining a light on the female primates. At that time male-dominated societies were regarded as the rule amongst primates. For example, in 1959 Irving De Vore published a study in which he established that a male pecking order was the key to baboons social behavior; to him the males were the most obvious sex to study, so he simply *did not see* what is now accepted — that males transfer between groups and that it is females which provide social stability in this species. Counter bias occurs in theory as well as observation: the relatively frequent cases of infanticide amongst Indian langurs, for example, was generally explained in terms of male sexual selection, as a means by which males bring females into heat; in 1970 Sarah Blaffer Hrdy demonstrated that females have reproductive strategies of their own by which, she argued, they use promiscuity to confuse paternity.

Historian of science, Donna Haraway, pinpoints the field of primatology as a crucial terrain for Western industrial society's account of 'the natural' — and, thus, one that is particularly contested. Its recent development has clearly been influenced by second wave feminism. In 1966, Thelma Rowell found no male domination in olive baboon societies in Uganda but explained this as an exceptional phenomenon contingent on the forest habitat; by the mid 1970s, she had greater confidence to take on theories of male domination. Haraway argues that American and British woman primatologists, like Rowell, who entered the field in the 1960s and 1970s "Changed the facts of nature by changing the visions of possible worlds." The various theoretical positions taken — by feminists and non-feminists alike — are all 'stories' about ourselves; primatology is thus "politics by other means."[3]

Feminists have highlighted just how subtly and subconsciously wider perceptions of gender can shape both what scientists find interesting to research, and how they 'see' the evidence in front of them. Thus, the very framing of sex differences research draws our attention away from the arguably much more significant area of sex similarities. Similarly, research on homosexuality (another contested area) has almost always focused on gay men — lesbians have remained largely invisible to science (as they have to other institutions), the assumption being that lesbian sex cannot be 'real' sex because it is not penetrative. A classic illustration of this masculine filter is provided by Emily Martin's delightful discourse analysis of scientific text and research articles on the subject of fertilization which portray the "egg as damsel in distress" and "sperm as heroic warrior to the rescue." Martin demonstrates that such language not only reflects and reinforces gendered stereotypes, it also blinkers how the researcher sees the process under study. Recent research has shown that the sperm is not all powerful and that the egg plays an active role, but even these findings are presented in gendered terms — eg., the egg "catches and tethers" the sperm.[4]

In sum, feminists have demonstrated that biologizing theories tend both to reflect and reinforce the sexism of the society in which they are produced. Scientists' own views are shaped by society at large, and will be evident in the theories and evidence they produce (especially where the subject matter relates to human behavior). So it is quite explicable that scientific theories will tend to support the dominant concerns and views of the time. Despite falling short of the ideal of 'objectivity,' the sexist research described here clearly strikes a chord in a society where sex differences are typically stressed as part of an ideology of male superiority and female inferiority. What does this critique mean for science? Is sexist science simply bad science which falls short of the principles of objectivity, or is objectivity more myth than reality?

The Myth of Value Free Knowledge

Three broad positions can be discerned within the feminist critiques of scientific knowledge: that sexist science is 'bad science' which can be corrected by the

FIGURE 3.1: THE NATIONAL PHYSICAL LABORATORY METROLOGY DIVISION. ONE OF THE TEST ROOMS FOR PRECISION MEASURING. © CROWN COPYRIGHT 1997. REPRODUCED BY PERMISSION OF THE CONTROLLER OF HMSO.

FIGURE 3.2: DIAN FOSSEY WITH A TWO-YEAR-OLD FEMALE GORILLA. WITH PERMISSION OF *NATIONAL GEOGRAPHIC MAGAZINE*.

elimination of bias; that sexist science is 'science as usual' because science is always value laden; and a potentially integrating position which holds that 'good science' is possible but only if scientists explicitly acknowledge the values they bring to it. The first and second of these positions were hotly debated within the radical science movement and the new sociology of science — two critical developments which emerged in the 1960s and 1970s, and together undermined the notion that scientific knowledge is objective. The third position was elaborated largely by feminist philosophers of science in the late 1980s.

Radical scientists stressed the links between science and ideology, for example in the case of race and IQ. However, there was no consensus on the extent of ideological influences on science — some adopted the position that all scientific knowledge is ideological, others that such knowledge is 'bad science' and that good, ideology-free science is possible. Although the new sociologists of science generally adopted a less explicitly political approach, their continuing project has been to demonstrate the socially constructed nature of scientific knowledge — that assumptions and evidence on which theories are based are contingent, and that observations are 'theory laden.' The sociology of scientific knowledge has been pivotal in this project; it adopts a relativist approach which is deliberately agnostic about the 'truth value' of any knowledge claims. Finally, both the radical scientists and the new sociologists of science highlighted the rhetorical importance of claims that science and of scientists are value free and objective. In their view, the fact that subjectivity and bias are a normal part of scientific practice is largely eclipsed by the widespread adherence to the notion of science and scientists as detached and disinterested, and so 'above' politics; and this notion in turn explains why scientific knowledge has so much authority in our society.

Many feminist critics were themselves active in the radical science movement, but became increasingly dissatisfied with the tendency towards abstract theoretical debates as divisions on ideology grew, as well as the apparent gender blindness of much of the political discussion. Feminists have also taken seriously the social constructivist analysis of the sociology of scientific knowledge, although they never really found a foothold in this field. One reason for this is that relativism presents a central problem for feminists: whilst it is extremely useful as a methodological stance — that one suspend judgement whilst analyzing knowledge claims — relativism often appears to be adopted by some as a principled, judgemental stance — that one should *never* adjudicate between knowledge claims. In recognizing the old equation between knowledge and power, we and other feminists want to be able to make judgements about competing knowledge claims in order to challenge and improve on sexist science; in Haraway's terms, we want an epistemology which allows for the possibility of distinguishing (and constructing) good stories from bad stories. In other words, the position that all science is value laden skirts too close to a kind of 'anything goes' political nihilism. On the other hand, the position that sexist science is bad science does not take account of the evidence that values necessarily enter science. So, many feminist theorists find themselves caught between a relativist insistence on the social construction of knowledge and a desire to hold onto some of the old standards of objectivity.

Feminist philosophers of science show that it is nevertheless possible to eshew the dichotomy between objectivity and relativism. Their epistemologies seek to steer a different course around these two poles by acknowledging the value-ladenness of science. For example, Sandra Harding argues for a practice of 'strong objectivity' in science which insisted that scientists be 'reflexive' about the subjective and wider social conditions shaping their knowledge claims. This is to be achieved through

> systematic understanding of powerful background beliefs [and] ... causal analyses of micro processes in the laboratory and macro tendencies in the social order which shapes scientific practice ... [this] permits a robust notion of reflexivity.[5]

In a similar vein, Haraway advocates the notion of 'situated knowledge', which

> requires that the objects of knowledge be pictured as an actor and agent, not a screen or a ground or a resource, never finally as slave to the master that closes off the dialectic in his unique agency and authorship of 'objective' knowledge. The point is paradigmaticaly clear in critical approaches to the social and human sciences, where the agency of people studied itself transforms the entire project of producing social theory. Instead, coming to terms with the agency of the 'objects' studied is the only way to avoid gross error and false knowledge of many kinds in these sciences.[6]

Haraway suggests that because agency is an inevitable part of knowledge production, it should be made explicit rather than be denied: this process, it is argued, produces more transparent, and therefore better, knowledge.

These potentially integrating positions are founded on new notions of objectivity. Fee, for example, argues that "objectivity is sufficiently vague to carry with it a multitude of meanings"; positive meanings include the

> constant process of practical interaction with nature; willingness to consider all assumptions and methods as open to question ... idea of individual creativity subjected to the constraints of community validation through a series of recognized procedures.[7]

Whilst wishing to hold on to these meanings, Fee rejects what she calls 'the hierarchy of distances' in objectivity, which is manifest in four ways: the treatment of the production of knowledge as separate from its social use; the separation between scientific rationality and emotion in the language of science; the distance between the subject and the object of study; and the view of science as separate and distinct from society. Helen Longino and Lynn Hankinson Nelson[8] favor similar reinterpretations of objectivity, based on a robust version of empiricism: i.e., one dependent on a commitment to critical evaluation of knowledge claims and beliefs based on the available evidence. Longino adds a reflexive twist to this position by arguing for 'contextual empiricism': scientists should allow their political commitments to guide their choice of particular models in science and not simply aim to uncover sexist bias.

These themes — combining reflexivity and rigor — have been increasingly important to feminists working in the social and life sciences, who have pointed to a number of changes which together would undermine the production of sexist science. One clear target is the field of biology and its disciplinary tendency to look for biologically determinist (and often reductionist) explanations of phenomena which (many would argue) are at least partly if not wholly socially based. Interactive models are seen as more appropriate, building on an interdisciplinary and holistic research approach. More generally, feminists point to the need for greater humility and honesty about the limits of both evidence and theory, alongside greater reflexivity about the personal and social values that enter science. Gendered assumptions and judgements should be made as explicit as possible, as should gendered language — scientists should recognize that readers do impute meaning to the written word. In this way, the possibility of alternative perspectives on a subject is made visible. Many feminists suggest, in addition, that the involvement of a wider range of social groups in science would lead to a wider range of perspectives and therefore theories.

We return to this last issue later; first we explore further criticism of conventional notions of objectivity in the light of the strong association between masculinity and objectivity in popular as well as scientific culture. Is this association also rhetorical (or stereotypical), and thus easily dismissed, or is there a sense in which men are intrinsically more objective? What does this mean for the conventional and revamped notions of objectivity discussed above?

Objectivity and Masculinity

Modern science was constructed as a specifically male enterprise. Francis Bacon and the founders of the Royal Society sought only male recruits to the new philosophy because they believed that women would be bad for science. In the words of an early Royal Society fellow, Joseph Glanvill, "True philosophy" could not progress where "the Affections wear the breeches and the Female rules."[9] Two centuries later, when women were beating a path into the medical profession, Dr Robert Christian of Edinburgh University pronounced that female medical practitioners would "be injurious to medicine as a scientific profession."[10] These beliefs were based on a thoroughly dichotomous and gendered worldview — inherited by the church (from Greek and pre-Christian beliefs) in the late thirteenth century, then adapted by Descartes and other mechanical philosophers — in which women were associated with nature, darkness, mystery, the body and emotionality, whereas men were associated with the heavens, light, enlightenment, the mind and rationality. The new philosophers believed that their science demanded a form of rationality — objectivity — that denies any emotional feelings for or identity with the object of study, i.e., nature. Women were deemed to be incapable of making this split on the grounds of their imputed carnal ties with nature and their intellectual inferiority to men. Significantly, such views were retained when so much other 'superstition and ignorance' was discarded.

According to Carol Merchant and Brian Easlea, the new philosophers' position hinged on the radically new attitude to nature introduced by the mechanical philosophers of the seventeenth century. Where previously nature was viewed as an organic entity, at once nurturing and dangerous, moved by 'spirits' or the 'unseen hand' of God, the mechanical philosophy sought to cast all spirits out of nature, rendering it mere inert matter, a machine, and relegating God to the one-off role of creation. As Lynn White has concluded, by divesting nature of its mysterious and creative powers, the natural philosophers "made it possible to exploit nature in a mood of indifference to the feelings of natural objects."[11] This shift was necessary to the Baconian program because the knowledge of the mechanisms of nature was not to be gained easily but required experimentation using technical instruments. There are chilling parallels (many of them explicit) with the contemporary witch trials in the gendered language used by Bacon and others to describe the new philosophers' relationship to the natural world. Witness the countless references to notions of 'putting nature on the rack' and 'penetrating her inner secrets and crevices.' Merchant and Easlea[12] further argue that turning nature into an inert machine opened the way for the implied 'rape of nature,' manifest in the destructive exploitation of the environment and the threat of nuclear annihilation (both by technological means) which face us now.

We view the gendered metaphors in the discourse of the natural philosophers as largely rhetorical, reflecting the stereotypes of the culture in which they were living; they cannot be imputed as necessarily reflecting the motives and attitudes of scientists today. However, whilst language is clearly more shaped by than shaping of social practice, it would be wrong to suggest that it has no material effects. Three significant things remain from the heritage of science's founding fathers: (i) the historically pervasive association between masculinity and science (and objectivity) runs deep still in popular perceptions; (ii) women have been very effectively excluded from science, in keeping with this discourse; and (iii), as Easlea and Merchant argue, many scientists do behave with 'apparent indifference' to both the natural world and the consequences of their work.

In seeking to explain this state of affairs, different feminists have tended to emphasize rather different aspects of the general relationship between gender and sex. One approach is to argue that science is masculine because it is done by men who bring particular psychological traits to it. Easlea, for example, attributes men's involvement in science, including nuclear weapons research, variously to a sense of sexual impotence or a desire to compensate for or appropriate female creativity. Building in part on his work, Evelyn Fox Keller draws dramatic parallels between scientists' quest for 'secrets of life' in the discovery of DNA and for 'secrets of death' in the design of the atom bomb.[13] She (and others) have concentrated on scientists' detachment from their subject of study, which for her is explained in terms of 'object relations theory' concerning masculinity. This theory stresses the psychoanalytic significance (in societies that sanction cross-gender behavior and where the primary parent is almost always female) of the separation which very young boys have to make between themselves and their mothers as they realize that she is of an 'other' gender; little girls it is argued do not separate so much and so do not have to deny or 'master' caring and emotional traits.

Feminist epistemologists in the standpoint tradition share the belief that masculinity and femininity are coherent but distinct categories, but explain gender differences in terms of the division of labor (drawing on Hegel) rather than early psychological development. Hilary Rose and Nancy Hartsock thus argue that men share a particular standpoint which is qualitatively different from that of women because they perform different forms of labor. Women's role as principle caretaker of the home and family — work involving what Rose calls "the unity of hand, brain and heart" — is held to afford them a very important vantage point from which to view the world. It suggests that women are less likely to adopt the same standards of detached objectivity that male scientists advocate (if not adhere to), and that a science based on women's standpoint would involve, "a practice of feeling, thinking and writing that opposes the abstraction of male and bourgeois scientific thought."[14]

In our view, these two positions share an implicit essentialism about gender. By this we mean theories which see masculinity and femininity as rather fixed and thus unchangeable categories explicable in terms of psychological or biological determinism. To be fair, Keller and the standpoint theorists are, like many feminists, implacably opposed to biological determinism; gender essentialism is most explicitly embraced by ecofeminism. This tradition emerged in the early 1980s and draws together environmentalism, feminism and women's spirituality movements, in campaigns against military technology and animal experimentation as well as environmental issues. Following the analysis of Merchant, ecofeminism largely accepts the Baconian dichotomies but inverts the values associated with them. So, men's (presumed) detached objectivity is seen as a negative trait largely responsible for environmental despoilage and the threat of nuclear annihilation. Women, on the other hand, are seen as necessarily closer to nature — by virtue of their involvement in childbearing and rearing and in consumption — and thus having a vital role to play in saving both humanity and the planet.

Each of the strands of feminism outlined here have very different answers to the question 'what makes men so different from women?' encompassing the psychological, material and spiritual realms as well as biology. However, all are concerned to establish the commonality of masculine and feminine traits to some degree, and all have sharply dichotomized notions of gender. We are left with a general concern that essentialist positions — whether implicit or explict — run the risk of naturalizing unequal relations between men and women, and so undermining the case for political change, in the same way.

We have three further problems with the general approach taken by these feminists. First, we are not convinced that men and women are really so divided along gender lines. From a postmodernist perspective, all of these theorists can be criticized for suggesting that women have more in common than they actually do, on the grounds that "women are too diverse in our experiences to generate a single cognitive framework."[15] Like Haraway, we are uncomfortable with the idea that all women share subjugated positions in society and hence a single vision of reality. Rather than there being any one standpoint from which to approach science, she argues that there are many, a flux of shared positions, or solidarities, amongst scientists.[16]

Second, we would argue that this approach ignores the possibility that stereotypes are at work in discussions about differences between men and women. Accordingly, it tends to play down or ignore the material benefits (for male scientists) to be gained from stereotypes about detached rational men and connected emotional women in a society where power attaches both to science and to the status of being objective. Maybe male scientists have appropriated the traits associated with being objective, in their rhetoric or their practice, because science seems to demand them and because it is a powerful social institution — not because these activities or traits are necessarily masculine.

Finally, feminists simply do not need to justify the rejection of these aspects of objectivity on the grounds that they are masculine. When objectivity is unpacked, as in the previous section, it becomes clear that indifference to the natural world and the consequences of scientific knowledge, and the practice of distancing the subject and the object of study, can be rejected on the grounds that they generate worse forms of scientific knowledge.

The difficulty with sharply dichotomized theories that men and women have such different ways of thinking and acting is further highlighted in the following section.

Do Women do Science Differently from Men?

If men are (historically at least) associated with the version of objectivity which is 'detached' from the object and denies the subjectivity of the scientist, the question is raised as to whether women practice (or would practice) science differently. This question is of more than academic interest: if women, in general, do or would bring a different approach to science, and if that approach is 'better' than the existing masculine one in terms of the epistemologies outlined above, then this represents an additional reason to dramatically increase the representation of women in science — additional, that is, to simple fairness. The question potentially links the liberal program to get more women into science with the radical program to transform the practice of science. Perhaps for this reason, or perhaps because its implicit essentialism finds wider resonance, the question has been a source of popular fascination in the 1990s. In a 1993 special feature on women in science,[17] *Science* journal gave the question center stage. Here as elsewhere, evidence that women do bring a different approach to the practice of science is taken to come from two sources: the field of primatology and the work of plant geneticist Barbara McClintock who won the Nobel Prize for the theory of genetic transformation (or 'jumping genes'). Nevertheless, there is considerable counter-evidence.

As noted earlier, the women who entered primatology in the 1960s and 1970s are frequently credited with 'canceling out' the bias of their male counterparts by shining a light on the female primates. But arguably their more profound contribution was to prompt a significant reorientation in both methods and understanding. Whereas previously the emphasis was primarily on theory, and observation was done mostly at a distance (from a jeep), the women entrants went to extraordinary lengths, virtually living amongst the apes or monkeys, to get to know and observe them at close quarters. As a result of their painstaking observation-based approach, they saw things which had never been seen before. Jane Goodall observed chimpanzees using tools made from twigs to extract termites from their nests, and so overturned dearly held views about what distinguished humans from other animals. More generally, she and others after her observed high levels of individual variation in primate behavior — a development which profoundly challenged the evolutionary emphasis of existing theory as well as furthering our understanding of these animals. One of the most highly cited articles in the field is a 1974 methodological paper by Jeanne Altmann on the importance of spending equal amounts of time observing each individual in a social group.

What appears interesting from a gender point of view is that these women legitimated the practice of establishing empathy with the object of study — and so contravened the stereotypically masculine, 'detached' version of objectivity. Male primatologists interviewed by the *Science* journalists readily identify a gender connection: Louis Leakey, who helped launch the careers of Goodall, Dian Fossey and Birute Galdikas, is reported to have "trusted women for their patience, persistence, and perception."[18] But this does not mean that men are incapable of such traits, or indeed of seeing the female object. Haraway has demonstrated that the role of the women pioneers in primatology is specifically anglo-saxon: in the 1950s, an all male team of Japanese primatologists studying rhesus macaques adopted similar methods as later female entrants in Britain and the US, and came to similar conclusions about the inadequacies of the male dominated model of primate social life.

On the surface, Keller's celebrated biography of Barbara McClintock also appears to support the stereotype of women having more empathy with nature. Keller argues that McClintock was a "philosophical and methodological deviant"[19] — philosophical because she challenged the 'central dogma,' that DNA shapes nature but is not shapeable, and methodological because of her "cultivated attentiveness" and "respect for difference."[20] Thus, in the course of six years work, she developed an almost kin like, mystical intimacy with the plants she studied, to the extent that she almost felt herself to be "in there amongst the genes":

> For her the smallest detail provided the key to the larger whole. It was her conviction that the closer her focus, the greater her attention to individual detail, to the unique

> characteristics of a single plant, of a single kernel, of a single chromosome, the more she could learn about the general principles by which the maize plant as a whole was organized, the better her 'feeling for the organism.'[21]

Keller suggests that McClintock's approach represents a guide to how science could be different — with its scepticism about hegemonic orthodoxy and its insistence on 'listening to the material.'

Although the McClintock story is frequently interpreted as evidence that women do science differently to men, Keller herself strongly refutes this notion. She suggests instead that McClintock may provide a model for a "gender free" science which (echoing Rose) renounces "the division of emotional and intellectual labor that maintains science as a male preserve."[22] But there is arguably a contradiction here, in light of Keller's wider views on gender. In particular, her insistence that the split between emotional and intellectual labor is precisely what keeps women out of science, implies that women are more likely to be the bearers of a gender free science that cultivates a more 'caring' and connected approach to the object.

Significantly, the perspective of women currently in science appears to refute this suggestion: most women scientists see no methodological differences between themselves and their male colleagues, and react strongly to suggestions that such differences exist, though some it is claimed believe privately that they do. It is widely acknowledged that women in science are constantly having to demonstrate that they can do science 'as well as the men'; emphasizing difference is therefore potentially to undermine one's already fragile standing.[23] In contrast to the methodological issue, however, many women scientists maintain that there is a difference in the way in which women and men approach social interactions in laboratory practice, with women placing more emphasis on cooperation and less on competition.

Conclusion: The Brockenspectres of Science

We can now return to the questions about sex and gender in science with which we opened this chapter. With respect to the first question, we have explored two important ways in which science is masculine — in the (masculinist) sexism of some scientific theories and practice, and in the historically based but still rhetorically powerful association between masculinity and objectivity.

Our second question — how much is this masculinity of science due to its being dominated by men? — obliges us to problematize the extent to men and women respectively share common traits. Clearly, we see it as a mistake to appeal to essential qualitative differences between men and women: stereotypes of masculinity are often as far removed from men's actual characteristics as the myth of value free knowledge is from the practice of science. To magnify the differences between men and women, as so many do, is to mask the complexities of and the differences within both masculinity and femininity and, potentially, to erect barriers to change. We would further argue that the masculine stereotype surrounding science serves a role which, to some extent at least, is supported irrespective of the fact that most scientists are men. The myth of value neutrality and detachment in science provides society with the resource of disinterested expertise, and provides scientists with the authority this status carries. It would take a sea change in attitudes to get scientists, the sponsors of science and the population at large to accept that a reflexive science could still be valid and useful — let along that it might be more rigorous and trustworthy than what we have now!

In short, then, we believe that the masculinity of this most powerful of social institutions is only in part a function of science being male dominated. On the other hand, there can be no denying — to answer our third question — that the masculinity of science also serves to reinforce the male dominance of science. Whilst scientific claims about sex differences in mental abilities have directly reinforced discriminatory practices and attitudes, the masculine image of science enters into gendered socialization at home and school in more subtle ways. Thus, socialization which encourages 'tinkering with machines' and abstract rationalistic thought in boys, and caring behavior and connected, emotional expression in girls, serves to reduce the likelihood of girls seeing science as a suitable or interesting avenue to pursue. Those women who do nonetheless opt for science, frequently report being treated as at best an oddity and at worst not quite a woman. They also experience working conditions similar to those of women in other male dominated areas: the excessively competitive careers structures; the exclusion from crucial informal networks; the long, antisocial hours which make it difficult to combine a career with family life; and so on. These are additional ways in which the practice of science is masculine, causing many women to abandon scientific careers.

Our final question is whether, if science were no longer male-dominated, the equal representation of women in science would in any way change its practice or products. The evidence on the impact of those women already in science is not convincing but may be, in the absence of wider changes in both science and gender relations, the question should be seen a necessarily hypothetical. After all, the argument presented above means that we are unlikely to see any significant increase in the number of women in science without an equally significant attack on the masculine image and practice of science — the two battles must go hand in hand.

It seems likely that in the short term at least the greater involvement of *feminists* will have a greater impact on the practice of science than that of women *per se.* Those feminist scientists who are trying to implement alternative approaches in science stress how difficult and slow a process this is. There are enormous institutional pressures on feminists, and any marginal groups within science, to conform in terms of the research methods and questions pursued — not least, in the lack of funding for alternative approaches. There is also strong ideological resistance to social constructivist approaches in particular since these appear to undermine so much of the authority of science. In this context, we would emphasize the real strength of the 'third position' advocated by feminists: that not throwing the baby — objectivity — out with the bath water — the rhetoric of value free knowledge — is indeed possible. This approach promises a better science precisly because it breaks out of and goes beyond the conventional dichotomy between relativism and objectivity — just as, we would argue, feminism promises a better society because it seeks ultimately to go beyond the constraining dichotomies of masculinity and femininity.

We have alighted on the brockenspectre as a useful metaphor for the perspective which feminism has brought to the question of sex and gender in science. Brockenspectre, or spectre of the brocken, is a term from meteorology which means the shadow of an observer cast by the sun on to a bank of mist; when seen from a hilltop, this phenomenon presents the illusion of a gigantic form on/in the cloud below. Metaphorically, the brockenspectre captures the relationship between the observer and the observed in science. It is the shadow cast on science by the subjectivity and situatedness of the observer producing it, and the hilltops on which scientific observers stand represent their epistemological position, including their social context or situation. Thus, there is a masculine shadow on science cast by the predominantly male population which produces it (evident in the sexist knowledge claims and symbolism of science). Paradoxically, feminists have exposed this shadow, but they have also cast shadows on the science they produce — other 'stories' — because science is always conducted from one vantage point or another.

The metaphor highlights important ambiguity in the relationship between sex and gender in science. The masculine shadow on science is linked to the material reality of a male dominated practice but, at the same time, has illusory qualities. Thus, science is not intrinsically or necessarily masculine but can have a larger than life appearance of being so: witness how the ideological aspect of the masculine shadow of science has been magnified — both by men scientists seeking to retain their social status, and by feminists seeking to challenge it. We ourselves remain ambivalent about the extent and significance of the illusion of science as masculine, though we are convinced that it is bad — both for women and for science — to seek to inflate it further. Our central project, reflected in the feminist epistemologies outlined here, is to expose the illusion — by challenging the rhetorical force of ideologies which view science as both value free and masculine, and by deconstructing the situatedness and subjectivity of the knower.

The metaphor has crucial implications for science itself since it tells us that science can never be 'shadow-free,' and that a whole range of vantage points (and thus shadows) are possible, because the practice of science is necessarily subjective and context dependent. Feminists' critiques of sexist biology obliged them to acknowledge that they too imprint their social location and politics on science, and to insist on such reflexivity in others. In addition to feminist and masculinist vantage points, we have shown that there are various feminist perspectives on science, and that there is no single standpoint which all women share. We believe strongly that this diversity is a strength not a weakness. Getting a greater variety of social groups to enter science will bring greater plurality in subjectivities and situations, and therefore in methods and theories — and this can only make for more rigorous science. Encouraging and respecting *difference* provides the most promising basis for a non-sexist and 'gender free' science: there should be many *more* 'shadows' on science!

REFERENCES

1. Harding, S. *The Science Question in Feminism.* (Ithaca, NY: Cornell University Press, 1986), p. 9.
2. Longino, J.I. and Doell, R. 'Body, bias, and behaviour: A comparative analysis of reasoning in two areas of biological science', *Signs: Journal of Women in Culture and Society* (1983), **9(2)**: pp. 206–27; Morell, V. 'Seeing nature through the lens of gender', *Science* (April 16, 1993), **260**: pp. 428–9.
3. Haraway, D. (1986) 'Primatology is politics by other means', in Bleier, R. (Ed.) *Feminist Approaches to Scheice.* (New York: Pergamon Press, 1986), p. 80 and p. 114; Haraway, D. *Primate Visions: Gender, Race and Nature in the World of Modern Science.* (New York: Routledge, 1989).
4. Martin, E. 'The egg and the sperm: How science has constructed a romance based on stereotypical male-female roles', *Signs: Journal of Women in Culture and Society* (1991), **16(3)**: pp. 485–501, p. 491 and p. 490.
5. Harding, S. *Whose Science? Whose knowledge? Thinking From Women's Lives.* (Milton Keynes, UK: Open University Press, 1991), p. 149.
6. Haraway, D. 'Situated knowledges: The science question in feminism and the privilege of partial perspective', *Feminist Studies* (1988),**14(3)**: pp. 575–99, p. 598.
7. Fee, E. 'Women's nature and scientific objectivity', in Hubbard, R. and Lowe, M. (Eds.) *Women's Nature: Rationalisations of Inequality.* (New York: Pergamon Press, 1983), p. 16.
8. Longino, H. *Science as Social Knowledge.* (USA: Princeton University Press, 1989); Nelson, L.H. *Who Knows; From Quine to a Feminist Empiricism.* (Philadelphia: Temple University Press, 1990).
9. Quoted in Easlea, B. *Science and Sexual Oppression.* (London: Weidenfeld and Nicolson, 1981), p. 70.
10. Quoted in Lutzker, E. *Women Gain a Place in Medicine.* (New York: McGraw Hill, 1969), p. 48.
11. White, L. *Medieval Technology and Social Change.* (London: Oxford University Press, 1962), p. 1205.
12. Merchant, C. (2nd ed), *The Death of Nature: Women, Ecology and the Scientific Revolution.* (London: Wildwood House, 1982); Easlea, B. *Fathering the Unthinkable: Masculinity, Scientists and the Nuclear Arms Race.* (London: Pluto, 1982).
13. Keller, E.F. 'From Secrets of Life to Secrets of Death' in Jacobus, M., Keller, E. & Shuttleworth, S. (Eds.) *Body/Politics Women and the Discourses of Science.* (London: Routledge, 1990), pp. 177–91.
14. Rose, H. 'Hand, brain, and heart' *Signs: Journal of Women in Culture and Society* (1983), **9(1)**: pp. 73–96, p. 73 and p. 88; Hartsock, N. (1983) 'The feminist standpoint: Developing the ground for a specifically feminist historical materialism', in Harding, S. and Hintikka, M. (Eds.) *Discovering Reality: Feminist Perspectives on Epistemology, Metaphysics, Methodology and Philosophy of Science.* (Dordrecht, The Netherlands: Reidel Publishing, 1983), pp. 283–310.
15. Longino *op cite* note 23, p. 111.
16. Haraway, D. 'A Manifesto for Cyborgs: Science, Technology and Socialist Feminism in the 1980's', *Socialist Review* (1985), **no. 80**: pp. 65–107; and *op cite* note 6.
17. 'Women in Science', *Science* (April 16, 1993), **260**: pp. 383–430.
18. Morell, V. 'Called "trimates," Three bold women shaped their field', *Science* (April 16, 1993), **260**: pp. 420–25.
19. *Ibid.*
20. Keller, E.F. *Reflections on Gender and Science.* (New Haven, CT: Yale University Press, 1985), p. 159, p. 164 and p. 163.
21. Keller, E.F. *A Feeling for the Organism: The Life and Work of Barbara McClintock.* (New York: WH Freeman, 1983), pp. 101.
22. Keller *op cite* note 21, p. 178.
23. Kerr, E.A. (unpublished) 'Feminising science: linking theory and practice'. (PhD Thesis, University of Edinburgh, 1995).

Chapter 4

Science, Political Power and the State

Dominique Pestre

Different social actors define the sciences differently. For some, they are systems of action which shape the material and social worlds; for others, they are systems of values which define ideals and norms; for others again, they are systems of representations underpinning positions of authority. In a high-energy laboratory like CERN, the physics practiced there is THE science. Seen as the leading edge of knowledge and understanding, it is defined as a system of propositions, describing the ultimate constituents of matter, which is confirmed or rebutted by experiments. For the director of an electronics company, science might be this system of statements, and he would certainly be willing to describe it as such at an official banquet, but it is above all a means to control and transform the material world, one means, amongst others, to impose an economic standard, to get around a law, to dominate a market, to beat his competitors, to make money — or to justify a new organization of work in his factory. For a graduate from the Ecole Polytechnique working for the French state at the beginning of this century, science was probably a general method, a tool for treating very different kinds of problems, social problems included. It was an approach which enabled him to find universal solutions which promoted 'the public good' that he often embodied, and which legitimized his claim to transcend particular interests that disrupted the perfectly managed society he dreamt of. Finally, for a qualified worker in a firm of the same period, science could, by the use of machines or the introduction of a new organization of the workplace, be a means of stripping him of his know-how, undermining his social status, destroying his culture, subverting his autonomy and submitting his working day to a rhythm which he found odious.

In this chapter I want to focus on the practice of the sciences in their interaction with the political world and the state. The sciences I study are sites of power, they contribute to the management of the state and society. Be they civilian or military, articulated around 'pure' or industrial science, they are an integral part of large techno-political systems.

A survey of the recent literature on the sciences in the twentieth century reveals four dominant trends. The first takes the sciences as systems of knowledge and tends to produce a history of major disciplinary developments, a history of ideas and concepts. Albeit essential, this will be of little concern in this chapter. The second approach considers the sciences and their associated techniques, as above all systems of practices. Accepting the intrinsic contextuality of scientific knowledge, it produces a history focused on the protocols and methods of proof, on the variety of the spaces of legitimation, but deals only marginally with the political apparatus. A third dominant tendency of contemporary historiography studies the science/technology/industry interface. It overlaps with industrial and business history, it is concerned with R&D and the modalities of innovation, but it intersects political questions primarily through public policy. More recently, the relations between science and the military have enjoyed a surge of interest, of which the central object has mainly been the post-1945 military/industrial/academic American complex.

Scientific relations with *la politique* are thus infrequently studied, and when they are, it is usually the national security aspect, or the management of the hard sciences for the benefit of national economies, that is of interest. The question of the relationships between science and the world of politics is far more wide-ranging however, and should include considerations on the nature of the scientific activity in its relationship to power for example, or on how science and scientific ideologies are used by the state to regulate the social. Since historical studies on these questions are few and far between, it appeared impossible for me to give a global overview of the issue here. Thus I propose a study in two parts. The first poses the question in its generality, but without giving a continuous historical narrative. The second focuses on a more delimited object: the relationship between the state and physicists and biologists throughout this century in Europe and the USA. Here the approach is more systematic.

Science and Political Power: Some Methodological Remarks

Scientific intellectuals and those who manage society at the political level, often have a rather similar image of the relations between science and the political system. It embodies the scientistic conception which has come to dominate Western societies for the last two to three centuries. It holds that science and the world of politics constitute two different, even antagonistic domains, two worlds with opposed logics and goals. Deeply rooted in social consciousness (in the sense that Louis Althusser gave to the expression: "spontaneous ideology of the savants") this image is crucial to the smooth functioning of both groups. The world of science is said to be the world of Reason, a world which enables one to settle issues with certainty. The political world has a less straight-forward logic and relies mainly on consensus established between conflicting interests.

One corollary is the image of science as pure by nature but polluted or twisted by 'material' interests or reasons of state. This too has a long history and regained popularity after World War II, particularly among physicists who worked on the bomb (it found expression in Oppenheimer's famous dictum that scientists had lost their innocence). It has contributed to posing the question in the very narrow mode of the personal 'responsibility' of the savant, the kind of ethical obligation which the scientist has (or does not have) in continuing research which he judges 'potentially dangerous for society.' That discourse rarely prevented physicists from working for the military however. A kind of 'mental compartimentalization,' to use Paul Forman' expression, enabled them to keep both attitudes in parallel without too much disquiet.

This image is of course to be found outside scientific contexts where it offers both an ideal of purity and certainty. It locates the perversion outside Science, in the economic or social system for example, and provides a neat and simple way of separating Good from Evil. The positive and negative values can easily be inverted however, and many people outside the scientific sphere clearly situate science, associated with technology, on the side of Evil and the impure. Marginal among scientists and politicians, and without doubt amongst the majority of those who esteem intellectual activity, the Faustian image of science, the image of the sorcerer's apprentice, of the pact with the devil, and of a science dangerous by nature, is rather commonplace among the remainder of the body politic.

Most of the time, science and the political world are then taken for granted and considered independent of one another. One way of better appreciating the pervasiveness of this image is to study the very extensive work done in the past forty years on the economics of technical change which postulates the separation between science, which is a public good freely available to all those who wish to appropriate it through publications, and the uses to which it is put, and which are informed by political, industrial or military logics. This split made it is easy for economists to conceptualize innovation and economic development and suggest how the independent variable that was science could be controlled and manipulated for other purposes. The problem is that knowledge is always deployed in very specific contexts. The characteristics of the spaces in which it is produced, whether it is a Bell laboratory, or an academic research center, profoundly determine its outlook. Unfortunately, as we know only too well, the effects of these beliefs on developing countries have been extremely painful throughout this century. Experience has proved rather quickly that science was only public in name and that it was just as costly to acquire as any other product or know-how.

I do not wish to deny the analytical value of separating the two notions of the Scientific and the Political, but this divide is at best a convenient tool and a first approximation. Science is always and already entangled with the state and political questions, and it always has been, and the question of power is always and already in the practice of science itself. For example, the Ecole Polytechnique in Paris was, from its creation at the end of the eighteenth century, an institution intrinsically linked to the state. It was, at the same time, the center of European science.

Similarly, the Physikalische Technische Reichanstallt in Berlin was a response, almost a century later, to the needs of industry and the German state at a time of imperial expansion. This did not prevent it being, in the same move, one of the citadels of scientific achievement. By definition, I would say. More generally, scientific intellectuals have often been associated with weapon systems, as with the management of the state and of production. They have contributed, and still do, to the definition of technical norms: norms specifying quality (for medicines, fertilizers or the color of fabrics); norms defining levels of socially acceptable security (for machines or nuclear reactors); norms and standards of measurement (which are essential for the entire industrial sector). They participate in the evaluation of risks (for the state, but also for insurance companies), in the elaboration of legislation as well as in the definition of the enabling instrumentation and institutions. By nature one might conclude, since science is a social institution, it is immersed in the political system, it is linked to it by a thousand networks.

Symmetrically, the fact that power over nature is at the very core of the modern scientific enterprise, the fact that the manipulation of the material world is intrinsic to science itself, and that it is even one of its defining characteristics, has often been shown, by feminist studies for example, and there is no need for me to belabor the point and demonstrate why scientific achievements are of interest to other social actors. These remarks stressing the interconnectedness between modern science and the state must not mask two points however. Very important differences are to found from one epoch to another, and I have no intention of obliterating the periodization so essential to the historian. World War II and the Cold War, for example, are important turning points in the relationship between science and political power. Secondly, this organic and long-lasting link between science and political management must not blind us to what scientists themselves say. It is extremely important for them to perceive what they do as autonomous and to claim a large degree of freedom for their practices.

If one starts from the idea that the democratic state is the forum in which the compromises needed to perpetuate the social order are negotiated, its management implies that an image of neutrality vis-à-vis interest groups be constantly recreated by words, by gesture or by symbolic action. This image of the neutrality of the state is sustained in two ways. It is first demonstrated by the democratic nature of the appointments procedures. The major decisions of civil society are taken during elections, and this is the way major interests express themselves. Through complex rituals codified in constitutions, a solution becomes a majority opinion. This way of legitimating the state and its social neutrality is nevertheless constantly short-circuited by another which is based on 'expert opinion,' on the authority of science pronouncements, and not on the opinion of a majority. In that case, the state no longer speaks on behalf of the majority mediated through a vote, its legitimacy no longer rests on the democratic compromise emerging from the ballot box — it speaks and acts in the name of statements which are universal truths. There is no 'choice of society' anymore, no particular interests to transcend. There is only the one best way, inevitable, surely painful for those excluded from its benefits, but the only possible road for society to follow. At least that is what experts are frequently ready to claim, and what governments are ready to accept when the solutions fit with their objectives.

In the exact and social sciences, the strength of experts comes from the fact that they are able to manipulate techniques and artefacts, to invent material and symbolic technologies which stand for the 'real world' out there and which allow them to act. More importantly however, it is the very idea of expertise, the conviction that such an expertise does exist, which is crucial. Experts might systematically disagree on concrete questions, as is too obvious, but that is of secondary importance. What counts is the certainty that social and natural facts can be assessed authoritatively by science. Be it in the state apparatus, in court or in front of a panel investigating an industrial desastre, what is important is the conviction that experts can establish facts without ambiguity, that they can be summoned to present the truth of things.

In Western societies in this century, science has thus become an authority to legitimate public action. Since science is a discourse that claims not to depend on partisan decisions, it enables one to 'technicalize' public action, to 'de-politicize' it, to render it impersonal, to bypass the democratic rules of accountability. This mode of action leads to an instrumentalization of politics through the use of specialists, it gives to political decisions the force of necessity, and it comes to substitute competence and technical knowledge for the affirmation of will and of values deliberately chosen. Backed by the rationalizing words of knowledge, some possibilities are excluded *a priori* in the name of criteria given as scientifically evident. The actions which follow are rarely submitted to public debate, a public deemed incompetent, anyway, notably in economic affairs, and the choices are made by those who 'know' in the best interests of all. At the extreme this way of doing 'scientific politics' converges with the notion that political practice can be reduced to the certitudes of a well managed leviathan state, or with the dream (or the nightmare) of the technocratic management of societies and of political engineering, two possibilities that have been ever present during this century in Western societies, as we know only too well.

The link between science and social management could take other forms, less tragic, more underground perhaps, but constraining all the same. The knowledge produced by scientists, and by intellectuals in general, contributes to defining the frameworks of social action, be they cultural, by inventing categories and institutions which become second nature to us and which trace the boundaries of what is thinkable and what is no longer doable, or embedded in our material world. Be it by words or by artefacts, by discourses or by techniques, the 'hard' and the social sciences construct the world in which we live.

The practice of social statistics provides a good example. Socioprofessional categories profoundly differ from one country to the next. Even when they are given scientific justification, these categories cannot be separated from complex social and political histories. In the statistics it publishes, the French Institut National de la Statistique et des Etudes Economiques (INSEE), for example, distinguishes the salaried work-force according to levels of education, and not to income. This practice seems to derive from the encounter of the left wing and egalitarian movements of the 1930s and 1940s with the engineers who were re-organizing the state apparatus after the war. Similarly we cannot easily find a parallel in French for the distinction between professionals and managers, which is important in the American system, nor can we easily translate the German category of 'Angestellte.'

Once the categories are established however, they can be strong enough to translate into new institutions, gain efficiency, and become even truer! In France, for example, the distinction between salaried and non-salaried workers, which has been present in all statistical data since the last third of the nineteenth century, has been used by unions, administrators and politicians. It has also found its place in the labor legislation. The categories of social scientific statistics thus imbue the social with an additional stability and durability because they generate social rules and new institutions, because they are inscribed, for example, in the collective deals (the *conventions collectives* in France) negotiated by social partners under the patronage of the state. Moreover, they give to different actors new modes of identification: to be a 'non-salaried worker' for example, to be 'head of the family,' or to work in the 'public sector.' In this very strong sense, science fabricates the social and the political.

My last example will be the 'scientistic' turn taken by the social sciences after 1945, notably in the context of the technocratic and technological optimism reigning in the United States. During the war anthropologists, sociologists, psychologists and linguists were mobilized to work alongside engineers on practical problems like the stress and fatigue of aircraft pilots, and to help with psychological warfare. After 1945 the approaches adopted in these studies were made the norms of 'good' practice in the social sciences. To be scientific in these domains meant to separate science from philosophy, and "[to solve] problems rather than [to] reflect on meanings." This also meant being operational and efficient. It meant being able to influence the world and to control it. Science, having passed its test during the war, showed how enormously useful it could be and it was only too clear that the 'humanities' and 'human sciences' had to relinquish what remained from their critical manners of thought. In short, they had to change to what was now the cardinal value: a genuine scientific approach and, if possible, one that was quantifiable.[1]

This attempt to renew the social sciences had an epistemology and a metaphysics, a social and a political value system. The epistemology was usually borrowed from the physical sciences, from electronic circuitry and from cybernetics, and it tended to be even more 'atomistic' and reductionist than before. It meant that the social was treated in terms of the psychology of the individual and not in terms of the group or its shared values. The theoretical posture was one of claimed neutrality, of claimed non-involvement with social and political 'ideologies,' a posture which sought its legitimacy through appealing to experiments (in political science as in psychology or sociology) and to a generalized will to quantify (for example in economics and even in history). De facto, the social and political value system was conservative and fought the war against communism both in theory and in 'practice.'

In short the specialists of the social sciences made themselves politically indispensable by instrumentalizing their work in favor of the state and the dominant social values, by paying homage to an image of science which the 'hard' scientists were militantly propagating in the same spirit, by redefining what counted as legitimate in their disciplines. Henceforth the social sciences would not be burdened with moral and metaphysical questions, and they would no longer be activities which could, for example, question the very idea that there existed an obvious distinction between the Scientific and the Political.

Research in the 'Hard Sciences' and the State, an Overview for the First Part of the Twentieth Century

The state has found itself endowed with two main functions in the twentieth century. One, classical and already discussed, is to maintain the social order, to find compromises in the face of competing claims and decide how to distribute the economic surplus. The second has been to take over the socialization of certain costs and risks deemed too onerous for private entrepreneurs. By way of example I would mention the reconstruction of part of the industrial and economic infrastructure of France by the state in the decade after the war and, of greater relevance to us here, the taking over of most policies of public health, of education, as well as the costs of most of scientific research and development. The corollary was an increasing role for the state in the expansion of scientifico-technical systems.

Wars are always decisive moments in the development of science and technology. The First World War confirms the rule. Since scientists had proven their effectiveness during the war with gas warfare, the detection of enemy batteries and submarines, but also the organization of production, there was subsequently a great demand for the proper integration of science into the national economy in all the belligerent countries.

The results of these demands differed considerably from one country to another. In France and Italy the calls were addressed to a state that was regarded as responsible for financing the sciences. Academics and industrialists lived in relatively distinct worlds (a situation reinforced in France by the split between universities and engineering schools), and the dominant system of innovation was not primarily in the integration of the research function in the firm. It consisted of two other practices. On the one hand, the ongoing integration of the knowledge of qualified workers and production engineers to improve production. The glass, aluminium and special steel industries, which in France were extremely competitive, show how efficient these practices could be. Secondly, it relied on the purchase of patents by firms and the external employment of *ingénieurs-conseils*. This was particularly the case in sectors like chemistry and the electrical industry where, internationally, innovation depended on the production of fundamental knowledge. The system, which had proven its worth in several sectors, was not fundamentally modified by the war.

In France and Italy the response of the state to these demands for greater involvement was slow and lukewarm. The financing of universities did not improve, their structures were not changed — and it was through the creation of national funds provided by the state that research was supported: the Consiglio Nazionale delle Riccerche in Italy, the Caisse Nationale des Sciences and the Centre National de la Recherche Scientifique in France. Prior to the second half of the 1930s the results were nevertheless rather mediocre. But even when substantial amounts of money were made available, as in France by the Popular Front government in 1936, no 'research economy' was really put in place.

For the bio-medical sciences, Great Britain is the exception: the war saw the creation of the Medical Research Council. The initial role of this council was the management of a national budget dedicated to the battle against tuberculosis, as well as the management of a bacteriological laboratory and the National Institute for Medical Research. The maintenance of the war budgets after 1918 transformed the council into a genuine research agency. Under the direction of the physiologist W. Fletcher, it initiated a policy which privileged biology (notably physiology and biochemistry) and favored university laboratories over hospitals.

The German case differs from French and Italian cases since there were many 'markets' for research before the war (businesses and foundations financed it in parallel to the state) and, in any case, Germany lost the war! Having no confidence in a Weimar Republic thought to be imposed by the winners, the university mandarins and the captains of industry were convinced that industry and science remained the only reliable forces which the country had at its disposal. They alone were the guarantors of national identity. This did not however mean that they wanted to bypass the state, as in the United States. As with the foundation of the PTR, or that of the Kaiser-Wilhelm Gesellschaft just before the war, the ideal remained that of *Selbstverwaltung*. In other words, science and industry would succeed much better if they were left free to arrange their affairs as they chose, but with the unfailing support and goodwill of the state.

The first result of the post-war campaigns in Germany was the establishment in 1920 of the Helmholtz Gesellshaft, a foundation involving industrialists in the Ruhr and electrotechnology companies in Berlin. Unlike what happened with chemistry, this foundation was a failure. Luckily for the physicists, who were directly concerned, another society soon followed, the Notgemeinschaft der Deutschen Wissenschaft. The main difference between the two societies was that the second was essentially financed by the federal government. Even so, several industrialists invested in research alongside the state, making a markedly more important contribution than that in France or Italy. These engagements were undertaken on the basis of deeprooted convictions springing from the experience of the years 1880 to 1914 when it was believed that science was Germany's trump card in economic competition and national rivalry. Nevertheless, there is an important difference with the United States, a difference which reveals that Germany was indeed a European country: the federal government was at the heart of the business, and its presence was considered a guarantee, and not as a thread to the neutrality and freedom of science.

In the United States, it was the relationship between the physical sciences and industry which was radically transformed by the war, and not the relationship with the federal government which restricted itself to relatively minor military and aeronautical affairs. A quarter of the physicists in the USA worked in industry in 1920. 1600 industrial firms claimed to have research laboratories in 1931 and 2000 said they had them in 1940. In the biological sciences, research institutions diversified through foundations, universities and agricultural colleges — but also through the Department of Agriculture with its network of agronomy stations. As is traditional, the federal state was markedly involved in agronomical matters. In industry, a new type of practitioner emerged, a "combination of physicist and industrial engineer." In universities, finally, scientists remained rather different from most of their European colleagues. Very schematically, I would say that the former were more often entrepreneurs building empires in a context were institutional and material achievement were an essential criterion for recognition while the latter more often claimed to be *intellectuels* or *savants.* The contrast could be illustrated by comparing the California Institute of Technology with its trio of founders (Hale, Noyes and Millikan) and the science faculty at the University of Paris.[2]

World War II, the Cold War, the Military-Industrial-Academic Complex and the Bio-Medical Complex

One of the most unambiguous changes in the twentieth century is the displacement, after 1945, of the center of gravity of the Western world to the United States. That is true for politics as it is for science. Of course the earth sciences in America were already very influential in the nineteenth century, but otherwise most scientific domains were dependent on Europe until the 1920s. After 1945, by contrast, the frontier and the norms of research were defined in the United States.

If the Second World War represented a major rupture in the relationships between science and politics, it was because this war was also a scientific and technological war, because it mobilized and trained a large number of scientists, engineers and technicians, and because it has never ended. Having become the incarnation of modern warfare due to the atomic bomb, radar, proximity fuzes, and also penicillin and oceanography, the sciences remained at the heart of the political system, at the nerve center of the military and industrial complex put in place during the cold war. Since the cold war was conducted by mutual displays of technological prowess, as well as by a number of contained local conflicts, to be sure, and because it supposed that the reserve army of scientists could be mobilized immediately in the case of conflict, the sciences continued to be of great interest to politicians and industrialists. From the war in Korea and the electronics of defense, from Sputnik, the space race, and the subsequent generalized studies of materials, the physical sciences and their associated techniques became a *sine qua non* of any military success.

The image of the US army distributing a miraculous medicine (penicillin) in Europe at the end of the war reminds us that the war also changed the character of biomedical research. Of course a mobilization of scientists for the health of the soldiers and the populace took place during the First World War (let us remember the image of Marie Curie and her daughter at the front, with their radiological apparatus), but the magnitude of the tasks undertaken during the Second World War on the fractionation of blood, vaccines, viruses, products against malaria, proteins and vitamins, radically changed the parameters of the problem — as did the systematic collaboration between industry and the state. One of the successes of the American OSRD was to launch large production programs and clinical trials, and to persuade industrialists to organize mass production. At the end of the war penicillin became an emblem of a new and efficient, technocratic and technicist, way to tackle disease. It represented an approach which required the labors of researchers, doctors and industry, the whole being financed and coordinated by government agencies. As a result, there was an acceleration after the war in the race to innovate in a fully expanded pharmaceutical industry carried along by the multiplication of antibiotics and their uses.

Contrary to those who see the Second World War as a parenthesis in an otherwise continuous history of the sciences (scientists went back to their research after a pause to do 'applied' work imposed on them by events) it is thus preferable to think of the post-war years as perfectly continuous with the war years. The process was surely first and foremost American, British and Soviet, since these countries immediately fought the cold war, but the economic and military constraints of the following decades, like the need for each scientist to remain in touch with the research frontier, meant that the new way of practising science was absorbed in continental Europe.

The relationship between the 'hard sciences' and the state changed drastically because scientific practices themselves were modified, because the war led to profound transformations in what it meant to be a scientist. In physics for example, instrumentation was no longer a mere means for doing experiments, but passed to the core of the research process and became at once its *raison d'être* and its output. Radar is a perfect example since it was at the origin of many research fields, like magnetic resonances and hertzien spectroscopy, after the war. Similarly, masers and lasers became interesting physics tools generating remarkable phenomena, like 'coherent' light and frequency standards, and were used for precision measurements and in weapons systems. In the 1950s and '60s, the driving force of instrumentation was reproduced in other fields than physics. The corollary of this evolution, which saw objects and know-how take the lead over 'abstract' knowledge, was the priority accorded to the practical control of phenomena over their 'understanding.'

Another symptom of this new definition of physics in the post-war period was the loss of prestige of the 'great philosophical questions,' so cherished in Bohr's institute in the 1920s and '30s, coupled with the extension of quantum techniques to all domains. Fundamental theorists were still essential, and they were highly respected, but they no longer had that mythical status which was accorded to the founders of quantum mechanics. They were also in minority with those (the 'phenomenologists') whose job it was to deal with the mass of experimental results produced in the laboratories. Seeking theories which were locally coherent and which could be immediately useful and produce numbers, their role was to display a practical efficiency. They thus participated in the development of a science which was increasingly integrated into its economic and political environment, and contributed to the multiplication of the sites where knowledge was produced. These were now the universities and the technical institutes, the national laboratories and the industrial laboratories (Siemens or General Electric), but also the myriad of small firms established as a result of government contracts. In all these places the differences between fundamental research, technical development and the invention of artefacts and techniques became increasingly blurred.

A similar development occured in the biological sciences. Contrary to what is commonly understood, molecular biology is not a theory which would have been later applied as genetic engineering. On the contrary, it was from its inception a 'biotechnology.' The phenomenon is connected to a change in the nature and role of instrumentation, and to the emergence of the biomedical complex. The emergence of a 'biomedecine' is undoubtedly the main feature of the post-war period. Medical organizations became the principal operators of research, and the legitimation of clinical practices depended increasingly on laboratory practices. One can speak here of a biomedical complex bringing together pharmaceutical industries, hospitals, universities and government agencies. After 1945 these institutions participated in all major national programs (from the war on polio to that on cancer), and their main targets were the fabrication, identification and selection of new chemiotherapeutic remedies.

The central role of instrumentation was clear from the place taken in all laboratories by ultracentrifuges, radioactive isotopes, electrophoresis equipment, electron microscopes or equipment for crystallography and X-rays. This instrumentation was not only a means to an end however, but contributed directly to the results which were produced. The standardization and commercialization of ultracentrifuges, for example, played a major role in the generalization of a vision of proteins and viruses as macromolecules. Similarly in the fifties, the biochemists multiplied the subcellular entities defined by the instrument's parameters (like the number of rotations per minute). Finally the influence of physics models was evident in theoretical matters. Cybernetics and works on control systems underpinned a conceptualization of biology in terms of information theory (genetic codes, signal transfer, translation, etc.) and 'feed-back' models were used to reformulate biochemical and physiological regulation.

Changes in the Social and Political Practices of Physicists and Biologists

It was not only the topics and the way of dealing with them that changed in the post-war period, but also the organization of scientific work, and the relationships between scientists and the Social and the Political. The term Big Science encapsulates some of the meanings. It refers to heavy equipment (reactors or accelerators, for example) whose use is shared; the 'science made by committees' (which will now decide who is permitted to experiment and under what conditions); work in teams (a collaboration at CERN in the 1980s could involve 400 scientists and last ten or fifteen years); national programs financed by contracts and aimed at resolving practical, military or technological, problems; and of course the state, the increasingly important if not exclusive supplier of funds. These structures, characteristic of large cold war programs, have had effects which have trickled down to 'little science' which was increasingly financed and assessed in the same way.

The new relations which the scientific elite had with the political world really emerged in the USA when they participated in the think tanks set up by the military in the late 1940s. In the fifties and sixties the most prestigious of these bodies was the Institute of Defense Analysis, a target for students opposed to the Vietnam war. The inheritors of the practices of Operations Research, notably developed to manage anti-aircraft defenses and to organize convoys in the North Atlantic in the United Kingdom during World War II, these institutions served to directly integrate scientists in the elaboration of major military and strategic planning.

In Western Europe, similar but less important structures were put in place, and, with the exception of Britain, later than in the USA. In France, for example, such a system was put in place only after de Gaulle's assumption of power. The creation, between 1958 and 1961, of the Délégation Générale à la Recherche Scientifique et Technique (an organism put in place by the state to orient, via targeted financing, the national research effort) and the Direction des Recherches et Moyens d'Essais (the supplementary body set up by the military) led to an unprecedented integration of scientific milieus into technological and military research. This relation remained less organic, or perhaps more hidden, for the bulk of European scientists than for their American colleagues. The main reason certainly was the greater diversity of political attitudes and the influence of leftist culture in scientific circles of many European countries.

Displacing the traditional role of agriculture, biomedical progress became the principal source of legitimacy for the biological sciences in the United States. The ceaseless increase of funds for research voted by Congress turned the National Institutes of Health into the leading research entrepreneurs ahead of the Department of Agriculture and foundations like the Rockefeller or the American Cancer Society. The success of antibiotics encouraged the celebration of continuous and regular biomedical progress, which diverted debate on a possible national health system. From the forties onwards discussions on national health insurance (a social and political question par excellence) increasingly gave way to the idea that support for research was the solution, the solution which would solve the political problem.

The combination of laboratory research and clinical practice went through a last transformation for which the political system was directly responsible. The best example of this was the normalization of screening procedures and clinical trials of chemiotherapeutics prepared by industry. The mass procedure called 'randomized and double blind' became the norm for clinical trials. It was centrally established by the state apparatus and supported by the experts who were statisticians. The rather strong doubts of some practitioners who were unhappy with methods of statistical aggregation which simply ignored their clinical know-how, were overcome, in the after-war period, by the organization of large-scale trials using already accepted antibiotics. Committees which brought together doctors, biologists, industrial experts and health administrators organized these trials. They chose the doctors, defined which patients were to be selected, and strictly limited the local autonomy of practitioners. Upstream of these trials, organized under the supervision of the accredited representatives of government agencies, a first selection of products was made from the thousands annually produced by the pharmaceutical industry. This selection was made, often also on the initiative of the state agencies, from living 'models' produced industrially and in series, like laboratory mice. In this way a techno-scientific complex unknown before the war was put in place, a complex largely regulated by the state apparatus.

The change in the relation between scientists and politics and public ethics was also profound. If before 'savants' were prone to present themselves as servants of truth and as bearers of humanist culture, which was the pre-war tradition in Europe, the scientists who started working during the cold war found themselves in a position which could not but force them to distance themselves from such ideals. Hitherto the image they had of themselves was that of the disinterested intellectual linked to a discipline that enriched culture. The newcomers saw themselves more often as professional scientists working in large systems determined by ponderous technical and social determinants which they were expected to make functional and efficient.

Before 1940, the rule was that the scientist of renown became a star in the cultural firmament. It was kind of an obligation to share a few philosophical writings or moral reflections. Mach, Planck, Poincaré, and of course Einstein and Bohr did so. The scientists often assumed the posture of the savant philosopher "who had the duty, or the psychological need, to put in place a coherent vision of the world." In the fifties and sixties, this kind of man became rarer. And although the change happened later in Europe, had less impact on the older generation, and influenced disciplinary fields in quite different ways, it became the most common norm. The best indicator of this change is the kind of 'testaments' which scientists now leave for future generations. These are increasingly hedonist biographies, but more rarely works of reflection.[3]

Conclusions

Two major institutional changes specific to Europe need to be given in conclusion. In the post-war period, the feeling that a huge gap had opened up vis-à-vis the United States led to a movement among European scientists in favor of transnational solutions. This movement materialized with the creation of CERN in the early 1950s. The CERN solution was often invoked in the subsequent decades. The model was either reproduced (as with ESRO, the European Space Research Organization or with EMBL, the European Molecular Biology Laboratory), adapted via the Europe of the Six (as with Euratom), or transformed into bi- or multilateral arrangements (as with the ILL, the Laüe-Langevin Institut in Grenoble, or the Franco-German satellites of the 1960s). In all these cases intergovernmental solutions controlled and financed by European states were a means to overcome 'the gap.'

At the level of individual countries the solutions found to the American challenge were often those of national centers, that is to say laboratories wanted and established centrally to tackle broad questions deemed to be in the national interest. In France the laboratories were often purely and simply creations of the state apparatus. The best known examples are of course in nuclear or telecommunications research, and industry became involved only in a second stage. In Britain, military establishments played a key role while in Germany the solutions often implied industry and the state(s) from the beginning. In any event, and this is the important lesson of the post-war era, these new arrangements led to the creation of complexes which bound the world of politics, the military, industrialists and academics more organically together than ever before.

References

What is said on the biomedical sciences is directly borrowed from a course given in July 1994 by Ilana Löwy and Jean-Paul Gaudillière during the summer school held at the cité des Sciences et de l'Industrie, Paris.

1. Heims, S.J. *Constructing a Social Science for Postwar America.* (Cambridge: MIT Press, 1993).
2. Weart, S. "The Physics Business in America, 1919–1940, A Statistical Reconnaissance" in N. Reingold *The Sciences in the American Context: New Perspectives.* (Washington: Smithsonian, 1979), pp. 295–358.
3. Holton, G. 'Les hommes de science ont-ils besoin d'une philosophie?', *Le Debat* (1985), 116–138.

Further Reading

F. Caron, 'La cacité d'innovation technique de l'Industrie française. Les enseignements de l'histoire', *Le débat* (Septembre–Novembre, 1997), **46**: 37–52.

A. Desrosières et L. Thévenot. *Les catégories socio-professionnelles.* (Paris: La Découverte, 1992).

Y. Ezrahi. *The descent of Icarus.* (Cambridge: Harvard University Press, 1990).

P. Forman. 'Into Quantum Electronics: The Maser as 'Gadget' of Cold-War America,' Paul Forman and José M. Sanchez-Ron, *National Military Establishments and the Advancement of Science and Technology.* (Dordrecht, Kluwer, 1996), 261–326.

R. Fox. 'France, Research, Education, and the Industrial Economy in Modern France,' in *The Academic Research Enterprise within the Industrialized Nations: Comparative Perspectives , Report of a Symposium.* (Washington D.C.: NSF, 1990), 95–106.

J.P. Gaudillière and I. Löwy. 'Disciplining Cancer. Mice and the Practice of Genetic Purity,' in J.P. Gaudillière and I. Löwy (eds.). *The Invisible Industrialist. Manufactures and the Construction of Scientific Knowledge.* (London: McMillan, 1997).

Steve Joshua Heims. *Constructing a Social Science for Postwar America.* (Cambridge: MIT press, 1993).

Dominique Lecourt. *Prométhée, Faust, Frankenstein, Fondements imaginaires de l'éthique.* (Paris: Synthélabo, collection *Les empêcheurs de penser en rond*, 1996).

Chandra Mukerji. *A Fragile Power, Scientists and the State.* (Princeton: Princeton University Press, 1989).

D. Pestre. *Physique et physiciens en France, 1918–194.* (Paris: Editions des Archives Contemporaines, 1984).

D. Pestre. 'The CERN system, its deliberative and executive arms,' in A. Hermann *et al.*, *History of CERN, vol. II.* (Amsterdam: North Holland, 1990), 341–415.

D. Pestre. 'Pour une histoire sociale et culturelle des sciences, Nouvelles définitions, nouveaux objets, nouvelles pratiques,' *Annales.*

Jacques Rancière. *Les noms de l'Histoire, Essai de poétique du savoir.* (Paris, Seuil, 1992).

S. Weart. 'The Physics Business in America, 1919–1940, A Statistical Reconnaissance,' N. Reingold (ed.), *The Sciences in the American Context: New Perspectives.* (Washington: Smithonian Institution Press, 1979), 295–358.

CHAPTER 5

Crafting and Disciplining the Tropics

Plant Science in the French Colonies

CHRISTOPHE BONNEUIL

"The means at his disposal, and the limited character of his needs," wrote Pierre Clerget in 1911, "prevents the 'savage' exploiting his environment abusively. Destruction starts with civilization [. . .]. But that is only the first step. Faced with the consequences of his abuse, the 'civilized' becomes aware of what he has done and embarks on 'rational' exploitation. The use of fertilizers, reforestation, fish breeding, the domestication of the ostrich, 'water policy,' all are steps back in time and simultaneously indicative of a better understanding of 'geographic realities.'" Thus, for this professor at the École Supérieue de Commerce de Lyon, promoting a "rational exploitation of the globe," colonization and its devastation was a fundamentally positive intervention to the extent that it led to a better control of the environment and to the progress of knowledge.[1]

The case of the plant sciences in the French tropical empire from the end of the nineteenth century will thus lead us to think of the colonialist phenomenon as a knowledge activity and to see science in the colonies as a process of control and of intervention. Michel Foucault made it clear that knowledge is not merely utilized for the exercise of power; one has recourse to mechanisms ("dispositifs") which fabricate knowledge and power, which render the natural and the social worlds at once more transparent, more amenable to control, and more productive.[2] He thought of the army, the school, the hospital and the prison as institutions equipped with these mechanisms, coproducing knowledge and domination effects which were then disseminated throughout an entire disciplinary regime. Science studies have recently described the laboratory as a central place where one forges locally a control of natural phenomena, as well as practices of inscription, of work discipline, an obsession with predictability, an ethics of precision and standardization which are then disseminated into our "calculative world," as Joseph Rouse suggested.[3] My concern here is thus the constitution of the tropics as a scientific object and their integration at the periphery of a calculative world.

I begin this chapter by describing the kind of orientalism which is at the origin of the idea of tropical nature, and, following Xavier Polanco, I suggest that a world-science was built in the same movement as a world-economy. I then go on to explore some important sites — botanical gardens, agricultural experimental stations, plantations — from which one can follow the constitution of the tropical space as a vast field of investigation, of intervention and of control.

The Invention of Tropicality

Climatic conditions and a common geological history gave the tropical landscapes a uniqueness which struck all travellers. This was a world without winter, a world in which the closeness to the equator implied fluxes of solar radiation, photoperiods, temperature ranges, and hydrological patterns which differed significantly from those in temperate zones, and had clear consequences on vegetative life. What is more, unlike the temperate zones whose vegetation was stripped by glacial phenomena in the quaternary period, tropical ecosystems were blessed with greater stability, from which their exceptional biodiversity.

The uniqueness of nature between the tropics of Cancer and Capricorn is not only to be traced to its natural history, however. It is also a product of Western culture. At the time of Pliny it was the world of the unknown and so of the extraordinary: the rich regions of India were populated with monsters, like the giant ants which hid gold in the heart of their anthills. Later, for Bernardin de Saint-Pierre, but also in Alexander von Humboldt's more instrumental "physical portrait," tropical nature was, above all, a world of beauty and heroism. It offered a limitless collection of new phenomena, and everything in it appeared to be more beautiful, more powerful, more colorful than in the temperate zones. Luxuriant and fertile, but also dangerous and fragile. Richard Grove showed that environmental concern and the naturalist project of a rational management of nature can be traced back to the eighteenth century in tropical islands. The disruptive colonial intervention, at first perceived in the limited space of islands, was to be experienced throughout the tropical world in the industrial age.[4]

Nature exuberant and fertile, savage nature that was to be civilized by the plough, to be disciplined by a scientific harnessing, nature as capital to manage or as sanctuary to preserve . . . we still need a comprehensive study on how the notion of tropicality was forged in the age of imperialism at the end of the last century and the beginning of this.[5] It is already clear however that representations were strongly influenced by the concerns and interests of successive Western observers. In the case of Benin in West Africa, Dominuque Juhé-Beaulaton's very detailed semantic analysis shows how the 1880s constitute a turning point in the Western perception of landscape. Whereas 'fields' used to occupy the first place, suddenly the 'forest' becomes "huge, virgin, continuous, impenetrable." As well as an inexhaustible reservoir to be exploited, it appeared indeed for the explorers and the military as a challenge ripe for conquest.[6]

Coining the term "virgin forest" was suggesting that large parts of the tropics remained to be used. For colonial scientists, the stress upon the specificity of the tropics went along with claims for autonomy and with discipline building. But defining an other world was also to postulate a complementarity, to naturalize an economic division of labor. Michael Osborne documented the initial dream of French military, engineers and the Société Impériale Zoologique d'Acclimatation to turn Algeria into a tropical colony on the model of the West Indies, that would not compete with metropolitan agriculture.[7] The sun in the Indies does not only shine for the Indies," said A. de Haulleville during the International Congress of World Expansion in 1905," and the inhabitants of the northern countries are entitled to the excess of light and heat which nature exudes there with a prodigenous generosity. In return the inhabitant of the tropics is entitled to the products coming from the cold zones. He is entitled to the iron which one finds there, and which one manufactures there, to the works of industry, of art and science, to all the advantages of religion and of civilization." So emerges the imperial dream of the rational exploitation of a planet with complementary resources, and of a science which will be the enlightened steward of the biosphere.

Exploration, Conquest and the Great Divide

The industrial revolution and the decline of the slave trade led, in the last century, to the expansion of European rule in the hitherto uncolonized tropical regions, and to the establishment of new ways of organizing and exploiting nature and human beings: a more systematic plunder of natural resources (minerals, ivory, gums and rubbers, wood), a new step forward in the plantation economy (notably in the Far East), as well as the orientation of indigenous farming towards cash crops farmers (particularly in West Africa). Place of birth and life, place of collective identity, nature was thus desanctified, pillaged, reorganized, so as to become simply a center for the production of resources needed by industrialized societies.

As European domination extended over the labor force and the natural resources in the tropics, a succession of voyages ensured an accumulation in the centers of a world-science of notes, of maps and drawings, of herborium species, of vegetable material, and allowed the elaboration of an hegemonic global knowledge. The perfectioning of the techniques of representation, of transport and of conservation ensured that vegetal materials and information were made sufficiently durable to cross oceans, while the systems of classification created an order which allowed accumulation and synthesis. The travelers undertook to collect names and certain local knowledge which were connected to new plants and which were incorporated at once into botanical science and into commercial activity, so annexing native knowledge into a comprehensive big picture. This process of integration was also a process of disqualification, constructing the great divide between beliefs and Western science.

Indeed, as imperialism advanced at the end of the last century the colonized societies, whose agricultural and botanical competence had often been praised in other ages, now found their knowledge and their relation to nature massively disputed by the colonizers engaged in a global struggle for resources. Thus Barret remarked that the African could not help the scientist lift the heavy veil that concealed nature's secrets because "he looks, but he barely sees, and having learnt nothing as he labours under the burden of his routine habits, he takes from the abundance of this rich milieu no more than that which his daily needs require. The black man 'does not know'! He is indifferent to the world around him, he is indifferent to himself."[8] In the view of the colonizers the fertility of the tropical territories was matched only by the laziness of the natives. Paul Leroy-Beualieu, principal French theoretician of colonialism claimed that the colonial movement "covered all the remaining open spaces of the globe, inhabited either by sleepy and languid peoples, or by tribes who were incoherent, devoid of any sense of progress and unable to exploit the regions where destiny placed them."[9] It was just this presumed capacity of the Europeans to know how to exploit rationally the tropical space which gave them the right to colonize. As Leroy-Beaulieu's English counterpart put it "The tropics will never be developed by themselves."[10]

Botanical Gardens and Experimental Gardens, the Laboratories of a Domesticated Tropicality

With the affirmation of the richness and fertility of the tropical zones, and with the urgent need to let tropical nature be exploited by the enlightened colonizers, a neo-physiocratic attitude, Richard Drayton suggested, took hold which was largely promoted by the scientists themselves and which created a demand for science.[11] Businessmen and administrators turned to botanists to assess the natural resources, to identify a plant, to determine the value of its product and the possibilities of growing it in this or that colony. Making inventories was part of the colonial culture.

In Britain the influence of the Royal Botanic Garden at Kew reached its apogee under the directorship of William Thistleton-Dyer from 1885 to 1903. It constituted then the center of a vast network for the exchange of plants, the central agency for information on the agricultural development of the colonies and the nursery for specialists sent abroad. The Botanische Zentralstelle für die deutschen Kolonien established in 1891 in the Botanical Gardens in Berlin played a similar role in the German empire. In France three institutions vied for the honor: the Institut Colonial of Marseille set up by botanists and businessmen in 1893, the Muséum National d'Histoire Naturelle in Paris, and the Jardin Colonial founded by the ministry of the colonies in 1899 near Paris. An École Supérieure d'Agriculture Coloniale, organized according to the French model of the 'Grandes Ecoles' was added to the latter in 1902.

Established at the beginning of colonial expansion, the gardens and experimental gardens were, in 1900, among the most important scientific institutions set up in the tropics. To the 28 establishments existing in 1880 in the British possessions, most of them created before 1820, were added 60 new gardens by 1900. Several dozens of new establishment were also set up in the French and German possessions between 1880 and 1900. In the Dutch Indies, where agricultural colonization already had a long history, the botanical garden in Buitenzorg was, at the turn of the century, a major center for tropical botany and ecology, and had created agricultural research stations for the main colonial products by means of contracts with the administration or with the planters. There were thus several hundred European botanists or botanist-gardeners working in the tropical periphery of the empires.[12]

Maxime Cornu, professor of horticulture in the Paris Museum trained most of the directors of the French experimental gardens overseas, selected among qualified gardeners from the École Nationale d'Horticulture de Versailles. These establishments varied considerably. Some were mainly designed to provide the Europeans with fresh vegetables while others were genuine botanical gardens comprising an herbarium, laboratories and a library. Some general traits can nevertheless be pointed to, deriving from the fact that the gardens constituted one of the main places for the taming of the tropical environment at the opening of the colonial era.

First and foremost, the systematic organization of the transfer of plants between the different parts of the planet, through the network of botanical gardens, profoundly redrew the agricultural and botanical map of the tropics. Well known examples are the rise of the hevea plantations in South East Asia after the transfer, via Kew, of seeds from Brazil; of the expansion of the cultivation of the cacao tree in West Africa, and the cinchona in the Dutch Indies. Landscapes were transformed. Large regions were reshaped by the introduction of exogenous vegetable elements. Gardens were the bridgehead of plantation agriculture. They were at once a showcase for the potential of each colony, the center for the distribution of plants to the settlers and demonstration plots for improved methods of cultivation.

Slowly breaking with the lamarckian conceptions which were so central and so specific to the French acclimatization movement, the systematic efforts at cultivating all kind of plants under the climate of the French tropical possessions (and the many setbacks) led to an appreciation of the limits of plant adaptation. Unlike the traveler who simply saw one place at one particular season, the gardener-botanist had to deal with the climate throughout the year. Ways of confronting nature sharply differed. In the controlled and filtered space of the gardens, the behavior of many different species, represented by only a few examplars, were placed under a kind of clinical scrutiny. It is not by moving through the field, but through a close and daily contact with — mostly imported — plants that gardeners contributed to a richer understanding of the climates, the phytogeography and the agricultural potential of diverse tropical zones, and to the establishment of suitably adapted agricultural and horticultural calendars.

Many historians have described botanical and experimental gardens as tools for the development of colonial agriculture and as important settings for the institutionalization of science overseas. But techniques for transporting plants and seeds and garden practices, and the local relations between scientific and technical staff in charge of the garden with the administrators, traders and the settlers remain overlooked. We need a better understanding of these vegetal "microworlds" in which nature was manipulated in the tropics.[13] It seems to me very fruitful to follow how objects, practices and ideas were crafted in the garden and made stable enough to spread out of its borders and to contribute at a macrolevel to attempts at reshaping tropical nature. In short we must understand gardens as laboratories for the control and 'civilization' of the tropics.

Because it was surely about civilizing. In the new territories "the European uprooted from his country attaches himself to his recent past with a kind of religious fervour (. . .) he faithfully protects his civilised customs in the midst of savagery."[14] Once the novelty of exotic foods had worn off nothing tasted better than a simple apple, and the French settlers tried to manufacture the Camembert. Brought to their knees by fever and the climate, the European inhabitants of the first colonial towns dreamt of the vegetables, the smells, the flowers, the colors of their native land. The experimental gardens were thus busy introducing the flowers the French liked most. The rose, which the colonial officer Pobéguin called "the French flower par excellence," or the carnation (*Dianthus caryophyllus*) were very prized.[15] In addition to these temperate plants, tropical or subtropical flowering plants, common in summer in the European gardens since the middle of the nineteenth century were also introduced, such as several species of dahlias or Geraniums (botanically *Pelargonium*). More rare in the metropole, the *Bougainvilea spectabilis* became one of the French colonizers' most valued ornamental plants in colonial outposts. Plants like *Lippia citriodora* and *Cymbogodon citratus* were also largely grown to repel mosquitos. Indeed, the maintenance of a 'civilized' life in the tropics required the appropriate plant environment. The activity of gardening, of central importance both to the military and to the missionary, was warmly recommended as a hobby to all the colonists for its moral value, no doubt to avoid the degeneration experienced by Ferdinand Bardamu as depicted by Celine's *Journey to the end of the night!*[16] It is by situating them in the European societies then being established in colonial settings, that we can understand the botanical gardens as institutions for both culture and cultivation.

The botanical and experimental gardens also affirmed the technical leadership of Western civilization. In the first phase of colonization experimental activity was directed primarily at establishing the superiority of the rulers' values and techniques for the management of nature, land utilization, the working and fertilization of the soil. Around 1900 Enfantin set about demonstrating, in the thin and light Senegalese soils, the virtues of planting in rows and of deep ploughing for the cultivation of the peanut. "We will penetrate further and further into those famous lands, still referred to as unknown on our present maps, with the help of our civilizing instrument par excellence, by which I mean the plough," he wrote.[17] One needed several pairs of oxen to pull a heavy plough but who cared. The results were there (staggering outputs of three to seven tons per hectare which no other trials could subsequently reproduce . . .). The governor general was enthusiastic about the superiority of European methods and expected increase in production. In Guinea in the same period, the director of the Jardin d'Essai in Camayenne complained that his African workers "were not familiar with the work of cultivation" to explain the difficulties he had had with making a clearing which he had ordered in the midst of the rainy season!

Having established scientifically that the native did not *know* how to use his environment, the European had to demonstrate how to harness nature properly. The garden, which by definition is an enclosed space, always echoes the image of a domesticated nature. Experimental gardens in the French colonies of tropical Africa, Madagascar and Indochina, had more precisely to dramatize the capacity of the white man to dominate and to order tropical nature, to civilize the bush. The experimental garden was the laboratory in which a teeming, disturbing, savage nature had to be tamed, ultimately disciplined, boxed in, remodeled according to the plans of a sophisticated exoticism. The gardens contained a rich variety of plant species, gathered from all over the globe, and requiring the most painstaking horticultural care, which displayed the multiple potential of the territory so as to attract investors. The avenues were carefully laid out according to the principles of garden architecture that Édouard André who had worked for the hausmannian transformation of Paris was teaching at the École Nationale d'Horticulture de Versailles. The grouping of the plants in patches was made according to their economic uses, rather than to the botanical classification or phytogeographical views. This functional approach expressed the ideal of the organization of nature as a reservoir of resources for industry and commerce. So was constructed in the space of the garden a kind of tropicality at once cosmopolitan and carefully controlled which would then leave its mark on the European plantations and on the colonial towns.

Conakry, capital of French Guinea, is an interesting and vivid example. In the 1890s Governor Ballay undertook to make of this village a modern colonial town and the major port of the Upper Guinea coast, to the disadvantage of Freetown. As in the experimental gardens, 'savage' vegetation had no place in the checkerboard cadastral plans of 1890. "Here the newcomers have let loose the fire and the axe. Their slogan was: flatten to the ground; their ideal was a 'champ de Mars'" complained a member of the Binger mission who passed through in 1892.[18] The military reference was not without foundation since the inauguration of the governor's residence in 1890 was accompanied by the staging of a display of colonial power in the tradition of the royal avenues: the 'place du Gouvernement' was covered with sand then planted with banana and pineapple trees in ordered and symmetrical arrangement. But colonial urbanism in Conakry repeated essentially the Hausmannian plan: a frame of the wide avenues bordered by two rows of trees on both sides, complemented by narrower arteries bordered by a single row. The vegetation had to respect the imperatives of geometry, aeration and hygiene, of order and of security. An experimental garden was established in 1897 in Conakry, headed by Paul Teissonnier. He supplied the Public Works officials with new species of avenue plants, in addition to the usual mango trees. The fine and elegant horsetail beefwood (*Casuarina equisetifolia,* from Australia), the siris tree (*Albizia lebbeck,* from the West Indies), the mahogany (*Khaya senegalensis,* from Senegal), the coconut palm, and the flamboyant tree (*Delonix regia* from Madagascar) with its scarlet inflorescence, were thus introduced from different parts of the world in the streets of the capital of French Guinea. This was imposed neglecting the African population's own preferences for and uses of certain trees. The police were also busy with the cattle, which roamed freely in the towns and were fond of the young trees. Teissonnier especially recommended the horsetail beefwood and the siris tree because they resisted the damage caused by the cattle.

These tropical yet new exogenous species defined a cosmopolitan exoticism which characterized colonial towns, and made them echo one another all over the tropics. They reinforced the image of modernity and urbanity of Conakry, which willingly presented itself between the two wars as the model colonial town. This vegetal organization also helped construct the tropical identity of Conakry, in contrast to Europe of course, but also as opposed to the scant vegetation found in the previous stop on the voyage, namely Dakar. It gave the town that picturesque stamp which ensured it first place among the West African ports of call.[19]

The town thus displayed a sophisticated and constructed tropically, worked out in the first instance in the experimental garden, that was an emblem of the colonial way of life worked out in the first instance in the experimental garden.

The Plantation as Laboratory

In a few cases the dream of a tropical nature entirely managed according to the settler's plan could be realized in viable estates. In the French empire the rubber plantations of southern Indochina provide the best example. In the 1930s Indochina's production of rubber, mostly concentrated in the hands of a few companies like *Terres Rouges* and *Michelin,* ranked fourth in the world and was the second most important export from the colony.

Alexandre Yersin, the director of the Pasteur Institute in Nha-Trang, was one of the very first rubber planters in French Indochina. In 1902, his plantation, which was managed by the agronomist Georges Vernet, was equipped with a chemistry laboratory. Just when extremely brutal tapping methods were practised in the plantations of Malaysia and Ceylon, Vernet became interested in the physiology of the production of latex which he saw as an ongoing function of plant life. He thus opted for moderate and daily tapping which kept the rubber concentration constant in the latex. He built a correlation table between the density, the temperature and the concentration of rubber. This tapping system required the daily measurement of the output of each worker (weight and density) in order to check the concentration of the collected rubber. This allowed Vernet to craft experiments comparing systems on the yield of trees under different tapping conditions, and to identify high-yield trees for seed selection. The same organizational strategy also enabled him to estimate (and reward or sanction) the activity of each tapper. "Since it is difficult for me to keep perfect watch on the workers spread throughout the plantation," explained Vernet in 1909, "one of them thought one day to take advantage of the situation by not tapping all the trees in his sector, replacing the latex he had not collected with water. This fraud was immediately revealed when the density of the latex was measured."[20] Vernet was thus not only interested in producing rubber but in organizing the plantation in such a way that the output of each element, be it human or vegetal, could be known, and compared and combined with that of another. In the inter-war years big estates applied this principle to thousands of hectares. Agronomical research stations dedicated to rubber were nearly inexistant in Indochina compared to Malaysia and the Dutch Indies. And yet, on the eve of the Second World War, the big Indochinese plantations were most successful in Asia. Younger than others, they had been deliberately equipped on a massive scale with the best performing clones which had been developed in the Dutch Indies in the 1920s. They were indeed sites of invention and intervention and constituted themselves as veritable experimental spaces.

In the space of the great plantation the control of the vegetal resonates with the discipline of the coolie. Wild animals were kept out and potential deserters from the contract labor force kept in its enclosure. Forest diversity was transformed into a homogeneous population of *Hevea brasiliensis*. In the thousands of hectares divided into divisions, blocks, sections and tasks, a single tree was present in a multitude of similar clones, each exactly the same habit, the same color of green, the same tapping incision made at the same level on the same face of the trunk, so that the (European) assistant manager could control, with one glance, the work that was done, like a general who sees just one head. Research directed to optimization of productivity, the effort made at standardization of methods and gestures, the quantitative measurement of different factors of output and costs, the ethic of precision, produced at once order, profits and knowledge. They made of the great plantation a laboratory, as well as a factory and a barrack. A human and vegetal microworld was constructed as a controlled system, with observable and manipulable elements, where intervention and inscription were aimed at producing signs, stable facts: the cheapest tapping system, the best weeding methods, the right distance between the trees, the workers to be rewarded or punished, the most productive clones, etc.

Scientific organization of tapping in the 1930s is worth exploring in further detail. The method of tapping one day in four on the entire circumference (denoted S.J/4), rather than one day in two along a semi-spiral (S/2. J/2) was developed in Malaysia by the Socfin company and in Indochina by the Terres Rouges group (Société de Plantations des Terres Rouges, Compagnie du Cambodge, Compagnie de Padang) which owned a quarter of the surface planted with hevea in this French colony. These two companies were controlled by the Franco-Belgian Société Financière des Caoutchoucs, one of the most important plantations-holding companies in the world. As tapping represented 70 percent of the costs of production, the aim in this period of crisis was to increase the productivity of the tapper without affecting the yield per hectare. The "full spiral," which reduced to 90 the number of days annually on which a tree was to be tapped, was put to the test in 1931 by François Gain of the technical services of Socfin in Malaysia. On a plot chosen for its homogeneity, one of his experiments compared "tasks" (being a group of heveas tapped during the morning by one coolie) in S.J/4 to "tasks" in S/2. J/2. As the worked proved to take about 30 percent more time in full spiral, the S.J/4 task was reduced to 250 trees instead of 350. The tappers' individual differences were compensated for by using them alternatively for the two kinds of work. Each task and its trees, each coolie and his bucket, were given a number. The weight of each bucket was measured each morning by an overseer before its contents were thrown into a vat for transfer to the factory. The overseer also took a sample from each with a pipette, of which 50g was coagulated, stocked, numbered and weighed after drying out. From this the daily production of dry rubber from each task was inferred. Gain obtained an increase of 37 percent in output per hectare for the full spiral, and of 95 percent in the productivity of the tapper.[21] Similar experiments by the research staff of the Terres Rouges led in 1936 to the standardized adoption of the full spiral on thousands of hectares, so allowing for a reduction by 30 percent of the number of tappers and a reduction in the supervision required.

Researchers from the technical departments of Socfin and Terres Rouges were not interested in the productivity of the trees per se, but only in its association with the tapper at work. They searched for the optimal relation to be set between the physiology of the trees and the tapper's activity. Basing himself on the methods of Taylor and Gibreth, Gain broke down the movements performed by the tapper while tapping the tree. His work led to the adoption of a sequence of standard movements. The generalization of clones ensured uniformity in the trees, and assisted this standardization of movements.

In the 1930s the demand for 'accurate tapping' emerged along with a device for measuring this accuracy. The control of tapping, initiated by the HAPM in Sumatra, attempted to quantify the quality of the work. For each worker, a few trees from his task was sampled from time to time and a different parameter measured: the depth of the notch (too superficial does not yield enough but too deep injures the tree), the consumption of the bark, the frequency of the wound injuries, the slope of the wound. For each parameter, a scale of points was set, proportional

Figure 5.1: Expedition of plants in wardian cases from the Jardin Colonial to overseas gardens. (no date). © CIRAD-Nogent

Figure 5.2: Practical training for packaging plants in wardian cases in the Colonial Agricultural College in Nogent (Ecole Supérieure d'Application de l'Agriculture Tropicale). (no date). © CIRAD-Nogent

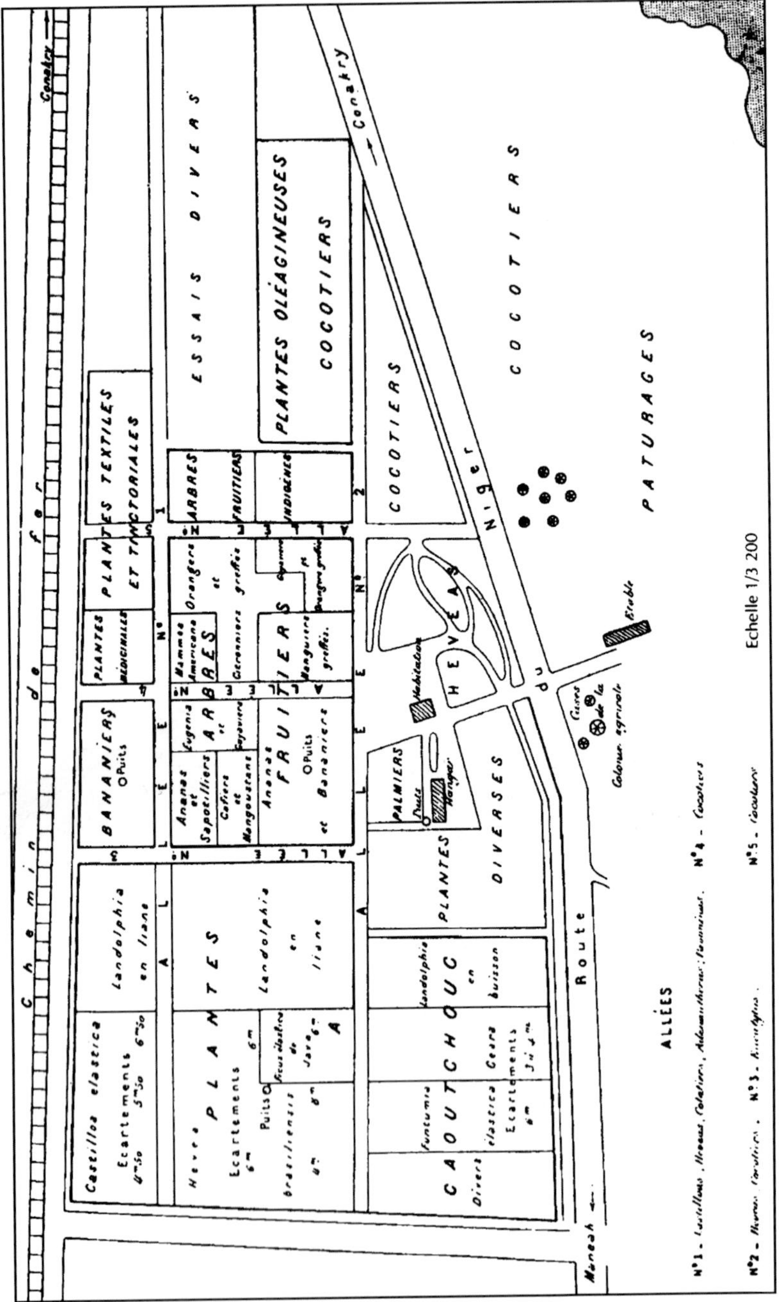

FIGURE 5.3: CONAKRY EXPERIMENTAL GARDEN (FRENCH GUINEA, 1906) Y. HENRY (GOUVERNEMENT GÉNÉRAL DE L'AFRIQUE OCCIDENTALE FRANÇAISE, INSPECTION DE L'AGRICULTURE), *RAPPORT AGRICOLE POUR L'ANNÉE 1906.* (PARIS, CHALLAMEL, 1907), PAGE 95.

FIGURE 5.4: BAMBEY GROUNDNUT EXPERIMENTAL STATION (SENEGAL). THE STAFF. 1930. © CIRAD-NOGENT

FIGURE 5.5: BAMBEY GROUNDNUT EXPERIMENTAL STATION (SENEGAL). A LAND TOTALLY CLEARED AND A PERFECTLY REGULAR PLANTING. 1930. © CIRAD-NOGENT

FIGURE 5.6: BAMBEY GROUNDNUT EXPERIMENTAL STATION (SENEGAL). 1930. A PERFECTLY REGULAR PLANTING. © CIRAD-NOGENT

FIGURE 5.7: SELECTING THE PARENT PLANTS FOR THE STRAINS "1931". BAMBEY GROUNDNUT STATION. 1930. © CIRAD-NOGENT

FIGURE 5.8: COMPARING THE STRAINS: "ROW VERSUS ROW" EXPERIMENT. © CIRAD-NOGENT

Figure 5.9: Sampling of 100m² in a large multiplication field of the station, To estimate the yield per hectare. Bambey. 1930. © CIRAD-Nogent

Figure 5.10: Weighing the harvest of these 100m². Bambey. 1930. © CIRAD-Nogent

Figure 5.11: Delivery of seeds from the station to the Société de Prévoyance of the province of Djourbel. 1929. © CIRAD-Nogent

to the difference from the norm, and the sum of these points was the overall mark the tapper was given for his work, and according to which a bonus could be indexed. The authority of Vietnamese overseers and European cadres in charge of the surveillance could thus also be measured. Along with other elements, this asserted the authority of research staff and the technical centralization within the group. Through the setting of standard methods and devices measuring distance from standards, the technical department of the Terres Rouges went about getting more reliable data on the conditions of production in each plot of the different plantations of the group (about 34000 hectares in 1942). This department, which was the brain of an extremely hierarchized organization, could document on a large scale the value of different clones or of different cultivation methods, and produced information to be quickly translated into directives.

Thus the tactics of keeping track of the workers (and the cadres) and of calibrating the movements of the tappers (and standardizing management) assured discipline and productivity and simultaneously allowed a better knowledge of the physiology and genetic features of the plant, its reactions to tapping and the parameters of productivity. The production of a homogeneous and cheap rubber, like that of a knowledge which was stable, relevant and applicable, depended on the imposition of a series of habits and practices on the worker's body.

From the Station to the Peasants' Field

In the tropics, European estates could compete only in a few cases with peasants. Only a small fraction of the peasants were bent to the discipline of industrial and scientifically organized agriculture. The question thus arose of supervising rural societies. At issue was no longer simply increasing the production of a microworld of men and plants but to intensify cash crop production of entire rural societies having their relation to nature, their knowledge and their own history and social organization. After a number of unproductive confrontations involving the failure of compulsory cultivation of new cash-crops and of European ploughing and manuring techniques, a looser form of guidance of the systems of production, usually by improving the seeds, was adopted by the colonial agronomists. The peasants who were asked to improve their production were still objects of agricultural experimentation and of administrative intervention. Their agricultural methods and their social and economic functioning were rendered less opaque for the colonizers.

Since the end of the last century, the peanut has been the major export of Senegal. Plant breeding got under way in 1924 in the agricultural experimental station in Bambey. At the request of the governor, the local administrators had the seeds collected from the "best looking plants" in the fields of the peasants. These were cultivated in the station along with those which the station agronomist had himself collected. Some high-yield plants were chosen as parent plants for crafting strains, following a pure line selection breeding strategy. Objects were thus removed, displaced from the peasants fields to the station, juxtaposed in rows to be compared under controlled conditions. These 'controlled conditions' were not already available, however. The station itself had to be carved out from a surrounding seen as disorganized, from which no stable fact could possibly emerge. It was with the organizing passion of Robinson Crusoe that the first experimenters enclosed, cut down and pulled out the trees, fragmented the land into plots according to a geometric plan, had templates made which could enable seeds to be sown at a standard depth and spacing, and had shot the birds which would introduce a non-controllable factor into the output. All these in the very middle of the Sereer region where the peasants maintained nothing less than a park of *Accacia albida* in the fields, so as to benefit from its fertilizing effects on the soil! The station was also a microworld, an arrangement whereby stable facts were produced by the analytical break down of the factors responsible for productivity. By neutralizing some of these factors by experimental strategies of control and record keeping, one was able to keep track of what had been done and of the successive steps in the experiment and of the genealogy of the plants. All of which demanded the discipline of researchers and African laborers.

Once pure and productive strains had been identified in the experimental plots of the station, they had to be multiplied on a larger scale, their yield per hectare had to be documented, and the plants which seemed suitable at the station but which were found to be less productive under the conditions of peasant cultivation had to be eliminated. Selection thus did not end at the gates of the station: at this stage the 'improved variety' did not yet exist. With the help of collaborative growers, that were usually chiefs and Mourid religious leaders, trials on a larger scale were initiated in the 1930s. Experimentation was then carried out in the entire Sengalese peanut belt, which became a field of intervention.

The success of this process demanded the imposition of a minimum number of conditions prevailing in the station. For instance the chiefs who accepted to run the multiplication of the new varieties signed a contract with the administration. Rewarded by higher prices, they had to plant all the seeds which were given them, and those alone, in plots chosen in advance with the agricultural agent, and they had to facilitate the supervision of the plots, to signal immediately the arrival of diseases or pests and to store the harvest separately. As in the station, it was particularly important to preserve the purity of the seeds to avoid contamination which would falsify further results. The agricultural agents went back and forth across the fields to make sure that the peasants had really sown the seeds given them, rather than getting rid of them or, when the improved variety had acquired a high reputation, selling them at a good price. It was not by chance that the region of the Terres Neuves was chosen to be the principal center for the distribution of the seeds. Very recently populated by the administration, its social structure was more easily bent to the imperatives of control, while its relative isolation inhibited the illicit circulation of the seeds which threatened the purity of the output.

The organization of large scale trial and distribution of the high-yield varieties were run by — and strengthened — the Sociétés Indigènes de Prévoyance. These provident societies which were set up by the administration from 1907 onwards to regularize production and limit rural debt by the public loan of seeds, became the real organizers of the rural economy in the context of the growing dirigisme of the 1930s. They held a monopoly in the seed business and they set themselves up as a sales cooperative (to the detriment of the private trading companies). The agents of the Department of Agriculture of the colony and of the Sociétés de Prévoyance policed the fields. With the help of the chiefs who were the relays of the administration also in the prevoyant societies, they ensured the distribution of the seeds, which were sometimes marked with methylene blue to better keep track of the losses. They centralized the harvest from a village, a canton, so enabling one to determine yields and took samples whose quality and oil content was determined in the station's laboratory. Equipped with this information, the geneticists guided from the station the distribution and the testing of different varieties throughout the whole country. In 1939, before the age of the green revolution, 24 percent of the peanuts grown in Senegal came from high-yield varieties (and about 50 percent in 1948).

From these processes grew knowledge of the genetics and the agroecology of the plant (what variety for what region), as well as the transparency of the peasant societies where one intervened, and the control of the colonial administrative apparatus on them. For the different social groups of Senegal, methods of cultivation, land and time use were better understood, and the department of agriculture could better forecast the output and quality of the crop. Experience was also gained on the ways of intervening in the peasant context, shaping practices and approaches in 'development.' Furthermore, the diffusion of improved 'Bunch' varieties (more compact than the 'Runner' generally grown) allowed the successful diffusion of the animal tracted hoe which had failed around 1900.

The rural world was thus adjusted, placed under scrutiny to render it more receptive to external constraints, to the practices and objects of the experimental station. It became more amenable to administrative action, more predictable. It was integrated at the periphery of our calculative world.

References

1. Pierre Clerget, *L'exploitation rationelle du globe.* (Paris: Doin, 1912), p. 6.
2. Foucault, Michel, *Surveiller et punir.* (Paris: Gallimard, 1975).
3. Rouse, Joseph, *Knowledge and Power.* (Ithaca & London: Cornell Univ. Press, 1987).
4. Richard H. Grove, *Green Imperialism, Colonial Expansion. Tropical Island and the Origins of environmentalism, 1660–1860.* (Cambridge Univ. Press, 1995).
5. M. Bruneau and D. Dory, *Les enjeux de la tropicalité.* (Paris: Masson, 1989), N.L. Stepan, "Tropical Nature as a Way of Writing', in Lafuente *et al.* (1993), 494–504; V.R. Savage, *European Impression on Nature and Landscape in Southeast Asia.* (Singapore: Singapore University Press, 1984).
6. Dominique Juhé-Beaulaton, "Environnement et exploration géographique de l'ex-Dahomay (Republique du Bénin) à la veille de la conquête coloniale", in Bruneau, Michel et Dory, Daniel (Eds), *Géographies des colonisations XVe-XXe siècles,.* (Paris: L'Harmattan, 1994), 289–314, quotation p. 300.

7. Osborne, Michael A., *Nature, the exotic, and the science of french colonialism.* (Indiana Univ. Press, 1994).
8. Dr. Paul Barret, *L'Afrique Occidentale, La nature et le Noir.* (Paris: Challamel, 1888), vol. 1, p. IX
9. Paul Leroy-Beaulieu, *De la colonisation chez les peuples modernes,* 5th edition, (Paris: Guillaumin et Cie, 1902), preface to the 5th edition, p. 1.
10. Benjamin Kidd, *The Control of the Tropics.* (London, 1898) quoted by Drayton (1993), p. 441, and discussed by Adas (1989), p. 218–220.
11. Drayton, Richard H., *Imperial Science and a Scientific Empire: Kew Gardens and the Uses of Nature, 1772–1903,* unpublished Ph.D thesis. (Yale Univ. Press, 1993).
12. Headrick (1988), Chapter 7: M. Worboys, *Science and British Colonial Imperialismm 1895–1940.* Unpublished Ph.D thesis, University of Sussex, 1979, Chap. 2: F.K. Timmler and B. Zepernick, German Colonial Botany', *Berichte Deutsche Botanishce Geselschaft* (1987), **100**: 143–58.
13. J. Rouse, *Knowledge and Power.* (Ithaca & London: Cornell Univ. Press, 1987), p. 102–105.
14. Dr. Paul Barret, *L'Afrique Occidentale, La nature et le Noir.* (Paris: Challamel, 1888), vol. 1, p. 373
15. Henri Pobéguin, *Essai sur la Flore de la Guinée Française.* (Paris: 1906), p. 120.
16. Louis-Ferdinand Céline, *Voyage au bout de la nuit.* (Paris: Denoël, 1932).
17. Archives Nationales du Sénégal. Fonds du gouvernemen Général de l'AOF R3 Rapport de Lucien Enfantin à M. le directeur de I'Intérieur, ler sept 1897.
18. Marcel Monnier, *France noire (Côte d'Ivoire et soudan).* (Paris, Plon Nourrit et Cie: 1894), p. 27.
19. Odile Goerg, Entre nature et culture: la végétation dans les villes coloniales (Conakry, Freetown), *Revue française d'Hist, Outre-Mer* (1996), **83**: 43–60.
20. Georges Vernet, *Organisation générale d'une plantation d'Hevea,* p. 2 (reprint from *J. Agri. Trop.* No. 96, 97 and 99 (1909)) I will look for the article reference.
21. Françcois Gain, *Contribution à l'étude technique et économique de la saignée de l'hevea en Malaisie Britannique.* (Nancy: G. Thomas, 1935), p. 101. This work was accepted as a Ph.D thesis in natural sciences at the University of Nancy.

Further Reading

Adas, Michael, *Machine as the Measure of Men, Science, Technology, and Ideologies of Western Domination.* (Cornell Univ. Press, 1989).

Boneuil, Christophe and Kleiche, Mina, *Du jardin d'essais colonial à la station expérimentale, 1880–1930.* (Ed du Cirad, 1993).

Bonneuil, Christophe, *Mettre en ordre et discipliner les tropiques: Les sciences du végetal dans l'empire français, 1870–1940,* Ph.D Thesis. (Univ. Paris 7, 1997).

Brockway, Lucille H., *Science and colonial expansion. The role of the British Royal Botanic Garden,.* (London: Academic Press, 1979).

Cittadino, Eugene, *Nature as the laboratory. Darwinian plant ecology in the german empire. 1880–1900.* (Cambridge: Cambridge Univ. Press, 1990).

Drayton, Richard H., *Imperial Science and a Scientific Empire: Kew Gardens and the Uses of Nature, 1772–1903,* unpublished Ph.D thesis. (Yale Univ. Press, 1993).

Foucault, Michel, *Surveiller et punir.* (Paris: Gallimard, 1975).

Headrick, Daniel R., *The tentacles of Progress. Technology tranfert at the age of imperialism, 1850–1940.* (Oxford Univ. Press, 1988).

Lafuente, A., Elena, A. and Ortega, M.L. (Eds.), *Mundializacion de la sciencia y cultura nacional.* (Madrid: Doce Calles, 1993).

Latour, Bruno, *La science en action.* (Paris: La Découverte, 1989).

Miller, David P. and Reill, Philipp H. (Eds.), *Visions of Empire. Voyages, botany and representations of nature.* (Cambridge University Press, 1996).

Osborne, Michael A., *Nature, the exotic, and the science of french colonialism.* (Indiana Univ. Press, 1994).

Polanco, X., *Naissance et développement de la science-monde.* (Paris: La Découverte, 1990).

Rouse, Joseph, *Knowledge and Power.* (Ithaca & London: Cornell Univ. Press, 1987).

CHAPTER 6

The Management of Society by Numbers

THEODORE M. PORTER

Since the late eighteenth century, social scientists and social reformers have generally considered that by relying on numbers they were following the model of the sciences of nature. This was at best only partly true. Administrative and mercantile uses of counts and measures go back at least as far as scientific ones. Book-keeping has as strong a claim as astronomy to be the prototype of quantitative reasoning, and the huge expansion of numerical information on populations and economies since about 1750 was largely directed by government officials with few if any scientific ambitions. For them, numbers were the quintessential form of neutral information, knowledge that could readily be communicated to great distances. Such information could also be easily summarized, combined, and rearranged, generally using operations no more demanding than the elementary operations of arithmetic. A tabular representation of births and deaths or commerce in each of the towns and provinces of a country, could be turned into a description of the whole simply by adding, supposing only that uniform categories had been used. Detailed verbal descriptions might be more informative, and of course in practice information would be reported in both verbal and quantitative form. But for the administration of large territories, it was convenient to rely on numbers whenever possible.

The relation between administrative and scientific quantification can by no means be reduced to a one-way borrowing of the methods and techniques of science. Some of the most fundamental ideas of statistics, for example, originated in the social sciences and in insurance studies rather than in mathematics, physics or biology.[1] Moreover, the push for quantitative rigor in accounting manipulations, social policy, and causal inference was not necessarily the result of a desire to become more scientific, but often was an adaptation to bureaucratic or political demands for orderly and non-subjective procedures. That is, strategies for managing populations and economies have also helped to define what it means to be scientific. In a variety of ways, then, the drive for social quantification in the twentieth century has been closely allied with a scientific impulse. The hugely increased role of numbers and of quantitative rules in the management of society is in its way as impressive and as important as is the well-known reliance on mathematics in science and engineering.

Numbers and Accountability

Financial accounts are in many ways the prototype for our modern reliance on numbers. Most people find them boring. We might call this the grey suit effect: that accountants are able to do much of their work without exciting interest or oversight, because they are presumed to be merely following technical rules. It is a remarkable achievement, when we consider that many well-educated and cultured people otherwise find money so fascinating that they can happily while away an evening discussing the appreciation in value of their houses or the malign consequences of public debt. Accounting numbers seem boring because their creative role is largely invisible to the public, and often even to accountants and business managers themselves.

In industrialized societies, almost everyone must become a part-time accountant in order to calculate taxes. The universality of accounts is often regarded as one of the costs of economic progress. Accounting, it is supposed, is the formal logic of capitalism. So, in an effectively discredited but still widespread view once urged by Werner Sombart, the birth of double-entry book-keeping in the trading cities of Renaissance Italy is often attributed to the needs of early European capitalism. A more recent example is the genesis of cost accounting and management accounting, conceived as a result of the emergence of large and stable joint-stock companies in America and Europe in the late nineteenth century. These arguments tend to presume that accounts are a necessary concomitant of economic growth and concentration, and that new methods were created to meet economic need.

There are two problems with this view of accounting. First, it discounts the possibility that new methods of recording and calculating may play an active role in the dynamic of business history. They were not merely created by the modern business corporation; in important ways they made it possible. Management accounting provided a system of 'centralized control and decentralized responsibility,' in which the activities of managers were not generally dictated or even observed in detail, but assessed periodically on the basis of their ability to generate profits. Top officials could not very well second-guess every decision made by their subordinates, especially since most problems and opportunities are inevitably understood better at the scene of the action than in the central office. Suitable accounts provided them with some information, but not too much. They also gave the middle managers who headed divisions and offices a basis for judging themselves, and in this way helped to create an identity for an 'organization man' who was yet far more than an automaton.

We need also to understand that the form of accounts was not shaped entirely or even mainly by the logic of capitalism or of large business organizations.

Accounting is tied up with public regulation, and is shaped in part by the state. Tax accounting provides the clearest example. It is dictated in extravagant and ever-increasing detail because people habitually exploit every loophole to minimize their tax obligations. Corporate financial statements are closely regulated in an effort to prevent them from degenerating into misleading puffery. Even accounts designed for internal business purposes have been shaped by governmental initiatives. Cost accounting, for example, originated in an effort by railroad companies in several countries to justify their charges in the face of public protest and political pressure, by linking rates to the expenses of providing the service.

In Britain, where there was relatively little regulation of railroads, cost accounting was one of the technologies that had to be perfected to help fight the First World War. Massive military procurement had upset markets, permitting some notorious profiteering. Factory laborers were asked in the name of patriotic duty to hold down their wage demands and to forego strikes. They were willing, but not if the consequence would be simply to enrich greedy investors. So the government began purchasing some of its supplies on the basis of cost plus a moderate profit. This was not at all a straightforward calculation, especially before it became a familiar one.[2] Companies also learned to use calculations of this kind to help set their own prices and as a tool for increasing efficiency, but there was always a political element mixed in as well. A recent critical study urges that 'management accounting,' the modern heir to cost accounting, has degenerated in the United States since its golden era in the first few decades of the twentieth century. Then, according to this argument, costing was worked out by engineers to improve operations, but the intrusions of government regulation and the misconceptions of business professors have turned it into an accommodation to public agencies and a dream of rigor, contributing rather little to the operation of the firm.[3]

So much government regulation takes place through accounts, that organizations such as universities and foundations have been compelled to adopt accounting procedures like those of corporations in the interest of proper regulation. Small businesses that could otherwise make do with elementary book-keeping have been forced by tax law to bring in accounting professionals. The management of society by numbers reshapes civil society, making it manageable. For this, it does not suffice that every individual and every organization keep elaborate financial records. Management by numbers tends also to make the accounts uniform and rigid. Across national boundaries, accounts are remarkably variable: basic entities such as profits and expenses are defined very differently. To permit such diversity within a country, though, would impede bureaucratic oversight. The precision and standardized form of accounts, in short, does not follow from their quantitative character and from the rules of arithmetic. They have been made rigorous and objective by the force of detailed regulation and in pursuit of regulatory ends.

It might seem that this has meant a massive expansion of the power of accountants, and in a way it has. When a Conservative government in the United Kingdom began to assert more direct authority over the National Health Service in the 1980s, it used expanded accounts to give business and financial specialists leverage over decisions that the physicians had previously been making on their own. But it should not be supposed that accountants are a distinctive species of humanity, which prefers always to seek understanding and formulate decisions based only on numbers. The leaders of American accounting when it was being organized as a profession, in the 1920s, claimed it was their job to make the financial condition of companies intelligible to stockholders and citizens, which could not be achieved merely by assembling a balance sheet. They insisted on their role as interpreters, what might now be called the hermeneutic function of accounts. They claimed, moreover, to be in possession of genuine expertise, which could never be reduced to the mechanical manipulation of numbers. Leaders of accounting continue, on occasion, to argue this way. The triumph of bare quantitative accounts over financial interpretation in the work of accountants was not their own doing, but rather that of government functionaries, who doubted that expert judgment could be distinguished from distortion and manipulation. The accounting standards boards that began in the 1930s to issue what became a mountain of regulations and definitions were set up within the accounting profession, but only because it was clear that the US Securities and Exchange Commission would draw up and issue standards if the accounting profession didn't.

STATISTICS

Accounts are generally denominated in money terms. Statistics can be registered in almost any unit, including money. Individual humans or human actions, such as marriages or murders, are common units. In general the objects counted are not literally the resources of the state, but official statistics are still closely allied to public accounts. The tallying of populations was often regarded in early modern times as a survey of resources. In the twentieth century, the massive expansion of statistical activity has occurred in tandem with the vastly increased scope of governmental economic and social planning. Even public accounting, the anticipation of revenues and expenditures, depends on a close monitoring of the economy. But the statistical regulation of life has gone much further, especially in the twentieth century. The mobilization of economies to fight wars or to combat depression led to new forms of quantitative social investigation.

Economic planning, as it developed in socialist countries and also in many capitalist ones was carried out using a language of numbers. In its more heavyhanded forms, as in the Soviet Union under Stalin, the state set heroic growth rates and then dictated the allocation of resources in order to meet the plans — aided, when necessary, by generous tallies. During the Second World War, the economies of the capitalist West were also managed in much detail by government planning agencies. In post-war France, 5-year plans were worked out rather through a dialogue of functionaries, industrial leaders, and labor representatives. These plans were to be 'indicative' rather than mandatory. In The Netherlands, an elaborate apparatus of economic modeling was used to formulate plans and determine the necessary tradeoffs. Such negotiations and models were informed by a detailed tabulation of resources. During the 1930s and 1940s, new methods of national accounting and of input-output analysis were put in place in much of Europe, North America and Japan. As nations were formed out of the European empires in Asia and Africa, they too attempted to promote economic growth through rational quantitative planning.

The practical, administrative side of such activity was often linked to the economic research of academic social scientists. The origins of modern econometrics, for example, involved well-known research economists working under the sponsorship of institutions that were closely tied to the American government without being part of it, especially the National Bureau of Economic Research and the Cowles Commission. The traditional subject matter of the econometricians had been business cycles, which by the late 1930s had become the problem of economic depression.[4] Political and administrative goals such as controlling inflation and stimulating growth provided the context for the emergence of many iconic numbers, including gross national product (or gross domestic product) and productivity growth rates, and of tools such as value-added accounting and planned programming and budgeting (PPB). This last scheme, put in place in the 1960s in the United States and somewhat later in various European countries, aimed to subject all public expenditures to a rigorous standard of rationality by determining the consequences of every expenditure. Academic economists were involved in the formulation of these tools, but this was intellectual labor linked closely to governmental purposes, and nothing like a pure research tradition.[5]

Clearly such statistics have to do with regulating social and economic life, not merely with describing it. It has only occasionally been noted, though, that even the collection and publication of statistics tends to alter or at least to crystallize what it sets out to describe. Before there were official crime rates, which began to be published in the 1820s, theft and murder were usually taken to be the bad acts of depraved individuals. Before unemployment began to be measured, around 1880, lack of work was attributed to laziness or sometimes to individual misfortune. Afterwards, these became social phenomena with social causes and, by implication, an important element of public responsibility. The very act of counting helped to make crime and unemployment into social problems.

Causes of death, similarly, were nebulous until resolute statisticians set out to tally them. Beginning in the late nineteenth century, international conferences on public health began working to establish a single classification of disease for the entire world, so that statistical records of mortality could be accurate and comparable. Causes of death present a thorny problem under the best of circumstances. Typically, a patient weakened by old age may become very sick with the flu and succumb in the end to pneumonia. That is, even where the cause of death is well understood, there may be serious difficulties in determining what should count as the cause — and hence, of coding. Often the cause of death is not so clear, and in most countries, autopsies are carried out only exceptionally. Finally, and perhaps most interestingly, understandings of disease and of death vary enormously from country to country. Even in a world that was reshaped by European colonialism and by its economic and intellectual traditions, medicine remains heterogeneous. It is scarcely possible to secure consistency of diagnoses without imposing some uniformity of training and of professional status on medical practitioners. International organizations responsible for collecting statistics lack the power to effect this. Hence the standardization of disease necessary for proper counting has remained highly incomplete.[6]

In other cases, the apparatus of statistics may be so closely tied to a regulatory function that the quantifiers help to generate or reshape the phenomena which they set out to describe. For example, occupational statistics have provided one of the most important agencies through which job classifications have become relatively uniform within countries. The rather different names in English, French, and German for professionals, *cadres* and *Angestellte*, emerged from a scene of political and ideological contestation. Or the statistics may form the site of intense political struggles about cultural identities and about what activities deserve to be counted and measured. Ethnic categorization has in the twentieth century been a political battleground, particularly in the United States. The very concept of 'ethnicity' came to be recognized in part through its institutionalization in the American census. Categories like 'Hispanic' and 'Chicano,' were advocated — and opposed — for political reasons. Once adopted by the statistical agencies, it became possible to say how many there were, how much they earned, where they lived, and how they were educated. All these collective statistical attributes gave ethnic and occupational groups an identity they had previously lacked. It remains for statistical and bureaucratic reasons very difficult to be of mixed ancestry, though this may soon change.[7]

Testing Minds and Bodies

Francis Galton set up an 'anthropometric laboratory' at the 1884 International Health Exhibition in London, where some 9000 visitors paid a modest fee to have various components of their strength, quickness, and intelligence measured. This was less a matter of administration than of curiosity, yet Galton conceived of the enterprise as a way of gathering information about the distribution of merit in society. He hoped to create a basis for deciding who should have children and who should not. For several decades afterwards, eugenics was one of the main incentives for the measurement of human attributes, and for the development of formal statistical methods.

Other ambitions for the management of human populations, though, have proved more enduringly influential. The broad movement in business administration known as scientific management was one of the most important. Measurement of productivity and the precise specification of tasks helped to shift control over production out of the workplace and into the offices of managers. From the beginning of the twentieth century, especially in the United States, systems of quantification were set up to determine the effects of diverse workplace arrangements, forms of supervision, and compensation schemes. The purpose was to increase efficiency and at the same time to reassure workers that when tasks were reorganized or new machinery introduced, their contributions to expanded production would be properly recognized and rewarded. In practice, employees often distrusted these measurement schemes, fearing that norms would be recalibrated as their own productivity increased, or that they would be held to the standard of the fastest workers. The experts in scientific management who worked out these experiments, in contrast, became convinced that the very act of experimenting produced a beneficial change, called the 'Hawthorne effect' after the factory where it was first identified. Workers, they supposed, became more efficient in response to the interest shown by managers.[8]

Mental measurement was also as much a part of the bureaucratic as the scientific world. Intelligence testing produced one of the most successful of the created quantitative entities of this century, IQ. Although it remains controversial, IQ is widely accepted as a precise measure of what we mean colloquially when we speak of intelligence. Its origins, like those of scientific management, date to approximately the beginning of the twentieth century. There was at the beginning a connection to eugenics, but the technology of mental testing was most closely associated with schools. Testing was linked to the view that choice of schools and rate of advancement should be tied more closely to individual merit and achievement, rather than being dictated by the age and social class of students. On this the American and British pioneers of standardized IQ testing, Lewis Terman and Cyril Burt, were in agreement. In 1917, during a rapid mobilization for war, the US Army was persuaded to give IQ tests to new recruits as an aid in picking out promising candidates to become officers. This was the first mass application of intelligence testing, a partial quantitative solution to the enormous bureaucratic problems of sorting millions of new recruits. Use of the tests, though, was not driven mainly by need. Entrepreneurial psychologists recognized in the sudden influx of new soldiers a large potential market for their methods, a chance to demonstrate their effectiveness, and an opportunity to assemble an unprecedented data base.[9]

American schools faced a comparable problem at about the same time. A huge expansion of elementary and then high schools in the late nineteenth and early twentieth centuries brought in a vast and socially mixed population of new students. Here again, the psychologists were not shy about advertising their wares. In this case, though, standardized tests were already being developed even before educational psychologists appeared on the scene with IQ measures. The efforts of the psychologists were welcomed as a solution to the problem of sorting the mass of new students. A new bureaucratic layer of principals and superintendents, mostly men, wanted some basis for checking and standardizing the assessments of ability and achievement given by a largely female teaching force. The move to increasing reliance on tests supplied by specialists was a natural continuation of trends toward an alliance of testing and sorting, and of the growing use of multiple-choice examinations by schools. Psychological testing was associated with centralization, changing the basis of evaluation in a way that made it unnecessary to be in the classroom, with the teachers, in order to assess outcomes. This enhanced the power of administrators, and helped them to define themselves as champions of science and objectivity. It also provided a basis for colleges and universities to compare students at local or elite schools with other applicants from far away.[10]

Objectivity and Public Policy

The claim or aspiration to objectivity in policy formation dates back at least to the Enlightenment, but the institutionalization of formal methods to guide administrative decisions is largely a twentieth-century achievement. These methods almost always have a legal dimension, the requirement to follow certain procedures, but the basis for choice — the bottom line — is generally quantitative. The prototype of quantitative decision-making is cost-benefit analysis. Economists urge that the comparison of costs and benefits is almost timeless, the natural way for profit-maximizing entrepreneurs and managers to make decisions. But at best this was almost always an informal way of thinking, rarely made explicit by the use of real numbers, until the creation of large corporations turned business decisions into a bureaucratic problem. There are further reasons to regard cost-benefit analysis as primarily a technology of government bureaucracies rather than of private enterprise, whether entrepreneurial or corporate. One is that business quantifiers seek only to predict expenditures and revenues. They have no need to convert into money terms benefits and disadvantages of a nonfinancial kind, such as environmental degradation or the saving of human lives. The other, which seems decisive, is that the codification of general rules for quantifying costs and benefits came entirely out of the public sector.

This last development is of much interest. It happened first of all in the United States, where it was associated most closely with public works. As early as 1902, a Board of Engineers within the US Army Corps of Engineers was required by Congress to certify as beneficial any water projects that the Corps proposed to undertake. During the first two decades of the century, the Corps almost never interpreted this law as requiring them to quantify anything, though of course budgeting procedures required estimation of costs for projects they favored. During the 1920s and 1930s, as a gesture of accountability, the agency formed the habit of listing enough expected benefits so that recommended projects would show an excess over anticipated costs. This was still very informal. When, in 1936, the

Congress wrote a cost-benefit standard into law, it was largely an act of penance. An embarrassing collection of manifestly uneconomic projects had been added during floor debate to the flood control bill of 1935. This legislative contrition might not have expressed itself so radically, were it not that the 1935 bill finally failed. Cost-benefit analysis was intended to force the Congress to respect some standard of procedural regularity, and at the same time to assure the public that tax money was being well spent.

The traditional looseness of cost-benefit calculation at the Corps persisted even after 1936, and the Flood Control Act was only one phase in the story of its formalization. More crucial was the controversy generated by the greatly expanded scale of flood-control and navigation work during the late Depression and especially during the post-war construction boom. The cost-benefit requirement gave opponents of Corps projects a more appealing basis for fighting them than a frank defense of their own interests. And while the Corps had a very close patronage relationship with Congress, some of its opponents were also very powerful. Private utility companies, for example, considered that the generation of power by flood-control and irrigation dams amounted to creeping socialism. They hired experts and undertook to show that government bureaucrats were unable to appreciate the complexities of electricity pricing. Similarly, railroad executives objected to government subsidized competition in the form of canals built at public expense. They argued that Corps methods were economically unsound, or were too loose, or were misapplied.

Such objections obliged the Corps to justify its procedures, and also to modify them so they could be defended. The really crucial incentive for the development of uniform, standardized, rationalized methods of cost-benefit analysis, though, came from interagency conflict. A variety of agencies were involved in water control and development. Two in particular, the Department of Agriculture and the Bureau of Reclamation, were often at odds with the Corps. Their battles almost always included a political dimension, as affected interests lined up behind their favorite agency. Agriculture and Reclamation, in response to pressures like those facing the Corps, had developed their own economic standards. Unfortunately, the methods of the three agencies were quite different, leading regularly to discrepant results. The resulting disputes could be very hard to settle.

The drive for uniform cost-benefit methods throughout the government arose as a response to these difficulties. A major effort in the late 1940s to negotiate shared rules was only partly successful. By the 1950s, the agencies had begun to rely increasingly on economic specialists to codify appropriate methods of cost-benefit analysis. In the early 1960s, cost-benefit became a recognized specialty within welfare economics, and soon the economists were promoting their techniques as the proper basis for decisions regarding highway construction, health policy, and educational 'investment' in a measurable entity that has become known as 'human capital.' In the 1980s, this form of economic analysis came almost to be fetishized, as for example in the Presidential executive order issued by Ronald Reagan which required all major new regulations in the United States to pass a cost-benefit test. Other countries, especially but not exclusively English-language ones, have been only slightly less fervent in the application of this form of quantification to public policy. It has in addition been used heavily by international bodies such as the World Bank, the International Monetary Fund, and the United Nations.

The idea behind this is partly of course an economic one, to make government more efficient. There was also a political dimension, at first liberal but then conservative. Liberals of the 1960s wanted to show that government projects were not pork-barrel largesse, but could be defended as economically rational. The strong advocacy of cost-benefit methods since the 1970s by Republicans in the United States and by Conservatives in the United Kingdom has been part of an effort to check or reverse the growth of government. There is clearly an element of technocracy, or administration by experts, in the growing influence of quantitative methods. But such tools are not unambiguously friendly to elite experts. Expertise means not simply the ability to apply difficult technical methods, but also, or mainly, the capacity to exercise judgment with wisdom and discrimination. An insistence on rigorous quantification generally implies a distrust of expert judgment, an insistence that it should be open to the public gaze and confined by rigorous or inflexible rules. The management of society by numbers tends not only to generalize away the individuality of the populations administered, but also to restrict that of the managers themselves.[11]

Inferential Statistics: Managing Science

'Statistics' means not only collections of numbers about societies and economies, but also a body of mathematical methods for analyzing those numbers. The modern tradition of mathematical statistics began in the early 1890s, and was associated with research on heredity, biological evolution, and eugenics. The biometric school was founded principally by Karl Pearson, a professor of applied mathematics in London. He was mainly interested in the analysis of observational data, especially large collections of biological measurements. His methods also formed the basis for psychometrics, the statistics of mental testing, and the analysis of economic data. Ronald A. Fisher, who worked at an agricultural station and then succeeded Pearson at University College, moved the field decisively in the direction of experimental design and the analysis of experimental data. Since the 1930s, Fisherian statistics has penetrated, or in many cases helped to constitute, many or most experimental sciences. Among the fields most dramatically shaped by it are agriculture, psychology, ecology, and medical therapeutics.

Mathematical statistics might be called an accounting of permissible belief. Fisher urged researchers to test candidate causes by formulating hypotheses of a negative kind, performing an appropriate experiment, and calculating to determine if the data refuted adequately the 'null hypothesis.' Thus one might divide an area of land into many plots, apply a certain fertilizer to half of them, chosen at random, then harvest and weigh, and analyze whether the resulting difference of mean yields is within the range that might result from chance. When the difference exceeds the calculated bounds of chance, Fisher's experimental protocol permitted a causal inference: the fertilizer was not merely correlated with, but caused, the increased yields. Suitably adapted, Fisher's methods could be used in almost any discipline in which experiments can be performed with numerous, randomized repetitions. This was of only modest consequence for fields like physics, where the standards of acceptable experiment and measurement had been evolving for two centuries and where tight experimental control was generally preferred to averaging over numerous repetitions.[12] In population biology, medical therapeutics, and the behavioral sciences, by contrast, Fisher very nearly defined what an experiment should be.

In medicine, for example, the statistical approach had long been tied to studies of public health. Fisher's methods did not apply readily here because experimental randomization was operationally difficult and ethically problematical. Austin Bradford Hill, the distinguished British pioneer of the randomized clinical trial, tried to adapt Fisher's methods in order to learn whether smoking causes lung cancer. Almost every comparison of smokers with nonsmokers, of which there were many beginning in the late 1940s, showed much higher cancer rates among the smokers. Hill tried to do better by identifying experimental and control populations that he took to be fully comparable. But it was impossible to randomize, since that would require a large number of experimental subjects willing to let their decision to smoke or not, over a very long term, be decided by the toss of a coin. Fisher, on this account, openly doubted that the observational results could justify a causal inference about the causes of cancer. Maybe it was a mere correlation, due perhaps to genetic differences between smokers and nonsmokers.

The randomized clinical trial was a tremendous feat of experimental organization. Patients had to be recruited and randomized. In most cases they were kept in the dark about whether they were receiving experimental treatments or placebos. Often the doctors too were not informed about their patients. In addition, they were required to follow a strict protocol of medical treatment, and then at the end to find that their judgment of its success was largely ignored in favor of instrumental measures. That is, the rules of experimental design and statistical analysis amounted to a strategy of coordinating large-scale therapeutic research in medicine by confining the discretion of doctors.

The pioneering instance of a randomized clinical trial was performed in Britain at the end of the Second World War to test the effectiveness of streptomycin in the treatment of tuberculosis. This form of medical experimentation was institutionalized, however, in the United States, where it provided a partial solution to the problem of regulating medical practice. Experimental trials for a new drug would normally be performed by the pharmaceutical company planning to produce and distribute it, obviously an interested party. The Food and Drug Administration could scarcely let the companies test their products in any fashion they saw fit. Instead, it specified the randomized clinical trial and a test of statistical significance as the proper way to assess clinical effectiveness. There were, of course, always departures from so strenuous a design, but these provided opportunity for doubt, and hence were to be minimized.[13]

Fisherian statistics, in short, came to be seen as a valuable tool for regulating drug companies and even doctors by defining what would count as acceptable science. There are other related cases, such as testing of food additives and workplace chemicals, where prescribed forms of statistical analysis came to be authorized legally and bureaucratically. In most sciences, the links to public administration are looser. Certainly there are some disciplines whose expectation of statistical significance testing arose from within, and in which public-policy aspects were decidedly secondary. But often it was more complicated. American research psychologists began to insist on experimental designs producing numbers that could be analyzed using broadly Fisherian methods during the 1930s and 1940s. This was intended to promote a respectable field of pure research, not to cultivate the links between the field and some applied domain such as psychiatry, clinical counselling, business management, or education. Ironically, advanced statistical methods had entered psychology through one of these applied fields, educational testing. Here again, the administrative use of numbers created an expectation that the science itself should also come up to the presumed standards of statistical rationality.

In these and other cases, the quantitative management of society has taken place through the management of scientific knowledge. There is something here akin to accounting and to cost-benefit analysis, a set of formal methods to reduce the play of discretion or to constrain the free exercise of scientific judgment. This cannot be explained as simply the natural outcome of the scientific disposition to quantify, for such methods were seized most eagerly, and often developed, by relatively low-prestige applied subdisciplines such as educational psychology and therapeutic medicine. From here they spread to the more 'basic' social and biomedical sciences, and eventually even to physical ones. That is, statistics entered scientific research as an element of a regime of regulation, arising partly from within the sciences themselves but also imposed or encouraged for bureaucratic reasons. Quantification is a tool of science that has been extended to society and government, but it is also an administrative strategy that has been applied aggressively to science.

References

1. Theodore M. Porter, *The Rise of Statistical Thinking, 1820–1900.* (Princeton: Princeton University Press, 1986).

2. Anne Loft, "Accountancy and the First World War," in Anthony G. Hopwood and Peter Miller (Eds.), *Accounting as Social and Institutional Practice.* (Cambridge: Cambridge University Press, 1994), 116–137.
3. H. Thomas Johnson and Robert S. Kaplan, *Relevance Lost: The Rise and Fall of Management Accounting.* (Cambridge, Mass.: Harvard Business School Press, 1987).
4. Mary Morgan, *The History of Econometric Ideas.* (Cambridge: Cambridge University Press, 1990).
5. Hopwood and Miller, *Accounting*; also Francois Forquet, *Les Comptes de la Puissance: Histoire de la comptabilité et du plan.* (Paris: Encres Recherches, 1980).
6. Geoffrey Bowker and Susan Leigh Star, "Knowledge and Infrastructure: Problems of Classification and Coding," in Lisa Bud-Frierman (Ed.), *Information Acumen.* (London, Routledge, 1994), p. 187–213.
7. William Peterson, "Politics and the Measurement of Ethnicity," in William Alonso and Paul Starr (Eds.), *The Politics of Numbers.* (New York: Russell Sage Foundation, 1986), p. 187–233; Alain Desrosières, *La Politique des grands nombres.* (Paris: La Découverte, 1993).
8. Richard Gillespie, *Manufacturing Knowledge: A History of the Hawthorne Experiments.* (Cambridge: Cambridge University Press, 1991).
9. John Carson, "Army Alpha, Army Brass, and the Search for Army Intelligence," *Isis* (1993), **84**, 278–309.
10. Kurt Danziger, *Constructing the Subject: Historical Origins of Psychological Research.* (Cambridge: Cambridge University Press, 1990).
11. Theodore M. Porter, *Trust in Numbers: The Pursuit of Objectivity in Science and Public Life.* (Princeton: Princeton University Press, 1995).
12. Stephen Stigler, *The History of Statistics: The Measurement of Uncertainty to 1900.* (Cambridge, Mass.: Harvard University Press, 1986).
13. Harry Marks. *The Progress of Experiment: Therapeutic Reform in the United States, 1900–1990.* (Cambridge: Cambridge University Press, 1997).

Further Reading

Boorstin, Daniel J., *The Americans: The Democratic Experience*, part 3. "Statistical Communities," (New York: Vintage, 1973).

Bowker, Geoffrey and Susan Leigh Star, "Knowledge and Infrastructure: Problems of Classification and Coding," in Lisa Bud-Frierman (Ed.), *Information Acumen.* (London: Routledge, 1994), p. 187–213.

Bulmer, Martin, Kevin Bales and Kathryn Kish Sklar, *The Social Survey in Historical Perspective, 1880–1940.* (Cambridge: Cambridge University Press, 1991).

Danziger, Kurt, *Constructing the Subject: Historical Origins of Psychological Research.* (Cambridge: Cambridge University Press, 1990).

Desorsières, Alain, *La Politique des grands nombres.* (Paris, La Découverte, 1993).

Fourquet, François, *Les Comptes de la puissance: Histoire de la comptabilité et du plan.* (Paris: Encres Recherches, 1980).

Gigerenzer, Gerd, Zeno Swijtink, Theodore Porter, Lorraine Daston, John Beatty, and Lorenz Krüger, *The Empire of Chance: How Probability Changed Science and Everyday Life.* (Cambridge: Cambridge University Press, 1989).

Gillespie, Richard, *Manufacturing Knowledge: A History of the Hawthorne Experiments.* (Cambridge: Cambridge University Press, 1991).

Hacking, Ian, "Making Up People," in Thomas C. Heller, M. Sosna and D. Wellbery (Eds.), *Reconstructing Individualism.* (Stanford: Stanford University Press, 1986), p. 222–236.

Hopwood, Anthony G. and Peter Miller (Eds.), *Accounting as Social and Institutional Practice.* (Cambridge: Cambridge University Press, 1994).

Johnson, H. Thomas and Robert S. Kaplan, *Relevance Lost: The Rise and Fall of Management Accounting.* (Cambridge: Harvard Business School Press, 1987).

MacKenzie, Donald A., *Statistics in Britain, 1865–1930: The Social Construction of Scientific Knowledge.* (Edinburgh: Edinburgh University Press, 1981).

Marks, Harry, *The Progress of Experiment: Therapeutic Reform in the United States, 1900–1990.* (Cambridge: Cambridge University Press, 1997).

Morgan, Mary, *The History of Econometric Ideas.* (Cambridge: Cambridge University Press, 1990).

Peterson, William, "Politics and the Measurement of Ethnicity," in William Alonso and Paul Starr (Eds.), *The Politics of Numbers.* (New York: Russell Sage Foundation, 1986), p. 187–233.

Porter, Theodore M., *The Rise of Statistical Thinking, 1820–1900.* (Princeton: Princeton University Press, 1986).

Porter, Theodore M., *Trust in Numbers: The Pursuit of Objectivity in Science and Public Life.* (Princeton: Princeton University Press, 1995).

Stigler, Stephen, *The History of Statistics: The Measurement of Uncertainty before 1900.* (Cambridge, Mass: Harvard University Press, 1986).

CHAPTER 7

Scientific Management and the Production Process

YVES COHEN

GENERAL OVERVIEW

Scientific management claims to make production practices scientific. In Germany this movement was called *wissenschaftliche Betriebsführung*, at least during its beginnings in the first half of the century. In France the phrase *organisation scientifique du travail* is preferred. Russia and Italy have translated the French terms. In other words the activity which Taylor publicly launched in 1903 with the publication of his *Shop Management* and above all in 1911 with the more well-known *Principles of Scientific Management* is about 'management' in the United States, about 'the leadership of firms' in Germany and about 'the organization of the workplace' in France, Italy and Russia. One could say that it is concerned with three domains which cover different kinds of practices. Let me only stress that in each country this first science of organization will be perceived differently, as the different languages both express and confirm. It is however the word 'scientific' that they all share.

The speed with which this new doctrine spread throughout the industrialized or industrializing world in the four years preceding the First World War, extending also to Japan and Russia, shows that it immediately touched a chord among businessmen and engineers. The accelerated growth in this period from Russia to the USA posed a number of entirely new problems concerning the organization of the workforce, the control of increasingly complex technologies, and the direction of firms. Taylor's doctrine differentiated itself from other contemporary movements seeking the more 'systematic' organization of the firm because it was more systematic and comprehensive and also, without doubt, by its claim to be scientific. Its major defining characteristic is not only that it proposed a quantification of human actions with a view to their material and temporal optimization, but that all quantification undertaken within the firm was now obliged to be based on, and derived from, the microscopic quantification imposed in the workplace (calculation of the cost price, control of throughput, salary system, conception of the conditions for entering new markets, etc . . .).

By virtue of the 'scientific' character of the 'system' the administration of the firm appeared to be 'objective.' A new breed of management specialists designated themselves the bearers of this new scientific mask, thus distinguishing themselves from the property holding entrepreneurs and their immediate delegates. The specialists invented their profession and, at the same time, the forms that this would take in its relationships to other jobs in the firm. One of these forms was the planning department, the specific place for this science of action which defined its aim as organizing workers' behavior and also that of other specialists, taking hold of the material processes of production and arranging them according to the relationships between different tasks, which now became 'functions.' It is a science which is immersed at the core of a constellation of actions to which it claims to be relevant, both as regards their conception and implementation and as concerns their mutual interactions. The innovatory regime which it deals with is neither at the level of the concept, nor at that of the object, but rather that of the human act and the relationships which it engenders.

The objectivity of a science aiming to administer action seemed to be the guarantor of the social peace sought in all industrialized countries, and thus the subjection of the workers movement. Scientific management also came to have a central role in the social discourse of the company owners who put it forward as a scientific and objective arbitrator of social conflict. The norms which it laid down had to be adopted by all actors. And indeed in the name of the universality of a single rationality, the major workers unions almost throughout the entire world would support this claim after the First World War.

The scope of scientific management was to be enlarged in two directions. Firstly, it quickly moved beyond the confines of the world of industry to conquer, under the influence of Taylor but above all a number of his disciples and competitors (Taylor died in 1915), a huge variety of human action from typing to domestic labor. The novel by the Russian Evgenii Zamiatin, *We*, published in English in 1924, shows how one can imagine extending Taylorism as far as love. Then, applying itself to action in the sense of social action, in hierarchical contexts, it would lay the foundations of another range of practices, namely those known as the sciences of organization. There it merged with other doctrines, at once competitive and complementary, like that of the Frenchman Henri Fayol on administration.[1]

Taylor's scientific management was extremely popular for sixty or seventy years, both as a point of reference and legitimation for a huge variety of organization and management practices, and as a foil for internal or external criticisms of the managerial movement. Today, at least in the most developed countries, only the second function remains, though both are very much on the agenda in the industrializing countries.

Even though it was not developed solely for big, mass production firms, from the First World War onwards scientific management in the industrial arena was, by and large, to be found in the large Fordian enterprise. The inter-war period was one of experimentation. After World War II, we find in all the major developed countries a stabilization and a normalization of procedures based on functionalism both in the organization of work and in the organization of the firm.

Once we begin to differentiate between countries' very different situations emerge. In Germany Taylorism merges with a movement for rational administration which accompanied the phenomenal acceleration of industrialization from its inception in the 1870s. The Germans were passionately attracted by *Taylorismus*, but it was not a passion which excluded alternatives. Its main rival was not only *Fordismus*. Both were immersed in a far broader movement for *Rationalisierung* which included, among other novel traits, a consideration of the relationships between firms.

After several attempts to implant Taylorism in tsarist Russia, the young Soviet Union undoubtedly became the country which made the most wideranging efforts to organize production scientifically. Taylor, Ford and also Fayol were the leading thinkers there in the 1920s. Industry was part of the state apparatus and the rationalization of the one was inseparable from the (frequent and vain) attempts to rationalize the other. The accelerated industrialization of the 1930s put an end to the efforts to introduce scientific management. Stalinist discipline was incompatible with sophisticated modes of organization and more direct hierarchical forms took their place. In a certain sense the authorities chose not to intervene at the heart of the labor process when it restricts all freedoms in all other areas of social life. The questions Taylor tried to answer in the 1890s are still very much alive in Russia though the conditions imposed by economic competition have changed radically.

The French experiment in 'scientific management' was steadily carried through, though rather less systematically than in Germany and less planned than in the Soviet Union. As in Germany, leading engineers tried to introduce a science of 'acts of production movements' into firms before the First World War, and factory owners were extremely impressed during the war and again when the economy revived in the 1920s. The main features of scientific management were understood by big, labor intensive firms by the end of the 1930s, just as mass production appeared on the horizon. After a delay imposed by the war, mass production was fully deployed from the 1950s onwards using methods borrowed from the United States until a serious crisis swept through the workplace in the 1970s.

Paradoxically, it is probably easier to write a history of scientific management as practised outside the United States than in the US. Taylorism is a clearly identifiable export throughout the period of international 'Americanization.' Furthermore America was exported along with material objects which embodied within themselves organizational practices for which they were conceived and produced. In the US, scientific management was a local product and so less visible, less easily detached from its context. There are many reasons for this. When Taylor's scientific management emerged, it did so in the context of a large number of consultants all proposing new modes of organization. This diversity was even to increase thanks to the opponents of the more orthodox approaches. After Taylor's death, scientific management as reflected in his official mouthpiece, the Taylor Society, changed into the Society for the Advancement of Management, and diverged from the letter of Taylorism. For example, it was no longer unacceptable to incorporate liberal policies of personnel training into a Taylorian reform of the production process. The identity of scientific management was blurred by a variety of organization systems proposed by an exponentially growing group of consultants. Other famous names emerged like Frank and Lilian Gilbreth, Emerson and Bedaux.

Taylorism also had another function. It was a fundamental point of reference against which other organizational movements could react and define their specificities. For example, we have the school of human relations developed by Elton Mayo and his colleagues at the Harvard Business School beginning in the early 1930s, which aimed to found a new theoretical and practical sociology based on a critique of the actions systematized by scientific management.

Two fundamental Taylorian ideas transcended the differences between his school and its diverse opponents: the material and temporal control of the workplace on the one hand, and the specialization of the task of defining norms, carried out by professionals in a quite separate department in the company, on the other. All theories born in the US after the Second World War and which claimed to be innovative in the field of scientific management kept these two characteristics which were essential to their scientific pretensions. It was the practice of the optimization of action which made it possible to establish a link between Taylorism and operational research and also some current research in artificial intelligence.

Why Scientific?

If we tackle the issue from a more phenomenological perspective, in what sense can we call scientific management scientific? And if it is a science, what is it a science of? What kind of science is it? When all is said and done, in what respect are the processes of production involved in and by such a science?

The first use of the term 'scientific management' can be dated quite precisely. Management as a science or scientific discipline, was born at the end of 1910. The term was not used by the founder of the discipline, Frederick Winslow Taylor, but was introduced to designate his work. It was used during a court case, and it was only a year later, in 1911, that Taylor published his highly successful *Principles of Scientific Management.* Its diffusion generated an efficiency craze which was remarked on by all his contemporaries.[2]

In 1910 a number of clients of a major railroad company brought a case against it, insisting that its fare increases were excessive. The company argued that it had had no choice, given the increase in salaries it had paid to its staff. One of the sitting judges, Louis Brandeis (the future Justice Brandeis), set out to prove that one could increase salaries along with productivity, so that one was not obliged to increase prices (nor the burden of labor). And he had the idea of basing his argument on Taylor's work. Taylor was already well advanced in his career as a consulting engineer. He had published numerous contributions to the very lively debate on organization and salary systems. His basic and pathbreaking book was *Shop Management*, published in 1903. A school had already been formed around him, bringing together enterprising, inventive but not necessarily faithful disciples. Taylor first received recognition by being elected chairman of the American Society of Mechanical Engineers. Judge Brandeis wanted to call some of Taylor's associates to the bar as witnesses. To ensure that their position was coherent, he gathered them together one evening to find a name for the doctrine that they were going to promote. Having suggested the 'Taylor system,' 'Functional Management,' 'Shop Management,' and 'Efficiency,' all agreed that, during the court case, the term 'Scientific Management' would be the most appropriate one to designate the system.[3]

It was because of the need to organize the public launch of a new discipline of industrial management that it was decided, under pressure from a biased judge, to label it scientific. To present it as scientific bestowed both unity and power on the discipline. At the same time, to call this discipline scientific was to ensure unhesitating agreement between its proponents. To baptise the management as scientific was thus to give it the best label one could imagine.

The choice was justified by the success which followed. Horace Drury remarked on the considerable increase in the number of publications on this topic once the label was forged, using data extracted from a bibliography at the Library of Congress in 1913. The adjective scientific apart, the fact that management itself had become a major issue in the economy also played an important role in this development. 1911 is also noteworthy for a similar event, namely a parliamentary commission of enquiry into the Taylorian system, demanded by the unions. There is a striking contrast between the small number of Taylor's works (two books and a few talks) and the enormous literature produced from 1910 onward on scientific management. The variety of possible interpretations was already in evidence, and this led to a huge variety of organizational practices. A discourse began to spread having its own logic, its authors, its modes of expression and diffusion, its societies, hierarchies and press. This discourse was both parallel, and tightly linked, to the organizational interventions of consultants or production engineers and with the conception and deployment of local and specific systems of organization.

Should the circumstances surrounding the designation of Taylor's system as scientific lead us to question its being considered a science? Certainly the discipline subsequently became sufficiently well-established under the labels 'scientific management' or 'science of management' for the original and rather fortuitous attraction of this designation to be regarded as a well-founded claim. On the other hand, I do not feel able to judge its appropriateness myself. One has no more reason to take this claim seriously than, symetrically, one has to take just as seriously the claims of those opponents who deny the scientific character of scientific management. That granted, if scientific management is scientific what is it about it that, from its point of view, makes it scientific? This is what I first want to explore.

What is it that is Scientific in Scientific Management?

Most works in American writing on scientific management which aim to show or to defend its scientific character add little to the first sentence in Taylor's *Shop Management*. This sentence states that "the writer's chief object in writing this paper is to advocate the accurate study of 'how long it takes to do work,' or Scientific Time Study as the foundation of the best management".[4] When the book was written, scientific management was not yet an issue, while shop management was already backed by 'scientific' studies. The precision of the formulation used in 1903 was not subsequently improved on except perhaps by Carl Barth, one of Taylor's closest and most faithful collaborators: "Our method is called scientific because it determines exactly — scientifically — the length of time in which a man can do a piece of work."[5]

Each of Taylor's phrases has been commented on so extensively, to the extent that some people have said that Taylorism has become a religion, that it is very difficult to add anything without risking ridicule. However, since I have undertaken to analyze Taylorism's claims to be scientific, something must be said. Firstly, the 'scientific study' has as its objective the better management of men at work. Put differently, a science of work was needed so as to manage work better, or again, the solution of problems in the process of production could not be left to the workers. They were the domain of management, and to achieve its objectives that management had to develop a scientific approach. The scientific nature of the study of work guaranted that the managers would come to control it. It was to ensure the power of the management over the work force that Taylor undertook to develop a new kind of scientific study: a science of power.

This has several corollaries. One had to educate experts in this scientific study: specialists who were independent of the workforce and the other technicians but also of the managers. Their task was to develop a specific knowledge of how to intervene in the production process. One had to create a *geographical* location where they could instal their own equipment. One had to establish them as an *organizational entity* in the vertical hierarchy and the circulation of the orders of production. Finally one had to define their *temporal* relationship to the production process. It was easy: they intervened before any job was undertaken. Their place geographically and in the organigram was labeled the *planning department*. This would be the operational base for the scientific study. This would be the entry point for studies of the workshop and for interacting with the masters and the workers.

What was the object of this new and situated scientific study? Taylor's and Barth's statements are clear: they are studying time. Never before had a scientific study of time with especially trained men been made. This also involved the associated material equipment, considerable organizational prestige, and a strong political and economic ambition, since the study of time was supposed to resolve all social questions. Of course Taylor was not interested in cosmic time. It was time at work; the time taken for actions in the production process. Scientific management thus presented itself as, above all, a science, ie., an activity of specialists, directed towards a study of the time embodied in movements within the production process. Simultaneously, the Taylorian system introduced into the business world this science, an organizational form whose vocation it was to take as its object all the actions being made in the world in which it found itself. By its very existence this science of time transformed the organization of the firm. Indeed nothing escaped its gaze.

A Science of What?

The wish to establish a science of human movements in the production process is surely very ambitious. This science has two dimensions.

Firstly, it has no unambigous object, no object with clearly defined boundaries. Work, the acts of production, are composites, mixtures, they include tools, manual and mental operations, material supports, written documents, machines, reading, talking, and so on. Bruno Latour would probably say that the object of this science is hybrid. I would rather describe this science of the time for action as a science of linkages. Firstly, there is the productive link between worker and machine. The language used in the texts on scientific management to describe this link shows this clearly when, for example, the author writes of "the efficiency of the worker and machine," a phrase which refers to a unity, or when, as is often the case, one refers to the problem of 'adapting' men to the machine or to their work.[6]

Scientific management, however, unites or bonds together more than just the couple man/machine. As one author put it around 1912, it could be described as "a scientific combination of brain and brawn."[7] More analytically, it is the links established between the environment put in place for the purpose of production (whatever the physical object involved: materials, machines, tools) along with a series of manual and cognitive operations and the behavior of men, the tools of management (graphs, tables, data . . .) as well with hierarchical and functional structures. A whole range of Taylorian innovations are the concrete manifestations of these links: techniques for routing and scheduling work, the systematization of the quality control of all steps in production, the introduction of instruction sheets for the operators, the setting up of a system to constantly calculate cost price, stock control of the items being produced, the functional division of management . . . The economist Laurant Thévenot has used the happy phrase 'investments in forms' to describe these innovations.[8] We are dealing then with the injection of science into the links between heterogeneous elements with a view to transforming them into operational time. We are also dealing with a science of action.

A Science of Action

A science of action seems to be a rather novel idea. Among the social sciences, for example, the nineteenth century saw sociology as a requisite condition for social action,[9] but the object of study was very clearly distinguished from systems of action as such. In medicine, by contrast, there was a close relation between knowledge and the act of healing. Medicine, however, remained part of what were called the sciences of nature. It did not deal with human relations. Scientific management tries both to understand and to intervene, and all knowledge is directed to this latter objective. And this so as to optimize human action during the production process. But it also changes all the human relationships associated with these actions. In these respects scientific management resembles psychoanalysis, another field which also emerged at dawn of the twentieth century as a science of action. Psychoanalysis also had a great "wish to be scientific," the double task of understanding and healing, and the hope of helping the patient to reform his relationships.[10]

In a science of action it is the same people who aim both to know and to act socially. The Taylorians in factories were obviously the representatives of the management. They were sent into the workshop to unravel the secrets of time which the workers jealously guarded. The workshop was the place from which data was extracted. The specialized department was the place were these data were elaborated according to more general rules, and shaped into written instructions. These instructions were thus materialized and had to be followed by both the shopfloor managers and the workers. They embodied both the authority of the owner and science. The process was constantly fueled by the recycling of information back to the planning department. "Information which (. . .) is supposed to be furnished by the knowledge of the workman or the gang-boss or foreman, is brought back to the planning room and becomes a part of the instruction card."[11] The knowledge of all the actors in the production process was only worthwhile to the extent that it had passed through the scientific mangle of the planning department.

It must also be said that this was a science of two kinds of action. It was firstly the science of the actions of those who carried out the cyclical task of extracting knowledge from the shopfloor, of elaborating the norms of work, and of re-inserting these now scientific norms into the workplace. Taylorians legitimated their activity in two different social worlds: one was that of the firm itself, from the line to the boardroom where they were expected to improve profitability. The other was that of their profession where their methods were circulated in the market of the new science (organizational literature, national and international professional conferences, establishment of consultancy firms . . .) In this market it was success in the factory, which was the primary, though not the only criterion for recognition.

Scientific management was also a science of the actions of its object/subject: workers (and others). Put differently, it was a science of the joint action of the human and the non-human, the direct source of all knowledges brought into play. This was action closely guided and regulated, where nothing could be left to chance, everything had to pass through the crucible of analysis. The always conflictual relationship between these two types of action could also be called Taylorism's analytical stage, again drawing a comparison with psychoanalysis.[12] This stage was localized at various levels stretching from the individual workplace to the factory, passing via units like the workshop. And it organized the conflict around an object: the chronometer.

The chronometer did not leave the workshop even after time and motion had been standardized in tables which were supposed to be internationally valid, and which began to spread towards the end of the 1930s and above all after World War II. Standardization did not relieve the organizers of firms of their workload.

The Mechanical Infrastructure

Taylor's system was not the only practical science of management competing for the crown in a broader movement for 'systematic management.' It did not acquire its influence only by declaring itself scientific in 1910. Nor was it only because it based its scientific pretensions on the invention of a planning department which was at once an identifiable place — like a laboratory — where the specialists of the new science were gathered together, an organizational cell, and a location temporally 'upstream' of the production process. Scientific management was able to distinguish itself from its competitors by its remarkable success with production techniques, and this success gave an added weight to its organizational innovation intended to free the production process from shirkers. For 26 years Taylor actually worked in a machine tool workshop to perfect the laws of metal cutting which he deemed necessary for the foundation of a new work organization. This done, he invented along with an engineering colleague a steel for high speed cutting tools whose effect on the cutting speed amazed all their contemporaries. This steel combined 18 percent tungsten, 6 percent chrome and 1 percent vanadium, and was quenched at 1200 degrees. Carl Barth, for his part, invented a slide rule to determine the optimal cutting speed. This ruler took account of no less than twelve variables identified by Taylor and relating to the machines, the tools, the materials . . . Taylor also made important advances in the heat treatment and use of the transmission belts on machines. Thus took the time to build up an impressive technical infrastructure which had the added advantage that it caught the eye of his contemporaries, who marveled at the red coloring of the metal offcuts and the note. High-speed steel tools helped build their faith in the organization.

Taylor's system was born in a mechanical workshop at a time when the mechanical was the paradigmatic technical domain, as physics was for the sciences. This played a decisive role in convincing the main French protagonists of Taylor's work, and above all Henry le Chatelier, polytechnician, chemist, inventor of a law of thermodynamic equilibrum, professor at the Collège de France, and principal non-American transmitter of Taylor's system. It was a very French kind of enthusiasm which swept over Le Chatelier when he saw the high speed steel tools operating at the Universal Exhibition in 1900. Le Chatelier translated *Shop Management* in 1907 and never wearied of bringing Taylor's ideas to France. Taylorism was not simply scientific for him; it was the model science. His preface to the French edition of *Principles* was the only foreign text included in the major bibliographic study of scientific management published in 1914.[13] Le Chatelier's contribution played an important role in the USA since it confirmed the scientificity of scientific management in the eyes of its partisans.

According to Le Chatelier industrial phenomena bring into play more variables than do natural phenomena, and he was stunned to silence by the twelve variables for cutting metals that were materialized in Barth's ruler. This proved that Taylor applied the scientific method, "so often scorned in the laboratories of pure science," more rigorously than the scientists themselves.[14] We find an echo of Le Chatelier's concern even in the pedagogic texts dealing with scientific management in France. Michelin, the tyre manufacturer, and one of the pioneers of the import of Taylorism, published a number of popularizing brochures towards the end of the 1920s. At the top of one of them we find this phrase: "Remember that the scientific method and Taylor's method which is the same thing — are based on analysis: begin, then, with a very small problem."[15]

Production Problems

Taylorism created a niche in the factory for specialists whose job it was to analyze and reform industrial processes. These processes joined materials and men: transformed materials, materials which transformed, and men and women. Scientific management was the science of the optimal time for these operations. That said, its statements were open to many different interpretations. What is Taylorism seen from the point of view of production?

One of the determinant criteria for adopting Taylorism was the magnitude of human work in the production process. The higher the cost of the basic work force with respect to the capital invested, the greater the interest shown for scientific management or for one of its variants of which there were many. The mechanical roots of Taylor's system also predisposed it to be taken up by factories which made a major use of machine tools. Offices, then being mechanized and feminized, were also one of its privileged sites.

The adoption of scientific management in automobile management was difficult in the USA. The automobile industry was new, and the major problems posed by the production of motor cars only became clear after the basic ideas of Taylor's system had been put forward. The process of production of the motor car is planned backwards from the final product, by the assembly, even before the single model and Ford's production line appear. Ford made the motor car the place in which many mechanical tasks which were by nature discontinuous were rendered fluid. The first management problem in automobile production is the control of a flux of tasks which result in the mass production of complex mechanical objects.

It is well known that Taylor's system plays no role in the conception of the assembly line which emerged in 1913. All the same Taylorian analysis is very much present therein. The first attempt at Taylorian management in the automobile industry occurs precisely with routing and scheduling, as was the case with textiles, where the machines imposed their rhythms on the worker. In the inter-war years the American automobile became a kingdom to be conquered for scientific management.[16]

The use of scientific management encouraged many production engineers to share Taylor's desire to reduce production costs scientifically by and for a better control of work. However the capital intensive industries, which are also those which are marked by continuous stream processes, like electricity, petroleum, chemistry, are not to be found among the pioneers of Taylorism. The attitude that derives from these sources is all the same instructive. The remarks of a Frenchman working in metallurgy reveal its principal traits.

Georges Charpy was a chemical engineer who managed a laminating and heat treatment factory. The opening lines of his 1919 article state that he "only appeals periodically to Taylor's ideas" and that the "organization of work is not a panacea: technological innovation, even reduced to its simplest forms, is dominant in most cases."[17] All the same his entire effort as an organizer of production was directed to finding ways to deprive the metal workers of all initiative in their jobs, a pure Taylorian project. For this, the laboratories embarked on an analysis of operations, and chose a solution 'as close as possible to the optimal.' These results were translated into "precise instructions, carried over into the workshop, and which left nothing indeterminate for the workmen." Cabins for guiding operation were installed on the shopfloor. They received instructions drawn up by the laboratory and sent out orders from afar, imposing a rhythm on the tasks of the workers. It was the metallurgical laboratory which here played the role of the time laboratory, the role played elsewhere by the planning department. It transformed a curve depicting the stages in heat treatment by regulating the ovens and the work rhythm of the laborer. We find the same aims and the same analytical approach as that of Taylor in the mechanical industry, the same transfer of information to a specialized department, and its recycling back to the workforce. Other highly capital intensive industries put similar systems in place to manage throughput.

The diffusion of Taylorian organization inside firms was a gradual process, and required learning whose duration and nature varied according to circumstances and economic and social choices. The coherence between different practices for managing flow systems and scientific management's overt ambition to generalize its approach has led to the identification of all the organization of work in a specific period with Taylorism. This label seems to sum up a general and clearly detectable mode of thinking about action.

Today the prevalent theme in industry is the search for ways to escape not only the division but also the opposition between planning and performing. This separation and opposition are attributed to Taylor as indicative of a modern respresentation of order as opposed to chaos and individual initiative. It was in the name of a deterministic science which refused uncertainty that one sought to discipline action. It is in the name of a postmodern science of uncertainty and chaos that one seeks today to liberate it: above all, perhaps, in the firm. But it is not at all obvious that the central demand for an efficiency based on the control of time has disappeared, even if Taylor's *name* has.

And the Other Sciences

At this point it might be useful to demonstrate how scientific management could serve as a stimulant for introducing other sciences into the factory. Le Chatelier has stated that "the essence of Taylor's system is the systematic application of the scientific method to the study of all industrial phenomena," and that "his point of departure is the belief in the *determinism*, ie., the existence of a necessary relationship between all material, intellectual or moral facts and work in industry." Once we adopt this approach it opens the doors of the firm to any industrial discipline which can present itself as a science, whether it concerns men, materials, machines and installations, numbers or prices. The claim has also a practical dimension to it.

With its laboratory which localized the practical science of time, Taylor offered a critical point of view from which to assess the introduction of any other discipline. Roughly speaking, Taylor's thesis was that everything was possible if work was organized, ie., if the workplace, the basic unit of the man-material link, was the object of the science of time.

Refined techniques of cost accounting, wage incentives, vocational guidance or welfare were put forward from the end of the nineteenth century to fulfil, wholly or partly, the objectives identified by scientific management. These can be summarized as the reconquering of work by the firm's management so as to assure social peace in the most efficient of all possible worlds. Submitting the workplace to scientific study, scientific management defined a point of entry into the process which was at once intelligible and coherent with the objectives of the firm, and around which all the rest could turn. Henceforth industrial accounting had a solid base, and could also be rendered predictive. Salary systems could be assessed with respect to different kinds of work (even if the differential system developed by Taylor himself was not always the preferred one). One could associate psychological traits of professional types to the particular posts. Welfare was included in a labor policy conceived from the point of view of a rational production process. All these efforts at practical compatibility were worked on during the three or four decades after World War I.

In short, into the confusion of industrial actions as seen by the rational engineer with an interest in production, and which is simply one kind of order that others manage (workers and classical management), Taylor inserts a kind of order based on the 'scientifically' established 'time needed,' an order which is always specific, and tied to places, countries, the kind of products produced, and the size of markets.[19]

This order is itself controlled by the functionalism which is part and parcel of Taylor's doctrine. Taylorian logic abstracts and simplifies actions and organizes their relationships. Any activity which is deemed not to be directly connected with the production process, like supplies, maintenance, control, etc., is functionalized ie., set apart from the central pole of the organization's function which is to produce. In other words the position of the planning department, on the one hand, and functionalism as a doctrine on the other, have their effects beyond the walls of the workshop. They define a hierarchical axis connected to the 'function' of production and at the periphery of which they situate all other practices. This is what happened with the 'functions' which were developed between the wars: personnel management, design, research, and industrial accounting.

Laboratory practices in the natural sciences were of course brought into play in an articulated arrangement of this kind, with the Taylorian science of time at its core. As for the social and human sciences, the deployment of scientific management provided them with an opportunity to become operational.

The liveliness with which an emergent industrial psychology criticized Taylorism before the 1914–18 war bears witness to the new discipline's interest in it. Hugo Münsterberg in the United States and Germany, but also Jean-Maurice Lahy in France were its main champions. Their criticism of a 'technicist' Taylorism in the name of another science, the psychophysiology of work which was not yet ergonomics, could only reinforce their own scientific pretensions. As in Taylorism, they based their scientificity on what they took to be the model of natural sciences, and used this to found their own science of the man/machine relationship. Their critique gradually changed into a dialogue by means of which scientific management and industrial psychology ended up being compatible with one another, though not without maintaining their original motives for being in disagreement of course. Their interaction lasted for decades taking different forms in different countries. From it would emerge the new field of 'psychotechnique' narrowly tied to the needs of industrial selection.[20]

In the interim scientific management became adept at articulating different branches of knowledge. Its demand that everything should be measured, made just as the drive for productive efficiency was getting under way, required that these knowledges be identified and evaluated (also economically) and that their circulation be regulated. It is not just a question of measuring, but of bending all kinds of sciences to efficiency; or industrial opportunity, in the sense which one finds in the French judicial system which enshrines the principle of the opportunity to institute legal proceedings but according to the requirements of the political powers that be.

This scientific management which absorbs and combines the sciences so as to suppress social conflict is not itself a science of conflict. Indeed it does not know how to handle conflict since its tries to eliminate it. It is not a neutral science. It takes sides with efficiency and the time for productive actions, seen as means to achieve industrial profitability. Nevertheless it was considered an arbiter by the unions, though it was never allowed to share the right to define the truth of working norms. The human dimension of the production process was in fact that which posed the most complex problems, even more complex than those of materials and machines. In trying to avoid conflict, scientific management looked towards the elimination of people from the work floor, and its aims are on the way to being achieved. Taylorian algorithms have become embedded in machines. The computer is the Age of Reason's long awaited marriage between electricity and organization.

However that may be, and whether one thinks that it is dead or that it survives, Taylorism is surely a scientific undertaking of the twentieth century. Correlatively, its questioning is coherent with debates on how science is practiced at this *fin de siècle*: is science something that is done in separate places, where its specialists gather together, or is it to be found in the integrating discourses of practices and results, of rules of thought and of material techniques?

References

1. Fayol, Henry. "Administration industrielle et générale," *Bulletin de la Société de l'Industrie minéerale.* (no. 3, 1916), followed by numerous reprintings.
2. Haber, Samuel. *Efficiency and Uplift: Scientific Management in the Progressive Era, 1890–1920.* (Chicago: University of Chicago Press, 1964).
3. Drury, Horace B. *Scientific Management: A History and Criticism.* (New York: Columbia University Press, 2nd ed. revised, 1918), pp. 251 (1st ed., 1915).
4. Taylor, Frederick. "Shop management", *Transactions of the American Society of Mechanical Engineers* (1903), **24**: pp. 1337–1456.
5. Barth, Carl G. *Scientific Management.* (Hanover (N.H.): Dartmouth College Conferences (Tuck School Conference), 1912), pp. 174–5.
6. Kendall, Henry P. "Management: Unsystematized, Systematized, and Scientific," in C. Bertrand Thompson (1914), pp. 113, 123.
7. Dodge, James Mapes, "A History of the Introduction of a System of Shop Management," *ibid.*, p. 231.
8. Thevenot, Laurent. "Rules and Implements: Investments in Forms," *Social Science Information* (1984), vol. 23, no. 1, pp. 1–45.
9. Savoye, Antoine, *Les débuts de la sociologie empirique.* (Paris, Méridiens Klincksieck, 1994).
10. Stengers, Isabelle. *La volonté de faire science. A propos de la psychanalyse.* (Le Plessis-Robinson Les Empêcheurs de penser en rond, 1992).
11. Kendall, 121.
12. Stengers, I. *op. cit.*, and also Yves Cohen, "Le vingtième siècle commence en 1900. Sciences, technique, action", *Alliage.* (no. 20–21, automne-hiver 1994), pp. 88–104.
13. Thompson, Bertrand. 1914.
14. Le Chatelier, Henry. *Science et industrie.* (Paris: Flammarion, 1925), p. 74.
15. *Sur le tas ou conseils pour débuter dans la méthode Taylor, Prospérité.* (2e année, suppl. au no. 7, décembre 1929), p. 1.
16. Charpy, Georges. "Organisation méthodique d'une usine métallurgique," *Bulletin de la Société d'Encouragement pour l'Industrie Nationale.* (mai–juin, 1919), pp. 572, 574.

17. *Ibid.*, p. 603–604.
18. Le Chatelier, Henry. *Frederick Winslow Taylor, 1856–1915.* (Paris: O. Doin et fils, 1915), p. 9 (emphasis by Le Chatelier).
19. Mottez, Bernard. *Systèmes de salaire et politiques patronales. Essai sur l'evolution des pratiques et des idéologies patronales.* Paris: CNRS, 1966), p. 143.
20. Rabinbach, Anson. *The Human Motor. Energy, Fatigue, and the Origins of Modernity.* (New York: Basic Books, 1990).

Further Reading

Taylor, Frederick W. *Scientific Management.* (New York, Harper and Bros, 1947).

Thompson, Clarence Bertrand. *Scientific Management. A collection of the more significant articles describing the Taylor system of management.* (Cambridge, Mass.: Harvard University Press, 1914).

Nelson, Daniel. *Managers and Workers. Origins of the New Factory System in the United States, 1880–1920.* (Madison: The University of Wisconsin Press, 1975).

Nelson, Daniel. *Frederick W. Taylor and the Rise of Scientific Management.* (Madison: The University of Wisconsin Press, 1980).

Nelson, Daniel (ed.). *A Mental Revolution: Scientific Management since Taylor.* (Columbus: Ohio State University Press, 1992).

Lamoreaux, Naomi R. et Raff, Daniel M.G. (eds), *Coordination and Information. Historical Perspectives on the Organization of Enterprise.* (Chicago: The University of Chicago Press, 1995).

Waring, Stephen P. *Taylorism transformed. Scientific Management Theory since 1945.* (Chapel-Hill-Londres: University of Carolina Press, 1991).

Fridenson, Patrick. 'Un tournant taylorien de la société française (1904–1918),' *Annales. Economies, Sociétés, Civilisations.* (vol. XLII, no. 5, Sept.–Oct. 1987), pp. 1031–1060.

Moutet, Aimée. *Les logiques de l'entreprise. L'effort de rationalisation dans l'industrie française de 1919 à 1939.* (Paris, Editions de l'EHESS, forthcoming).

Kleinschmidt, Christian. *Rationalisierung als Unternehmensstrategie. Die Eisen- und Stahlindustrie des Ruhrgebiets zwischen Jahrhundertwende und Weltwirtschaftskrise.* (Essen: Klartext, 1993) (there is no overall work on Germany).

Beissinger, Mark R. *Scientific Management, Socialist Discipline, and Soviet Power.* (Cambridge: Harvard University Press, 1988).

CHAPTER 8

The Discovery of Growth

Statistical Glimpses of Twentieth-Century Science

SUSAN E. COZZENS

Over the twentieth century, every aspect of the shape and size of science has changed. Not only has the list of countries involved in the scientific effort expanded; the countries themselves have changed their boundaries, their names, their independent or colonial status. Two imperial eras have come and gone — those of Western Europe and Eastern Europe — and with them patterns in scientific relations that reflected and reinforced larger geopolitical forces. The intellectual map of science has likewise been transformed, through differentiation within broad disciplines established in the nineteenth century; through the emergence of major new conglomerations; and in recent years through the blurring of disciplinary boundaries themselves. While the categories of institutions that house science have endured, the institutions themselves have evolved differently in different national contexts and under the influence of a shifting set of patronage relations: academies have faded; in some areas institutions based on private philanthropy have been overshadowed, while in others private funding behemoths have emerged; an evolving set of industries has invested and disinvested in research capacities, and extended their influence into the academic sphere; and areas of government research have waxed and waned, as a shifting set of nations has shouldered responsibility for war, the environment, and other global problems.

The appearance of statistics that chart the growth of science mirrors the fundamental institutional processes that characterize the century, and in particular the absorption of science into the various world wars. Fascination with numbers that describe science, with the scale and anatomy of the enterprise, has emerged over the twentieth-century in particular times and places. The growth of science, as a problem to be studied in its own right, was discovered in post-Sputnik America, for example. In telling the history of twentieth-century science through its statistics, we learn most not from the numbers themselves, but rather from asking who collected them, for what purposes, and how they chose their categories.

A full history of science statistics in the twentieth century would be enlightening, but has not yet been written. In the first sections of this chapter, I imagine what science statistics might have looked like before the Second World War, and in the second section, I sketch only the broad outlines of its post-war development. Mid-century theories of the limits to growth in science, although they were futuristic when they were introduced, are predicting more and more features of the contemporary scene as the century draws to a close. The third section of the chapter uses those theories to highlight selected features of the development of twentieth century science, and the final section articulates the questions they raise for conditions in science as the twenty-first century begins.

The Evolution of Science Statistics

What appears at first to be a commonplace observation is in fact significant: At the turn of the century, no one was interested in developing statistical descriptions of science, at either national or world level. Especially outside the United States, national governments ran many research institutions, but they had not yet developed the sense of government as the patron of science that gave birth to later statistical efforts. A much larger share of the research enterprise was privately supported in 1900 than is true today, a fact that contributed to the scattering, rather than gathering, of information about research as a whole. Industrial research had scarcely been invented, and would not have been seen by contemporaries as significant enough to count. At the world level, only the British Empire had the global identity and communication network to study the distribution of research efforts in different countries, but it did not do so. In short, the scale and distribution of effort in science were not problematic for the official record-keepers of the first half of the twentieth century.

Let us therefore invent a turn-of-the-century statistician to attempt, fictionally, to construct a rudimentary science indicators system. The raw materials were available, but highly scattered. Most of her task would focus on Europe and the United States, and could start in the higher education system, where the research role became professionalized in the late nineteenth century. The universities of Europe were well-established, and those of the United States were growing, and she could track the founding of chairs in the new science disciplines. The formation of universities on the Western model would form a convenient index for the diffusion of European learning styles into the countries of the periphery, following the process of colonization and quasi-colonization over the preceding two centuries. To name just a few examples, the Mexican School of Mines had been formed in 1792, only two years after its more famous counterpart in Paris; and in 1804, Mexico City was a scientific center comparable to Philadelphia or Boston.[1] In Australia, the University of Sydney was formed in 1852, and Melbourne in 1855.[2] In Japan, the predecessor institution to Hokkaido University was formed in 1876, and the University of Tokyo in 1877.[3]

For non-university institutions, our fictional census-taker would face the problem of setting an operational definition for a scientific institution — a problem also faced by later science statisticians. In her case, the task would be complicated by the fact that the realm she was trying to describe was incompletely and unevenly professionalized. Both professional associations and private research institutes entered the scene in the late nineteenth century,[4] and with a well-placed network of correspondents, our fictional statistician might be able to enumerate them. A rough, retrospective estimate based only on institutions that survived to the late twentieth century indicates that in 1900, European institutions outnumbered those in the United States by about two to one, and that less than ten percent of the world's scientific associations and research centers were located outside Europe or the United States. South Africa, for example, had only 32 such institutions before 1900, including institutes, universities, observatories, parks, game preserves, museums, professional associations, and journals.[5]

Turning from institutions to people, our fictional figurer would face similar issues of definition. Some help might come from membership numbers for learned societies, but they would have to be interpreted carefully. Present-day observers of such societies are quick to point out that at the turn of the century they included amateur as well as credentialed scientists. Perhaps the best estimate of the size of the technical workforce in the British colonies, for example, is the membership of colonial learned societies, which stood at about 5000 in 1900. But no more than 100 of these members were likely to have been active scientific researchers. In 1900, 35 percent of the members of the Royal Society of New South Wales, for example, lacked "scientific backgrounds."[6] In Britain itself, the scientific workforce numbered only about 2400 at the turn of the century. This retrospective estimate includes those employed in industry, secondary schools, universities and primary schools, government, and elsewhere.[7]

University faculties might provide the most dependable counts, but again in a quite decentralized manner. In Japan at the turn of the century, there were only 33 chairs in all academic fields, with only five of those in the physical sciences.[8] The European systems were much larger; for example, German universities had 2667 members of the academic staff at that time.[9] University degrees provide another source of estimates of the total size of the trained workforce, as well as an indication of the contrast in size between scientific core and periphery nations. US universities awarded 382 Ph.Ds in all fields in 1900, of which perhaps two thirds were in the sciences.[10] In contrast, Quebec and Ireland graduated only about 100 scientists each over the entire nineteenth century.[11]

While it is tempting to leave the fictional statistician at her desk at the turn of the century, sending letters to the four corners of the globe to gather counts of institutions and scientists, for our purposes it is more useful to ask her to revisit her task on the eve of the First World War, and then the Second. We had better give her a few assistants, however, since growth in all the aspects of science she is measuring had been rapid. For example, by 1913 South Africa had doubled its institution list. The scientific workforce in the UK had tripled, and was to triple again by 1939.[12] For Japan, the numbers were up ten-fold over the university counts for 1900, and showed much greater diversity in employment. A third of the 387 Japanese research scientists in 1913 were located outside universities, 30 at government laboratories, 15 in higher technical schools, and 70 in medical schools.[13] At the same time, arch-rival Russia claimed (retrospectively, in the 1950s) to have had about 4000 scientists working in 289 scientific institutions.[14] US doctorates in all fields increased by 40 percent between 1900 and 1914. In the sciences between 1920 and 1940, they grew by a factor of six.[15] Overall, the data on institutions that survived to the last decade of the century indicate that about three times as many professional associations and research centers were formed between 1900 and 1940 as had been formed previously, with only about 5 percent of the new institutions located outside Europe and the United States.

Again, like those who followed her in the last half of the twentieth century, our fictional statistician would need to be wary of characterizing change in the science system purely in terms of quantitative expansion in the categories she established at the turn of the century. The most important shifts she observed in 1913 and 1939 were perhaps the qualitative changes within the categories. For example, an early manifestation of the self-conscious role of a government in stimulating research appears in Japan in 1918 in the form of the Science Research Grants Program.[16] Later activities of just this sort would be reflected in a large share of national science statistics. The Japanese program had in fact been formed to counteract perceived inequities in government funding for research institutions — again, a theme that would appear many times in the last half of the century in the statistics on institutional and geographic distribution of research support.

Likewise, new national roles for science were reflected in the formation of Japan's National Research Council (NRC) in 1920.[17] Bartholomew describes the motivations: ". . . the council's establishment helped liberate Japanese science from its obsession with Germany. The NRC was created as an affiliate of the International Research Council (IRC). Since this agency had been founded as a device for isolating German science, members of the Japanese scientific community were obliged to evaluate carefully their place in the international community of science."[18] That international community, with its inherent tension between conflict and cooperation, would also eventually generate volumes of cross-national statistics and comparative standards.

Finally, on the list of institutions of early twentieth century science our fictional statistician might notice the gradual emergence of industrial research, and its symbiotic relationship with government science. In the West, the pharmaceutical firms of Lilly, Upjohn, and Smith, Kline, and French were assembling research staffs in the period before World War I,[19] while in the East, Takeda and Sankyo Pharmaceuticals formed research divisions in 1915 and 1916.[20] In the United States, General Electric was helping to support research at Harvard, and AT&T endowed research at MIT,[21] while the Japanese Institute for Metals Research, a government laboratory, was in partnership with Westinghouse Corporation from 1917 into the 1920s.[22] The US National Bureau of Standards was set up in 1901 to establish uniform industrial standards not only of weight, but also of chemical and electrical quantities.[23] The Australian Council for Scientific and Industrial Research Organization was founded in 1916,[24] and stayed largely an industrial service organization until the Second World War.[25] Relationships like these between public and private research activities were to become items of enduring interest in the indicator systems that were developed later.

Cold War to Cold Shoulder

Like so many features of present-day discourse on science policy, centralized statistics on research were an outgrowth of the Second World War, a sign of the new scale of government involvement in the support of research. Some observers, however, trace the origins of national-level statistics on science to earlier lobbying efforts on the part of scientists. The lobbying genre made its appearance in Britain in the 1930s, where Julian Huxley attempted to calculate the total cost of research carried out in Britain, and J. D. Bernal published figures on the number and distribution of scientists, their sources of support, and their contributions to the scientific literature.[26] The equivalent lineages in the United States culminated in Vannevar Bush's *Science: The Endless Frontier*, which recommended the formation of a National Science Foundation (NSF). The report included a table with annual figures from 1920 to 1944 on expenditures for science in various categories of research institutions, as compared with national income. Careful study of Bush's footnotes, however, reveals most of the numbers to be extrapolations from surveys in the 1930s.[27]

When the Foundation was finally established in 1950, its organic act gave it a mandate to collect data on research and development activities in the United States. Two types of numbers on science appeared early, and have persisted as the mainstays of science statistics systems, in the United States and elsewhere: personnel data, and data on expenditures for research and development. The personnel systems have their origins in registries of scientific personnel started during the war, and continued afterwards, to increase readiness for any future technologically-based conflict. An early example appears in Australia, the *Science on Service* directory published in 1943.[28] (Figure 8.1) The first comprehensive scientific personnel counts in the United States also came from a registry, and are available for the early 1950s.

The Korean and Cold Wars kept the interest in so-called manpower issues lively throughout the 1950s in both the United States and Europe. In 1951, for example, the US National Security Resources Board issued "Plans for the Development and Use of Scientific Manpower." Beginning with the statement that "In terms of gross numbers of men, the United States is inferior to its potential enemies,"[29] the aim of the plan was to make best use of technical talents within the services, on the

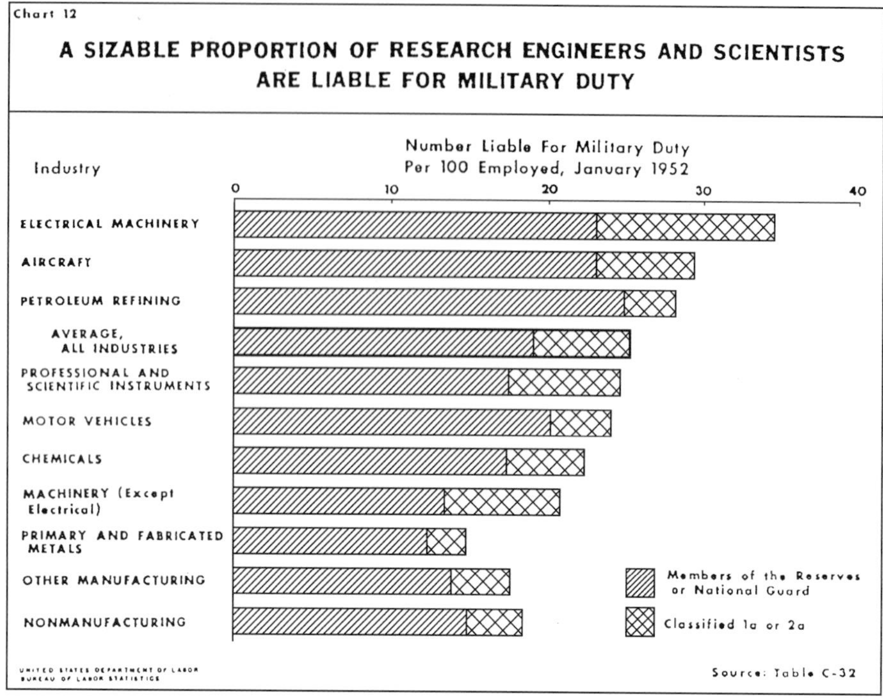

FIGURE 8.1: US OBSERVERS IN 1953 WERE CONCERNED ABOUT THE ELIGIBILITY OF RESEARCH SCIENTISTS AND ENGINEERS FOR ANOTHER DRAFT.

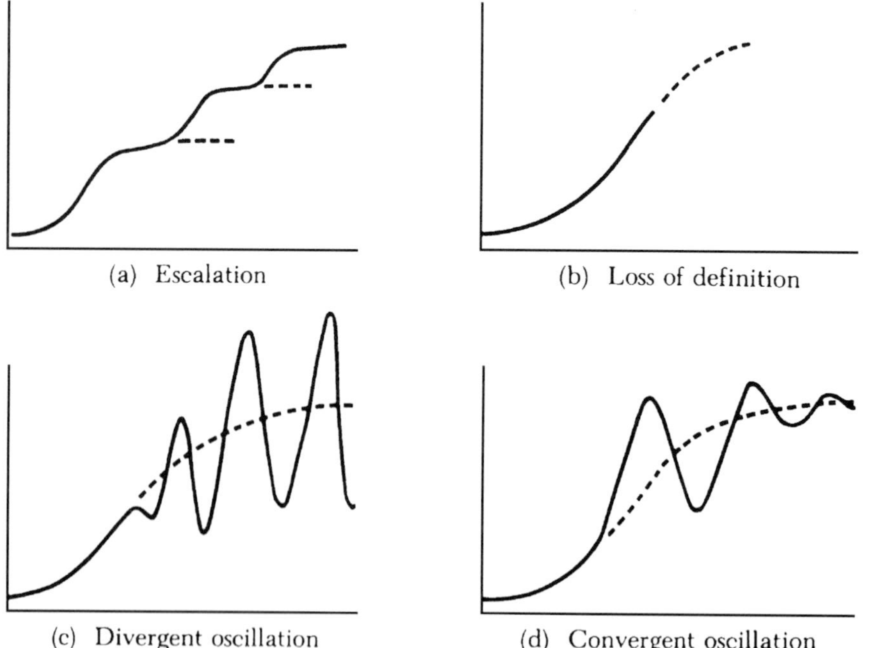

FIGURE 8.2: IN 1963, DEREK PRICE CHARACTERIZED THE END OF GROWTH FOR SCIENCE WITH THESE DIAGRAMS. (FROM PRICE, 1963).

assumption that they would not be exempted from conscription. Cold War issues appeared in the analyses of manpower in the Communist states. A Rand study in 1957 of "The Growth of China's Scientific and Technical Manpower" compared China's college enrollments in technical fields with those of the US, USSR, and India, calling attention to potential weakness on India's part because of its underinvestment in scientists and engineers.[30]

An awakening to the importance of industrial research and development appears in the early post-war interest taken in the topic by the US Bureau of Labor Statistics, in cooperation with the Department of Defense, in a study of research scientists and engineers employed in defense-related companies.[31] Close to 100,000 researchers worked in the nearly 2000 companies in the study, and about half were working on federally-financed projects. Observers were clearly worried about the possible impact on the civilian economy of a renewed draft. For example, one figure caption in the report read, "A sizable proportion of research engineers and scientists are liable for military duty" (Figure 8.1). As many as 35 percent of the research engineers and scientists in defense-dependent industries such as electrical machinery, were eligible to be called up in time of war.[32] The survey yielded the estimate that two-thirds of the nation's research and development expenditure was for work done in industry's laboratories, although a large portion of that total came from the federal government.

NSF's early statistical efforts concentrated on a different part of the system: the universities, for which NSF had been given responsibility. The earliest NSF funding data are for 1950–51 — an achievement which is particularly impressive given that the Foundation was not really operational until 1951.[33] Paralleling the growth of government research investment, several other industrialized countries started science-statistics systems in the 1950s, including Japan, Canada, the United Kingdom, the Netherlands, and France; and by 1960, most of them were engaged in this kind of activity.[34] In Australia, for example, the first available comprehensive data for expenditure on research are for 1958/59, and show governments to be responsible for more than 80 percent of the national research expenditure.[35]

The international organizations that sprung up after the war played an active role in developing science statistics systems. The Organization for European Economic Cooperation (OEEC), which was created in 1948 to deal with the US on the distribution of Marshall Plan aid funds, started a scientific manpower program in 1958, with the United States as full participant. Post-Sputnik competition was an explicit motivation.[36] The OECD became a particularly active player in the science statistics area. In 1962 a consultant prepared the "Proposed Standard Practice for Surveys of Research and Development," and a working group adopted it in 1963. In honor of the location of the meeting, the standards became known as the "Frascati Manual."[37] The United Nations Educational and Scientific Organization (UNESCO) has also collected and published statistics on science and technology on a world basis since 1966.[38]

Collecting data in a standard way in many countries has been seen as particularly important so that data can be used within various national contexts to justify research expenditures — the aforementioned lobby function that traces back to the 1930s. Particularly useful for these purposes has been the percent of Gross Domestic Product devoted to research and development (GERD, or Gross Expenditure on Research and Development) — a latter-day cousin of the table on government expenditures in *Science: The Endless Frontier.* Since its invention, this statistic has appeared in thousands of science policy statements, generally with the admonition that a particular country's share is not as high as one of its neighbors' and that government funding for science therefore needs a boost. Only in the no-growth 1990s have analysts begun to recognize that the statistic rises either when the numerator increases (i.e., through greater expenditure) or when the denominator drops.

The 1976 edition of the "Frascati Manual" begins, "Statistics of R&D are presently limited to the inputs" — that is, to the staple figures on people and funding.[39] But as early as the 1960s, the limitations of these types of data for understanding the dynamics of a rapidly-growing science system were apparent. In 1970, a committee of the National Science Board (the National Science Foundation's governing body) began developing the idea of a more analytical quantitative report on US science and engineering.[40] Their work eventually inspired the 1972 volume of *Science Indicators,* a compendium of existing statistics on science and technology and a platform for developing new ones. They hoped the new volume would produce assessments of the effects of science and technology, and synthesis of a variety of information sources. The volume has instead been successful as a reference source, and has been copied in a dozen other countries, and recently the European Community.

Science and Engineering Indicators, as the volume is now called, still contains about half input data — the funding and personnel series that stretch back to the 1950s.[41] Data on international linkages and links among government, academic, and industrial research have also expanded over the years, and over the 1980s the volume expanded to include much more information on science education and literacy. The connection between science and the economy has been of continuing interest, but on an unstable conceptual base. The indicators in that area — trade and innovation statistics, for example — show the highest rate of turnover.[42]

The US indicators volume pioneered the collection of data on public attitudes toward science, which is now gathered on a cross-national basis. Ironically, given the origins of the statistical series themselves in the response to war, the attitudes data show clearly the ambivalence of the American public toward the use of science in war. Well over 60 percent of the public judge that science and technology have had a positive impact on standards of living, public health, general working conditions, and enjoyment of life, but less than half think they have helped world peace.[43]

THE DYNAMICS OF GROWTH

In the 1970s, as the profusion of data on science became available, a burst of scholarly effort was devoted to thinking through the assumptions and methods that underlay indicator systems. Historians, economists, information scientists, statisticians, and sociologists collaborated in discussions around "the metric of science."[44] Elkana placed this effort in the context of a "growing faith in creating equivalents of the established economic indicators for other areas of social behavior," and as a "continuation of the great information-gathering tradition of Western Civilization."[45]

The dean of this movement was Derek deSolla Price, an historian of science who, since the early 1960s, had been practicing scientometrics — that is, postulating quantitative laws about the development of science. Growth was Price's special obsession; in this he was a product of his times. In his 1963 classic *Little Science, Big Science,* he himself used hand-gathered data, but the popular impact of his quantitative style was surely magnified by the new statistical tools that the 1950s had made available.

Price's central hypothesis was that exponential growth had characterized science throughout its history. He compared the rapid increase in science and engineering doctorates and total scientists and engineers (numbers that double every 10 to 15 years) to slower growth in the general population (doubling every 40 to 50 years), and came to his famous conclusion:

> It is clear that we cannot go up another two orders of magnitude as we have climbed the last five. If we did, we should have two scientists for every man, woman, child, and dog in the population, and we should spend on them twice as much money as we had. Scientific doomsday is therefore less than a century distant.[46]

A growing social system, he observed, shows one set of dynamics, but a social system in the process of stabilizing is likely to have quite a different atmosphere. Unless the pattern of growth is transformed through a change in organization that allows a new spurt, systems are likely to go into a saturation phase characterized by wild fluctuations around the maximum (Figure 8.2). Price's work was aimed at issuing a general warning about this impending change in the character of science, especially in the countries of the mature science core. He observed that the overall growth curve for science is the composite of many particular ones. Nations that have been active longest and have the largest investments in science were likely to be closer to the saturation point for growth than those that are just entering the system, according to his theory. Thus the crisis of saturated growth may hit in the United States and Europe while other nations are still rapidly increasing in scientific strength. Japan was his prime example of a catch-up system.

Over the last two decades, the statistics on growth in the numbers of scientists and engineers relative to the general workforce have shown some of the patterns Price predicated (Figure 8.3). In the United States and several major European

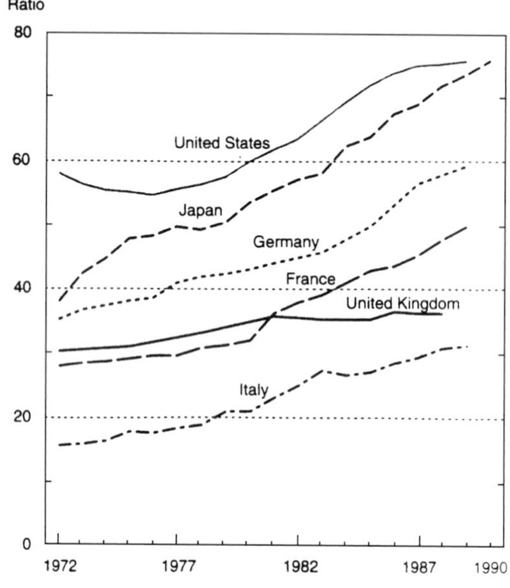

FIGURE 8.3: THE REST OF THE WORLD IS CATCHING UP WITH THE UNITED STATES, AS PRICE PREDICTED. (FROM NATIONAL SCIENCE BOARD, 1994).

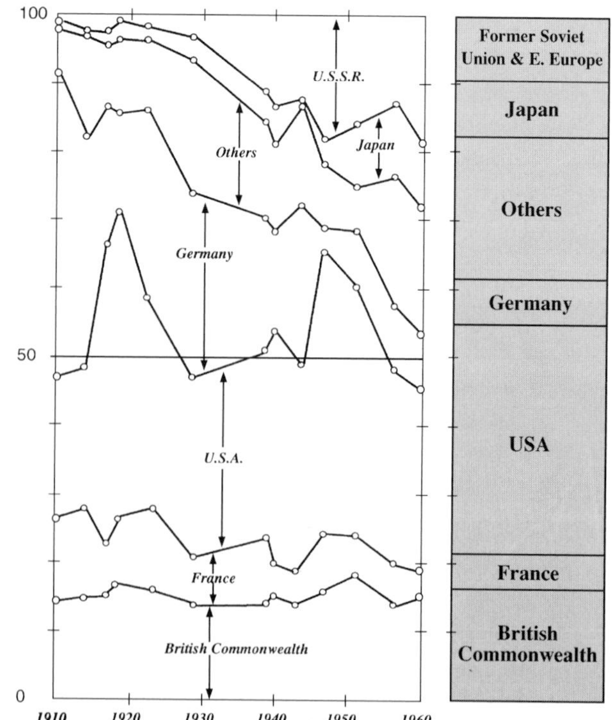

FIGURE 8.4: ON THE LEFT, HAND-CRAFTED DATA FOR CHEMISTRY IN THE EARLY TWENTIETH CENTURY (FROM PRICE, 1963). ON THE RIGHT, MACHINE-GENERATED COUNTS FOR THE EARLY 1990S (FROM NATIONAL SCIENCE BOARD, 1994).

nations the percentage of the labor force in technical occupations increased steadily and at about the same rate over the 1970s and 1980s. (The United Kingdom is an exception, stabilizing at an apparently pre-mature, low level in 1980.) The comparable Japanese figures, however, increased much more quickly, and overtook the United States at the top of the scale in 1990. The populations of scientists and engineers in other Asian nations are increasing even faster.[47]

Price was fond of using scientific literature as another indicator of this uneven growth pattern. More recent data support him in the notion that growth has slowed in the core but continues in the periphery. Scientific and technical periodicals in the United States, for example, are growing much more slowly than the total world stock.[48] In the specific case of chemistry, modern bibliometric sources show a similar pattern to the one Price explored in his hand-crafted data (Figure 8.4). Price's graph of the change in distribution of chemical literature between 1910 and 1960 revealed the growth of the contribution from the USSR and Japan, displacing German strength from the early part of the century. Data for 1990 continue to show the dominance of the US and Europe in the literature of the field, with a slight trend toward continued Japanese relative growth and US relative decline.[49]

The scientometricians of the 1960s were fond of combining the two staples of science statistics, personnel and funding data, to formulate laws linking the newest institutional feature, government funding for research, to aspects of the growth of science. Price observed that in the 1950s, the costs of science were rising as the square of the number of scientists — a pattern that creates obvious and immediate pressures for a slowing of the underlying growth.[50] While the 1950s numbers fit this hypothesis well enough for Price's purposes, figures for later decades have not followed the pattern (Figure 8.5). The number of scientists has grown steadily at about three percent per year, but R&D expenditures per research scientist and engineer have been stable since 1975. United States spending almost equals that of the European in this area ($144,000 per person in 1989 in Europe, as compared to $137,000 in the United States). Asian investment, however, shows again the catch-up pattern, with the newly-industrializing nations moving even faster on this front than the Japanese.[51]

In the last section, I noted the emergence of government statistics on industrial research spending in the early 1950s, as a symptom of the technical nexus between defense needs and the economic interests of manufacturing firms. Further elaboration of the science-government-industry relationship has been a dominant feature of the institutional growth of late twentieth-century science. Especially in this area, international comparative data has allowed the construction of various alarming scenarios on the part of national science advocates. An example is the relationship shown in Figure 8.6, taken from the US *Science Indicators* volume, which depicts US industrial R&D spending in stagnation and the Japanese catching up fast.[52] As with Vannevar Bush's excursions into science statistics, this graph uses extrapolation strategically.

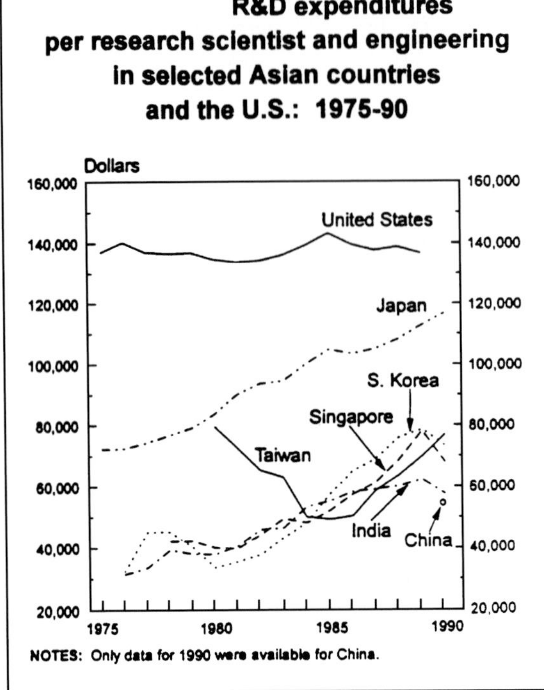

FIGURE 8.5: NO GROWTH FOR THE UNITED STATES IN THE LATE TWENTIETH CENTURY, WHILE ASIA GROWS APACE (FROM NATIONAL SCIENCE BOARD, 1994).

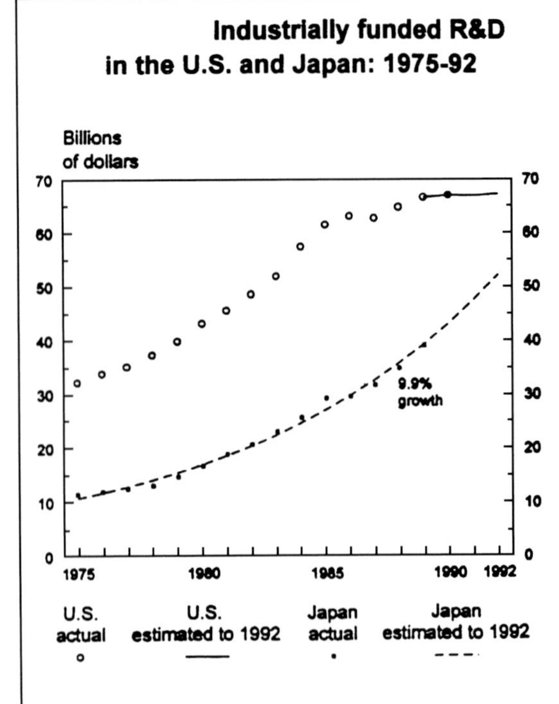

FIGURE 8.6: A FAVORITE STRATEGY IN STATISTICAL VISUAL RHETORIC: OUR COMPETITORS ARE CATCHING UP (FROM NATIONAL SCIENCE BOARD, 1994).

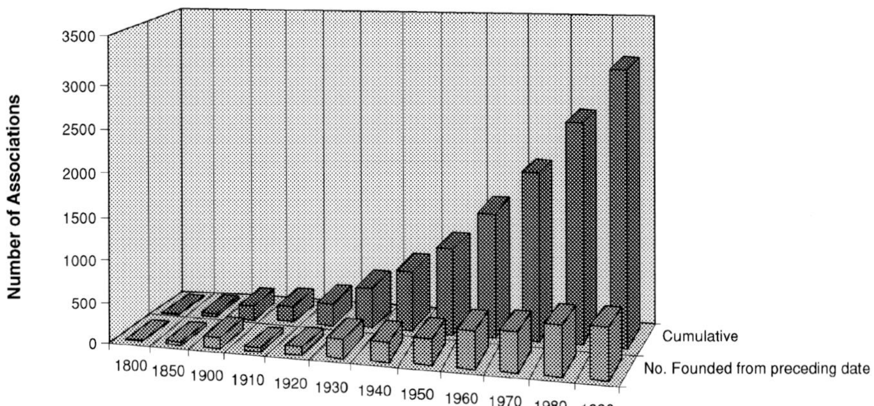

FIGURE 8.7: FOUNDING DATES OF PROFESSIONAL AND TECHNICAL ASSOCIATIONS, CUMULATIVE AND DECADE-BY-DECADE (SAMPLE DRAWN FROM DRESSER, 1994).

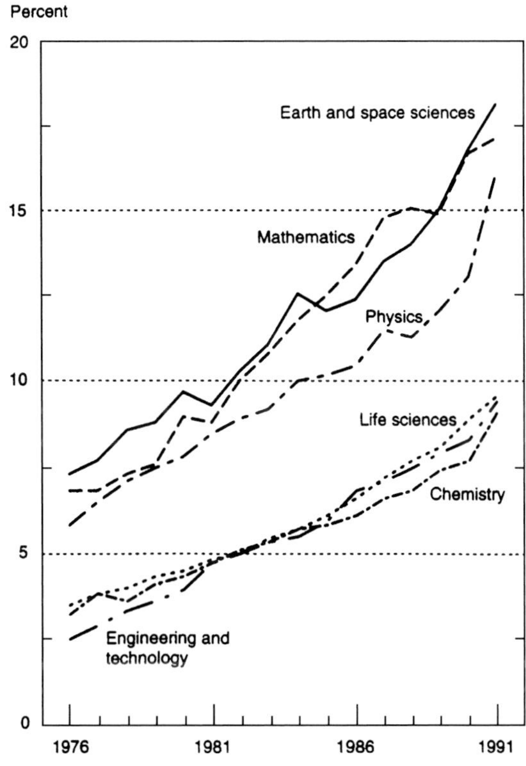

FIGURE 8.8: CO-AUTHORSHIP ON THE RISE (FROM NATIONAL SCIENCE BOARD, 1994).

Price and others in the 1960s saw an inevitable connection between growth and differentiation in science: the more science grew, the more it would differentiate. Larger disciplines would subdivide; new journals would be formed in proportion to the number of active researchers, etc. The data of the late twentieth century indicate, however, that these early hypotheses underestimated the integrative forces that work against pure differentiation within the research community. Price's concept of the "invisible college," for example, a set of about 100 scientists who follow each other's work, was based on the notion of a limit to the amount of information an individual scientist could absorb. The number of journal subscriptions per scientist, however, has been rising since 1970, from an average of about 9 to close to 12.[53] Have researchers learned how to absorb information more effectively? The rise of large, multidisciplinary journals also suggests greater integration as a counter-balance to differentiation: subscriptions to large journals (those with over ten thousand subscribers) rose monotonically between 1970 and 1985, while subscriptions to smaller journals have stayed stable.[54] Finally, while the number of professional associations continues to rise (a sign of disciplinary differentiation), the rate of increase has been stable in the 1990s, after large jumps in the 1960s and 1980s (Figure 8.7). As with journals, the average membership of the largest associations is probably growing, and they may be forming more internal sections, but the pure ability to maintain coalitions indicates that integrative forces are stronger than the scientometricians of the 1960s predicted.

Indeed, networking and collaboration, rather than differentiation, seem to be the watchwords of social structure for science as it moves into the twenty-first century. Our fictional turn-of-the-century statistician would scarcely have thought to count collaborations in the sciences, so dominant was the mode of individual investigators in their own laboratories, albeit in some cases surrounded by students. Yet collaboration rates have grown explosively in the sciences in the late twentieth century. Institutionally-collaborative articles increased by fifty percent as a share of the total between 1973 and 1984,[55] and articles co-authored between US academic and industrial collaborators went up by half between 1981 and 1991[56] (Figure 8.8). International collaboration tripled between 1976 and 1991, with every broad field of science showing an increase of some size.[57]

Power and Politics: Saturation Science

Our statistical survey thus brings us back to the theme with which we started. The twentieth century was a time of great change, not only within the intellectual core of sciences, but also in the institutional features that science statistics capture. As we enter the twenty-first century, we find a scientific community not only orders of magnitude larger, but also more consolidated in many ways than at the turn of the last century. The global interactions of science are stronger, as evidenced by international programs and collaborations. Institutional boundaries in the patronage and performance of research are blurring; universities work more closely with industry, industry works more closely with government, government is the mainstay of universities, etc. At the level of laboratories and individual researchers, the integrative forces are even clearer. Laboratory benches are linked to microscopes and supercomputers throughout the world, as research operations themselves take place over the networks. Electronic communication far outpaces paper publication, and makes collaboration with anyone, anywhere — not to mention collection of science statistics from anyone, anywhere — a reality.

The frenetic pace of interconnectedness may give a feeling of acceleration and growth, but at the same time, the signals Price described for saturation science are appearing. Resource allocation in at least some mature scientific countries has indeed been fluctuating wildly, and the need to set priorities among fields, another of Price's prognostications, is a widely-established fact. If scientists learned arrogance when their institution was a rising star, they must now learn a new lesson, as Price so aptly described in 1963:

> The new state of scientific maturity that will burst upon us within the next few years can make or break our civilization, mature us or destroy us. In the meantime, ... we must look for considerable assumption of power by responsible scientists, responsible within the framework of democratic control and knowing better how to set their house in order than any other men at any other time.[58]

REFERENCES

1. Chambers, D.W., "Period and Process in Colonial and National Science" in Reingold and Rothenberg, *Scientific Colonialism* (1987), p. 301.
2. Inkster, I., and Todd, J., "Support for the Scientific Enterprise" in R.W. Home, *Australian Science in the Making* (1988), p. 111.
3. Watanabe, M., *The Japanese and Western Science* (1988), p. 1.
4. Watanabe M., *The Japanese and Western Science* (1988), p. 1. Inkster, I., and Todd, J., "Support for the Scientific Enterprise" in R.W. Home, *Australian Science in the Making* (1988), p. 111.
5. Brown, A.C., *History of Scientific Endeavour in South Africa* (1977), p. 70.
6. Inkster, I., and Todd, J., "Support for the Scientific Enterprise" in R.W. Home, *Australian Science in the Making* (1988), p. 114.
7. Edgerton, this volume.
8. Batholemew, J., *The Formation of Science in Japan* (1989), p. 196–7.
9. Ben David, J., *The Scientist's Role in Society: A Comparative Study* (1971), p. 129.
10. Jarrell, R.A., "Differential National Development and Science in the Nineteenth Century: The Problems of Quebec and Ireland" in Reingold and Rothenberg *Scientific Colonialism* (1987), p. 343.
11. Edgerton, this volume.
12. Bartholemew, J., *The Formation of Science in Japan* (1989), p. 165.
13. Krementsov, this volume.
14. US Bureau of the Census, Historical Statistics of the United States, Colonial Times to 1970 Part I (1975), p. 386.
15. Bartholemew, J., *The Formation of Science in Japan* (1989), p. 7.
16. Bartholemew, J., *The Formation of Science in Japan* (1989), p. 7.
17. Bartholemew, J., *The Formation of Science in Japan* (1989), p. 254.
18. Swann, J.P., *Academic Scientists and the Pharmaceutical Industry: Cooperative Research in Twentieth Century America* (1988), p. 22–23.
19. Bartholemew, J., *The Formation of Science in Japan* (1989), p. 230.

20. Kevles, D., *The Physicists: The History of a Scientific Community in Modern America* (1971), p. 100.
21. Bartholemew, J., *The Formation of Science in Japan* (1989), p. 246.
22. Kevles, D., *The Physicists: The History of a Scientific Community in Modern America* (1971), p. 66.
23. Smith, C., "Saving the Archives of Scientific and Industrial Research" in Kirsop and Borchardt *Some Sources for the History of Australian Science* (1988), p. 15.
24. Home, R.W., *Australian Science in the Making* (1988), circa p. 220.
25. Thackray, A., "Measurement in the Historiography of Science" in Elklana *et al.*, *Toward a Metric of Science* (1978).
26. Bush, V., *Science — The Endless Frontier* (1945: Reprinted 1980), p. 86.
27. Home, R.W., *Australian Science in the Making* (1988), p. 231.
28. US National Security Resources Board, p. 2.
29. Ikle, F.C., *The Growth of China's Scientific and Technical Manpower* (1957), p. 54.
30. Wood, H., *Scientific Research and Development in American Industry: A Study of Manpower and Costs* (1953).
31. Wood, H., *Scientific Research and Development in American Industry: A Study of Manpower and Costs* (1953), Chart 12.
32. The 1950–1 data are cited in US Bureau of Labor Statistics, p.1.
33. OECD The Measurement of Scientific and Technical Activities ("Frascati Manual") (1976), p. 9.
34. Johnston, R., and Buckley, J., "The Shaping of Contemporary Scientific Institutions" in R.W. Home, *Australian Science in the Making* (1988), p. 375.
35. US Bureau of Labor Statistics p. 1.
36. OECD The Measurement of Scientific and Technical Activities ("Frascati Manual") (1976), p. 10.
37. OECD The Measurement of Scientific and Technical Activities ("Frascati Manual") (1976), p. 11.
38. OECD The Measurement of Scientific and Technical Activities ("Frascati Manual") (1976), p. 15.
39. Based on Cozzens, S.E., *Science Indicators: Description or Prescription, Report to the US office of Technology Assessment.* (1990). Contract Number N3–2315.0. Available from National Technical Information Service, PB 91-166-611.
40. Cozzens, S.E., *Science Indicators: Description or Prescription* (1990), p. 8.
41. Cozzens, S.E., *Science Indicators: Description or Prescription* (1990).
42. National Science Board 1993, Figures 7–12.
43. Elkana, Y., *et al.*, *Toward a Metric of Science: The Advent of Science Indicators* (1978).
44. Elkana, Y., "Political Contexts of Science Indicators" in Y. Elkana, *et al.*, *Toward a Metric of Science: The Advent of Science Indicators* (1978).
45. Price, D.J. De Solla., *Little Science, Big Science* (1976), p. 19.
46. National Science Foundation Asian Report.
47. King, D.W., MacDonald, D.D., and Roderer, N.K., *Scientific Journals in the United States* (1981), Figure 2.4.
48. Data from National Science Board 1993.
49. Price, D.J. De Solla., *Little Science, Big Science* (1976), p. 92.
50. National Science Foundation Asian Report.
51. National Science Foundation Asian Report.
52. King, D.W., MacDonald, D.D., and Roderer, N.K., *Scientific Journals in the United States* (1981), Figure 2.8.
53. King, D.W., MacDonald, D.D., and Roderer, N.K., *Scientific Journals in the United States* (1981), Figure 4.2.
54. *Science and Engineering Indicators 1987*, Tables 4–36 and 5–23.
55. National Science Board 1993, p. 429.
56. National Science Board 1993, p. 426, Figure 5–15.
57. Price, D.J. De Solla., *Little Science, Big Science* (1976), p. 115.

CHAPTER 9

Academic Research, Technical Change and Government Policy

KEITH PAVITT

ANALYTICAL PERSPECTIVES

Scientific discovery, invention and innovation have become increasingly professionalized in the twentieth century, with the emergence of expensively equipped and professionally run laboratories in universities and business firms as the main centers of research activities. In the advanced OECD countries, business-funded research and development (R&D) now typically amounts to 1–2 percent of Gross Domestic Product (GDP), and academic research between 0.2–0.4 percent of GDP, more than 80 percent of which is typically funded from government sources.

The importance of basic research as an input into technological practice has been recognized for a long time. In the eighteenth century, Adam Smith had already identified (in Chapter 1 of *The Wealth of Nations*) what we would now call professional scientists as one of the three major sources of technical change:

> All the improvements in machinery, however, have by no means been the inventions of those who had occasion to use the machines. Many . . . have been made by the makers of the machines, when to make them became the business of a peculiar trade: and some by . . . those who are called philosophers, or men of speculation, whose trade is not to do anything but to observe everything: and who, upon that account are often capable of combining together the powers of the most distant and dissimilar objects. . . . Like every other employment . . . it is subdivided into a number of different branches, each of which affords occupation to a peculiar tribe or class of philosophers; and this subdivision of employment in philosophy, as well as in every other business, improves dexterity and saves time.[1]

In the first half of the nineteenth century, A. de Tocqueville[2] predicted that the contribution of basic research to economic and industrial change would grow in importance, that practice and theory would develop together, and that public support would be necessary for the full development of the latter. And in the second half of the century, Karl Marx argued that technological development in the capitalist system stimulated basic science by generating resources, problems, data and instruments for basic research activities.[3]

Since then, it is tempting to argue that the quality of analysis about the economic usefulness of basic research has declined. Many academic scientists and economists have a great deal to answer for, in perpetrating oversimplified and potentially misleading models of the contributions of basic research to technology. The scientists tend to emphasize the so-called linear model: 'basic researchers produce ideas and discoveries, then technologists apply them,' thereby reinforcing the notion of basic research as the main source of the technical change necessary for improved living standards. In a similar vein, the economists treat the output of basic research as potentially useful 'information' — costly to produce, but virtually costless to reproduce and re-use, and thereby justifying government subsidy as a public good.

Neither of these models is devoid of empirical substance. Over the past 100 years, basic research in physics, chemistry and now biology has produced discoveries and techniques that have had revolutionary effects on technology.[4] And governments in all the advanced market economies recognize the central importance of public funding for basic research. However, the models also lead to some awkward questions. If the results of basic research are costless to transmit, why can not any one country simply obtain discoveries and ideas from other countries without paying for them? If basic research is useful, why are published results of basic research cited so infrequently in patents protecting technological inventions?[5] Why cannot basic researchers be oriented towards having *useful* ideas and making *useful* discoveries? On the other hand, why do large firms like ICI, Phillips and Hitachi publish hundreds of journal papers annually, when they should (at least according to the economists' model) keep them secret and appropriate the results for their own commercial advantage? And in this era of globalization, why do many large firms say that local universities should continue to concentrate on high quality basic research that is in principle available to everybody?

Technology as 'Applied Science' or 'A Capacity to Solve Complex Problems?'

These puzzles are resolved, once we recognize that technology is both more than — and different to — 'applied science.' Certainly, both bodies of knowledge emerge mainly from experimental activities in laboratories, and are performed by qualified scientists and engineers. Certainly, some large firms make considerable expenditures on basic research, publish many scientific papers, make major scientific discoveries and sometimes their scientists even win Nobel Prizes.

However, in spite of these similarities, important differences remain. Academic research is mainly basic research: business research is mainly the development and testing of prototypes and pilot plant. Academic institutions dominate in the publication of scientific papers, and business firms in the granting of patents. And despite examples of spectacularly close links between basic research and technology (e.g., biotechnology), basic research builds mainly on basic research (scientific papers cite other scientific papers much more frequently than patents); and technology builds mainly on technology (e.g., patents cite other patents much more frequently than scientific papers).[6]

These differences reflect the *purposes* of the two activities. One of the main purposes of academic research is to produce codified theories and models that explain and predict natural reality. To achieve analytical tractability, this requires simplification and reduction of the number of variables (e.g., 'Under laboratory conditions . . . ,' 'Other things being equal . . .'). The main purpose of business research and development is to design and develop produceable and useful artefacts. These are often complex, involving numerous components, materials, performance constraints, interactions and research fields, and are therefore analytically intractable (i.e., theory is an insufficient guide to, and predictor of, practice). Knowledge is therefore accumulated through trial and error. As a consequence, the methodologies of 'experiments' in the two types of laboratories are often very different. According to one eminent engineering practitioner

> . . . we construct and operate . . . systems based on prior experiences, and we innovate in them by open loop feedback. That is, we look at the system and ask ourselves 'How can we do it better?' We then make some change, and see if our expectation of 'better' is fulfilled . . . This cyclic, open loop feedback process has also been called 'learning-by-doing,' 'learning by using,' 'trial and error,' and even 'muddling through.' Development processes can be quite rational or largely intuitive, but by whatever name, and however rational or intuitive, it is an important research process . . . providing means of improving systems which lie beyond our ability to operate or innovate via analysis or computation.[7]

Tacit knowledge is of central importance in such learning processes, in deciding which component of the system to change, in interpreting the consequences of change, and in deciding what to do next. As we shall now see, academic and business research should be seen as overlapping and interacting systems, with the former augmenting the capacity of the latter to solve an increasing range of complex problems.

How Academic Research helps Technological Problem-Solving

A variety of empirical studies have traced how academic research contributes to technological problem-solving.[5,8–18] The main channels and mechanisms can be summarized as follows.

- **Useful knowledge inputs**, where academic research leads directly to prospects of application (e.g., X-rays, lasers).
- **Engineering design tools and techniques**, including modeling, simulation and theoretical prediction have become central features in the design and testing of complex technological systems. These methods are often developed in academic engineering departments, and sometimes generate research programs in related and more 'academic' disciplines like mathematics.

- **Instrumentation** — products like cathode ray tubes were first developed as laboratory instruments, and techniques developed in academic laboratories are centrally important in today's biotechnology.
- **Trained scientists and engineers** are considered by many business practitioners as the primary benefit of academic research, since such training brings with it skills that can be applied well beyond the scope of the specific subjects of postgraduate research.
- **Background knowledge** — industrial practitioners are often less interested in the contents of the published papers of academic researchers than in the tacit skills and experience that underlie them. Scientific publications by industrial and other practitioners are partly devices for signaling and identifying potentially relevant expertise to the academic community.
- **Membership of national and international professional networks** — trained scientists and engineers bring to technological problem-solving 'knowledge of knowledge' — in other words, membership of often informal networks that enable them to draw on the knowledge of other scientists and engineers, including those beyond national boundaries.

Table 9.1 helps us understand how these dimensions of academic research contribute to engineering practice. It reproduces Vincenti's categorization of different types of engineering knowledge, and of the activities that generate them. From this, it emerges that academic research can in principle contribute to all categories of engineering knowledge. Some contributions will be *direct*, when academic research leads to applicable discoveries, engineering research techniques and instrumentation. Others will be *indirect*, when academic research training, background knowledge and professional networks contribute to a business firm's own problem-solving activities: in particular, the experimental engineering research, design practice, production and operation that will be mainly located there. One danger in both analysis and policy-making is that excessive attention will be given to the *direct* contributions of academic research to technology, to the neglect of the *indirect* contributions that are often more highly valued by the practitioners themselves.

Persistent Differences amongst Research Fields and Industrial Sectors

Mapping the links between academic research and technology is further complicated by the considerable differences amongst research fields and industrial sectors in the nature of the links between the two, reflecting long-standing differences in their origins and development.[4] Tables 9.2, 9.3 and 9.4 give quantitative evidence reflecting these differences in the USA — a country that is large and technologically advanced, and which also has excellent statistics on scientific and technological activities. From previous international comparisons, it is reasonable to assume that broadly similar patterns and trends exist in other technologically advanced countries.

TABLE 9.1: SOURCES OF DIFFERENT CATEGORIES OF ENGINEERING KNOWLEDGE

Knowledge-Generating Activities	Knowledge Categories					
	Fundamental Design Concept	Criteria and Specifications	Theoretical Tools	Quantitative Data	Practical Considerations	Design Instrumentalities
Transfer from Science			X	X		
Invention	X					
Theoretical Engineering Research	X	X	X	X		X
Experimental Engineering Research	X	X	X	X		X
Design Practice		X			X	X
Production				X	X	X
Direct Trial (including operation)	X	X	X	X	X	X

Source: Vincenti[40]

TABLE 9.2: INDICATORS OF RESEARCH FIELDS' LINKS WITH INDUSTRY IN THE USA

Field	Transfer of Academic Knowledge<--------------->Transfer of Trained Problem-solvers			
	1. University Share of All Publications (percent)	2. Share of Industry Publications Co-authored with University (percent)	3. Industry's Share of all Publications (percent)	4. Share of Doctoral Scientists & Engineers in Industry (percent)
Mathematics	91.6	49	3.0	20.4
Clinical Medicine	69.4	45	5.1	}
Biomedical	75.9	40	5.7	} 26.0
Biology	75.7	45	3.4	}
Chemistry	72.2	24	16.2	} 52.0
Physics	65.6	31	16.0	}
Engineering & Technology	59.8	26	24.4	56.8
All Fields	**70.5**	**35**	**8.9**	**36.0**

Source:			
Column 1	National Science Board[41]	Appendix Table 5–25	Data for 1981–91
Column 2	Op. Cit.,	Appendix Table 5–26	Data for 1991
Column 3	Op. Cit.	Appendix Table 5–25	Data for 1981–91
Column 4	Op. Cit.	Appendix Table 3–14	Data for 1991

Table 9.2 shows some striking differences between the research fields of mathematics, the life sciences, and the physical sciences plus engineering, in the channels between academic research to industrial practice. Broadly speaking, mathematics and the life sciences rely more heavily on the transfer of *knowledge* from universities to business practitioners, since relatively high shares of publications and doctorates are located in universities, and industry's share of publications is low and highly dependant on joint authorship with university researchers. The physical sciences and engineering, on the other hand, tend to transfer *trained problem-solvers,* and this is reflected in industry's relatively high shares of both publications and doctoral employment, and by its relatively low dependence on joint authorship with universities.

In addition, the studies by Narin and Olivastro,[5] Klevorick *et al.,*[13] and Katz *et al.,*[19] show that the industrial research competencies the life sciences, as reflected in publications, tend to be concentrated in a few industries (particularly pharmaceuticals) and the hospital sector. The physical sciences, engineering and mathematics, on the other hand, have their competencies and publications spread over a wide range of industries. From this, we may conclude that the practical impact of academic research in the life sciences tends to be more *direct and concentrated,* whilst that in the physical sciences and engineering tends to be more *indirect and pervasive.*

Table 9.3 shows that equally big differences emerge amongst major industrial sectors in the nature of the impact of academic research on technological development. At one extreme, the chemical industry makes direct intensive use of academic knowledge (as reflected in both patenting by universities, and by its own patent citations to journals), performs high levels of research itself, and employs a relatively high proportion of scientists compared to engineers.

At the other extreme, the transport sector hardly cites the journal literature at all in its patenting activities, but performs high levels of development activities, and employs relatively large numbers of scientists and (above all) engineers. Thus, if chemicals can be described as a *research*-intensive industry, transportation can equally be defined as a *development* (or *engineering*)-intensive industry. The machinery sector resembles transportation in employing large number of engineers, but not in formal research or development activities. The electrical and electronics sector is intensive in both research *and* development activities.

These different patterns reflect each sector's historically distinct technological trajectory.[4,20] In the non-electrical machinery sector, technological development activities have generally been pragmatic, and business firms have tended to be small, specialized and therefore without distinct research and development activities. Their high level of technological activity — reflected for example, in the continuing high share of mechanical inventions in total patenting — is better reflected in their total employment of scientists and engineers than in their R&D activities.

TABLE 9.3: INDICATORS OF MANUFACTURING SECTORS' LINKS TO ACADEMIC RESEARCH IN THE USA

Manufacturing Sector	Percentage Share of Manufacturing Sectors in:				
	1. Patents Granted to Universities	2. Patent Citations to Journals	3. Business Funded Research	4. Business-funded Development	5. Employment of Scientists & Engineers (percent Engineers)
	Direct Use of Academic Knowledge<---------->In-House Problem-Solving with Academically Trained Staff				
Chemicals	55.8	51.2	35.4	16.4	13.9 (44.9)
Electrical & Electronic	15.2	17.3	27.4	35.0	15.4 (91.3)
Instruments	6.0	14.8	9.0	9.4	13.5 (88.5)
Machinery	3.4	5.3	3.8	5.5	16.8 (90.8)
Transport	0	0.4	12.1	26.9	23.5 (88.2)
Other	19.6	11.0	12.3	6.8	16.9 (n. a.)
Total	**100.0**	**100.0**	**100.0**	**100.0**	**100.0 (78.8)**

Source:				
	Column 1	Rosenberg and Nelson[16]	Table 6	(data for 1990)
	Column 2	Narin and Olivastro[5]	Table 2	(data for 1975–80)
	Columns 3 & 4	National Science Board[41]	Appendix Table 4–30	(data for 1991)
	Column 5	Op. Cit.,	Appendix Table 3–1	(data for 1992)

In the chemical, electrical and electronics sectors, major innovations have often emerged from R&D activities in large firms that rely heavily on advances in basic science. Their reduction to operating practice has required the establishment of new engineering disciplines.[16] This helps explain these sectors' relatively high shares of patent citations to the journal literature, of both research and development activities, and of total employment of qualified scientists and engineers.

However, in the automobiles, aerospace and other transport equipment sectors, major innovations have not been based on advances in basic research, but on the mainly incremental improvements in the design and operation of complex systems and of their major components and materials. This explains the low rate of citations to research literature in patents granted, the low intensity of research but high intensity of development activities, and the high intensity of employment of qualified scientists and engineers.

This rich and varied set of relations between academic research and technological knowledge is further complicated by substantial changes over time in the technological opportunities being opened up by radical improvements in the underlying knowledge base, especially in biotechnology and information technology. Some of these are reflected in Table 9.4, which shows that US science and engineering employment has been increasing most rapidly for life scientists (with many ending up in the chemical industry), and computer specialists. At the same time, science and engineering employment has been increasing more rapidly in non-manufacturing than manufacturing industry, and is now larger. The most rapid increases have been in financial and computer services, where computer specialists account for more than half of science and engineering employment.

This is convincing evidence of the emergence of a major new technological trajectory, with a new locus (the service sector) and a new style of technical change (software). However, we should not automatically assume that it will resemble earlier science-based industries (chemicals and electronics), with their strong, direct links into basic science, high levels of research and development, and the dominance of large firms. For example, computer and engineering services in 1991 accounted for only 4.2 percent of total company funded R&D, compared to 8.2 percent of science and engineering employment. This structure is closer to that of machinery, with small firms — without the formally designated R&D departments common in large firms — supplying specialized software to large operators of complex information systems in such sectors as finance and distribution, and with design knowledge accumulated mainly through problem-solving rather than from the direct application of academic research.[21]

Some Implications for Theory and Policy

A number of implications for theory and policy emerge from the nature of these links between academic research and technological application.

TABLE 9.4: THE GROWTH OF US SCIENCE AND ENGINEERING EMPLOYMENT IN LIFE SCIENCE, COMPUTING AND SERVICES

Growth of Industry Employment of Scientists and Engineers: 1992/1980	
All Fields	**1.44**
Amongst which: Life Sciences	3.12
Computer Specialists	2.03
Manufacturing Sectors	**1.30**
Non-Manufacturing Sectors	**1.69**
Amongst which: Financial Services	2.37
Computer Services	4.10
Industries' percentages of Business Employment of Scientists and Engineers, 1992 (Percentage that are computer specialists)	
Manufacturing	**48.1** *(10.9)*
Non-Manufacturing	**51.9** *(23.7)*
Amongst which: Engineering Services	9.1 *(3.2)*
Computer Services	8.3 *(51.8)*
Financial Services	6.1 *(58.5)*
Trade	5.2 *(25.5)*

Source: National Science Board[41] Appendix Table 3–1

The main economic benefits of basic research are not easily transmissible information or ideas and discoveries available on equal terms to anyone anywhere in the world. Instead, they are various elements of a problem-solving capacity, involving the transmission of often tacit (i.e., non-codifiable) knowledge through personal mobility and face-to-face contacts. The benefits therefore tend to be geographically and linguistically localized. This important conclusion has recently been confirmed in empirical studies by economists, sociologists and bibliometricians.[11,22–25]

They are unlikely to be modified either by the increasing globalization of business activities, or the increasing use of information technology. Whilst they are often participants in global research networks, large firms continue to concentrate their R&(especially)D activities in relatively few countries, given the advantages of geographic concentration in the development and launching of major new products and systems.[26] Similarly, information technology cannot dispense with personal contact as the main means of developing and exchanging tacit knowledge.

Although academic research has attributes of a *public* good (i.e., the low cost of transmitting the information output; no wear and tear on re-use), it is certainly not a *free* good (i.e., intelligible and applicable without cost to the user). Countries and companies can benefit academically and economically from basic research performed elsewhere, only if they belong to the international professional networks that exchange knowledge. This requires (at the very least) high quality foreign research training, and (at the most) a strong world presence in basic research.[11, 27]

Business firms can and do capture some of the economic benefits — mainly in the form of tacit knowledge — from their own investments in basic research. However, since codified knowledge leaks out from business firms, and industrial R&D workers can and do move from one company to another, business firms cannot capture all the economic benefits of their own investment in basic research. This means that — left to itself — the market would under-invest, which is why governments in all advanced market economies spend substantially on basic research.

The academic engineering disciplines are more than 'applied sciences,' since they involve the design, development and operation of complex systems, through the integration of knowledge and skills from a variety of scientific disciplines. As such, they generate academic research both within the engineering disciplines (e.g., improving methodologies and techniques in design), as well as in related disciplines (e.g., from aeronautical engineering to asymptotic perturbation theory).[16]

Countries with world class technology support domestic world class science, not as a form of conspicuous intellectual consumption, but a necessary long-term investment. World class technology requires high quality researchers, who are skilled in the latest research techniques and instrumentation, and are also knowledgeable of state-of-the-art advances in other parts of the world. This is precisely what world class basic research provides. It explains why countries with industries which spend high proportions of Gross Domestic Product on R&D activities also have high quality basic research measured through the number of citations per published paper.[28] It also explains why Germany overtook the UK in research quality in the 1980s, and why Japan is catching up fast.

Economists often criticize the funding of basic research in developing countries, by saying that useful results can easily be obtained from other countries, and sociologists do likewise by saying that the developing world's basic research is of low quality, and has no impact on the world's scientific community. On the basis of the above analysis, neither of these criticisms is well founded. On the contrary, there is a strong economic case for newly developing countries to invest in basic research, in order to create a technological problem-solving capacity, even if the research is not initially at the world frontier. In the earlier stages of development, priority should probably be given to the engineering disciplines and those necessary to exploit local natural resources, to the establishment of basic research capable of plugging into the world system, and to support for some of the best students for post-graduate education in the technologically advanced countries.

In this context, the successful processes of technological catch-up in East Asian countries have been misinterpreted. Their investment in academic engineering has always been substantial, and so has the commitment by the four tiger countries (Hong Kong, South Korea, Singapore and Taiwan) to training post-graduates in the technologically advanced countries.[29,30] The early Japanese investment in academic research has not been fully reflected in the international scientific literature, probably because it contained a large imitative element, and because rest of the world did not read Japanese. A spectacular example of the latter occurred in the mid-1980s, after the discovery of buckminsterfullerene (the 60-carbon atom molecule — C_{60}). It then emerged that a young Japanese researcher at Kyoto University had predicted the existence of the C_{60} molecule in a book published in 1971: the text was in Japanese and therefore completely ignored by the rest of the scientific world.[31,32]

Conclusions (and Unsolved Problems)

Two reasonably firm conclusions emerge from the above analysis. The first is that the generous state support for academic science in the twentieth century is not given solely — or even mainly — for the pursuit of knowledge for its own sake. There is a strong and justified expectation of economic, social and other benefits. The second (paradoxically) is that these benefits are more likely to be achieved when academic researchers are:

- left to choose their research topics on the basis of inherent intellectual interest;
- judged by their peers;
- closely linked to post-graduate training;
- part of national and international networks of researchers with similar fields of interest;
- linked with technological practitioners, particularly through the provision of trained scientists and engineers.

But these guidelines are insufficient for a full understanding of the purpose and procedures of public support for basic research. Two problems remain. First, how to allocate resources for academic research amongst competing fields of science and engineering. Second, how to ensure an effective matching between the supply of academic research skills and knowledge, on the one hand, and the demand for such skills from business and other practitioners, on the other.

Judgement by peers is useful in decisions to allocate resources amongst academic research in the fields of, say, mechanical engineering, astronomy and molecular biology. Yet such choices must be made, and are becoming more difficult and acrimonious, as academic research budgets grow more slowly than in the past, and sometimes not at all. Governments and national research councils have therefore been experimenting with a variety of methods to improve the allocation of resources amongst fields. Each have revealed difficulties and drawbacks.

Attempts to allocate resources on the basis of *expected economic and social returns* turn out to be relatively straightforward and useful at one level, and complex and dangerous at another. At the level of broad fields, many governments and research councils (and large firms) are capable of identifying the fast changing fields that almost by definition are of potential long-term economic and social interest. In the 1980s, they usually identified information technology, bio-technology and new materials as fields where research competencies should be strengthened, and greater funds allocated at the margin. Some research councils have also reallocated resources amongst fields in the light of changing geo-political and economic circumstances: in the UK, for example, the proportion of funds devoted to Big Science (radio astronomy and high energy physics) has progressively reduced, and that to academic engineering progressively increased, as the UK has increasingly recognized itself as a medium-sized European power rather than a large world power.

Similar processes have led to changes in broad research priorities in all countries, but they have the drawback of being highly political, adversarial and controversial. In an attempt to make decision-making more 'rational,' some governments and research councils have put in place procedures to integrate into resource allocation forecasts of economic and social usefulness at the more detailed level of research sub fields, programs and projects.[33] But these are potentially misleading, given the documented inability to predict accurately the technical and commercial success of research projects, especially those that are long-term or revolutionary.[34] Corporate practitioners deal with the problem by beginning with broad and inexpensive research programs aimed at reducing uncertainties and creating research competencies.[35,36] Similar policies at the national level are likely to be more efficient than attempts to 'pick winners' at the level of projects and programs.

In addition, the increasingly used distinction between 'strategic' (i.e., potentially useful) and 'blue-sky' (i.e., mostly useless) research is spurious, since it confuses the interests of the researcher (e.g., an organic chemist, whose university research is purely curiosity driven and therefore 'blue-sky') with the motives of the agency providing the funding for the researcher (for whom curiosity-driven research in organic chemistry will continue to generate useful discoveries, skills and techniques in the future, just as it has done in the past, so that the very same research is considered 'strategic').

Finally, problems in public policies for academic science arise when the quality and quantity of academic research is not matched by the quality and quantity of economic and social demand for it. The post-war experience of technologically successful countries like Germany and Japan confirm the observations and predictions of de Tocqueville and Marx, namely, the strong expansion of high quality corporate R&D creates the conditions (political and financial) for high quality academic R&D.[28]

But the reverse turns out not to be true: the expansion of high quality academic research does not necessarily create the conditions for high quality corporate R&D. In certain countries, the high quality of academic research has not been matched by growth in demand from the economic and social spheres. The most extreme cases were the centrally planned countries of Central and Eastern Europe, where there were virtually no links between often quite basic research and very poor industrial practice[37] and where science and technology systems are now going through revolutionary changes.[38]

The same holds in a much less extreme form in the UK and US, where the quality of world class academic research has not been matched by a high volume of corporate R&D. In both cases, this has led to bouts of what can be called 'techno-nationalism,' namely, complaints that foreign countries have been exploiting — even stealing — their discoveries and inventions, and to calls for some sort of 'protection' of indigenous academic research from the prying eyes of foreigners. It has also resulted in calls for academic research to become more 'market-led.' In the UK, there is at least one case where a high quality academic program (in solid state physics) has been asked by the research council that is funding it to change its focus and priorities to match those of less than world class British firms in the field.[39]

It is not too difficult to see how such a process could lead to cumulative decline in the quality of both academic and corporate research. It also illustrates nicely how and why national policies in academic research cannot be divorced from the policies and practices of the corporate sector in applied R&D, where the means of government influence are far more indirect, complex and controversial.

References

This chapter is based on research undertaken in the Centre for Science, Technology, Energy and Environment Policy (STEEP), funded within SPRU by the Economic and Social Research Council (ESRC). It has benefited from comments on an earlier draft by Diana Hicks, Sylvan Katz and Ben Martin.

1. Smith, A. (1776). *The Wealth of Nations.* (London: Dent, 1910).
2. Tocqueville, A. de (1840). *Democracy in America.* (New York: Vintage Classic, 1980).
3. Rosenberg, N. "Karl Marx and Economic Role of Science", *Perspectives on Technology.* (Cambridge: Cambridge University Press, 1976).
4. Mowery, D. and N. Rosenberg, *Technology and the Pursuit of Economic Growth.* (Cambridge: Cambridge University Press, 1989).
5. Narin, F. and Olivastro, D. "Status Report: Linkage between Technology and Science", *Research Policy* (1992), **21**: 237–249.
6. Price, D. de Solla, "Is technology historically independent of science? *Technology and Culture* (1965), **6**: 553–568.
7. Kline, S. "A Numerical Measure for the Complexity of Systems: the Concept and Some Implications", Report INN-5, Dept. of Mechanical Engineering, Stanford University, (1990).

8. Gibbons, M. and Johnston, R. "The Roles of Science in Technological Innovation", *Research Policy* (1974), **3**: 220–242.
9. Faulkner, W. and Senker, J. *Knowledge Frontiers.* (Clarendon Press, Oxford, 1995).
10. Hicks, D. "Published Papers, Tacit Competencies and Corporate Management of the Public/Private Character of Knowledge", *Industrial and Corporate Change* (1995a), **4**: 401–424.
11. Hicks, D. "A Morphology of Japanese and European Corporate Research Networks", *Research Policy* (1995b), (forthcoming).
12. Lyall, K. *The 1993 White Paper on Science and Technology: Realising our Potential or Missed Opportunity,* M.Sc. Dissertation, Science Policy Research Unit, University of Sussex (1993).
13. Klevorick, A., Levin, R., Nelson, R and Winter, S. "On the Sources and Significance of Inter-industry Differences in Technological Opportunities", *Research Policy* (1995), **24**: 185–205.
14. Mansfield, E. "Academic Research underlying Industrial Innovations: Sources, Characteristics and Financing", *Review of Economics and Statistics* (1995), **77**: 55–65.
15. Pavitt, K. "What makes Basic Research Economically Useful?", *Research Policy* (1991), **20**: 109–119.
16. Rosenberg, N. and Nelson, R. "American Universities and Technical Advance in Industry", *Research Policy* (1994), **23**: 323–348.
17. Rosenberg, N. "Critical Issues in Science Policy Research", *Science and Public Policy* (1991), **18**: 335–346.
18. Rosenberg, N. "Scientific Instrumentation and University Research", *Research Policy* (1992), **21**: 381–390.
19. Katz, S., Hicks, D., Sharp, M., Martin, B. and Ling, N. *The Bibliometric Evaluation of Sectoral Scientific Trends: Final Report,* ESRC Centre on Science, Technology, Energy and the Environment Policy, Science Policy Research Unit, University of Sussex (1995).
20. Freeman, C., Clark, J. and Soete, L. *Technology and Unemployment.* (London: Pinter, 1982).
21. Torrisi, S. (1994) *The Organisation of Innovative Activities in European Software,* D. Phil. Dissertation, Science Policy Research Unit, University of Sussex.
22. Jaffe, A. "Real Effects of Academic Research", *American Economic Review* (1989), **79**: 957–970.
23. Hicks, D., Izard, P. and Martin, B. "A Morphology of Japanese and European Corporate Networks", *Research Policy* (1995), **23**: 359–378.
24. Katz, S. "Geographical Proximity and Scientific Collaboration", *Scientometrics* (1994), **31**: 31–43.
25. Narin, F. (1992), "National Technology has Strong Roots in National Science", *CHI's Research,* 1. Haddon Heights, New Jersey.
26. Patel, P. "The Localised Production of Global Technology", *Cambridge Journal of Economics* (1995), **19**: 141–153.
27. Callon, M. "Is Science a Public Good?", *Science, Technology and Human Values* (1994), **19**: 395–424.
28. Patel, P. and Pavitt, K. "National Innovation Systems: Why they are Important, and how they might be measured and compared", *Economics of Innovation and New Technology* (1994), **3**: 77–95.
29. Kim, L. "National Systems of Industrial Innovation: Dynamics of Capability Building in Korea", in Nelson (Ed.), *Op. Cit.* (1993).
30. Hou, C-M. and Gee, S. "National Systems supporting Technical Advance in Industry: the Case of Taiwan" in Nelson (Ed.), *Op. Cit.* (1993).
31. Kroto, H. and Walton, D. *Personal Communication,* University of Sussex (1992).
32. Yoshida, Z. and Osawa, E. *Aromaticity* (in Japanese). (Kyoto: Kagakudojin, 1971), p. 174–178.
33. Office of Science and Technology *Progress through Partnership: Report from the Steering Group of the Technology Foresight Programme, 1995.* (London: HMSO, 1995).
34. Freeman, C. *The Economics of Industrial Innovation.* (London: Pinter, 1982).
35. Mitchell, G. and Hamilton, W., "Managing R&D as a Strategic Option", *Research-Technology Management* (1988), **31**, pp. 15–22.
36. Miyazaki, K. *Building Competences in the Firm: Lessons from Japanese and European Optoelectronics.* (London: Macmillan, 1995).
37. Hanson, P. and Pavitt, K. *The Comparative Economics of Research, Development and Innovation in East and West: a Survey.* (Chur.: Harwood Academic Publishers, 1987).
38. Pavitt, K. "Transforming Centrally Planned Systems of Science and Technology: the Problem of Obsolete Competencies" in D. Dyker (Ed.) *The Technology of Transition.* (Budapest: Central European Press, 1997), (forthcoming).

39. Aldhous, P. "An Industry-Friendly Science Policy", *Science* (July 29, 1994), 596–598.
40. Vincenti, W. *What Engineers Know and How they Know it.* (Baltimore: Johns Hopkins Press, 1990).
41. National Science Board-National Science Foundation *Science and Engineering Indicators, 1993.* (Washington: US Government Printing Office, 1993).

FURTHER READING

Caron, F., Erker, P. and Fischer, W. (Eds.), *Innovations in the European Economy between the Wars.* (Berlin: de Gruyter, 1995).

Dosi, G., Giannetti, R. and Toninelli, P. (Eds.), *Technology and Enterprise in a Historical Perspective.* (Oxford: Clarendon Press, 1992).

Freeman, C., Clark, J. and Soete, L. *Unemployment and Technical Innovation: a study of long waves and economic development* (London: Pinter, 1982).

Hounshell, D. and Smith, J. *Science and Corporate Strategy: Du Pont R&D, 1902–1980.* (New York: Cambridge University Press, 1988).

Leonard-Barton, D. *Wellsprings of Knowledge: Building and Sustaining the Sources of Innovation.* (Boston: Harvard Business School Press, 1995).

Mowery, D. and Rosenberg, N. *Technology and the Pursuit of Economic Growth.* (Cambridge: Cambridge University Press, 1989).

Nelson, R. *National Innovation Systems: a Comparative Analysis.* (Oxford: Oxford University Press, 1993).

Patel, P. and Pavitt, K. "The Wide (and Increasing) Spread of Technological Competencies in the World's Largest Firms: A Challenge to Conventional Wisdom" in *The Dynamic Firm,* A. Chandler, P. Hagstrom and O. Solvell (Eds.). (Oxford: Oxford University Press, 1997), (forthcoming).

Reader, W. *Imperial Chemical Industries, A History.* (Oxford: Oxford University Press, 1975).

Reich, L. *The Making of American Industrial Research.* (Cambridge: Cambridge University Press, 1985).

CHAPTER 10

Science and the University

Patterns from the US Experience in the Twentieth Century

ROGER L. GEIGER

Universities have been aptly called 'the home of science' — places where scientists are not only trained but where they also conduct research to advance knowledge.[1] This arrangement originated when nineteenth-century German universities, inspired by the Humboldtian ideal, forged a lasting link between teaching and research. In the twentieth century, the mantle of leadership in academic science passed to the universities of the United States, which became the most prolific generators of scientists and scientific knowledge. Even there, however, the role of universities in science has remained circumscribed. Only half of the basic research in the United States (measured by expenditures) is performed in universities. The education of scientists too, particularly at the postdoctoral level, can occur almost anywhere cutting-edge research takes place. The university system and the system of science might be represented as partially coextensive spheres that interact somewhat differently in each country. The role of universities is thus an important variable in the history of science.

Two historically-minded sociologists have provided valuable perspectives on this topic. Joseph Ben-David, seeking a sociological theory of scientific growth, posited that the vitality of science as an autonomous subsystem required the establishment of secure social roles for scientists. Since the modern era saw this accomplished most powerfully in universities, Ben-David's writings analyzed the ascendancy of German universities in the nineteenth century and American universities in the twentieth. The key in both cases was a condition of de-centralized competition accompanied by considerable intellectual freedom. Ben-David illuminated particular factors that allowed the US to surpass Germany in scientific productivity by about 1930: the American graduate school was better able to accommodate both advanced training and research; American departments could encompass greater and more diverse subject specialization than the German pattern of chairs and institutes; and American universities derived additional vitality through their multiple ties with external constituencies.[2]

Burton R. Clark has approached this topic from the opposite direction, constructing a sociology of higher education that emphasizes the centrality of knowledge and knowledge growth. He has suggested that the Humboltian ideal ought to be reformulated for current purposes as the "research-teaching-study nexus," the terms and effectiveness of which vary for each national system. The imperatives of scientific research and of mass higher education generate centrifugal forces that strain the nexus, but it can be insulated and preserved through differentiation within and across institutions. The distinctive US pattern of the 'vertical university' has allowed this nexus to prevail with great autonomy at the level of the graduate department. Across the system, the competitive arena in which they operate has produced a hierarchy of 'research universities' which concentrates scientific resources at a limited number of institutions. In Clark's formulation, nevertheless, the vitality of the research-teaching-study nexus remains comparatively and historically problematic.[3]

The complementary perspectives of Ben-David and Clark go far to elucidate the intersection of science as an investigative activity with universities as educational institutions. They leave implicit, however, one of the driving forces of twentieth-century science — the direct support of academic research by external sponsors. This chapter proposes a larger framework that incorporates institutional encouragement for scientific research with the nature and consequences of two kinds of external support. This scheme will be used here to describe the evolution of universities and science in the United States in the twentieth century. However, this general framework might be applied to any country.

Support derived directly from the educational mission of the institution will be examined first. The creation of a scientific role for university teachers, as emphasized by Ben-David, is perhaps the central institutional contribution; but the provision of the infrastructure that sustains study and learning constitutes support for the advancement of knowledge flowing directly from the university's mission of educating students.

The disinterested patronage of science is examined next. Science and learning have been objects of philanthropic patronage since at least medieval times. Today modern governments all take some responsibility for these activities. At issue here are forms of support provided chiefly to advance scientific knowledge, regardless of longer-term ulterior considerations. Situations in which special interests are foremost fall into the next category.

The third source of support can most readily be identified as coming from consumers of research. Once again, these consumers might be public or private entities, but the key consideration is that their support for science is offered, and accepted, with explicit recognition of the funder's special interest in the results.

These latter two forms of support, considered as ideal types, have rather different implications for university science. Patronage is generally supportive of the whole scientific process. This process might be conceptualized rudimentarily as possibly having four desirable outcomes:

1) confirming hypotheses (finding what one is looking for);
2) serendipitous findings;
3) expanding the expertise of the investigator(s);
4) training students to be future scientists.

All of these results have the effect of advancing the scientific enterprise, and thus are worthy objects of patronage. Consumers of science often place great value on process outcomes, especially those concerning scientific personnel. However, their primary concern tends to be with the anticipated findings of research. These different orientations produce different dynamics.

The process outcomes of scientific research are highly congruent with the mission of universities — so much so that most governments support academic research for precisely this reason. Over time, however, the supply of scientists eager to perform research outstrips the demand generated by public and private patronage. Rationing becomes an inescapable issue. Distribution of research funds on the basis of merit through peer review is considered the most legitimate way to deal with this problem, but there are myriad complications in implementing this approach. The essential point, nevertheless, is that the patronage of science is inherently scarce and that forces choices to be made.

The demand for research generated by the interests of consumers would seem in theory to have greater expansionary potential, particularly since World War II. Consumers, however, make sophisticated decisions about what research to perform themselves and where to purchase complementary research. Universities, for their part, make judgements about the nature of the research they agree to conduct and the terms of trade for supporting it. There is consequently a high variability, nationally and internationally, in the extent to which universities perform research for consumers.

University research thus appears to depend on resources derived from three dynamic processes, operating to some extent independently of one another and varying considerably from country to country.

Universities and Educational Support for Science

American colleges in the nineteenth century were inherently teaching institutions. Although some included affiliated professional schools, their fundamental mission was to provide four years of unspecialized, pre-professional education emphasizing classical languages and literature. The colleges nevertheless sought to employ professors in the principal fields of science. Some of these individuals used the advantages of their positions for fruitful scientific work. However, the vast majority fit the description of Joseph Lovering, longtime Hollis Professor of Mathematics and Natural Philosophy at Harvard, who "felt no more called upon to extend the domain of physics than as a preacher he would have felt obliged to add a chapter to the Bible."[4] In the last quarter of the century, this situation was altered by developments in the universities and in science.

Three institutions spearheaded the emergence of the American university. The opening of Johns Hopkins University in 1876 signaled the institutionalized cultivation of German-style erudition in the United States. Financed by what then seemed like a bountiful endowment, the university consciously dedicated itself to faculty scholarship and graduate education. Hopkins soon challenged Harvard and its reformist president, Charles W. Eliot (1869–1909). Eliot forced the adoption of an elective system, which allowed the faculty to teach specialized courses on an advanced level and at the same time justified the addition of able scholars. The faculty in Arts and Sciences alone grew from 23 to over 150 as America's oldest and wealthiest university also became its most scientifically accomplished. Harvard was closely rivaled by another upstart. The University of Chicago opened in 1892, but with the ambitions of president William Rainey Harper setting the course, and the benefactions of John D. Rockefeller billowing the sails, it quickly became the country's second vessel of scholarship and research. During the 1890s other universities joined the race, and research became an indelible mission of the country's finest universities.[5]

Coeval with the emergence of universities was the formation of formal disciplinary organizations. Both professional associations and scholarly journals were formed, jointly or separately, for all the principal disciplines in the last two decades of the century. By 1900, journals were being founded for secondary specialties like astrophysics, physical chemistry, and microbiology. Johns Hopkins and Chicago, in particular, actively encouraged the founding of journals and associations, but once in existence these organizations represented an autonomous system of professional evaluation and recognition. Operating in conjunction with a system of universities eager to employ the most talented scientists, the reward system of science was translated directly into institutional rewards. Most importantly, scientific leaders were now assured of being given the best available opportunities to conduct research and set the research agendas for their fields.

That these opportunities were found largely in universities was strikingly confirmed by an effort in 1906 to identify the '1,000 Leading American Men of Science.' Sixty percent held academic posts and two-thirds of those were at the leading 13 universities. The same 13 employed the majority of the top 200 scientists. By the opening of the twentieth century the United States looked to the universities, principally in their educational role, to support and sustain the cream of American science.

For universities, this responsibility had two distinct facets. Their general prosperity as educational institutions gave them the wherewithal to sustain a research role, but to do so also required internal adjustments to their teaching mission.

The 15 institutions that might be called research universities at this juncture were for the most part the largest and wealthiest in the country. Although a few had moderate enrollments (Johns Hopkins, Princeton, MIT, Stanford) these fifteen together enrolled one of every five students at the beginning of the twentieth century. Their size was principally due to the multiple functions they embraced. Besides collegiate and graduate instruction in arts and sciences and the traditional professions, each offered degree programs in a variety of semi-professions like engineering, education, mining, and dentistry. Despite wide variations in endowments for private universities, and in state support for public ones, the research universities in the first two decades of the century primarily translated large enrollments into large faculties capable of covering the proliferating base of academic knowledge. But in addition, large size helped to cover the equally crucial overhead expenses for libraries, laboratories, and facilities.

After 1920, the paths of public and private institutions diverged. State-supported universities continued to grow, and their numerous undergraduates continued to defray the overheads for faculty research. The private universities, however, made a strategic decision to limit enrollments and to improve institutional quality. These choices were undoubtedly influenced by the interest and generosity of alumni, who gave favored institutions — especially Harvard and Yale — enormous sums during these years. These gifts were motivated predominantly by sentimental feelings toward undergraduate education. But quality was an inclusive notion: the private universities became more selective in choosing their students; they upgraded student residences and enriched extracurricular life; and in their zeal to obtain the finest faculty, they employed outstanding scientists and provided them with appurtenances for research. These last efforts were strongly affected by a confluence of interest with external agencies, as will be seen below. But the inherent wealth of the research universities in endowment and physical assets resulted primarily from solicitude for their educational mission.

Internally, universities faced the problem of making the faculty role not merely a potential scientific role, but an actual, necessary one. This process had begun in the last decades of the nineteenth century with the spread of the elective system and graduate education, both of which permitted professors to teach advanced specialties. A pronounced decrease in teaching loads at major universities aided as well. After 1900 it nevertheless became apparent that further steps were needed to allow faculty substantial time free for research. Sabbatical leaves (the seventh year off at half salary) became an accepted practice at these universities. Probably more consequential was the use of graduate students as research and teaching assistants. First instituted at Harvard in 1899, this practice provoked controversy for nearly a decade before its obvious advantages overwhelmed all opposition. In addition, universities experimented with different assignments for 'research men' and 'teaching men.' Pure research professorships were even tried, but they did not fit well with the emerging egalitarian academic department. Early in the century, then, the time and the expectations for research were embedded into faculty positions at major universities. The rising direct costs of research, however, were not yet factored into this equation.

Recognizing this need, universities envisioned that each institution would develop its own specific research funds, preferably in the form of endowments. Harvard received some such funds; and the presidents of Cornell and Wisconsin exhorted benefactors to endow research at their respective universities. Aside from medical research, this rarely happened. State universities budgeted small amounts for revolving research funds, but scarcely enough for such needs as radium for physics experiments, sophisticated instrumentation, or research expeditions. The financial structure of the American university was erected around its teaching mission and could provide little for the direct support of research. By World War I, this limitation was the bottleneck of academic research.

Patronage of Science and University Research

Science and learning have been consistent objects of disinterested support in the modern era, sometimes from the gifts or bequests of individuals and sometimes from the patronage of governments or sovereigns. Given the pragmatic and populist tone of American democracy, virtually all support for pure science for most of US history came from private benefactors. Colleges and universities before 1900 at least shared in this bounty along with learned societies and, in the case of Joseph Smithson's bequest, the federal government. The donations of Boston merchants in 1844 founded the Harvard Observatory. Although tinged perhaps with a vested interest in navigation, this was nevertheless the first gift to spawn a permanent academic research unit. For the remainder of the century, telescopes and museums were the most conspicuous results on campuses of the patronage of pure science, but much everyday research depended on a regular flow of subscriptions and small gifts.

In 1900, nevertheless, it was by no means apparent that universities were the best place for philanthropists to invest in research. Andrew Carnegie and John D. Rockefeller faced such a decision at this juncture, and both chose to create free-standing research institutes. The creation of the Rockefeller Institute for Medical Research (f. 1901) and the Carnegie Institution of Washington (f. 1902) reflected a view that university professors were still too encumbered by teaching responsibilities to be entrusted with the responsibility of advancing science.

Less than two decades later the same situation arose once again, but now the balance of scientific strength had tilted in favor of universities.[6] In the intervening years, much of the vast Carnegie and Rockefeller wealth had been given to philanthropic foundations with broad mandates. The Carnegie Corporation (f. 1911; $125 million) aspired "to promote the advancement and diffusion of knowledge"; and the Rockefeller Foundation (f. 1913; $182 million) sought "to promote the well-being of mankind throughout the world." In addition, American participation in World War I had fostered cooperative efforts among academic, industrial, and federal scientists. In the aftermath, these foundations took an interest in accelerating the development of American science. The original impulse of the Rockefeller

Foundation once again was to establish an independent institute for the physical sciences, but leading scientists now recommended that science be supported within universities instead.

The Rockefeller Foundation awarded a grant for the creation of postdoctoral fellowships, administered through the National Research Council (NRC). The fellowships in themselves bolstered academic science by providing promising young scientists the opportunity for intensive research and study. The grant also set a precedent for foundation support for science mediated by an independent scientific body (the NRC). The Carnegie Corporation, for its part, made grants to develop the NRC, to create the Stanford Food Research Institute, and to help found the California Institute of Technology. Despite these promising initiatives, philanthropy met only a small portion of the needs of academic science in the early 1920s.

This situation was soon changed through the exertions of two heads of Rockefeller trusts. Beardsley Ruml became director of the Laura Spelman Rockefeller Memorial in 1922 and soon committed that foundation to an extensive program of building the social sciences in American universities. The following year Wickliffe Rose assumed the directorship of both the General Education Board and the International Education Board. Despite the titles of those trusts, Rose believed his greatest contribution to education would be to raise the quality of American science. He consciously supported the strongest universities, vowing "to make the peaks higher." Both men in fact proceeded in the same manner. After carefully assessing needs and potentialities during their initial tenures, they made grants at an accelerating pace until 1929, when their organizations were incorporated into the Rockefeller Foundation. All told, Ruml granted nearly $20 million to academic social science; and Rose channeled roughly $27 million for science in American universities. These funds produced an extraordinary burst of prosperity at a time when universities were already enjoying nearly optimal conditions. It also produced permanent gains for academic science.

Foundation patronage during the 1920s consciously bolstered process outcomes. Some of the most productive grants were undoubtedly those given for fellowships through the NRC. Small grants and fellowships were also administered by the Social Science Research Council. Given the style of Ruml and Rose, much support took the form of massive grants to the leading private universities. These grants erected science buildings, provided ongoing research funds, and created new research units like the Yale Institute for Human Relations and the Oriental Institute at the University of Chicago. Yet another important effect was indirect — altering university behavior toward research. For example, when a Ruml grant forced Princeton to raise $2 million in matching funds to expand research in mathematics and physics, the university was obliged to educate Princeton alumni about the importance of research. In a more blatant case, a rebuff from the Rockefeller Foundation caused MIT to reevaluate its policies. It hired Princeton physicist Karl Compton to be its new president and placed greater emphasis on basic research.

The onset of the Depression brought an end to the boom of the late 1920s, yet American universities remained largely dependent on philanthropy for external support of basic research. The number of foundations actually increased, but the amounts they could grant stagnated at best. Support for medical research remained fairly strong, but funding for the social and natural sciences ebbed. When possible, scientists stressed medical applications for their research in order to obtain funding. This relative scarcity brought one important change in the relationship between foundation patrons and university scientists.

When the Rockefeller Foundation had to limit its support for natural science, it opted to concentrate on strategic areas where potential breakthroughs might occur. The director, Warren Weaver, selected 'experimental biology' with the expectation that recent advances in the physical sciences might soon produce breakthroughs in biology. To achieve this result, however, research projects had to be carefully evaluated by the foundation staff. Weaver thus became a "manager of science."[7] Instead of giving 'fluid' research funds to universities to spend as they wished, the Foundation now gave grants for specific research projects, selected for both scientific excellence and relevance to foundation goals. The implications of the project system were profound, even if they would not be realized for some time. Now the grantors had to possess scientific expertise (something Ruml and Rose never claimed) in order to make informed judgements about competing proposals. Grantees would be chosen for the potential contribution of their research, rather than the reputation of their university. Thus, university scientists were in competition with one another for the wherewithal to conduct research. This was still a policy of supporting best science, but now scientists from any university might benefit.

The foundations that became the principal patrons of academic science in these decades were motivated by ulterior goals of serving mankind, but the means they chose was support for basic, not applied, research. The achievement of American science under these conditions was remarkable. According to Ben-David, the ascendancy of American science was accomplished by 1930 — before the forced emigration of European scientists. In atomic physics, the most competitive international field, Americans had moved to the forefront by the early 1930s. Three Nobel Prize-winning discoveries were made in 1932 alone, as Harold Urey at Columbia identified deuterium, Carl Anderson at Caltech established the existence of the positron, and E.O. Lawrence at Berkeley achieved one-million electron volts with his latest cyclotron.[8] Conditions during the 1930s also encouraged the training of many more American scientists, as well as drawing distinguished immigrants. These accomplishments belonged to the predominant, pure-science sphere of university research. It was accompanied, however, by another vital sector, also privately funded, that was shaped by serving consumers of science.

University Research for Consumers of Science

The US government, which had eschewed the role of science patron in the nineteenth century, played a key role in imposing a practical role on American universities. The Morrill Act of 1862 offered a land grant to each state to support a college teaching, among other subjects, agriculture and the mechanical arts. The knowledge base for a useful agricultural curriculum, however, was largely lacking. In 1887, the colleges joined with agricultural interests to secure passage of the Hatch Act, which gave $15,000 annually to each state for an agricultural experiment station. These research units did not have to be connected with the land-grant universities, but in fact they were. Just as American universities were able to add separate scientific units like observatories or museums, they could also accommodate externally supported units intended to serve farmers. With their funding augmented by the Adams Act (1906), the agricultural experiment stations proved quite successful in undertaking basic and applied research and disseminating their findings locally through agricultural extension and nationally through the Department of Agriculture.

A nexus between universities and industrial consumers of research developed in engineering, also largely in land-grant institutions. Before World War I, MIT was almost alone in cultivating close ties with industrial sponsors.[9] Instruction in electrical engineering began there in 1882, and by the new century it had developed close ties, accompanied by occasional support, with the major high-technology firms of the day — American Telephone and Telegraph and General Electric. Actual research contracts appeared in the next decade. Priority here belongs to the Research Laboratory of Applied Chemistry, founded in 1908 explicitly to provide contract research services to industry.

In the aftermath of the Great War, the MIT pattern of conducting research for industry was widely imitated. For a time it seemed that Congress would create 'engineering experiment stations' on the model of the Hatch Act. Several state universities had already launched such units and, despite the failure of federal legislation, more continued to be formed. By 1937, engineering experiment stations were operating in 38 universities. The 1920s nevertheless seem to have been the heyday for this kind of cooperation. In 1929, MIT's Division of Industrial Cooperation and Research conducted $270,000 of industrial research, while the Department of Engineering Research at the University of Michigan contracted for $300,000. This type of research, at universities as well as at independent laboratories, covered the gamut from routine testing to basic research, but the former predominated. Such narrowly focused activities supported the work of engineering faculty and their students, as well as helping to purchase sophisticated equipment. With the growth of industrial research laboratories, however, large firms seem to have brought proprietary research in-house, and the demand for testing and analysis in universities shrank. But industrial patronage of academic research took other forms.

Industrial firms that conducted their own research clearly had the greatest interest in university science. After World War I, firms created fellowships that supported graduate students and also brought them to work in industrial labs. Prominent professors in strategic areas were tapped as consultants, a role that often expanded into direct support for their university laboratories. These forms of interaction were manifest first in the chemistry, electricity, and petroleum industries, but by the end of the 1930s the pharmaceutical industry may have developed the closest links with academic science.[10] Caltech succeeded in establishing cooperative arrangements with firms to share large-scale equipment. All these activities were monitored and encouraged by the National Research Council, which for a time attempted to broker such relationships. Performing research for self-interested clients was nevertheless viewed with suspicion throughout much of the academic community.

The major foundations, in particular, believed the most proper university role to be the conduct of disinterested basic research. Their influence was largely responsible for the reorganization of MIT in 1930, which curtailed its direct services to industry. Direct research services in general were gradually withdrawn from academic settings. For example, the Mellon Institute, which had been loosely associated with the University of Pittsburgh, became independent in 1927, and MIT closed its Research Laboratory of Applied Chemistry in 1934.

More controversy was generated by the issue of patents in biomedical science. University scientists were generally not expected to patent discoveries that might benefit mankind, but patents sometimes helped to bring a product to market in a safe and controlled manner. Universities holding such patents, however, were vulnerable to criticism for stifling further research or for squandering proceeds on attendant legal fees. The most famous case of the period occurred when Harry Steenbock awarded the patent for Vitamin D to the non-profit Wisconsin Alumni Research Foundation, with the income designated to support further research at the university. A boon for science, this arrangement was nevertheless attacked through the 1930s for almost every aspect of its commercial involvement.[11]

During the inter-war years a privately funded university research system emerged in the US that drew upon the resources of both philanthropy and private industry. It became increasingly apparent by the late 1930s, though, that philanthropy lacked the wealth and industry the incentive to sustain this system adequately — let alone realize its burgeoning potential. Scientists clearly foresaw the necessity of eventual government support, but they failed to anticipate whether federal support would resemble the pattern of altruistic patronage or interested purchase.

The Federal Government: Patron or Consumer of Research?

World War II brought federally sponsored research to campuses in a rush. By 1944, American universities were performing roughly three times the volume of pre-war research, virtually all of it under government contract. To take stock of this situation and adapt it to peacetime conditions was the object of Vannevar Bush's famous report, *Science: The Endless Frontier.*[12]

Its central argument was that the federal government should become the patron of basic research, largely in universities, in order to generate the fundamental knowledge that subsequently would be applied to industry, medicine, and national defense. Such research was ultimately vital to the national interest, but to safeguard the autonomy of science he recommended that it be supported as disinterested patronage. So important was this last point to Bush and the generally conservative scientific leadership that an independent 'national research foundation' was proposed, where scientists alone would determine the most meritorious projects. Bush envisioned a 'division of natural science' as the largest unit of the foundation, followed by divisions for scholarships and fellowships, medical research, and national defense. In the post-war reorganization of federal science, however, a different pattern emerged: disinterested patronage was swamped by federal consumers of science.

Essentially, federal research was continued, despite some name changes, in its same wartime channels. The armed services all simply continued their research programs. Large laboratories, like the Johns Hopkins Applied Physics Lab and the Jet Propulsion Lab at Caltech, persisted in loose affiliation with their academic hosts. MIT, where radar research had been concentrated, reorganized these activities into a more diffuse laboratory for research in electronics. Each service devised its own research organization. The most important of these was the Office of Naval Research (ONR), which took an extremely broad approach to its mandate. It perceived the importance of establishing long-term relationships with scientists in key fields, rather than focusing on immediate problems. As a result, ONR became for a time the chief patron of basic research in universities, and as such a new departure in the research system. In medical research, wartime contracts were transferred to the National Institutes of Health where they became the kernel of an external grants program that would grow enormously in the following decades. Research linked with nuclear physics was confided to the Atomic Energy Commission, the lineal descendent of the Manhattan Project, and thus implicitly linked with the development of nuclear weapons. University research thus entered the post-war era invigorated by an influx of federal dollars which, whether for basic or applied research, was proffered at the behest of mission agencies. Eager at this early stage to win the confidence of academic scientists, most agencies were highly supportive of the science process.

Bush's national research foundation fell victim to political wrangling, chiefly as a result of its author's determination to insulate it from presidential authority. A weaker version was finally created in 1950 — after the post-war federal research system had already assumed the shape described above. The National Science Foundation assumed the one unfilled role, what its historian has called the federal "patron for pure science."[13] Separated from the spheres of application that had been emphasized in the Bush Report, pure science held little attraction for Congress, and NSF received scant funding for the rest of the decade. Research spending by the defense agencies, on the other hand, boomed as a result of the Cold War, but also became far more focused on enhancing military technology.[14]

In a huge military R&D budget, universities shared in the approximately five percent devoted to basic research. By the 1950s, the armed services had developed fairly clear policies toward academic research. In certain fields defense interests were paramount — oceanography, fluid dynamics, the upper atmosphere. Here, military funding simply dominated research in all settings. In fields like solid-state physics, meteorology, computing, and statistics, defense interests were shared by many sectors. The services accordingly sought to stay abreast of all advancements through active programs of research. In still other fields, the military maintained only 'listening-post' activities, staying alert for possible breakthroughs with implications for defense. As extensive as this coverage was, much of academic science remained subject to the vagaries of federal 'consumers.' By the mid-1950s, academic scientists were venting their frustrations with this situation — dependance on federal support for research with inadequate provision for disinterested science or university needs.

American scientists prevailed on this occasion, thanks largely to the Soviet Union. The launch of Sputnik in October 1957 galvanized a national effort in education and science to reestablish American preeminence. For the first time, the federal government wholeheartedly embraced the role of patron of science. The mission of the National Science Foundation was finally validated. Its support for academic research increased from $16 million in 1958 to $201 million in 1970. The National Institutes of Health, benefiting from the post-Sputnik mood and Congressional favor for biomedical research, saw its grants grow from $72 to $615 million. Research support from the defense establishment tripled during these years, but it was far overshadowed by the rise in civilian support. Moreover, direct support for research was accompanied by a number of programs that strengthened the capacity of universities to perform research. During this 'golden age' that lasted through most of the 1960s, academic research was more generously treated and was less accountable to sponsors than at any other time.

A Pluralist System, Perpetually Evolving

As the term 'golden age' implies, the conditions of the 1960s could not endure.[15] Some characteristics of the decade represented extremes for academic research: the predominance of basic research, dependance on federal support, federal assistance for institutions, and — non-quantifiable — an ivory-tower mentality concerning the university's research role. In other respects, the system that emerged from that tumultuous decade endured for the remainder of the century: the respective roles of NSF and NIH have persisted and universities have performed roughly half of the nation's basic research since that time. Nevertheless, the

American system of academic research has not previously and is not likely in the future to assume a fixed form. Depending as it has upon three dynamic bases of support, developments in one area perturb the larger system, with implications for behavior all around.

The federal government became the chief patron of academic science in the 1960s, but ever since that role has been clouded with ambiguity. As indicated above, scarcity and choice inevitably overtook scientific patronage. In the US, these conditions are confronted with appeals to 'relevance.' Congress and the President began demanding practical results from the government's huge investment in basic research even during the 'golden age.' The resulting tension was particularly awkward for NSF, the patron for pure science, which has been compelled to embrace relevance while still keeping a free hand to sustain the best science on its own terms. In the 1980s, for example, engineering was elevated to a status commensurate with science, and enhancing the country's economic competitiveness became part of its official mission.

Academic research supported by NIH — one half of the federal total — has always had a dual personality. For the most part NIH has funded basic biomedical research, but its justification has always been to conquer disease and improve health. Behind this rationale its research grants grew inexorably from the 1950s to the 1990s. This investment has yielded spectacular results — in scientific knowledge and improved medical care. However, the largest patron of basic research also spawned an enormous community of researchers who now clamor for continual support despite a constraining fiscal environment. Additional federal patronage reaches universities from NASA, for whom the cultivation of space sciences contributes to public awareness and interest in the agency's mission. At the Department of Energy, a half century of development in high-energy physics has seen the purest academic research confounded with pork-barrel politics.

The overweening presence of the federal patron had the effect of displacing former science patrons. In 1960 the Ford Foundation provided more support to universities than NSF, but before that decade ended it had withdrawn entirely from supporting academic science. Philanthropic foundations have nevertheless continued to play a vital role, particularly in the social sciences, by assisting neglected areas. Basic research in education, for example, is largely beholden to the Spencer Foundation. In a rather different case, the Howard Hughes Medical Institute, now the world's largest foundation, supports biomedical research chiefly in university settings.

The rise of the federal patron in the 1960s also had the effect of pressuring consumers of academic research. Those years constitute a relative nadir for industrial involvement with universities. This situation was exacerbated by the prevailing ivory-tower outlook. Administrators, students, and often scientists too, argued that university research should be confined to basic science and related to education.

Congress, for its part, deemed that consumption and patronage should not be mixed. Legislation in 1970 instructed the Department of Defense to fund only research in universities directly related to its mission. It has largely adhered to the role of a consumer since that time, despite subsequent invitations to resume a larger role. Defense agencies nevertheless continue to dominate academic research in certain areas like computing and artificial intelligence.

For industry, on the other hand, a rapprochement with academic research began to form in the late 1970s and blossomed in the 1980s. The emergence of biotechnology presented an irresistible paradigm for the close interconnection of basic research and commercially valuable discoveries. Combined with the insatiable appetite of universities for research support, a revolution ensued in attitudes and behavior. 'Technology transfer' became the watchwords of the decade, justifying an increasing volume and variety of university-industry relationships (often encouraged with public subsidies). These ties now account, directly or indirectly, for approximately 10 percent of academic research. Future predictions are for a greater orientation toward 'generic research' — in areas like materials research and biotechnology that underpin future commercial technologies. Thus, a sophisticated emphasis on consumer-oriented, and largely consumer-supported, research is foreseen as the engine of growth for university science.

The universities' educational support for science has nevertheless been a crucial and at times contested dimension of academic research. Once the magnitudes of post-war federal research support were apparent, it became clear that scientific excellence would in one sense pay for itself through external grants. Universities consequently mirrored or at times exaggerated the reward system of science by aggressively competing for the services of eminent scientists. They thus excelled in creating the scientific roles that Ben-David found crucial to scientific advancement. In other respects, this quest carried internal costs borne in varying degrees by the university: building and equiping laboratories, sustaining graduate education, or creating endowed chairs. Particularly since 1970, universities have often been hard-pressed to maintain the infrastructure that cutting-edge research demands. As multipurpose institutions, they became vulnerable to criticism when research in graduate departments appeared to overshadow their other missions. This clearly occurred in the 1960s, when the boom in research presented an easy target for student wrath. Something similar occurred in the 1990s as well, as research universities have been repeatedly criticized for being tied too closely to the reward system of science. Thus, in Clark's terminology, research-teaching-study nexus in the vertical university, despite its great effectiveness in nurturing scientific advancement, has been subject to periodic challenges.

The pluralism of the American system of academic research has produced extraordinary vitality, inexorable growth, and laudable scientific achievements. However, that very pluralism generates inherent and inevitable conflict: patronage is never sufficient; consumer-driven science carries innate risks; and limited educational resources are claimed for numerous tasks. For these very reasons, then, the evolution of academic research is a continuing process.

References

1. Dael Wolfle, *The Home of Science: The Role of the University.* (New York: McGraw-Hill, 1972); Roger L. Geiger, "The Home of Scientists: A Perspective on University Research" in *The University Research System: The Public Policies of the Home of Scientists*, Bjorn Wittrock and Aant Elzinga (Eds.). (Stockholm: Almqvist & Wiksell, 1985), p. 53–74.
2. Joseph Ben-David (1974); and *Scientific Growth: Essays on the Organization and Ethos of Science*, Gad Freudenthal (Ed.). (Berkeley: University of California Press, 1991), p. 97–186.
3. Burton R. Clark (1995); and *The Research Foundations of Graduate Education: Germany, Britain, France, United States, Japan.* (Los Angeles: University of California Press, 1993).
4. Samuel Eliot Morison (Ed.), *The Development of Harvard University Since the Inauguration of President Eliot, 1869–1929.* (Cambridge: Harvard University Press, 1930), 227.
5. Kohler, Robert. *Partners in science: foundations and natural scientists, 1900–1945.* (Chicago: University of Chicago Press, 1991), pp. 265–302, pp. 330–57.
6. Kevles, Daniel J. *The physicists: the history of a scientific community in modern America.* (New York: Random House, 1979), pp. 222–35; Heilbron, John L., and Robert W. Seidel. *Lawrence and his laboratory: a history of the Lawrence Berkeley Laboratory.* vol. 1. (Berkeley: University of California Press, 1989).
7. Noble, David. *America by design.* (New York: Oxford University Press, 1977); John W. Servos "The Industrial Relations of Science: Chemical Engineering at MIT, 1900–1939," *Isis* (1980), **71**: 531–49.
8. Swann, John P. Academic scientists and the pharmaceutical Industry: cooperative research in twentieth century America. (Baltimore: Johns Hopkins University Press, 1988),
9. Charles Weiner, "Patenting and Academic Research: Historical Case Studies," *Science, Technology, & Human Values* (Winter, 1987), **12**: 50–62.
10. Vannevar Bush, *Science — The Endless Frontier.* (Washington, D.C.: National Science Foundation, 1960 [1945]). The following is drawn from Geiger (1993).
11. J. Merton England, *A Patron for Pure Science: The National Science Foundation's Formative Years, 1945–57.* (Washington, D.C.: 1982).
12. Leslie, Stuart W. *The Cold War and American science: the military-industrial-academic complex at MIT and Stanford.* (New York: Columbia University Press, 1993); Geiger, Roger L. *Research and relevant knowledge: American research universities since World War II.* (New York: Oxford University Press, 1992).

Further Reading

Ben-David, J. *The scientist's role in society: a comparative study.* (Chicago: University of Chicago Press, 1984 [1971]).

Bulmer, Martin. *The Chicago School of Sociology: institutionalization, diversity, and the rise of sociological research.* (Chicago: University of Chicago Press, 1984).

Clark, Burton R. *Places of inquiry: graduate education and research in modern society.* (Los Angeles: University of California Press, 1995).

Geiger, Roger L. *To advance knowledge: the growth of American research universities, 1900–1940.* (New York: Oxford University Press, 1986).

Geiger, Roger L. *Research and relevant knowledge: American research universities since World War II.* (New York: Oxford University Press, 1993).

Heilbron, John L., and Robert W. Seidel. *Lawrence and his laboratory: a history of the Lawrence Berkeley Laboratory.* vol. 1. (Berkeley: University of California Press, 1989).

Kevles, Daniel J. *The physicists: the history of a scientific community in modern America.* (New York: Random House, 1979).

Kohler, Robert. *Partners in science: foundations and natural scientists, 1900–1945.* (Chicago: University of Chicago Press, 1991).

Leslie, Stuart W. *The Cold War and American science: the military-industrial-academic complex at MIT and Stanford.* (New York: Columbia University Press, 1993).

Noble, David. *America by design.* (New York: Oxford University Press, 1977).

Oleson, Alexandra and John Voss (Eds.), *The organization of knowledge in Modern America, 1860–1920.* (Baltimore: Johns Hopkins University Press, 1979).

Swann, John P. *Academic scientists and the pharmaceutical Industry: cooperative research in twentieth century America.* (Baltimore: Johns Hopkins University Press, 1988).

Veysey, Laurence. *The emergence of the American university.* (Chicago: University of Chicago Press, 1965).

CHAPTER 11

Science, Scientists, and the Military

EVERETT MENDELSOHN

At 02:45 hours on the sixth of August 1945 three combat planes, a bomber and two escorts took off from Tinian Atoll in the South Pacific, six and a half hours flying time from the main Japanese islands. They reached the main island at approximately eight a.m. local time and while the two escorts dropped back, the lead plane, a B-29, the Enola Gay (named after the pilot's mother Enola Gay Haggard) went on by itself. At 08:16:02 the payload exploded at just about nineteen hundred feet above the Shima Hospital in Hiroshima. It was a uranium bomb, 'little boy,' with an equivalent yield of approximately twelve thousand five hundred tons of TNT. The population of the city at the time was between two hundred and eighty and two hundred and ninety thousand civilians and some forty three thousand military personnel. Seventy eight thousand people were killed at once by blast and fire; thirty seven thousand were subsequently missing. By the end of 1945 a total of one hundred and forty thousand were dead and at the end of a five year period, some two hundred thousand deaths were directly attributable to the bombing. As an analyst of the United States Army Institute of Pathology put it 'little boy' produced casualties including dead six thousand five hundred times more efficiently than ordinary high explosive bombs. The press release from the White House in Washington, at mid-day August 6, 1945 (local time) called the bombing "the greatest achievement of organized science in history."

In 1979, the British-born physicist, Freeman Dyson, in his book, *Disturbing the Universe,* mused about what was still happening:

> Nuclear explosions have a glitter more seductive than gold to those who play with them. To command nature to release in a pin point the energy that fused the stars, to lift by pure thought a million tons of rock into the sky-those are exercises of human will which produce an illusion of illimitable power.

HISTORIOGRAPHIC QUESTIONS

How should we express the relationship between the military and science in the course of the twentieth century? Were the physicists and the bomb an accidental or aberrant example of the sciences' encounter with the military? Were the bomb, radar, and other World War II developments examples of military needs, problems, and resources, seducing an innocent science, or innocent scientists? Or were scientists and science driving the formation of a new military in areas of theory, strategy, and practice? Of course neither science nor the military were static during the twentieth century, the years since 1900 have been tumultuous in many areas. While the relationships between the sciences and their applications have always existed, the distinction between the applied sciences and many basic sciences has become increasingly blurred as advanced technologies and scientific practices occupy overlapping terrains, sharing instruments, modes of work, and on many occasions, personnel as well. As Walter Millis, the military historian, made the point in his book, *Arms and Men*, World War I was responsible for the "mechanization" of war, implying that technology had become the decisive factor. World War II, in contrast brought a "scientific revolution" to war. What did this transformation mean for science and scientists on the one hand and for those responsible for the practices of war, the military, on the other? Can we discern the means by which this transformation occurred?

War has always involved technologies, their use and often their development. Whether it be clubs, ram-rods, canons, swords, or rifles, the use of instruments and applied technique, was a very visible feature of the battle field.

The critical point however, is that during the twentieth century many of the technologies employed were intimately science-related, especially the cutting edge technologies.

While the history of twentieth-century warfare has had many practitioners, historians of science have devoted surprisingly little attention to the interaction of war, the military and the sciences. With some notable exceptions (for example, A. Hunter Dupree and Daniel Kevles) earlier historians of science seemed markedly nervous in examining this connection. Even when the problem was addressed, the historians seemed to be over protective and did not want to find too close an identification between the military, the practice warfare, and science for which the traditionally accepted image was one of a normative neutrality and a resistance to entangling alliances. Alex Roland has noted that it seemed as though only advocates of war and the military took up writing analytical histories of the science/military relationship. He added that it was as though to write a history of the plague one had to support the plague and its affects. In consequence there have been no sustained, comparative histories of the science/military relationship in which the theories, the practices, the institutions, the practitioners, the politics, and the moralities have come under sustained scrutiny. There are clearly political as well as intellectual reasons for this distance. Not surprisingly it was with the end of the cold war that historians seemed liberated enough to begin this task.

What elements should be part of such histories? The method adopted in this chapter will be to examine selected episodes and events through the century; to identify organizations that became part of the military/science interaction, both those that were permanent, and those that were transient. In addition I will examine the new forms established especially in the post World War II 'cold war' era. I will move in roughly chronological fashion through the wars of the twentieth century: World War I, World War II, and the Cold War, leaving unanswered the question of whether the end of the Cold War will in some way alter or even sever the relationship of mutual embrace that has come to mark the science/military interaction. The role of practitioners, particularly among the scientists will be examined pointing to scientists as patriots on the one hand and to their role as dissidents on the other with some attempt to disaggregate where appropriate the 'foot soldiers' in the scientific community and the leadership or establishment. The interactions between the individual scientist and organized science and its corporate structures will force greater attention to the issue of actual agency as compared to passive involvement. At the same time there is an evaluative agenda; what has been the effect of the interaction with the military on the cognitive agenda, the nature of laboratory and institutional practices, the moral and political life of the scientist and the institution, the structure of the organizations of scientific activity and not least the dramatically changing nature of the funding of the sciences? Looked at from the point of view of the military, there is the increasingly close involvement with the sciences and the effects on the development of strategy and tactics, the nature of weaponry, the command structures and organization of forces, and the internal and external reliances created by the continuing close-working relationship.

World War One

World War I has been identified as the first 'post industrial revolution' war. That is, it is seen to be a war of industrial production of ordnance, ammunition, and machines. As Alex Roland put it, "the Germans were never really defeated in the field; rather they ran out of the fodder of war . . ."[1] This was also a war in which new technologies and new scientific practices played increasingly significant roles, albeit not decisive ones. The submarine, the machine gun, poison gas, the tank, radio, and aircraft were introduced during the war and played mixed, but generally important, roles.

There are several important debates on the role of science which are worth examination. Daniel Kevles, in several articles and in his book *The Physicists* challenged what had become the conventional wisdom that World War I was really the chemists' war. He juxtaposes two events which took place in April and May of 1915:

> Late one April afternoon in 1915 thick yellow clouds of chlorine gas rolled toward the French line on the Belgium front at Ypres, and soon hundreds of men were choking and vomiting and dying. A few weeks later, a German U-boat (submarine) sank the Lusitania.

He then goes on to cite the contemporary assessment made by the Harvard classicist Roy K. Hauk, that the new sciences had been expected by many to bring the world closer together. But not only did the sciences not prevent war, he argued, rather they enabled humans better to "destroy each other on land . . . *assassinate* under the sea, . . . defile the air with zeppelins."[2] Kevles argues that it was the submarine, and not poison gas that was the most important technology.

In the first quarter of 1917, German U-boats sank one million three hundred thousand tons of allied and neutral shipping with numbers rising each month. Some relief was gained by sending boats in convoys, but there was clearly a desperate need for submarine detection and anti-submarine weapons. At the height of the new efforts, Robert Millikan, the physicist, working at a shore-side research station at Nahant, just north of Boston, claimed that detection was nothing more than "a problem of physics pure and simple." He was referring to the development of acoustic listening devices. Since physicists were being organized for war work by the National Research Council, the active arm of the National Academy of Sciences they requested financial support, received it and quickly established the New London (Connecticut) Experimental Station. This station was created only through "Millikan's dogged insistence upon a role in submarine research for academic physicists; it included thirty two professors, a large plant of laboratories and test facilities, three submarine chasers, three yachts, a precious destroyer and more than seven hundred enlisted men."[3] This sizeable project was conceived by the scientists and organized largely outside of the military command structure. Among the scientific and advanced technology efforts undertaken during the war, submarine detection was probably the most effective. It helped shape the outcome of the war by protecting supplies arriving in the ports of the United Kingdom.

Nevertheless, the most notorious scientific contribution to the first World War was the introduction of poison gas. While obviously a different kind of weapon, the pattern of its development and use is very similar to that of submarine detection, particularly in so far as it involved organized science and the participation of scientists. The recent detailed history written by L.F. Haber, the son of Fritz Haber the chemist responsible for the major developments of poison gas in Germany during the first World War, presents in great detail the manner in which modern chemical warfare was integrated into military practice. In the years 1914 and 1915 both the French and the British were actively exploring a variety of tear gas 'options' for wartime use; although in all probability they were not used on the battle field. The Germans, having heard reports of these efforts, and suspecting that additional moves would follow, escalated efforts of their own. Walter Nernst, the physical chemist, and Karl Duisberg, a leader of the German chemical industry, attended a meeting called by the German military high command. The result was an examination of numerous chemicals and gasses which did have irritant qualities but seemed in actuality ineffective. From the German perspective however, there was a shortage of all the key elements: shell cases, propellants, explosives, as well as the heavy howitzers and canons needed to deliver chemical warheads. The Germans recognized that a novel solution would be required. The chemist Fritz Haber fairly rapidly came forward with a new proposal. Tear gas, he pointed out, was useless when utilized on the small scale; instead what was needed was 'a gas cloud.' He proposed using xylyl bromide filled shells which would be fired from grouped mortars. The initial response was that his solution was impractical and the military turned out to be quite resistant to his ideas. He revised his plan, proposing instead chlorine which would be expelled from gas cylinders to form a cloud. This would drive the enemy out of the trenches into the open where they could become easy targets for more traditional weapons. Haber, who at the time served in the military as a non-commissioned officer in the reserves, was promoted to the rank of captain, and at the age of forty six, put in charge of German gas warfare. As his son L.F. Haber recounts "in Haber the OHL (high command) found a brilliant mind and an extremely energetic organizer determined and possibly unscrupulous." Did the development of this weapon represent a violation of the Second Hague Convention? Haber claimed that he was 'never consulted' about this legal issue and believed that his superiors would only have engaged in developing those weapons permitted by international law.

Poison gas was rushed into battlefield use with only minimum prior experiment, and with potential secondary and tertiary implications military and legal almost totally sidelined. On April 22, 1915, on the Belgian front at Ypres, Field Marshall Sir J.P.D. French described what occurred:

> Following a heavy bombardment the enemy attacked the French division at about five p.m. using asphyxiating gases for the first time. Aircraft reported at about five p.m. thick yellow smoke had been issuing from the German trenches between Langemarck and Bixschoote. What follows almost defied description.[4]

Some six thousand cylinders released approximately one hundred and fifty tons of chlorine. The gas cloud advanced slowly one mile to the west and although it became somewhat thinned out it provided a physical and perhaps equally importantly a psychological shock. The Franco-Algerian soldiers in the trenches were engulfed by the gas. They choked, they suffocated, they went into spasms. Those who still could, ran, but the gas followed. The front collapsed in the face of a series of further attacks. Poison gas had appeared successful. The Allies claimed that fifteen thousand troops had been wounded and five thousand killed, though historians judge the numbers to be inflated. There were also German casualties because the movement of the gas was by no means completely reliable.

Otto Hahn, a rising star of German physical chemistry, was urged by Haber to come to the front as an 'observer, and he took command of a machine gun station. James Franck, the physicist, and his colleague Gustaf Hertz, were also recruited by Haber to join the poison gas program. The research itself was conducted at the prestigious Kaiser Wilhelm Institut for Physical Chemistry and Electrochemistry in Berlin.

When the United States entered the war in 1918, American chemists also became involved in making poison gases. A recent biographical study of James Bryant Conant, the organic chemist from Harvard University, and later the University's president, recounts how he joined the gas making effort. He took leave from Harvard to become part of a unit working at the American University in Washington DC, which later became the Chemical Warfare Service. There is no indication that any of the scientists who joined the American effort had any moral qualms. Conant himself recalled later that the morality of poison gas was rarely addressed; after all, he noted, the Germans had used it first. Several hundred chemists were recruited to take part in the research, and they established the typical community conducting dances, organizing a glee club, as well as basketball and baseball leagues. In September 1917 Conant branched out and began work on the new and much deadlier mustard gas. Advancing the US efforts emerged as his highest priority. The Americans quickly designed and built a manufacturing plant capable of producing thirty tons of mustard gas a day. Its first use against the Germans came in June of 1918. Conant followed through with the next step as well and helped to develop the significantly more deadly lewisite (dichloro [2chlorovynyl] arsine) an arsenic base for chlorine compounds. It was both more lethal than mustard gas and would dissipate more rapidly which meant that once it did its work, the battle field could be occupied by advancing troops. Fortunately, the war ended before the gas was used. Ironically, the manufacturing plant was located in a suburb of Cleveland, Ohio.

The bright, young Conant was in many ways a figure representative of many in the scientific community at the time. He willingly took on what he referred to as the "highly unattractive task" of producing poisons. He considered the job to be a real 'chemical challenge.' Many years later, in his autobiography *My Several Lives* (1968), Conant still defended the logic of what he had done, despite his knowledge that chemical or gas warfare had since been outlawed. "To me, the development of new and more gases seemed no more immoral than the manufacture of explosives and guns." In his own direct way, he gave the justification which was to be used by others when inventing new weapons. "I did not see in 1917 and did not see in 1968 why tearing a man's guts out by a high explosive shell is to be preferred to maiming him by attacking his lungs or skin." And what of the problems in civilian areas which could be dusted with poison gas if the wind changed? In this Conant indicated that he was 'old fashioned' and assumed that the users did not want to kill civilians. This same problem of course followed Conant into the Second World War when he took on the role as a key advisor in making the atomic bomb and later in the Cold War years as a senior member of the advisory group for the hydrogen bomb project.

Fritz Stern, the historian of Germany, as he evaluated Fritz Haber's role in the wartime making of poison gas might well have been talking of Conant and other chemists on the Allied side:

> Like others later he was a scientist who under the pressures of war developed a new weapon, untroubled by its consequences, anticipated and unanticipatable. Haber was above all concerned with the effectiveness of the new weapon; science, he once said belonged to humanity in peace time and to the fatherland in war. He looked for a weapon that would break the decimating stalemate, that would bring an early, victorious end.[5]

If internationalism had often been singled out as a preeminent value of the sciences, it was scientific nationalism as demonstrated during the First World War that clouded the earlier vision. Scientists readily enlisted in the national service and willingly took on tasks that brought the sciences themselves face to face with fundamental professional and ethical queries. The scientists proposed solutions, developed new organizational forms, new technologies, and outlined new tasks in which their work would be invaluable. In the United States, the Chemical Corps, the Gas Service, the Signal Corps, the Air Service, and the Corps of Engineers, all enlisted scientists for the war effort. As George Ellery Hale, the prominent American astronomer and scientific organizer put it, war and the involvement of scientists "forced science to the front."[6]

In the fall of 1914 German scientists produced a *Manifesto of German University Professors and Men of Science* (*The Manifesto of the Ninety Three*). In this document, they defended the German invasion of Belgium, including the burning of the library at the university in Louvain. They proclaimed German militarism to be the spearhead of civilization and saw the involvement of scientists and intellectuals as a high task. The *Manifesto* carried the signatures of Fritz Haber and many others including Ernst Haeckel, Walther Nernst, Wilhelm Ostwald, Max Planck, Wilhelm Roentgen, and Wilhelm Wein. All were members of the scientific elite and a number were current and future Nobel Prize winners. Through their manifesto they wanted to repudiate the charges of German atrocities made by the Allies and, by implication argued for German innocence, blaming all the misfortunes on Germany's enemies. But as Fritz Stern has lamented, "intellectuals everywhere joined in this chorus of hatred and the cry for blood."[7] In Britain, the Fellows of the Royal Society proposed that all German and Austrian foreign members be dismissed. The French Academy of Sciences dropped all the signers of the *Manifesto* from their rolls and Emile Picard, a past president of the Academy, called for the full ostracism of German scientists. However, there was some dissent; not all scientists joined the nationalist cause. Albert Einstein remained an unbeliever and helped shape a counter manifesto which demanded that a just peace be made without territorial annexations. In November 1914, together with nine other German scientists, he formed the *Bund Neues Vaterland*, something akin to a German Fabian Society. The existence of an alternate German voice indicates that choice was possible, that the 'times' were not totally determinant. In the aftermath of the war,

perhaps having short-lived benefit, probably had less beneficial long-term consequences. By the end of the war and demobilization very little was left of these scientifically oriented military services, and many of their efforts and organizational structures had to be recreated for the Second World War.

One of the questions that historians have asked on a number of occasions is just how effective science and the new technologies were in the First World War. Paul Koistinen has argued strenuously that it was not the researchers but the managers and engineers, the 'doers' who were the greatest contributors. It has become apparent that the potential of the new sciences and technologies were not fully exploited, not for lack of technical competence, but rather because a doctrinal basis, or a clearly understood strategic role in the military system did not exist. This is particularly the case with aircraft. Furthermore adequate preparations had not been made to exploit the opportunities which the new technologies and weapons, like the tank and gas created once the initial surprise was over and no one had faced the problem of how to deal effectively with the counter measures which the other side rapidly undertook. Nor was attention given to production and resupply as the chemical munitions were used up or tanks became incapacitated. Similarly, the radio which represented a technological breakthrough of great magnitude in communications, while used well at sea, was not yet reliable on the land battlefield. In trench warfare, the runner, the oldest of the modes of communication, was often preferred by military commanders for reliability. Similarly, the automobile with its internal combustion engine, a major new technology with far reaching military possibilities, took second place to the horse in the moving of ordnance, weapons, and supplies at the front.

The First World War as compared to earlier wars on the European continent, was fairly widespread and intense, and while not quite the 'total war' that marked the Second World War, it did involve large-scale national participation and a command economy.

Science and technology were not decisive in the outcome of the First World War, but the war itself did have a significant affect on science and technology. During the war, governments created scientific and technical organizations and institutions, some of which survived the end of the war though in reorganized forms. Governments, recognizing their potential became permanent promoters of science and technology and steadily moved to integrate them into the state system, often in the area of military-related research and development. Several examples drawn from the United States' experience serve to strengthen this point.

The National Advisory Committee for Aeronautics (NACA), which was established on the eve of the US entry into the First World War in 1915, influenced in part by the war already underway in Europe, had as its mandate the "scientific study of the problems of flight with a view to their practical solution."[8] While the NACA was not a military institution, it was established within the Naval Appropriations Bill of 1915. There were military members on its main committees and certainly the primary government interest in aviation at that time was focused almost exclusively on military issues.

In the civilian sector, George Ellery Hale, the distinguished astronomer and one of the early 'influentials' in American science, created an active research arm for the National Academy of Sciences. The National Research Council (NRC), as it was named, came to straddle every scientific constituency. During the war itself, the National Research Council programs were militarily controlled. The National Defense Advisory Commission (NDAC), established to link science and technology to the mobilization of the economic and industrial sectors for war production, showed the manner in which the military, the industrial, and the science-technological became linked. Koistinen in his study of science military relations identifies the NDAC as the direct predecessor of what President Dwight D. Eisenhower was to identify in his 1961 farewell address as the 'military-industrial-complex.'

Industrial research as an important segment of the US scientific establishment, can be traced to the war years. The effort was marked by cooperative research on a large-scale with groups working together for quick solutions, where specialists rapidly crossed lines of their disciplines and expert knowledge. In the aftermath of the war there was a revulsion against everything that was linked with it and in turn, as the US Congress became isolationist, there was a sharp decline in Army and Navy funding. Research structures were seriously undermined and the wartime National Research Council actually liquidated itself and adopted a peacetime coloration relying initially exclusively on private funding sources, largely the philanthropic foundations. The military research agencies themselves suffered even more severely. Scientists who had held military commissions rapidly left the services and the military in turn was left with very few officers who had research backgrounds.

The French experience was different. Initially it was slow to involve scientists in the war effort; only after the gas attacks on the Belgian front at Ypres was this rectified. Paul Painleve, the mathematician and political activist, led the effort to establish a Directorate of Inventions Concerning National Defense; he was joined by colleagues like Perrin, Borel and others. The agency helped to reshape the French artillery and to give substance to the anti-submarine efforts. But this was largely a body of scientists and one of their own creation. It was the extension of previous links. The war took its toll on the ranks of France's trained scientists; of the one hundred and sixty one graduates of the Ecole Normale Superieur classes 1911–1913, sixty one were wounded and eighty one killed. Another wartime agency, the Service Scientifique de la Defense Nationale emerged from the war, after a series of changes, as the National Office of Scientific and Industrial Research and Invention and was placed under the Ministry of Education. Then it was combined in 1922 with the Caisse des Reserches Scientifiques which created a body representing the first version of Big Science. The two bodies were installed in a 'science city' at Bellevue which has been referred to as a giant scientific and technical complex. During its existence, it carried over its early military linkages and received large governmental subsidies. This office itself was abolished in 1938 with the creation in its place of the Institut de la Recherche Scientifique Appliquee a la Defense Nationale. During the same year the Centre Nationale Recherches Scientifiques (CNRS) was established.

In all, it can be said that the legacy of the organization for the application of science and technology to military pursuits during the First World War was mixed. Not one of the major military powers developed a clear policy for the role of science and technology in the military during either peace or wars of the future. This meant the reinvention on the eve of World War II of most of the agencies linking science and the military.

World War Two

In his foreword to one of the earliest histories of science in the Second World War (James Phinney Baxter, *Scientists Against Time,* 1947), Vannevar Bush refers to it as "the history of a rapid transition, from warfare as it has been waged for thousands of years by the direct clash of hordes of armed men, to a new type of warfare in which science becomes applied to destruction on a wholesale basis. It marks, therefore, a turning point in the broad history of civilization."[9] Bush, who presided over the American involvement of science and scientists in the wartime efforts was full of praise for the heroic contributions that were made and was fully aware of the application of scientific knowledge to a range of new technologies made possible by this large-scale enlistment of science in the war effort. Bush alludes to the fact that the scientists and engineers rapidly adopted "a new and strange way of life and set of human relations." Their motivation, he was certain, was a response to the recognized scientific competence of the enemy, and the desperately short time in which the transformation of weaponry and warfare had to be carried out. He was proud of the fact that in America "there were no disloyalties, or so few as to be negligible." He applauded the fact that toward the end of the war there emerged a full partnership of scientists, engineers, industrialists, and military men "such as was never seen before." The catalogue of new weapons, or very significantly enhanced devices, is impressive: radar, and radar counter-measures; antisubmarine warfare; aerial warfare; rockets, fire control, and proximity fuzes; new explosives and propellants, smoke incendiaries and flame throwers; and of course, the atomic bomb. These make up the headings of Baxter's book. Many of these new weapons and devices have been, in the years since the initial history, the subject of important detailed historical studies. Alongside these significant achievements,there were the range of developments in military medicine and associated fields: new anti-malarial drugs and work with blood and blood substitutes; the discovery of the medical utility of enhanced penicillin; and the creation of insecticides (eg., DDT) and rodenticides. In addition there was an important transformation of the way in which scientists worked with the war system: the development of Operations Research which significantly enhanced the role of scientists in both designing the questions to be asked and the ways in which answers would be found, from anti-submarine strategies to the creation of the early radar networks which protected the British Isles.

THE ATOMIC BOMB

When J. Robert Oppenheimer came to MIT in 1947 to deliver the Arthur D. Little Memorial Lecture, the scientist who presided over the making of the atomic bomb was in a reflective mood:

> Despite the vision and far seeing wisdom of our wartime heads of state, the physicists felt a peculiarly intimate responsibility for suggesting, for supporting and in the end in large measure for achieving the realization of atomic weapons. Nor can we forget that these weapons as they were in fact used, dramatize so mercilessly the inhumanity and evil of modern war. In some sort of crude sense which no vulgarity, no humor, no overstatement can quite extinguish, the physicists have known sin, and this is a knowledge which they cannot loose.[10]

Although some of Oppenheimer's compatriots objected and claimed they had no sense of guilt, what Oppenheimer was driving at was the intimate role that scientists, drawn in large numbers from university laboratories, played in the vision and ultimate realization of this most destructive weapon.

The 'vision' goes back to the earliest days of work on radiation preceding even the First World War, when scientists like Pierre Curie, Ernest Rutherford, Frederick Soddy, recognized the potential destructive capacity locked in the atomic nucleus. The image, and in fact the name 'atomic bomb,' entered public discourse in H.G. Wells' novel *The World Set Free* (1914) published on the very eve of the First World War. He was reflecting in fiction, the fear of many physicists, of the extraordinary destructive potential of an atomic bomb. Robert Millikan, an establishment physicist, was highly critical of those who "have pictured the diabolical scientist tinkering heedlessly, like a bad small boy with those enormous stores of atomic energy and some sad day touching off the fuse and blowing our comfortable little globe to smithereens" (1930). Even Rutherford, by 1933 one of the 'statesmen of science,' at the London meeting of the British Association For the Advancement of Science claimed that "he who talks about atomic energy on a large-scale is talking moonshine." It was this debate that caught the eye of the wandering Hungarian physicist Leo Szilard who, being an outsider and contrarian, attempted to think the problem through at the theoretical level and even took out a patent on a process that he believed would work. He chose the atom beryllium and then thought of indium, but neither of these gave the results he was looking for: a nucleus which would emit two neutrons when bombarded by a single neutron. Szilard even wanted to construct an apparatus to test all the elements to see if he could find one which would give him explosive results.

In the mid to late 1930s three different groups in Europe were examining the implications of neutron bombardment of atomic nuclei. Enrico Fermi in Italy with his collaborators; Marie Curie and her son-in-law, Pierre Joliot, working at the Radium Institute in Paris with Halban and Kowarski; and Otto Hahn, the physical chemist at his laboratory in Berlin, working together with Lise Meitner and later Fritz Strassman. The scientific breakthrough came on the nineteenth of December 1938 when Hahn, who had been an early student of Rutherford, found unexpected results from the neutron bombardment of uranium — a finding they claimed "contradicts all previous experiences of nuclear physics."[11] Hahn immediately communicated the news directly to Lise Meitner then in exile in Sweden since as an Austrian Jew she had had to flee Germany in 1938 after the German takeover of Austria that year. The interpretations came rapidly from an international group of physicists who quickly became aware of the potential for creating a chain reaction starting with the bombardment of the uranium nucleus.

The story of the next steps has been told in detail during the past decades, but several items are worth emphasizing. It was physicists who saw the explosive implications and it was physicists who pushed the project forward after briefly considering in 1939 the idea of keeping the work secret and perhaps achieving a moratorium, at least on publication, if not on further experimentation. The fact that the initial discovery was made in a German laboratory was of great concern, particularly to those emigre physicists who were working primarily in England and the United States. The involvement of the United States government at the highest level was judged to be necessary if significant research was to get underway. Leo Szilard, having moved to the United States, and working in concert with a number of other physicists, convinced Albert Einstein, recognizably the greatest living scientist, to write an appeal directly to President Franklin D. Roosevelt, a task Einstein undertook on August 2, 1939. The letter, drafted by Szilard, and received and read by Roosevelt personally, recounted the scientific work that had been taken and indicated that the work which had been done to date could lead to the construction of extremely powerful bombs of a new type. "A single bomb of this type, carried by a boat and exploded in a port might very well destroy the whole port together with some of the surrounding territory." Einstein urged that a concerted effort be made by various agencies of the US government, and that experimental work be speeded up to examine the potentials. To underscore the urgency, Einstein closed his letter by noting that Germany had stopped the sale of uranium from Czechoslovakian mines and that the physicist son of the German Under Secretary of State, von Weizsacker, was a member of the Kaiser Wilhelm Institute in Berlin where the work on uranium was being done. The fear of German scientific success was a strong motivating force.

While initial steps were taken by the US government in 1939, and additional scientific information was compiled and analyzed by physicists, the pace of development of a potential atomic weapon was slow. The German invasion of the Soviet

Union in June 1941 caused President Roosevelt to act more forcefully, and by November of that year, he initiated a project of intensive work. A month later the United States itself joined the Second World War following the Japanese attack on Pearl Harbor.

The story of the success in designing, building, testing, and using the atomic bomb is a fascinating mix of scientific and engineering ingenuity which had to be matched by major industrial activity and a large scale organization of the scientific community away from traditional laboratories and into government establishments closely supervised and facilitated by the military. Scale is important to recognize alongside organization. The US Army Corps of Engineers established the so-called 'Manhattan District' to oversee the bomb project while the task of constructing an atomic pile and the achievement of a sustained chain reaction were undertaken by a relatively small group of physicists working under the sports stadium at the University of Chicago. It achieved very rapid success under the leadership of Enrico Fermi, who in 1942 notified his command in Washington with the coded message "the Italian navigator has landed."

At Oak Ridge Tennessee, close to the large electrical supplies produced by the Tennessee Valley Authority, the Corps of Engineers, working with several large industrial contractors, built two enormous plants to separate isotopes of uranium necessary for bomb construction. The center at Oak Ridge quickly became the fifth largest city in the state of Tennessee. At Hanford Washington, on a site of six hundred and seventy square miles, the plutonium separation took place in close proximity to the Columbia River and the Grand Coulee Dam, another source of large quantities of electrical power. The scale of the efforts, the willingness to expend unprecedented amounts of money (the total project cost $2 billion) and the willingness to create redundant systems left its mark on all those who participated in the project.

The bomb itself was assembled at Los Alamos, New Mexico in a remote site on a mesa not far from Santa Fe. Perhaps in recognition of the concentration of physicists involved, the University of California was contracted to operate the laboratory which opened in the spring of 1943 and quickly became the best equipped physics laboratory in the world. A cyclotron was brought from Harvard, two van de Graf generators were shipped in from the University of Wisconsin, a Cockcroft-Walton high voltage machine was brought from the University of Illinois and freight loads of apparatus came from the recently shut down Princeton laboratory of the Radiation Project. The scientific director was J. Robert Oppenheimer, a forty one year old Harvard-educated theoretical physicist, then on the staff of the University of California. The key explosives expert was George Kistiakowsky, a Russian emigre, who had been chief of the Explosives Division of the Office of Scientific Research and Development and took over the same role at Los Alamos. The military director, the head of the Manhattan District of the US Corps of Engineers, General Leslie Groves, presided over the basic administrative and security aspects of the whole project. Scientists from Britain and refugees from Germany complemented the staff of American physicists, chemists, and engineers who had been drawn in secret from hundreds of university centers across the United States. In less than four years from its initiation, an atomic explosive device was built, one instrument tested at Alamagordo New Mexico on July 16, 1945, and an uranium bomb and a plutonium bomb dropped over the cities of Hiroshima and Nagasaki, Japan on August 6 and August 9. US Secretary of War, Henry L. Stimson described what had occurred as "the greatest achievement of the combined efforts of science, industry, labor, and the military in all history."[12]

A mission led by Samuel Goudsmit, and Boris Pasch, code named ALSOS, had closely followed the advancing Allied forces as they entered Germany and secretly rounded up the German physicists, who they suspected were working on an atomic bomb. The Germans were taken to Farm Hall, an English estate, where they were interogated and secretly taped. The question that emerged then, and only in 1992, when the full tapes were released, could be answered is why the Germans failed to build a nuclear bomb of their own? Fission was a German discovery. The German physics team' though somewhat weakened by the expulsion of the Jews, was very strong. Several myths were propagated by participants like Werner Heisenberg, which claimed that attempts were held back by moral scruples, that the scientists wanted to maintain control and that they steered research away from the bomb. The new evidence makes it clear that it was not theoretical misunderstandings, nor moral ojections that account for the failure. The German scientists had concluded, in 1941, that a bomb could not be built before what they expected to be the rapid German victory in Europe and the end of the war. The leadership concurred and Albert Speer, the new Ordnance Minister agreed that the large scale use of resources was not warranted. The project continued at a much slower pace while increased efforts were directed to rockets. There is an ironic tone to the exchange between the young Carl Freidrich von Weizsacker and Otto Hahn caught on the Farm Hall tapes:

> von Weizsacker: I believe the reason we didn't do it was because all the physicists didn't want to do it, on principle. If we had all wanted Germany to win the war we could have succeeded.
>
> Hahn: I don't believe that, but I am thankful we didn't succeed.[13]

When the German scientists heard the news on August 7, 1945 of the Hiroshima bombing there was a mix of chagrin, depression and relief.

Radar

Baxter relates the story of a visit of the physicist I.I. Rabi to the US Senate shortly after the war. Senator J. William Fulbright asked, "Is radar in the same class with atomic energy? Is it of a similar nature as far as its implications in warfare are concerned?" Rabi, who had worked on both projects replied, "yes and no. Radar represents an extension of man's senses and power. He sees further. He sees more clearly. He measures distance more accurately . . . Then you apply it to something. It extends your senses in dropping the atomic bomb, it extends your senses in guiding the missile . . ."[14] While the invention of radar never gained the same notoriety as the atomic bomb, its utility in warfare and in civilian technology is substantially greater. Like the bomb its story is very much a physicists' story. The broad theory behind it is relatively simple, projecting radio waves and monitoring the echoes. The ingenuity came in the application; creating a device that could operate at the most appropriate wavelengths.

Laboratory experimentation had begun in the 1920s and by 1939 there were active projects in France, Germany, Great Britain, and the United States each working independently and secretly. Although there were many, sites not much money had yet been appropriated. The name for the new technology was given by the US Navy, an acronym for radio detection and ranging. It was in Great Britain that the most intense efforts were underway. Fearing a German air attack if war were to breakout, the British military began recruiting scientists as early as 1934 to work on air defenses. By December 1935 the Air Ministry had established the first operative radar system. Five radio locating stations were built on the east coast of England. By August 1937, fifteen additional stations were approved for the east and south east coasts. Beginning Easter 1939, a twenty four hour watch was being manned: the so-called Chain Home station network. These installations, still using relatively primitive long wave detection played a decisive role in Britain's crucial battle against German bombing in the early years of the war.

While the British went on to make a series of additional advances in adapting radar to night fighter control, anti-submarine warfare and blind bombing, the most important innovation that they were responsible for was the invention of the resonant cavity magnetron, the instrument that became the heart of all radar. It moved radar from its early two meter wavelength to the centimeter range permitting much more accurate locating. Inspite of the fact that the US was technically not yet at war, intense cooperation had begun shortly after the German campaigns of 1939. A group of scientists led by Sir Henry Tizard brought the magnetron model to the US in 1940. The British were anxious to see radar development accelerated as the 'Battle of Britain' depended on it. One American commentator called it the "most valuable cargo ever brought to our shores." The American scientific team was ready to utilize the British breakthrough and they undertook a high priority work under the aegis of the National Defense Research Committee to develop a microwave airborne intercept system. Research began on the 10th of November, 1940 at the Massachussets Institute of Technology in a unit soon to be named the Radiation Laboratory, the 'Rad Lab' as it became known. The physicist Lee DuBridge was brought from Rochester University to lead the group of some thirty physicists. By the Spring of 1941 the group had constructed a useable device, although this was after the British had beaten back the German air attacks and broken the German air superiority.

The Rad Lab, as well as continuing into second and third generation radar systems adapted the device for air to surface ship (submarine) detection, and extended the technology to the proximity fuze, a small tough radar that could be placed in an artillery shell and cause the shell to explode at a determined distance from its target, eg., an airplane. Radar was applied to bombing accuracy through target identification in night and poor visibility making possible the massive strategic bombing which marked the final years of the war in Europe and the move to intensive bombing of Japan. Loran, the long range navigation instrument was another important outgrowth of the radar program.

The Rad Lab rapidly grew in size and several of its operations moved to other sites in California and England. The budget for 1942 was running at $1,150,000 per month and the staff had grown from 450 to almost 2,000 by the year's end. At the peak of its operations nearly 5,000 staff were at work in or under contract to the Rad Lab. They undertook not only design but the construction of prototypes and they followed the devices into the field. By the end of the Second World War over $3 billion was spent on radar research and equipment. Industry was strongly involved and worked out a procedure where industrial staff were sent to the Rad Lab for training and orientation. Over 100 industrial firms were engaged in radar work with prime work being done by such industry giants as Sperry, General Electric, Westinghouse and Philco.

Operations Research

P.M.S. Blackett, the British physicist and one of the founders of Operations Research (OR) explained what was involved:

> Many war operations involve considerations with which scientists are especially trained to compute, and in which serving officers are in general not trained. This is especially the case with all those aspects of operations into which probability considerations and the theory of error enters . . . the scientist can encourage numerical thinking on operational matters, and so can help avoid running the war by gusts of emotion . . .[15]

The sense of applying rational, science-based procedures to complex problems had occupied Blackett and a number of other leftwing scientists in the face of the economic and social turmoil that came with the great Depression of the 1930's. Marxist as well as socialist humanists were involved in a number of efforts to replace what they saw as chaotic social decision making. They believed scientists had a special knowledge and therefore responsibility to act. One of their number, only half in jest, suggested replacing the House of Lords with a Senate of Scientists. Quite a few of these scientist activists were among the early members of the Operational research units established on the eve of Britain's entry into the war.

Operational Research, as it was called in Britain, was closely linked to the efforts to build a protective radar net around the East and South of Great Britain. Sir

Henry Tizard, the chemist was the founder but within a short time Blackett emerged as the leader and the group came to be known as 'Blackett's circus.' The original team included three physiologists, two mathematical physicists, an astrophysicist, an army officer, a surveyor, a general physicist and two mathematicians. In its earliest incarnation the group of scientists examined the way in which radar was being deployed and how effectively the information was being utilized. Initial judgments suggest that the use of radar increased the effectiveness of the British defense by a factor of ten and that operations research added an additional factor of two to the efficiency. The name reflected the expected site of action of the teams: in the operations rooms themselves, using all the data, signals, charts and combat reports, normally available to the service personnel. As Blackett put it, the "scientific analysis, if done at all, must be done in or near the operations room."

In the United States, operations research began in late 1941 as the Antisubmarine Warfare Operations Research Group. By mid-summer 1943 Philip Morse of MIT was the head, with William Schockley of Bell Labs as research director. A group of forty-four was chosen with the expected mix of scientists with the addition of fourteen actuaries and an architect. For the US the primary concern was German submarines and antisub warfare became the initial focus of the American group.

Several analysts concurred in Vannevar Bush's assessment that the primary role of operations research during the war was to help the services make the most effective use of the forces and weapons that they had rather than engage in research on new weapons. By working in close proximity to the operations themselves, their recommendations could be followed at once. One point that has been stressed in the evaluation of OR is that the physicists had a particular outlook. As Fortun and Schweber put it in their recent study, for the physicist **rationality = computability**.[16] Another team of British authors noted that OR reduced war to a rational process. They compared this to the romantic view of war held by Hitler. "Systematic scientific work on known weapons paid larger and quicker dividends. It beat Hitler."[17] Both sets of analysts note that the German system was hierarchical, preventing collaboration between the armed forces and the scientists which as a group was seen as rational and egalitarian not hierarchical.

Operations Research flourished for a period after the war. Several of its major practitioners joined together to organize the Operations Research Society of America and in 1952 they established a journal devoted to the field. Several text books were published and steps taken to develop university courses. In the early 1950s the field was still quite influential with one report noting that the Walt Disney organization called in an OR team to help them decide where in Southern California their new enterprise, Disneyland, should be located. But the influence of OR as a specific form did not last long. The scientific advisory activities for the military took on other organizational forms, and systems engineering and scientific management in the civilian and military industrial sectors developed their own styles, theories and practices.

In Britain it was explicitly politics that did in the OR units. The leading practitioners, like Blackett, Zuckerman, and Bernal were too far left in their politics for the new conservative government which took power as the Cold War was getting under way. Fortun and Schweber identify another element that may have initially given them prominence but ultimately lessened their appeal: hubris. They cite an author who identified "the arrogant attitude . . . of the physicists who inherited the mantle of scientific leadership coming out of the war [and] made them believe [that] they could address all technical problems without needing assistance from other scientific disciplines." This was the attitude embedded in Alvin Weinberg's call for the application of a 'quick technological fix' to complex social problems.

World War II was different from World War I and science and scientists emerged in new roles in the war system and military organization. The full impact of the scientific and industrial revolutions was felt at almost every level of combat activity from the frontline to strategic planning. Scientists and the managers of science played key roles as advisors at the and new agencies were established to project science forward and avoid it being lost in organizational conservatism. Although it can still be argued that the most widely used technology was the internal combustion engine (land vehicles, tanks, airplanes, and submarines) it was new technologies, the product of intense scientific activity, like the atomic bomb and radar which added a critical new edge to war fighting. The large scale involvement of scientists (willing involvement it should be noted), in turn transformed science. There were dramatically larger budgets for science coming from government; the projects envisioned were of significantly increased scale; there was a rapid acceleration in change in the material culture and organization of the laboratory, initially most notably in physics; the sciences grew dramatically in size and importance in universities; and of deep cultural significance the scientific community found itself in a sustained and very close relationship with the military, which was enhanced, but still willingly accepted, during the Cold War.

The Cold War and the Sciences

As he prepared to leave office in January 1961, President Dwight D. Eisenhower addressed the American people by radio and television. While crediting the armed forces with a vital role in keeping the peace, he turned to a new problem: "Our military organization today bears little relation to that known by any of my predecessors in peacetime, or indeed by the fighting men of World War II and Korea." A transformation had occurred with the formation of "an immense military establishment" coupled with a permanent very large arms industry. This he identified as "the military-industrial complex." Although recognizing the need for its creation, he expressed concern for the implications of the magnitude of its influence (economic, political and spiritual) that was felt in every part of the nation. He was further worried about the acquisition of "unwarranted influence," whether sought or unsought by this new configuration of power. But Eisenhower went a critical step further in identifying the source of the changes that had transformed the nations military-industrial posture, "the technological revolution of recent decades." He spelled out what he meant:

> In this revolution research has become central; it also becomes more formalized, complex, and costly . . . Today the solitary inventor, tinkering in his shop, has been overshadowed by the task forces of scientists and laboratories and testing fields. In the same fashion the free university, historically the fountainhead of free and scientific discovery, has experienced a revolution in the conduct of research. Partly because of the huge costs involved a government contract becomes virtually a substitute for intellectual curiosity. For every old blackboard there are now hundreds of new electronic computers.[18]

The new military, he claimed, was a scientific and technically shaped military. He was ambivalent about the outcome, but also worried about the process and the participants. "Yet, in holding scientific research and discovery in respect, as we should, we must also be alert to the equal and opposite danger that public policy could itself become a captive of a scientific-technological elite."[19] Like many on the political left, this Republican president was concerned about the core democratic processes of the society that could be overwhelmed by the knowledge and technique, often out of range of those without scientific education, necessary for the military enterprise of the new era.

The idea of a knowledge elite, including the scientists, was not Eisenhower's alone. Five years before Eisenhower sounded the alarm, C. Wright Mills, a sociologist from Columbia university, had identified and criticized the cluster of military-industrial and political leaders and powerbrokers who were coming to wield excessive control of American life. In his book *The Power Elite*, 1956, Mills explicitly examined the role of the scientists and technologists. Although he did not give them the central place, and relegated them to the role of 'technical lieutenants of power,' he saw danger in the new position of scientific experts in league with industry and the military. Historians of the nuclear test ban debate tell us that Eisenhower was frustrated by his inability to secure a weapons test ban, feeling he had been close to negotiating a treaty only to have it aborted by a group of scientists headed by physicist Edward Teller and their close allies in segments of the military services. They identified complex and somewhat arcane scientific and technical arguments against the reliability of a test ban treaty.

The Cold War was 'officially' inaugurated in 1947 by the British Prime Minister Winston Churchill in a speech at Westminster College in Fulton, Missouri, a site chosen by President Harry S. Truman near his own home base. Churchill proclaimed that an 'iron curtain' had been pulled down across Europe and that Soviet power must be resisted and overcome. In actuality science had been deeply implicated in the Cold War beginning in the final years of World War II. As the new generation of revisionist historians have shown, at least one of the deciding factors governing the decision to use the atomic bomb on Japanese cities was an effort to 'manage' the Soviets in the post-war world. The Soviet leadership had not been told about the bomb's development and were excluded from any discussion about its use. Although they probably had received some hints from Klaus Fuchs who later admitted spying for them, the first hard knowledge the Soviets had came from the news announcements on the day of the Hiroshima bombing. In its way the bombing, represented to a significant extent the first 'shot' of the Cold War. The subsequent failure to achieve an atomic weapons control pact added to the tensions growing between the former allies. As US Secretary of War , Henry Stimson, put it in a September 1945 memo, ". . . I consider the problem of our satisfactory relations with Russia as not merely connected with but as virtually dominated by the problem of the atomic bomb."[20] Secretary of State James Byrnes, a hardliner, saw the US monopoly of the bomb as an important asset in dealing with the Soviets. The question was how long could the monopoly last , how long would secrecy be maintained. Stimson was blunt, "we don't have a secret to give away — the secret will give itself away." The major industrial concerns Union Carbide, Dupont, Eastman Kodak who managed the Manhattan District facilities and isotope separation, advised as early as June 1945 that it would take no more than five years for the Soviets to build their own bomb, judging it to be largely an industrial and manufacturing problem. But it was the advice of General Groves which was taken. He assured all willing listeners that a prolonged US monopoly, for as long as ten to twenty years, was possible.

The Soviets for their part came to understand the full import of the bomb only after the explosion at Hiroshima. As David Holloway demonstrates in his recent study Stalin's crash program to build a bomb relied on the existence in the Soviet Union of strong schools of physics and radiochemistry and a competent engineering structure.[21] Because Soviet nuclear scientists had conducted significant research toward a chain reaction between 1939–1941, they were able to make good use of the intelligence they received from Britain and the US. The shock to American security planners came on September 3, 1949 when the airborne data they collected gave unmistakable indications of fission products from the explosion of a plutonium bomb on 29 August! In just four years the Soviets had broken the atomic monopoly. Although Stalin remained suspicious of most scientists, physics came to occupy a privileged position and gained a degree of intellectual autonomy. Holloway draws an interesting conclusion: "The Soviet nuclear project shows not that science and totalitarianism are compatible, but that totalitarian regimes have to allow some zones of intellectual autonomy in the society if they are to reap the benefits of science." The physicists in turn gave to Stalin the knowledge and technique to make the Soviet Union a powerful state with a powerful and competitive military apparatus.

The subsequent forty years involved the Soviet Union and the United States in an expensive and at times dangerous arms race, which was in very real terms a science/high technology race. President Truman rejecting the counsel of his scientific advisers, responded to the Soviet a-bomb by following the advice ofthe Cold War scientists, led by Edward Teller, and launched an urgent effort to build a hydrogen bomb. As successfully tested in 1952 it was some 1,000 times more powerful than the bomb dropped on Hiroshima. The Soviets reciprocated and two years later exploded their own fusion weapon.

The race to develop missile delivery systems for nuclear warheads was based in part at least on the work of German rocket experts who were brought to both the US and the Soviet Union immediately following the Second World War. Werner von Braun and a team of 120 of the core personnel from Peenemunde were reassembled in the US and by 1946 hard at work at White Sands, New Mexico, and later Huntsville, Alabama.

The Soviets who had rounded up their own share of German rocket experts housed them at a site north of Moscow. The results were startling and one year after the Germans' arrival the Soviets fired a remanufactured V-2 from a test area near Stalingrad. But the Soviets did not make as good use of the Germans as did the Americans. Russian rocket experts gained all the information they could and went on independently, with great success. In October, 1957 Sergei Korolev's Sputnik satellite was launched, once again startling the American strategists. The American presidential campaign of 1960 was partly conducted over John Kennedy's claim that the US had fallen behind during the Eisenhower years and that a 'missile gap' existed. When elected the new young president launched his plan to land a human on the moon before the end of the decade; another presidential initiative rejected by his science advisers but carried out none the less. The pattern of move and counter move, new weapon countered by newer counter weapon fueled the arms race and kept armies of scientists and engineers at work on weapons development. Scientists became used to getting what they wanted with few if any budget constraints.

The Soviets handled the years of preparing for confrontation by creating a command economy (at great cost to overall economic health, it turned out) and diverted to the science-military-industry sector what was called for. The US adopted its own version when it secretly adopted a national security policy directive committing the country to a form of permanent mobilization (NSC-68), its own form of command economy. Institutions as well were transformed. Stuart Leslie in his valuable study, *The Cold War and American Science*, recounts a visit in 1962 to MIT made by Alvin Weinberg, director of the Oak Ridge National Laboratory, when he quipped that it was becoming increasingly difficult "to tell whether the Massachussets Institute of Technology is a university with many government research laboratories appended to it or a cluster of government research laboratories with a very good educational institution attached to it."[22] By the late 1960s MIT ranked 54th among defense contractors sitting between missile giant TRW and Thiokol Chemicals. Its contracts for 1969 topped $100 million. Numerous other educational institutions, while not as deeply involved as MIT, mimicked its military arrangements. As Philip Morrison the MIT physicist feared, the military would end up buying American science and engineering "on the installment plan."

In addition to the universities as centers where science-military cooperation took place another class of institutions was established: 'think tanks.' They linked scientists and experts from other disciplines to the military to provide expert advice on the emerging new areas of security. RAND (research and development) was the first established directly out of Air Force interests. It was followed rapidly by other centers, for example the Hudson Institute, Heritage Foundation, American Enterprise Institute, Georgetown University's Center for International and Strategic Studies and other specific groups like Project Charles in the 1950s and 1960s and the Jason Group, (the 'fair haired superbrains') who, among other things designed the electronic battlefield for Vietnam in the 1960s. The think tanks were often used as revolving doors for scientists and 'defense intellectuals' as they moved into and out of government. Among other innovations they were home to General Graham's 'High Frontiers,' the predecessor to Star Wars in 1980. The MX missile and the MIRV missile concepts were also among the weapons systems and strategic concepts born in these science-military groups.

Commentators have pointed to the deep inter-penetration that began during the second World War and consolidated during the Cold War of the military and the sciences and shaped the thought and practices of both. As the Cold War wound down in the late 1980s the modes of work and thinking that have become so much a part of scientific organization remained intact with new linkages between the sciences and civilian sector industries, especially in biomedicine and technology. And the former independence and open critical spirit so often mentioned in the histories? All that seems left is the 'illusion of autonomy.'

References

1. Alex Roland. "Science and War," *Osiris*, new series (1985), vol. 1, p. 262.
2. Daniel J. Kevles. *The Physicists, the History of a Scientific Community in Modern America.* (New York, 1971, 1979), p. 102.
3. Ibid., p. 126.
4. Ouoted in James G. Hershberg. *James B. Conant, Harvard to Hiroshima and the Making of the Nuclear Age.* (New York, 1993), p. 41.
5. Fritz Stern. *Dreams and Delusions, the Drama of German History.* (New York, 1987), p. 42.
6. Kevles. *The Physicists*, p. 138.
7. Stern.
8. Roland. "Science and War," p. 263.
9. James Phinney Baxter, 3rd. *Scientists Against Time.* (Boston, 1946 [pb,1968]), p. xv.
10. J. Robert Oppenheimer. "Physics in the Contemporary World," in *The Open Mind.* (New York, 1955), p. 88.
11. Cited in Richard Rhodes. *The Making of the Atomic Bomb.* (New York, 1986), p. 255.
12. Cited in Baxter, p. 438.
13. Jeremy Bernstein (ed.). *Hitler's Uranium Club, The Secret Recordings at Farm Hall.* (New York, 1996), pp. 129–30.
14. Cited in Baxter, p. 136.
15. Cited in Baxter, p. 404.

16. Michael Fortun and Sylvan S. Schweber. "Scientists and the Legacy of World War II: The Case of Operations Research (OR)," *Social Studies of Science*, vol. 23, 1993, p. 625.
17. Cited in Fortun and Schweber, p. 625, from J.G. Crowther and R. Whiddington, *Science at War.* (New York, 1947).
18. Dwight D. Eisenhower. "Farewell Radio and Television Address to the American People, January 17, 1961," *Public Papers of the President of the United States, Dwight D. Eisenhower, 1960–61.* (Washington, DC, 1961), pp. 1035–40.
19. Eisenhower, p. 1037.
20. Gregg Herken. *The Winning Weapon, The Atomic Bomb in the Cold War, 1945–50.* (New York, 1980), p. 23.
21. David Holloway. *Stalin and the Bomb, The Soviet Union and Atomic Energy, 1939–1956.* (New Haven,1994).
22. Stuart W. Leslie. *The Cold War and American Science, The Military-Industrial-Academic Complex at MIT and Stanford.* (New York, 1993), p. 14.

FURTHER READING

Gar Alperovitz, *The Decision to use the Atomic Bomb and the Architecture of an American Myth*, New York, 1995.

Ronald W. Clark, *The Rise of the Boffins*, London, 1962.

Ludwig F. Haber, *The Poisonous Cloud: Chemical Warfare in the First World War*, Oxford, 1986.

Guy Hartcup, *The Challenge of War: Britain's Scientific and Engineering Contribution to World War II*, New York, 1970.

R.G. Hewlett, O.E. Anderson, *The New World 1939/1946: A History of the United States Atomic Energy Commision*, vol. 1, State Park, PA, 1962.

Lillian Hoddeson, Paul Henriksen, Roger Meade, Catherine Westfall, eds., *Critical Assembly: A Technical History of Los Alamos during the Oppenheimer Years, 1943–1945*, Cambridge, 1993.

Paul A.C. Koistinen, *The Military-Industrial Complex: A Historical Perspective*, New York, 1980.

Everett Mendelsohn, Merrit Roe Smith, Peter Weingart, eds., *Science, Technology and the Military*, Sociology of the Sciences, vol. XII, 1&2, Dordrecht, 1988.

Michael Neufeld, *The Rocket and the Reich, Peenemunde and the Coming of the Ballistic Missile Era*, Cambridge, MA, 1995.

F. Ordway, III, M.R. Sharpe, *The Rocket Team: From the V-2 to the Saturn Moon Rocket*, Cambridge MA, 1982.

Carrol W. Purcell, Jr., ed., *The Military Industrial Complex*, New York, 1972.

Richard Rhodes, *Dark Sun: The Making of the Hydrogen Bomb*, New York, 1995.

Merrit Roe Smith, ed., *Military Enterprise and Technological Change. Perspectives of the American Experience*, Cambridge, MA, 1985.

Irvin Stewart, *Organizing Scientific Research for War*, Boston, 1948.

Mark Walker, *Nazi Science: Myth, Truth and the German Atomic Bomb*, New York, 1995.

Spenser R. Weart, *Scientists in Power*, Cambridge, MA, 1979.

Figure 11.1: Nuclear weapon of the "Little Boy" type, the kind detonated over Hiroshima, Japan, in World War II. The bomb is 28" in diameter and 120" long. The first nuclear weapon ever detonated, it weighed about 9000 pounds and has a yield equivalent to approximately 20,000 tons of high explosive. © Imperial War Museum

Figure 11.2: From the highest building in Hiroshima, ratings view the devastation; cranes of the Mitaukishi's shipyard show on the horizon. © Imperial War Museum

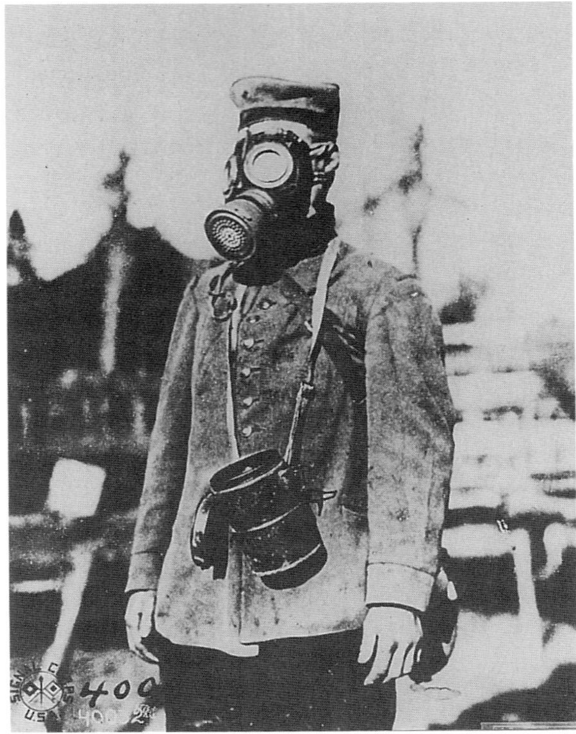

FIGURE 11.3: GERMAN SOLDIER WEARING A GAS MASK.
© IMPERIAL WAR MUSEUM

FIGURE 11.4: BRITISH TROOPS WEARING VARIOUS TYPES OF RESPIRATORS. DECEMBER 1915.
© IMPERIAL WAR MUSEUM

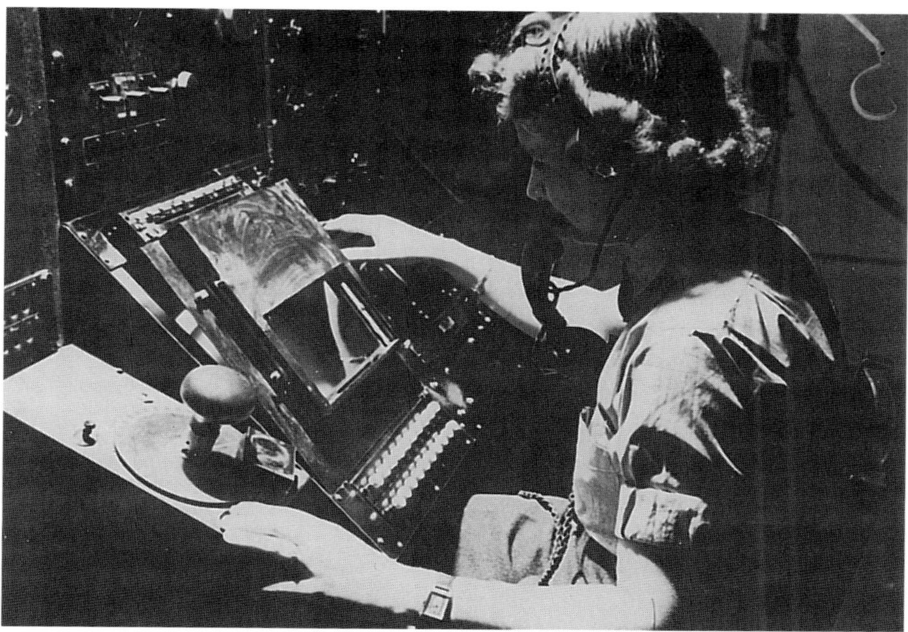

FIGURE 11.5: A W.A.A.F. RADAR OPERATOR PLOTTING AIRCRAFT ON THE CATHODE RAY TUBE. © IMPERIAL WAR MUSEUM

FIGURE 11.6: THE VARIOUS AERIALS EMPLOYED BY A CHAIN HOME LOW STATION, WHICH WERE DESIGNED TO DETECT LOW-FLYING AIRCRAFT. © IMPERIAL WAR MUSEUM

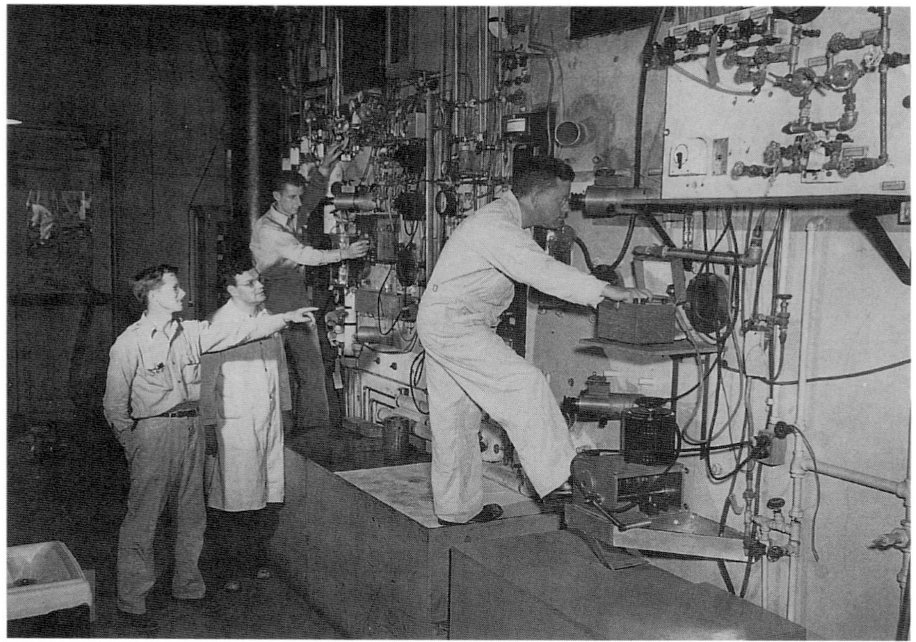

FIGURE 11.7: ATOMIC ENERGY HARNESSED. OAK RIDGE, TENNESSEE. HERE IS THE REMOTE CONTROL AND VIEWING EQUIPMENT. © IMPERIAL WAR MUSEUM

FIGURE 11.8: ATOMIC ENERGY HARNESSED. OAK RIDGE, TENNESSEE. © IMPERIAL WAR MUSEUM

CHAPTER 12

Innovation and the Modern Corporation

From Heroic Invention to Industrial Science

W. BERNARD CARLSON

If you ask the typical schoolchild of the 1990s where new technology comes from, he or she would probably conjure up the image of a scientist in a white lab coat, huddled over a computer terminal or microscope. If asked where this scientist worked, the child might tell you that his or her laboratory was part of some giant corporation. Central to this image of technological innovation is science and big business.

If, however, you were to go back and ask a schoolchild in the 1890s about new technology, they would probably tell you all about Thomas Edison or a similar heroic inventor. The image they would offer would be of an eccentric individual who through some mix of genius, luck, and hard work created revolutionary new technologies such as electric lights, automobiles, or motion pictures. The child might reveal that the heroic inventor was independent of big companies and free to pursue whatever problems struck his fancy. In contrast to the twentieth-century image of technological innovation, this nineteenth-century vision highlights individual genius and small-scale organizations.

As mirrors of popular culture, children often reflect profound economic and cultural changes. This chapter explores how this profound change in nature of technological innovation took place. How and why did business firms stop depending on heroic individuals for new products and instead come to employ professional scientists? When did firms establish in-house R&D laboratories? As firms hired scientists, how did they expect those scientists to employ science to develop new products? Did managers think that scientists would draw on their ability to break down problems, their experimental skills as well as their knowledge of theoretical science? And once in place, how did R&D labs evolve and take on new missions?

These questions are not new. Over the past thirty years, historians of technology, business, and science have extensively studied the origin and operation of research laboratories. They have generated a number of first-rate case studies of individual laboratories in major American companies such as General Electric, AT&T,

Du Pont, RCA, and Alcoa. Detailed and nuanced, these studies generally recount how R&D laboratories fared once they were established; they focus on how research managers and scientists carved a niche for themselves in the world of American big business. What these studies typically do not cover is the 'prehistory' of R&D — namely how firms secured new products and processes prior to the arrival of professional scientists. This neglect of the prehistory is unfortunate in that historians often overlook the creative ways firms secured innovation before the advent of R&D labs and scholars miss some of the factors leading up to science-based R&D. All too frequently, the coming of R&D is presented as a discontinuous phenomenon; as firms took up new science-based technology (such as electricity or chemicals), they were suddenly obliged to hire scientists and bring scientific research inside their organizations. Such explanations have a strong flavor of technological determinism, in the sense that inevitable technological change brings on inevitable social change in the form of the new R&D laboratory. What is lost is a perspective on why individuals and companies may have chosen to replace inventors with professional scientists in response to organizational needs.

Consequently, this chapter begins by addressing the 'prehistory' of industrial research, tracing the evolution of inventors as a source of new technology and how they came to play a special role in the US economy.[1] One could write an equally interesting essay on the 'prehistory' of R&D in Europe focusing on the early efforts of German dye manufacturers or French scientists. However, in the interest of telling a coherent story of the gradual development of R&D laboratories, I have chosen to devote this chapter to tracing American developments. I then discuss how inventors came to work inside firms and how for a time they were able to provide firms with a steady supply of new products, exploring how different factors — the size of companies, the growing supply of scientists, and sudden competitive threats — led corporate managers to replace inventors with professional scientists. Following this, the chapter examines several innovations at General Electric, Du Pont, and AT&T from 1900 to 1950 to reveal how scientists carved a niche for themselves by helping firms diversify and by undertaking fundamental research. I conclude with a brief discussion of R&D since 1955 and the significance of a historical perspective for how we think about innovation today.

Inventors in History

Since the Italian Renaissance of the fourteenth and fifteenth centuries, individuals who created new devices have called themselves inventors. These first inventors often began as artists, because the development of a new machine was enhanced by the ability to sketch and build models. Like artists, inventors claimed that their ability to create new technology was based on deep personal knowledge and that inspiration frequently came in a flash — the Eureka moment. Hence, if one wanted an invention, one had to let the inventor pursue his craft freely. While inventors sought a social role of intellectual freedom, they nevertheless realized that they could only build their creations by aligning themselves with powerful patrons.

Throughout the Commercial and Industrial Revolutions of the next three centuries, numerous individuals created new goods, machines, and industrial processes and styled themselves inventors. These individuals frequently began as craftsmen, and their ability to envision and fashion new machines was grounded in their profound understanding of a craft. The early contributions of inventors to advancing industry and commerce was sufficiently clear to monarchs and governments that they began rewarding inventors with patents which gave them exclusive control over their inventions. The first patents for inventions were awarded by the Republic of Florence in 1421, and the first British patent law was passed in 1623.

Just as creative technologists in eighteenth-century England called themselves inventors, so ambitious technologists in colonial America did the same. As early as 1641, American inventors petitioned colonial governments for patents. By the end of the American Revolution in the 1780s, the English Industrial Revolution was well underway, and the American Founding Fathers fully appreciated the value of stimulating technological change. As a result, when they framed the Constitution, one of the specific powers delegated to the new Federal government was the creation of a patent system. Early American inventors frequently developed new machines for processing agricultural products; for instance, Oliver Evans introduced an automated flour mill in 1790 and Eli Whitney patented his cotton gin in 1794.

During the first half of the nineteenth century, numerous inventors contributed to the creation of American industry. Spurred not only by the patent system but by first-hand experience of using machines in trade or farming, Americans readily imagined new inventions. As one European visitor remarked, "there is not a working boy of average ability in the New England states . . . who has not an idea of some mechanical invention for improvement . . . by which, in good time, he hopes to better his position, or rise to fortune and social distinction."[2] Two prominent inventors, Robert Fulton (the steamboat) and Samuel F. B. Morse (the electric telegraph), were trained as artists in England; modeling their inventive careers on their artistic careers, they struggled to support themselves exclusively by invention. More typically, however, inventors frequently developed only one or two new devices which they then put into manufacture themselves or sold to eager entrepreneurs. For instance, a young Philadelphian, Matthias Baldwin, designed a new locomotive in 1830. Finding no one willing to build it for him, he set up his own company which became the leading manufacturer of locomotives in the United States for the next 80 years.

While various individual inventors made their way in the antebellum American economy, their efforts were nonetheless circumscribed. Firms in this period were generally small partnerships and lacked substantial capital. Most industries were marked by sharp price competition, which forced businessmen to avoid the long-term investment necessary for improving technology. While businessmen were willing to purchase patents from inventors and put them into use, they generally kept inventors at arm's length, reluctant to employ them or subsidize their development costs. For example, after accidentally discovering the process of vulcanizing rubber in 1838, Charles Goodyear spent an additional five years and $50,000 perfecting and patenting his process. Even though he was able to sell the rights to his patent in both Europe and America, he nonetheless died in 1860 with debts of nearly $200,000.

The Golden Age of Heroic Invention

In the decades after the American Civil War, this pattern remained true in many American industries — inventors were a major source of new technology but not integrated into business firms. For instance, Wilbur and Orville Wright initially developed the airplane without any assistance from business. However, it was the telegraph and electrical manufacturing industries that created a new situation, a brief "Golden Age" for individual inventors.[3]

Based on the inventions of Morse, Charles Wheatstone, and William Cooke, it is perhaps not surprising that the telegraph industry should have become a hotbed of heroic inventors in the 1870s. However, it was not the heroic origins of this industry that determined the frenzy of individual activity in the 1870s, but the appearance of the giant Western Union Telegraph Company. Although Morse's original invention was promoted by a host of small, regional telegraph companies in the 1850s, it became clear to the industry's leaders by the 1860s that the telegraph would only flourish if one system connected cities and villages throughout America. Under the leadership of Hiram Sibley and then William Orton, Western Union gradually created a nationwide system by absorbing its competitors and building the first transcontinental line in 1861.

However, no sooner had the company achieved national dominance in 1867 than it had to fight off rival networks and governmental interference. The first threat came from Wall Street. Western Union had expanded rapidly by erecting lines along railroads and placing telegraph offices in railway stations. This meant that as new transcontinental railroads were built, railroad financiers could create their own telegraph network, compete with Western Union, and attempt to gain control of the telegraph giant. Jay Gould pursued this strategy twice, unsuccessfully in 1874–8 and successfully in 1879–81. A second challenge for Western Union came from Washington. As one of the first monopolies, Western Union was attacked by politicians and reformers as a threat to American democracy. Reformers feared that this corporate giant would raise prices and not serve the public interest. Critics were concerned that Western Union had access to market information as well as private business messages, and that the firm could use this information to ruin individual businessmen and manipulate stock and commodity markets in its favor.

In responding to these threats from Wall Street and Washington, Western Union employed various tactics (price competition, political lobbying, and hostile takeovers) and in this turbulent environment technological innovation came to play a new and important role. To maintain its dominant position, Western Union needed to adopt new inventions that would permit it to operate more efficiently. At the same time, the challengers — financiers and reformers alike — also realized that innovations might be used to gain a foothold in the industry. As the *Telegrapher* observed in 1875,

> improved apparatus has become of vital importance, and, consequently, telegraphic inventors who, for some years past, have been regarded as bores and nuisances, suddenly find themselves in favor, and their claims to notice, recognition and acceptance, listened to with respectful attention. All parties are now desirous of securing the advantages which may be derived from a development of the greater capacity of telegraph lines and apparatus. The fact has become recognized that the party which shall avail itself to these most fully will possess a decided advantage over its competitor or competitors.
>
> That this state of telegraphic affairs affords the opportunity for the inventive talent and genius of the country which has hitherto been wanting, is unquestionable.[4]

By the mid–1870s, the combination of Western Union's dominance and the possibility of a rival network created a unique market for telegraph inventions. There was a strong demand for 'blockbuster' inventions which could be used by Western Union or its challengers, and this demand prompted dozens of ambitious men to turn their attention to developing improved devices and entirely new systems. Typical of these inventors was Alexander Graham Bell who started inventing after reading a newspaper story about Western Union purchasing a patent for a duplex (two-message) telegraph from Joseph Stearns for $25,000 in 1872. While over 400 individuals secured patents for telegraph inventions between 1865 and 1880, the most successful inventors were men such as Thomas Edison and Elisha Gray who had established themselves as telegraph equipment manufacturers. Edison was especially skilled at developing new systems for both Western Union and Wall Street financiers, and he used this patronage to leave manufacturing and build an "invention factory" at Menlo Park, New Jersey in 1876.

Although Menlo Park is frequently regarded as the ancestor of modern R&D labs, one should be careful about claiming such institutional lineage. Edison was able to build and operate Menlo Park because Western Union contracted with him to develop several new devices, including a telephone to compete with the new Bell Telephone Company. In this sense, Menlo Park was one of the first facilities to harness technological innovation to corporate strategy. However, Menlo Park was not integrated into the Western Union organization in the way that later labs were. Indeed, Western Union seems to have viewed technological innovation as a risky and expensive proposition, and they chose to minimize their risk by supporting an outside research facility. Thus, Western Union chose to secure innovation through contracts and a strategic alliance with Edison — not unlike semiconductor firms in the 1980s who supported innovation through the Sematech consortium. From Edison's standpoint, not being tied to Western Union was equally desirable since it permitted him to move into new fields, as he did with electric lighting in 1878. Given its contractual relationship with Western Union, Menlo Park is better seen as the ancestor of R&D consulting organizations such as Arthur D. Little or the Battelle Institute.

For Edison, Menlo Park was an ideal creative environment, and during his seven years there (1876–1883), he turned out a series of spectacular inventions — an improved telephone, the phonograph, and an incandescent lighting system. Skillfully playing up images of the romantic genius for newspaper reporters, Edison gave the American public a highly individualistic myth of technological innovation which perhaps served as an antidote for the realities of the expanding, impersonal organizations (corporations, government agencies, and universities) that were coming to dominate American culture. Edison's success at Menlo Park stimulated other inventors — such as Nikola Tesla, Edward Weston, and Reginald Fessenden — to set up their own independent laboratories in the 1880s. Even today, American inventors and scientists frequently invoke Menlo Park as the inspiration and model for how they organize their creative efforts.[5]

Inventors and the Corporation, 1880–1900

As inspiring as they might be, Edison and Menlo Park were soon surpassed by other individuals and institutions. Both inventors and businessmen realized that the real challenge in introducing new technology lay not with idea generation (research), but with working out the details of manufacturing and marketing (development). While idea generation could take place away from the firm, effective development had to be done inside the firm where one could match the characteristics of a new invention with the company's resources. Consequently, as the high technology of the 1880s — electric lighting — took shape, inventors such as Charles Brush, Edward Weston, and Nikola Tesla chose to locate themselves inside new manufacturing companies. Even Edison came to understand the need for this transition. While he used Menlo Park to generate and incubate new ideas for electric lighting, when it came time to convert his incandescent lamp into a commercial lighting system he abandoned Menlo Park and moved his research efforts to the Edison Electric Lighting Company in New York City.

Representative of this new trend of inventors moving into companies was Elihu Thomson. A chemistry teacher from Philadelphia, Thomson was fascinated by the arc lighting systems he saw while visiting Paris in 1878. On his return to the States, he and Edwin J. Houston began developing their own system for lighting factories and shops. However, Thomson soon realized that while he could invent ingenious devices, he knew little about manufacturing and marketing his system. Consequently, he allied himself with several different groups of entrepreneurs who provided the funds and expertise needed to commercialize his system. After two unsuccessful attempts, Thomson finally found the right set of backers among the shoe manufacturers in Lynn, Massachusetts. Led by Charles A. Coffin, the shoemakers were familiar with marketing since they had developed techniques for selling shoes throughout the United States. Moreover, Coffin was able to secure capital from industrial financiers in nearby Boston and develop new arrangements for extending credit to the newly established utility companies.

Under Coffin's leadership, Thomson was able to concentrate on inventing, and the Thomson-Houston Company grew rapidly. Because many towns and cities in America rushed to create their own local utility companies, there was tremendous demand for electric lighting equipment, and in response, Thomson developed a wide range of new products (dc and ac incandescent lighting systems, motors, streetcars, and meters). The company built a huge factory complex in Lynn, and by 1891 was employing 2400 workers. To reach customers throughout the United States and the world, Thomson-Houston established sales offices in major cities and had a large force of salesmen. To help the new utilities set up their systems, Thomson-Houston had a construction subsidiary as well as a staff of engineers at the Lynn plant. These many facets of the electrical manufacturing business meant that the Thomson-Houston Company soon came to have a complex management structure and by 1891, it was capitalized at $10.5 million. For Thomson, the rapid growth of the firm meant that there was a steady demand for his talents as an inventor; new products were needed to reach new markets and compete effectively with the electric manufacturing companies established by Edison and George Westinghouse.

What makes Thomson-Houston significant is that it was more effective than its competitors at coordinating the basic tasks of product innovation, manufacturing, and marketing. During the late 1880s, it took over all of its smaller competitors, and in 1892 merged with Edison General Electric to form the General Electric Company.

Thomson demonstrated to Coffin and the other managers at Thomson-Houston and GE that not only could inventors work within the firm but that new products were essential for rapid corporate growth. By the early 1890s, the leaders of GE had come to realize that the size of their firm (in terms of money invested, plant capacity, and organizational complexity) was such that they could no longer hold inventors at arm's length; in order to protect their huge investment, product innovation had to be brought within the firm. With Thomson as an employee, GE had product innovation inside the firm and they supplemented Thomson's abilities by hiring several other creative inventors and engineers, including Charles Steinmetz.

The Limits of Invention in the Giant Firm

While the size of the firm prompted GE's managers to support product innovation, size nonetheless interfered with the process of developing new products. As the company became larger, with more factories, departments, committees, and employees, it became increasingly difficult for creative individuals like Thomson or Steinmetz to secure and coordinate the resources they needed to develop new inventions.

This problem is illustrated by Thomson's experience with developing high-efficiency engines and automobiles in the mid-1890s. During this period, the US economy was experiencing a severe depression, and GE's primary customers, utility companies, were unable to purchase new equipment. In response, GE developed more efficient generators and lamps which permitted utilities to make money by lowering operating expenses. GE sought to improve not only its generators but also the engines used to drive them, and the firm asked Thomson to develop a simple engine that could be used in small central stations and isolated plants. Because orders were down for big generators, GE officials hoped that the manufacture of engines might utilize the idle capacity of their large factory in Schenectady, New York.

Thomson quickly realized that an automobile would be an excellent way to test a small engine. For an engine to be successful in an automobile, it would have to be lightweight, simple, and easy to operate. If he could produce an automobile engine with those characteristics, then Thomson figured that the same engine would be excellent for powering generators in small stations lacking highly trained attendants. With the growing popularity of the bicycle as a form of individual transportation, it seemed highly desirable to create a self-propelled vehicle. Along with Edison, Henry Ford, Hiram Maxim, and others who took up the challenge of developing a practical automobile, Thomson and Coffin sensed that the success of the bicycle indicated a huge market for a horseless carriage.

After investigating electric motors and internal combustion engines, Thomson chose to focus on a steam-powered vehicle. For that vehicle, Thomson designed his "uniflow" engine, which achieved improved thermal efficiency by exhausting cool steam at the end of the stroke through a special set of exhaust ports. By August 1898 the steam vehicle was operational, and Thomson's assistant, Hermann Lemp, demonstrated the vehicle's practicality by driving from Lynn to Newburyport and back, a distance of 25 miles. Coffin was sufficiently impressed that he encouraged Thomson and Lemp to begin planning for production. In May 1899 they began work on a new and lighter design that was to be "complete and perfect in all parts; in other words to reduce the carriage to a standard article, as if we were building an arc lamp or dynamo for reproduction."[6]

GE, however, chose not to put the Thomson-Lemp steam automobile into production. After consulting with Thomson, Coffin and the company's patent attorneys concluded that they would not be able to secure adequate patent coverage. Although Thomson and Lemp had filed patent applications for details of the vehicle and its engine, it became clear that the company would not be able to assemble a group of patents that would prevent other firms from entering the automobile field. Full-fledged production of vehicles would require a substantial investment by the company, and Coffin believed that it was too risky to make that investment if the company could not control the field. Equally important, in 1898 GE's core business had begun to recover. The company was receiving new orders from utility companies for equipment, thus eliminating the need for a new product to employ the underutilized plants.

In the course of the automobile project, Thomson became frustrated with how the company handled new product development. The engine and automobile work had gone slowly because of the existing organizational arrangements. To build and test his engines and automobiles, Thomson had to have different parts made by workers in both the Lynn and Schenectady factories, and then assembled at his laboratory in Lynn. He also had to coordinate with several different groups within the company, such as the Manufacturing Committee and the Patent Department, and these groups did not always cooperate. By September 1899, Thomson concluded that what was needed was an organizational change, and he wrote to Coffin

> that it has grown upon me strongly within the last four or five months that what is needed is a department at the Works especially for the development of this kind of machinery [i.e., engines]. We should have men and machinery wholly devoted to work in this field — together with the automobile field — and they should be separated out as it were in a building or department by themselves. As it is, the work is scattered and partly done in one place and partly in another, and it is almost impossible to force it along at the rate required. I find it extremely difficult with the work scattered as it is, to impress upon the men the necessity of saving time or to get a proper appreciation of the value of time in the development of new work. Things move at an exasperatingly slow rate, and the only cause for it that I can discover is the lack of concentration in one place of draftsmen, men and tools.[7]

What Thomson wanted was a department isolated from manufacturing operations, staffed by specialists, and equipped with the necessary machine tools. Although he did not suggest that scientists be hired, what is more important is that he wanted individuals "wholly devoted to work in this field," that is, specialists. Significantly, Thomson proposed a research department as a way of coordinating resources and expediting the innovation process. He had clearly demonstrated that new products could be developed at GE, but because it was such a large and complex organization, he was not able to control the innovation process and deliver new products in a timely fashion. If the firm was to succeed in using new products to gain a competitive advantage, Thomson realized that it would need a new institution suited to the scale of the firm: the R&D department or industrial laboratory.

The First R&D Laboratory

Although GE did not go into the automotive field and did not create the research department he proposed, Thomson discussed his concerns about the organizational arrangements for new product development with the company's other major innovator, Steinmetz. Working first in the Calculating Department at the Schenectady plant, and then in a laboratory at his boardinghouse, Steinmetz had applied his mathematical skills to improving the efficiency of ac generators, transformers, and motors. By the late 1890s, Steinmetz had become worried that GE's carbon-filament lamp was about to be overtaken by several new and more efficient lighting devices: the Welsbach gas mantle, the Hewitt mercury-vapor lamp, and the Nernst metallic-filament lamp. Aware that those devices had been invented by men familiar with electrochemistry, Steinmetz proposed in July 1897 that the company establish a chemical laboratory where those devices could be investigated. Although his first proposal was ignored by GE officials, Steinmetz repeated his request in early 1899, and he enlisted the support of the vice president for engineering, Edwin Wilbur Rice, and the chief patent attorney, Albert G. Davis.

In September 1900, Steinmetz, Thomson, Rice, and Davis succeeded in convincing the company that a research laboratory should be established to investigate and develop new products. To head the new laboratory, the company hired Willis R. Whitney, a professor at the Massachusetts Institute of Technology who had earned his Ph.D in chemistry at the University of Liepzig under Wilhelm Ostwald. Whitney's mission was to develop immediately a metallic filament lamp, and to fulfill this mission, he was given a laboratory at the Schenectady plant in 1901 and an annual budget of $15,830. Whitney gradually built up his staff to 45 scientists and technicians by 1907.

In his proposal for the new laboratory, Steinmetz emphasized that a chemist should be hired to apply chemical theory and laboratory techniques to the development of new lighting devices, but it was Thomson who suggested that the laboratory might have a broader mission of pursuing fundamental research; as he wrote to both Steinmetz and Rice,

> it does seem to me that a company as large as the General Electric Company should not fail to continue investigating and developing new fields; there should, in fact, be a research laboratory for commercial applications of scientific principles, and even for the discovery of those principles.[8]

In espousing both the application and discovery of scientific principles, Thomson was implementing the values he had acquired while teaching chemistry in Philadelphia in the 1870s. In his high school valedictory address, Thomson had spoken eloquently of how science might help the manufacturers of Philadelphia. Having come to a position of authority with a major manufacturer, he helped create an institution that embodied the marriage of science and industry.

GE created this new entity, the research laboratory, as a result of three factors. First, the company was confronted by an immediate competitive threat. If it did not acquire a new high-efficiency incandescent lamp, it was likely to lose a significant portion of the lamp market to Westinghouse and European lamp manufacturers. To protect its substantial investment in technology, capital, plant, and a skilled workforce, GE had to respond to this threat.

GE had two choices as to how it could respond. Like Westinghouse which bought the patents for the Nernst lamp, GE could have purchased patents from outside inventors. The other choice was to develop a new lamp in-house. GE chose the latter alternative because Thomson and Steinmetz had demonstrated that innovation could take place within the firm. Because of his numerous inventions at both Thomson-Houston and GE, Thomson had shown the potential of new products for capturing new markets and enhancing the firm's position. Hence, a second factor contributing to the creation of the research laboratory was that GE had an established tradition of in-house product innovation.

But it was not enough to have a tradition of innovation. A firm must also have a structure that permits the coordination of people and resources necessary for developing innovations. Thomson's recent experience with automobiles revealed that GE's size and structure were impeding product innovation, thus suggesting that a new kind of laboratory was needed. In order to develop competitive products in a timely fashion, it would be necessary to concentrate resources in a single department. Consequently, a third factor leading to the industrial research laboratory was the gap between the tradition of product innovation and the existing organizational arrangements; because all the activities related to innovation could be performed in the new research laboratory, it was hoped that the new lab would fill this gap.

Although they began in 1900 with enthusiasm and high expectations, Whitney and his team found it extremely difficult to develop a better lamp, and in 1906, GE was forced to buy the German patents for manufacturing a tungsten-filament lamp. The lab's first success came only in 1907 when William Coolidge demonstrated how tungsten could be made ductile and hence shaped by machine into lamp filaments.

Given that Whitney and GE lab were unable to contribute any immediate results, why did GE support the lab for the first six years? There are several reasons why, once established, the lab survived. First, besides engaging in basic research, Whitney made sure that his chemists provided the company with a range of services. Along with developing new products for different departments, Whitney and his scientists consulted on production problems, tested materials, and designed pilot plants. The lab actually manufactured some specialty items, such as carbon resistance rods and tungsten contacts, which were used by other parts of the company. And Whitney made sure that his staff filed patents which GE used defensively to protect its existing product line and offensively as bargaining chips in negotiating with rivals. By performing multiple tasks for the firm, Whitney secured funds and support which could be used to subsidize fundamental research.

While GE executives and senior engineers valued these services, they also came to value the laboratory as part of a broader strategy of minimizing risk and uncertainty. As the business historian Alfred D. Chandler, Jr. has argued, giant corporations such as GE grew and survived by performing a wide range of tasks relating to production and distribution. To reduce costs and eliminate uncertainty, firms frequently integrated backward toward their sources of raw materials and forward toward the customer. In pursuing this vertical integration, managers generally chose to bring activities inside the firm rather than to depend on outside suppliers; only by having key functions inside the company did they feel it was possible to minimize risk and protect their large organizations. Given this general business strategy, it is not surprising that some managers of technology-oriented firms brought one of the key inputs, research and product innovation, inside the firm. By generating its own new products and patents, a firm ensured a regular supply of these inputs which it could direct toward increasing productivity and efficiency. Unlike Western Union in the 1870s which was comfortable in contracting with Edison, GE in the 1900s felt that it could only protect itself by fully integrating innovation into its corporate structure. In this sense, the creation of the industrial research laboratory was part of the early twentieth-century trend in the American economy toward minimizing risk by bringing key functions inside the firm.

But why did big firms like GE invest in a *scientific* laboratory? Why not hire more talented inventors like Thomson and Steinmetz? Here the answer is both economic and cultural. From an economic standpoint, one difference between the 1870s and the 1900s was a change in the supply of scientific manpower. In the 1870s, only a handful of American universities offered advanced research degrees in the sciences, and like Whitney, those few Americans wishing to become research scientists went to Germany to study physics or chemistry. Yet by 1900, American universities had undergone a profound expansion, particularly in scientific research. Thanks to private philanthropy and the Federal land grants to state colleges, American universities now trained hundreds of Ph.D-level scientists each year. In fact, George Wise has suggested that the supply of scientists probably exceeded the demand for science professors, and this situation led some scientists around the turn of the century to seek careers in industry.[9] Hence, given the growing supply of scientists, it made sense for managers at GE and other large companies to hire scientists and not inventors for product innovation.

But there were also cultural reasons for choosing scientists over inventors. Inventors generally explain and legitimate themselves by claiming that they possess unique personal knowledge (genius) and skills. The basis of their expertise is personal and idiosyncratic. If inventors actually invent based on a eureka moment, then their work is fundamentally discontinuous and unpredictable. Who knows when the muses will speak? Given this rhetorical stance, inventors were not especially appealing to managers trying to minimize uncertainty and protect companies capitalized for tens of millions of dollars. Yes, a genius like Steinmetz can do great work, but should one bet the company on him?

Instead, along with other attempts to rationalize their organizations, corporate leaders turned to scientists who promised to produce new technology in an efficient and predictable manner. Central to the rhetorical stance of the new industrial scientists of the twentieth century were promises of predictability. and continuity. A central characteristic of science was its claim to be able to predict the behavior of natural systems; if this was generally true of science, then the process of applying science to industrial problems should be predictable as well. Moreover, by taking a team approach to solving problems by breaking down complex problems into a series of routine experiments, scientists promised managers that they would get results sooner or later. By promising to be predictable and continuous, industrial scientists spoke a language that made sense to managers struggling to protect big firms in the face of uncertainty. The rhetorical position of the industrial scientist was perhaps best summed up by Carl Duisberg, director of the research laboratory at the German chemical company, Bayer; when describing the routine nature of inventing new dyes, he loved to point out that in his lab, "Nowhere any trace of a flash of genius."[10]

From Defense to Diversification

During its early years the GE industrial research lab fulfilled several needs of the company, but it did not produce any major breakthroughs. Whitney and his team of scientists defended GE's position in the incandescent lamp market by gradually improving the tungsten filament lamp. All this changed, however, in 1909 when Whitney hired a new young chemist, Irving Langmuir, and assigned him to study why light bulbs acquired over time a coating on the inside. Langmuir's work led GE into the world of electronics, and permitted Whitney to define the lab's role in terms of helping GE diversity into new markets.

Langmuir had beem awarded his Ph.D from Gottingen University in Germany where he had studied the behavior of incandescent filaments under the direction of Walther Nernst, the inventor of the metallic filament lamp. To take advantage of Langmuir's background, Whitney initially assigned him to study the basic physical and chemical processes taking place in the ductile tungsten filament lamps, but in 1911, Whitney specifically asked Langmuir to study the Edison effect and lamp blackening. In 1880 Edison had observed that the inside of his lamps became black by what were apparently particles of carbon. After some experimentation, Edison decided that these particles were being discharged because there was a current flow from the negative side of the hot carbon filament to the neutral or positively charged interior surface of the bulb. Following Edison, Lee De Forest and Andrew Fleming had used this effect to develop the first vacuum tubes for detecting radio waves, but scientists were debating what caused the effect. Determined to resolve the scientific controversy and make a name for himself in science, Langmuir pounced on this problem. He found that a hot filament discharged a stream of electrons when the electrons encountered "a sort of subatomic traffic jam just outside the filament surface."[11] This traffic jam was called the 'space charge effect,' and it had been first observed for positive ions by C. M. Child at Colgate University. By using the space charge effect, Langmuir was able to not only account for the electron discharges he was measuring on the benchtop but could also improve the design of vacuum tubes and incandescent lamps.

While Langmuir thought about his discoveries in terms of scientific papers, Whitney and another GE manager, Laurence A. Hawkins, were quick to see the commercial advantages of his work. Hawkins in particular saw the connection between Langmuir's filament work and vacuum tubes, and he arranged for Langmuir to work with Ernst Alexanderson, GE's leading radio engineer. As a result, GE soon acquired key patents on radio tubes, which it used during and after World War I to secure a strong position in the new field of radio. Meanwhile, Langmuir convinced fellow researcher William Coolidge to apply his findings about the space charge effect to improve X-ray tubes. Coolidge used this knowledge in 1913 to a create a tube which reliably generated more powerful rays. Although GE was initially reluctant to produce the new Coolidge tube on a large scale, it proved to be highly popular with doctors and hospitals, and it led GE to diversify and commit substantial resources to developing X-ray equipment.

The development of both vacuum tubes and the Coolidge X-ray tube marked a new phase not only at the GE lab but in American industrial research generally. Prior to 1913, industrial research labs had been established primarily to protect the company's existing market position by securing patents and improving existing product lines. The mission of the GE lab was to defend the company's electric lighting business by improving lamps. Yet through these new radio and X-ray tubes, the lab permitted GE to diversify and move into new markets. As a result, business leaders came to see industrial research as both a defensive and offensive tool for corporate strategy.

At the same time, Whitney and other industrial research leaders realized that Langmuir's work marked a new approach to product development. Previously, they had assumed that professional scientists would improve existing products by using a combination of experimental skills, and teamwork. Langmuir, however, was one of the first researchers who got ahead by converting a practical problem (blackening lamps) into a scientific problem (why did a hot filament discharge electrons?) and then translating his scientific knowledge (the space charge effect) back into new products (better vacuum tubes). This blend of the practical and abstract deeply impressed Whitney and other lab directors, one of whom called this approach "pioneering applied research."[12] Langmuir continued to successfully mix practical problems with scientific research, and in 1932, he was awarded the Nobel Prize in physics. With this award both GE management and the public were convinced of the value of industrial research.

World War I and the Spread of R&D

Prior to World War I, only a few large firms followed General Electric's lead and established in-house research laboratories. In most cases, companies re-invented the idea of a research laboratory in response to their own peculiar, immediate circumstances. For instance, American Telephone and Telegraph established its research branch in 1907 when its chief engineer, J.J. Carty, came to the conclusion that he needed to hire physicists to develop a new repeater (i.e., amplifier) for the company's first coast-to-coast long-distance line. Likewise, George Eastman at Kodak authorized the creation of a laboratory in 1912 in order to conduct research and secure such broad patent coverage that it would discourage other companies from moving into the photographic field.

It was World War I, however, that fully convinced both American managers and the public of the value of scientific research. Prior to their entry into the war, Americans were simultaneously impressed with the role science seemed to play in the industrialization of Germany and frightened by America's dependence on Germany for chemicals, dyes, pharmaceuticals, and optical glass. During the war, American scientists were mobilized by the government, and they worked on a variety of defense problems. The American public was favorably impressed with the way in which collaborative scientific research improved radio and anti-submarine defenses, and these achievements convinced many Americans that science was essential to industrial and social progress.

Following World War I the number of research labs grew rapidly. By 1931, 1600 companies reported laboratories employing 33,000 scientists, engineers, and technicians. By 1940, more than 2000 firms reported R&D departments employing 70,000 people.[13] Although inventors such as Edwin Armstrong continued to come up with new products (FM radio), the transition from heroic inventor was more or less complete by the start of World War II. By then, few American managers looked to "flashes of genius" as a source of new products. The new paradigm was exemplified by chemist Wallace H. Carothers inventing nylon at DuPont in the 1930s.

The Shift to Fundamental Research

The paradigm embodied by Carothers marked an additional step in the evolution of R&D in American business. Unlike other industrial scientists who were hired by firms to apply their benchtop skills to practical problems, Carothers was hired by Du Pont to conduct fundamental research, identical to the theoretical investigations which scientists were pursuing in American universities. While Du Pont executives initially assumed that fundamental research would not necessarily result in profitable new products, Carothers' work did yield two new products, neoprene (artificial rubber) and nylon.

Du Pont had established its first R&D facility, the Eastern Laboratory, in 1902 to improve the manufacture of high explosives. Like the GE lab, the Eastern Laboratory grew into the Chemical Department before World War I in order that the company might diversify into new areas such as dyestuffs, celluloid, paints, and artificial leather. By 1927, Du Pont was employing 850 people and spending $2.2 million on research, but its research director, Charles Stine, believed that the company should take a more radical approach to R&D. Rather than have professional scientists apply the results of pure science to industrial problems, why not permit scientists to generate new science in the company's lab?

In proposing to the top management of Du Pont that it should fund fundamental research, Stine offered four reasons. First, he thought that the company would gain in prestige and 'advertising value' by being able to report that its scientists were publishing papers. These public relations concerns were probably not frivolous since a portion of the American public saw Du Pont solely as a munitions company that had made fabulous profits from the carnage of World War I. Second, Stine argued that interesting scientific research would make it easier to recruit and retain first-rate Ph.D chemists who might prefer academic careers. Third, he anticipated that results from fundamental research could be useful in bartering with other companies for patents and proprietary information. And only fourth did he suggest that fundamental research might lead to new products.

After another round of negotiations, Du Pont's executive committee authorized Stine to spend $250,000 annually on fundamental research. With these funds in hand, Stine built a new laboratory which was soon dubbed 'Purity Hall,' and he began recruiting chemists from academia. Among Stine's first hires was an organic chemist, Wallace H. Carothers who was an instructor at Harvard University.

Carothers initially resisted Stine's offers, fearing a loss of intellectual freedom and worrying that his "neurotic spells of diminished capacity" [what is now called depression] might be a handicap in a corporate environment.[14] However, after receiving reassurances that fundamental research meant pure science, Carothers joined the company in 1928 as head of a new group investigating long-chain molecules or polymers. Just as Langmuir had been attracted by the controversy surrounding electron discharges from hot filaments, so Carothers was excited by the controversy surrounding the nature of polymer molecules. While some chemists thought that polymers were held together by the same forces that operated in smaller molecules, others thought that these large molecules involved some other kind of forces. Carothers resolved this controversy by building long-chain molecules, one step at time, employing well-understood reactions which used acids and alcohols to form esters. In the course of this research, Carothers and his team not only laid the foundation for our modern understanding of polymers, but in 1930 they also discovered two valuable materials, artificial rubber or neoprene and a strong manmade fiber which came to be called nylon.

Under Stine, Carothers' principal obligation was to publish papers about his results, but shortly after the discovery of neoprene and the new fiber, Stine was replaced by a new director of research, Elmer K. Bolton. Bolton had worked his way up the ranks in Du Pont by converting laboratory research on dyestuffs into commercial products "in the shortest time with the minimum expenditure of money."[15] Consequently, Bolton had little patience with Stine's ideas about fundamental research, and in response to the Great Depression, Bolton reorganized the groups in 'Purity Hall.' Bolton believed that new products could be developed faster by combining fundamental and applied research in single teams, and in 1933 he asked Carothers concentrate his group on developing nylon as a commercial fiber. Carothers did so, but at the personal cost of new bouts of depression. In 1937, just a few weeks after the basic patent for nylon had been filed, Carothers committed suicide.

During the 1940s, nylon came to be used in women's stockings, reinforcement cords in automobile tires, rope, and a variety of industrial applications. According to David A. Hounshell and John K. Smith, "[n]ylon became far and away the biggest money-maker in the history of the Du Pont Company."[16] Based on its commercial success, Du Pont invested heavily in R&D in the 1950s in the hope of getting similar winning products. In doing so, the company assumed that fundamental research would automatically lead to revolutionary products, but the story of Carothers and nylon should be read as a cautionary tale. Nylon came about only because of several lucky organizational developments. On the one hand, Stine had to create a positive environment which would attract a talented chemist like Carothers and which would permit him to do creative research. On the other hand, Carothers would never have converted the fiber he discovered in 1930 into the commercial product nylon without pressure from Bolton. Without the right blend of creative freedom and practical considerations, fundamental research in corporate labs will not yield new 'nylons.'

Bell Labs and the Development of the Transistor

The importance of blending creative freedom and practical considerations in R&D can also be seen in the development of one of the most famous products of industrial research, the transistor. Invented in 1947, the transistor was the first multi-purpose semiconductor device and it led to the establishment of the modern semiconductor electronics industry. Often touted as the product of theoretical solid-state physics, the transistor is better seen as the result of a mix of physics, business, and hands-on skill.

The transistor was invented at Bell Laboratories, the research arm of AT&T. Established in 1925 as a separate subsidiary, Bell Labs combined AT&T's older Research Department with scientists and engineers from Western Electric, AT&T's manufacturing arm. Bell Labs quickly became the largest corporate R&D lab in the United States, and by the late 1940s, it was employing 5700 people, of whom over 2000 were professional scientists and engineers.[17]

Like other corporate labs, Bell Labs pursued several missions, including solving manufacturing and operation problems, securing patents, and conducting research in areas which would affect the future of telecommunications. One of the future issues which worried Bell Labs was the growing size and complexity of telephone exchanges. By the mid 1930s, the Director of Research at Bell Labs, Mervin Kelly, was becoming especially concerned that as exchanges grew, the mechanical relays used as switches would have to be replaced with electronic devices. While vacuum tubes would be faster than the mechanical relays, Kelly was concerned that tubes would burn out too quickly and draw too much power. Would it perhaps be possible, wondered Kelly, to develop an entirely new type of electronics?

Kelly's concerns were temporarily set aside during World War II, during which Bell Labs worked on a wide variety of military projects, particularly the development of radar. Because radar utilized microwaves which could not be detected by vacuum tubes, scientists at Bell Labs, MIT, and Purdue investigated semiconductor materials such as germanium and silicon in order to develop new detectors using point-contacts. In these investigations, Bell Lab scientists learned how to make two kinds of semiconductor materials (n-type and p-type) by deliberately doping germanium or silicon with traces of other elements. To understand these new materials, researchers drew on the new field of solid-state physics, which used theories and discoveries about electrons and atomic structure to understand the nature of all kinds of materials.

As World War II came to an end, Kelly was anxious to capitalize on the solid-state physics expertise that Bell Labs had acquired, and so in 1945, he ordered the creation of a solid-state physics sub-department. The mission of this sub-department was to obtain

> new knowledge that can be used in the development of completely new and improved components and apparatus elements of communications systems . . . There are great possibilities of producing new and useful properties by finding physical and chemical methods of controlling the arrangement and behavior of the atoms and electrons which compose solids.[18]

To lead this new group, Kelly selected a chemist, Stanley Morgan and a physicist, William Shockley. Kelly also assigned two more top-notch physicists, Walter Brattain and John Bardeen to the team. While Brattain was an experimental physicist who had worked in the Vacuum Tube Department since 1929, Bardeen was a theoretical physicist who had just joined Bell Labs.

Under direction from Shockley, Brattain and Bardeen focused their efforts on finding a semiconductor device which could amplify signals and thus serve as a possible replacement for vacuum tubes. Shockley first had them try to build a field-effect amplifier by creating a 'junction' of p-type and n-type semiconductors. Shockley theorized that when a current was placed across the junction, the induced charge carriers would be free to move, increase the conductivity of the device, and hence boost the signal. When these experiments failed, Bardeen suggested that they did not fully understand what was happening to electrons on the surface of the semiconductor, and he developed a theory to explain what was happening. To test this theory, they needed to place two point contacts on the surface of a slab of germanium, and Brattain devised a way to do so by drawing on work he had previously done with point-contact rectifiers. Because Bardeen's theory predicted that the contacts had to be only a few microns apart, Brattain fashioned two contacts by wrapping a plastic triangle with gold foil, cutting a small slit at the apex, and filling the gap with wax. In December 1947, Brattain, Bardeen, and Shockley found that this device could amplify signals. After filing patent applications, Bell Labs announced this invention in June 1948, calling the new device a transistor. Transistor research continued at Bell Labs, culminating in Shockley's development of a successful field effect transistor in 1951. The following year, Bell Labs began offering seminars on this new technology and licensing the manufacture of transistors by other companies. For their pioneering research, Brattain, Bardeen, and Shockley shared the 1956 Nobel Prize in Physics.

R&D Since 1955

Emboldened by the success of nylon and the transistor, many American firms invested heavily in R&D in the 1950s. By 1955, total R&D expenditures had risen to $6.1 billion.[19] Like their predecessors at GE, managers during this period continued to see investment in science as part of a risk-averse strategy. As John K. Smith has observed, "if basic science was the seed of new technology, then the entire innovation process could be contained within the firm; reliance on unpredictable outside sources of technology was no longer necessary."[20] American firms found further incentives as a result of antitrust litigation. During the New Deal, the Federal government had attacked a number of many firms (including both AT&T and Du Pont), and it continued these investigations. In this political environment, American firms often found it preferable to develop new products inhouse and avoid acquiring new technology through corporate acquisitions or cooperative arrangements.

Not only did the Federal government shape corporate R&D through antitrust litigation, it also influenced it through defense spending. During the Cold War, the Federal government spent hundreds of millions of dollars not only on complex new weapons systems but also on developing new manufacturing techniques. Significantly, the Federal government often financed the risky and expensive commercialization phases of a new technology; for instance, Bell Labs was only able to perfect manufacturing techniques for silicon transistors with support from the Pentagon. One commentator estimates that during the fifties and sixties that one-half to two-thirds of the funds for R&D came from Washington.[21] To date, historians have only begun to investigate how Federal support changed corporate R&D.

In investing in R&D, American companies employed thousands of Ph.D scientists and built elaborate research 'campuses.' At these new facilities, scientists were granted a large degree of autonomy, in the belief that such freedom had been the crucial ingredient in the development of nylon and the transistor. And yet despite ample funds, new facilities, and unprecedented freedom, scientists at the major corporate labs came up with few major breakthroughs from the 1950s to the 1980s. Indeed, the major blockbuster innovations in this period — such as the integrated chip, the personal computer, and the birth control pill — were developed and introduced by small, start-up firms. But before celebrating small start-up firms in the electronics industry as entrepreneurial wonders, we should remember that these small firms were often highly dependent on military funding and on Bell Labs for providing information and personnel.

As a result of these disappointments, American firms pursued several different strategies for technological innovation in the 1980s and early 1990s. Some firms have sold off R&D facilities. For instance, shortly after buying RCA in the late 1980s, GE sold RCA's famous Sarnoff Laboratories to the RAND Corporation in 1989. Responding partly to a relaxation of anti-trust litigation during the Reagan and Bush administrations, firms have found it expedient to pool their resources and expertise in research consortia. In order to develop new integrated circuit chips, in 1987 a number of semiconductor firms created Sematech. Finally, some firms have found that the only way to develop new products rapidly and successfully is to bypass their existing R&D organization. Despite having an extensive R&D facility in Yorktown, New York, International Business Machines (IBM) was only able to develop its own personal computer by creating an entirely new and independent division in Boca Raton, Florida. While it is difficult to predict whether any of these trends will come to define American R&D in the last quarter of the century, these developments clearly show that it is an ever-changing institution, evolving in response to changes in the economic and political environment.

CONCLUSION

In this chapter, I have examined the role of science in the modern corporation by tracing the transition from the heroic inventor to industrial scientist. By examining technological innovation and scientific research in a series of individual firms, we can glean several lessons about how and why science came to be a part of big business in America.

One lesson concerns the nature of the transition from the individual inventor to the R&D laboratory. Despite corporate mythology and older historical scholarship, the development of the first industrial research laboratories was not a sudden, discontinuous event. Corporate managers did not simply realize one day that they needed to bring product innovation in-house and under control. Instead, the process was much more evolutionary. Beginning in the 1870s with the telegraph and electrical industries, firms first created alliances with inventors and then brought them into the firm as employees. During a brief 'Golden Age,' many inventors did much creative work inside companies. However, as firms grew, it became more difficult for individual inventors to preside over the creative process from start to finish, and at General Electric, this problem led Thomson and Steinmetz to think about creating a new institution, the research laboratory. Hence, rather than creating a stark 'before' and 'after' scenario with inventors constituting the 'dark' ages of innovation and scientists representing the 'modern' world, it is more useful to see the transition from inventors to industrial scientists as being a more gradual, process in which inventors played various roles in companies.

Another lesson of this transition concerns technological determinism.[22] Prior histories have tended to assume that the 'science-based' technologies of electricity and chemistry could only be developed by the application of science. Hence, these histories present the coming of R&D as an inevitable event. Once the infant electrical and chemical industries were established, so the argument runs, it was only a matter of time before the industrialists would have to hire scientists to develop better products. In direct contrast to this deterministic view, I have suggested here that new product development could be done by either inventors or scientists, and that firms chose to hire inventors or scientists for a variety of reasons. In particular, managers shifted from inventors to scientists in part because the image and rhetoric of science appealed to managers intent on protecting their organizations by minimizing risk. Inventors fell by the wayside because they came to be perceived by managers as an unpredictable source of innovation.

As a further challenge to technological determinism, I would suggest that the exact timing of the appearance of the first research laboratories had more to do with the nature of business organizations than with the inherent content of the technology. The technology of incandescent lamps had not become so complex that it could only be solved by turning to science; rather, the size and complexity of the GE organization had reached the point where innovation could only be done by creating a new kind of laboratory. To be sure, there was a genuine commercial threat before the company, but the significant factor contributing to the formation of research laboratory was the fact that the existing arrangements for product innovation had become unsatisfactory, leading Thomson and Steinmetz to think about a new institutional configuration.

A final lesson is that the rise of R&D in the twentieth century cannot simply be seen as a straightforward transfer of scientific knowledge into a business setting. The evolution embraces a series of changes involving corporate strategy, structure, teamwork, and knowledge. Though often celebrated as the success story of theoretical science generating revolutionary new technology, the case of the transistor summarizes many of the changes which came to make R&D a powerful source of innovation in the United States. While solid-state physics was a necessary ingredient, there first had to be a strategic context in which this knowledge could be utilized. Just as Western Union had turned to technological innovation to maintain its dominant position in the telecommunications industry in the 1870s, so AT&T supported technological innovation at Bell Labs in the 1940s to protect its monopoly position. Next, not only had there to be a tradition of product innovation but also a place in the organization where creative work could be done. Just as Thomson-Houston had permitted Thomson to work freely in the Model Room in the 1880s, so AT&T established Bell Labs as a free-standing organization in the 1920s. In this way, innovators had their own physical and intellectual space but at the same time their work was tied to the needs of the firm. In this new organizational space — Bell Labs — the research managers pushed ahead by breaking down complex problems into smaller, more routine tasks and organizing a teams of investigators like that of Brattain, Bardeen, and Shockley. As we have seen, the team approach was pioneered by Edison at Menlo Park and perfected by Whitney and Stine at GE and Du Pont in the 1910s and 1920s. Of course, the Bell group drew on what they knew, solid-state physics, to develop the transistor, but it is interesting to note that they moved back-and-forth from the practical goal (getting an amplifier) to a scientific controversy (what was happening on the semiconductor surface) to a new theory (Bardeen's model of surface charges) to finally a new device (the transistor). In this zig-zag course, the Bell team behaved much as their predecessors Langmuir and Carothers had done. Like Langmuir and Carothers, the Bell group did not do 'pure' academic research but instead pursued a research program which blended creative freedom with practical needs. While the creative freedom and 'pure' aspects of R&D are often celebrated, the success of R&D has nonetheless frequently depended upon researchers and managers achieving a delicate balance between creativity and practicality.

Although there have been some memorable failures (such as RCA's videodisc), American firms have by and large successfully balanced creative freedom and practical needs. In so doing, they have converted technological innovation from a chance and random event to a reliable component of corporate strategy. From Edison's Menlo Park to recent research consortia such as Sematech, managers have viewed technological change as a valuable tool and they have strived to create better organizational arrangements for the efficient production of new ideas and goods. American corporations created successful institutions for technological innovation in response to a particular set of intellectual, organizational, and economic forces in the early twentieth century, and the story of the transition from heroic invention to industrial science reveals much about the challenges of linking the appropriate knowledge and people to the technical and organizational problems at hand.

REFERENCES

This paper was prepared with the support of the Bankard Fund for Political Economy at the University of Virginia.

1. John J. Beer, "Coal Tar Dye Manufacture and the Origins of the Modern Industrial Research Laboratory," *Isis* (1958), **49**: 123–31; George Meyer-Thurow, "The Industrialization of Invention: A Case Study from the German Chemical Industry," *Isis* (1982), **73**: 364–381; and Henk van den Belt, and Arie Rip, "The Nelson-Winter-Dosi Model and Synthetic Dye Chemistry" in W.E. Bijker, T. Pinch, and T.P. Hughes (Eds.), *The Social Construction of Technological Systems.* (Cambridge: MIT Press, 1987), pp. 135–58.
2. Quote is from George Wallis, an English engineer and manufacturer. See Marvin Fisher, *Workshops in the Wilderness: The European Response to American Industrialization, 1830–1860.* (New York: Oxford University Press, 1967), p. 48.
3. Thomas P. Hughes *American Genesis: A Century of Invention and Technological Enthusiasm. 1870–1970.* (New York: Viking-Penguin, 1989), pp. 13–95.
4. "The Progress of the Telegraphic Contest," *The Telegrapher* (30 Jan. 1875), **11**: p. 28.
5. Kenneth A. Brown, *Inventors at Work: Interviews with Sixteen Notable American Inventors.* (Washington: Tempus, Redmond, 1988).
6. Thomson to Coffin, 11 May 1899, Letterbook 4/99–7/1900, p. 107, Elihu Thomson Papers, Library of the American Philosophical Society, Philadelphia.
7. Thomson to Coffin, 12 September 1899, Letterbook 4/99–7/1900, pp. 371–4, Thomson Papers.
8. Thomson to Steinmetz, 24 September 1900, Thomson collection, GE Hall of History, Schenectady, New York.
9. George Wise, "A New Role for Professional Scientists in Industry: Industrial Research at General Electric, 1900–1916.", *Technology and Culture* (1980), **21**: 408–29.
10. Quoted in van den Belt and Rip, p. 155.
11. George Wise, *Willis R. Whitney, General Electric, and the Origins of US Industrial Research.* (New York: Columbia University Press, 1985), p. 175.
12. This term was used by Charles Stine at Du Pont in the 1920s. See John Kenly Smith, Jr., and David A. Hounshell, "Wallace H. Carothers and Fundamental Research at Du Pont.", *Science* (2 August 1985), **229**: 436–42, p. 436.
13. Kendall Birr, "Industrial Research Laboratories" in N. Reingold (Ed.), *The Sciences in the American Context: New Perspectives.* (Washington: Smithsonian Institution Press, 1979), pp. 193–208, p. 199.
14. Quoted in David A. Hounshell, David A. and Smith, John Kenly Jr., *Science and Corporate Strategy: Du Pont R&D, 1902–1980.* (New York: Cambridge University Press, 1988), p. 230.
15. Smith and Hounshell, "Carrothers and Fundamental Research", pp. 439–440.
16. Hounshell and Smith, *Science and Corporate Strategy*, p. 273.
17. Ernest Braun and Stuart Macdonald, *Revolution in Miniature: The History and Impact of Semiconductor Electronics*, 2 ed. (New York: Cambridge University Press, 1982), p. 33.
18. Quoted in Dirk Hanson, *The New Alchemists: Silicon Valley and the Microelectronics Revolution.* (New York: Avon, 1983), p. 74.
19. Birr, p. 202.
20. Smith, p. 128.
21. Birr, p. 202.
22. Merrit Roe Smith and Leo Marx, *Does Technology Drive History? The Dilemma of Technological Determinism.* (Cambridge: MIT Press, 1994).

FURTHER READING

Birr, Kendall. *Pioneering in Industrial Research: The Story of General Electric Research Laboratory.* (Washington: Public Affairs Press, 1957).

Carlson, W. Bernard. *Innovation as a Social Process: Elihu Thomson and the Rise of General Electric, 1870–1900.* (New York: Cambridge University Press, 1991).

Chandler, Alfred D. *The Visible Hand: The Managerial Revolution in American Business.* (Cambridge: Harvard University Press, 1977).

Fagen, M.D. *A History of Engineering and Science in the Bell System: National Service in War and Peace, 1925–1975.* (N.p.: Bell Telephone Laboratories, 1978).

Fleming, A.P.M. *Industrial Research in the United States of America.* (London: His Majesty's Stationery Office, 1917). Reprinted New York: Arno, 1972.

Galambos, Louis. "The American Economy and the Reorganization of the Sources of Knowledge" in A. Oleson and J. Voss (Eds.), *The Organization of Knowledge in America, 1860–1920.* (Baltimore: Johns Hopkins University Press, 1979).

Graham, Margaret B.W. *RCA and the VideoDisc: The Business of Research.* (New York: Cambridge University Press, 1986).

Graham, Margaret B.W. and Pruitt, Bettye Hobbs. *R&D for industry: A Century of Technical innovation at Alcoa.* (New York: Cambridge University Press, 1990).

Hoddeson, Lillian. "The Discovery of the Point-Contact Transistor," *Historical Studies in the Physical Sciences* (1981), **12**: 41–76.

Jenkins, Reese V. *Images and Enterprise: Technology and the American Photographic Industry, 1839 to 1925.* (Baltimore: Johns Hopkins University Press, 1975).

Misa, Thomas J. "Military Needs, Commercial Realities, and the Development of the Transistor, 1948–1958," in M.R. Smith (Ed.), *Military Enterprise and Technological Change.* (Cambridge: MIT Press, 1985).

Pretzer, William S. *Working and Inventing: Thomas Edison and the Menlo Park Experience.* (Dearborn: Henry Ford Museum and Greenfield Village, 1989).

Reich, Leonard S. *The Making of American Industrial Research: Science and Business at GE and Bell, 1876–1926.* (New York: Cambridge University Press, 1985).

Sturchio, Jeffrey L. "Chemistry and Corporate Strategy at Du Pont," *Research Management* (Jan–Feb 1984), **27**: pp. 10–18.

Wise, George. "Science and Technology" in S.G. Kohlstedt and M.W. Rossiter (Eds.), *Historical Writing on American Science: Perspectives and Prospects.* (Baltimore: Johns Hopkins University Press, 1985), p. 229–33.

Chapter 13

The Transformation of the Pharmaceutical Industry in the Twentieth Century

Louis Galambos and Jeffrey L. Sturchio

Pharmaceutical research and development is big business. In 1995, the global industry spent about $34 billion on the discovery and development of new medicines. US firms alone spent $14.4 billion on research and development (R&D) that year, and by 1996 the figure was nearly $15.8 billion. Over the past fifty years, investment in R&D has given good returns, both for the companies involved and for society. This has been particularly true for US pharmaceutical companies which have introduced about three-fifths of all of the new drugs approved for therapeutic use around the world. In the process, healthcare professionals have acquired many new weapons to prevent or fight diseases, ranging from infections and cancer to congestive heart failure and influenza. This is one of the success stories of twentieth-century science and technology. However, scholars know more about eighteenth-century drug jars than they do about the relations between industry, academic research, and government that have led to most of these innovations.

During the past century, the sources of innovation in this industry have been decisively transformed. Scientific, economic, and political developments have interacted to reshape both the institutional and cultural setting. Expanding urban markets enabled some of the manufacturing pharmacies of the late nineteenth century to grow in scale and scope, gradually evolving into the specialized, integrated pharmaceutical manufacturing firms of the 1930s and 1940s. This institutional change coincided with a shift from research conducted outside the firm to the emergence of modern, corporate R&D organizations which sparked a new wave of innovations: the vitamins and sulfonamides in the 1930s and the steroids, antibiotics, and cardiovascular products in the 1940s and 1950s.

Links between business, government, and a wide range of professional organizations sustained innovation in the United States and the leading nations in Europe after World War II. During the post-war era, these nations constructed diverse infrastructures for conducting basic biomedical research, training new cadres of researchers (with particular emphasis on the emerging disciplines of enzymology, molecular genetics, rational drug design, and clinical research), and expanding public sector markets for health care. These developments supported the transformation of the leading pharmaceutical firms into the multinational enterprises that lead the industry today.

Many of these changes reflect general trends in industrial development during these decades. The pharmaceutical industry thus provides unusually rich sources of information on the process of innovation in the twentieth century. Much of the science that has contributed to the industry's development is well-documented, as are the various government programs that have grown up around this industry. We know less about the private firms than we do about the public dimensions of the industry, but in recent years, a number of far-sighted businesses have begun to make available research materials on their development and to sponsor studies of their evolution.

As yet, however, these rich materials have not produced a consensus about the sources of innovation or even about the proper approach to this subject. Many of the studies focus on the contributions of individuals or on the intellectual aspects of discovery. Economic studies use a different framework entirely, as do the histories of business and public health. Frequently, these historical volumes substitute the dynamic organization for the dynamic individual as the central moving force in the industry's evolution. Many historians of technology have in recent years begun to shift their attention from the heroic inventor to the manner in which creative individuals interact with their changing social and technical contexts. There are in addition important studies of national, institutional and economic settings and their impact on innovation across the entire range of a country's industries.

While all of these approaches to innovation have produced useful information and analyses, we have adopted a somewhat different strategy. We reappraise the sources of innovation by focusing on the relationships between developments at three different institutional levels in modern society: the national level; the network level; and the level of the individual, formal organization.

At the national level, we look for innovations that were substantial enough to have some impact on the aggregate measures of economic performance. This is the approach pioneered by Joseph A. Schumpeter, the father of entrepreneurial studies, and adopted by most analysts working with national income statistics.

We devote considerably more attention, however, to the network level of analysis. By networks, we mean the loosely integrated, informal organizations that are joined by a common interest (in the germ theory of disease, for instance, or virology) and are characteristic of the biomedical sciences. Networks of this sort exist in numerous other areas of modern life (in the academic disciplines and professions, for example), and they frequently impinge on the process of innovation. At the network level of analysis, innovation is any action that is new to the common interest of the network and that advances that interest. The networks we examine consist of individuals (researchers and public health officials, for example) and a variety of organizations: private profit-making organizations (for instance, pharmaceutical companies); private nonprofits (professional institutions and their publications); and such public institutions as the US National Institutes of Health (NIH).

Seen from the network level, the process of change is non-linear. It normally follows a cyclical or wave-like pattern, with a slow buildup to a high level of innovations, followed by a long, slow decline in entrepreneurial activity associated with that particular network. Thus the network associated with the original work on the germ theory of disease led to a major wave of innovations in serum therapy during the 1890s and early 1900s. This cyclical pattern of change has important implications, which we discuss, for organizations in the pharmaceutical industry and for the national institutions involved with innovation in the medical sciences.

The third level of our analysis focuses on the particular organization: the firm, the public health organization, the regulatory agency. These are formal organizations, and at this micro-level of analysis, we define innovation as any action that is new to an organization, that is employed in an on-going manner, and that yields an advantage.

We use three business organizations as examples and focus primarily on vaccine and serum antitoxin development at Merck & Co., Inc., at Sharp & Dohme (which merged with Merck in 1953), and at the H.K. Mulford Company (which Sharp & Dohme acquired in 1929). We examine innovation in vaccines, which are preventive medicines that use whole live, altered or killed microbial pathogens to induce relatively long-term immunity against a disease. We also look at serum antitoxins that treat such infectious diseases as diphtheria by employing antitoxins from the blood of an animal (for example, a horse) deliberately infected with the microbes. We are thus concentrating on biologicals (biochemical substances used in the treatment or prevention of disease), which is only one part of the much larger industry involved with pharmaceuticals (all chemical substances used in this manner). By exploring the rich history of biologicals, we can cover roughly a century of change in the relationships we discuss and in the sources of innovation.

Each of these firms attempted to maintain creative relations with the research networks, their executives and scientists considered advantageous. Indeed, the ability to interpret signals from the relevant networks accurately and to manage interactions with other key actors astutely are two of the most important predictive factors for the sustained success of any of these organizations. That was why corporate and research leaders encouraged active participation in these networks: to contribute new knowledge; to recruit new talent; to monitor new trends in conceptual approaches and technological tools; and to secure access to external innovations through consultancies, licenses, and broader strategic alliances.

We describe and evaluate the manner in which the industrial scientists went about this task and also consider the role that company culture played in sustaining innovation. Although the firms were in competitive situations that shaped their behavior, they also collaborated extensively with government researchers and officials, with university scientists, and at times with other pharmaceutical firms. It is impossible to understand the history of vaccine and serum antitoxin development at Mulford, Sharp & Dohme, or Merck without understanding the cooperative aspects of these network linkages. In this industry the networks have always had international dimensions, but in recent decades the process of innovation and the entire industry have become increasingly global in scope.

While the evolution of biological research at Mulford or Merck is not necessarily typical, it is representative of the critical role of innovative networks in transforming the pharmaceutical industry in the twentieth century. This interpretive framework can be applied as usefully to innovation at Glaxo and Wellcome in the United Kingdom; at Bayer and Hoechst in Germany; at CIBA-Geigy, Roche, and Sandoz in Switzerland; or at Abbott, Lilly, Pfizer, SmithKline, Squibb, or Upjohn in the United States. As R&D has increasingly become a central competitive aspect of the industry, the maintenance of creative links with diverse innovative networks has become the single most important element shaping success in the global pharmaceutical industry.

Mulford and the Bacteriological Cycle

In the late nineteenth century when the H.K. Mulford Company first became involved in biological production, a relatively new innovative network was just taking shape, built around new concepts of the bacteriological origins of disease. This network of scientific, public health, and private institutions was international in scope and the central ideas and new therapies were European (largely French and German) in origin. While it was very loosely integrated, the network was held together by shared ideas, some consensus about leadership, and a dominant ideology framed in terms of scientific progress. Communications were frequently personal, but there were elaborate channels for professional discourse in journals such as the *Deutsche Medicinische Wochenshrift*, the *Annales de l'Institut Pasteur*, and the *Bulletin of the Johns Hopkins Hospital*. This particular network had quickly spread to the United States, even though America was still an underdeveloped nation in medical science and pharmaceuticals.

To understand why it spread so quickly in the United States between 1890 and the First World War, we need to look at certain basic characteristics of the nation. At the turn of the century, it was just becoming the leading industrial producer in the world. Rapid economic expansion, population growth, and urbanization had created unusually large markets for pharmaceutical and biological products. The country's complex, decentralized array of public health and university institutions facilitated change, as did the existence of a small number of wealthy foundations prepared to support research in the new medical sciences. The United States did not have the kind of strong public support for research at the national level that it would later have. But state and local institutions, like the New York Health

Department, played an important role in fostering such innovations as the use of serum antitoxin to treat diphtheria.

Two of the private organizations that were to take advantage of this network of institutions, and to make commercial use of the new diphtheria therapy were Parke-Davis and Company in Detroit and H.K. Mulford Company of Philadelphia. Both firms established laboratories in 1894-95 and worked with public health authorities as they acquired the ability to produce and standardize the new antitoxin. While Parke-Davis had the advantages of scale and scope, Mulford was the first to bring the antitoxin to market in the spring of 1895. Subsequently Mulford applied the same therapeutic principles to the development of a variety of antitoxins and vaccines. The company brought out a tetanus antitoxin (human and veterinary), smallpox and rabies vaccines, and serums for anthrax, dysentery, brucellosis (also called at that time "melitensis," or undulant or Malta fever), meningitis, pneumonia, and streptococcal infections.

In the years prior to World War I, Mulford also developed a variety of "bacterins" and "serobacterins" that were advertised as "sensitized bacterial vaccines." Bacterins employed killed bacteria (antigens) in an effort to produce active immunity by stimulating the body of the recipient to form antibodies. The serobacterins combined antigens and the antibodies derived from an immune serum with the objective of providing both short-term and long-term therapeutic action against a specific disease. Biologicals of this sort were developed for treating everything from acne and hay fever to cholera and typhoid fever.

During this burst of innovation in biologicals, Mulford had only a small number of competitors in the US market. In 1912, for instance, there were six firms producing antityphoid vaccine. In addition to Parke-Davis, they included the National Vaccine and Antitoxin Institute; Lederle Antitoxin Laboratories; the Cutter Laboratory; Burroughs, Wellcome & Co.; and the Swiss Serum and Vaccine Institute, whose American agents were the Pasteur Laboratories of America.

All of the companies in this industry produced a number of products whose efficacy was not substantiated by reliable clinical evidence. In these early days of bacteriology and immunology, the knowledge base was so narrow and the techniques of clinical testing so rudimentary that no one had much reason to be restrained in experimenting with such new products as bacterins. There was a theoretical basis for their use in the pre-war years. Bacterins were thought to produce an increase in the 'opsonins,' serum substances that prepared invading bacteria for their engulfment by white blood cells, that is, for 'phagocytosis.' In later years this theory would be questioned, and in the late 1930s the use of the therapy would decline (only to be revived again after new forms of bacterial vaccines were developed in the seventies). But at the turn of the century, these therapies were used extensively by physicians who had few alternatives.

It was not until 1902 that the United States established any sort of national regulatory agency. Even then, the powers of the US Public Health Service were circumscribed. The Hygienic Laboratory of the Service's Division of Pathology and Bacteriology inspected the facilities of organizations making these products, and was empowered to test the serums and vaccines for purity and strength. The Service could suspend an organization's license when there were problems with contamination. But the federal government only began to monitor and control the efficacy of pharmaceuticals and biologicals in 1962.

Mulford achieved competitive advantage by dint of its early entry into biologicals and by establishing a strong reputation for the quality of its products and the scientific standards of its facilities. The firm employed well-established bacteriologists (including Dr. Joseph James Kinyoun, former head of the Washington laboratory of the US Public Health Service) to ensure that its procedures were up-to-date. Its reputation eased relations with the government's new regulatory body and became the leitmotif of the firm's advertising and its organizational culture.

Mulford's innovations included the establishment of a substantial capacity for production and a distribution network that used both branch offices and agencies throughout the United States and overseas. Mulford did not make any unique contributions to the basic or applied science of antitoxins and vaccines. Moreover, the firm's system of batch production employed standard laboratory techniques on only a slightly enlarged scale, and there were thus few opportunities to achieve economies of scale. Nevertheless, its production and distribution facilities were significant additions to the nation's nascent biological industry. These aspects of the business were especially important in a country that was as large as the United States and that had such a variety of public health institutions. By 1918, the company had offices or depots in fifteen cities and was also selling its biologicals in nine other countries. During the First World War, Mulford was a major supplier of diphtheria antitoxin to the US military.

This network's immediate economic impact on the nation (à la Schumpeter) was not particularly important. The entire pharmaceutical industry was only a small part of the chemical sector, and biologicals were only a small part of pharmaceuticals. The total value of all fine chemicals, which included pharmaceuticals, was less than 10% of the value of chemical products in 1909.

Nonetheless, the spread of this network of ideas, institutions, and scientific leaders from Europe to the United States and the elaboration of the network through the activities of such firms as Mulford and Parke-Davis were important. The development of the bacteriological network altered American medical science and education in dramatic ways. Bacteriology became the basis for a more scientific approach to the understanding and treatment of disease. This transformed such leading institutions as the Johns Hopkins Hospital, the Medico-Chirurgical College in Philadelphia, and the New York City Health Department. The Health Department in New York established a Division of Pathology, Bacteriology, and Disinfection to conduct research in this new field, and other public health organizations followed suit. Gradually, these innovations reshaped medical practice and laid the foundation for a therapeutic revolution that following the Second World War would make pharmaceuticals and biologicals one of the nation's important growth industries.

By the First World War, however, the pace of innovation in Mulford's network, and in the firm itself, had slowed substantially. It had accelerated quickly in the 1890s, continued at a high rate well into the early 1900s, and then produced very few new products or ideas thereafter. The industry leveled off, entering a relatively stable phase. This cyclical pattern would be repeated. Understanding these patterns, and responding creatively to them, would become a major problem for the managers of private firms and public institutions alike as they attempted to keep their organizations on the cutting edge of industry and biomedical science.

Sharp & Dohme and the Next Cycle

The second chapter in our narrative of innovation provides an excellent example of the problems private firms encountered when two such cycles overlapped. Here we focus on the Sharp & Dohme Company, which purchased H.K. Mulford Company in a merger consummated only months before the US economy began to collapse into the Great Depression. Sharp & Dohme was a Maryland-based pharmaceutical firm with a well-developed distribution system but virtually no experience in biologicals, and with limited research capabilities. In 1930, the company employed only three full-time scientists.

During the early years of the depression, Sharp & Dohme marked time. But in the mid-thirties the company set out to transform its R&D and product line. Under new professional managers, the company increased the number of full-time scientific personnel to 140 by 1952. Between 1940 and 1952, the research budget (which sometimes included medical division funds) jumped from $184,000 to $1,589,000, a figure comparable to the R&D expenditures for such British firms as Glaxo and Wellcome. By way of comparison, Merck & Co., Inc., had an R&D budget of $865,000 in 1940 and $5,525,000 in 1952. As this contrast suggests, Sharp & Dohme was still lagging behind the frontline of research-oriented firms in the US industry, but it had by the early fifties clearly enhanced its research capabilities.

Sharp & Dohme achieved its major goals in R&D, while taking advantage of only one of the two new networks taking shape in medical science. In the thirties and forties, the company followed the industry's leaders by emphasizing the network forming around advances in medicinal chemistry. Sharp & Dohme concentrated with considerable commercial success on research on sulfonamides and formulations of penicillin. In the wave of enthusiasm for the revolutionary antiinfectives and antibiotics associated with this network, many thought there would no longer be any need for vaccines and serum antitoxins. Indeed, one American firm withdrew two effective pneumonia vaccines because it thought there would be no future demand for them. Thus, Sharp & Dohme devoted only limited attention and research to the second network emerging at this time: virology.

FIGURE 13.1: THIS ADVERTISEMENT APPEARED IN THE *BULLETIN OF THE MEDICAL AND CHIRURGICAL FACULTY OF MARYLAND* IN FEBRUARY OF 1912. COURTESY MERCK & CO., INC., WHITEHOUSE STATION, NEW JERSEY, U.S.A.

FIGURE 13.2: AN ADVERTISEMENT FOR MULFORD'S DIPHTHERIA ANTITOXIN (COURTESY WILLIAM H. HELFAND). COURTESY MERCK & CO., INC., WHITEHOUSE STATION, NEW JERSEY, U.S.A.

FIGURE 13.3: THE MULFORD LABORATORIES, C. 1920. COURTESY MERCK & CO., INC., WHITEHOUSE STATION, NEW JERSEY, U.S.A.

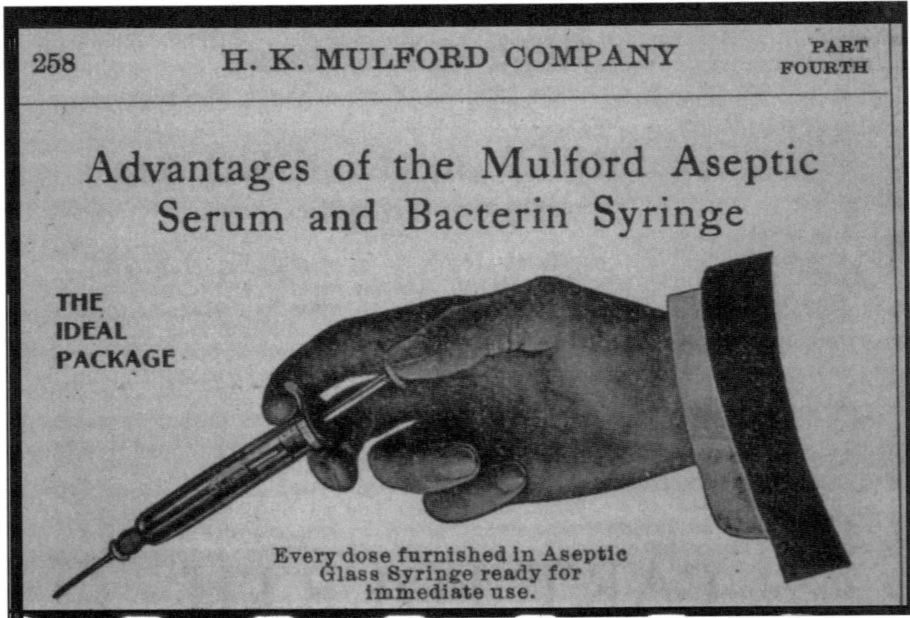

FIGURE 13.4: AN ADVERTISEMENT FOR THE MULFORD SYRINGE. COURTESY MERCK & CO., INC., WHITEHOUSE STATION, NEW JERSEY, U.S.A.

Since 1898 . . . This Expert Has Produced Smallpox Vaccine Mulford

HERE Dr. William Franklin Elgin, Director of the Mulford Smallpox Vaccine Laboratory, and his assistant are grinding vaccine calf lymph. Dr. Elgin has probably made more smallpox vaccine than any other person in the world—almost one hundred millions of vaccinations since 1898! Under his direction, the Mulford Biological Laboratories have pioneered in the production of smallpox vaccine—from the old-fashioned dry ivory points, dry glass points, glycerinized glass points, to the Mulford Improved Capillary 'Tube-Points' (scarified-applicator, Mulford).

"FOR THE CONSERVATION OF LIFE"

FIGURE 13.5: SHARPE & DOHME KEPT THE MULFORD ORGANIZATION INTACT. THIS ADVERTISEMENT APPEARED IN THE 1944 SHARP & DOHME SALES CATALOGUE. COURTESY MERCK & CO., INC., WHITEHOUSE STATION, NEW JERSEY, U.S.A.

Ampoules Are Subjected to Rigid Bacteriological Control

SHARP & DOHME ampoule solutions are prepared with all the care of biologicals . . . with sterility established by actual test before being placed in stock. In fact, the entire filling and testing processes are conducted by our Mulford Biological Laboratories operating under government license. This includes a seven-day incubation test for sterility. A special release must be obtained from the Biological Laboratories, in addition to the approved chemical analysis of the Analytical Control Laboratory, before ampoules can be marketed. Sterility and safety are assured.

"FOR THE CONSERVATION OF LIFE"

FIGURE 13.6: A SHARPE & DOHME ADVERTISEMENT STRESSING THE FIRM'S QUALITY CONTROL. COURTESY MERCK & CO., INC., WHITEHOUSE STATION, NEW JERSEY, U.S.A.

Famed "Park 8" Strain of Corynebacterium Diphtheriae . . . Used Here Since 1894

FIRST to produce Diphtheria Antitoxin commercially in the United States, the Mulford Biological Laboratories today continues to develop its diphtheria products from this famous "Park 8" strain of Corynebacterium diphtheriae.

In the illustration above, pure cultures of diphtheria bacilli are being transferred to Fernbach flasks where they grow more readily because of the larger surface. Growth takes place only on the surface, the toxin produced during growth dissolving in the medium. Later, by filtration process, all bacteria are removed, leaving the clear amber liquid which is termed diphtheria toxin.

"FOR THE CONSERVATION OF LIFE"

FIGURE 13.7: FIFTY YEARS AFTER DR. PARK GAVE MULFORD ITS ORIGINAL STRAIN OF DIPHTHERIA BACILLI, SHARPE & DOHME WAS FEATURING "PARK 8" IN ITS ADVERTISING. COURTESY MERCK & CO., INC., WHITEHOUSE STATION, NEW JERSEY, U.S.A.

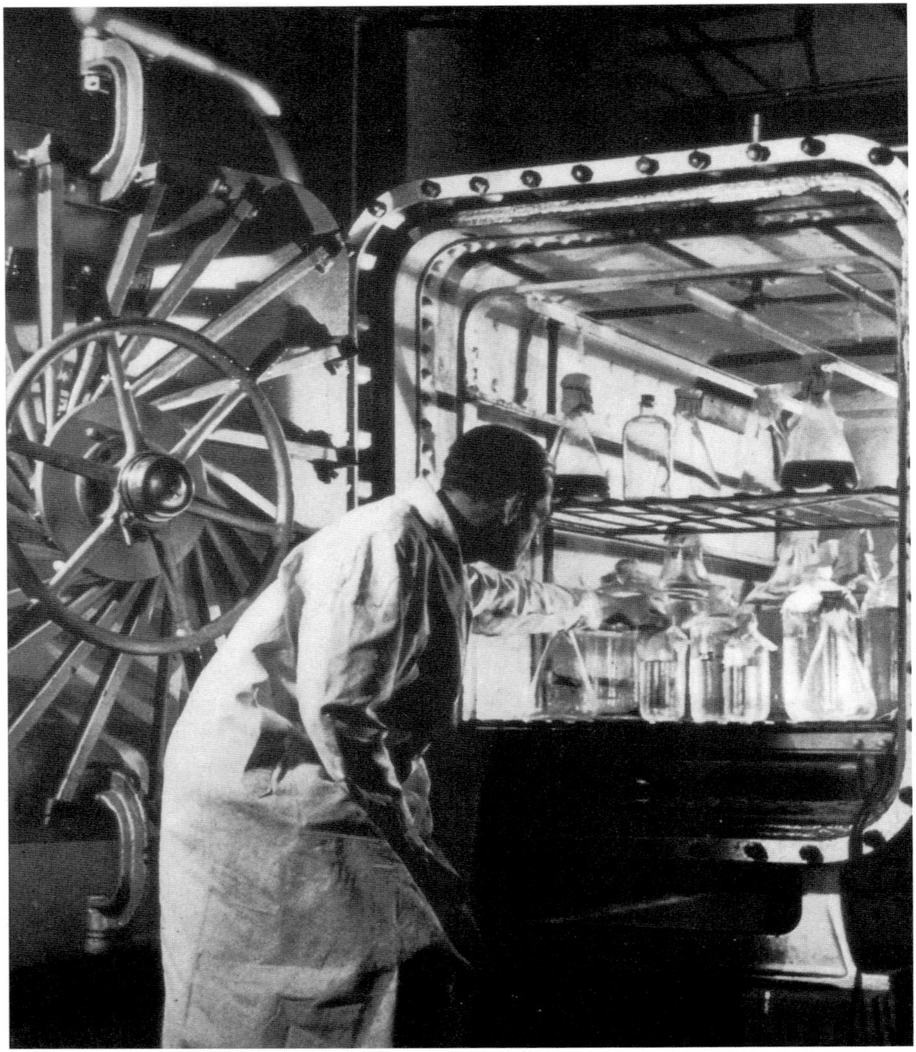

FIGURE 13.8: SHARPE & DOHME'S BIOLOGICAL LABORATORY IN THE 1930S. COURTESY MERCK & CO., INC., WHITEHOUSE STATION, NEW JERSEY, U.S.A.

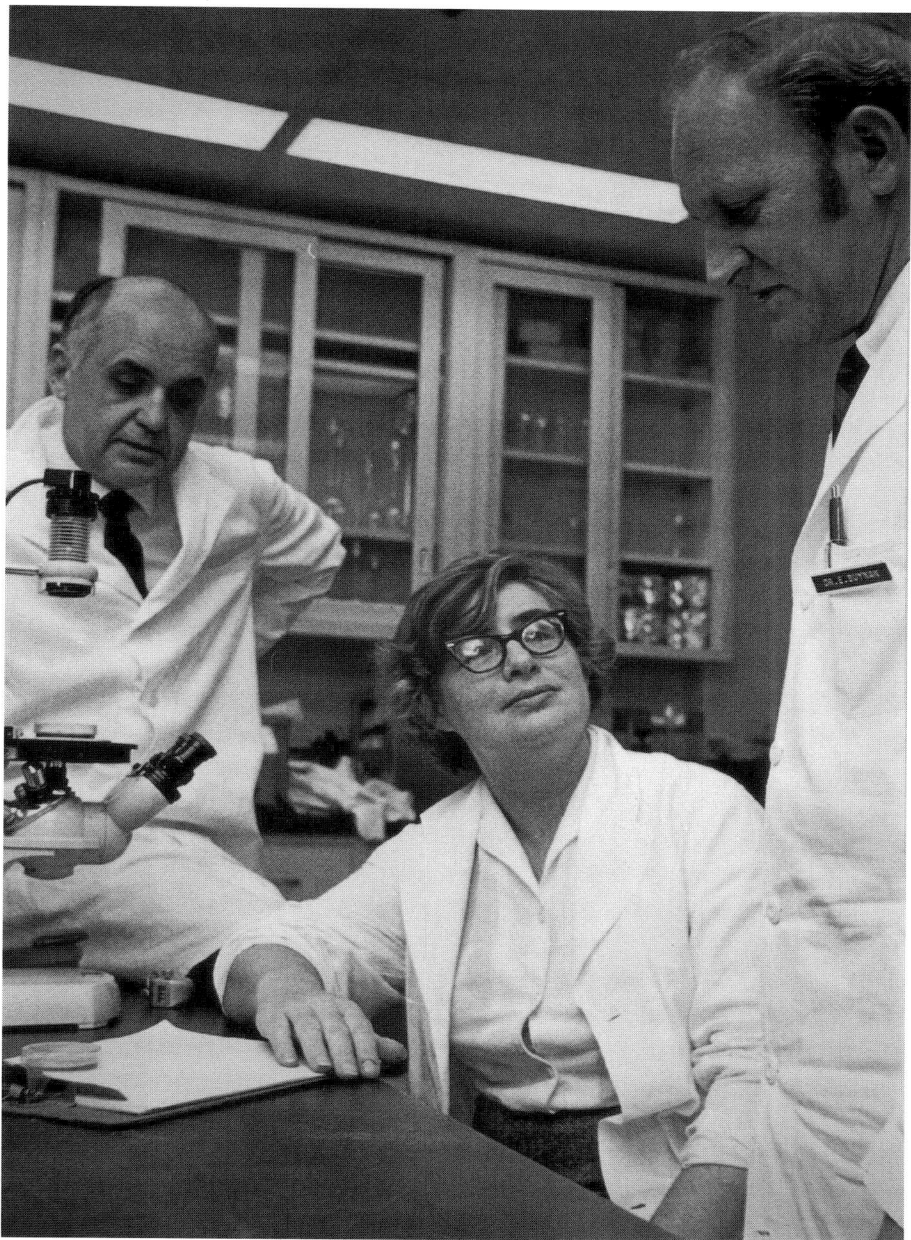

Figure 13.9: Left to right: Dr. Maurice R. Hilleman (Executive Director of Virus and Cell Biology Research), Dr. Beverly Jean Neff (Senior Research Virologist), and Dr. Eugene B. Buynak (Director of Viral Immunology Research) review test results of Merck's vaccine against Marek's disease. Courtesy Merck & Co., Inc., Whitehouse Station, New Jersey, U.S.A.

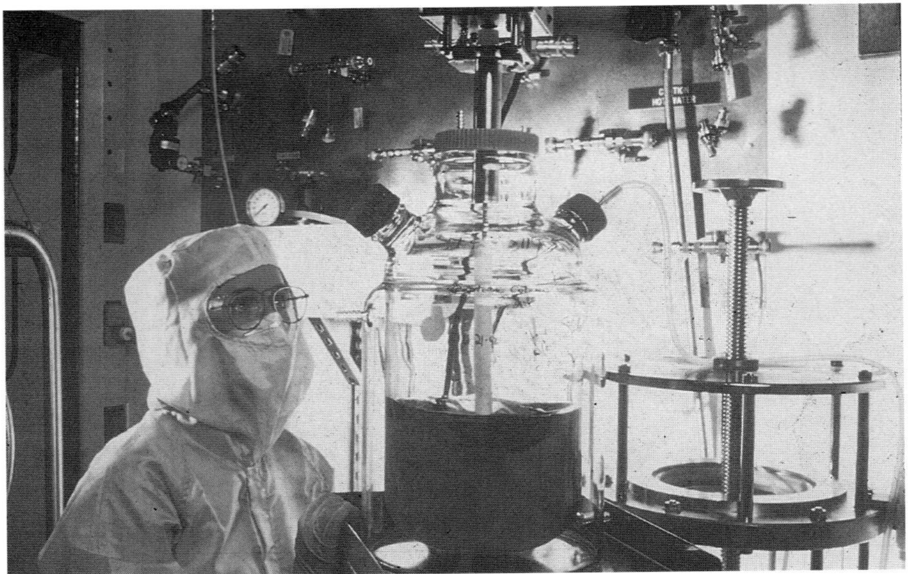

FIGURE 13.10: ENGINEER JULIE CAMBURN ENTERS DATA ON MERCK'S HEPATITIS A VACCINE. COURTESY MERCK & CO., INC., WHITEHOUSE STATION, NEW JERSEY, U.S.A.

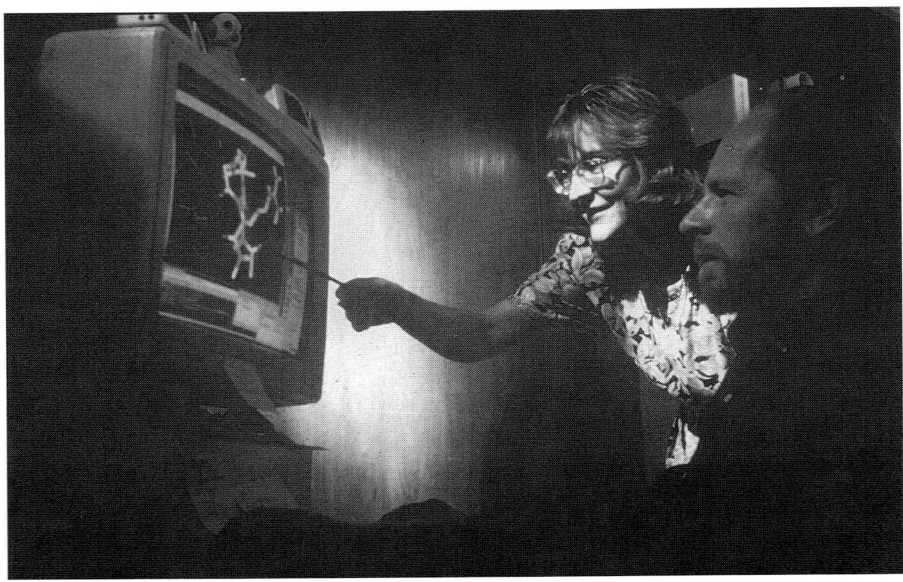

FIGURE 13.11: DR. DENNIS UNDERWOOD AND DR. KRISTINE PRENDERGAST OF THE MERCK RESEARCH LABORATORIES USE COMPUTER MODELING TO DEVELOP FUTURE VACCINES. COURTESY MERCK & CO., INC., WHITEHOUSE STATION, NEW JERSEY, U.S.A.

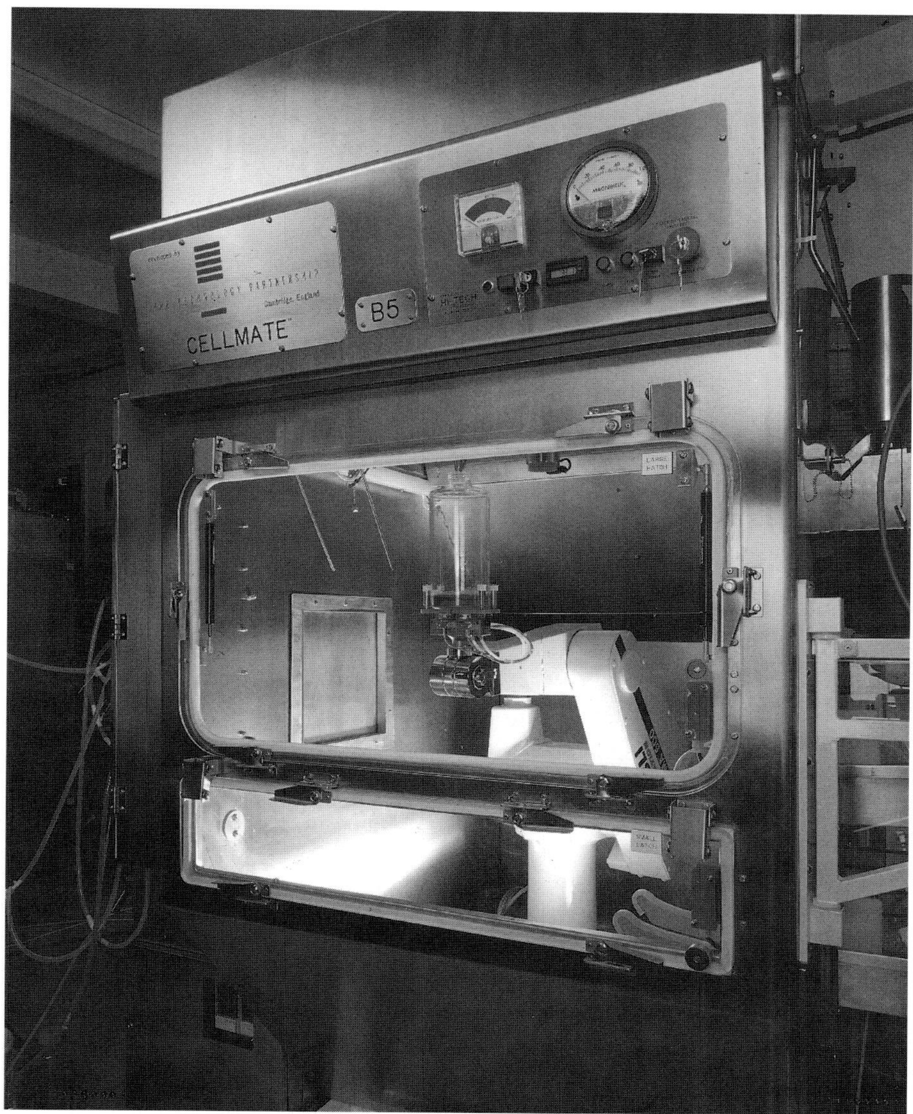

FIGURE 13.12: ROBOTS USED AT MERCK & CO., INC., TO PRODUCE THE VACCINE AGAINST CHICKEN POX. COURTESY MERCK & CO., INC., WHITEHOUSE STATION, NEW JERSEY, U.S.A.

When Sharp & Dohme's biological laboratories finally developed a new vaccine, it was at the behest of the United States Army during World War II. This episode was symptomatic in several ways of how the new virology-centered network was constituted and would evolve during the post-war era. National institutions such as the US Army and later the NIH played a far more important role than they had during the bacteriological cycle earlier in the century. Wartime exigencies brought important advances in antibiotics (penicillin), antimalarials, and other areas of pharmaceutical and biological research. The involvement of academic, government, and industrial leaders in the mobilization of medical research also seemed to offer a useful lesson to those concerned about the allocation of national resources. As a result of the wartime experience, national organizations provided more of the funding for research and later became far more important as regulators and distributors of the industry's products. This was the case in Europe as well as the United States.

Sharp & Dohme became directly involved with the virology network after US Army scientists produced a new influenza vaccine. The Army was concerned about the potential for an epidemic. The memory of what had happened in the great pandemic of 1918–1919 (when between 21 and 25 million had died, over half a million in the United States) lent urgency to their efforts.

Building on the progress that had been achieved in identifying the strains of influenza virus and charting their epidemiology, US Army researchers developed (1943) a formalin-inactivated influenza vaccine. Subsequently the military contracted with several firms, including Sharp & Dohme, to provide the services with this new bivalent (types A and B), 'killed-virus' vaccine. Working under wartime pressure, Sharp & Dohme was able to provide the Army with 20,000 shots that were used successfully during a 1944 epidemic.

Although the government licensed the firm to sell the product for civilian use, Sharp & Dohme was neither able to create an adequate market for the product nor improve it substantially in the immediate post-war period. The vaccine's major problem was that it caused fevers or other reactions. Its 'reactogenicity' was a product of the relative impurity of the vaccine, and neither Sharp & Dohme nor the other companies involved were able to solve that problem.

They might have been able to work out a solution to this technical problem and others if they had devoted more resources to virology, but as late as the early 1950s, companies such as Sharp & Dohme had not developed significant capabilities in this emerging field of science. Important advances were being made during these years, and the opportunity was there. The electron microscope, which had been developed prior to the war, became widely available in the 1940s, and this instrument at last enabled scientists to see the viruses they were studying. As of 1949, they could also propagate viruses in cell culture, outside of a living host, and during the fifties acquired a new understanding of how DNA governs and viruses transform the cell. The result was a burst of scientific research on the prevention of viral diseases. But to capitalize on these important innovations, companies like Sharp & Dohme needed to make substantial investments in biological research.

At this time, Sharp & Dohme placed its investment dollars elsewhere. It did not exploit the opportunity the Army had provided in order to establish a strong position in the virology network ahead of the rest of the industry. It opted instead to remain focused on organic chemistry and in particular on fermentation, areas of research in which companies like Pfizer, Eli Lilly, and Merck & Co., Inc., were discovering new antibiotics, steroids, and vitamins. Glaxo made a similar decision at this time. Sharp & Dohme's laboratories began to conduct research on kidney functions and continued to give substantial attention to its established business in blood products, while keeping its vaccine operations on hold.

It is not at all clear that a company the size of Sharp & Dohme could have afforded to maintain active links with more than one network. That was all that Mulford had been able to achieve at the turn of the century. The price of entry into a new network (like the one emerging in virology) was relatively high: the company would need scientists who were 'at the tip' in this new area of research; new facilities; science administrators to guide the internal efforts; and science 'diplomats' to maintain relations with other institutions in the network. In 1950, the firm spent slightly over $1.2 million on research expenses and was building a new $4 million research center. That year, R&D expenditures were 2.8 percent of net sales (comparable to Glaxo and less than Merck) and 22.6 percent of net income. A strong commitment to virology would probably have required Sharp & Dohme to double its research budget. That would have reduced the income per common share from $4.27 to $3.30: hardly a welcome outcome for the firm's stockholders.

Thus Sharp & Dohme let the opportunities in virology slide past. Not until it merged (1953) with Merck & Co., Inc., a substantially larger company, would the combined enterprise make the heavy investments needed to become a leading innovator in vaccines.

Merck Establishes a Leading Position in the Virology Network

This was a marriage of opposites. The Rahway-based Merck & Co., Inc., was the nineteenth century offspring of a German fine-chemical producer, and it still retained some of the values and operations inherited from its parent firm. An independent, American-based company since 1919, Merck had maintained a strong reputation for the quality of its reagents, coal-tar derivatives, narcotics, iodides, and other medicinal products. Since the 1930s, the company's R&D had also acquired a substantial reputation for cutting-edge research in organic chemistry.

During the Great Depression, Merck transformed itself into one of the industry's most innovative organizations. In 1929 the firm recruited Randolph Major, a young organic chemist from Princeton University, to lead its new R&D program. He in turn recruited Karl Folkers from Yale (where he had done postdoctoral work with

Treat Johnson, an eminent biochemist), Max Tishler from Harvard (where he had studied with E.P. Kohler and J.B. Conant, two of the leading *organiker* of the day), and Lewis Sarett from Princeton. Together these men and their associates generated a significant body of original research on the vitamins, sulfonamides, antibiotics, and antiinflammatories, culminating in the 1940s with Sarett's synthesis of cortisone and Folkers' isolation of vitamin B_{12}. By the early 1950s, Merck was a large (6,400 employees in 1952) and successful (sales of over $160 million that year) fine-chemical manufacturer with a company culture that emphasized science-based innovation.

Still, the firm's future was in danger. Merck sold most of its products in bulk to pharmaceutical manufacturers, who distributed the drugs under their own trademarks. Most of the sales were to a limited number of major firms such as Eli Lilly & Co., Upjohn Co., and Sharp & Dohme. In the forties and early fifties, several of these customers began to produce their own fine chemicals, threatening to close off some of Merck's most important markets. To solve that problem, Merck merged in 1953 with Sharp & Dohme, a pharmaceutical company with major capabilities in marketing to US physicians and hospitals. This merger (downstream, toward the wholesale and retail customers) enabled Merck to protect its markets and, after the two organizations were successfully consolidated, to acquire an even larger market share in the United States and abroad.

The combined enterprise was large enough (total sales in 1956 were over $172 million) to sustain the investments needed to promote research related to several networks in medicinal science, including virology. For three years prior to the merger, Merck had employed Richard Shope, one of the leading scientists in the field, to help establish a program for growing viruses in cultures consisting of different forms of tissue (including human cells). But Shope had quickly tired of commercial research, and between 1953 and 1957, the Merck, Sharp & Dohme Research Laboratories (MSDRL) made only moderate progress in virology, a discipline now attracting tremendous attention because of the vaccines for poliomyelitis. Using tissue cultures, Dr. Jonas Salk and his colleagues at the University of Pittsburgh produced (1954) a vaccine that was effective against all three types of the virus. Merck was one of the companies that contracted to produce polio vaccine for the subsequent trials, but refused to release its product. Even though the virus in the vaccine was inactivated with formalin (it was a 'killed-virus' vaccine), MSDRL was concerned about its safety. It was 1956 before Merck was prepared to release its first shipment of the Salk polio vaccine.

The following year, the company moved decisively to upgrade its virology research. MSDRL started by hiring the scientist/science manager/science diplomat it needed to establish fruitful ties with the work being done in this fast-changing, very complex international network. Dr. Maurice R. Hilleman was, Merck's executives thought, "a great scientist *and* a great mover." Hilleman had been at the Walter Reed Army Institute of Research in Washington, D.C., serving as Chief of the

Department of Respiratory Diseases. The Institute was a major center for research in virology, and Hilleman had spent ten highly productive years under the tutelage of Dr. Joseph Smadel, who headed the Division of Communicable Diseases. During those years, Hilleman had been a codiscoverer of the respiratory viruses now known as adenoviruses, had devised diagnostic tests for and defined the clinical features of the diseases caused by three forms of the virus, and had developed an effective killed-virus vaccine. In addition to this breakthrough in pediatric and military medicine, Hilleman had done work of fundamental importance on the manner in which influenza viruses change their antigenic form. During his tenure, Walter Reed had become a major center for the study of influenza epidemiology. Although diplomatic language was not his common mode of communication, Hilleman was certainly well qualified to play the special role that Merck had in mind. For one thing, he had experience in the pharmaceutical industry, having worked for E.R. Squibb & Sons before going to Walter Reed.

Hilleman quickly reoriented the virology program at MSDRL. He eliminated an effort to find antiviral agents through traditional chemical research and concentrated most of his group's resources on a major exploration of three childhood diseases. Building on the tissue-culture techniques developed by the Nobel Prize winner Dr. John F. Enders and his colleagues, the Merck team was able by the early 1960s to produce a measles vaccine suitable for clinical trials. Hilleman's vaccine used an attenuated (that is, weakened, not killed) form of the virus. In this case and others, Merck worked closely with Dr. Joseph Stokes, Jr., of the Children's Hospital, Philadelphia, in conducting the necessary clinical studies. In 1963, after substantial developmental research on control procedures and safety tests, the US government's Division of Biologics Standards (DBS) issued Merck a license for *Rubeovax*, the first innovation to emerge from its revitalized virology program.

The second was a mumps vaccine on which the firm's research was of central importance and the Hilleman contribution particularly crucial. In this case Hilleman's daughter, Jeryl Lynn, provided the mumps virus needed to start the process of attenuation by successive passages through tissue cultures. By 1965, the product was ready for clinical testing and then for large-scale field trials. Two years later, the government licensed this live-attenuated virus vaccine, which Merck marketed under the tradename of *Mumpsvax*.

In 1969, the US government licensed Merck's third vaccine, *Meruvax*, for rubella (German measles). This particular virus was not isolated until 1962, and during the following three years, a major epidemic prompted significant public, governmental, and commercial interest in developing an effective vaccine. Anticipating that another epidemic would follow in the 1970s, Merck and the DBS cooperated in solving the problems of attenuating the live virus so as to reduce its 'reactogenicity' to acceptable levels. Clinical tests in 1968 indicated that the vaccine was both safe and effective, and the company had 600,000 doses ready for distribution when the DBS issued its license. Two years later, the firm had on the market a combination of these three vaccines, *M-M-R*, which provided protection against all three diseases with one injection.

Between 1957 and 1971, Merck's virology program had established an important position in the national and international networks that emphasized the development and distribution of new vaccines. In contrast to Mulford's earlier cycle of innovation, Merck was now making significant, original contributions to the basic science, as well as the applied science and technology of vaccine production, and the distribution system for these products. By making the substantial investments required to internalize skills in vaccine development and clinical research, Merck had developed new capabilities that promised to pay dividends for many years to come.

Distribution, however, was a growing problem. In the United States (which used about one-half of the world's supply of vaccines) large public immunization programs became a major element in the market following the successful polio campaign in the fifties and sixties. From the perspective of the private firms that supplied the vaccines, these programs simultaneously introduced a new element of monopsony and greater liability. Profit margins were under increasing pressure at a time when heavier investments in research and development were needed to remain competitive. In Merck's case, the company was also funding emergency programs like those which accompanied the periodic influenza epidemics and pandemics in the 1960s and 1970s, including the troubled episode with swine flu vaccine in 1976.

These kinds of economic pressures transformed the US industry in the seventies. A number of major vaccine producers, including Pfizer and Eli Lilly, left the market entirely, leaving Lederle, Parke-Davis, Wyeth, and Connaught as Merck's major competitors. Although Merck was now the largest vaccine producer in the world, it was still unable to penetrate European markets. They were dominated by European companies that were favored by their national governments and health programs. The big three were Pasteur and Mérieux in France and Behringwerke in Germany, followed by a number of firms which produced primarily for their national markets: Burroughs-Wellcome and Glaxo in the United Kingdom; Sclavo in Italy; and RIT (acquired in 1968 by Smith, Kline and French) in Belgium.

Despite the economic problems of the seventies, Merck continued to invest in its virology program while pressing forward with important innovations in several other human therapies and animal health. By the middle of the decade, the firm was posting annual sales of about $1.5 billion and was spending over $122 million a year on research and development. It was by that time one of the top five pharmaceutical companies in the world (in sales). Merck was maintaining and improving its ties with a number of networks (including those involved with cardiovascular and antiinflammatory research) something neither Sharp & Dohme nor Mulford could have financed. Meanwhile, it was exploring to good effect new ways of preventing bacterial infections through immunization. Out of this research came vaccines (using the polysaccharide capsule of the bacteria) against meningitis, pneumococcal pneumonia, and *Haemophilus influenzae* type b.

New Networks and Global Transformations

While to all outward appearances, Merck in the early seventies was a successful, growing firm, and MSDRL was an outstanding R&D organization, the company was on the edge of another difficult transition. The forty-year reign of medicinal chemistry and pharmacology was coming to an end. In 1969 Max Tishler had handed the reins of MSDRL over to another organic chemist, Lewis Sarett, who presided over the early phases in the development of several promising new compounds. But none were blockbuster commercial successes. The heyday of medicinal chemistry had passed.

The industry was entering a new era, one in which networks integrated by the biochemistry of the cell and later by genomics and biotechnology would produce most of the breakthrough drugs. The United States was pumping hundreds of millions of public dollars into the basic research that promoted these new networks, both of which had experienced rapid expansion during the sixties. At a time when many US firms in other industries were failing to make the transitions that would enable them to deal effectively with intense global competition, Merck once again acquired the new leadership and research strategy it needed. It introduced a new style of targeted biochemical research under the leadership of Dr. P. Roy Vagelos, the president of MSDRL from 1976. Targeted R&D, which was grounded in microbial biochemistry and enzymology, looked for specific molecules that would interfere with mechanisms of disease by preventing one of the biochemical transitions essential to the disease process. Traditionally, most drugs had been discovered by random screening of chemicals to determine if they showed activity against a disease. While biochemical targeting did not entirely supplant screening, the new strategy occupied more and more of the laboratories' time and support. Under Vagelos the efforts of all of MSDRL's research teams were focused on the most promising therapies, and coordination between research and marketing was strengthened.

Merck's vaccine program might well have been terminated during this transition. As we noted before, margins in vaccine markets were tight, liability was a problem, and the new biochemical research based on enzymology was unusually promising. The global vaccine business only amounted to about $200 million dollars a year in a pharmaceutical industry that was on the verge of developing single products (for example, *Vasotec,* a Merck product to treat high blood pressure, or *Zantac,* a Glaxo product for ulcers) that would have annual sales of over a billion dollars. An economic analysis conducted in the late seventies concluded that "no company not in the vaccines business would (logically) choose to get in. Most companies currently in it would get out if they could, or had the courage to, or had other options."

MSDRL had "other options" that were very attractive, but Vagelos protected the vaccine program and used it to establish links with other new networks. Biochemical targeting, combined with insights from structural biology and computer modeling, was now known as rational drug design, and it had become the dominant approach to R&D. But the next cycle of innovation, Vagelos thought, would be based on newly emerging networks in genomics and rDNA technology. One means of establishing a presence in those networks was by keeping the vaccine program, a decision based more on medicine and science than economics. That decision emphasized long-term prospects more than short-term profits, and in this regard, his strategy was consistent with the dominant culture at Merck.

With MSDRL continuing to provide substantial support for virology, Hilleman's team was able to bring to a successful conclusion an extended effort (1968 to 1981) to develop a vaccine for hepatitis B. Since they derived the virus from plasma, however, *Heptavax B* could never provide a solution to the global problems created by this disease. The firm could not locate a sufficient supply of plasma to satisfy the need, nor could it bring the price down to a level that would permit extensive vaccination where it was most needed. Meanwhile, the beginnings of the AIDS pandemic generated fears that any plasma-based product might be dangerous.

Vagelos then turned to the new networks for assistance. Since MSDRL lacked the kind of scientific capabilities it needed in the fields of molecular genetics and rDNA technology, the company worked out a collaborative research program to employ recombinant DNA technology to produce hepatitis B antigen. Merck collaborated with Dr. William Rutter of the University of California: with Chiron, a small biotech firm; and with a University of Washington research team. Meanwhile, MSDRL built up its internal capabilities in these new fields of science and technology. Leadership was provided by Vagelos and Dr. Edward Scolnick, an eminent molecular geneticist recruited from the NIH. Scolnick became president of MSDRL (1985), when Vagelos became the firm's Chief Executive Officer. The initial product (1986) from their program was *Recombivax HB*, the world's first vaccine for human use made using recombinant technology.

In order to build on this scientific foundation, the company organized a new Merck Vaccine Division in 1990, led by Dr. R. Gordon Douglas, Jr., a leading infectious disease specialist recently recruited from the Cornell University Medical College. The new division tightened the coordination between research, marketing, and manufacturing. By that time, the field of competitors in vaccines had again narrowed significantly. Now there were only four leading companies in the international market. The leader in global sales was Mérieux, which had absorbed Connaught BioSciences and established an alliance with Pasteur Vaccins; followed by Merck; Lederle Praxis Biologicals; and SmithKline Beecham. These four firms now did over 70 percent of the world's business in vaccines.

The Merck Vaccine Division responded to this situation and to the need for additional combined vaccines by establishing a series of strategic alliances and joint ventures. These included alliances with Institut Mérieux and its US subsidiary,

Connaught Laboratories Limited; with Pasteur Mérieux Serums et Vaccins; and with Commonwealth Serum Laboratories in Australia and New Zealand. In this and other regards, the global network in virology/vaccines has been restructured, changing dramatically the for-profit, corporate components and the processes of innovation in this sector of preventive medicine.

By pulling through the economic crisis of the seventies and making the transition to the new genomic network, Merck and its three major global competitors were positioned to make full use of the unusual opportunities for innovation being generated in the nineties by molecular genetics and rDNA technology. Now viruses could be engineered to carry a portion (a peptide) of a pathogen, thus stimulating the host to develop a broad immunity to the disease. While efforts to use the new technology to produce an AIDS vaccine have been unsuccessful because of the mutability of the human immunodeficiency virus, several new approaches to vaccine development appear to be opening a front for innovation even broader than the one stemming from the virology of the 1950s and 1960s.

Innovation in Firms, Networks and National Systems

The pharmaceutical firms that have been successful innovators in this field have maintained close links between industry, academe, and government. This has been as true in France, Germany, and Britain as it has in the United States. These organizations have read the signals coming from their relevant scientific networks and responded to the cyclical patterns of change by periodically finding new leaders and adopting new strategies of growth. This was true in the 1890s, when the focus was on the germ theory of disease; in the 1930s, when Merck, Lilly, Squibb, and Abbott all reorganized their research efforts to take advantage of new networks in organic chemistry and pharmacology; and in the 1970s, 1980s and 1990s when hundreds of biotech startups and the established pharmaceutical firms began to exploit the emerging networks in rDNA technology and genomics.

Organizational cultures and leadership played important roles in these transitions. Successful pharmaceutical companies developed cultures emphasizing scientific innovation and took a long-term perspective on performance. Success over the long-term also depended upon the presence of astute managers who could mediate the many links between the firm's internal capabilities and the relevant networks. Where firms have failed to respond creatively to changes in their networks, an active market for corporate control has provided an effective mechanism for transferring resources: thus Sharp & Dohme acquired a lagging Mulford; and Merck upgraded its virology research after merging with Sharp & Dohme.

At the national level of analysis, our study indicates how important public investment in basic science can be to the process of innovation in the private, for-profit sector. The post-World War II national system in the United States clearly promoted a high level of successful innovation in this industry. Support for research and professional training spurred network expansion, providing new opportunities for firms able to develop the appropriate leadership and internal capabilities. The public investments paid off over the long-term even though no single innovation had an impact on the national economy that would excite the followers of Joseph A. Schumpeter. Nevertheless, those firms that successfully navigated through what Schumpeter called the "storm of creative destruction" became major national assets in the global pharmaceutical industry.

Over the past century, a series of loosely integrated networks associated with bacteriology, virology, enzymology, rDNA and genomics, were the most decisive institutions shaping innovation in pharmaceuticals and biologicals. Each of these networks evolved along a cyclical path that created first unusual opportunities and then problems for both the public and the corporate organizations involved. In vaccines, networks not directly involved with biologicals also influenced the process of innovation by providing alternative investments in the public as well as private sectors. By the 1990s, only a few of the world's major pharmaceutical firms had the resources necessary to continue promoting innovation in biologicals while exploiting the much larger economic opportunities provided by the networks relevant to pharmaceutical innovation.

FURTHER READING

R.P.T. Davenport-Hines and Judy Slinn, *Glaxo: A History to 1962.* (Cambridge: Cambridge University Press, 1992). Rolv Petter Amdam and Knut Sogner, *Wealth of Contrasts: Nyegaard & Co. — A Norwegian Pharmaceutical Company, 1874–1985.* (Oslo: Ad Notam Gyldendal, 1994). Jeffrey L. Sturchio (Ed.), *Values & Visions: A Merck Century.* (Rahway: Merck & Co., Inc., 1991).

Robert Ballance, János Pogány, and Helmut Forstner, *The World's Pharmaceutical Industries: An International Perspective on Innovation, Competition and Policy.* (Cheltenham: Edward Elgar, 1992).

David L. Cowen and William H. Helfand, *Pharmacy: An Illustrated History.* (New York: Abrams, 1990).

Harry F. Dowling, *Fighting Infection: Conquests of the Twentieth Century.* (Cambridge: Harvard University Press, 1977).

Louis Galambos, with Jane Eliot Sewell, *Networks of Innovation: Vaccine Development at Merck, Sharp & Dohme, and Mulford, 1895–1995.* (New York: Cambridge University Press, 1995).

Alfonso Gambardella, *Science and Innovation: The US Pharmaceutical Industry During the 1980s.* (Cambridge: Cambridge University Press, 1992).

Jonathan Liebenau, *Medical Science and Medical Industry: The Formation of the American Pharmaceutical Industry.* (Baltimore: Johns Hopkins University Press, 1987).

Jonathan Liebenau, Gregory J. Higby, and Elaine C. Stroud, eds., *Pill Peddlers: Essays on the History of the Pharmaceutical Industry.* (Madison: American Institute of the History of Pharmacy, 1990).

Richard Nelson, ed., *National Innovation Systems: A Comparative Analysis.* (New York: Oxford University Press, 1993).

Stanley A. Plotkin and Edward A. Mortimer, Jr., eds., *Vaccines.* (Philadelphia: W.B. Saunders Company, 1994).

David Schwartzman, *Innovation in the Pharmaceutical Industry.* (Baltimore: Johns Hopkins University Press, 1976).

Arthur M. Silverstein, *A History of Immunology.* (New York: Academic Press, Inc., 1989).

Jane E. Smith, *Patenting the Sun: Polio and the Salk Vaccine.* New York: William Morrow and Company, 1990).

John P. Swann, *Academic Scientists and the Pharmaceutical Industry: Cooperative Research in Twentieth Century America.* (Baltimore: Johns Hopkins University Press, 1988).

CHAPTER 14

The History of Electronics

From Vacuum Tubes to Transistors

JOHN PETER COLLETT

Few other fields could serve better to illustrate the many-faceted nature of twentieth-century science and its complex inter-relationship with technology than the field of electronics. In a standard dictionary, electronics is defined as "the study, design and use of devices that depend on the conduction of electricity through a vacuum, gas or semiconductor."[1] The key word in this definition is "devices." The history of electronics is about the development of such devices and the core inventions on which they depended, the electron tube (or valve) and the transistor. Electronics is a prime example of the 'science-based' technologies that characterize modern industrialized society, and its origin and subsequent development has depended on the development of a range of scientific disciplines. However, when it comes to distinguishing the technology part of the history from the scientific, we soon run into trouble. In electronics, science (if we define it as an activity aimed at explaining in theoretical terms the properties of nature) is inextricably linked with technology, when we define this as an organization of knowledge and skills aimed at solving the practical undertakings of mankind. Scientific understanding has sometimes been ahead of, and sometimes lagged behind the technical devices that have been developed. Indeed, electronics can only be explained as the outcome of particular forms of organizing the development of knowledge that exceeds the categories of science and technology.

Electronics is often described as a revolutionary technology. It has in the course of the twentieth century profoundly changed the every-day life of practically everyone, through the development of radio and television. The development of the electronic computer has revolutionized industrial production as well as public and private administration. Electronic devices have brought about fundamental changes in warfare. As the basis of information technology, electronics has been crucial to a transformation of the modern world, often conceived as comprehensive enough to qualify as a shift from Industrial Society into Information Society. As a by-product, it has brought about a revolution in the way in which scientific research is conducted, through providing new research instruments ranging from the electron microscope to the artificial satellite, and it has created a new field of computer science.

None of these electronic devices existed at the beginning of the twentieth century. They derive from two basic inventions, the electron tube and the transistor. The first of these inventions date from the beginning of the century: J.A. Fleming's diode from 1904; Lee de Forest's triode from 1906–07; the transistor, from the mid-century through the discovery by a group working at Bell Laboratories made public in 1948. From a technological point of view, the development of electronics thus falls in two separate epochs, which coincide with a periodization according to scientific criteria: the first epoch based chiefly on the principle of thermionic emission of electrons in a vacuum; the second on the conductive properties of certain materials called semiconductors.

The electronics technology of the two epochs offered both possibilities and constraints. Its revolutionary character derives from the way the possibilities have been exploited and the constraints overcome. But in order to illuminate this, I shall relate the development of electronics to its social, economic, political and legal context.

The Era of the Individual Inventor: The Invention of the Electron Tube

The point of departure would be the last decade of the nineteenth century, when wireless communication, the embryo of electronics, was established. The possibility of transmitting and detecting signals via electromagnetic waves had been proven through experiments undertaken in the years 1884–93 by Heinrich Hertz. At that time, Hertz was working in the institutional setting of a German university. His experiments were part of physics research, aiming at experimentally testing the theories concerning the existence and properties of electromagnetic waves advanced by James Clarke Maxwell several decades earlier. Hertz showed no interest in pursuing the potential practical use of his findings but, as was required by the academic scientific community in which he worked, published his discoveries in the appropriate periodicals which were available to the general public.

The way in which Hertz' experiments were turned into technical devices was through the intervention of a group of individuals belonging to a category of people central in the development of nineteenth-century technologies — the inventors. Inventors were the mediators between science and technology, and cannot easily be pigeonholed in either category. They were familiar with the world of academia and were often professionally linked to universities, and their inventions could be the outcome of theoretical research. However, they were aiming at a use of their findings that exceeded contributing to the extension of human knowledge. An inventor would be a person who would apply for a patent, a legal protection of his right to the benefits deriving from the application of his particular finding in a commercial setting. Even when working outside of the academic world, inventors were able to grasp opportunities that science unfolded, by exploiting the free flow of information from academic research. The early development of radio communication provides us with a clear example. The twenty-year old Italian Guglielmo Marconi read about Hertz' experiments in an electrical journal and immediately started developing equipment for radio communication in his parents' house. Three years later, in 1897, the British Marconi Company was founded with the aim of commercially exploiting the principle of wireless telegraphy, with Marconi's patents on various devices as its main asset.

The attempt to develop commercially a system for wireless telegraphy spurred a race among numerous inventors who shared Marconi's belief in wireless communications as a field of great economic potential, and who had access to the same scientific and technical information. A host of patents were registered in Europe and in the United States by inventors and companies who wanted to have their share of a promising market. Among these was a young American, Lee de Forest, holding a Doctor's degree from Yale University, who in 1901 founded the De Forest Wireless Telegraph Company. Its aim was to compete with Marconi's American subsidiary, established in 1899, by developing its own system for ship-to-shore communication.

A working wireless communications system would consist of a range of equipment enabling the generation, emission and reception of signals. It would be a technical system with several components. While most of the inventors had rushed to file patents covering separate components, Marconi had from the outset aimed at getting a functioning system and had secured rights to all its necessary parts. In this, he was only matched by the Germans, where the imperial government had intervened to end conflict over patent rights between the two competing systems and had enforced the amalgamation into one (Telefunken). Marconi's aim was to control the technology of spark telegraphy, and in this his strategy was successful in so far as no competitor was able to put up a working system without infringing at least one patent. This was what happened to de Forest's company. In 1911, it was taken over by Marconi after having lost a case over patent infringement. Lee de Forest had already been fired from his own company in 1906.

De Forest and other American radio pioneers were inventors in the tradition of the American inventor-entrepreneurs, matching inventive genious with commercial entrepreneurship. They were emulating Alexander Graham Bell and Thomas Alva Edison. Edison, with his research laboratories functioning as patent factories, pouring out patent applications over a wide range of appliances, was obviously a source of inspiration and imitation. However, the radio pioneers were never able to copy the success of their eminent predecessors in building complete technical systems. They failed in the field of spark telegraphy, and they would fail again in the efforts to construct a working system of continuous-wave radio.

The thought of replacing the established spark telegraphy with a system of wireless communication using continuously emitted waves started as an engineer's dream at the beginning of the century. Such a system would overcome the obvious shortcomings of spark telegraphy, and, notably, open the road to wireless transmission of the human voice. Several solutions were suggested to the problem of generating and detecting continuous-wave signals. One if these was the 'audion,' a device patented by de Forest in January 1907. The following month, de Forest formed the De Forest Radio Telephone Company, so there could be no doubt about his own conviction that his new device would make the dream of radio voice transmission come true. However, he would not have been able to foresee the revolutionary potential of his invention, described by I.I. Rabi in 1945 as "so outstanding in its consequences that it almost ranks with the greatest inventions of all time."[2]

De Forest predicted the audion's use as a detector of radio signals. It was the first three-element vacuum tube or triode and was based on the same principles as the diode, the first and simplest type of electron tube, developed for the Marconi company by J.A. Fleming in 1904. It was based on the 'Edison effect' discovered by T.A. Edison in the early 1880s, by which he had shown that a current would flow from a heated element in a light bulb to a cold electrode. Edison had taken out a patent for this in 1884 as a possible way of regulating and measuring the flow of electric currents. Otherwise he had found no use for it. Fleming, who had been working with Edison, took up the idea as a means to ameliorate the detecting device of the wireless telegraphy system, commonly regarded as the weakest point of the system.

The devices which the principal inventors designed functioned in ways that could only be partially explained in theoretical terms. Though the principles of conductivity had been subjected to academic research — through his experiments with Cathode rays, J.J. Thomson was able to explain the Edison effect as thermionic emission of electrons in 1897 — the lack of full understanding was a drawback as was the inability to produce a reliable triode. In 1912 de Forest and his business associates were facing fraud charges, the audion being described by the prosecutor as a "device [that] had proven worthless."[3]

The first problem for de Forest and the other American radio inventors was their inability to assemble a working system containing all necessary components. The social and legal setting was barring this through patenting practice. By the outbreak of World War I, patent litigation had created a complete stalemate. The United States Navy, when ordering radio sets for its ships, found that there was no single company which "possessed basic patents sufficient to enable them to supply, without infringement, . . . a complete transmitter or receiver."[4]

The individual inventors had been instrumental in exploring the possibilities of radio-wave communication and a flow of inventions had followed. However, they were unable to continue their inventive activity into the construction of complete working systems. Constraints imposed by patenting were partly responsible for this, but there were also constraints imposed by lack of capital, and the facilities and staff necessary for turning the patented devices into functioning instruments.

The Era of the Industrial Laboratory: The Struggle for Control of Radio

The years immediately before World War I saw the eclipse of the individual inventor as driving the development of electron tube technology. Their role was taken over by new actors: the large industrial research laboratory. It was in this setting that the potential of the electron tube was developed. Simultaneously, the difficulty of constructing a technological system of radio communication was overcome by the emergence of a powerful protagonist — the state at war.

The industrial research laboratory was a result of the emerging complex technological systems and a means by which large companies could exert the control over them. The companies whose research laboratories would contribute most to the development of electronics in the electron tube era — the General Electric Company (GE) and the American Telephone & Telegraph Company (AT&T) — were working in the field of lightning and communications, respectively. Both were recent technologies, based on late nineteenth-century inventions in electricity. GE was the commercial continuation of Thomas Edison's activity, while AT&T was the outgrowth of Alexander Graham Bell's invention of the telephone. Both companies were controlling large technical systems serving huge markets, and their control rested for a large part on the exploitation of basic patents. With work going on outside the companies, and with academic research in the field providing would-be competitors with free information, the companies had reason to fear losing control if competing inventions were patented by others. The companies established their in-house research laboratories as a counter-measure. These were to serve as 'patent factories'; aiming at developing patentable devices that would serve to strengthen control over their technologies.

The two companies entered the radio field and took up research on electron tubes. GE had started doing radio research in 1903 and became interested in the development of electron tubes as a means of improving its high-frequency alternator. In 1912, the laboratory started work de Forest's audion. In the GE laboratory, there was both theoretical knowledge updated by academic research on conductivity, and expertise on the design and production of vacuum tubes, which constituted the basis for GE's production of electric bulbs. The industrial laboratory could thus provide the combined scientific and engineering capability that was to prove necessary to develop the first electronic devices beyond the level of curiosities. The GE engineers realized that the functioning of the triode demanded a vacuum, and the company was able to produce vacuum tubes with the necessary accuracy to assure a length of use many times longer than the first pitiful prototypes of the audion. GE's research also led to the development of the triode as an oscillator. Through these developments, the triode emerged as the key component of a functioning voice-carrying radio system. At the same time, AT&T acquired the audion patent rights from the unfortunate de Forest, and the company's laboratories started studying its potential.

Unlike GE, AT&T was not interested in the triode as a radio component, but as a device for amplifying telephone signals. AT&T exerted a hegemony on the United States telephone network and had a keen interest in strengthening control of long-distance telephony. The scientifically skilled staff at AT&T's Bell Laboratories conducted the research necessary to grasp the principles of the electron tube. The aim of the laboratories was to control technology. An important way to obtain and to strengthen this control was through the understanding of the basic principles of the technology. Only by assuring a grasp of the basis of the technology could the company be sure to foresee developments that could endanger its market position.

The research undertaken in GE and Bell Laboratories would inevitably lead to further court battles over patent infringement. The stalemate was complete when a United States court decision in 1916 ruled that Marconi's Fleming patent on the diode and de Forest's triode patent were mutually infringing, and both parties were denied the right to produce their tubes. The stalemate was broken by the needs of war. In need of radio equipment for the United States Navy, the government decided to lift all restrictions on the use of patents relating to radio.

The subsequent development of electron tubes during World War I under the state of emergency declared by the United States government was crucial in the development of the voice-carrying radio as a functioning system, and it was decisive for assuring the position of the electron tube as the hegemonic component in transmitters and receivers. It was driving out alternative technical solutions, like the arc and alternator as wave generators and the crystal receiver. The massive demand by United States government for electron-tube radio equipment had turned the electron tube into a standardized mass-produced device, as part of standardized radio receivers and transmitters. At the end of the war, these developments had created both a large production capacity for tubes and radio sets, and a large potential market of ex-servicemen with war-time experience in operating radio sets.

The United States government continued its concern for radio developments into peace-time. Facing the possibility a foreign private company (Marconi) obtaining a de facto monopoly on a system for inter-continental radio connection, the government decided that this would be contrary to the nation's vital strategic interests. In order to avoid this, it arranged for the holders of the different radio patents to form a consortium — the Radio Corporation of America (RCA) — and to enter an intricate agreement on the cross-licensing of patents. The aim of these arrangements was to enable the creation of an American radio communication system marketed by RCA using equipment manufactured by GE, while assuring AT&T's control of its core markets.

The RCA agreement was an effort by the companies holding the patent rights in radio technology to maintain control. In one respect, the effort was fruitless as radio was soon transformed into an activity unforeseen by the contracting parties. By 1920, Westinghouse, the third partner of the arrangements, started operating a radio 'station,' serving the 'amateurs' that had purchased radio receivers. Radio broadcast was born and was soon to form the basis of a radio production which was one of the fastest-growing industries in the world. Sales of home radio receivers in the United States reached 2 million by 1925, 10.5 million by 1939. Europe was soon catching up, establishing broadcasting services in the early 1920s. By the end of 1929, 3 million listening licenses had been granted in the United Kingdom and approximately the same number in Germany.

This rapid growth in the output and sales of radio sets was accompanied by a succession of product innovations, which perfected the radio as a means of receiving undistorted spoken and musical entertainment in the home. RCA was forced by market demand and by government anti-trust measures to grant licenses to many producers of radio and receiving tubes. This, in turn, led to fierce competition and price cutting that allowed a further expansion of the market. Nevertheless, tube production at the end of the inter-war years was still dominated by a few companies in the United States, Europe and Japan. These were large integrated companies, producing tubes in-house as components and parts of equipment they themselves were manufacturing and selling. Through patents, price agreements and market sharing these manufacturers had been able to establish a fairly stable oligopoly and deter the destructive effects of competition.

The struggle for control of the radio market had given rise to further innovative activity. Competition spurred innovation, yet, at the same time, contributed to relative conservatism. Patent litigation was used to keep new entrants out of the field and stabilize the competitive position between the oligopolists. No less than 1500 infringement suits were recorded in the United States concerning radio patents in the years up to 1941, the bulk of these after the introduction of radio broadcasting. These figures demonstrate that the holders of the patents were scrupulously defending their position, and doing what they could to keep radio as a technical system firmly within the boundaries defined by the patents they were holding. Consequently, innovation in the radio industry concentrated on incremental perfection of the system rather than on radical changes in technology. Radio as a means of mass communication and as a household commodity was in rapid growth, while at the same time radio as a technical system was stabilized. Radio sets were becoming an increasingly standardized mass-produced consumer good. Electron tube producers controlled radio and linked it to the basic technology they controlled. The exception that proves the rule was the invention of frequency modulation (FM) which was made outside of the radio industry and was much resented by the insiders. It was virtually forced upon the United States radio industry through government intervention.

The major radio companies, especially RCA, preferred to concentrate their innovative efforts in the 1930s on the development of television, where the potential return on investments in basic technology was regarded as more promising than in the highly competitive radio manufacturing industry. Innovating activity was abandoning a stable technology and heading for a technology still in the making. The forecasts were fully justified. Although interrupted by World War II, research on television was to bring about a revolutionary new technical system of mass communication, which was also based on the use of electronic devices; the cathode-ray tube constituting the key element. In its development from initial uncertainty as to the choice of technical principles, and a proliferation of inventors advocating alternative solutions, into a phase of massive research in industrial laboratories, and then to a striving for stabilization in an oligopolistic setting, the television story is in many ways a replay of the radio story.

While broadcast radio was on its way to stabilization, much effort was expending in exploring the possible use of electron tube technology for other purposes. Whereas radio development in the hands of the oligopoly could hardly be expected to yield more than continued perfection of the established technology, the other applications of electron tubes were to provide the foundations of a further technological revolution by rapidly challenging the limits of the tube technology and thus leading to a search for alternatives. Two such applications were to be particularly significant. The first was in long-distance telephony. The other was in the development of a system for detecting aircraft and ships — the radar.

The Search for Alternatives: The Beginning of Semiconductor Research

By the 1930s Bell Laboratories had developed into the largest industrial research laboratory in the world, and also won recognition for important scientific work beyond the practical demands of the mother company. For example, in 1937, Clinton J. Davison, a Bell Labs scientist, was awarded the Nobel Prize in Physics.

By 1936, the Bell director of research, Marvin Kelly, had come to the opinion that in the telephone network of the future, the existing mechanical connections (relays) would represent a bottle-neck halting further expansion, and believed that they should be replaced by equipment based on electronics. How clearly articulated this belief was at the time, has been debated. Nevertheless, Kelly decided to set up a research program to find an alternative to the vacuum tube, and, specifically, to find solutions that would eliminate the tubes' obvious shortcomings — the large power they demanded and the heat they generated. The alternative approach was to develop equipment based on the conductive properties of certain solids called semiconductors. The scientist that was assigned to carry out Kelly's program was William Shockley.

The research project was of an undeniably long-term character and involved great risks. However, the Bell Labs scientists were able to draw upon theoretical and experimental research that had been carried out on the properties of semiconducting materials for several decades. In fact, the use of semiconductors in radio communications preceded the introduction of the electron tube. In the

Figure 14.1: Marconi Wireless Station. The Lizard. 1901. © GEC Marconi

Figure 14.2: Marconi Wireless Station. Cape Cod, USA. 1901.© GEC Marconi

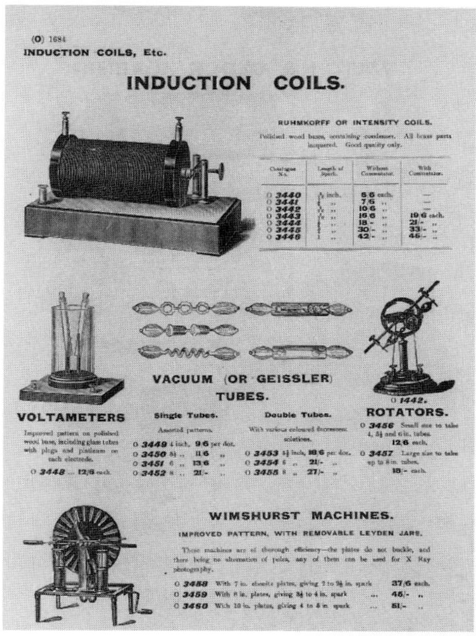

(O) 1684
INDUCTION COILS, Etc.

INDUCTION COILS.

RUHMKORFF OR INTENSITY COILS.

Polished wood bases, containing condenser. All brass parts lacquered. Good quality only.

Catalogue No.	Length of Spark.	Without Commutator.	With Commutator.
O 3440	[illegible] inch.	6/6 each.	—
O 3441	[illegible] ,,	7/6 ,,	—
O 3442	[illegible] ,,	10/6 ,,	—
O 3443	[illegible] ,,	16/6 ,,	19/6 each.
O 3444	[illegible] ,,	18/- ,,	21/- ,,
O 3445	[illegible] ,,	30/- ,,	33/- ,,
O 3446	1 ,,	42/- ,,	45/- ,,

VACUUM (OR GEISSLER) TUBES.

VOLTAMETERS

Improved pattern on polished wood base, including glass tubes with plugs and platinum on each electrode.

O 3448 ... 12/6 each.

Single Tubes.

Assorted patterns.

O 3449 4 inch, 9/6 per doz.
O 3450 5½ ,, 11/6 ,,
O 3451 6 ,, 13/6 ,,
O 3452 8 ,, 21/- ,,

Double Tubes.

With various coloured fluorescent solutions.

O 3453 5½ inch, 18/6 per doz.
O 3454 6 ,, 21/- ,,
O 3455 8 ,, 27/- ,,

O 1442.

ROTATORS.

O 3456 Small size to take 4, 5½ and 6 in. tubes. 12/6 each.

O 3457 Large size to take up to 8 in. tubes. 15/- each.

WIMSHURST MACHINES.

IMPROVED PATTERN, WITH REMOVABLE LEYDEN JARS.

These machines are of thorough efficiency—the plates do not buckle, and there being no alternation of poles, any of them can be used for X Ray photography.

O 3458 With 7 in. ebonite plates, giving 2 to 2½ in. spark ... 37/6 each.
O 3459 With 8 in. plates, giving 3½ to 4 in. spark ... 45/- ,,
O 3460 With 10 in. plates, giving 4 to 5 in. spark ... 51/- ,,

Figure 14.3: Induction Coils "GEC Catalogue 1905–6". © GEC Marconi

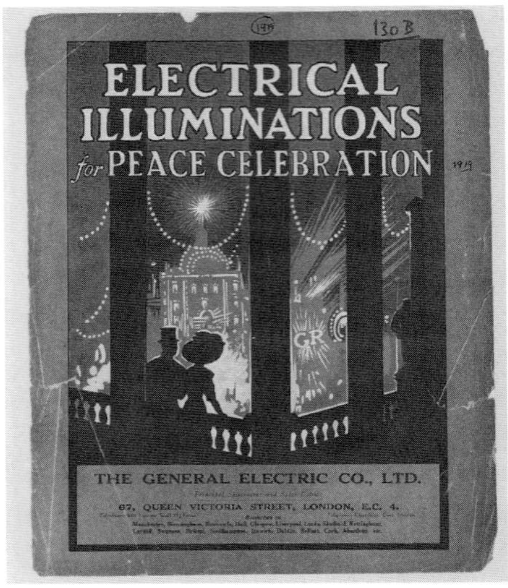

FIGURE 14.4: FRONT COVER "ELECTRICAL ILLUMINCATIONS 1919 FOR PEACE CELEBRATIONS". © GEC MARCONI

FIGURE 14.5: AMPLIFYING VALVES. LEFT TO RIGHT: V 24 C. 1922 (1ST HIGH FREQUENCY VALVE); S 625 C. 1926 (1ST PRODUCTION SCREEN-GRID VALVE); VMS 4 C. 1933 (VARIABLE GAIN VALVE); Z 63 1940S (INTERNATIONAL BASE PIN LAYOUT-OCTAL); Z 77 C. 1955 (SIMILAR FUNCTION TO Z 63 BUT MORE EFFICIENT). © GEC MARCONI

FIGURE 14.6: HIRST RESEARCH CENTRE (GEC) WEMBLEY. © GEC MARCONI

FIGURE 14.7: 500KW DEMOUNTABLE CONTINUOUSLY-EVACUATED THERMIONIC VALVE INSTALLED AT RUGBY IN 1932 FOR THE G.P.O.; THE DISCOVERIES EMBODIED IN THIS VALVE MADE LONG DISTANCE RADAR TRANSMISSION POSSIBLE.
© GEC MARCONI

FIGURE 14.8: TRANSMITTER/RECEIVER IN SHORT "SHETLAND" AIRCRAFT 1154/1155.
© GEC MARCONI

FIGURE 14.9: GEC "MUSIC MAGNET 4" RADIO.
© GEC MARCONI

FIGURE 14.10: TYPE 158 MODULATOR, 1947.
© GEC MARCONI

era of spark telegraphy, the usual detector of radio signals was the 'cat's whiskers' (now called a point-contact crystal diode). The crystal detector was a typical product of the inventive early radio days, and was not to be given any theoretical explanation until the 1930s, by which time it had been replaced in radio communications by tubes. Compared with the electron tube, the cat's whiskers suffered from a decisive short-coming: it could not be used as an amplifier.

As a field of academic research, the study of the properties of semiconductors advanced during the first decades of the twentieth century. This advance had mostly been of a theoretical character, and was related to basic physics research. With Max Planck's formulation of the Quantum theory in 1900, and Albert Einstein's application of this theory to explain the photoelectric effect, the emission of electrons from solids was a phenomenon attracting theoretical interest. German academic researchers in particular, had made contributions to the theoretical explanation and classification of semiconductors.

When the Bell Laboratories turned to the study of semiconductors as a possible way of replacing the vacuum tube, they were by no means the only ones interested in turning the character of semiconductors to practical use. The aim of such efforts was mostly to construct a solid-state amplifier. Several unsuccesful attempts was made at accomplishing this. One of these was the attempt at Bell Labs, made by Shockley and Walter Brattain in 1938, to produce a semiconductor amplifier using copper oxide. Brattain has downplayed the originality of the idea: "Anybody in the art was aware of the analogy between a copper oxide rectifier and a diode vacuum tube and many people had the idea of how do we put in a grid, a third electrode, to make an amplifier."[5]

In hindsight, it is easy to conclude that the failures were due to lack of material of appropriate quality as well as a lack of theoretical understanding. In general terms, the basic principles of semiconductor theory were fairly well known. It was recognized that semiconductors could be changed from insulators to conductors, or vice versa, by exposure to other materials. However, the distance from theory to practice was still long, and there was no known way of manufacturing semiconducting material of the necessary purity to make a working device possible.

With the outbreak of World War II, the research on semiconductor amplifiers at Bell Labs was discontinued. The United States war effort required a massive mobilization of the scientific expertise. Brattain and Shockley were transferred elsewhere to do research on submarine detection. Other parts of Bell Labs were put at the disposal for the program for development of radar.

Intervention of the State at War: Research on Radar

Radar had been one of the fields where the application of electron tubes had been attempted. The use of reflected electromagnetic waves for the detection of a moving object at a distance had already in 1900 been suggested by Nikola Tesla. In the inter-war years, research was well on its way in many countries on making workable systems for the detection of ships and aircraft. These projects were mostly funded by military authorities. By 1939, Germany, the Netherlands, the United Kingdom and the United States already possessed working radar systems, while parallel work was going on in France, Italy and Japan.

The development of radar was to demonstrate the constraints of the electronic tube technology. In order to make the radar systems as efficient as possible, the researchers wanted to construct oscillators generating signals at a very high frequency. New types of tubes were constructed. The cavity magnetron developed in the United Kingdom, and the United States invention, the klystron, were oscillators capable of generating microwave signals which would be reflected at a much higher resoulution of the target than signals at lower frequencies.

The use of microwave radar signals highlighted a shortcoming of electron tubes: they could not be used as detectors of microwave signals. The reason for this was weaknesses inherent in the physical principles of electron tubes, one of which was that the transit time of electrons between the electrodes in the tube introduced unacceptable time-delays at such high frequences. This led to an intense search for alternative detectors in order to make radar systems operative. The alternative was found in the point-contact crystal rectifier. With the coming of the war, large-scale research was started in Germany, and elsewhere in order to produce large quantities of point-contact diodes of constant quality.

The joint United Kingdom-United States effort in the development of radar was an undertaking of giant dimensions. The Radiation Laboratory at the Massachusetts Institute for Technology was established under government auspices to act as a co-ordinating agency for work carried out at universities, government and private industrial laboratories. It was aimed at developing radar equipment that would give the Allied a decisive advantage against the Germans and Japanese, irrespective of cost. The program is estimated to have cost $2.5 billion; more than the Manhattan project to build the nuclear bomb.

One important outcome of the war time radar research effort was the capacity to produce semiconductor material (silicon and germanium) of very high purity. This was to be crucial for the later developments of semiconductor electronics. Without the methods developed during the war, it would not have been possible to obtain material of a sufficient purity to make a semiconductor amplifier possible.

The Birth of the Transistor — The Outcome of a Unique Institution

With the demobilization of research after World War II, Bell Labs was able to take up its interrupted project for a solid-state amplifier. In the summer of 1945 a sub-department for solid state physics was established. In the reorganized group, Schockley and Brattain were joined by John Bardeen whose speciality in academic research was the theory of metals. The subsequent story is one of continued failure in reaching the defined goal for the project, but also, as an unforeseen outcome, the spectacular success in the invention of the point contact transistor. This breakthrough, established just before Christmas 1947 and made public by Bell on the 30 June 1948, marked the beginning of the era of solid state electronics.

The Bell Labs team did not achieve what they had expected. Shockley had expected to make use of the principle later known as the field-effect transistor. This was what theoretical studies had suggested. By then, theory could reasonably well explain what would happen in the body of semiconducting materials. In practice, however, constructing such a device still proved extremely difficult. Unexpectedly, the group found phenomena related to the surface of semiconductor materials that turned out to be of very great importance. In December 1947 Bardeen and Brattain demonstrated that electrical amplification could take place on the surface of a semiconductor material (they used germanium) by closely spacing two wire electrodes. The effect was unexpected and could not be theoretically explained. This was a cause of embarrassment to the Bell Labs. The invention had to be kept secret for several months, as it was impossible to file a patent application without the establishment of the precise functioning.

The invention of the point-contact transistor was epoch-making. Yet, the first transistor was hardly more than a laboratory curiosity, and it was a challenge of great dimensions to turn it into a working device. Yet, by the beginning of 1948, the principle of solid-state electrical amplification had been proved. What had previously been a project of a long-term and high-risk character could now be turned into a goal-oriented program for the development, production and sale of amplifiers. The Bell Labs created a new organization for carrying on this project and assured the necessary expertise and resources.

It had taken a unique research institution to create the transistor. Bell Labs was the world's largest privately-owned industrial research laboratory. In the early 1950s, its staff numbered 6,900, of whom 2,500 were engineers and scientists. The staff was multi-disciplinary and offered a blend of theoretical and practical skills that turned out to be decisive for the development of the transistor. There was room for academically-oriented fundamental research. However, as an industrial laboratory attached to the AT&T and Western Electric, its aim was still to produce technical innovations of economic interest to the owners. This implied a multi-disciplinary goal-oriented working style that differed from traditional academic institutions. It also assured funding of greater continuity than was possible in most universities.

In fact, a university research group was AT&T's nearest rival to the invention of the transistor. Under the war time mobilization of science for the development of radar, a number of university departments were assigned projects to examine the properties of semiconductors. The Physics Department at Purdue University, under the leadership of Karl Lark-Horovitz, took up work on germanium. Their research resulted in important new insight into its conductive properties and of the theory of semiconductors in general. The Purdue group was also working actively on the development of functioning radar equipment. Had the group taken its practical work further, it might have arrived at the construction of the transistor before the Bell group. However, with the demobilization of the scientific war effort, the Purdue group withdrew from practical experiments and returned to the the traditional academic pursuit of fundamental knowledge. Ralph Bray, who with another graduate student, Seymour Benzer, was to continue the Purdue work on conductivity of germanium, later said that his interest was "in explaining the effect that I was seeing, but I wasn't particularly interested in the idea of a solid-state triode."[6] Their experiments were close to those later undertaken at Bell Labs. However they lacked the incentive to continue in the direction of constructing electronic devices.

Theoretical insight in the properties of semiconductors had been important for the invention of the transistor, but it had not been enough. The first phase of transistor research was characterized by "a curious blend of abstract quantum mechanics and cut-and-try tinkering."[7] The successfull outcome was dependent on the research style of the industrial laboratory, where theory and practice were closely coupled and the former did not necessarily take precedence over the latter. This was another reason why the transistor was not likely to emerge from a university department.

However, the subsequent story of the transistor shows the importance of a theoretical approach to semiconductor electronics. Shockley maintained his conviction that the field-effect transistor would be better than the clumsy point-contact transistor and pursued his idea of a junction transistor. This time, theory preceded practice. Shockley formulated the precise theory of the junction transistor two years before such a device was produced, in 1951.

The distance from invention to practical application of the new electronic device proved longer and more cumbersome than its sponsors in AT&T and the Bell Labs had probably foreseen. The public presentation in 1948 aimed at marketing the new device as a revolutionary replacement of the electron tube. Scientifically, it was greeted as a real breakthrough and earned Brittain, Bardeen and Shockley the Nobel Prize in Physics in 1956. From an industrial and technical point of view, however, the response ranged from enthusiasm to outright scepticism.

Bell stressed the transistor's potential for miniaturization and mass-production. However, it took time to realize this potential. Production of transistors in Western Electric did not really start until 1951. The following year, Bell started campaigning for the transistor, inviting firms to a symposium where the new invention was presented. Part of the campaign was the promise of liberal licensing of patent rights to manufacturing firms that might be interested.

The openness demonstrated by AT&T's leadership, and the eagerness to open access to the transistor technology, was in striking contrast to the early history of the vacuum tube. The reason for this difference is partly AT&T's cautious use of its position as a de facto telephone monopoly. It was important for the firm to maintain an image of a supplier of a public good, showing willingness to share its technology with all potential users. Also, the transistor was not developed as a crucial part of a technical system which the inventing company was anxious to control. On the contrary, it was a discrete device in search of a system where it might be used.

It was not easy to find commercial use for the transistor in its initial phase. It was publicized as a replacement for the vacuum tube, but in practically all appliances it was less suited than its established rival. The first transistors were more expensive than vacuum tubes. They were less reliable. Only in a few appliances would there be full use of its practically sole advantage over the vacuum tube, namely, the small size. The first commercial use of any importance was in hearing aids. In memory of Alexander Graham Bell's deafness, Bell relinquished its rights to licensing fees for the use of transistors for such aids. However, the first transistorized aids were of limited benefit to their users. They contributed to the reduction of their size and weight, but they made more noise than electron tubes and were more expensive.

In the competition with electron tubes, the transistor was for a long time hampered by the double handicap of poor performance and high cost. Not even its inventing firm was able to find much use for it. Not until well into the 1960s were transistors widely used in the telephone network.

Nevertheless, the production and sales of transistors and other semiconductor components showed rapid growth during the 1950s. By 1956, world sales reached $115 million. Part of the explanation is found in the gradual dissemination of transistors into consumer products, as transistor performance was improved by further innovative activity. From 1952, Western Electric was able to produce Shockley's junction transistor which was an important step forward from the first point-contact transistor. Drawing upon expertise in metallurgy and other disciplines, competing firms were also making important contributions. Notably, GE developed the alloy junction transistor (also in 1952). Such transistors could operate at a higher frequency and with higher currents. Also, with the alloy process, mass-production of transistors was made possible.

The production of transistors was taken up, through AT&T's liberal licensing policy, by a number of manufacturing firms.In the first stage, these firms were largely the same firms that formed the oligopoly controlling production of vacuum tubes. Both European and Japanese vacuum tube-producers rapidly took up the production of transistors as a supplement to their product range. Gradually, transistors were replacing or rather supplementing tubes in consumer products like portable radios and television sets.

The consumer market was of marginal importance, however, to the development, in the 1950s, of the transistor as a technical device. The forces that brought this about, which were to propel semiconductor electronics into the forefront of a technical revolution, are to be found in the United States military establishment and its procurement policy.

The Era of the Military-Industrial Complex: Semiconductor Electronics Brought to Triumph

The initial development of the transistor had taken place without defense funding. Bell Labs policy was to avoid defense contracting, and it was able to fund from its own resources the approximately $1 million spent on transistor development from 1945 to 1948. However, Bell was well aware of the potential military interest in its invention, and took pains to demonstrate the first transistor to military authorities before it was announced to the general public.

Unlike the lukewarm reception given the transistor in industrial circles, the military responded with enthusiasm. The Army Signal Corps Engineering Laboratory immediately established its own transistor group, and in 1949 Bell signed its first contract for a study of transistor applications with both the United States Navy, Army and Air Force as partners.

The military authorities were able to spot the advantages that transistors potentially could have over electron tubes. The crucial point was their small size and small power consumption. The transistor fitted into a program of miniaturization of military equipment that had begun in 1930s.

Throughout the 1950s the United States defense agencies were able to pour millions of dollars into the further development of the transistor and for a massive build-up of manufacturing capacity. The total amount spent between 1955 and 1961 is estimated at $66.1 million. The reason for this is to be found in developments at the highest political level.

The mobilization of science and industry that had taken place during World War II had ended with the Japanese surrender. It was generally regarded as an extraordinary measure dictated by the emergency situation. In the early 1950s, however, this situation changed. With the Cold War, and, specifically, the Korean War, the need for a build-up of the nation's military capability again became a matter of national prirority, and the defense agencies were given a free hand to establish close alliances with research laboratories and industrial firms. The United States military became the sponsor of technological developments in fields that were targeted as of vital strategic interest.

Military needs dictated choices. The military needed equipment of high performance under extreme conditions. They preferred standardization to custom-making, and they were only marginally concerned with cost. Industry responded by providing solutions in the form of the silicon transistor (1954) which could operate at temperatures much higher than the previous germanium transistors, and was preferred in aircraft and guided missiles, despite the much higher initial cost.

In 1960 a decisive breakthrough was made in the techniques for producing transistors, through the introduction of the planar process. This is a process whereby a layer of an electrically stabilizing oxide is thermally grown on the surface of a semiconducting material. Through holes etched in this layer suitable impurities is diffused into the semiconductor to produce the desired regions of opposite polarity. It produced components with a flat, or planar, surface with increased reliability. With this process, which rapidly became dominant in the industry, the problem of large-scale mass-production of transistors was solved. As the planar process was better suited for silicon than germanium, it entailed the standardization of the silicon transistor. The transition from germanium to silicon was an important step in the direction of making the transistor a robust mass-produced device suited for a variety of purposes.

Both the silicon transistor and the planar process represented a shift in the semiconductor manufacturing industry. They were not introduced by established electronics firms that were producing transistors alongside their manufacturing of electron tubes. The silicon transistor was the invention of Texas Instruments; the planar process was developed by Fairchild Semiconductor. Both firms were newcomers, with no history in electronics of the tube era and no bonds to an established market for consumer goods.

It was an important step in the development of the transistor that producers outside the established oligopoly emerged as major producers, taking a lead in product and process innovation. Transistor development was impeded as long as the transistor remained a mere by-product of vertically integrated producing firms whose main interest lay in tubes and products in which tubes were used. In the mid–1950s, sales of tubes were still more than ten times greater than that of transistors.

By the late 1950s, the transistor was breaking out of the grasp of the established tube producers. This was a result of a change of policy of United States military authorities. They now favored the new entrants dedicated to the technology that the military wanted. This policy was reinforced by United States anti-trust policy, which in 1956 forced Bell to confine its production of transistors to its own use and the military market. The shift in government policy coincided with a tendency of skilled scientific workpower in the big electronics firms to leave their companies for the smaller companies or to form their own.

Unlike the aftermath of World War I and World War II, the close alliance of defense, science and industry was not loosened after the Korean War. The arms race during the Cold War established a lasting and powerful pact in what President Dwight D. Eisenhower in 1960 was to label the 'military-industrial complex.' The complex comprised science as one of its corner-stones, and the forging of the transistor into a high performance mass-produced industrial product was one outcome of this alliance.

Increased reliance on missiles in the United States defense played an important part in the continued quest for components of ever smaller size and weight; greater reliability and robustness to meet the extreme demands of use in rocketry. The space race that was spurred by the 'Sputnik shock' from 1957, and the subsequent determination of the United States government to place a man on the moon, gave a further impetus to continued miniaturization of electronic devices.

The Age of the Digital Computer: Electronics as the Basis of Information Technology

The integrated circuit marked a milestone in the continued miniaturization process. Throughout the 1950s, transistors were discrete components that had to be linked in circuits. With the integrated circuit (developed by Jack Kilby of Texas Instruments in 1958–59), several components could be linked together on the same piece of semiconducting material. The planar process of production made this invention technically and commercially viable, and from 1963 integrated circuits were taken into use for appliances where small size and weight were crucial. The first application was, like transistors, for hearing aids. Very soon, however, the military took over as the biggest and practically only client — the initial high cost of integrated circuits barring customers other than those who demanded minimum size and weight regardless of cost. Like the discrete transistor, the integrated circuit was carried through the process of development from prototype to commercial product by military demand and through military funding. In the case of integrated circuits the sponsor was a large missile program.

The large upswing in the production and sales of the United States semiconductor producers that had been spurred by military demand, came to an abrupt end in 1961, when military demand leveled out. The industry was facing overcapacity, with fierce competition leading to depressed prices and dwindling profits. The civilian market with a preference for low cost exposed the industry to competition from abroad. Japanese producers were penetrating the United States market from the low end, notably with cheap transistorized radios and television sets that United States producers had regarded with disdain.

The integrated circuit presented itself as the product that could save the United States transistor industry from commercial disaster. Furthermore, it strengthened the United States industry as the world leader in semiconductors, placing it well ahead of competitors in Europe and Asia. The way to success was through an alliance with a powerful partner, which this time was the electronic computer.

Machines for automatic calculations had been a dream for mathematicians and accountants for centuries. All sorts of mechanical devices had been tested. The idea of using electronic devices for a calculating machine had been realized as part of the World War II mobilization of science for war. An Electronic Numerical Integrator and Computer (ENIAC) was made for the United States defense authorities at the Moore School of Engineering at the University of Pennsylvania in 1945. It was a typical example of a device that would not have been constructed but for the state of national emergency. With its 18,000 tubes it was the largest electronic mechanism seen until then, consuming 174 kWatt. Most tube experts gave it little chance of success.

The subsequent history of computers cannot be told here. The important point in our context is the emergence of the electronic digital computer, which stood out as the leader over rival computer designs after several years of uncertainty.

Military demand was again decisive for this outcome. Electronic digital computers offered the high speed and reliability that was required for specific defense purposes. Miniaturization of electron tubes helped reduce size and power consumption, but the computer was one of the appliances that demonstrated their limitations. Tubes burned out; the energy consumption was still significant; the machines were slow.

Towards the end of the 1950s, experimenters started using transistors in computers. Initially, computer scientists were sceptical. Transistors were regarded as unreliable and unable to handle large power or operate at the high temperature in a computer. Through experimental programs where the United States government was heavily involved in initiatiting and funding, the transistor found its way into the computer.

The transistorized digital computer became a commercial product in the late 1950s, with machines produced by newcomer firms like Control Data and Digital Equipment Corporation. However, it was when the established computer manufacturer International Business Machines released its IBM–360 in 1965 that the digital computer made its breakthrough. The computer left the world of scientific institutions and military establishments and became a tool for industry. The number of computers in use in the United States trebled from 1964 to 1968 and by that year had reached 70,000.

As a commercial product for a mass market the computer relied heavily on the use of integrated circuits. They made possible the larger-scale production of reliable computers at a reasonable cost. The further development of computers was, conversely, the driving wheel behind the subsequent evolution of semiconductors and a seemingly unending development towards further integration and miniaturization. Large-scale integration (LSI) started with the introduction of the microprocessor (by Intel) in 1971. LSI in its turn has led to very large-scale integration (VLSI) and the development of ever more powerful components, performing ever more complex functions.

With the microprocessor, a computer was available as a single piece of silicon. This enabled the computer to break the confinements of professional use and entered the mass market in the form of personal computers and a range of consumer products. Through the alliance with the digital computer the transistor technology was brought to fully unleash its revolutionary potential. Brought up by military demand and government funding, the transistor technology had grown up through the integrated circuit and the planar process and was able to find an ideal marriage partner in the computer. The computer addresses a combined professional and mass market which seems both endless in its size and in demand for continued innovations.

Computer technology has penetrated nearly all sectors of modern society. It has provided man with powerful tools for communication and control. Information technology has developed as a means of combining the two. Science has provided a much broader basis for information technology than the hardware technology which has been treated in this chapter. Software technology also draws upon a multitude of disciplines.

Science is not only at the base of information technology. Science is integrated in this technology in a way that makes it extremely difficult to separate the two. The transition from tubes to semiconductors created a new type of scientist-engineer. Indeed, the industrial engineer became a scientist. When the transistor challenged the tube, there was a fundamental change in engineering. The new technology could not be handled through empirical experience, but demanded theoretical calculation. The engineer resembled the university scientist, both in the style of work and the style of life. Engineers challenging company authority and moving to other firms have been instrumental in the 'perennial gales of creative destruction,' to use Joseph Schumpeter's famous phrase, that have been characteristic of the industry. The university scientist has at the same time become like the engineer, and there has been a wide exchange of personnel between industry and academia. California's 'Silicon Valley' with its numerous firms established on the fringe of Stanford University symbolizes this merger of technology and science.

The history of electronics is a history of how scientific knowledge has been used as a means of creation and of control. The electron tube was invented in order to break up an existing monopolistic technical system. In the hands of an oligopoly of large firms, the tube in turn became an instrument for the maintenance of control of a new large system, the radio. This oligopoly was broken through the introduction of a new radical innovation, the transistor. What started as a search for a means to ameliorate the weak part of the system, led to a total reconfiguration of the system which had been established, as well as the opening of new vast perspectives of which we not yet have seen the limits.

The history illustrates the strength of the forces behind the development of modern technology: on one hand, the potential of scientific creativity, on the other hand, the enormous power of large industrial corporations with research laboratories as one of their instruments. Furthermore, the history shows that the combination of forces that have made the development of electronics possible, is for a large part enforced through the intervention by the state, in situations of national emergency during periods of war or preparations for war. War has helped unleash the potentials of new technology by breaking down constraints on its development and bringing about new societal settings where innovations were made possible. Only situations of national emergency have allowed the breaking up of forces that have impeded the development of technology or the combination of separate forces into new alliances which have brought about innovations.

In this way, the history of electronics is a forceful demonstration of how creation and destruction seem to be inextricably linked in all human activities, regardless of time or space.

REFERENCES

1. The New Penguin Dictionary of Electronics. (Penguin Books, 1979), p. 129.
2. I.I. Rabi. "The Physicist Returns from the War". (Atlantic Monthly, Oct. 1945), p. 109.
3. G. Archer. *History of Radio to 1926.* (American Historical Company, 1939), p. 110; quoted in W.R. Maclaurin, *Invention & Innovation in the Radio Industry.* (New York: Macmillan, 1949), p. 84.
4. *Memorandum of Commander Loftin.* (U.S. Navy, 1919), quoted in W.R. Maclaurin, op.cit., p. 105.
5. E. Braun and S. Macdonald. *Revolution in miniature,* 2nd edition. (Cambridge: Cambridge University Press, 1982), p. 37.
6. E. Braun and S. Macdonald, op.cit., p. 38.
7. T.J. Misa. "Military Needs, Commercial Realities, and the Development of the Transistor, 1948–1958," in M.R. Smith (ed.), *Military Enterprise and Technological Change.* (Cambridge, Mass: MIT Press, 1985), pp. 253–287.

FURTHER READING

H.G.J. Aitken. *Syntony and Spark: The Origins of Radio.* (Princeton: Princeton University Press, 1976).

H.G.J. Aitken. *The Continuous Wave: Technology and American Radio 1900–1932.* (Princeton: Princeton University Press, 1985).

J.R. Beniger. *The Control Revolution. Technological and Economic Origins of the Information Society.*

E. Braun and S. Macdonald. *Revolution in Miniature. The History and Impact of Semiconductor Electronics.* Second edition. (Cambridge: Cambridge University Press, 1982).

J. Brooks. *Telephone: The first Hundred Years.* (New York: Harper & Row, 1976).

Electronics. Special Commemorative Issue on the History of Electronics. (April, 1980).

T. Forester (ed.). *The Microelectronics Revolution.* (Oxford: Blackwell, 1980).

T. Forester (ed.). *The Information Technology Revolution.* (Oxford: Blackwell, 1985).

P. Griset. *Les Revolutions de la communication.* (Paris: Hachette, 1991).

T.P. Hughes. *Networks of Power: Electrification in Western Society 1880–1930.* (Baltimore: Johns Hopkins University Press, 1983).

T.P. Hughes. "The Evolution of Large Technological Systems" in W.E. Bijker, T.P. Hughes and T. Pinch (eds.). *The Social Construction of Technological Systems.* (Cambridge Mass: MIT Press, 1987).

S.W. Leslie. *The Cold War and American Science. The Military-Industrial-Academic Complex at MIT and Stanford.* (New York: Columbia University Press, 1993).

W.R. Maclaurin. *Invention & Innovation in the Radio Industry.* (New York: Macmillan, 1949).

F. Malerba. *The Semiconductor Business. The Economics of Rapid Growth and Decline.* (Madison: The University of Wisconsin Press, 1985).

S. Millman (ed.). *A History of Engineering and Science in the Bell System: Physical Sciences (1925–1980).* (Murray Hill, New Jersey: Bell Laboratories, 1983).

T.J. Misa. "Military Needs, Commercial Realities, and the Development of the Transistor, 1948–1958" in Merrit Roe Smith (ed.) *Military Enterprise and Technological Change: Perspectives on the American Experience.* (Cambridge Mass: MIT Press, 1985).

A.H. Molina. *The Social Basis of the Microelectronics Revolution.* (Edinburgh: Edinburgh University Press, 1989).

P.R. Morris. *A History of the World Semiconductor Industry.* (London: Peter Peregrinus, 1990).

L.S. Reich. *The Making of American Industrial Research. Science and Business at GE and Bell, 1876–1926.* (Cambridge: Cambridge University Press, 1985).

S. Sturmey. *The Economic Development of Radio.* (London: Gerald Duckworth, 1958).

Chapter 15

Science in Public Policy

A History of High-Level Radioactive Waste Management

Frans Berkhout

Science plays many roles in public policy, but its most pervasive applications have been in the field of health and environmental regulation. Scientific instruments are frequently required to detect environmental pollutants, scientific evidence and reasoning is used to define the hazard represented by pollutants, and scientific claims typically underpin the formation and enforcement of regulations which control environmental hazards. However, 'regulatory science' is frequently characterized by uncertainty and disagreements, partly because, as Jasanoff makes clear, its goal is to provide 'truths' relevant to policy, rather than 'truths' of originality and significance which is formally the aim of research science. While scientific claims are essential to the legitimacy of many regulatory decisions, the true importance of science in informing those decisions is often ambiguous. This gap between the perceived (or ideal) role for science in public policy ('speaking truth to power'), and actual experience has been the focus of much debate.

In the public policy arena, science is put to work in a variety of ways. It may be used directly: in analysing and understanding problems; instrumentally to assess alternative ways of dealing with them; and rhetorically to justify decisions. Science may also be used indirectly as an instrument of policy: to delay policy; as an instrument in the centralization of decision-making; as a diplomatic tool in the defence of sovereignty; and as a means of clarifying the conflicting interests of major participants in a policy-making process. In the process of serving policy interests, scientific activity is itself transformed.

Regulatory science agendas are, to a significant extent, not internally generated, but determined by the interests of dominant groups in the policy-making process. The relative independence of science agendas will depend on the nature of policy which is being informed, and on the 'style' of policy-making (whether adversarial or consensual, for instance). Typically, science plays a role which is subservient to the resolution of what frequently become social and political conflicts. This is an uncomfortable position for scientists who habitually seek to extend the range of their autonomy.

A history of science in public policy is therefore a history of several different types of science: 'core' science which is able to control its own research agenda, 'mandated' science in which control of agendas has been to a greater or lesser extent ceded to policy-making or other institutions, and 'useful' science which is typically integrative and multidisciplinary science organized to resolve technical solutions to social needs.

The boundaries between these different contexts of science are blurred and often contested. Collingridge and Reeve have argued that this problem is so acute that scientific expertise can *never* be of use to policy. They suggested that scientific knowledge will always be either 'overcritical' (partisan and divergent), or 'undercritical' (determined from the outset by élites).

In public and contested roles, the authority of science and its success in persuading public opinion about the correctness of a decision, or in calming fears, is always under threat. This threat comes not only from traditional processes of scientific dispute such as falsifying evidence or new theories, but also from political and industrial sources. Technology policy decisions are usually regarded as redistributing risks and benefits between groups in society. Concomitantly, there are usually also perceived to be losers, economically, culturally or symbolically. It is in the interest of these groups to dispute all promotional claims, including those made on the basis of 'core' science. The deconstruction and reconstruction of scientific claims is therefore characteristic of science in public policy. This process may take place in many different institutional contexts and in each one the conditions of authority and legitimacy are different. Science is employed in a multitude of loosely linked discourses as a resource. It is therefore not 'disinterested,' but strategically used as a currency with some value in disputes which have become increasingly public in the last three decades of the twentieth century. The problem of 'risk' is often central to these disagreements.

To a large extent, the effectiveness of science in regulatory decision-making therefore depends on the resolution of disputes about equity and fairness. Resolving these disputes is a political process involving many interest groups and institutions, but one with a tight reciprocal relationship with scientific institutions, knowledge and argument. For instance, the estimation of a risk of human exposure to man-made radiation some hundreds of years into the future requires an extensive architecture of scientific reasoning and problem-solving. Some commentators argue that the evolution of underpinning 'core' science is related to the evolution of interests around the policy problem, setting in place assumptions and styles of analysis which may, in turn, channel policy in particular directions.

Typically there are also aspects of the policy problem itself which are intractable to resolution with reference to science alone. For instance, the definition of what level of risk from man-made radiation would be 'acceptable' (now or in the future) clearly depends on some form of agreement and consent which is generated in the wider political sphere. The critical problem for policy dealt with in this chapter

— whether a method of radioactive waste disposal is acceptably safe — therefore has both technical and political elements which are inter-related and cannot be decomposed into separate categories. Alvin Weinberg characterized this type of problem as 'trans-scientific.'

In this chapter I present an historical survey of what is perceived as one of the most intractable of environmental problems — the long-term isolation of high-level radioactive wastes. In particular I will assess how ideas of technical control over these wastes — the underlying objective of safety — have evolved over the past fifty years. Science and its institutions have played key roles in defining the hazard represented by these wastes and in proposing and validating solutions to the problem of control. Despite this investment and the emergence at various times of agreement within technical communities about the safety of a disposal approach, there has been a continuous dispute, more or less public depending on the circumstances, over the credibility of scientifically-based prescriptions, and more broadly over the legitimate reach of technical analysis. Critics have often argued that technical definitions of problems are used purposely to limit the extent of public disclosure. Science itself is therefore used as a device of control.

For these reasons, we may see a history of disputes about the control of radioactive wastes as demonstrating some general features about the uses of science in public policy. The chapter begins with a discussion of the critical problem for policy — what we describe as the problem of defining a threshold of control. A basic element of this problem is the question of the human health hazard represented by ionising radiations, and this is discussed in the next section. Following a brief discussion of the nature of radioactive wastes, the rest of the chapter is given over to an historical review of debates about the isolation of radioactive wastes from the human environment. I argue that a linear, sequential history of ideas about the control of these wastes cannot be written, partly because the final outcome is not yet known. Although one approach — deep geological disposal — has attracted more official institutional support than others, the technical consensus built around it has periodically been destabilized and its long-term viability still hangs in the balance in most countries.

The Threshold of Control

All industrial processes produce residuals which are treated as wastes. Some of these wastes may be hazardous to people or to other forms of life, and a proportion may be retrieved and reused. Wastes are often disorganized mixtures of many different materials which are hazardous along several dimensions — they may be toxic, carcinogenic, noxious, ozone depleting and so on. Because of their compositional variety, the definition of wastes is frequently problematic. Radioactive wastes represent a comparatively simple type of waste because only one form of hazard — ionizing radiations — is usually considered important, and in almost all cases the only dimension of hazard considered is the risk to human health. This relative clarity has concentrated regulatory and scientific efforts, but also served as a beacon for controversy.

Two characteristics of radioactive wastes have shaped ideas about their control: the emission of ionizing radiations is a spontaneous, uncontrollable physical process; and the decay of radioactivity means that the relative hazard of radioactive wastes declines through time. It is difficult to alter the physical nature of radionuclide atoms so as to reduce their radiotoxicity or hasten their decay. The hazards represented by radioactive wastes were from an early stage seen to be especially challenging and of global significance. For epistemological, institutional and economic reasons, the main objective of radioactive waste management became its isolation from the human environment.

What form would this isolation take? In principle it would be possible to maintain continuous institutional control until all the wastes had decayed to a level where they could be safely dispersed directly into the environment. This has been standard practice with many shorter-lived radioactive wastes. For longer-lived wastes containing materials like plutonium-239 with a half life of 24,000 years, it has been argued that institutional control would need to be preserved for periods before they could be emitted into the biosphere. This option of 'monitored and retrievable' storage has been systematically rejected by industrial and scientific interests over the past 50 years. They have argued that final disposal can be safe and that there are strong ethical objections to passing on custody of toxic wastes to future generations. There has recently been a reawakened interest in the indefinite storage option, especially amongst environmental groups who want to avoid sanctioning disposal programs.

Almost all national radioactive waste policies today assume final disposal. This would occur in two distinct phases: a first stage in which direct institutional control is maintained and a second stage when institutional control is irrevocably relinquished. During this second stage some combination of engineered and natural barriers would secure the continued isolation of radioactivity from the human environment until it had substantially decayed. There are two approaches to this problem: the elimination of radioactive wastes from the earth's environment (such as launching it into space or transforming it into shorter-lived products); or its engineered emplacement in an isolated part of the earth.

The critical policy question is how to draw the boundary between direct, institutional control (monitored and reversible) and indirect, engineered control (not monitored, and essentially irreversible). This we call the 'threshold of control.' On one side control is exerted directly, on the other control is indirect and reversibility is not a readily available option. Science's main function is to assist in establishing a politically sufficient agreement over the adequacy of indirect control and so to help authorize the boundary which is drawn between the two forms of control. Science therefore has a clear industrial and political task, and its institutional forms, agendas, practices and dissemination will be shaped by the contingencies associated with that task. Many procedures are adopted from 'core' science, but the orientation of problem-solving is deeply ensnared in social and political objectives which are not for the research scientist to determine. What leads science on are not problems raised in the light of other science, but by problems raised in the erection of what Jasanoff has called a 'serviceable truth.' She describes a 'serviceable truth' as a 'state of knowledge that satisfies tests of scientific acceptability and supports reasoned decision-making, but also assure those exposed to risk that their interests have not been sacrificed on the altar of impossible scientific certainty.'

Broadly speaking, four interests have a stake in the threshold of control: the polluter; the 'public' (frequently also represented by proxy organizations); the state; and science itself. Producers of nuclear wastes — primarily the nuclear weapons and nuclear power producers — have an interest in disposing of the problem because it represents an economic and political cost. As long as critics can argue that no 'solution' has been found for the isolation of radioactive waste, the industry will continue to be vulnerable, no matter how firm the technical consensus that long-term isolation can be safely done. Lined up against them in most settings, but varying in power and effectiveness, are environmental organizations representing components of lay public opinion. These organizations have an *a priori* sceptical attitude to claims of safety by the industry, an attitude embedded in their public appeal and political positions. Criticism and opposition to waste storage and disposal plans is underpinned by an awareness that opposition to nuclear power as a whole will benefit from continued obstruction of waste storage and disposal projects.

The third block of interest is government. In the role of both policy-maker and regulator, government departments have acted as rule makers and final arbiters about nuclear safety. In principle nuclear regulators are independent bodies with the objective of protecting public welfare. Under the rubic of sustainable development the rights of future generations have achieved greater prominence. The reality of state interests is usually more complex. Environmental policy does not stand alone, but is always in some relation to energy, industrial, economic and other policy objectives in government. Radioactive waste policy has been strongly affected by the degree of state interest in promoting nuclear policy, and in particular in promoting nuclear fuel reprocessing. States with strong nuclear programs have been weakly committed to setting a threshold of control; while states without interests in promoting nuclear power have been moderately committed to a threshold of control and have faced fewer obstacles to the implementation of policy.

Lastly, research science itself has an interest in staking out and defending problem areas as its own. Engagement in policy science disputes frequently leads to a threat to the legitimacy of scientists as problem areas are invaded by new entrants who may have little standing within scientific communities, but who are nonetheless able effectively to deploy scientific claims within the policy arena. This process of the 'opening up' of science has led to deep critiques of the composition and practices of some scientific communities and these have damaged the authority of science as a whole.

Ideas about the threshold of control have therefore been constructed through a combination of industrial, environmental, state and scientific interests. Knowledge has not accumulated within an autonomous research program but has been shaped by the action of particular interests outside the science system. In the development of engineering concepts for waste isolation the process of knowledge accumulation has frequently been hampered by the absence of opportunities for validation through testing. A great weight has therefore been placed on computer-based models, only very partially underpinned by empirical data. The indeterminacy of many of the results of these simulations has proved fertile ground for political-technical dispute. A wide arena of debate could be established by the interests involved because the critical problem — is it safe and will it be safe forever? — cannot be definitively resolved.

Within this interrelated set of shaping forces the role of organized environmentalism has often been important, although its influence should not be overstated. In countries where state commitment to a threshold of control was weak these groups often pushed at an open door. In these cases it was often easier for government and industry to adopt a policy of waiting than of pressing on with a policy that mattered little to them. In countries such as Germany during the 1970s and 1980s, where a stronger commitment to a threshold of control existed, the successes of political environmentalism have paradoxically been much smaller, even though their activities have generated far more heightened disputes. It is arguable that the British environmental organizations, generally seen as relatively weak, have been highly effective in obstructing government radioactive waste policy.

The Institutional Setting of Radiological Protection

The biological effects of ionizing radiation were first reported by Minck and Lister in 1896. In a presidential address to the British Association for the Advancement of Science, Lister discussed the exciting prospects for medical radiography using newly discovered X-rays, while warning that "If the skin is long exposed to their action it becomes very much irritated, affected with a sort of aggravated sun burning. This suggests that their transmission through the human body may not be altogether a matter of indifference to internal organs!" Precautions such as lead shielding soon became commonplace in laboratories and hospitals using X-ray machines and these reduced the numbers of reported radiation injuries to the skin (erythemas).

By the 1920s a second more disturbing set of effects began to be observed with the deaths of radiologists and their patients from blood diseases and cancers with long latency periods. This prompted a new round of protection efforts and the publication of recommendations by medical and radiological societies. By 1928 a widely accepted set of safety guidelines had been agreed by a new scientific organization, the International Congress of Radiology (ICR), meeting in Stockholm. In 1934 the ICR accepted Mutscheller and Sievert's argument that tolerable doses of radiation be based on the threshold erythema dose (the radiation dose producing skin reddening). More by accident than design, both had arrived at an estimate of a tolerance level of radiation exposure of one tenth of an erythema dose per year (typically expressed in roentgens — the unit of radiation exposure agreed upon in 1928 by the ICR).

The institutional form of an international radiation protection community was set in place by these events. Although much has changed since the early 1930s, the process by which scientific opinion is formed has changed little. The state of knowledge about the health effects of radiation, and judgements about acceptable levels of risk are periodically reviewed by a small self-selecting group of scientists (there are currently 13 members on the full Commission, with American and British members dominating) in the International Commission on Radiological Protection (ICRP, renamed 1950). Their views and recommendations are published in the Annals of the ICRP, and major revisions of recommended doses to workers and the public have been accepted more or less unaltered by most national governments since the late 1940s. Only the US National Council on Radiological Protection (NCRP) and the United Nations Scientific Committee on the Effects on Atomic Radiation (UNSCEAR) have formed competing concentrations of scientific authority.

The stability of the ICRP's independence and authority up to the very recent past requires some explanation. One observation is that authority in science is *given* as well as taken. The ICRP has derived its scientific authority from its membership, while its political authority has been bolstered by governments for whom the committee continued to be an attractive unified source of expertise. Near universal application of the ICRP's recommendations have avoided relativistic comparisons being made between standards adopted in different countries. For the nuclear industry, in its continual search for legitimation, clearly stated recommendations from ICRP reduced regulatory uncertainty to a minimum.

On the preliminary question of the health effects of ionizing radiation therefore, the codification of scientific opinion, and its transmission into regulatory policy has been highly concentrated and uniform for about 70 years. In this field of regulatory science prevailing industrial and state interests have supported a unified structure of science advice production and acted on the advice which was produced. The robustness and authority of the structure contrasts with, for instance, the frequently ignored advice provided by the World Health Organization on food safety. Unlike some fields of regulatory science, radiological protection has adjusted to new science relatively quickly, leading periodically to more or less uniform lowering of dose limits applied through national regulations.

Figure 15.1: Typical example of encapsulated Magnox swarf in concrete matrix. © British Nuclear Fuels plc.

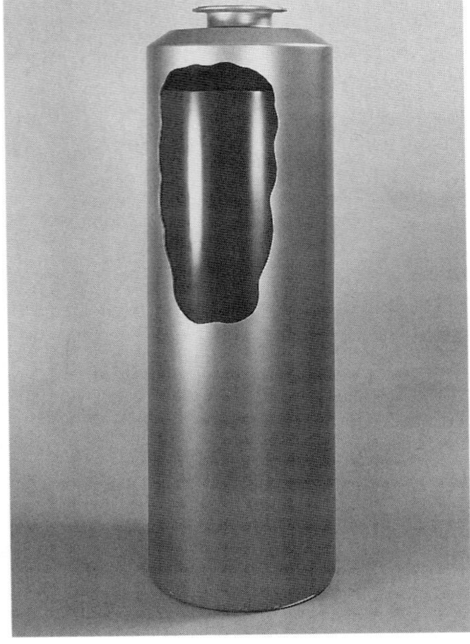

Figure 15.2: Sectioned vitrified waste container. © British Nuclear Fuels plc.

Figure 15.3: Storage Pond in (THORP) Thermal Oxide Reprocessing Plant Receipt and Storage. © British Nuclear Fuels plc.

Figure 15.4: Flask handling operations in the THORP Receipt and Storage Facility inlet cell. © British Nuclear Fuels plc.

Figure 15.5: The Vitrified Product Store built to hold containers of highly active vitrified waste, cooled by natural air convection. © British Nuclear Fuels plc.

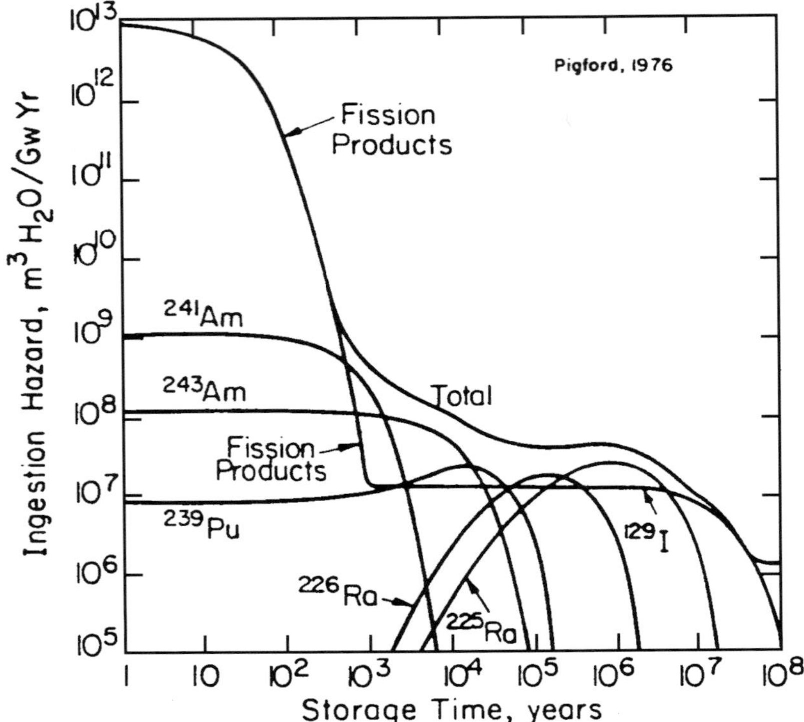

FIGURE 15.6: PRINCIPLE CONTRIBUTIONS TO THE INGESTION HAZARD INDEX AT HLW FROM THE REPROCESSING OF A URANIUM FUELED LWR AS A FUNCTION OF TIME. SOURCE: T.H. PIGFORD AND J. CHOI, "EFFECT OF FUEL CYCLE ALTERNATIVES ON NUCLEAR WASTE MANAGEMENT," PROC. OF THE SYMPOSIUM ON WASTE MANAGEMENT, TUCSON, ARIZONA, ERDA, CONF-761020, 1976.

TABLE 15.1: RADIATION PROTECTION STANDARDS: 1934–PRESENT

RECOMMENDED LIMITS FOR ANNUAL WHOLE BODY DOSES TO WORKERS AND INDIVIDUAL MEMBERS OF THE PUBLIC

DOSE FIGURES ARE IN MILLISIEVERTS

Year	Worker	Public
1925	700	
1934	250	
1941		30
1949	150	
1953		15
1966	50	
1977		5
1990	20	1

The evolution of recommended dose limits is shown in Table 15.1. Although dose limits by themselves are no longer regarded as a suitable guide to action, their development demonstrates a common feature of science in regulation — new knowledge and reduced social tolerance of risk with growing affluence will force acceptable levels of exposure to hazards down over time. The experience that accumulated knowledge about pollutants and their effects will typically reveal higher risk factors underlies much of the debate over the safety of industrial activities. Indeed, this has led to the articulation of the precautionary principle in environmental management.

Over the past twenty years, the primary cause of falling dose limits has been the analysis of mortality data from four groups of people exposed to radiation: atomic bomb survivors at Hiroshima and Nagasaki (the most important); radiation therapy patients (ankylosing spondilytis and cervical cancer); patients diagnosed using radiation (X-rays); and workers who have had occupational exposures radiation. The most recent revisions of the ICRP's recommendations were made in 1990 when the risk of a fatal cancer for all ages was increased from 1.25E-2 per Sievert (Sv) to 5E-2 per Sv (i.e., the risk of death in any year following an exposure of one sievert of radiation would be five percent). The risk of severe genetic effects now stands at E-2 per Sv. This revision was influenced especially by an influential reassessment of atomic bomb dosimetry in Hiroshima and Nagasaki involving a sample of 75,400 people in those cities. Risk estimates based on the epidemiological data from Hiroshima in particular needed to be altered as more became known about the radiation yield of the unique weapon dropped on the city.

Radioactive Wastes: Defining Hazards

Definitions of radioactive waste hazards have drawn strongly on radiation protection but with far less clear-cut results. The most significant waste stream in the nuclear fuel cycle is irradiated fuel discharged from nuclear reactors once its fissile content becomes depleted. Fissioning of uranium and plutonium in the reactor produces heat and leads to the production of a suite of new isotopes including both lighter (fission and activation products) and heavier (transuranices, including plutonium) isotopes. Of these only about 20 are significant. Fission products typically have shorter half-lives (averaging about 30 years) and emit more energetic radiations (beta and gamma radiation), than longer-lived transuranics (alpha radiation).[1] The irradiated fuel may be chemically reprocessed to enable the separation of plutonium and residual uranium. High-level waste (HLW) occurs in two forms: irradiated nuclear fuel; and the first cycle raffinate produced in the chemical reprocessing of irradiated fuel. HLWs in the latter form are initially stored as aqueous liquors.

Significant quantities of spent fuel and high level waste have been generated in about sixty countries in plutonium production, power and research reactors. A large quantity of these fission product and transuranic wastes have been separated in reprocessing in the past, although in future the majority will be kept in the form of stored irradiated fuel.

There is no simple way of representing the hazard of radioactive wastes, but one commonly used definition is the radiotoxicity index. This takes as a starting point regulatory limits for the intake of radiation by adult humans. Making standard assumptions about the intake of water by adults, these limits can be transformed into maximum permissible concentrations for radionuclides in water. For a given mixture of nuclides it is therefore possible to calculate how much water would be needed to dilute it to drinking water quality. A radiotoxicity index for HLW from the reprocessing of light-water reactor uranium fuel is shown in Figure 15.6. The main aim of this type of figure, the product of computer models developed in the early 1970s, is to give some impression of the time-scale over which control may need to be imposed on these wastes. After 1000 years about a cubic kilometer of water would still be required to dilute the waste generated by a year's operation of a single large nuclear power station. This graph is still used pervasively by the nuclear fuel reprocessing industry to show how gross radiotoxicity will be reduced by the separation of plutonium from the HLW waste stream. A quite different representation is used by those who advocate the direct disposal of irradiated nuclear fuel.

The Isolation of High-Level Wastes: A History of Approaches

I have argued above that the critical problem for radioactive waste policy has been to create the political and technical conditions under which direct management control over wastes can be relinquished. These conditions are both technical and political. Agreement, or 'closure' therefore has more onerous demands, it requires both scientific closure and the suspension of related disputes concerning public welfare and national policy. In the absence of social agreement, technical concepts have suffered from continual threat and periodic destabilization. The history of high-level waste management is a history of one approach — geological disposal — being established as a reference concept or 'ideal' for dealing with the problem of long-term containment, but continually coming under threat from alternatives because of the lack of social agreement over its soundness. This appears to be true across a wide range of very dissimilar political systems. Countries as culturally and politically diverse as Sweden and South Korea have experienced profound political dislocation over radioactive waste disposal policy.

The conditions for the formation of broad-based social agreement vary from country to country and evolve through time. Governments always have a primary role to play as the principal regulator, and in some countries through being directly responsible for waste management (as in the United States and Germany). Governments are usually responsible for articulating a waste policy and for setting safety criteria. In this role they are responsible for defining the technical conditions for the threshold of control they have received advice from an increasing variety of sources. Much of this advice is based on scientific claims. Each claim will need to compete with other scientific claims and will be balanced against further claims which have no foundation in science. Science is one resource amongst many, used to both inform decisions and to justify them.

The procedural conditions for establishing legitimacy are also mostly set by governments. The fiercest and symbolically most important disputes generally arise over the licensing of new facilities such as an underground rock laboratory. Disputes are often conducted within formal settings such as quasi-judicial public inquiries, expert reviews and in legal proceedings. The procedural setting has been shown to have an influence on final outcomes; with some authors arguing that more open, less conflictual settings lead more quickly to the construction of coalitions of agreement. Recourse to law is traditional in other settings, and this has tended to lead to more formal regulatory systems, greater opportunities for opponents to challenge decisions, and more 'friction' in the creation of social agreement. Beyond these institutionalized settings, the more diffuse influence of public opinion may at critical moments have a strong impact on decision-making by industry and government.

Given contingency in the conditions of consent formation and the fluctuation of the political and industrial context of these disputes, it is impossible to write a linear, sequential history of ideas about the control of radioactive wastes. Different ideas have competed for ascendancy at any moment, as they do today. Moreover, the shape of ideas is country specific — because of historical commitments to nuclear power, resource (technical and environmental) constraints, and the national specificity of debates about policy. Nonetheless, it is possible to pick out common dominant tendencies in the many disputes which have taken place, and to distinguish the emergence of new ideas which find common currency. An international (Western) history of core ideas about the control of high level wastes can therefore be written, even if these are overlapping and heterogenous at any particular moment. The history of ideas presented here illustrated with specific cases. Most of these examples come from America and Europe. This may give a skewed picture since we are now learning that ideas and practices in the Soviet Union were often different to those in the West.

Operational Management in Atomic Weapons Programs

High-level radioactive wastes were first produced in the atomic bomb projects of the 1940s in the United States, the Soviet Union and the United Kingdom. During this phase radioactive waste management was the responsibility of the military and was a part of normal operational activities at nuclear sites. De la Bruhèze has shown that each production site in the United States bomb program dealt with wastes differently. Each adopted different definitions, treatment, storage and disposal strategies. Some sites chose a cautious approach emphasizing control, others (especially those in the West such as Hanford) opted to disperse certain wastes into the ground and river courses. The basic approach to high-level wastes in the

United States was to store them untreated in large cooling tanks. Some were experimentally and accidentally permitted to dry out through evaporation. This approach led elsewhere to at least one serious accident — the explosion at Kyshtym in Russia in 1957 — and to numerous leaks and an enormous clean-up problem whose dimensions only became apparent in the late 1980s.

How can we explain this cautious approach to the management of high level wastes. The strongest explanation appears to be the pre-war emergence of new evidence about the hazards of radiation exposure in the radium dial painting industry and the desire to protect workers in the nuclear industry. Radium intake limits were set by the US National Bureau of Standards in 1941, although no attention was paid to the disposal of radium components. The radium limits led to stringent safety precautions in the Manhattan Project. After the war military interest in radiation effects grew as assessments were commissioned on the lethality of radiation doses from atomic weapons and on the conditions under which troops could operate in contaminated theatres of war. When the Soviet Union exploded its first nuclear weapon in 1949 the problem of organizing a civil defense against nuclear attack provided a further stimulus for radiation biology research in the United States and elsewhere.

In 1947 the US National Council on Radiation Protection (formed in 1946) set tighter permissible doses for workers. This new move was motivated mainly by emerging concern about genetic effects of radiation, based on the work by Muller on genetic effects in mice and fruit flies. New research suggested that there was no safe dose of radiation. A shift of emphasis in radiation protection took place in the 1940s from the control of 'appreciable' injuries to minimizing injuries which could not easily be quantified or studied through direct observation of individuals. Risk estimates came to be based instead on epidemiological studies with far more complex and problematic models of causality. The risk of genetic damage to human populations appears to have been a primary influence on the storage approach to high-level waste management. Other interests also played a part. Production sections at some weapons laboratories saw the wastes as a potential source of heat and valuable minerals. Security interests were also served. By holding wastes on-site secrecy was maintained by avoiding the need for new controls to be imposed beyond production sites.

The 1947 dose limits raise two important points about the constraints placed on regulatory science. First, without a safe dose of radiation a certain number of injuries were deemed acceptable whatever dose limit was set. The question of what was acceptable has typically been seen as a matter for social and political, not scientific, judgement. To deal with this problem the concept of 'as low as reasonably achievable' (ALARA) was developed in the late 1950s. This widely used concept states that a dose limit represents a limit of acceptable risk and that actual doses from industrial activities should be as far below this as possible given technical and economic considerations.

Second, possible economic impacts on the nascent atomic energy industry were taken into account in setting the 1947 limits. The ICRP's decision to adopt the NCRP limits in 1953 followed industry acceptance of the limits. Industry is often deeply involved in the environmental rule-making. The standard model which assumes that regulations are derived scientifically and imposed on industry has been shown to be at odds with reality. Creation of new regulations and their implementation is a more or less complex process involving regulatory agencies, other governments departments, the affected industry and other interested parties. Economic and industrial considerations always play a major role in judgements about the stringency and monitoring of regulations.

Debates about Final Disposal: Dispersion or Containment

By the mid-1950s tank storage of high-level wastes was regarded as expensive and ultimately unsafe. An alternative to storage was also deemed necessary by those who had begun to plan for civil applications of atomic power. At the seminal 1955 United Nations conference on Peaceful Uses of Atomic Energy in Geneva, Wolman and Gorman, two influential American delegates, argued that: "The disposal of reactor and fuel processing wastes will be one of the major controlling factors determining the extent of the use of power reactors competitively with other sources for energy production." Both American and British commentators of the time were already aware of the critical role public opinion would play in the choice of long-term isolation strategies.

Formal discussions within groups of government-appointed experts began in the US and in Britain on the long-term disposal of radioactive wastes. The first great issue to be resolved was whether to disperse high-level wastes in the oceans, or whether these wastes should be confined and contained. This debate was effectively won by the mid-1950s by a group of health physicists and academic geologists who showed that the dispersion of fission product wastes produced by forecast power reactor programs would require a significant proportion of the world's water resources.[2] Perceived uncertainty about the physical behavior of the oceans, the transmission of radioactivity through marine food chains and the problem of how to develop an effective dispersal technique were also factors in the debate. This was resolved differently in the Soviet Union where dispersion of high-level wastes (both deep geological injection and dispersion into surface lakes) was chosen.

By the time the Hess and Wolman Reports (1953/55) and the Key Report (1959) were published in the United States and Britain respectively, containment had been accepted as the basic objective of waste management. The favored approach in the United States was geological disposal in mined cavities although the Key Report limited itself to the statement that high-level wastes should be dumped in "inaccessible parts of the earth." The first appraisal of geological disposal was published by the US National Academy of Sciences in 1957 which focused on the suitability of different rock types. It concluded that salt would be the best medium for waste disposal because it was homogenous, dry and had the characteristic of 'creep' causing fractures which do occur in the rock body to be 'self-healing' (rather like highly viscous treacle). Various alternatives such as injection into abandoned oil and gas fields through deep wells were also considered and actively supported by the Oak Ridge National Laboratory.

Work on ocean disposal of packaged wastes was also launched, enthusiasm generally being more muted in the United States than in Britain. The international implications of sea disposal were clearly apparent. The 1958 Law of the Sea contained specific reference to disposal of radioactive wastes and authority was given to the newly formed International Atomic Energy Agency (IAEA) to study the question. This work promoted the idea that the disposal of radioactive wastes should be internationalized. Many commentators have argued, especially in the mid-1970s, that the global scope of radioactive waste hazards means that there would be benefits for states in scientific collaboration, technical standardization and in even in cooperative waste disposal.

While a high level of collaboration developed in 'radioactive waste disposal science' beginning in the 1970s, and the process of setting international safety standards gained momentum during the 1980s after the Chernobyl accident, the idea that one country should accept the radioactive wastes of others has not yet proved to be politically viable. By the end of the 1950s, the national geological waste repository concept had been established as the single accepted solution to HLW isolation. This was a 'weak' closure of a policy debate since little industrial or state interest existed at the time in developing the concept further. The question of how to establish a threshold of control in practice could not yet be embraced since all technical efforts were devoted to the task of developing and industrializing nuclear technologies. Up until the 1970s radioactive wastes were a minor and unfashionable concern within the nuclear power community.

The Emergence of Public Concern: The Fallout Debate

Public interest in radiation in the environment was given substance for the first time by a debate about the health effects of fallout from atmospheric nuclear weapons testing which began in 1954 and lasted for nearly ten years. An open scientific controversy about the genetic and health effects of low level radiation proved to be a turning point in public attitudes to radioactivity. Its general effect was to compound public confusion and fear. The previous dichotomy in public discourse between embracing the positive therapeutic and other benefits of the atom while fearing the potential consequences, was replaced in many countries with a more limited, negative discourse about radiation. Cold War images of hidden threats and global catastrophe now became deeply identified with radiation.

In early 1954 about 300 people, among them a group of Japanese fishermen, were accidentally contaminated with fallout from a US hydrogen bomb test in the South Pacific. The incident aroused Japanese indignation and speculation that atmospheric testing might injure the health of people in both countries. These fears were dismissed by the head of the US Atomic Energy Commission (AEC), Lewis Strauss who was, in turn, challenged by a community of academic geneticists, among them A. H. Sturtevant, Genetics Chair at Caltech, and Linus Pauling.

A controversy emerged about the status of human genetics and especially about whether data from animal models could be extrapolated to estimate human genetic effects. The AEC argued that doses caused by fallout were small compared to background radiation, whereas Sturtevant and his colleagues argued that any dose above the minimum would be expected to cause genetic mutations. The main result of this debate, effectively concluded with the publication of a National Academy of Sciences (NAS) report in 1959, was to confirm the genetic risks of radiation. More importantly however, a second NAS committee on radiation pathology highlighted the link between strontium-90 ingestion and the blood cancer, leukaemia. E. B. Lewis' work on the induction of leukaemia published in 1957 suggested that there was a linear relationship between dose and effect, and therefore that no threshold dose existed for somatic radiation injury. He also showed that the risks of leukaemia were higher than for genetic damage.

Lewis' results were not finally accepted until the mid-1960s, but the key result of this debate was that it led to a widespread public experience of exposure to man-made radiation. Measurable levels of fallout radiation began being monitored in the early 1950s all over the United States and were reported in the Press. By 1959, record levels of Strontium-90 was found in milk, wheat and soil in Minnesota and the Dakotas. Following the nuclear weapons testing moratorium fallout levels began to decline, only to rise again with the resumption of atmospheric testing in 1961. Precautions were taken in some American states such as the removal of cows from grazing and the discarding of milk for direct consumption (the threat to milk supplies also played an important symbolic role in public perceptions of the Windscale reactor fire in the UK in 1957). There was much dispute about the necessity of these measures, but the authority of the AEC and of groups advising it had been critically damaged.

Pressure for a Demonstrated Solution

Very little progress was made in radioactive waste management on either side of the Atlantic between the early 1960s and the mid-1970s. A plan developed by the Oak Ridge National Laboratory to build a geological repository at Lyons, Kansas was abandoned due to local public opposition and the discovery of geological problems at the site, but elsewhere there was very little activity on disposal. Nuclear programs were not restrained by an absence of waste policies. Irradiated nuclear fuel and untreated HLW was stored at nuclear sites with slow progress being made towards the development of a glassification technology to immobilize HLW liquors. In Britain, France and the United States development of technology to vitrify (glassify) HLWs became a substitute for planning towards disposal, especially in the growing amount of public information literature which the nuclear industry began to produce.

In the mid-1970s the problem of radioactive waste management suddenly reappeared as a major issue for nuclear policy. National policies were quickly articulated and, during a period of intense public interest and policy debate, the geological disposal option was both further elaborated and challenged by a number of more exotic alternatives. There appear to be four reasons why this happened. First, fears of energy scarcity and new projections for the growth of nuclear power reopened the question of what would eventually happen to wastes. In several countries new nuclear development (including Germany, Japan, Sweden and the UK) was made conditional upon a 'demonstration' of long-term high-level waste isolation. This was frequently tied to an industrial policy objective to establish a domestic reprocessing industry. Institutional momentum for scientific and engineering research was therefore provided by the urgent pursuit of wider national industrial policy priorities.

Second, there was a need for nuclear industries and governments to compete against a mobilized anti-nuclear movement which argued that no solution to waste disposal was available. These groups had emerged at the end of the 1960s as part of the cultural revolution of that decade and had become organized around an anti-nuclear program by the early 1970s, especially in the United States and West Germany. The programmatic coherence of these groups continues to be dependent on their anti-nuclear stance.

Third, the mid-1970s were a period during which the United States attempted to reassert control over the international nuclear fuel cycle, partly out of concern for the proliferation of nuclear weapons. This was resisted by European and Asian countries in a dispute contested primarily within the International Nuclear Fuel Cycle Evaluation (INFCE) based at the IAEA in Vienna between 1977 and 1980. A key aspect of the technical work of INFCE was to investigate the security and environmental aspects of alternative fuel cycles (reprocessing vs once through, uranium vs. thorium), and this work led to the first systematic scientific research into high-level waste disposal.

This intensive period of activity led to a number of seminal outcomes. The first was that executive responsibility for waste disposal was formally allocated, usually to government disposal agencies, although in some countries (Germany and the UK) to nuclear utilities and fuel cycle companies. In all cases governments took responsibility for setting policy for the disposal of radioactive wastes, usually with the argument that this was necessary to protect the public good. In taking these roles, conflicts of interest frequently arose, since governments became both disposers of waste and regulators. This dual role became most vulnerable in countries where governments were still deeply committed to nuclear energy, and where radioactive waste policies were therefore seen as self-serving instruments of energy policy.

The second outcome was a formalized disposal concept which came to be known as the multi-barrier approach. This stressed the importance of many redundant engineered barriers in containing radionuclides within a repository environment for as long as possible, but accepted that over the periods of hundreds of thousands of years some release of radioactivity would be anticipated under both exceptional and normal scenarios. Simple heat and mass diffusion models were developed for conceptual repositories, supported by new investments in laboratory and *in situ* research into the transmission of radionuclides through hydrogeochemical environments.

The first systems assessment was the Kärnbränslesäkerhet (KBS) project, of the safety of a geological repository completed cooperatively by the Swedish nuclear utilities in 1978 under extraordinary conditions. Over a period of about a year the question of whether radioactive waste could be disposed of with 'absolute safety' came to dominate Swedish national politics. In 1976 one of the first acts of a new anti-nuclear coalition government led by Thorbjörn Fälldin, was to make new reactor operating licences conditional on a demonstration of safe radioactive waste disposed. A rudimentary safety case using published literature was quickly drawn up and this was reviewed through several rounds by national and international experts. There was little agreement about whether the findings represented adequate proof that safe disposal had been demonstrated. In lieu of agreement, one question therefore emerged as being crucial — whether a site with the characteristics assumed in the KBS dispersion model could be found in Sweden.

A search process was launched for a site with an 'unfractured rock mass' similar to that assumed in the KBS study. Conditions for judging whether a site at Sternö in southern Sweden met this criterion were initially set by the leading coalition party, but disagreements over these led to the demise of the government. With a new minority government in power, the nuclear safety inspectorate, was given the task of deciding on the suitability of Sternö. Following a new review of the evidence, the board of the safety inspectorate, which contained several political appointees, could agree only that investigations at the site *did not disprove* the availability of a suitable site in Sweden. In highly contentious circumstances, the government awarded an operating licence for a new nuclear reactor. Successor safety studies (KBS2 and KBS3) abandoned the unfractured rock mass hypothesis and sought instead to develop models of migration through fractured masses which could be adapted to site-specific conditions. Each site came to be seen as being different, and an engineered repository system and its safety assessment would need to be adapted to these specific local hydrogeochemical conditions. The role of generic safety cases was therefore thrown into doubt.

What stands out from the KBS experience is the failure of research and formal review procedures to yield definitive technical agreements which could then be transmitted into the political process. In the complex, multidisciplinary arena of repository safety cases closure of technical disagreements has often proved elusive. Few strategies exist for overcoming the 'hypothetically' of safety cases which make forecasts for the behavior of an engineered system over many thousands of years. Only partial tests exist to check results produced by simulation models; partial because only one component of the system can usually be tested, and partial because these tests will all be over comparatively short time periods compared with the operating life of a repository. The traditional engineering response, to build in redundancy, is also only a partial solution where very high standards of confidence are required. An important response to these details was the development of underground rock laboratories: Asse in Germany (operating since 1965); Stripa in Sweden (operating since 1976); Grimsel in Switzerland; Mol in Belgium; with further laboratories planned in France, Britain, Japan and Sweden. Germany is the only country to have funded a repository science research program since the beginning of its nuclear program. Research was begun on 'natural analogues' to high-level waste repositories. At the best known of these, the Oklo uranium deposit in Gabon, spontaneous chain reactions in several 'reactor zones' were first described in 1975. Uranium and fission products were shown to have migrated only a few metres into the surrounding altered rock. The status of analogue research was uncertain for a long period in the 1970s and 1980s, mainly because it could not yield specific answers to the question: Would disposal be safe here? Analogue research has again found favor more recently as a complement to modeling approaches.

By the late 1980s, despite the growing confidence in technical communities in the efficacy of the multi-barrier approach, less emphasis was being laid on natural barriers in many countries. Greater weight came to be placed on engineered barriers because these were regarded as 'testable' and therefore easier to demonstrate in public 'deconstructions' of safety cases. The adoption in the Swedish KBS concept of a copper canister with a predicted life of about 100 thousand years is a clear example. This engineering approach was resisted by geochemists and geologists who remained more critical of the capacity of engineered barriers. The general problem of the right balance between engineered barriers and natural hydrogeochemical barriers was left unresolved. In the safety case for the Yucca Mountain repository in the United States for instance, a rapid breach of engineered containment is assumed.

The Proliferation of New Concepts

A second result of the crisis of the mid-1970s was the proliferation of new waste isolation concepts. These included alternative repository sites (sea bed and icecap disposal) and new concepts to eliminate wastes from the earth environment (space disposal and transmutation). Just as the land disposal option was being elaborated and tested, a new set of competing concepts were launched. The many disputes over repository safety generated political and funding opportunities for alternatives, but the eventual outcome was to further undergird the geological disposal approach. The major problem for alternatives was not that they did not appear conceptually attractive or feasible, but that the conditions for demonstrating their safety were even more onerous — technically and institutionally.

The story of the sea disposal concept illustrates this. Sea dumping of relatively low level activity wastes was began in 1946 by the United States and by the 1960s it had been adopted by the Soviet Union and several European countries. In 1967 sea dumping operations by member states of the Organisation for Economic Cooperation and Development (OECD) were coordinated within the OECD Nuclear Energy Agency to overcome criticism that discrete national dumping programs could lead to a cumulative hazard to fisheries and fish consumers. Concerted oceanographical research into the sea disposal of high-level wastes was launched in 1973 by the United States, even though these wastes had been placed on the 'black list' of prohibited wastes by the London Dumping Convention (LDC) in 1972. Britain was a keen supporter of this work owing to the UK Atomic Energy Authority's continuing ambivalence about securing a suitable land disposal site. On-going dumping operations and plans to extend them to new wastes became highly contentious during the late 1970s. Greenpeace began a campaign of harassment of sea dumping in the Atlantic in 1979, and this was followed in 1983 by the adoption by the national parties to the LDC of a voluntary moratorium on dumping pending the results of a scientific review of safety.

In the event two sets of reviews were carried out, the most important by the Intergovernmental Panel of Experts on Radioactive Waste Disposal (IGPRAD) which finally reported in 1993. IGPRAD reviewed not just technical questions such as risk assessments for sea disposal of radioactive wastes, but also wider legal, political and social aspects. Arriving at scientific agreement proved to be impossible due to the entrenched positions of several Nordic countries under political pressure at home to resist a resumption of sea disposal. Although IGPRAD concluded that no technical case had been made for significant radiological impact being caused by the sea dumping of low and intermediate level wastes, the final report of the group was ambiguous. More weight was attached to the trend over the past 20 years towards the prohibition of sea disposal on a regional basis, and a challenge to the right of states to use the high seas for activities which might lead to pollution. In 1993 the LDC prohibited the disposal of all radioactive wastes at sea.

In the late 1970s large scientific, institutional and political investments were made in radioactive waste isolation in Europe and the United States. National disposal policies were created for the first time in the late 1970s. Work on safety models for repositories developed from 'conservative' deterministic assessments to more probabilistic assessments which drew on a better understanding of radionuclide mobilization and migration processes. These assessments provided strong confirmation that under normal conditions very small doses of radiation would result from the deep land disposal of high-level radioactive wastes. A new set of competing isolation concepts were also generated and assessed, but generally failed to attract sufficient institutional support as governments became increasingly invested in geological disposal under pressure from both the nuclear industry in search of a 'solution' and from environmental critics.

Risk Perception and the Retreat from the Threshold

Despite an almost universal agreement in technical communities about the feasibility of safe geological disposal, doubts about its political acceptability persisted and in some places grew. Where they had been initiated (Germany, the United States) repository programs faced delay and obstruction by state governments which could use their opposition to extract other concessions from central government and from environmental organizations. Controversies persisted in countries like Sweden where the state had withdrawn its interest in promoting the nuclear industry and where government decision-making is characterized by openness, and in countries like France where the state retained a strong interest in promoting nuclear power together with a tradition of centralized decision-making power. The institutionalist critique, which argued that the resolution of technical policy disputes was to be found in the 'democratization' of decision-making, has therefore not been borne out in practice. There is now broad concern about 'paralysis by analysis.'

Two broad explanations of this outcome are possible, one cognitive; the other cultural. By the late 1970s the persistence of opposition to nuclear power, and radioactive waste disposal projects in particular, led to research into the sources and structures of human risk perception. How could the dread of radiation be explained in advanced industrial societies where industrial risks were commonplace and generally accepted? An initial response was to argue that people were misinformed, but that this could be put right through better information and education. Research within psychology was also launched which analyzed responses to different types of risks according to criteria such as whether they were voluntary or led to catastrophic consequences. This work on the cognition of risks was criticised for ignoring the 'risk receiver.'

Cultural theorists have concentrated on the symbols, codes and practices through which understandings of danger, illness and death are transmitted. Authors such as Mary Douglas and Brian Wynne have highlighted the role of social relations in the construction of understandings about risk, and demonstrated that a wide range of public perceptions about the nuclear industry have fed through to public mistrust of its handling of radioactive wastes. The nature of radiation risks and the fear of cancer may play only a small part in the responses to nuclear power which stem more from a suspicion of anonymous and powerful institutions of state. In the Habermasian sense, the wider legitimation problems of the nuclear industry and of states intervening to support it will taint the reception of scientific advice such as an assertion about the safety of radioactive waste disposal. In open, public disputes about waste disposal it is not possible to disentangle scientific or technical judgements from other judgements and perceptions about standing, welfare and equity.

In response to these deep-seated political obstacles to waste disposal and the political judgement that the social strife it generates is not sustainable, the 1990s have seen a renewal of interest in radical alternatives in some countries. We may be entering a new period in which the status of geological disposal as the benchmark of national policy again comes under serious threat. The first of these threats is from the concept of 'retrievability' which has been discussed since the late-1970s. In its simplest form this means that wastes are placed in secure storage, either above or below ground, and monitored, with the option of retrieving them if a problem arises, or if a better approach is developed. In the extreme version of the new isolate, contain, monitor (ICM) policy adopted in the Netherlands, institutional control over all hazardous wastes would be extended indefinitely.

Although the Dutch are unique in having formally adopted this approach, ICM has been the *de facto* policy in most other countries. Stress incrementalism and the need for public participation, leading to a conscious social shaping of technology. There is a common recognition that scientific assessments change over time, social tolerance to risk changes, and that social consent is constructed. Intergenerational equity has also been invoked more strongly under the rubric of sustainable development. The results are ambiguous, policy communities arguing for producer responsibility critics contending that the risk of serious long-term environmental burdens is unacceptable.

The other retreat from the objective of long-term isolation has been in the renewed interest in partitioning and transmutation (P&T) as a significant option in some countries, France and Japan in particular. P&T was first developed in the early 1970s and gained considerable attention during the INFCE process as a means of eliminating plutonium. Under this concept long-lived radionuclides are fissioned in an accelerator or a fast neutron reactors to form shorter-lived fission products. The timescale over which isolation would need to be guaranteed would therefore shrink, although because the process could not be perfectly efficient the need for long-term disposal would probably not be eliminated. There are three main reasons for current interest in P&T: the failure of fast reactor programs leaving large numbers of reactor physicists unemployed; the need to redefine the rationale for reprocessing and a genuine concern that final disposal may never be acceptable to a significant number of people.

CONCLUSION

I have argued in this chapter that the main framework of ideas about the control of radioactive waste had been established by the late 1950s in the US and Europe. The 'critical problem' was identified as the choice between the dispersion of man-made radiation into the environment and its isolation. If isolation was chosen then the question was whether to isolate under direct institutional control (evoking problems related to the longevity, competence and trustworthiness of social institutions) or under indirect engineered control based on untestable forecasts of safety.

Isolation quickly emerged as the approach preferred by the nuclear industry and government for high-level wastes (although not for all species of radioactives wastes) because of the scale of the hazard they represented — the first environmental hazard of perceived global significance — and because uncertainty about the final health technical communities. The longevity of waste actinides persuaded these same communities that direct institutional control was not viable in the long run. Geological disposal, especially in dry environmental like salt, was established as the preferred route, although little progress was made toward acting on the concept. Public mistrust and the need for secrecy played a part in reinforcing technical caution. These factors were generally absent in the Soviet atomic weapons programme and this may explain the preference for 'dilute and disperse' solutions to high level waste management at the two main plutonium production sites of Chelyabinsk and Krasnoyarsk.

Planning for final disposal in the West did not begin until expansive plans had been laid for nuclear power in the early 1970s. Then it became necessary in many countries (especially those where nuclear growth was linked to progress on radioactive waste management) for governments and industry to demonstrate that a containment approach was 'technically feasible.' A move towards what we may characterize as 'strong' closure on the question of long-term repository safety therefore began. The key scientific and technological effort here was not in the waste handling or repository, but in the assessment techniques, instruments and computer models which were the basis for risks assessments for repositories. Opportunities also existed during this period for a set of technical alternatives to geological disposal to be developed. Closure was therefore disrupted right at the point when the dominant institutional interests in closure were strongest.

Following this period of policy formation and disruption, nationally-based geological disposal was re-established as the preferred route and the main contender (sea disposal) was abandoned by all states even for low activity wastes, against scientific advice. The need to demonstrate long-term repository safety forced many disposal agencies (whether state- or industry-run) to begin the assessment of real sites. This ignited public interest and opposition to waste policies, forcing governments less committed to establishing a threshold of control to retreat from policy positions, often with the claim that more research was needed. Scientific research has therefore become an instrument of a policy of delay. Repository projects in many countries were redefined as research projects.

Nuclear fear has not been 'normalized.' Mistrust of assertions of safety as well as less choate equity considerations have meant that implementation of disposal policies has been seriously delayed. The ultimate viability of projects such as Gorleben, Sellafield and Yucca Mountain is still in question and their survival will depend more on ethical, political and economic, than on scientific considerations, even if scientific rationales are still for decisions given. Under these conditions of uncertainty and drift, the idea of final disposal is again being challenged from two opposite directions: the 'retrievability' option argues for an abandonment of disposal in favour of indefinite storage under direct, monitored control; 'partitioning and transmutation' advocates argue that long-term containment can be avoided through the destruction of long-lived waste nuclides.

Further Reading

U. Beck, *Risk Society: Towards a New Modernity.* (London: Sage, 1992).

U. Beck. *Ecological Politics in an Age of Risk.* (Cambridge: Polity, 1995).

F. Berkhout, *Radioactive Waste: Politics and Technology.* (London: Routledge, 1991).

R. Brickman, S. Jasanoff, and T. Ilgen, *Controlling Chemicals: The Politics of Regulation in Europe and the United States.* (Ithaca: Cornell University Press, 1985).

C. Caufield, *Multiple Exposures: Chronicles of the Radiation Age.* (London: Penguin, 1989).

N.A. Chapman and I.G. McKinley, *The Geological Disposal of Radioactive Waste.* (Chichester: John Wiley, 1987).

D. Collingridge and C. Reeve, *Science Speaks to Power: The Role of Experts in Policymaking.* (London: Frances Pinter, 1986).

M. Douglas and A. Wildavsky, *Risk and Culture: An Essay on the Selection of Technological and Environmental Dangers.* (Berkeley: University of California Press, 1982).

A. Giddens, *Modernity and Self-Identity.* (Cambridge: Polity, 1991).

International Atomic Energy Agency, *Report on Radioactive Waste Disposal.* (Vienna: Technical Reports Series no. 349, 1993).

S. Jasanoff, *The Fifty Branch: Science Advisers as Policymakers.* (Harvard University Press, 1990).

M. Schwarz and M. Thompson, *Divided We Stand: Redefining Politics, Technology and Social Choice.* (New York: Harvester Wheatsheaf, 1990).

United Nations Scientific Committee on the Effects of Atomic Radiation (UNSCEAR), *Sources, Effects and Risks of Ionizing Radiation, 1988 Report to the General Assembly.* (New York, 1988).

CHAPTER 16

From Eugenics to Genetic Manipulation

DANIEL J. KEVLES

The word 'eugenics' was coined, in 1883, by the English scientist Francis Galton, a cousin of Charles Darwin and a pioneer of the mathematical treatment of biological inheritance. Galton took the word from a Greek root meaning 'good in birth' or 'noble in heredity.' He intended the term to denote the 'science' of improving human stock by giving the "more suitable races or strains of blood a better chance of prevailing speedily over the less suitable." The idea of eugenics dated back at least to Plato, and discussion of actually achieving human biological melioration had been boosted by the Enlightenment. In Galton's day, the science of genetics had not yet been invented. Nevertheless, Darwin's theory of evolution taught that species did change as a result of natural selection, and it was well known that by artificial selection farmers and flower fanciers could obtain permanent breeds of plants and animals strong in particular characteristics. Galton thus supposed that the race of men could be similarly improved using two complementary approaches — by getting rid of the 'undesirables' and multiplying the 'desirables.' Through eugenics, he proposed, mankind could take charge of its own evolution.[1]

Galton's eugenic ideas took popular hold after the turn of the twentieth century, developing a large following in the United States, Britain, Germany, and many other countries. The backbone of the movement was formed of people drawn from the white middle and upper middle classes, especially professional groups. Its supporters included prominent laymen and scientists, particularly geneticists, for whom the science of human biological improvement offered an avenue to public standing and usefulness. Eugenicists declared that they were concerned with preventing social degeneration, which they found glaring signs of in the social and behavioral discordances of urban industrial society — for example, crime, slums, and rampant disease — and the causes of which many attributed primarily to biology.

Eugenics entailed as many meanings as did terms such as 'social adequacy' and 'character.' Indeed, eugenics mirrored a broad range of social attitudes, many of them centered on the role in society of women, since they were indispensable to the bearing of children. On the one hand, positive eugenicists of all stripes argued against the use of birth control or entrance into the work force of middle class women on grounds that any decline in their devotion to reproductive duties would lead to 'race suicide.' On the other hand, social radicals appealed to eugenics to justify the emancipation of women, contending that contraception would enable women to divorce sexual pleasure from reproduction and thus allow them to approach child-bearing with a purely eugenic interest. Yet what tied eugenicists of all stripes together was an absorption with the role of biological heredity in shaping human beings. A number of eugenicists, notably in France, assumed that biological organisms, including human beings, were formed primarily by their environments, physical as well as cultural. Like the early nineteenth century biologist Lamarck, they contended that environmental influences might even reconfigure hereditary material. However, most eugenicists in most places, particularly in the United States, Britain, and Germany, believed that human beings were determined almost entirely by their 'germ plasm,' their inheritable essence, which was passed on from one generation to the next and which overwhelmed environmental influences in shaping human development. Their belief was reinforced by the rediscovery, in 1900, of Mendel's theory that the biological makeup of organisms was determined by certain 'factors,' which were later identified with genes. Human beings, who reproduce slowly, independently, and privately, are disadvantageous subjects for genetic research. Nevertheless, since no creature fascinates us as much as ourselves, efforts were mounted and institutions established in the early twentieth century to explore human inheritance, especially eugenically relevant traits.

In the English-speaking world, one of the most prominent of these institutions was the Galton Laboratory for National Eugenics, at University College London, under the directorship of the statistician and population biologist Karl Pearson. An adamant anti-Mendelian, Pearson probed the hereditary underpinnings of traits by calculating correlations among relatives or between generations for the frequencies of occurrence of different diseases, disorders, and behaviors, but the approach that dominated eugenic science in most laboratories was Mendelian evaluation — the analysis of phenotypical and family data to account for the inheritance of a variety of medical afflictions and social behaviors in genetic terms. Mendelism dominated the work of the Eugenics Record Office, which was affiliated with, and eventually became part of, the biological research facilities that the Carnegie Institution of Washington sponsored at Cold Spring Harbor, on Long Island, New York, under the directorship of the biologist Charles B. Davenport. Eugenic science of a strongly Mendelian bent was also institutionalized in Germany beginning in 1918, with the establishment of what became the Kaiser Wilhelm Institute for Research in Psychiatry. The institutionalization continued with the creation, in 1923, of a chair for race hygiene at Munich, to which the biologist Fritz Lenz was appointed; and with the founding, in 1927, of the Kaiser Wilhelm Institute for Anthropology, Human Heredity, and Eugenics in Berlin, which was directed by the anthropologist Eugen Fischer, a conservative nationalist Catholic who then headed the Society for Racial Hygiene.

Eugenic research included the study of medical disorders — for example, diabetes and epilepsy — not only for their intrinsic interest but because of their social costs. A still more substantial part of the program consisted of the analysis of traits alleged to make for social burdens — traits involving qualities of temperament and behavior that might lie at the bottom of, for example, alcoholism, prostitution, criminality, and poverty. A major object of scrutiny was mental deficiency — then commonly termed 'feeblemindedness' — which was often identified by intelligence tests and was widely interpreted to be at the root of many varieties of socially deleterious behavior. Typically for eugenic scientists, Charles B. Davenport concluded that patterns of inheritance were evident in insanity, epilepsy, alcoholism, 'pauperism,' and criminality. Such findings were widely disseminated and made their way into middle-class culture. A chart displayed at the Kansas Free Fair in 1929, purporting to illustrate the "laws" of Mendelian inheritance in human beings, declared, "Unfit human traits such as feeblemindedness, epilespy, criminality, insanity, alcoholism, pauperism, and many others run in families and are inherited in exactly the same way as color in guinea pigs."[2]

It was Davenport who helped introduce Mendelism into the influential studies of 'feeblemindedness' that were conducted by Henry H. Goddard, the psychologist who brought intelligence testing to the United States. Goddard speculated that the feebleminded were a form of undeveloped humanity: "a vigorous animal organism of low intellect but strong physique — the wild man of today." He argued that they lacked "one or the other of the factors essential to a moral life — an understanding of right and wrong, and the power of control," and that these weaknesses made them strongly susceptible to becoming criminals, paupers, and prostitutes. Goddard was unsure whether mental deficiency resulted from the presence in the brain of something that inhibited normal development or from the absence of something that stimulated it. But whatever the cause, of one thing he had become virtually certain: it behaved like a Mendelian character. Feeblemindedness was "a condition of mind or brain which is transmitted as regularly and surely as color of hair or eyes."[3] According to later studies by Goddard and others, it also occurred with disproportionately high frequency among lower-income and minority groups — notably recent immigrants in the United States from Eastern and Southern Europe.

Like eugenic scientists elsewhere, many American eugenicists held different national groups and 'Hebrews' to represent biologically different races and express different racial traits. Davenport found the Poles "independent and self-reliant though clannish"; the Italians tending to "crimes of personal violence"; and the Hebrews "intermediate between the slovenly Servians and the Greeks and the tidy

Swedes, Germans, and Bohemians" and giving to "thieving" though rarely to "personal violence." He expected that the "great influx of blood from Southeastern Europe "would rapidly make the American population "darker in pigmentation, smaller in stature, more mercurial . . . more given to crimes of larceny, kindapping, assault, murder, rape, and sex-immorality."[4] Such observations were based upon crude, often anecdotal anthropological data, but after World War I, they obtained quantitative authority in the results of the IQ tests that were administered to the thousands of draftees in the US Army.

Robert Yerkes, the head of the testing program, and others claimed that the tests were almost entirely independent of the environmental history of the examinees, and that they measured 'native intelligence;' but the tests were biased in favor of scholastic skills, and test performance thus depended on the educational and cultural background of the person tested. A post-war testing vogue generated much data concerning the 'intelligence' of the American public, yet the volume of information was insignificant compared with that from the wartime test program, which formed the basis of numerous popular books and articles about intelligence tests and their social import. According to a number of popular analyses of this data, almost four hundred thousand draftees — close to one-quarter of the draft army — were unable to read a newspaper or to write letters home. Particularly striking, the average white draftee — and, by implication, the average white American — had the mental age of a thirteen-year-old.

The psychologist Carl Brigham, one of the wartime Army testers, extended the analysis of the Army data in 1923, in his book *A Study of American Intelligence.* The Army data, Brigham said, constituted "the first really significant contribution to the study of race differences in mental traits." Brigham found that according to their performance on the Army tests the Alpine and Mediterranean "races" were "intellectually inferior to the representatives of the Nordic race." He declared, in what became a commonplace of the popular literature on the subject, that the average intelligence of immigrants to the United States was declining.[5] The average intelligence of black Americans appeared to be just as low as most white Americans had long liked to think it. The Army test data, and various test surveys disclosed that blacks accounted for a disproportionately large fraction of the feebleminded. The Army tests also appeared to indicate that the average black person in the United States had the mental age of a ten-year-old.

Clearly a variety of causes, particularly the poor education of many of the Army test takers, might have accounted for the results. Yet the test data further convinced many Americans not only that mental deficiency was genetically determined but that so was intelligence. White college students scored very well on the alpha tests, and so did high school students from Anglo-Saxon or white-collar homes. These results were taken to mean that gifted students came from homes that, in the words of one educator, "rank high racially, economically, intellectually, and socially."[6] Various psychologists were quick to point out that opening up avenues of opportunity to the children of lower socioeconomic groups probably made no sense; they did not have the IQ points to compete. President George B. Cutten of Colgate University took the Army test results as a starting point to attack the democratization of higher education and wondered aloud in his inaugural address whether democracy itself was possible in a country where the population had an average mental age of thirteen.

The IQ test results reinforced the overall eugenic perception that 'racial degeneration' was occurring in the United States, just as eugenicists elsewhere believed it was occurring in their own countries. Eugenicists fastened on British data which indicated that half of each succeeding generation was produced by no more than a quarter of its married predecessor, and that the prolific quarter was disproportionately located among the dregs of society. Before the war in the United States, leading eugenicists had warned that excessive breeding of the lower classes was giving the edge to the less fit. The growth of IQ testing after the war gave a quantitative authority to the eugenic notion of fitness: The vogue of mental testing not only encouraged fears regarding the "menace of the feeble-minded"; it also identified the source of heedless fecundity with low-IQ groups, and it equated national deterioration with a decline in national intelligence.

Eugenicists in a number of countries, perceiving such social problems as dire, urged that their putatively objective human-genetic knowledge should be used to solve them. They offered the expertise available in eugenic research institutions to state and national governments for the formation of biologically sound public policy, advising interference in human propagation so as to increase the frequency of socially good genes in the population and decrease that of bad ones. The interference was to take two forms: One was 'positive' eugenics, which meant manipulating human heredity and/or breeding to produce superior people. The other was 'negative' eugenics, which meant improving the quality of the human race by eliminating or excluding biologically inferior people from the population.

An interest in positive eugenics was particularly strong among environmentalists like the Lamarckians in France, who contended that more attention to factors such as nutrition, medical care, education, and clean play would, by improving the young, better the human race. Some urged that the improvement should begin when children were in the womb through sound prenatal care. The pregnant mother should avoid toxic substances such as alcohol. She might even expose herself for the sake of her fetus to cultural enrichment, including attendance at fine plays and concerts. But where Mendelian eugenics was strong, little attention was paid to positive eugenics, although eugenic claims did figure in the advent of family-allowance policies in Britain and Germany during the 1930s, and positive eugenic themes were certainly implied in the so-called 'Fitter Family' competitions that were a standard feature of the eugenic programs that were sponsored at a number of state fairs during the 1920s in the United States. These competitions were held in the 'human stock' sections of the fairs. At the 1924 Kansas Free Fair,

FIGURE 16.1: EUGENICS SOCIETY ARCHIVES. SA/EUG/G.43, POSTER: 'ONLY HEALTHY SEED MUST BE SOWN!' BY HAYWOOD (PRINTER?), NORFOLK, N.D. (C. 1935?). WELLCOME INSTITUTE LIBRARY, LONDON.

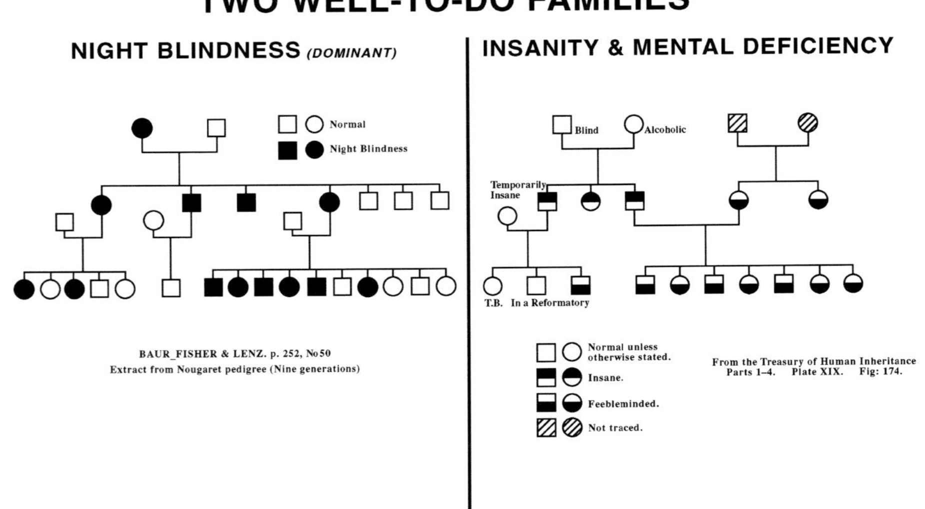

FIGURE 16.2: EUGENICS SOCIETY ARCHIVES. SA/EUG/G.34, CALICO POSTER: TWO WELL-TO-DO FAMILIES: NIGHT BLINDNESS, INSANITY & MENTAL DEFICIENCY. WELLCOME INSTITUTE LIBRARY, LONDON.

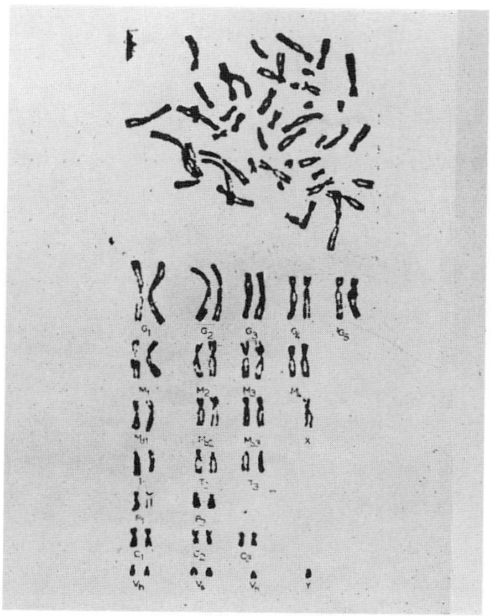

Figure 16.3: Karyotype of a person with Down's syndrome, showing an extra chromosome no. 21 (the third "Vu" chromosome in the bottom row).

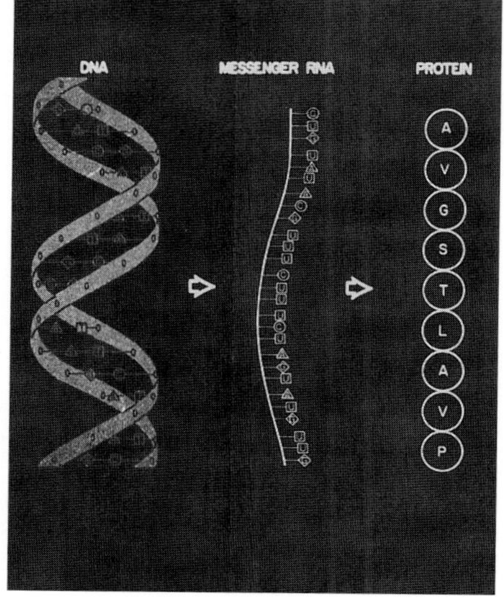

Figure 16.4: A schematic for how DNA codes for the cellular creation of a protein.

Figure 16.5: Winner of a Fitter Family Competition, Texas State Fair in the 1920s, in the Large Family Class.

FIGURE 16.6: THE COPPER MEDAL, AWARDED TO OUTSTANDING INDIVIDUALS IN THE FITTER FAMILY COMPETITION.

FIGURE 16.7: EUGENIC HEALTH EXHIBIT, KANSAS FREE FAIR, 1920S.

winning families in the three categories — small, average, and large — were awarded a Governor's Fitter Family Trophy, which was presented by Governor Jonathan Davis, and "Grade A Individuals" received a medal that portrayed two diaphanously garbed parents, their arms outstretched toward their (presumably) eugenically meritorious infant. It is hard to know what made these families and individuals stand out as fit, but some evidence is supplied by the fact that all entrants had to take an IQ test — and the Wasserman test for syphillis.

Much more was urged for negative eugenics, notably the passage of eugenic sterilization laws. In Britain, despite the pleas of eugenic activists, no such measures ever passed; by the 1930s, the activists were reduced to urging the legalization of voluntary sterilization. But in the United States by the late 1920s, some two dozen American states had framed compulsory eugenic sterilization laws, often with the help of the Eugenics Record Office, and enacted them. The laws were declared constitutional in the 1927 US Supreme Court decision of *Buck v. Bell*, in which Justice Oliver Wendell Holmes delivered himself of the opinion that three generations of imbeciles were enough. The leading state in this endeavor was California, which as of 1933 had subjected more people to eugenic sterilization than had all other states of the union combined.

The most powerful union of eugenic research and public policy occurred in Nazi Germany. Much of the eugenic research in Germany before and even during the Nazi period was similar to that in the United States and Britain, but during the Hitler years, Nazi bureaucrats provided eugenic research institutions with handsome support and their research programs were expanded to complement the goals of Nazi biological policy, exploiting ongoing investigations into the inheritance of disease, intelligence, and behavior to advise the government on its sterilization policy. Fischer's Institute, the staff of which included the prominent geneticist Otmar von Verschuer, trained doctors for the SS in the intricacies of racial hygiene and analyzed data and specimens obtained in the concentration camps. Some of the material — for example, the internal organs of dead children and the skeletons of two murdered Jews — came from Josef Mengele, who had been a graduate student of Verschuer's and was his assistant at the Institute.

The Hitler regime promulgated a eugenic sterilization law in 1933. Although partly modeled on the law in California, it went far beyond any American sterilization statute in several respects, but especially in making sterilization compulsory for all people who suffered from alleged hereditary disabilities ranging from feeblemindedness to physical deformities that were grossly offensive. The law was administered by hundreds of Hereditary Health Courts established to adjudicate the German procreational future. Within three years, German authorities had sterilized some two hundred twenty-five thousand people, almost ten times the number so treated in the previous thirty years in the United States.

Sterilization was only the beginning of the Nazi eugenic program. The Nazi state established a variety of national and local incentives to encourage the fecundity of couples whose offspring were likely to be a credit to the *Volk*. Moreover, as the Hitler regime turned ever more overtly against Jews, Nazi racial and Nazi eugenic policies, which in the beginning were not notably racial, increasingly merged. After the Nuremberg Laws of 1935, marriages between Jews and non-Jews were prohibited. In 1939, the Third Reich moved beyond sterilization to inaugurate euthanasia upon certain classes of the mentally diseased or disabled in German asylums. Among the classes were all Jews, no matter what the state of their mental health. The euthanasia program provided the model that led to the death camps: People designated for euthanasia were herded into rooms disguised as showers, where they were gassed.

Although during the 1930s eugenic sterilization rates rose considerably in the United States, there and in Britain, scientific opinion turned increasingly against eugenics, partly because of its association with the Nazis, partly because of the scientific shoddiness that colored its theories of human heredity. Eugenic science was attacked for its neglect of polygenic complexities — the dependence of a trait on many genes — in favor of reliance on single-gene explanations. It was also indicted for its racial and class bias, its incautious speculations, and its disregard for how social and cultural environment might shape behaviors such as prostitution, not to mention performance on intelligence tests.

However, the eugenic idea remained tantalizing to some scientists and drew an exceptionally talented cadre into human genetics, including the British scientists Ronald A. Fisher, J.B.S. Haldane, Lancelot Hogben, and Julian Huxley and the American Hermann J. Muller. One might call them 'reform eugenicists' because, unlike their predecessors, they held that any eugenics must be free of racial and class bias and must also be made consistent with the development of a sound science of human genetics.

Partly to emancipate the field of human genetics from a prejudicial eugenics, the new students of human heredity preferred to search for well-defined, sharply segregating traits as immune as possible both to uncertainty in identification and to environmental influence. They thus welcomed with particular enthusiasm the rapidly increasing knowledge of the human blood groups, seven of which were known by the early 1930s. The blood groups displayed patterns of inheritance that seemed to conform to Mendel's laws. Being readily identifiable, they also might provide precise and universal genetic markers, presumably located at the same chromosomal place in most individuals, relative to which the genes for other traits might be linked.

Eugenicists placed high hopes in such linkage studies, thinking that tying traits to markers like the blood types would be promising for eugenic prognosis. They had long been stymied by the problem of tagging the carriers of single genes for recessive traits, which were not expressed until — too late from a eugenic point of view — they joined homozygously in offspring. Linkage studies might reveal that a deleterious recessive gene occurred on the same chromosome as did one of the blood groups; it would not be necessary to know which chromosome was involved to spotlight someone found to have that blood group as a probable carrier of the recessive. Similarly, if the gene was a dominant that expressed itself later in life — for example, the gene for Huntington's disease — people fated to contract it could be advised before they had children of the chance of transmitting it to their offspring and they might then refrain from reproduction. A good deal of effort was spent, especially in England, on the search for linkages, but by the late 1940s, none was found between, on the one side, the blood groups or any universal, non-sex-linked character and, on the other, any type of genetic disease or disorder.

The disappointing results dashed hopes for eugenic prognoses of reproductive outcomes — and also for the emerging field of genetic counseling. The field, which began to develop in the decade after World War II, aimed to provide prospective parents with advice about what their risk might be for bearing a child with a genetic disease. Without linkage indicators, genetic counselors had nothing to offer couples except a calculation of odds on the likelihood of bearing a child with such a malady. Still, some geneticists sought to turn the practice of genetic counseling to eugenic advantage — to reduce the incidence of genetic disease in the population, and by extension to reduce the frequency of deleterious genes in what population geneticists were coming to call the human gene pool. To that end, some claimed that it was the counselor's duty not simply to inform couples about the possible genetic consequences of their union but also to instruct them whether or not to bear children at all. During the 1950s, however, the standards of genetic counseling turned strongly against the offering of eugenically oriented advice — that is, advice aimed at the welfare of the gene pool rather than that of the family. The new standards had it that no counselor had the right to tell a couple not to have a child, even for the sake of the couple's own welfare. In the wake of the Holocaust, eugenics had become something of a dirty word.

Yet at the very time that eugenics fell completely out of fashion, genetic research was raising the curtain on a new, potentially revolutionary era in the control of heredity, including the human variety. Rapid progress in human cytogenetics — particularly the recognition in 1959 that Down's syndrome arises from a chromosomal anomaly — soon made prenatal diagnosis possible, with the option of abortion for women at risk of giving birth to children with severe chromosomal disorders. The unveiling of the structure of DNA, in 1953, opened the door to the discovery of how genes actually control the development (and misdevelopment) of organisms. By the mid-1960s, it was understood that genes embody a code, written into their chemical structure, that instructs the cell what specific proteins to manufacture. Finding proteins associated with diseases made it possible to identify flaws in DNA that generated illness and to detect disease genes in recessive carriers and fetuses homozygous for illnesses such as Tay-Sachs disease and sickle-cell anemia. The working out of the genetic code inspired Galtonian visions. As early as 1969, Robert Sinsheimer, a prominent molecular biologist at the California

Institute of Technology declared that "for the first time in all time, a living creature understands its origin and can undertake to design its future" — and, in consequence, might eventually control its own evolution.[7]

Sinsheimer's vision may have seemed utopian, but during the quarter century since his pronouncement, the pace of advance in molecular genetics appeared to be enlarging the power of genetic manipulation at a dizzying pace. The invention of the technique of recombinant DNA enabled scientists to snip a gene from one species and insert it into the genome of another — which they promptly did. They genetically engineered bacteria, plants, and animals, creating transgenic organisms by inserting foreign genes into fish, cows, fowl, and mice. Analysis of the human genome revealed that it contains numerous markers — precisely what the reform eugenicists of the 1930s had hoped to find with the blood groups — that could be linked to disease genes. At the end of the 1980s, the United States led the scientific nations of the world in establishing the Human Genome Project, which aims to establish a complete marker map of the human genome and obtain the DNA sequence of all the 100,000 genes estimated to comprise the human genetic complement.

In the 1990s, many biomedical scientists found the medical and therapeutic prospects of genetics exhilirating — but many lay observers regarded them with apprehension. A number deplored the creation of transgenic species, calling the practice an attack against the sanctity of life and deploring such interference with God's designs. A clear subtext of the critique was that if genetic manipulation of animals could be done today, it would surely be attempted with human beings in the future — and not necessarily with benign intent. In April 1991, an exposition opened in the hall atop the great arch of *La Defense,* in Paris, under the title: *Life in a Test Tube: Ethics and Biology.* The biological exhibits included displays about molecular genetics and the human genome project. The ethical worries were manifest in a catalogue statement by the writer Monette Vaquin that was also prominently placarded at the genome display:

> Today, astounding paradox, the generation following Nazism is giving the world the tools of eugenics beyond the wildest Hitlerian dreams. It is as if the unthinkable of the generation of the fathers haunted the discoveries of the sons. Scientists of tomorrow will have a power that exceeds all the powers known to mankind: that of manipulating the genome. Who can say for sure that it will be used only for the avoidance of hereditary illnesses?[8]

Vaquin's apprehensions have been echoed frequently by scientists and social analysts alike. They indicate that the shadow of eugenics hangs over any discussion of the social implications of the Human Genome Project in particular but generally over all considerations of the genetic manipulation of human beings. Indeed, since the opening of the DNA era, observers have wondered whether new genetic knowledge will be deployed for positive eugenics, for attempts to produce a super race or at least to engineer new Einsteins, Mozarts, or athletes like Martina Navratilova. The apprehensions are not entirely unfounded. In Singapore in 1984, Prime Minister Lee Kwan Yew deplored the relatively low birth rate among educated women, contending that their intelligence was higher than average and that they were thus allowing the quality of the country's gene pool to diminish. Since then, the government, embracing a crude positive eugenics, has adopted a variety of incentives — for example, preferential school enrollment for offspring — to increase fecundity among such women and provided similar benefits to their less educated sisters who would have themselves sterilized after the birth of a first or second child.

Many commentators — for example, the late Nobel Laureate biologist Salvador Luria — have cautioned that the human genome project is likely to foster a revival of negative eugenics. The technology that permits the identification of disease genes either directly or through their association with markers might be applied to single out fetuses, children, or parents who will become diseased or disabled or presumptively 'anti-social.' The state might then intervene in reproductive behavior so as to discourage the transmission of these genes in the population. Indeed, in 1988, China's Gansu Province adopted a eugenic law that would — so the authorities said — improve "population quality" by banning the marriages of mentally retarded people unless they first submit to sterilization. Since then, such laws have been adopted in other provinces and have been endorsed by Prime Minister Li Peng. The official newspaper *Peasants Daily* explained, "Idiots give birth to idiots."[9]

Economics may well prove to be a powerful incentive to a new negative eugenics. In July 1988, the European Commission proposed the creation of a human genome project for the European Community, calling the venture a health measure under the title "Predictive Medicine: Human Genome Analysis." Its rationale rested on a simple syllogism — that many diseases result from interactions of genes and environment; that it would be impossible to remove all the environmental culprits from society; and that, hence, individuals could be better defended against disease by identifying their genetic predispositions to fall ill. According to the summary of the proposal: "Predictive Medicine seeks to protect individuals from the kinds of illnesses to which they are genetically most vulnerable and, where appropriate, to prevent the transmission of the genetic susceptibilities to the next generation."[10] In the view of the Commission, the genome proposal, which it found consistent with the Community's main objectives for research and development, would enhance the quality of life by decreasing the prevalence of many diseases distressful to families and expensive to European society. Over the long term, it would make Europe more competitive — directly, by strengthening its scientific and technological base but also indirectly, by helping to slow the rate of increase in health expenditures.

The emphasis on 'preventive medicine' raised a neo-eugenic flag in the minds of many members of the Parliament, notably Benedikt Härlin, a Green Party member from West Germany, where sensitivity to anything that might smack of eugenics was high. In a hard-hitting report, Härlin reminded the Community that in the past eugenic ideas had led to "horrific consequences" and declared that "clear pointers to eugenic tendencies and goals" inhered in the intention of protecting people from contracting and transmitting genetic diseases. The application of human genetic information for such purposes would almost always involve decisions — fundamentally eugenic ones — about what are "normal and abnormal, acceptable and unacceptable, viable and non-viable forms of the genetic make-up of individual human beings before and after birth." The Härlin report also warned that the new biological and reproductive technologies could make for a "modern test tube eugenics," a eugenics all the more insidious because it could disguise more easily than its cruder ancestors "an even more radical and totalitarian form of 'biopolitics.'"[11]

Memories of Nazi eugenics have also been forceful in debates over human genetic issues in the United States, where the focus on such diseases or disabilities is said to cast a shadow over the worth of people who suffer from them. The attitude that a newly conceived child with such an affliction merits abortion has been atacked as stigmatizing the living who have the ailment. Protests have come from individuals and families with diseases such as cystic fibrosis and sickle cell anemia, but especially from the handicapped and their advocates. Barbara Faye Waxman, an activist for the disabled who herself has a neuromuscular impairment, criticizes her fellow workers in a Los Angeles Planned Parenthood clinic for displaying "a strong eugenics mentality that exhibited disdain, discomfort and ignorance toward disabled babies."[12]

The more expensive health care becomes, the greater the possibility that taxpayers will rebel against paying for the care of those whom genetics dooms to severe disease or disability. To be sure, the more that is learned about the human genome, the more will it become obvious that we are all susceptible to one kind of genetic disease or disability; we all carry some genetic load and are likely to fall sick in one way or another. Since everyone is in jeopardy of genetically based illness, then everyone would have an interest in a well-financed public health program — national health insurance — and everyone would have a stake in extending its benefits universally. However, not everyone's genetic load is the same; some are more severe and costly than others. It is likely that, on grounds of cost, even a national health system might seek to discriminate between patients, using the criterion of how expensive their therapy and care might be. Public policy in all the Western industrial countries might feel pressure to encourage, or even to compel, people not to bring genetically affected children into the world — not for the sake of the gene pool but in the interest of keeping public health costs down.

All this said, however, a number of factors are likely to offset a scenario of socially controlled reproduction let alone a revival of a broad-based negative eugenics. Analysts of civil liberty know that reproductive freedom is much more easily curtailed in dictatorial governments like that of the Nazis than in democratic ones. Eugenics profits from authoritarianism — indeed, almost requires it. The institutions of political democracy may not have been robust enough to resist altogether the violations of civil liberties characteristic of the early eugenics movement, but they did contest them effectively in many places. The British government refused to pass eugenic sterilization laws. So did many American states, and where they were enacted, they were often unenforced. It is far-fetched to expect a Nazi-like eugenic program to develop in countries with strong protections of civil liberties. If a Nazi-like eugenic program becomes a threatening reality in any country, that country will have have a good deal more to be worried about politically than just eugenics.

What makes contemporary political democracies unlikely to embrace eugenics is that they contain powerful anti-eugenic constituencies. Awareness of the barbarities and cruelties of state-sponsored eugenics in the past has tended to set most geneticists and the public at large against such programs. Most geneticists today know better than their early twentieth-century predecessors that ideas concerning what is 'good for the gene pool' are highly problematic. Then, too, handicapped or diseased persons are politically empowered, as are minority groups, to a degree that they were not in the early twentieth century. They may not be sufficiently empowered to counter all quasi-eugenic threats to themselves, but they are politically positioned, with allies in the media, the medical profession, and elsewhere, including the Roman Catholic Church — long a staunch opponent of eugenics — to block or at least to hinder eugenic proposals that might affect them.

The power of anti-eugenic groups in the West was manifest in the European Parliament's response to the European Commission's proposal for a human genome project. Opposition was directed not at the human genome project as such but, following Härlin's lead, at the way the Commission had framed it as a project in predictive medicine. In mid-February 1989, the Parliament modified the genome proposal in numerous respects, including the complete excision from the text of the phrase "predictive medicine." Collectively, the modifications were mainly designed to exclude a eugenically oriented health policy; to prohibit research seeking to modify the human germ line; to protect the privacy and anonymity of individual genetic data; and to ensure ongoing debate into the social, ethical, and legal dimensions of human genetic research.

The changes drew support not only from the Greens but also from conservatives on both sides of the English Channel, including German Catholics. The Parliament's action prompted Filip Maria Pandolfi, the new European Commissioner for Research and Development, to freeze indefinitely Community human genome monies in early April 1989. The move was believed to be the first by a commissioner to block one of Brussels' own technological initiatives. Pandolfi explained that time for reflection was needed, since "when you have British conservatives agreeing with German Greens, you know it's a matter of concern."[13] The reflection produced, by mid-December 1989, the adoption of a Modified Proposal from the European Commission that accepted the thrust of the amendments and even the language of a number of them. There would be a three-year program of human genome analysis as such, without regard to predictive medicine. The program committed the Community in a variety of ways — most notably, by prohibiting human germ line research and genetic intervention with human embryos — to avoid eugenic practices, prevent ethical missteps, and protect individual rights and privacy. Several percent of the genome budget would be devoted to research in related ethical and legal issues, a policy that James D. Watson, the first director of the genome project in the United States, had already laid out for the American effort.

The eugenic past is prologue to the human genetic future in only a strictly temporal sense — that is, it came before. Of course, the imagined prospects and possibilities of human genetic engineering remain tantalizing, even if they are still the stuff of science fiction, and they will continue to elicit both fearful condemnation and enthusiastic speculation. However, the near-term ethical challenges of the human genome project lie neither in private forays in human genetic improvement nor in some state-mandated program of eugenics. They lie in the grit of what the project will produce in abundance: genetic information. They center on the control, diffusion, and use of that information within the context of a market economy.

The advance of human genetics and biotechnology has created the capacity for a kind of "homemade eugenics," to use the insightful term of the analyst Robert Wright — "individual families deciding what kinds of kids they want to have."[14] At the moment, the kinds they can select are those without certain disabilities or diseases, such as Down's syndrome or Tay-Sachs. Most parents would probably prefer just a healthy baby, if they are inclined to choose at all. But in the future, some might have the opportunity — for example, via genetic analysis of embryos — to have improved babies, children who are likely to be more intelligent or more athletic or better looking (whatever those comparative terms mean).

Will people pursue such opportunities? Quite possibly, given the interest that some parents have shown in choosing the sex of their child or that others have pursued in the administration of growth hormone to offspring who they think will grow up too short. Benedikt Härlin's report to the European Parliament noted that the increasing availability of genetic tests was generating increasingly widespread pressure from families for "individual eugenic choice in order to give one's own child the best possible start in a society in which heredity traits become a criterion of social hierarchy." A 1989 editorial in *Trends in Biotechnology* recognized a major source of the pressure: "'Human improvement' is a fact of life, not because of the state eugenics committee, but because of consumer demand. How can we expect to deal responsibly with human genetic information in such a culture?"[15]

The torrent of new human genetic information poses challenges to systems and values of social decency. Employers have sought to deny jobs to applicants with a susceptibility — or an alleged susceptibility — to disorders or illnesses arising from features of the workplace. Life and medical insurance companies want to know the genomic signatures of their clients, their profile of risk for disease and death. Even national health systems might choose to ration the provision of care on the basis of genetic propensity for disease, especially to families at risk for bearing diseased children. The eugenic past has much to teach about how to avoid repeating its mistakes — not to mention its sins. But what bedeviled our forebears will not necessarily vex us, certainly not in the same ways. In human genetics as in so many other areas of life, the flow of history compels us to think and act anew — not about eugenics but about the control of human genetic information by geneticists, the media, insurers, employers, and government.

References

1. Francis Galton, *Inquiries into the Human Faculty.* (London: Macmillan, 1883), pp. 24–25; Karl Pearson, *The Life, Letters, and Labours of Francis Galton* (3 vols. in 4; Cambridge: Cambridge University Press, 1914–1930), IIIA, 348.
2. Daniel J. Kevles, *In the Name of Eugenics: Genetics and the Uses of Human Heredity.* (Cambridge, MA: Harvard University Press, 1995), p. 62.
3. Henry H. Goddard, *Feeble-mindedness: Its Causes and Consequences.* (New York: Macmillan, 1914), pp. 4, 7–9, 14, 17–19, 413, 504, 508–9, 514, 547.
4. Charles B. Davenport, *Heredity in Relation to Eugenics.* (New York: Henry Holt, 1911), pp. 216, 218–219, 221–22.
5. Kevles, *In the Name of Eugenics*, pp. 82–83.
6. *Ibid.*, p. 83.
7. Robert Sinsheimer, "The Prospect of Designed Genetic Change," *Engineering and Science* (April 1969), **32**:, p. 8.
8. *La Vie en Kit: Éthique et Biologie.* (Paris: L'Arche de la Defense, 1991), p. 25.
9. *The New York Times* (15 August 1991), p. 1.
10. Commission of the European Communities, *Proposal for a Council Decision Adopting a Specific Research Programme in the Field of Health: Predictive Medicine: Human Genome Analysis (1989–1991)* (20 July 1988), COM (88) 424 final-SYN 146, pp. 1, 3.
11. European Parliament, Committee on Energy, Research, and Technology, *Report . . . on the Proposal . . . Decision Adopting a Specific Research Programme on the Field of Health: Predictive Medicine: Human Genome Analysis (1989–1991)*, pp. 23–28.
12. *The New York Times* (4 July 1991), p. 12.
13. *London Financial Times* (5 April 1989), BioDoc: documents on biotechnology, European Economic Community, DG-XII, Brussels.
14. Robert Wright, "Achilles Helix," *The New Republic* (9 & 16 July 1990), **203**: 27.
15. European Parliament, Committee on Energy, Research, and Technology, *Report . . . on the Proposal . . . Human Genome Analysis (1989–1991)*, pp. 25–26; John Hodgson, "Editorial: Geneticism and Freedom of Choice," *Trends in Biotechnology* (Sept. 1989), p. 221.

Further Reading

Adams, Mark B. (Ed.), *The Wellborn Science: Eugenics in Germany, France, Brazil, and Russia* (New York: Oxford University Press, 1990).

Carol, Anne. *Histoire de L'Eugenisme en France: Les Médecins et la Procréation XIXme-XXme Siècle.* (Paris: Seuil, 1995).

Cook-Deegan, Robert. *The Gene Wars: Science, Politics, and the Human Genome.* (New York: W.W. Norton, 1994).

Gould, Stephen Jay. *The Mismeasure of Man.* Rev. (Ed.). (New York: W.W. Norton, 1996).

Kevles, Daniel J. *In the Name of Eugenics: Genetics and the Uses of Human Heredity.* (Cambridge, MA: Harvard University Press, 1995).

Kevles, Daniel J. and Leroy Hood (Eds.), *The Code of Codes: Scientific and Social Issues in the Human Genome Project.* (Cambridge, MA: Harvard University Press, 1992).

Kitcher, Philip. *The Lives to Come: The Genetic Revolution and Human Possibilities.* (New York: Simon and Schuster, 1996).

Lemaine, Gérard and Benjamin Matalon. *Hommes supérieurs, hommes inférieurs.* (Paris: Armand Colin, 1985).

Ludmerer, Kenneth. *Genetics and American Society: A Historical Appraisal.* (Baltimore: Johns Hopkins University Press, 1972).

Müller-Hill, Benno. *Murderous Science: Elimination by Scientific Selection of Jews, Gypsies, and Others, Germany, 1933–1945.* (New York: Oxford University Press, 1988).

Nelkin, Dorothy M. and Susan Lindee. *The DNA Mystique: The Gene as a Cultural Icon.* (New York: W.H. Freeman, 1995).

Proctor, Robert N. *Racial Hygiene: Medicine Under the Nazis.* (Cambridge: Harvard University Press, 1988).

Schneider, William H. *Quality and Quantity: The Quest for Biological Regeneration in Twentieth Century France.* (Cambridge: Cambridge University Press, 1990).

Searle, G.R. *Eugenics and Politics in Britain, 1900–1914.* (Leyden: Noordhoff International Publishing, 1976).

Soloway, Richard A. *Demography and Degeneration: Eugenics and the Declining Birthrate in Twentieth Century Britain.* (Chapel Hill: University of North Carolina Press, 1990).

Weindling, Paul. *Health, Race and German Politics between National Unification and Nazism, 1870–1945.* (Cambridge: Cambridge University Press, 1990).

Weiss, Sheila Faith. *Race Hygiene and National Efficiency: The Eugenics of Wilhelm Schallmayer.* (Berkeley: University of California Press, 1987).

Wexler, Alice. *Mapping Fate: A Memoir of Family, Risk, and Genetic Research.* (New York: Times Books, 1995).

CHAPTER 17

In the Name of Science

BERNADETTE BENSAUDE VINCENT

Popularization of science can mean many different things, depending on who is dealing with it. Historians of mentalities are concerned with the images of science circulating in various popular cultures. Historians of the press, publishing, radio broadcasting or television would include the popularization of science within the development of mass communication. Historians of science are more inclined to consider the emergence of the popularization of science as a result of the specialization of science. Since scientific knowledge is increasingly specialized, complex and esoteric, it is assumed that mediations are required in order to bridge the gulf between a small elite of learned scientists and the mass of other citizens. The notion of a gulf between science and the public, widely shared by academic scientists and science journalists, thus appears as a prerequisite underlying the whole development of science popularization. Since the emphasis of this volume is the history of science, this chapter will be mainly concerned with the latter view of science popularization. Privileging this approach, however, does not mean that the evolution of science popularization will be described as a necessary consequence of the advancement of science.

The growth of popular scientific press occured in the nineteenth century when science became intensively professionalized with academic curricula, chairs, learned journals and disciplinary societies. It should be noticed, however, that defining science popularization as a kind of side-effect of a universal process of professionalization implies a tacit and largely un-noticed premise: that each advance of knowledge widens the circle of people who do not know that promoting the advancement of science is also *de facto* generating more ignorance. It means that the so called 'gulf' is expanding and proliferating. To the great traditional 'gulf' between the scientist and the layman, a series of smaller gaps should be added between groups of experts sealed off in their narrow specialties. Thus the history of science should be complemented by a history of the advancement of ignorance and a history of the attempts to maintain public knowledge.

Historians of science, usually more attracted by the great discoveries of a few leading figures have paid little attention to popular science writers. The few historical studies of popularization follow the main lines of disciplinary boundaries and are usually devoted to the public reception of established doctrines, like Darwinism or Relativity physics. As long as popular works are considered as a by-product of the academic production of science, they are not worth investigating for an understanding of the development of scientific knowledge.

Because of this relative neglect, materials on most national cases is too scarse to allow a deep understanding of the circumstances which prompted the development of popularization in local contexts. This chapter, based on a few British, French and American cases, can only suggest a more complex and interactive view of the joint development of academic science and popular science.

Mass Communication

In 1867, Louis Figuier, a prolific French scientific writer wrote: "Science is a sun: everybody must move closer to it for warmth and enlightenment." "Everybody" had to be interested in science. Such was the credo and the program underlying all of the nineteenth-century developments of science popularization. Whereas in the eighteenth century the public of science was enlightened 'amateurs' who attended public lectures of chemistry and electricity and occasionally cultivated science as a leisure activity in their elegant 'cabinets,' in the nineteenth century a mass consumption of science developed.

The rapid growth of a cheap press in the nineteenth century, as a result of technological advances such as rotary systems, made the diffusion of science into the public at large a major objective. Hundreds of books, journals and magazines aimed to place science within everyone's reach. Some of them are still published today like *The Scientific American,* founded in 1845 or the British Weekly *Nature* founded in 1869.[1] Indeed, science was an integral part of the development of a popular press in Europe. The presence of science in social life was increased considerably during the second half of the nineteenth century by the World Exhibitions. At regular intervals they attracted millions of visitors, bringing together industrialists and workers, experts and amateurs, over a period of six months. It was progress that was invariably celebrated with technology displayed at the forefront of civilization in the huge, ever-bigger, ever-higher, machine galleries, in the enormous steam-engines, and later in the magics of the electrical palace at the Paris Exhibition, in 1900. Though science as such remained inconspicuous in the exhibitions, it was omnipresent in the accompanying rhetoric, being praised as a precondition of technological progress and prowess. While propagating no more than a bare minimum of knowledge, the World Exhibitions had clear educational intentions and, in fact, conveyed strong images of science and technology.

The creation of such public spaces as science museums, botanical and zoological gardens, observatories . . . had also been encouraged by the private mass-consumption of popular science periodicals and books at home. From cheap booklets to heavy dictionaries, the nineteenth century inaugurated all kinds of printed publications. Some of them were modest, in their intellectual ambitions, and their prices. On the assumption that scientific knowledge was useful and even indispensable in everyday life, they were thoroughly practical and included natural science as well as agriculture, medicine, hygiene, rural economics and new technologies. Other magazines, edited by active scientists such as Alexander Humboldt who founded *Kosmos* in Germany (1845), and Normann Lockyer editor of *Nature* in Britain (1869), provided a broad view of the advances in natural sciences. Intellectual magazines like the *Quarterly Review* (1866) *Revue des Deux Mondes* (1830), and *Die Deutsche Rundschau* (1874) presented science as a cultural activity and debated its role in modern society and civilization.

A wide range of books and serial publications were sold to suit all tastes, all classes, all conditions: manufacturers, farmers, teachers, women, children, gentlemen . . . . By the end of the nineteenth century popular science had become a lucrative business, managed by prosperous publishers. The circulation of such books was obviously dependent on the spread of literacy into the lower urban and rural classes i.e., on the development of educational systems.

An unexpected decline of popular science literature can be noticed at the turn of the century in such countries as Great Britain, Germany, Italy and France. By 1900, when the Fairy Electricity was celebrated at the Paris exhibition, the public's confidence in science as providing the key solution to all kinds of human problems seemed to have seriously declined. The repetitive celebrations of progress no longer appealed to the public, and the 'value of science' had become a matter of debate in the last decade of the nineteenth century, especially in France where a controversy over the 'bankruptcy of science' was widely echoed in public opinion.

However the tremendous growth of mass communication of science during the nineteenth century resulted in a definite image of the public. Popular science becoming a profitable market good, one of the many mass productions which developed in the late nineteenth century — at least in industrialized countries — the public of science can be adequately considered as 'consumers' of popular science. When the notion of a 'gulf' was used, it was mainly meant as a gap to fill in the market place, with a periodical addressed to a specific category of public between low culture and upper culture, for instance. The social division between scientists and the lay public induced by the professionalization of scientific work throughout the nineteenth century, was undoubtedly reinforced and stabilized by the emergence of the category of 'science consumers.'

This does not mean that the distinction between knowledge producers and knowledge consumers was clearly established. There was no consensus on the role of popular science press and its publications. Should they mirror academic science or promote amateur-practice of science? In the early twentieth century, popular science did not necessarily mean 'popularized science.' Rather, popular science and academic science formed two distinct networks. The network of professional scientists emerging through international conferences and academies was reflected or copied by a network of popular science writers, popular observatories or botanical gardens, of popular magazines and publishers exchanging articles, printing plates, compilations and translations, observations or specimens. Both of these networks were international, sometimes they were distant more often they interacted.

A Cultural Offensive

While we still use the phrases 'popular medicine' and 'pop culture,' in the twentieth century, the words 'popular science' are not supposed to refer to any thing other than the image of science as reflected by vehicles of pop culture such as commercial advertisements, best-seller fictions or television serials.[2] Instead the term 'popularization' has prevailed in English language and the more problematic rather inelegant term 'vulgarization' in the French language. The notion of a 'popular science,' a science distinct from that of professional scientists, taught in our schools, is no longer accepted. Any alternative science, any practice that would not been regulated by the current norms of the scientific community, is labeled a 'pseudo-science.' Science itself is one, unique. If professional scientists hold the monopoly of true, valid statements, everyone else — whatever his or her capacities — is 'the public.' What was the role of science communication in this process?

In the aftermath of World War I, a number of star scientists were promoted in the mass media. The image of science as pure and disinterested sacrifice was embodied in Marie Curie and Einstein. The journalists paid little attention to the subject of their achievements and often humorously distorted them. In the case of Marie Curie, the purpose was less to put science in 'everyone's reach,' than to raise public funds for research on radium. The public was supposed to retain one message: that science is cultivated by exceptional individuals, living on a separate planet, in a pure world of spiritual values. "Extraordinary intelligence, persistence, foresight, and modesty" are the major features emerging from a survey of the images of scientists in the American popular press.[3] This image of 'great scientists,' whose ethos compensates for the acceptance of public ignorance, is closely connected to the growing prestige of the Nobel Prizes in the press, following their creation in 1901.

Although purity and detachment were already common virtues attached to the popular image of the savant in the nineteenth century, they were given a new meaning after the poisonous clouds of the war. The use of chemical weapons and the mobilization of scientists in the war efforts on both sides deeply affected the image of a wonderful age, still prevailing in 1913, when even artillery could be described as an instrument of peace "provided it was more and more scientific."[4]

The effects not only on the public at large but in the scientific communities as well were ambivalent. On the one hand, the war increased the social visibility of science in the newspapers and reinforced the image of scientists as public servants. Popular science magazines systematically combined patriotic propaganda and scientific information. But after the war, the image of science bringing death and destruction raised doubts about the legitimacy of scientific endeavor and the dangers of uncontrolled progress.

This "crisis of faith," which sometimes turned into "a revolt against science,"[5] was followed by a number of initiatives for disseminating science to the public at large through various media such as magazines, exhibitions, encyclopedias, movies and books. It was during the inter-war period that science popularization was established as a public institution in many countries. Most initiatives continued the nineteenth-century genres of popular science and the tradition of popular series of cheap science books, relying on a supposed 'public hunger for knowledge' flourished. In the United Kingdom, for instance, Benn's Sixpenny Library competed with Routledge's "Introductions to Modern knowledge." In the thirties the publisher Allen Lane launched the Penguin editions, which sold between 50,000 and 60,000 copies. The commercial success of the Penguin editions relied on the opening up of distribution outlets beyond the normal booksellers. Though the publishers were bold in commercial enterprise, this 'revolution' was hardly an intellectual one. There was a rather strict conformism with respect to the contents of the books which relied on prestigious and classical authors such as Julian Huxley, James Jeans or the French popular enthomologist Jean-Henri Fabre.[6]

Science museums also boomed in the thirties. Three opened in the USA: the Chicago Museum of Science and Industry in 1933; Philadelphia's Franklin Institute of Science, in 1934; and the New York Museum of Science and Industry which opened at the Rockefeller Center in 1936. In France, the Palais de la découverte opened in 1937, the Musée de l'Homme in 1938 and a Popular Arts and Crafts Museum, planned in the thirties, eventually opened in the 1960s.

The campaign for improving the public image of science involved more and more people. The American case is extremely interesting in this respect because, in stark contrast to France where big projects were supported by the state, it shows a variety of initiatives from different social groups. Chemical companies, like Du Pont which met serious problems in its public image after selling explosives to the Allies during the First World War, sponsored a number of popular exhibits. The chemists' 'crusade' culminated in the famous Philadelphia display of 1937, called after Du Pont's new slogan "Better Things for Better Living . . . Through Chemistry."[7]

In addition to economic interests, academic strategies also played a part. In 1919, the American Chemical Society set up a "News Service" to keep the public informed of chemical developments and more seriously to restore public confidence in chemistry. Another agency, the "Science Service," was created in 1920 by L.W. Scripps, a newspaper magnate with the support of the American Association for the Advancement of Science (AAAS), the National Academy of Sciences and the

National Research Council. The Science Service was intended to disseminate information through a weekly Science Newsletter and radio broadcasts, and also helped to train science journalists. After a formative period of a few years, it was so successful in its encouragement of science in the media that daily newspapers hired their first full time science editors.[8] It was as a result of a close interaction of journalists and scientific institutions that science journalism emerged as a profession in the USA. In the 1930s a small group of full time science writers, supported by the AAAS, created the National Association of Science Writers, a professional association formed to legitimize and protect the activity of its members in their relations both to publishers and to scientists.

While science writers and journalists were promoting their professional interests, working scientists did not retire from the public stage. On the contrary, scientific communities, who felt weakened by the war and threatened by the economic depression, were an integral part of the public communication of science. In their efforts to recruit young scientists and repair the losses of the war, to raise public funds and build up institutions for research, they needed to address the public. Scientists took an active part in the emergence of radio broadcasts in the 1930s, organizing whole series of programs of the most recent scientific developments, such as the 'big-bang' theory. They were also involved in the field of International Exhibitions.[9] Supporting the Chicago Century of Progress Exhibition was the National Research Council. As early as 1927 Georges E. Hale, an astrophysicist, suggested that the Chicago Fair should exhibit "the services of science to humanity during the past hundred years." Visitors were welcomed in a Hall of Science, designed as a temple, a place to worship science as the highest and purest spiritual value. The slogan at Chicago in 1933 — "Science discovers, industry applies, man conforms" — clearly suggested a linear process of development with science as the prime mover. The cultural offensive of American scientists, culminating in the New York Fair in 1938, was clearly intended to restore public confidence in scientific expertise and large corporations.

The Paris World Exhibition, "Sciences et arts," opened in 1937 with a Palais de la découverte. Designed by the French physicist Jean Perrin as a kind of cathedral of science, the Palace of Discovery was intended as a recruiting station for scientific research. Without any concern for preserving collections, the Palais was focused on contemporary discoveries with working experiments, interactive displays and emotional ambiances. The emphasis was on the pursuit of knowledge for its own sake. Science was depicted as a purely investigative and exploratory activity, a promised land which brought public welfare and technological inventions, 'in addition' to its purely disinterested aims. In Perrin's view, pure science was more fruitful than applied research, at least in the long term. In showing the beauty of science and scientific research, in celebrating the life of heroic figures of scientists, the Palais was intended to attract young visitors into the exciting adventure of scientific discovery.

EXHIBITS AND DEMONSTRATIONS (continued).

Room 20. Demonstration of " Daylight " Lamps, using special glass bulbs.

Room 24. **Primary Battery Development.** A laboratory-factory.
The components of a primary battery are exhibited.

Room 25. **Primary Batteries.**
The testing of primary batteries and their characteristics — Constant temperature room.
Physico-Chemical Measurements.
Apparatus for studying the deposition of films on wires.—Viscosity Meter.

Room 26. **Photometry of Life Test Lamps.**
Cubical Integrating Photometer.
Colour match method of measuring filament temperature.

Room 27. **Life Test of Lamps.**
Racks for the life testing of electric lamps of both experimental and standard types.
The systematic proving of the quality and the life of Osram Lamps is undertaken here.

Room 29. **Standard Photometry.**
This room is not yet equipped.
Demonstrations are given of lamps and luminous devices using the electric discharge in gases.
Different types of neon and other discharge lamps.
Trigger device using a neon lamp.
Discharge tube oscillograph.
Special Spectroscope Lamp with intense Neon Glow (made for Messrs. Adam Hilger.)

Administration Section.—(South Block).
This part of the building is open to general inspection.
In the library (ground floor) is demonstrated the Laboratory General Information Index and the arrangements for the circulation of periodicals in the Laboratory.
On the first floor adjoining the drawing office is demonstrated the process for reproduction of typed matter and drawings.

OPENING

OF THE

RESEARCH LABORATORIES

OF

THE GENERAL ELECTRIC Co., Ltd.

WEMBLEY - - - FEBRUARY 27th, 1923.

Opening Ceremony.

2.45 p.m. to 3.45 p.m.

In HIGH TEMPERATURE LABORATORY (*at present unequipped*).

Chairman - - MR. HUGO HIRST.

FIGURE 17.1.

EXHIBITS AND DEMONSTRATIONS.

(The rooms are enumerated below in the order of the Route).

Room 1. **Vacuum Physics Laboratory.**
Demonstration of the principle of the Gasfilled Lamp—Langmuir's discovery of the constant thickness conduction layer.
Typical evacuation apparatus and demonstration of the operation of the diffusion pump.
Wheatstone Bridge Method of determining high vacua.
Apparatus for X-Ray analysis of Crystal structure, and X-Ray Spectroscopy.

Room 6. **Lamp Development**—A laboratory-factory.
Demonstration of the various processes in the making of electric lamps by automatic machinery.
Glass Blowing.

Room 5. **Valve Development**—A laboratory-factory employed in connection with Research Work for the M.O. Valve Co., Ltd.
This installation is not complete. Pump tables for making thermionic valves in the laboratory are exhibited.
A High Tension Electrostatic voltmeter.
X-Ray screening.

Room 3. **Electrical Laboratory.**
Demonstration of high tension rectified current with applications to corona discharge and the electrical precipitation of fumes. (Lodge Cotterill process).
The Photoelectric cell and its use for the measurement of the blackening of lamp and valve bulbs.
Constant ratio current divider for measuring very small direct currents.
Hay Sullivan alternating current bridge.
Telephone Repeater Tests and Measurements.
Thermionic Valve characteristic testing table.
Thermionic Valve Life Test arrangements for the product of the M.O. Valve Co's Works. (Installation not complete.)
The dull emitting filament—demonstration of properties.

Room 2. Demonstration to be given at 4.30 p.m. of a simple method for the analysis of very small quantities of gases.
(This demonstration is not of a popular character, and is of interest only to Physicists and Chemists.)

Room 7. **Substation for Laboratory** supplies—Electric Plant—Generators and Switchboards—Vacuum Pumps.

Room 8. **Metal Workshop.**
Room 9. **Wood Workshop.**
All the special apparatus and equipment for the Laboratories is made in these shops.

Room 10. The Leesona automatic coil winding machine.
Apparatus for weighing rapidly a few milligrammes.

Rooms 11, 12 & 13. **Microscopy.**
Method of preparation of specimens for microscopic examination. Handling of very fine filaments.
Exhibition of photomicrographs of copper clad nickel steel wire for electric lamp seals, and samples at various stages of manufacture. Measurement of coefficient of expansion. Exhibition of photomicrographs of tungsten filaments—showing crystal growth and single crystal wire.

Room 16. **Refractories.**
This laboratory is not yet equipped for its proper work.
Demonstration of a method of examining the flow of glass in machine feeding devices.
Measurement of coefficient of expansion of glass.
Hilger strain viewer for detecting strains in glass.
Method of detecting the composition of bubbles in glass.

Rooms 18a, 18 & 19. **Tungsten Wire.** A laboratory-factory.
It is not easy for the visitor to see this process in proper sequence.
Tungsten Oxide Powder is prepared in Room 19. It is reduced to metal powder in Room 18, pressed into ingots in Room 18A, sintered in Room 18A, swaged into wire of 0·8 m/m. diameter in the gallery above (up staircase from corridor), and drawn down into fine wire in Room 18.

(Continued overleaf.)

FIGURE 17.1: (CONTINUED) EXHIBITS AND DEMONSTRATIONS AT THE OPENING OF THE RESEARCH LABORATORIES, G.E.C. © GEC MARCONI

ALL STAIRCASES FROM THE MAIN CORRIDOR LEAD TO THE GALLERY,
WHERE REFRESHMENTS WILL BE SERVED.

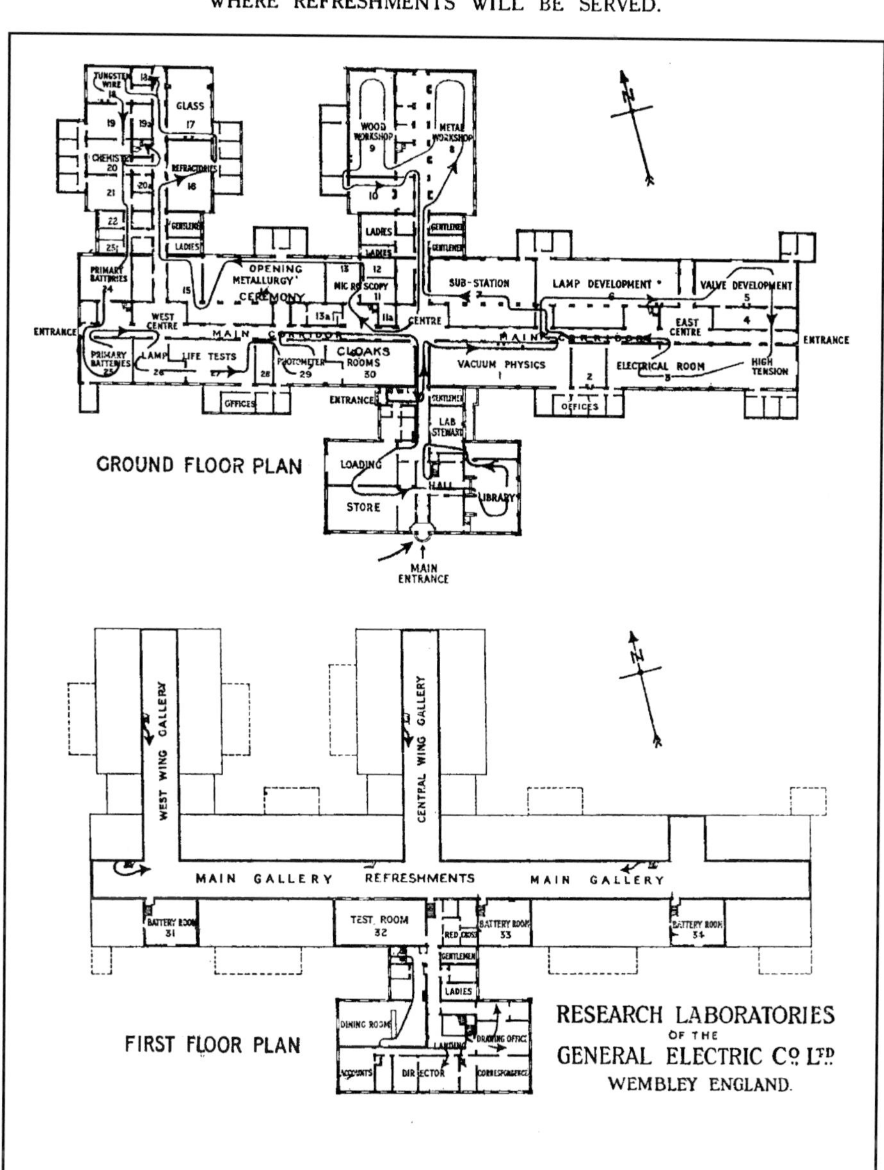

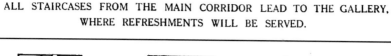

FIGURE 17.2: GROUND PLAN OF G.E.C. © GEC MARCONI

At the same time as Perrin was addressing the public, he was trying to persuade the French government to create a National Center for Pure Scientific Research. Both Perrin in France and Hale or Robert Millikan in the USA illustrate how addressing to the public can promote scientific disciplines and professional interests. Promoting science to the public and promoting scientific research as a profession could be combined in the same campaign.[10] Paradoxically, the popularization of science played a key-role in the professionalization of science.

The Scientist's Responsibility

In a sense, the emphasis on 'pure' science in the exhibitions of the 1930s was also a process of purification of science. Those individual scientists, who accused science of having 'prostituted itself' to war, were extremely concerned with science popularization. Paul Langevin, for instance, urged his fellow scientists to devote themselves to peace movements as well as popularizing tasks. "It is the duty of those who are called the elite to contribute to the treasure shared by all manhood, and to dedicate themselves to popularization."[11] In his own public lectures, Langevin presented science as a cultural activity and praised it as the noble product of human reason not only bringing technological benefits but useful also for combatting the raise of fascism. The notion of a 'gulf' was mainly used in the thirties to deplore the fact that social justice was left far behind the advancement of science and technology. Langevin assumed that it was the social responsibility of scientists to fight the delay of social progress, by placing moral and social values on equal footing with technological development through publications and conferences.

The social implications of science for the public were also the focus of the popularizing activity of a number of marxist British scientists like J.B.S. Haldane, J.D. Bernal, Lancelot Hogben and Hyman Levy. They felt it was their responsibility as scientists to deal with social problems. Their model was the Soviet Union's efforts to spread science among industrial workers and to involve scientists as decision makers in economic and social matters. With vigorous support of politically active journalists such as Richtie Calder and G.C. Crowther, they started a movement outlining the social functions of science and calling for an alliance of manual and intellectual workers. The 'social relations of science' movement initiated a specific style of popularization. It was nothing like a report on recent scientific achievements. Scientific knowledge was presented in a broad socio-historical approach with a clear social and political agenda. This program was extremely successful, from an editorial point of view, even with disciplines like mathematics which were usually considered less popular. Hogben's *Mathematics for the Million*, published in 1936, went through seven printings in the first six months and was followed by *Science for the Citizen* in 1938.[12]

Science and Common Sense

In the 1920s, relativity theory and quantum mechanics seemed to challenge all popularizing enterprises. Repeatedly, physicists said that their new concepts were counter-intuitive and distant from all 'common sense' understanding. The entry 'popularization' in the *Encyclopédie française*, written in the 1930s, referred to the new physics and shaped a reflexive view in three stages. In the eighteenth century, there was simply a difference of 'styles' between the scientist and the layman. In the nineteenth century it evolved into a difference in 'languages' requiring a 'translation.' In the twentieth century there was a difference of 'worlds.'

If there is an incommensurability between science and common sense, science cannot be 'translated' without being deeply altered; non-specialists will never find out what it means. Science and 'opinion' are two different incommensurable worlds and popularizing is an impossible task:

> Science, in its need to provide completeness, as well as in its fundamental character, is totally opposed to the concept of opinion. If it happens that science confirms an opinion about a specific point, then this is for other reasons than those which formed that opinion, in the sense that opinion in law is always wrong. Opinion is the outcome of bad thinking, of no real thought: it translates needs into knowledge. (...) The scientific spirit doesn't permit us to have an opinion about questions we don't understand, about questions we cannot formulate clearly. Above all one has to know how to present the problems.[13]

It was in the thirties that Gaston Bachelard developed his notion of a radical break, an epistemological rupture. Bachelardian epistemology implied that it was impossible to communicate the meaning of science without leading the mind beyond epistemological obstacles such as early intuitions, primary observations, practical interests. Popularization must involve the "formation of a scientific spirit," in other words, it has to become a full education in science.

While these difficulties seemed insurmountable, a new wave of public communication of science was initiated. Physicists like Einstein himself with Infeld, Arthur Eddington, James Jeans, Paul Langevin, Charles Nordmann and Louis de Broglie devoted a great deal of time to spreading the new theories to the learned public through lectures and papers and through wide circulation books. As early as 1911, the relativity theory was presented and discussed at the International Conference of Philosophy in Bologna. In France, the new theories of physics became a classical topic for historians and philosophers while most French physicists still ignored them. To be sure, crossing the line between science and humanities was an indirect way of convincing their fellow physicists. It was more than a campaign of persuasion, however. Addressing different kinds of specialists was decisive in developing the new concepts. The quantum physicist Niels Bohr, for example, refined his notion of "complementarity" through lectures addressed to biologists, anthropologists and psychologists . . . As he emphasized the limits of our understanding of nature, he also tried to extend his notion of complementarity to many other fields and to find a unifying principle of human knowledge.

A broad cultural program lay behind the emergence of a cross-disciplinary communication of science. Aspirations to a new synthesis of knowledge beyond the specialization of scientific investigation had already prompted a number of nineteenth century popularizing enterprises by Hemholtz and Du Bois-Reymond for instance. In 1900, the synthesis of knowledge motivated the creation of a Centre international de synthèse, a private foundation founded by Henri Berr, a French historian. In the 1920s and 1930s, he managed to bring together the leading figures from the natural sciences and humanities from all over Europe. He published the proceedings of the annual conferences and also encouraged the publication of a cross-disciplinary vocabulary of science.

The intense activity of science communication during the inter-war period did not result in the emergence of a definite genre of popularization. So great was the variety of styles in science communication in the mid-century that no standard figure of the author of popularization was shaped. Neither the scientific organizations nor the science writers' associations overtook the tradition of the great men of science addressing the public at large. Professional scientists and professional science writers or journalists were often allied in their efforts to win the support of industrials, of private foundations and state agencies. The territory of science communication thus remained opened to various social groups.

Out of a considerable diversity of scientific writings a rather uniform and standard image of science prevailed. The utilitarian ideology underlying most of the nineteenth century science media was superseded by the image of science as a pure and autonomous activity whose technological 'retombées,' or by-products, were profitable to society. The common image of science as a neutral activity bringing social benefits had been re-established through the cultural offensive of the inter-war period. The only exception was Lyssenko's repeated campaigns in 1936, 1939 and 1948 for spreading Michurinist biology in the Stalinian period which eventually reinforced a clear distinction between science popularization and political propaganda among scientists.

Shaping science as a liberal art, independent from social and state pressures, with an ethos of its own and its own esoteric language, was certainly a sure way of enhancing its prestige among the public and of favoring the image of scientists as spiritual guides of mankind. But the renewal of scientism thus generated was also a source of hostility especially in the fascist regimes. This ambivalence was the focus of Robert K. Merton's 1938 famous paper on "Science and the Social Order." Merton emphasized the conflict between the ethos of science and of other social institutions. In advocating pure and disinterested science, scientists claimed that science was independent from the state or from any other social power. Scientists themselves, Merton depicted them as the guardians or lieutenants of a tenuous and uncertain fortress, threatened by emerging new popular mysticisms.[14]

Information-Persuasion

In analyzing the case of science under the Nazi regime, Merton argued that the attacks on the autonomy of science were in a sense encouraged by science itself, because esoteric languages allowed new mysticisms:

> The modern scientist has necessarily subscribed to a cult of unintelligibility. The result is an increasing gap between the scientist and the laity. The layman must take faith in the publicized statements about relativity and quanta or other such esoteric subjects (. . .) To the public mind science and esoteric terms become indissolubly linked. The presumably scientific pronouncements of totalitarian spokesmen on race or economy or history are for the uninstructed laity of the same order of the announcements concerning the expanding universe or wave mechanics. (. . .) Partly as a result of scientific advance, therefore, the population at large has become ripe for new mysticisms clothed in apparently scientific jargon.[15]

After the Manhattan Project of the Second World War, the "tenuous and uncertain fortress" was a powerful organization whose methods and efficiency were able to reshape the world. The impact of science, as revealed by Hiroshima, and expressed by the state and military budgets for basic research, could have altered the image of science as a neutral and 'pure' activity. If the image survived nevertheless, it was the result of a huge campaign of public communication, which involved the scientific community, science journalists, and government agencies as well.

One important task was to convince the public that atomic energy was useful in peacetime. President Eisenhower's "Atoms for peace" initiative was widely echoed in many countries. The "Atomium," symbol of the World Exhibition held in Brussels (1958), during the Cold War, was an instrument of propaganda for nuclear energy. The message conveyed by the Hall of Science was that science was universal, that it was intrinsically cooperative in its methods and radically distinct from all its applications.[16] Atomic Energy Commissions, in various countries, developed a policy of public relations contrasting with the secrecy of their results. Many research agencies and scientific organizations also initiated campaigns to convince the public that the methods used during the Manhattan Project would help solve social and material problems. Beyond their own institution, beyond the national research policy, scientists were all campaigning 'in the name of science.' Science was idealized, being depicted as a rational and context-free enterprise.

Retrospectively, it is clear to us that in the 1950s, popularizing essentially meant persuading the public of the social value of science. A critical view of the United States after World War II led to the conclusion that "the term public understanding of science became equated with public appreciation of the benefits that science provides to society."[17] However persuasion was never at the cost of information. On the contrary, in this period the unsensational, factual, serious style of popularization flourished. The hundreds of small, cheap volumes of *Que sais-je*, launched by Philippe Angoulevent and the Presses universitaires de France, in 1949, and offering an "unbiased" presentation of science, even at the cost of being unattractive, have been extremely successful for a half century or so.

There were two important assumptions in such publishing initiatives. The popularizers displayed a deep conviction that more information about science would automatically improve the public's attitude toward science. There was a large consensus between science writers' associations, research agencies, scientists and even commercial publishers on this point. To provide 'factually correct information' was their common aim. There were no doubts that factual information would have the 'magic power' of attracting greater support for science.

It was also assumed that there was a social demand for information about science meeting the supply of information. Public surveys were conducted in the United States by the Survey Research Center in the 1950s (before and after the launching of Sputnik) which concluded that the public wanted more science news. The creation of the Air & Space Museum could be seen as an answer to such demand and a support for the US Space program.[18] What kind of demand was it? In fact, editorials and speeches make it look as if the 'public demand' was more like the scientists' plan for public support. Because scientific research is to a very large extent state funded, it is dependent on taxation. Therefore, disseminating research results amongst the public was an important way of maintaining confidence.

Proletarians of Knowledge

In the 1950s and 1960s the 'gulf' became a topical subject of public discussion. The social fabric and the integrity of culture were threatened by the tremendous advancement of science. The large audience of C.P. Snow's famous pamphlet *The Two Cultures*, published in 1959 clearly indicates the importance of the notion of a gap. Over the past thirty years, this booklet had been referred to in many various circumstances and for various purposes, thus showing the remarkable flexibility of the notion of "gap."[19] Certainly, Snow's purpose was to prompt interactions between distinguished Cambridge scholars in science and humanities rather than disseminate scientific knowledge to the public at large. In view of the many attempts at dialogue between science and the humanities in the inter-war period, this pamphlet can be read as a kind of nostalgia of the thirties. However it has mainly been used as the missionary document of a new age of cross-disciplinary communication, as a program to rescue modern civilization.

The gap between science and the public was not only considered as threatening the integrity of culture but also as dangerous for democracy because, without being properly informed, citizens could not discuss the social consequences of science and consequently exercise their civil rights. These common views undoubtedly helped science writers and journalists present themselves as gatekeepers of democratic societies and civilization. In 1965, an editorial of the French magazine *Science et Vie* defined the task of the popularizer in these terms:

> Entre le savant et l'homme de la rue, le vulgarisateur propose ses images, ses analogies, ses simplifications, traduisant pour le plus grand nombre ce que font les avant-gardes. Son métier, qui exige les qualités les plus rares, est ainsi devenu - parfois contre le gré des savants - une fonction majeure des sociétés modernes. Pour que son rôle satisfasse en même temps l'homme des sommets et la multitude grouillant dans les vallées, il faut

que s'équilibre deux poussées inverses: celle de l'isolé, du conducteur ne connaissant que les visages du premier peloton et celle des masses exigeant la popularisation des idées, l'application des découvertes. Un débat risque donc d'opposer les aristocrates du savoir aux prolétaires de la connaissance ... Et ce débat est d'autant plus dramatique qu'un pays, mieux équipé en hommes et en instruments, surpasse les autres dans les sciences et les techniques. Le clivage entre groupes sociaux risque d'être tragique ...[20]

Although the 'gulf' provided professional science journalists with a relevant definition of their status as bridge-builders, an ambiguity remained concerning their loyalties. Their self-perception as "the proletarians of knowledge," as opposed to the "aristocracy" of knowledge producers betrays potential conflicts between two different concepts of popularization. Once it is admitted that disseminating scientific information is the full responsibility of working scientists, the science writer or journalist is only considered a substitute for the scientist who has no time to address the public because of the professional pressures of specialization and competition. His task is simply to accommodate or 'translate' the scientific languages in order to maintain the social order. When science communication is considered a form of journalism, the science writer regards himself first of all as a member of this profession to which he gives almost all his allegiance and energy, even at the cost of defying the scientific authorities. The ambiguity was recognized in France by the creation of two distinct associations. The *Association des journalistes scientifiques de la presse d'information*, created in March 1955, emphasized the function of mediation. It aimed to promote accurate information through regular contact with scientific research organizations, and defend the freedom of journalists in the face of secrecy, whether private or official. The other association, *l'Association des écrivains scientifiques de France* (AESF), created in 1950 under the aegis of UNESCO, stressed the specificity and the dignity of science. Its mission was to "indicate the right way to the media which shape the public's minds." More militant than professional, this association is greatly involved in the defense of rationality against superstition and the 'tide of pseudo sciences' in the mass-media.

There were, and indeed there still are many complaints from scientists that science journalists did not correctly report on their work, and complaints from science writers that scientists did not like to cooperate. But these superficial conflicts did not alter the coalition of interests between science mediators and scientists. They agreed not only on the image of the 'gulf,' the existence of a social demand for science information, but in the assumption of a passive audience which inspired the so-called 'diffusionist' model of popularization. The view of the public as a simple receiver of scientific messages was behind most of the references of a 'gulf' between science and the public. In order to bridge the distance between the area of production and the area of 'consumption' of science, only a one-way flow of information from the summits of knowledge to the lowest cultures was envisaged. In other words the process of diffusion was modeled after educational activities.

Science for the People

However, new perspectives on the relationship between science and the public have been suggested by Hannah Arendt. She viewed the gulf as a process of self-exclusion of scientists from civil society. Because of the adoption of their esoteric languages of mathematical symbols, which can no longer be translated into natural language, scientists have no reliable political judgement.[21]

This philosophical critique of science was echoed by self-criticism of science. Two magazines with nearly the same title denounced the technocratic uses of science as an authority legitimizing economic directions. In the USA, the bi-monthly *Science for the People* was created by an association of "Scientists for Social and Political Action" (SSPA) which later extended to include engineers. Their political action went from public debates to strikes in research laboratories or industrial companies. A British magazine *Science for the People* was also created in 1969 by the "British Society for Responsibility in Science." It was a mixed association at the onset but it became more and more radical throughout the seventies.

Fighting the belief in the neutrality of science was the prime motivation which prompted a new form of popularization challenging the idea of a passive audience in favor of an active assimilation or rejection of scientific information. A number of Dutch radical scientists initiated the Science Shops in 1977.[22] Based on the respect of popular cultures, the Science Shops were small and flexible centers scattered in urban areas intended to facilitate interactions between scientists and the public. While trying to meet public demand, the 'shop-keepers' quickly realized that the questions the public had about science were by no means scientific. They had to answer them through a process of subtle reformulation which was rather different from the conventional one-way communication of science, where scientists answer their own questions. From Netherlands, the Science Shops movement spread in Belgium and France in the early 1980s. But the activity of these small independent centers relying only on volunteers, the Science Shop did not last very long.

From this experience, however, a new concept of "scientific culture" emerged. One key idea was that science communication was indispensable not so much to fight public ignorance but rather to prevent the isolation of scientists from culture.[23] In order to "mettre la science en culture," a number of centers of science and culture were created in the 1980s. Spreading scientific information was no longer the main objective. Rather it was one of the possible means to facilitate confrontations and debates about science. More recently, the debates about nuclear plants, nuclear tests and epidemics have prompted a number of counter-expertises challenging the views held by the scientific establishment.[24] The near future will tell us whether it would be a new function for science mediators to assume a role of judge between scientific milieux and the public.

Alternative science communication programs have been reinforced, in the 1980s, by sociological studies of science. The self-image of mediators has been described as an ideology legitimizing their social role of 'bridge-builders,' and disguising the fight for power within and outside the scientific community.[25] The 'gulf' between the lay public and the working scientists bridged by the 'mediator' has been also questioned. Considering the great variety of scientific writings — from reports to journals and magazines — the view of a continuous process of reformulation of scientific statements seems more adequate.[26]

Did such criticisms and alternatives condemn all conventional attempts at disseminating scientific knowledge? While the 1970s were an 'era of suspicion' with regard to all kinds of popularization, they were followed by a re-emergence of the "Public Understanding of Science" movement, institutionalized in the 1980s. The one-way educational perspective prevailed again. It relied on the assumption, based on a number of public surveys of a high demand for scientific informations relating to health, the environment, energy and space exploration.[27] Public surveys more concerned by the level of public knowledge of science have led to the notion of 'scientific literacy' (a rate defined by 7/10 correct answers to questions on specific scientific issues) and to repeated complaints about the lack of basic scientific knowledge. These surveys provide a better understanding of the public attitude to science. However, the popularization of science conceived of as an auxiliary of the educational system, is necessarily a one way communication, under the control of scientific expertise. In other words, Public Understanding of Science tends to ignore the controversy initiated in the nineteenth century about the function of the science mediator as echo of academic science or promoter of alternative popular knowledge.

The controversy cannot be denied and the question of the 'public demand' is still a matter of debate. A new style of science museums has emerged during the past decades, modeled on amusement parks. Through hands-on exhibits, interactive displays, video and computer games and additional activities like workshops, clubs, and summer camps, they reinvigorate the great tradition of amateur science. Completely separated from the conservatory functions of earlier science museums, these science centers are extremely successful in terms of numbers of visitors.[28]

Do they fulfill their objectives? This is a more complex question. Their goal is broadly educative, but informal education aims at increasing the interest for science rather than at a better understanding of science. The spontaneous enthusiasm for knowledge, presupposed by nineteenth century popularizers now seems so problematic that museums have to appeal to the public's desires and emotions. The 'need for scientific information' is felt as a constraint linked to problems raised by technological developments rather than a spontaneous desire. Scientific communication has to rely on the idea that the easier the access to scientific information, the greater the demand for more. The brochure advertising the Fleet Center in San Diego said "We want our center to give you a feast of science and to keep you hungry for more."

Up to now the success of the numerous and various enterprises in science communication more or less relied on the tacit assumption of a close link between two public attitudes: because the general public was using more and more science in daily life through technological environment it was expected to consume more and more science through magazines and museum visits. Science-user and science-consumer were supposed to be one and the same phenomenon.

This close association seems more questionable today. It may be a consequence of the evolution towards black-boxed technological items which can be used without any knowledge of the scientific processes embedded in them. Consequently we do not even know that we do not know. If the public is not in a position of 'desiring' knowledge, the main task of science communication in the future might be to revive the desire for knowledge.

Conclusion

In conclusion three major points can be stressed. There is nothing like a linear process of development of the popularization of science responding either to the specialization of science or to a public demand. It is a complex and multidimensional phenomenon which has periods of expansion and relative decline. The development of science communication in the twentieth century mainly rests on initiatives of scientific communities in response to social, political and cultural contexts. The various images of science are never definitively outmoded or obsolete. Like the layers of a collective imagination, they can be reinvigorated according to circumstances. This leads us to question the standard universal explanation-legitimation of the emergence of popularization as a necessary consequence of the 'advancement of science.' Rather it invites us to search for local and cultural explanations.

The question 'who should be in charge of the public communication of science' is still open. The professional status of the mediator promoted in mid-century remains controversial, uncertain, and extremely fragile. One important result of this brief historical survey is to deconstruct two long-lasting myths among the actors of science communication. The indifference of working scientists towards the mass-media, and the increasing and irresistible contamination of scientific milieux by the sensationalism of the mass media.

Such myths rested on the assumption that scientists and mediators are the only actors on the stage of popularization. The historiographical perspective suggests that a wide variety of actors are involved in the popularizing enterprises. Not only cognitive interests are at stake but disciplinary and professional strategies as well, and local or state policies, industrial, commercial and financial interests ... and the most silent actor, the audience. On the attitude of the audience, one point, at least, can be made. In spite of the increasing importance of professional scientific communities, the social division between science producers and passive science consumers is not immobile and can be revised at any moment.

REFERENCES

1. Special centennial issue of *Nature* (1 November 1969), **224**: 423–452.
2. Basalla Georges "Pop Science: The depiction of Science in Popular Culture," in G. Holton & W.A. Blampied, *Science and Its Public: The Changing Relationship*, Boston Studies in the Philosphy of Science. (Dordrecht: Reidel, 1976), p. 261–278.
3. La Follette, Marcel *Making Science our Own: Public Images of Science 1910–1955.* (Chicago: University of Chicago Press, 1990).
4. Gabriel Lippman in *La Science et la Vie,* T.1, (1913), p. 104.
5. Daniel J. Kevles, *The Physicists: The History of a Scientific Community in Modern America* (New York, Vintage Books, 1979).
6. Ring Katy, *The Popularization of Elementary Science Through Popular Science Books c. 1870–1939,* PhD Diss. (University of Kent at Canterbury, 1988), p. 268–270.
7. Rhees David "Corporate Advertising, Public relations and Popular Exhibits," in Schoeder-Gudehus (Ed.) *Industrial Society and Its Museums, 1890–1990.* (Harwood Academic Publishers, 1993), pp. 67–76.
8. Ehrardt George R., "The background, formation and early years of the national Association of Science Writers." Paper presented at the 1st Annual UNC-Charlotte History Forum, 1989; *Descendants of Prometheus: Popular Science Writing in the United States, 1915–1948,* PhD. Dissertation. (Duke University, 1993).
9. Robert W Rydell, "The Fan Dance of Science. American World's Fairs in the Great Depression," *Isis* (1985), **76**: 525–542.
10. Jean Perrin, *La Science et l'espérance,* Paris, 1936.Eidelmann Jacqueline, "Politique de la science ou politique de l'esprit? Genèse du palais de la découverte," in T. Shinn, R. Whitley (Eds.), *Expository Science: Forms and Functions of Popularisation.* (Dordrecht, Boston: Reidel, 1985), p. 195–207.
11. Paul langevin, "L'avenir de la culture," Madrid 1933, see also Bensaude-Vincent B. & Blondel C, "Deux stratégies divergentes de vulgarisation: Georges Claude et Paul Langevin," in *Cahiers d'Histoire et de philosophie des sciences,* no. 24 (1988).
12. Katy Ring *The popularization of Elementary Science Through Popular textbooks c. 1870, c.1939,* Ph.D. Thesis. (Canterbury: University of Kent, 1988).
13. Bachelard Gaston, *La Formation de l'esprit scientifique.* (Paris: Vrin, 1938), in ed. 1970, p. 14.
14. Robert K. Merton, "Science and the Social Order" (1938) in *The Sociology of Science: Theoretical and empirical investigations.* (Chicago & London: The University of Chicago Press, 1973), p. 254–266.
15. *ibid.* p. 264.
16. Brigitte Schroeder-Gudehus, David Cloutier "Popularizing Science and technology during the Cold War: Brussels 1958," in Robert W. Rydell, Nancy E Gwinn, *Fair Representations. World's Fairs and the Modern World.* (Amsterdam: VU University Press, 1994), pp. 157–180.
17. Lewenstein B. "The meaning of 'public understanding of science' in the United States after World War II," *Public Understanding of Science* (1992), **1**: 45–68.
18. Alex Roland "Celebration or Education? The Goals of the U.S. National Air & Space Museum," in Brigitte Schroeder-Gudehus (Ed.), *Industrial Society and its Museums.* (Harwood Publishers, 1993), p. 77–90.
19. C.P. Snow, *The Two Cultures.*
20. Labarthe, A., "La démocratie du savoir," *Science et Vie* (May 1965), No. 572, p. 56–58.
21. Hannah Arendt, *The Human Condition* (1958), Preface.
22. John Stewart, Véronique Havelange, "Les boutiques de sciences en France," *Alliage* (1990), No. 1, p. 97–103.
23. Levy-Leblond J.M. *Mettre la science en culture* (Nice: ANAIS, 1986), p. 56.
24. Bryan Wynne "Sheep farming after Chernobyl: a case study in communicating scientific information," *Environment Magazine* (1989), **31 (2)**: pp. 10–15 et 33–39. L. Wilkins, P. Patterson (Eds.) *Risky Business: Communicating Issues of Science, Risk, and Public Policy.* (New York: Greenwood Press, 1991).
25. Jurdant B., *Les problèmes théoriques de la vulgarisation scientifique* . (Strasbourg: Thèse de l'Université de Louis Pasteur, 1973); Roqueplo P. — *Le partage du savoir; science, culture et vulgarisation.* (Paris: Seuil, 1974).

26. Whitley, R. "Knowledge producers and knowledge acquirers: Popularisation as a relation between scientific fields and their publics," in Shinn T. et Whitley R. (Eds) *Expository Science: Forms and Functions of Popularization.* Reidel: Sociology of Science Yearbook, 1985), p. 3–28. Hilgartner, S., "The dominant view of popularization: conceptual problems, political uses, *Social Studies of Science* (1990), **20**: 519–539; Myers G. *Writing Biology: Texts in the Social Construction of Scientific Knowledge.* (Madison: University of Wisconsin Press, 1990).
27. Bruce Lewenstein, "Science and the Media," in S. Jasanoff *et al.*, eds, *Handbook of Science Technology and Society.* (Thousand Oaks, Calif.: Sage, 1995), p. 343–360.
28. W. Orchinson "Introducing the Science Centrum: A new Type of Science Museum, *Curator* (1984), **27**.

Further Reading

Collins H.M.: "Certainty and the Public understanding of Science: Science on Televison," *Social Studies of Science* (1987), **17**: p. 689–713.

Roger Cooter, Stephen Pumphrey, "Separate Spheres and Public Places: Reflections on the History of Science popularization and Science in Popular Culture," *History of Science* (1994), **32**: 237–267.

Yves Jeanneret, *Ecrire la science. Formes et enjeux de la vulgarisation.* (Paris: PUF, 1994).

Sheets-Pyenson, S.: 'Popular Science Periodicals in Paris and London: the Emergence of a Low Scientific Culture,' *Annals of Science* (1985), **42**: p. 549–72.

Lewenstein, B. "A Survey of activities in Public Communication of science and technology in the United States," in B. Schiele (Ed.) *When Science Becomes Culture.* (Boucherville, Québec: University of Ottawa Press, 1994), p. 119–178.

Union centrale des arts décoratifs, *Le livre des expositions universelles, 1851–1989.* (Paris, 1983).

Schroeder-Gudehus B., Rasmussen A., *Les fastes du progrès, Guide des expositions universelles.* (Paris: Flammarion, 1992).

CHAPTER 18

Science Fiction and Science in the Twentieth Century

KARLHEINZ STEINMÜLLER

THE SPIRIT OF AN AGE

When in 1957 Charles P. Snow formulated his theory of the two cultures and wrote about a gulf of incomprehension between science and the humanities, there already existed a kind of bridge: science fiction. Usually disregarded by serious scholars and of course ignored by Snow himself, science fiction has become a unique medium for discussing science and technology, their prospects and hazards, and more generally their social and cultural impacts. As an integral part of postmodern culture, science fiction has penetrated all fields of the media landscape: fiction, comic books, movies, even plays and musicals. Science fiction themes and images surface sometimes quite unexpectedly in everyday life, in TV commercials and video clips, not to speak of computer games. Battery powered aliens, monsters and robots have nearly expelled the traditional fairy tale characters from children's rooms. Data highway enthusiasts use science fiction jargon and imagery to depict their 'cyberspace' visions. A generation ago, the race to the moon was at least partly initiated by the dreams of early science fiction writers and readers. For the public, technology is science fiction come true. For many scientists and engineers science fiction provides the imagery of their visions.

It is therefore no exaggeration to regard science fiction — as Michel Butor did in 1953 — as the mythology of the modern, scientific age, or to claim — as the historian Michael Salewski does — that science fiction is itself the *Zeitgeist* of the age.[1]

Within the framework of this volume, a study of science fiction may provide insights into the popular perception of science and the picture of the scientist, insights into the ideology of progress as well, and — since science fiction is mostly technology fiction — into the hopes and fears connected with technology. Science fiction is an unexplored source for the history of science, reflecting (sometimes in a distorted way) its development, pivotal themes and achievements, and shaping the guiding images of the scientific-technological community.

BACK TO BACON AND PROGRESS

When Hugo Gernsback coined the term "science fiction" in 1926 as a selling label for his pulp magazines, he was well aware that he was christening a field of literature with a long tradition, going back perhaps to the fantastic deeds of Gilgamesh or the legend of Daedalos. Many famous pieces of literature have been claimed as ancestors of science fiction, among them Francis Bacon's *Nova Atlantis* (1629) and the third voyage from Jonathan Swift's *Gulliver's Travels* (1726). Both are literary reactions to the rise of experimental science during the renaissance. But whereas Bacon describes a technological utopia, Swift is bitterly satirical. Bacon founded his utopia on the *scientia nova*, methodical experimentation in laboratories, and portrayed an association of scientists, which could well have served as a model for the later Royal Society. Swift parodied the activities of this same institution as useless and unpractical project-mongery and ridiculed the bold theories and visions of new materials and technical appliances. Up to now science fiction oscillates between the two extremes: the Baconian vision and the Swiftian criticism.

The next crucial date in the pre-history of science fiction is Louis-Sebastien Mercier and his novel *L'An 2440* (1771), perhaps the first piece of future fiction with a fixed date. In contrast to the early utopians, who set their model state in an *ou-topos*, a place not known, Mercier introduced time as an utopical dimension: good governance, industry and industriousness will bring about times of luck and prosperity and, amongst other things, balloon airlines to China. After Mercier, writers no longer established their visions in a place nowhere, but in a time somewhen. Not uncharted lands but the near and distant future became the realm to be conquered by science fiction.

During the nineteenth century science fiction was shaped by one dominant conceptual framework: the ideology of progress and the underlying positivist philosophy. The idea of the steady improvement of technology and the perfection of social institutions combined well with the concept of evolution. Writers like Edward Bulwer-Lytton (*The Coming Race*, 1872) and Edward Bellamy (*Looking Backward 2000–1887*; 1888) dreamed of a technocratic or socialist future, powered by electricity, equipped with a nearly perfect command of all natural forces, with new means of communication, new weapons, new remedies for diseases, and sometimes inhabited by superhuman beings. Darwinism, taken to its logical conclusion, implied an imperceivable but constant evolution of the human race, with either its genetic degradation due to unnatural living conditions of the lower classes, or its rise to "minds immeasurably superior to ours."

The latter phrase is a quotation from H.G. Wells' *The War of the Worlds* (1898), and refers to the brains of Martians. The idea behind that novel however is not just evolution but the plurality of inhabited worlds. This philosophical concept had seemed a practical possibility (even a challenge to communication engineers!) when Giovanni Schiaparelli proclaimed in 1877 the discovery of channels on the surface of Mars. But Wells' novel has another, darker background: the tide of future war novels, which flowed shortly after the German-French war of 1871, and which expressed the British fear of an invasion and more generally the fear of the next great war, fought not by men but by machines, more disastrous than any war in history and perhaps initiating the end of civilization.

JULES VERNE CONTRA H.G. WELLS

Jules Verne (1828–1905) and H.G. Wells (1866–1946), the founding fathers of modern science fiction, represent two different approaches: the extrapolative and the speculative.

For Verne and even more for Hetzel, the editor of the *Magasin d'éducation et de récréation* in which most of Verne's "extraordinary voyages" were originally published, SF was a means of popularizing science, of educating the public and informing it of the immediate prospects of technology. The heroes of Verne are appropriate for this task: the scientist-engineer (*l'ingenieur-savant*), the explorer, and the industrialist. Working along the lines established by their real counterparts, great inventors, engineers, geographers, aeronauts, they design more powerful vehicles, more perfect means of communication, more destructive explosives, only rarely resorting to imaginary principles. It is therefore not surprising, that most of Verne's predictions have been realized. Among them there are even several tongue-in-cheek predictions, such as the replacement of a worn-out stomach by a new one, an anticipation of organ transplantation, in the story *A Day in the Life of an American Journalist in the Year 2889* (1889). Well-fitted illustrations convincingly contributed to the atmosphere of scientific realism — even in the case of such unusual phenomena as weightlessness aboard a cosmic projectile, which Verne attributed only to the gravitationally neutral point between Earth and Moon (*Around the Moon*, 1869).

Like Jules Verne, writers of 'hard SF' try to stick to the possible, or at least scientifically feasible. They draw their material from science journals or (like Verne) from recent patents. Some even calculate the orbital parameters of their planets, the necessary dimensions of their spaceships or, say, relativistic effects near black holes, or they have — like Verne — a friend to do the job.

H.G. Wells followed another line. His scientific romances transgress the established principles of science: he speculates about a drug which accelerates the human metabolism, a substance which screens gravity, a time machine. He was not much interested in the gadgetry of tomorrow: he muses about evolution, the conquest of space, machine warfare and their implications for society, culture, and man. According to Well's credo, science and technology will completely reorder human society, and affect every aspect of life. Therefore the world has to be run by technical experts.

When Jules Verne looked through Wells' *First Men in the Moon* (1901), he was shocked:

> We do not proceed in the same manner. It occurs to me that his stories do not repose on very scientific bases. No, there is no *rapport* between his work and mine. I make use of physics. He invents. I go to the moon in a cannon-ball, discharged from a cannon. Here there is no invention. He goes to Mars [sic!] in an airship, which he constructs of a metal which does away with the law of gravitation. *Ça, c'est très joli,*" cried Monsieur Verne in an animated way, "but show me this metal. Let him produce it.[2]

Both traditions, the Vernian extrapolative one, and the Wellsian speculative one with its philosophical and socio-cultural orientation, are still present in science fiction.

One Century of Science Fiction: An Overview

The new century made a headstart in SF with biology, relativity, wireless communication and radioactivity — quite parallel to the rediscovery of Mendel's laws, to Einstein's theories, to Marconi, to Becquerel and Soddy. Hugo Gernsback's classical novel *Ralph 124C 41+* (1911–12) gives a good example: video telephones, automatic restaurants, solar energy plants, treatment of diseases with radio-isotopes, space flight, hypnopaedia (learning by hearing while asleep), in short: a glamorous high-tech life in a society, which is by itself more or less organized along scientific principles (for an overview on SF in the twentieth century see Table 18.1).

In the twenties — during the 'pulp era' — SF magazines like *Amazing Stories, Science Wonder Stories,* most of the American ones originally edited by Gernsback, were filled with the exploits of space flight, robots and monsters and mad scientists. In addition to narrative, many magazines had a 'science section,' where the science questions of their mostly juvenile readers were answered. That tradition continued throughout the forties and fifties, when German emigree Willy Ley had his own columns in *Astounding Science-Fiction, Amazing Stories,* and *Galaxy Science Fiction* and became known as "the man who explains science," a part which was also played by biochemist and SF celebrity Isaac Asimov.

By propagating science and expounding its basic terms, science fiction, even in its pulp variety, prepared the field of rocketry. According to Arthur C. Clarke, who invented the communication satellite before becoming famous as a SF writer, many leading rocket engineers and astronautically inclined scientists received their "initial infection" from SF:[3]

> Popular attention began to be focused on rockets from 1925 onward, with the appearance of serious technical literature and the rise of small experimental groups in Germany, the USSR and elsewhere. For a long time, fact and fiction were inextricably entangled; many of the pioneers were writers, and used their pens to spread the news that space travel need no longer be fantasy.[4]

K.E. Tsiolkowski was one of these pioneers. He tried to communicate his ideas of the conquest of space by multi-stage liquid-fueled jet devices through didactic

SF, for example *Dreams of Earth and Sky* (1895) and *Outside the Earth* (1916). With the advent of cinema, space flight movies reached a broader audience than fiction, be it pulp or not. Rocket pioneers were employed as consultants to ensure technical accuracy, as Hermann Oberth did for Fritz Lang's *The Woman in the Moon* (1929), and Tsiolkowski for the Russian *Cosmic Voyage* (1935). The German motion picture company UFA even enabled Oberth to construct the most advanced liquid-fuel rocket of that time: simply for promotional purposes. (Alas, the rocket failed.) After World War II space flight movies were used rather commonly to promote national astronautic programs. But generally the link between space fiction and real space flight loosened after Stanley Kubrick's *2001. A Space Odyssey* (1968, based on a story by Clarke), and after the Apollo missions to the Moon (see Table 18.2).

Besides hopes and aspirations, SF also reflected an epoch's fears and concerns. The Great War had shaken the conviction that mankind could benefit from scientific and technological progress. It had reduced soldiers to machines fighting with other human and mechanical machines. At the same time, the assembly line reduced workers to machine parts. Could machines not rebel against man and substitute him? As early as 1920, the Czeck writer Karel Čapek imagined that possibility in his play *Rossum's Universal Robots*, coining, by the way, the term "robot" for what we today call androids (made not of mechanical but biological parts). Fritz Lang's movie classic *Metropolis* (1926) relates the enthusiastic and disturbing vision of an industry producing unprecedented wealth and consuming its blue collar workforce. However only Aldous Huxley transposed the principle of the assembly line to the production of human beings. *Brave New World* (1932) takes the vision of an utopian high-tech world state (as prophezied by H.G. Wells) and turns it into its opposite: dystopia. Technocratic control of life from birth out of the bottle to death in an euthanasia center. Pursuit of happiness achieved through manipulation of emotions by drugs and the media. Like most utopias, *Brave New World* may also be defined by what it abolishes: all atavistic traits in man and society, motherhood, individual choice, war . . . Whereas in the nineteenth century a scientific government, a society run by technocratic principles (be it socialist or liberal ones) was regarded as the up-to-date way out of social clashes, corruption, bad governance and waste effort, it became during the twentieth century synonymous with manipulation and suppression of the individual. Before Huxley, Yevgeny Zamiatin had imagined a technocratically controlled future state, where men were reduced to numbers. Although Zamiatin's novel *We* (1924) should mainly be understood as a reaction to the Bolshevist revolution in Russia, it also criticizes the futurist avantgarde which proclaimed 'machine art' and a rationalist, Taylorist[5] organization of life. Orwell's *Nineteen-Eighty Four* (1948), the third major dystopia of this century, added manipulation of history, of language, of thinking and feeling to the totalitarian nightmare vision, but without explicitely denigrating an underlying technocratic or rationalist world-view.

FIGURE 18.1: *METROPOLIS* 1926. MENNINGEN: FILMBUCH SCIENCE FICTION, 1980, S. 99.

FIGURE 18.2: ANTHONY FREWIN: ONE HUNDRED YEARS OF SCIENCE FICTION ILLUSTRATION, JUPITER BOOKS: LONDON, 1974

FIGURE 18.3: BRIAN ASH: THE VISUAL ENCYCLOPEDIA OF SCIENCE FICTION, PAN: LONDON & SYDNEY, 1977

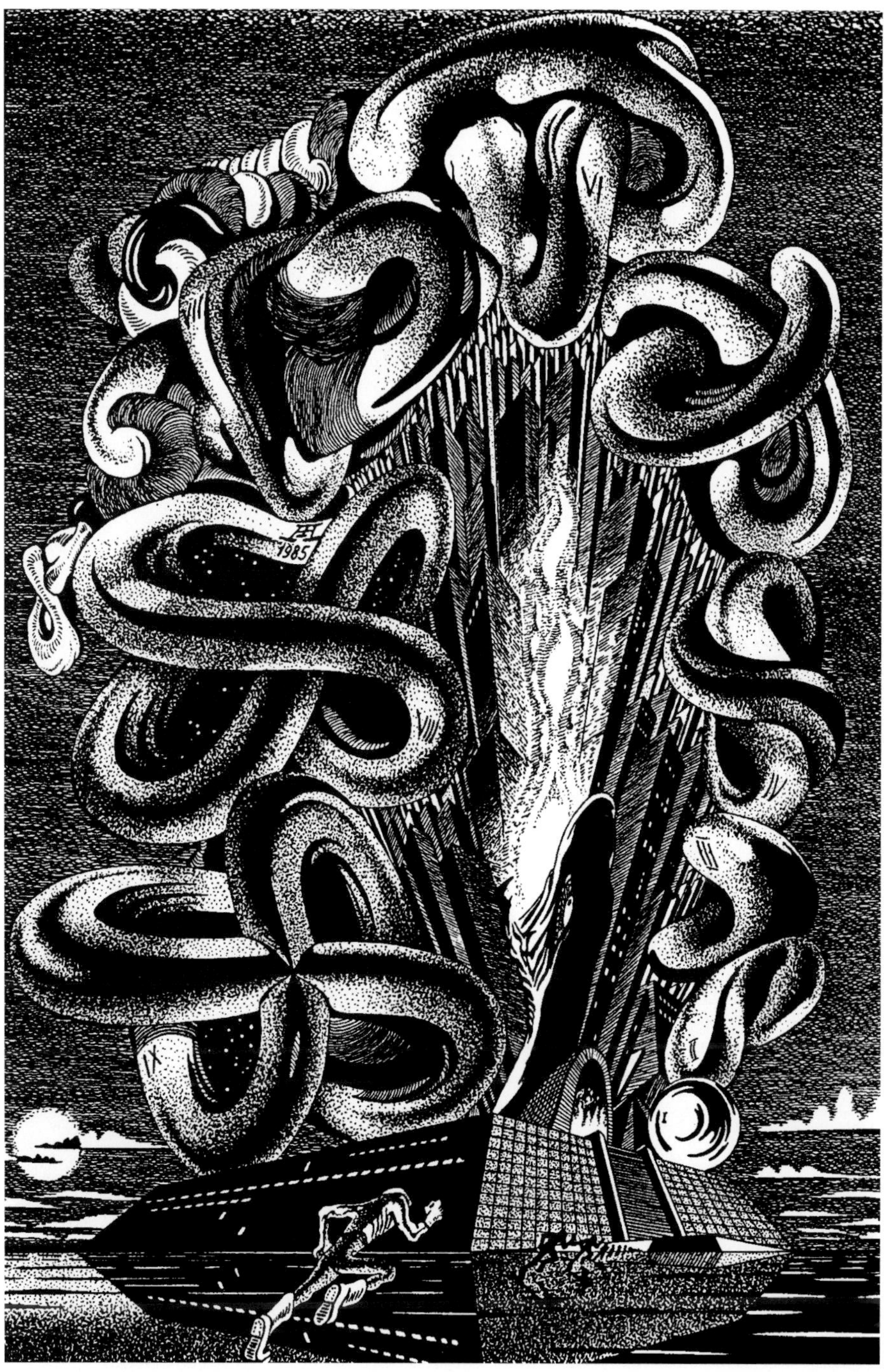

Figure 18.4: Topology of Manifolds, 1985. These convoluted curlicues of tubings define a space above a ziggurat-like habitat, set on a reflective sea of glass on which humanoid creatures run about seeking shelter. The image demonstrates the process of turning a sphere inside out. Through a series of transformations, the sphere's inner and outer surfaces will eventually change place. © Anatoly Fomenko.

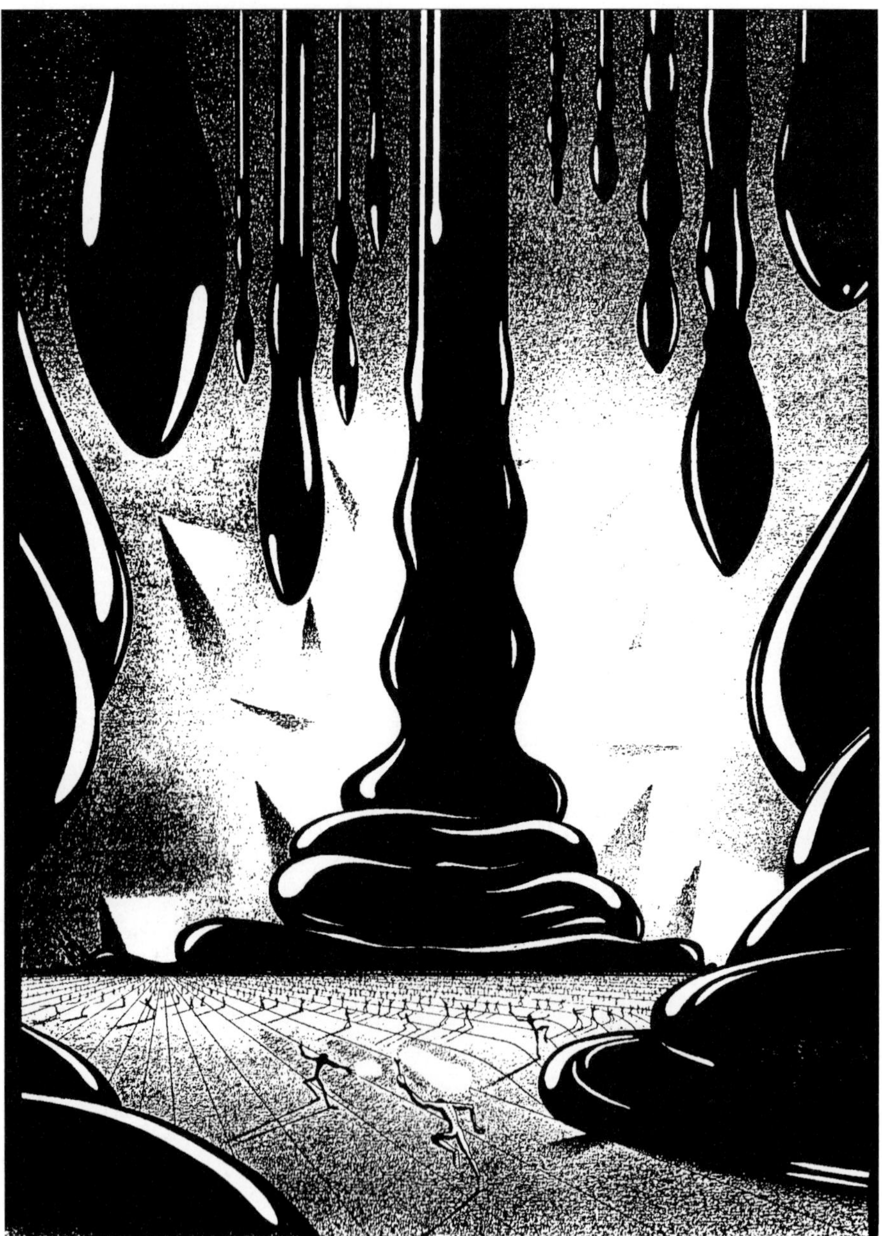

FIGURE 18.5: HOMOTOPY AND VICOUS LIQUID, 1987. A PUDDING-LIKE LIQUID POURS OUT OF THE SKY INTO AN ENOURMOUS SPACE WHERE PEOPLE RUN RANDOMLY ABOUT. THE UNDERLYING THEME IS HOMOTOPY, THE WAY IN WHICH AN OBJECT IS CONTINUOUSLY DEFORMED WITHOUT BREAKING. © ANATOLY FOMENKO.

TABLE 18.1: ONE CENTURY OF SCIENCE FICTION
AN OVERVIEW

Year	Mutations of SF	Highlights
1900	Scientific Romances	Wells: The Time Machine (1895)
	Mars Novels	Laßwitz: On Two Planets (1897)
	Future War Novels	Wells: The War of the Worlds (1898)
		A Voyage to the Moon (1902, movie)
		Verne: Master of the World (1904)
		Forster: The Machine Stops (1909)
1910		Gernsback: Ralph 124C 41+ (1911)
		Doyle: The Lost World (1912)
		Burroughs: Under the Moons of Mars (1912)
		Wells: The World Set Free (1914)
		Lindsay: A Voyage to Arcturus (1920)
1920		Capek: R. U. R. (1921)
		Tolstoy: Aelita (1922)
	Pulp Magazines	Zamiatin: We (1924)
		Metropolis (1927, movie)
1930		Stapledon: Last and First Men (1930)
		Huxley: Brave New World (1932)
		E.E. Smith's Lensmen Series (1934-48)
	Space Operas	Flash Gordon (1936, comic strip)
	SF Comics	Lewis: Out of the Silent Planet (1938)
1940		van Vogt: Slan (1940)
		Asimov: Foundation (1942-50)
		Orwell: 1984 (1948)
1950	'Golden Age' of SF	Asimov: I, Robot (1950)
	Outer Space	Bradbury: The Martian Chronicles (1950)
	Nuclear Energy	Clarke: Childhood's End (1953)
	Monster & Disaster	Miller: A Canticle for Leibowitz (1955-57)
	Movies	Yefremov: The Andromeda Nebula (1957)
		Blish: A Case of Conscience (1958)

TABLE 18.1: (CONTINUED) ONE CENTURY OF SCIENCE FICTION
AN OVERVIEW

Year	Mutations of SF	Highlights
1960		Lem: Solaris (1961)
		Boulle: Planet of the Apes (1963)
		Herbert: Dune (1965)
	Star Trek Serial	Brunner: Stand on Zanzibar (1968)
	New Wave SF	2001 (1968, movie)
	Inner Space	Dick: UBIK (1969)
1970		Niven: Ringworld (1970)
		Strugazki: Roadside Picknick (1972)
	Feminist SF	Komatsu: Japan Sinks (1973)
		LeGuin: The Dispossessed (1974)
		Haldeman: The Forever War (1974)
	Cyberpunk SF	Star Wars (1977, movie)
1980	Cyberspace	Benford: Timescape (1980)
		Aldiss: Helliconia (1982-85)
		Gibson: Neuromancer (1984)
		Atwood: The Handmaid's Tale (1985)
		Cherryh: Cyteen (1988)
1990		Crichton: Jurassic Park (1990)

TABLE 18.2: SPACE FLIGHT AND SCIENCE FICTION
A CHANGING RELATION
"ASTRONAUTICS IS THE ONLY SCIENCE THAT CAN BE SAID TO HAVE BEEN NOT JUST SHAPED, BUT KEPT ALIVE, BY WRITERS AND ARTISTS." (RANDY LIEBERMANN)

Stage		
Age of Visionaries	Space Flight	Uses of Science Fiction
	Space flight theoretically possible	Early space fiction: – (satirical) space utopias – popularizing astronomical knowledge (copernican world view)
	Technical principles of space flight known since Tsiolkowski, Oberth, Goddard	*Space flight as a technically feasible endeavor since Jules Verne*
Age of Rocket Pioneers	Technical feasibility known to a circle of pioneers Exaggerated hopes of pioneers Limitations of technology and costs of realization not yet experienced Scientific community still hostile	SF promotes guiding image of 'space flight by jet propulsion' Recruitment of young space scientists/rocket technicians Popularization of aims and means of space flight to the broad public – fund raising
	End of pioneer age with A4	SF as popular science – touchstone: scientific accuracy
Age of Implementation	Wide-spread implementation beyond original fields of application Necessary expenses and limitations known	Growing gulf between SF and astronautics: – SF goes literary and loses interest in space flight technology – experts pushed back by lack of scientific accuracy in SF SF as a means of public relations

During the thirties, nuclear energy slowly but surely took hold of science fiction, going back to Wells' *The World Set Free* (1914) with its anticipation of nuclear chain reaction. This was one of the correct and even influential anticipations of SF, at least through the person of Leo Szilard, who recollects

> In the spring of 1934 I had applied for a patent which described the laws governing such a chain reaction. This was the first time, I think, that the concept of critical mass was developed and that a chain reaction was seriously discussed. Knowing what this would mean — and I knew it because I had read H.G. Wells — I did not want this patent to become public.[6]

John W. Campbell presented nuclear physics in several stories — even scientific details like isotope separation. As editor of the magazine *Astounding Stories* (from 1937) Campbell inspired other writers to follow. Immediately after World War II visions of nuclear power, combined with 'the conquest of space,' were common fare in science fiction. But after Hiroshima, 'energy freedom' suggested the horrors of nuclear holocaust. SF writers tried to conceive the inconceivable, a post-doomsday world. Catastrophes and monsters crept into cold war fiction, and the cold war movie. Characteristically, monsters like Godzilla or Tarantula are either created by scientists or awakened by atomic explosions. In addition, cold war SF expressed the fears of communist subversion. When McCarthy's Committee for Unamerican Activities went witch-hunting among scientists and artists, evil aliens, often disguised as humans, took the role of 'commies' for example in movies like *Invasion of the Body Snatchers* (1955).

Nevertheless, the forties and early fifties became the 'golden age' of science fiction, with the appearance of writers like Asimov, Bradbury, Clarke, Dick, Heinlein, Sturgeon, van Vogt . . . etc., and with a market favorable for magazines and pocket books. Following the expectations of that time, science fiction was full of innovations, new visions, and a sense of wonder. Cybernetics, game theory, even general semantics (not to speak of all kinds of extra-sensory perception) made their way into science fiction. What SF had failed to predict — computers — now became a central topic. Computers that aided scientists, computers that ran government, computers that fought wars, computers that composed poetry. Again SF missed the possibility of personal computers, the next stage of development, perhaps because it was too concerned with the mythical qualities of the dominating, main frame, big brother computer.

The sixties brought about the creation of the two most popular SF series, which were both space operas: the German magazine and pocket book series *Perry Rhodan* and the American TV serial *Star Trek.* Their initial success was due to the pro-space climate of the decade, with the Apollo missions, and the 'race to the moon.' Possibly, their lasting success is at least partly due to the nostalgic appeal of these space visions.

In Eastern Europe and the Soviet Union, science fiction gradually freed itself from the straight jacket of 'socialist realism,' the official literary doctrine, which had confined it to the propagation of scientific knowledge and especially the alleged achievements of socialist technology: not completely unlike some utilitarian views of SF in the West. During the Khrustchev era Ivan Yefremov, a paleontologist of reputation, opened the way to a renaissance of Soviet SF with his disputed far-future novel *The Andromeda Nebula* (1957), which overcame the tight near-future time frames and thematic restrictions. The sputnik — shattering the American conviction of technological superiority — inspired writers in the Soviet block. Science and technology in the hands of the people would help to establish utopia on Earth; and only a peaceful united mankind would succeed in the mastering of nuclear energy and the conquest of space. SF had to play a role in that vast endeavor: to lead the reader to the foremost outposts of science, and spread enthusiasm among the young technical intelligentsia. Criticism of science, gloomy or dystopian visions of the future would hamper progress. Mostly willingly, writers paid lip-service to the official image of a bright communist future for all mankind. Nevertheless imaginative writers — like Stanislaw Lem in Poland or Arkady and Boris Strugatski in the USSR — would not stick to established guidelines and used science fiction as a medium for speculation on the philosophical basis and ethical implications of science.

With the counter-culture movement of the 1960s, with flower-power and LSD, with anti-establishment insurrection and the revolution on the campus, a new kind of science fiction emerged. It was called 'new wave,' a term borrowed from the *nouvelle vague* in French cinema and applied originally to a special brand of British SF written by authors like Brian W. Aldiss and James G. Ballard and published in the magazine *New Worlds* (edited from 1964 by Michael Moorcock). The proponents of that movement abjured hard science and technology (which belonged to the establishment). They combined a rejection of outer space with stylistic experiments, sometimes outright surrealism. Inner space, the psychological realms within the human mind, were to be explored. Drugs, religion, sex, in fact the breaking of as many taboos as possible became their focus, and entropy their guiding metaphor for social and intellectual decay. The 'new wave' moved quickly to the US, where Harlan Ellison edited an anthology with the tell-tale title *Dangerous Visions* (1967), one of the first and most influential original anthologies (of previously unpublished stories).

At the same time — well before the Club of Rome — science fiction developed a broad understanding of ecological and environmental issues; quite contrary to contemporary visions of an affluent, prospering high-tech society, but at least partly inspired by futurological writers like Paul Ehrlich (*The Population Bomb*, 1968). Risks of nuclear industry, pollution of the environment, dangerous substances in food or the general shortage thereof, and last but not least overpopulation became important topics. Even before Ehrlich, Harry Harrison described in *Make Room! Make Room!* (1966) an overpopulated world, where human corpses are 'recycled' into food. One of the most complex visions of a future world, hit by overpopulation, environmental problems and torn by ethnic clashes is John Brunner's monumental novel *Stand on Zanzibar* (1968), compelling also for the use of reportage techniques and futuristic jargon. Later Brunner took the message from Alvin Toffler's *The Future Shock* (1970) and created the picture of a society without any stable social relations, ruled by the media and a combination of perverted futurological Delphi techniques and Gallup polls (*The Shockwave Rider*, 1975). In his vision of the communication revolution and all-embracing computer networks he anticipated computer viruses, Trojan horse programs and software bombs.

During the late sixties and early seventies women writers, like Kate Wilhelm, Joanna Russ, C.J. Cherryh, finally made their mark. It has even been argued that feminism rescued SF from decline. Some like "James Tiptree jr" successfully exploited the established topics and conventions, others brought feminine concerns or a definite feminist touch into SF, or developed non-technical visions of an alternative but not necessarily matriarchalic or feminist society. One of the richest novels of that period is Ursula K. LeGuin's *The Dispossessed. An Ambiguous Utopia* (1974). In this novel LeGuin compared two societies on a planet and its moon: an anarchist and a capitalist one. The hero is a convincingly imagined theoretical physicist, who travels from one place to the other, at home in neither.

Visions of a post-industrial or information society abound in the science fiction of the seventies, whereas the fears of a big brother computer who is in control of everything receded even before the Orwellian year of 1984. At that time SF was well in its most recent wave of ingenuity, cyberpunk, combining (as the Reagan era might have suggested) high-tech with 'low life' in decaying slum areas afflicted by all the vices of modern civilization: violence and organized crime, drugs, and pollution. It has, like much earlier science fiction, a definite anti-capitalist touch, taking for granted that nation-states or international organizations will not be the big players of the future, but rather greedy and ruthless profit-chasing megacorporations that are beyond any democratic control and make up their own perverse laws. Virtual realities, computer simulated worlds, into which you can 'plug in' by means of special hardware (interfaces), had been introduced by writers well before, for example by St. Lem, Ph. K. Dick, Daniel F. Galouye or the German writer Herbert W. Franke in the fifties and early sixties; but only now, with the rise of video and computer games, they have become a major subject in SF, reflecting also postmodern philosophy and its 'deconstruction' of reality.

William Gibson's novel *Neuromancer* (1984), which marks the beginning of the cyberpunk wave in SF, combines all the characteristic elements, the "cyberspace" (a term for virtual realities allegedly invented on a meeting of American SF writers in 1982) decadent and rotten city agglomerations (called "sprawl"), hackers ("console cowboys") and artificial intelligences as protagonists. Humanity itself will be transformed by technology. Electronic devices are invading the human body, genetic engineering will overhaul it from the very core. Thus we end up, at the close of the century, in the anticipations of Bruce Sterling and other writers with man turning into alien, with Wellsian "minds immeasurably superior to ours," trans- and ahuman beings.

At this *fin de siècle,* scientific and technological progress no longer implies the steady rise of mankind to cultural perfection.

'What if . . .' — Science Fiction as Thought Experiment

According to Moskowitz, science fiction can be defined as a

> . . . branch of fantasy identifiable by the fact that it eases the "willing suspense of disbelief" on the part of its readers by utilizing an atmosphere of scientific credibility for its imaginative speculations in physical science, space, time, social science, and philosophy.[7]

The general principle of that imaginative speculation has often been characterized by the phrase 'What if . . .' What if machines could be made more intelligent than men? What if a plastic-eating microbe escapes from a laboratory? What if the internal combustion machine had never been invented? What if interstellar space travel or time travel were feasible? Sometimes the question is 'How could . . .' How could a sustainable economy based mainly on solar energy work? How could I survive in a global ecological disaster? How could we inform our distant descendants, perhaps living in a new medieval age, of the hazards of nuclear waste deposits?

Following these questions, SF can be understood as a kind of thought experiment similar to thought experiments in science. The experimenter — the writer — begins with a hypothesis and sets up initial conditions. Following the inherent logics of these conditions (i.e., the plot) he derives some results, perhaps surprising ones, as in pointed short stories with twisted or double twisted endings. Use of imagination is as central to the fictional thought experiment as to the scientific one, with the difference that the imagination of a writer is not controlled by scientific, methodological constraints, but by aesthetic, narrative principles. Characteristically, the writer does not look for the most plausible outcome of the experiment but for the most striking, most dramatic one. Perhaps the most profound reason why so many scientists feel attracted by science fiction, is that — beyond the methodological restrictions of science (enforced by social conventions!) — SF opens up vast opportunities for a playful manipulation of scientific concepts, for speculations on alternative laws of space and time, on more than two genders or on changed sexual roles, on machine self-reproduction and last but not least on cunningly devised political and sociological models. For a concerned scientist like Leo Szilard, SF was even a means to elaborate and promulgate models of mutual nuclear deterrence (*The Voice of the Dolphins and Other Stories,* coll. 1961).

In some ways, SF is a quasi-scientific and, like science itself, a collective enterprise. Like scientists, SF writers take notice of their colleagues' work and results; they borrow the fundamental concepts from previous generations of writers. They elaborate on and transform these concepts, apply and test them in new situations, and add new ideas. They regularly share even the technical terms introduced by others, to which the readers are accustomed and which are mostly coined to resemble true scientific terms. Thus, SF has its own topical lineages and traditions and undergoes — like science — phases of accumulative growth (for example during the pulp era) and of deep paradigm shifts, like the one when environmental concerns transformed its overall image of the future.

Of course, most science fiction is superficial adventure and its cognitive content may be questioned. But any piece of fiction, which wants to qualify as science fiction, has to introduce at least one deviation from our common empirical world (called 'the novum' in SF theory), and to work under the conditions imposed by that deviation. Even the dullest space opera is based on more or less elaborated assumptions on space flight and draws its 'legitimation,' its "willing suspense of disbelief" from science, which will, perhaps one day, make these inspiring dreams of interstellar battleships and the annihilation of whole planets come true.

The more interesting examples of science fiction follow the line of real thought experiments and ask questions which could challenge science too. Take for example the possibility of duplicating persons, be it by means of 'beaming' a person to two places or by some other way of recording and reconstructing the physical (and/or informational) structure of a person.[8] Could we have more than one copy of a person? What would become of the individual self? What does this mean for the philosophical notion of personality? What are the juridical consequences? Among others, Stanislaw Lem has tackled questions like these. In his novel *Solaris* (1961), all human efforts to communicate with an intelligent ocean on a distant planet fail, perhaps due to a lack within our scientific reference system. The ocean itself sends 'messengers' to the observing station: simulacra of dead persons taken from the memories of the observers. The epistemological riddle turns into personal tragedy.

Science fiction, used in this way, can prompt what Darko Suvin calls "cognitive estrangement": fantastic imagining in the service of knowledge, not as vehicle for escapism.[9] However, one should not mix up cognitive value with prediction. The principal question 'What if . . .' does not aim at predictions, but implications of a presupposed novum. SF, from that point of view, comes close to a kind of fictional technology assessment. Or, as Fred Pohl put it: "A good science fiction story should be able to predict not the automobile but the traffic jam."[10]

Science and Science Fiction: An Irksome Relation

Science fiction is controversial. It has been blamed for different, even opposing sins: propagating an elitist, technocratic, even authoritarian world view, giving a distorted picture of science, idealizing or charicaturing it, transgressing the borders between science and pseudo-science. It has been hailed as the only kind of literature which really matches an era of science and technology, and it has been scolded for its irrationalist beliefs in science. As early as 1953 Philip Wylie asked: "Does science fiction owe anything to the exalted standards of science itself? In short, does science fiction augment or aberrate human sanity in this age?"[11] — Science fiction as a pro-science enterprise and science fiction as a contra-science enterprise . . .

Of course, all that is true. SF has committed all the sins it has been blamed for, and it has deserved all the praise given to it. As a literary genre with some thousands of first editions per year there is a novel or a story that corresponds to all of these claims.

SF writers use science in many ways, as foreground, background, context or subject of their stories. According to Lambourne *et al.*, one should differentiate six different roles, which science can play in SF:[12] Science may provide the information necessary to describe a real, but relatively unfamiliar, environment. Similarily an imaginary environment may be constructed as consistent as possible with established facts and principles. Thus the ecology of the planet Dune from Frank Herbert's novel of that name (1965) or the seasons on Brian Aldiss' *Helliconia* (1982–85) are designed with the utmost scientific care.

The writer may also use a piece of scientific information as the basis for a puzzle which frequently follows the structure of detective stories. Take Larry Niven's story *Neutron Star* (1966) in which two scientists investigating a neutron star in a fly-by mission are found crushed to a bloody pulp. The hero discovers that this frightful accident is due to the enormous tidal forces of the star even within the dimensions of a spaceship. At that time, neutron stars were still a theoretical concept, the first astronomical observations were made in 1967.

Very commonly science is utilized to justify the existence of devices or processes. As an example, Michael Crichton's novel *Jurassic Park* (1990) — and consequently the movie — are based on genetic engineering. Crichton does even more. He uses the scientific process itself as a credible scientific setting to his novel *The Andromeda Strain* (1969, filmed 1971), where scientists analyse with extreme bio-hazard precautions harmful extraterrestrial spores. Nearly all SF uses science at least peripherally, to justify a device or process, or to provide a generally 'scientific' background.

Some of the examples quoted are 'hard,' or 'hard core,' SF. In contrast to 'soft' SF, this subgenre tries to follow as closely as possible the established facts of hard, natural science and technology. Writers of this kind of SF, like Jules Verne, look for plausible extrapolations; they stick as far as possible to scientific accuracy which implies that they do not use scientific concepts and technical devices as metaphors (as it characteristically happens in 'new wave' SF). Writers like Crichton or David Brin often incorporate realistic pieces of science shop talk into their novels. Unfortunately, some 'hard SF' tends to get infected by the 'technosyndrome': cardboard scientists mumbling polysyllabic neologisms involved in the technical solution of purely technical problems.

Not uncommonly, science is used only verbally: to provide the atmosphere of credibility, necessary for Moskowitz's "willing suspense of disbelief." Even worse, SF tends to mix up real and fake science. Some writers propagated (from conviction or only for the sake of the plot, it does not matter) 'hollow Earth' theories, Velikovsky's cosmogony, Hubbart's dianetics or popular UFO lore.[12] Thus they gave pseudoscientific concepts, like telepathy or giant 'mutants,' the decorum of science — undistinguishable in the minds of readers not scientifically trained. Furthermore, SF has sometimes willingly or unwillingly depicted science as a new magic, which solves in due time all problems and which works without any negative side effects. Especially during its 'golden age' SF gave rise to expectations of a wonderful 'world of to-morrow' which real science could never fulfil (see Ill. 4). But, one can argue, SF only borrowed its verbal and visual rhetorics from science and took for granted what many scientists, engineers, and politicians promised of the coming age of affluence.

Naturally, most readers know that time travel is an impossibility, that space flight faster than light is not feasible, but a slight doubt, an inextinguishable expectation of the big scientific breakthrough (proclaimed by SF) remains. There are, alas, fields where the demarcation line between the possible and the impossible is not as distinct as in basic physical questions. Think of the recording and replay of personality, of technical immortality . . . Perhaps, science fiction has the strongest impact on the public conception of future scientific achievements and future hazards in these fields.

Frankenstein, Einstein and the Scientist in Science Fiction

Science fiction cannot be expected to give a realistic image of science and scientists. For the ordinary reader, tedious lab work, nightly calculations, boring committee meetings, the frustrating fight for funding, even problems with inappropriate equipment are of little interest. Readers expect and writers prefer the eccentric researcher, the extraordinary experiment, the spectacular discovery, the shocking application. The image of the scientist in SF is full of misconceptions and stereotypes. Geniuses and mad scientists displace less colorful and characteristic types. Crackpots and eggheads abound. Glamorizing and demonizing science; SF also glamorizes and demonizes the scientist.

This image of the scientist reveals a double origin: the savant (searching wisdom in old scriptures) and the sorcerer commanding magic, white and black. Quite commonly, SF stories are plotted along the lines of Goethe's *The Sorcerer's Apprentice* who is unable to control the forces he has called forth. Within SF, this kind of plot goes back to Goethe's contemporary, the young Mary Shelley, whose novel *Frankenstein, or The Modern Prometheus* (1818) became the inspiration of a whole subgenre of SF/horror movies — and of stories on genetic engineering.

Victor Frankenstein, M.D., personified the aspirations and the hubris of science, challenging the 'natural order' of things. The scientist in the role of a demiurge, imitating the act of creation, that image fits well into the conflicts between science and religion in the nineteenth century and it fits well into the strained relation between science and ethics in the late twentieth century.

Frankenstein, perhaps the paradigm of the mad scientist, was copied many times, e.g., in Fritz Lang's movie *Metropolis* (1926), where a scientist with the looks of a cliché alchemist creates a robot in the likeness of a beautiful woman. But never, not even during the pulp era of the early twentieth century, did vicious crackpot scientists outnumber their sane colleagues. The 'professor' in SF, savant and wizard in one person, was to explain the intricate wonders of science to the laymen heroes and readers. He was the man of fundamental science, the theoretician and researcher, as well as the man of applied science, the inventor. He always had a useful new gadget at hand; or he quickly invented it during a crisis. Dr. Zarkov from the *Flash Gordon* comics (from 1934 onwards) is a case in point.

Contradicting the sociology of real science, fictional conventions helped the isolated scientific hero to survive the beginnings of Big Science, even the Manhattan project. Only rarely did writers mould their protagonists after the thousands of scientists involved in the construction of the atomic bomb. Albert Einstein, the outstanding visionary mind (at that time already rather secluded from the scientific community), remained their model; and he embodied the social responsibility of science. Perhaps the best example for this high regard of a scientist's social role and integrity is given in the movie *The Day the Earth Stood Still* (1951). Here, an extraterrestrial emissary tries to contact the leading figures of the world to warn them about the cosmic law of peace or destruction. Unsurprisingly there is no politician to receive the message — only a physicist, working alone on his home blackboard, summons his fellow scientists from all over the world to listen to the messenger.

Since the late fifties scientists have been less commonly represented in SF. This may be due to a decline of the professional status and the social significance of scientists. In a study of scientists in SF, Patrick Parrinder argues that a change in writers' careers may be an additional reason. Whereas in the 'golden age' SF writers were characteristically science graduates, the more recent generation has (with exceptions) received little or no scientific education.[13] One could debate whether this was effected by the growing reputation of SF as literature.

Today scientists are frequently represented as a faceless force, as anonymous as scientific-technological progress by itself. Parrinder concludes, that the heroes of cyberpunk SF in the eighties and nineties are not researchers producing knowledge, but persons who control information: computer scientists and hackers who make sure that knowledge remains unregulated and potent.

Science Fiction at the End of the Century

At the close of the century science fiction resembles a large supermarket, offering anybody what he or she wants. Space operas for 'trekkers,' BattleTech for weapon junkies, 'hard SF' and 'soft SF,' cyperpunk for computer kids, optimistic and pessimistic visions, feminist futures and macho ones, all kinds of mixtures with horror and heroic fantasy, high-tech nightmares and high-tech utopias — with a distinct preponderance of dystopian (or at least ambiguous) anticipations. Philip Wylie's question whether SF has strengthened or weakened human sanity cannot be answered for the whole of the genre.

During the last decades, the association of science fiction with science has diminished. Science fiction has lost most of its belief in and enthusiasm for science; its conviction that science and technology will initiate a better future. But science fiction only shares a common scepticism ...

References

1. Butor, M. "La crise de croissance de la science-fiction" (1953), in Michel Butor, *Essais sur les modernes.* (Paris, 1964), p. 233; M. Salewski, *Zeitgeist und Zeitmaschine. Science Fiction und Geschichte.* (München, 1986).
2. Interview with Jules Verne in *T.P.'s Weekly* (9 October 1903), quotation according to P. Parrinder (Ed.), *H.G. Wells. The Critical Heritage.* (London, Boston, 1972), p. 101f.
3. Clarke, A.C. "Science Fiction: Preparation for the Age of Space", in R. Bretnor (Ed.), *Modern Science Fiction. Its Meaning and Its Future.* (New York, 1953), p. 202.
4. Clarke, A.C. "Space Flight — Imagination and Reality", in A.C. Clarke, *1984: A Choice of Futures.* (New York, 1984), p. 107.
5. See Cohen's chapter in this volume.
6. Quoted from Weart, S.R./Szilard, G.W. (Eds.), *Leo Szilard — His Version of the Facts. Vol. II.* (Cambridge, 1978), p. 18.
7. Moskowitz, S. *Explorers of the Infinite. Shapers of Science Fiction.* (Westport/Conn., 1974), p. 11.
8. Moravec, H. *Mind Children. The Future of Robot and Human Intelligence.* (Cambridge/Mass., 1988).
9. Suvin, D. *Metamorphoses of Science Fiction: On the Poetics and History of a Literary Genre.* (London and New Haven, 1979).
10. Quoted from Lambourne, R./Shallis, M./Shortland, M. *Close Encounters? Science and Science Fiction.* (Bristol and New York, 1990), p. 27.
11. Wylie, Philip. "Science Fiction and Sanity in an Age of Crisis", in: Bretnor (1953), p. 230.
12. Lambourne *et al.*, (1990), pp. 39–48.
13. Parrinder, Patrick. "Scientists in Science Fiction: Enlightenment and After", in R. Garnett/R.J. Ellis (Eds.) *Science Fiction Roots and Branches. Contemporary Critical Approaches.* (Houndsmill and London, 1990), p. 62–64.

Further Reading

Pamela Sargent, *Women of Wonder* (1975), *More Women of Wonder* (1976), *The New Women of Wonder* (1978).
Aldiss, B.W./Wingrove, D. *The Trillion Year Spree.* (London, 1986).
Ash, B. (Ed.), *The Visual Encyclopedia of Science Fiction.* (New York, 1977).
Barron, N. *Anatomy of Wonder. A Critical Guide to Science Fiction.* (New York and London, 1995).
Clarke, I.F. *The Tale of the Future.* (London, 1972).
Clarke, I.F. *Voices Prophesying War. Future Wars 1763–3749.* (Oxford and New York, 1992).
Clute, J./Nicholls, P. (Eds.), *The Encyclopedia of Science Fiction.* (London, 1993), augmented edition as CD-ROM. *The Multimedia Encyclopedia of Science Fiction.* (Danbury/CT, 1995).
Goswami, A./Goswami, M. *The Cosmic Dancers. Exploring the Physics of Science Fiction.* (New York etc., 1983).
Gunn, J. (Ed.), *The Road to Science Fiction* (four volumes), (New York and London, 1977–82).
Moylan, T. *Demand the Impossible. Science Fiction and the Utopian Imagination.* (New York, 1986).
Pringle, D. *The Ultimate Guide to Science Fiction. An A–Z of SF Books.* (London etc., 1990).
Sadoul, J. *Histoire de la science-fiction moderne, 1911–1984.* (édition révisée et complétée, Paris, 1984).
Slusser, G.E./Rabkin, E.S. (Eds.), *Hard Science Fiction.* (Carbondale and Edwardsville, 1986).
Suvin, D. *Positions and Presuppositions in Science Fiction.* (London, 1988).
Warrick, P. *The Cybernetic Imagination in Science Fiction.* (Cambridge, 1980).

Chapter 19

Seeing and Picturing

Visual Representation in Twentieth-Century Science

Martin Kemp

Modern science is for the most part a highly visual affair, and never more so than in the age of computer graphics. An astonishingly wide range of instruments, technologies and media are used to 'see' effects, including means for rendering visible those phenomena that cannot normally be seen by the human eye. Even emissions beyond the scope of the eye, such as the sound waves used to survey the ocean floor, can be used to construct 'pictures' which are designed to be understood through sight. In exploring theories and expounding results, there are few sciences which do not resort to visual demonstrations at some stage, and many place illustration at the heart of their systems of exposition. We only need open a typical issue of *Nature* or *Science* to see a brilliant array of visual material, not only in the articles but also in the advertisements. Much of what we learn about science as an activity — its practitioners, its equipment, its buildings — is acquired from various forms of visual representation, most notably from the photographic media. The popular image of the scientist, whether the obsessed boffin in a white lab coat working to combat a deadly virus or the lunatic genius bent on making Frankenstein's monster, comes to us in visual form, not least through the screens of cinemas and television sets.

This chapter is concerned with the ways that science is pictured in relation to mechanisms for 'seeing' — taking picturing in its broadest sense to embrace any form of giving visual expression to the content of science whether by direct representation or by the artificial construction of an image intended to convey specific information. In this sense we will not simply be thinking of pictures in terms of those we see in an art gallery — though a few of the images can be considered pictures more or less in that manner — but as any visual transcription of what claims to be how things 'look,' even when it is machines that do the looking and drawing for us. In some instances, as with multi-dimensional geometry or gas chromatography, we are required to think of looking in a rather different way from our normal idea of using the light which is entering the eye to reconstruct the three-dimensional array of objects in the space around us.

The images will be of two main types. The first are those which depict the subject-matter of scientific investigation, that is to say the phenomena which are being studied or the theories to be expounded. Such depictions range from the direct representation of, say, the plumage of a bird, to a visual record in a cloud chamber of the track of an atomic particle which cannot in any sense be seen in a particular place at a particular time like a normal object in our visual field. The second type of image concerns the iconography of science, that is to say those images which convey an idea of science as a pursuit. The simplest of the images are portraits of famous scientists, such as adorn journals and provide frontispieces for biographies. The more complex depictions show science actually happening, most typically in a laboratory with teams of scientists and technicians operating intricate banks of equipment — though in truth such images are more often posed for the sake of the photograph than showing what actually occurs in the normal way. It is this more complex kind of image that has nourished the film-makers' cliché of the agitated genius surrounded by forests of wires, tubes, bubbling stills of technicolor liquid, beeping electronic devices and gaseous emanations.

The two kinds of image might seem on the surface to be very different — the one concerned with the internal, technical exposition of scientific matters, and the other interesting from the human point of view but essentially trivial in terms of genuinely scientific content. However, as we have increasingly come to recognize that science is a social activity, and that the content of scientific thought can have a period style no less than the buildings in which science occurs, so the two types of image may be seen to be intimately interconnected. If the former images occupy far more space in this chapter than the latter, this is a reflection of the greater range of types and medium involved and the complexity of the visual issues of what is represented — the second type is dominated by the photographic media — rather than an indication that depictions of the business of science are secondary in understanding science in the twentieth century.

Means, Ends and Styles

No era has ever been more visual than the late twentieth century. Science and technology have invented an array of visual media of unparalleled complexity, potential and geographical extension. Science itself has not been slow to exploit the range of visual devices, both as tools for investigation and to create new ways of demonstrating ideas and results in compelling manners. The public brandishing of state-of-the art visuals has become one of the prime means for the broadcasting of discoveries and promoting the status of the sciences. We so accept the role of the visual in the publications of modern science that it comes as a surprise to find that there is nothing inevitable about illustrated science. In mediaeval science — up to around 1450 — texts of many of the sciences which we would assume require illustration, such as medicine, botany, optics and physics, were at best sparsely illustrated, and many of those illustrations were not intended to convey technical information about the science in the text. What has survived of the scientific legacy from classical antiquity also suggests that visual exposition generally played a secondary role. In anatomy there was even a positive proscription which viewed illustration as a dangerous distraction from understanding about 'real' things. The rise of illustration as a major tool of science in the European Renaissance depended upon the revolution of the means for depiction — most especially the invention of perspective in the fifteenth century — and upon the invention of the printed book with printed illustrations, but the new tools of depiction only assumed power for science in the context of a reformed agenda for those disciplines which were concerned with the direct observation and recording of natural appearance. The modes of representation in twentieth-century science are very much the heirs of the Renaissance revolution, and the dominant conception of space in computer imaging, with its three spatial co-ordinates and the depiction of forms as seen from a single viewpoint (even if it can now be made to move), is essentially the same as that of a fifteenth-century picture in perspective. When geometers have attempted to step beyond the parameters of the familiar three dimensions, in representing three- or n-dimensional bodies — they have inevitably been locked into the invention of artificial conventions which signify the extra dimensions within the practical limits of the three.

The greatest difference between the visual qualities of twentieth-century science and what went before is the new predominance of the representation of things that are technically invisible to the human eye. There were, of course, earlier instances where technologies had rendered visible the previously invisible, most notably the telescope and microscope in the seventeenth century. And these earlier instruments raised the key question that arises with later systems of artificial seeing, namely whether the qualities of what appeared in the instruments depended more on the nature of the instrument itself than upon what was really 'out there.' This question could of course be asked of the eye, with the decoding system of the brain, no less than of the instruments. But the use of an instrumental system of perception on top of the natural one served to double the problems of whether what we think we see accurately reflects some external reality. The kind of 'visual' systems which are characteristic of the twentieth century are those which can record rays which lie immediately outside the visible spectrum, such as ultra-violet or infra-red, or which use different kinds of emission, such as sound, heat, X-rays, radio waves and electrons. There are also a host of techniques for getting the phenomena to describe themselves, by leaving some kind of trace which can be seen by the aided or unaided eye. Atomic physics has relied very much on the setting up of systems for the visual recording of tracks along which the particles have moved in tiny instants of time. Also characteristic of the visual in the twentieth century, particularly in its final years, has been the extension of the visual means for depiction beyond a static image on a flat surface to embrace cinematographic motion and the use of stereoscopic images and virtual reality to achieve three-dimensional illusions.

Where science uses non-visible emissions to construct a visual image, the artificial system is designed to deliver codes of representation — such as lines, shadows, textures and perspectival space — which have proved efficacious in visual communication over the centuries. In sonar scanning, for example, a machine is set up to generate the sound emission at frequencies of up to 500 KHz, as the equivalent of the light we wish to use to view the sea bed. As a speaker (Quentin Higget) said at the British Association in 1994, "It's bird song we're trying to see." The emissions impinge on those surfaces directly exposed to the source, and the rays are reflected back towards a robotic receiving device (the 'eye' of the 'organism') which is linked to the decoder (the 'brain') which in turn drives the depicting mechanism (the 'draftsman's hand'). All these stages are active, in the sense that the mechanisms are intentionally designed to achieve something specific and are switched on to perform their assigned functions. Not the least important of the stages is the decoding, in which the 'noise' in the system — the often muddled array of unwanted information which tends to conceal the clarity of desired perception — is filtered out, most often nowadays by computer programs. This stands in an analogous relationship to the mechanisms by which the human brain filters the myriad stimuli impinging on our senses to achieve a selective and coherent perception of whatever we are striving to see at a particular moment. The selective and intentional nature of any process of filtering inevitably raises questions about whether we are simply seeing what we want to see.

The high degree of direction and intentionality in the system, in which ends and means are locked together, often has the consequence that the means of visual representation becomes the automatic and unquestioned mode of communication between scientists, and more radically conditions the subjects and methods of the investigation. Thus a model of a complex organic molecule, whether constructed from the old-fashioned rods and balls or realized in animated form in stereoscopic computer graphics, embodies important aspects of the theoretical constructs which lie behind research programs and achieves a level of reality in communication in scientific journals, books and in teaching such that it becomes part of the mind-set of visualization which determines the trajectory of research inquiries. Some representations, such as Niels Bohr's famous 'planetary' visualization of the atom as a nucleus surrounded by a girdle of electrons, achieve such a compelling visual status that they retain their grip after they have lost their ostensible status as accurate depictions of the physical set-up.

The range of representations with which we are concerned cover the full range from direct depiction to the most abstract of conceptualizations. At the former extreme we have a photograph or direct depiction 'from life' of something visible, which stands as a particular kind of record of what the eye might actually see in front of the object. Such naturalism is always selective to a greater or lesser degree, and every depiction stands somewhere on a relative scale between detailed realism and schematization. For instance, a painting of birds in a habitat stands closer to the realistic end than the depiction of silhouettes for field recognition, but both convey usable visual information about what we can expect to see. At a remove from normal seeing are images (photographic or drawn) derived from optical instruments which extend the power of sight, generally by high magnification. At a further remove on the road to visual contrivance are those depictions formed by non-visible emissions. One of the characteristics of images which arise outside the parameters of normal sight is that they frequently require a second layer of visual interpretation, using forms of labeling or diagrammatic exposition which depend heavily upon prior knowledge. Even more artificial are those systems which transcribe data into visible form by some kind of mapping, such as graphs, bar charts and so on, and graphic means for the represention of such statistics as mean temperatures or population densities on maps of the surface of the world. At the most conceptualized extreme are those graphic representations which give visual form to theoretical propositions or abstract concepts, such as evolutionary trees or the diagrams which purport to explain multi-dimensional geometry. The more abstract and conceptual a system becomes the more likely it is to feature as a dynamic part of the scientist's processes of speculation, either as a form of experimental visualization within the mind or as actually drawn on a piece of paper, blackboard or computer screen in the graphic form of 'thought experiments,' as often exemplified in Edison's sketches.

A convenient way of giving some visual substance to the range of representations and, as it happens, to the notion of visual style in science, is the poster issued in 1994 by the journal *Nature* to celebrate '125 years of publishing excellence.' (Figure 19.1). A small photograph at the top left of the inaugural issue in 1869 immediately announces in its subtitle its visual ambitions as "a weekly illustrated journal of science." A series of collaged images scattered across the surface of the poster (using computer design procedures) announce some of the landmark discoveries and give some idea of the variety of visual techniques involved in different branches of science over the period 1925–1985. The discovery of Australopithecus, one of the 'missing links' of popular evolutionary theory, is celebrated with a photograph of the skull. (Figure 19.2a). Chadwick's demonstration of the existence of the neutron in 1932 is signalled by a computer depiction, using the bright colors and rendered surfaces which have become so popular in the late twentieth century. (Figure 19.2b). Almost inevitably, Watson and Crick's double helix for DNA is given star billing, again in a modern computer simulation. (Figure 19.2c). The discovery of quasars in 1963 is conjured up by suggestively mysterious images derived from the emissions reaching us from outer space. (Figure 19.2d). The widely applicable technique of magnetic resonance imaging is colorfully represented by an MRI scan of the human brain, selecting the type of scan which has most caught the public imagination. (Figure 19.2e). Similarly dramatic in a public context is the demonstration of the dangerous hole in the ozone layer revealed in the ultraviolet survey by the satellite Nimbus 7.

(Figure 19.2f). Finally, we are given a neat geometrical diagram of Carbon 60, the 'first fullerine' or 'Bucky ball,' (Figure 19.2c) so named because its semi-regular configuration had been already adopted by the American architect, Buckminster Fuller, in his futuristic geodesic domes.

The range, from a veridical photograph of a seen object to a geometrically conceptualized model of something that can only be reconstructed from data provided by modern techniques for seeing the unseen, neatly encapsulates the variety of visualization and representation in modern science. It is also characteristic that none of the images have been allowed to stand in the form they originally appeared. The images of the neutron and double helix are far removed in technique from those in the original publications, and all the images have been processed by the designer of the poster to look as if they are mingled in a disorderly manner with chemical stains and some more regular patches of indeterminate identity. Discerning the message of the poster, like the deciphering of any visual image, involves an interaction between the intentions of those responsible for the image and the spectator's proclivities. My overall impression, which is likely to be reasonably close to what was intended, is of successive visual revelations of previously unseen wonders, broadcast through self-consciously high-tech design and printing techniques, and an overall sense of dynamism — of science as process and progress. By contrast the front page of the first issue (Figure 19.2h) — with the 'rustic' type face of branch-like letters for 'NATURE' and its sober listing of the contents — speaks of the visual and intellectual styles of Victorian science. The old *Nature* would sit as happily in the wood-paneled library of a Victorian Gothic laboratory building as the bright, computer-graphic cover of the new *Nature* belongs with the metallic shelves and computer terminals of a late twentieth-century university building. Science clearly has a style, at any given period, and that style speaks volumes about the intellectual and social climates in which science is practised.

Supplementing the Eye

Virtually from its inception in 1839 photography was hailed as a tool of science, as well as an 'art' or an adjunct to art. William Henry Fox Talbot, the inventor of the negative-positive process, saw its potential, not only as a recording device but also as a way of directly reproducing images seen in an optical microscope, much as is still current today when very high magnifications are not required. As soon as exposure times were reduced by new photographic technologies, it also became possible to 'freeze' motions too rapid to be detected by the unaided eye. In 1878 Eadweard Muybridge's famous sequential photographs of a horse running radically altered our conception of a horse's gait and brought into question earlier modes of pictorial representation. Perhaps the nicest of all the later revelations were the studies of the beguiling plasticity of splashes published by A.M. Worthington in 1908, later extended radically by the 'instantaneous' images produced in Harold Edgerton's "Strobe Alley" at MIT, which froze such rapid motions as humming-bird wings and speeding bullets.

Another extension of sight within the visible spectrum has involved the enhancing of stimuli too slight to be registered by the human eye. A spectacular instance is the visual output of Voyager 2 in 1989, which was equipped with a camera that could take pictures with 1/900 of the amount of light impinging on earth while traveling at a speed of 27 km per second, which were then transmitted from this robotic eye to earth on radio signals of astonishing slightness. The resulting images of Neptune, synthesized from trillions of pieces of electronic data can give the effect of what it might be like if we could actually see the planet from a seat in Voyager, in the form of a computerized mosaic. The data can also be used to generate so-called 'false color images' which convert information about, say, the composition of the gaseous mantle around Neptune, into visual form. To some extent, the designation 'false-color images' — which indicates that features in bodies of data are signalled by the arbitrary assigning of a different color to each of them — is a misnomer, since the principle is no essentially different from that of the human eye. Our eye and perceptual system take in stimuli of light at different wavelengths and differentiate them in terms of what we call color, but the object we see as colored is not literally of such-and-such a color. Indeed, eyes of other organisms are adapted to encode stimuli in a different way and presumably do not see the same colors as we do. Thus the assigning of colors to such properties as the reflectance of different gases, zones of temperature or levels of neurological activity in brain scans is not so much 'false' as relying on an encoding through color, analogous to that accomplished by our eyes. Often the choice of color is not actually arbitrary in terms of how we view color in our normal visual world. Zones of 'hot' activity are characteristically designated as red or orange, while the darker and cooler hues are used to express lesser concentrations and energies.

Other phenomena result in the modification of light within the visible spectrum which cannot be seen because the eye is not attuned to their reception. For instance, the rapid passage of an object through the air (or *vice versa*) generates wave patterns which result in changed refractive indices for the disturbed air currents. There are physical methods for making the patterns of flow visible, as by the injection of a tracer substance into a fluid, but these inevitably affect the properties of the fluid to some degree. Non-disturbing methods, which use techniques for the registration of the optical effect of the currents, have been developed. The Mach-Zender reference-beam interferometer, which exploits the interference of a reference beam and a refracted beam, presents flow patterns with considerable clarity and often of great beauty (Figure 19.3). Interference phenomena have also played a significant role in optical mineralogy, in which polarized filters or nicol prisms (calcite crystals) in a microscope are combined ('crossed') to produce interference figures which reveal fundamental properties of the minerals under investigation. In such methods, we are seeing optical effects, but contriving both the properties of light and the receptor in such a way that a form of seeing is achieved beyond the unaided phenomena and unaided eye.

The orthodox spectrum of sight can also be extended at each end through the recording of infra-red and ultraviolet emissions. It was the ultraviolet light from the sun reflected back through the earth's atmosphere to orbiting satellites in the 1970s that alerted us to the hole in the layer of ozone that protects us from ultraviolet radiation. The information could be and was demonstrated in graphs, but it has been the colored 'picture' constructed from the data that has proved really potent. Working at the other margin of our visual range, the multispectral scanner used in the Landsat satellites to survey the surface of the earth supplement wave lengths from selected slots within the visible spectrum with ones in the near infrared band. Using a mirror which oscillates at thirteen times a second, and detectors tuned to different wavelengths, the scanner supplies lines of data at the rate of millions of digitized components per second. The images which are synthesized from the data (the 'sensory impulses' within the device) give a computed picture of the earth's surface, but not just in the optical-photographic sense, since the wavelengths indicate a whole series of activities and qualities in the landscape, such as the moisture content of the soil or the vigor of growing crops. The color does not automatically correspond to our way of translating wavelength into the Newtonian spectrum. Thus, for instance, fully-grown plants appear most vividly in the infrared range, and are according represented as red in the depictions rather than green. The 'false-color' images are, again, not so much literally false, but correspond to an artificial 'eye' rigged to translate visual data into different coloristic conventions from our eye.

Such images are also termed 'thematic,' since they illustrate selected themes, but the human system is also rigged to be thematic, though in practice it can adjust the themes to which it attends at any given moment with remarkable fluidity. The 'rigging' of the image affects every stage of the process, from the conception, design and calibration of the 'eye,' though the analytical filters which strain out unwanted sets of data, to the visual conventions of the final output, which finally has to work with what our eye can use and is accustomed to using. The underlying conception behind the rigging is never a neutral matter. It is determined by a series of intersecting intellectual, human, social, economic and political values. The launching of satellite, involving huge slices of state and/or corporate expenditure requires more than scientific curiosity, and the decision of what to seek — data for military surveilance, evidence of crop vigor, earthquake activity, etc., — is determined by a variety of human interests. In this, mechanical or electronic seeing is no different in principle from the partiality of human seeing. It is partial, but within the fields of partiality aspires to a non-arbitrary conveying of information to a knowing viewer.

Alternative 'Lights'

Almost any emission which passes through space or phenomenon which manifests itself in a differentiated manner in space and time and which can be detected and recorded may be used to build up a picture in terms of a map or landscape of objects or their effects. The most spectacular debut of one of the most familiar of the twentieth century techniques, occured at the very end of the last century, when Wilhelm Röntgen registered what was seen on a photographic emulsion as the result of X-rays passing through the hands of Frau Röntgen and himself (Figure 19.6). His curious discovery was to revolutionize medicine and to have enormous implications in the fundamental sciences of the arrangements of atoms and the distant cosmos. Subsequently, emissions of almost every conceivable type have been used as a surrogate for visible light, including sound waves (normal and ultrasound), radio waves, electro-magnetic forces, electrons, positrons, gamma rays and even gravitational attraction. Sometimes the emissions are deliberately generated and controlled by the investigator, while in others, such as radio waves emanating from distant stars, the source is given and lies outside our control.

Perhaps the easiest to grasp, since it resides within our normal sensory compass, is the use of sonar techniques to map features which are inaccessible to visible rays of light. Since no light effectively penetrates below a depth of 200 metres in the sea, an alternative method needs to be found to look at the bottom of the oceans from the surface. In the 1940s and 1950s, adapting technology which had been devised to detect submarines in the Second World War, sound waves were bounced of the sea bed to a tuned receiver which could register the distance the reflected ray had travelled. If the scan is taken at an angle — a 'side-scan' — a series of highlights and shadows are generated in terms of sound. Bruce Heezen, Marie Tharp and Maurice Ewing used sonar scans of the Atlantic to posit radically new views of the configuration and dynamics of the sea bed (Figure 19.4). The immediate transcription of the reflected waves is not into visible form, but consists of a series of measurements. These can be translated into various visual forms, ranging from silhouetted profiles of sections of the bed to the underwater equivalent of bird's eye maps (fish's eye maps?) in which a physiographic drawing renders the topography in much the same way as has been done with land masses since the Renaissance.

The form of sonar used to survey the ocean bed is supplemented for deeper structures by seismic surveying, in which lower frequency sound, generated by percussive methods (such as explosions), achieve a penetration to underlying layers of rock and are selectively reflected back to a series of detectors. The raw images, as translated into visual traces of the layered reflections, require substantial filtering to eliminate the background noise which tends to obscure the main features, and each method of filtering itself produces a depiction which needs to be understood in terms of the parameters of its own procedures. As always, when dealing the three-dimensional features, if a perspectival image can be synthesized — with all the artifice that it necessarily involved — the understanding of both structure and likely process is enormously enhanced. However, even heavily filtered and synthesized images make heavy demands upon the knowledge, experience and judgement of the geologists, since the investigator needs to know what kind of thing can be seen in what is an enormously mediated representation in relation to what might be the actual features below the earth.

The perception of other forms of emission lying definitely outside our sensory range have played a reforming role in a wide range of sciences. Indeed, there is probably no science concerned with the direct scrutiny of the material properties of things — however remote — that has not been radically affected. It is not practical to survey the whole field in this chapter, and therefore my tactic will be to take two contrasting examples. The first concerns the scanning of the human body using specially generated emissions under controlled conditions, while the second concerns the detection of naturally occurring radiations from some of the most distant bodies known to us.

X-rays in their original form gave entirely fresh insights into aspects of the interior of the body, but only into a few selected aspects, depending on the relative opacity of the various substances of the internal organs. Bones showed up well, but blood vessels barely at all. One procedure is to rely upon selective injection of an X-ray opaque material to provide the necessary contrast, which can further be enhanced by the manual or digital subtraction of unwanted background information, but the amount of spatial information in the image is still limited, since forms at different depths are inevitably overlaid in the see-through view. A more radical extension of the use of X-rays to picture the internal topography of the whole body has been provided by computer tomography (CT), in which an X-ray fan beam is received by detectors, and subjected to digitization and image processing to produce complete sectional images of designated thickness, which in a sequential series present a remarkably complete picture of the configuration of organs and may reveal abnormalities. A related technique to produce sectional slices is magnetic resonance imaging (MRI), which exploits a strong magnetic field and radio frequency to detect hydrogen protons in fat and water. Like CT, the technique is dependent for its new power (and indeed for being 'visible' at all) on the analytical and synthetic power of the modern computer as the 'brain' of the perceiving mechanism.

Both CT and MRI scans are essentially concerned with revealing morphology, while the other major new system for the scanning of brains, positron emission tomography (PET), involves registering the levels of activity in different sites. The basic notion of injecting a short-lived radioactive substance into the bloodstream dates from the late nineteenth century, but only with modern computed tomography has it been possible to build a usable picture of the relative densities of positron emission from the cell nuclei. The levels of activity are conventionally encoded in terms of the range of a normal spectrum, with higher ('hotter') activity signaled by red and the lower by violet, with white and black reserved for the highest and lowest extremes. The technique has obvious implications for the study of the localization of brain functions, but it has also been yielding data about irregularities in brain function, such as those caused by tumors. The plotting of psychological abnormalities is an altogether more problematic area, both with respect to the parameters of the interpretation of the tomographic artefacts and of the 'normal' and 'abnormal' brain. Interpretations of the nature of brain function and of particular conditions are not dictated by the image; the image is a tool which yields varied results in different hands according to the skills and preconceptions of the users.

Over the years, refinements in optical telescopes, most notably the development of giant reflecting telescopes and telescopes mounted on space vehicles, have resulted in stunning images of distant features, but there are limits to what can be achieved using the visible spectrum. The wavelengths beyond either end, the infra-red and ultra-violet, have both been used to supplement optical telescopy, the latter most spectacularly in the Hubble space telescope, which has the largest mirror ever located in space. Although designed with faulty geometry, its defective 'eyesight' was corrected by astronauts in an adventurous space walk, and is now providing huge bodies of data which can be synthesized into stunning images.

It had been noted in the 1930s that the Milky Way seemed to emit radio waves, but it was not until the next decade that radio astronomy became a serious tool for the investigation of very distant bodies. It subsequently proved especially useful in building maps of carbon monoxide concentrations, which are taken as markers of star formation. The analogous possibility that X-rays might be used for astronomical investigation arose in 1960, when it was demonstrated that the sun emitted X-rays which could be detected by a rocket outside the earth's atmosphere. Two years later it was demonstrated that other bodies also emit X-rays suitable for telescopic observation. The ability to send robotic eyes outside the protective screen of the earth's atmosphere has also opened up the field of gamma-ray astronomy, which is able to detect some of the highest energy processes in the most distant reaches of the universe. Depending on the quality of the data and the processes to which the image is subjected, the resulting 'pictures' can assume a striking resemblance to seen landscapes, such as the pictures of Venus synthesized from the billions of bits of radar and other data transmitted by the Magellan space probe in 1991. (Figure 19.5). There are clearly heavy preconceptions at work in the construction of such an image, not the least of which are conventional ideas about the aesthetics of landscape. A series of 'beautiful' observations have been synthesized to give 'beautiful' results. The production of such images is clearly satisfying in itself, but they also serve to provide spectacular demonstrations in the popular media which can help defray public doubts about the huge costs involved.

Too Small to be Seen

Using high magnification to reach to the most distant tracts of the universe has been complemented by techniques of magnification which have enabled us to see down to atomic level and indeed even within the tiny structures of the atom itself.

The most productive techniques for 'seeing' below the levels of resolution of the optical microscope have been the electron and scanning tunneling microscopes, X-ray diffraction crystallography, and cloud and bubble chambers.

In the basic technique of transmission electron microscopy, which was first used in the 1930s, the image acquires contrast from the scattering of the electrons by denser portions of the sections, with little of the differential absorption which provides so much assistance in our perception of light impinging on or passing through surfaces. Notwithstanding the fundamental differences, the results are processed to give the appearance of optically seen images, (Figure 19.7) sometimes like tiny landscapes, most effectively when the contrast in examining materials of low mass density is enhanced by the oblique deposit of a very thin layer of metal on the irregular surface. A specialized variant is the scanning tunneling microscope, which uses a tip so fine that its point is comprised of just one protruding atom. When the end atom is moved to within a few atomic diameters, electrons 'tunnel' into the sample, and the resulting current can be used to provide a vivid image of the atomic topography of the surface in three dimensions and with good depth of field. In a sense, we are now able to see what such surfaces 'look' like at the atomic level, but in another way this idea is misleading, since we are contriving techniques to construct images in terms of what we mean by the 'look' of things as experienced in our normal visual range. This is not to say that the images are false or arbitrary, and they have great utility when certain kinds of information about surfaces is required, but they are based upon highly selective perceptions of what comprises an atomic surface.

In the earliest years of the century the question was not how we could see what atoms and molecules look like, but whether they were real or simply useful theoretical constructs. The crucial proof came not so much from measurements and calculations, valuable though these were, but from when they could be 'seen' to exist, or rather, when visual traces of their presence became undeniable. The work of Perrin, Svedberg and Rutherford drew much of its conviction from the visual records of what was happening in their experiments. By 1911 the cloud chamber, invented by Charles Wilson, was being used to plot the paths of single charged particles as recorded by tracks of water droplets. Like its later cousin, the bubble chamber (which contains a superheated liquid) the images present the most visually compelling representations of charged particles as they perform their geometric dances under the influence of a magnetic field (Figure 19.8). In a bubble chamber, tens of thousands of photographs may be analyzed, using more than one camera to achieve a three dimensional picture of the tracks. Fundamental discoveries in particle physics, including the detection and often exotic naming of new particles, have depended heavily on the propensity of the particles to produce their own characteristic signatures in the lines they draw in cloud and bubble chambers during minute increments of time. The signatures themselves only describe in a very selective manner some of the traces of what happens when particles collide, exchange quanta and so on. Diagrammatic representations of a different kind are required to show what is believed to be happening in terms of 'deep structures.' The most effective of these representations have been Feynman diagrams — named after their inventor, Richard Feynman — in which graphic conventions such as arrows and zig-zags act to encode in visual form the assumed pattern of physical events.

The technique which achieved most in giving visual evidence of the nature of atomic and molecular structure in solids arose in 1912 when Max von Laue decided to use naturally occurring crystals as gratings for the study of X-rays. The result was an interference pattern which could be recorded on a photographic plate. The image is not a photograph of the interior structure of the crystal, but a record of the interruption on a photographic emulsion of the X-rays diffracted by the electron clouds around the nuclei of the atom . To translate this image into a three-dimensional arrangement of the atoms in the unit cell requires the co-ordination of a series of measurements of intensities with a theoretical model of the likely structure. Initially the results, such as those of the Braggs in Leeds, concentrated on the confirmation of the supposed models for crystalline substances such as diamond, but increasingly X-ray diffraction was establishing unexpected structures. The photograph which played perhaps the most dramatic role of all in modern science was that of deoxyribose nucleic acid (DNA) by Rosalind Franklin. With such complex organic molecules, the image requires an extremely high level of visualization before it can be translated into a hypothetical model of the three-dimensional structure, and we should not be surprised when alternative models are proposed. This is just what happened when model builders in Britain and America worked with the first photographs of DNA. It was access to the photographs of Franklin that permitted Watson and Crick to formulate their famous double helix scheme for the fundamental genetic components of living organisms.

Modeling and Mapping

On the face of it, the building of a model of a phenomenon that cannot be seen and is thus inaccessible for direct depiction and the making of a map as a way of plotting features that can be directly seen may seem to be contrasting activities. However, in as much as the model and the map both take a carefully selected body of data and achieve a carefully directed representation of those features deemed to be most significant for the particular exercise, the model and the map share much in common. This is particularly true when we make maps that are not wholly concerned with topography, and embody other data such as population density, temperature or rainfall. It is also particularly true of some of the more schematic modeling of molecules, in which complex three-dimensional structures are represented by chemical abbreviations and lines representing bonds.

The diagrams of molecular structures that we now use are the ancestors of systems developed in the nineteenth century to express, in diagrammatic form, what was known about the bonding of various substances. As atoms increasingly became knowable as real items and as such techniques as X-ray diffraction began to provide ways of looking inside molecular structures, so more data became available for the construction of increasingly complex models in three dimensions. When Watson and Crick used Franklin's image to hypothesize the double helix configuration for DNA, they exploited one of the then current methods for modeling, which is basically a spatial rendering of the conventional diagram of bonds in an organic substance. Watson and Crick were not of course claiming that the DNA molecule would look like their model if we could see it. They were demonstrating the underlying linkage and distribution of the component parts in the most visually effective way available to them. The problem is, however, that in the minds of the public, and to some degree the minds of scientists, the compelling visual nature of such a model can easily assume a level of reality beyond the modelers' intentions.

Later modeling systems have both recognized the conventionality of the representations and striven for greater realism. At the conventional end are the ancestors of the attenuated rod and ball method, in which it is the relative positions of the atoms and their bonds which are the focus of attention. The more linear the representation, the more possible it is to see within and through the structure to gain some sense of its spatial disposition. Towards the realistic end are space-filling systems, in which the globes representing the atoms assume sizes equivalent to the space they are actually thought to occupy. It has been found that retaining the perfect globe shape for all the atoms falsifies the distances between their centers, and more recent models have used fusion (resulting in Van der Waals surfaces at the interfaces) to provide more accurate dimensionality. The beguiling computer images, for which tailor-made programs are available, show complex arrays of brilliantly colored balls, gently shining, like a piece of constructivist sculpture made from smooth plastic. Such images, particularly of the double helix, have become one the most compelling icons of late twentieth-century science, not only in the popular imagination but also in the pages of articles and advertisements in journals.

With respect to the 'appearance' of a molecule, the models are highly contrived, since their design is directed to represent certain properties with which the scientist wishes to work. A major concern of modern biochemistry is how compounds might be developed that can lock on to viruses, most urgently in the case of HIV, and the elaborate illusionistic models which can be produced by computer, which now involve animation, rotation, stereoscopic viewing and even virtual reality, serve as powerful tools in the investigation of how molecules or parts of molecules might fit together. A model is necessarily a mental construct, but the conventions do mean that the model does not contain a realistic representation of particular

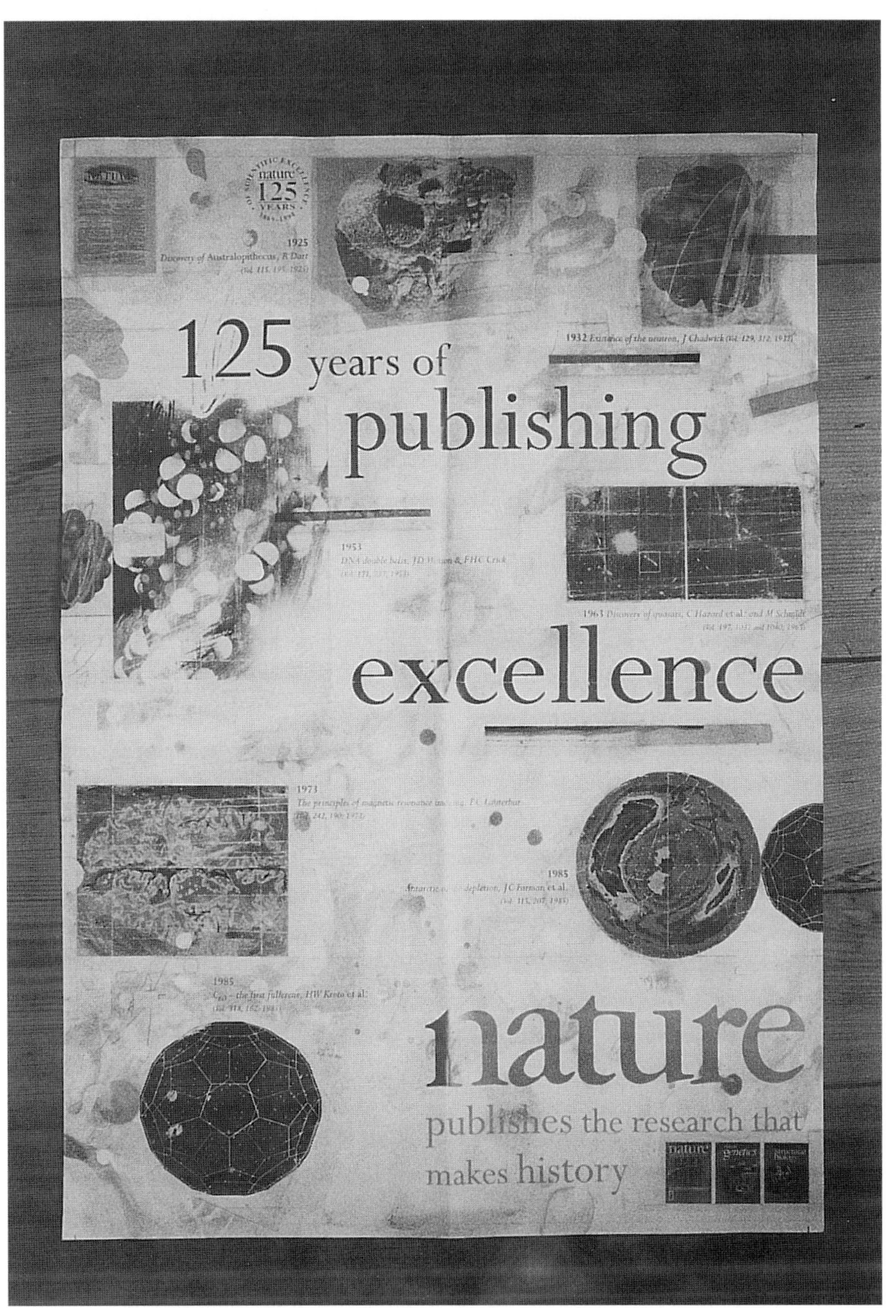

FIGURE 19.1: POSTER TO CELEBRATE 125 YEARS OF *NATURE*, 1994.

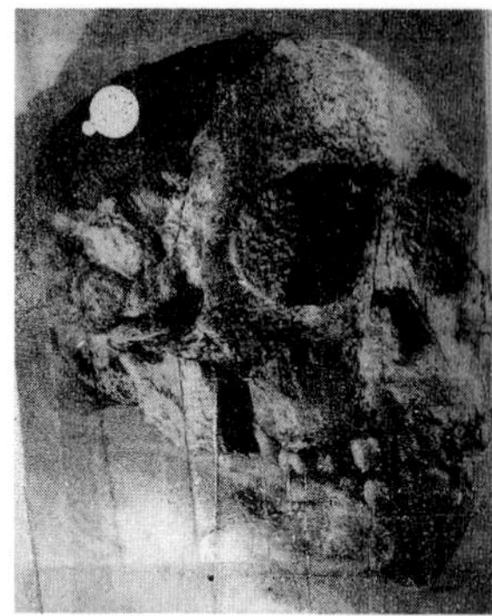

FIGURE 19.2A: SKULL OF AUSTRALOPITHECUS, DETAIL OF FIGURE 19.1.

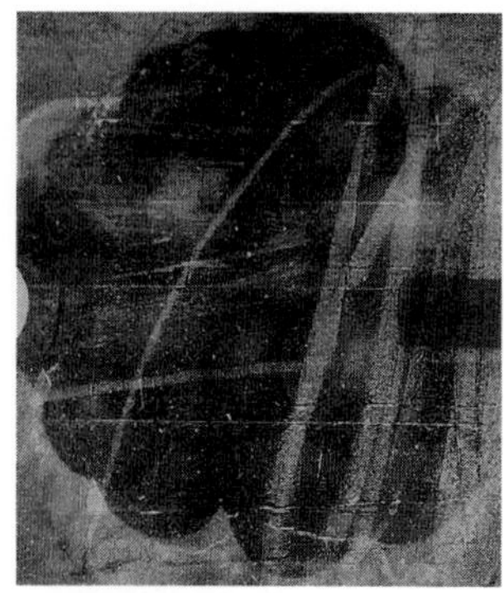

FIGURE 19.2B: THE NEUTRON, DETAIL OF FIGURE 19.1.

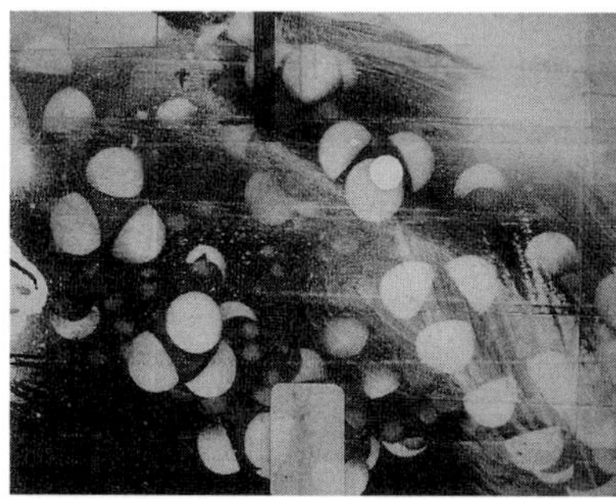

FIGURE 19.2C: DNA, DETAIL OF FIGURE 19.1.

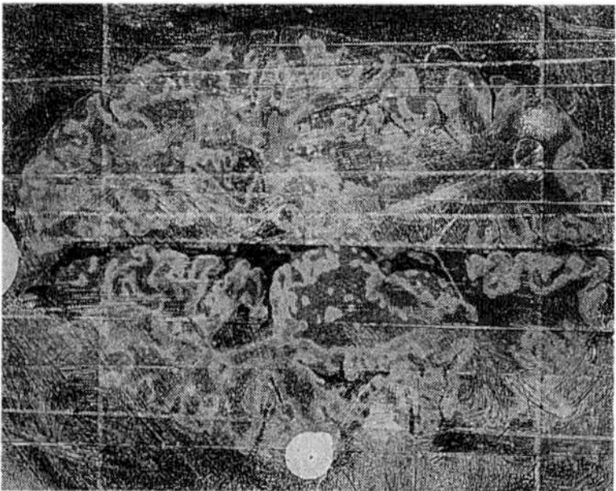

FIGURE 19.2D: QUASARS, DETAIL OF FIGURE 19.1.

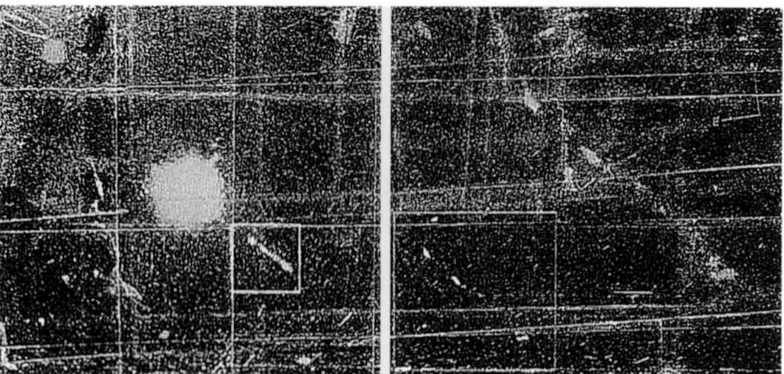

FIGURE 19.2E: MRI SCAN, DETAIL OF FIGURE 19.1.

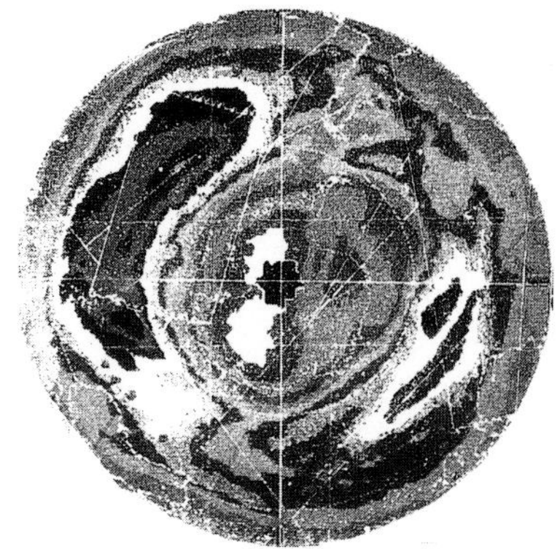

FIGURE 19.2F: THE OZONE LAYER, DETAIL OF FIGURE 19.1.

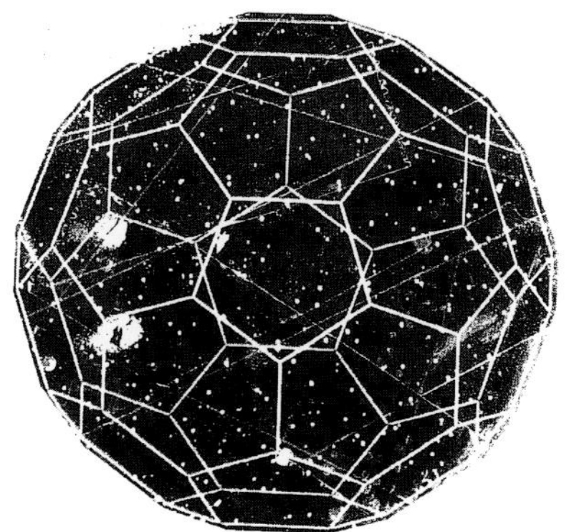

FIGURE 19.2G: CARBON 60, DETAIL OF FIGURE 19.1.

NATURE

FIGURE 19.2H: COVER OF THE FIRST ISSUE OF *NATURE* IN 1869, DETAIL OF FIGURE 19.1.

FIGURE 19.3: INTERFERENCE PATTERN OF SUPERSONIC FLOW AROUND AN AIRFOIL, TAKEN WITH A MACH-ZENDER INTERFEROMETER.

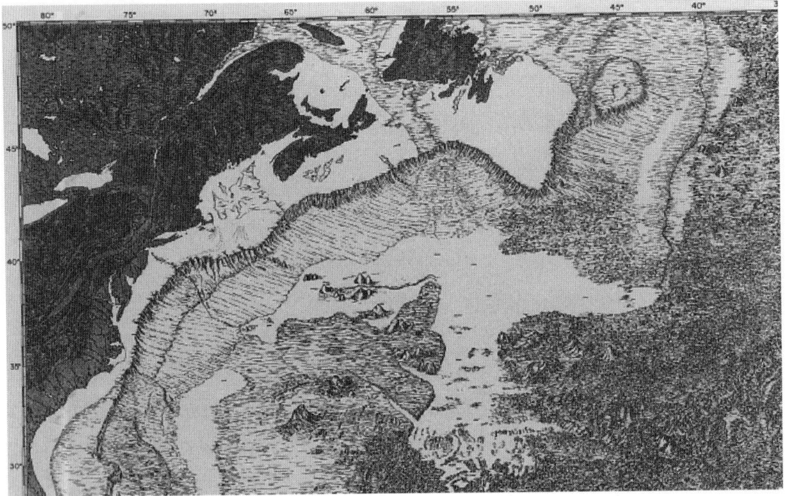

FIGURE 19.4: PHYSIOGRAPHIC DIAGRAM OF THE NORTH ATLANTIC, BY BRUCE HEEZEN AND MARIE THARP.

FIGURE 19.5: MOUNTAINS OF MAXWELL MONTES ON VENUS, FROM THE MAGELLAN SATELLITE, 1991.

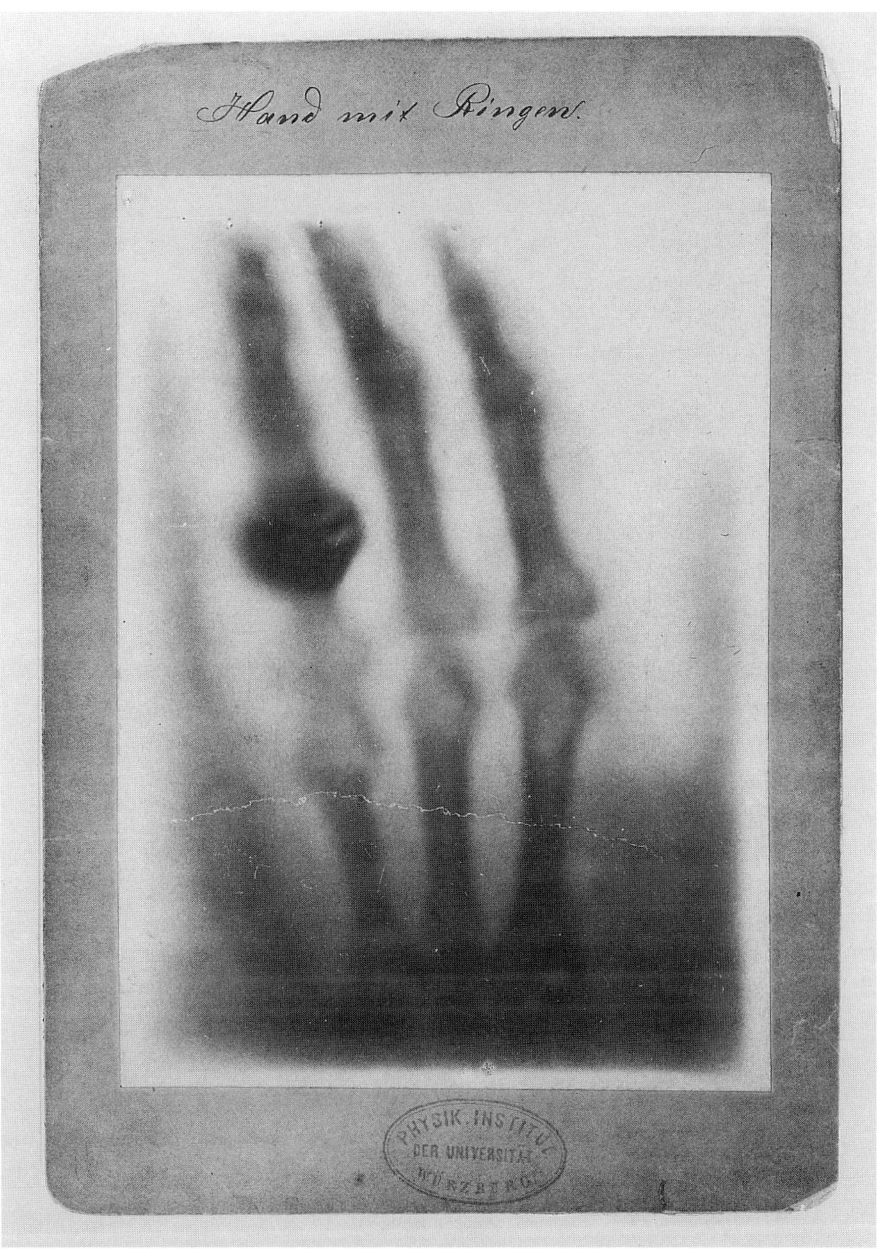

FIGURE 19.6: RADIOGRAPH: PROBABLY THE HAND OF FRAU ROENTGEN WITH A RING; BY W.K. ROENTGEN, 22 DECEMBER 1895. S 0326

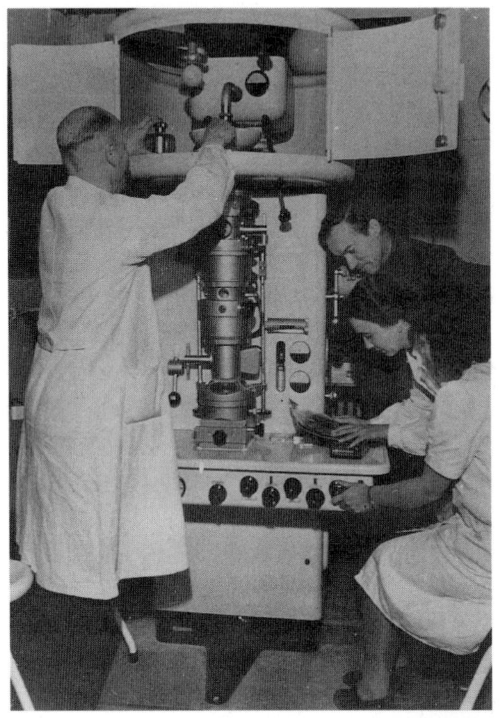

FIGURE 19.7A: TWO IMAGES OF AN *ELECTRON MICROSCOPE*.

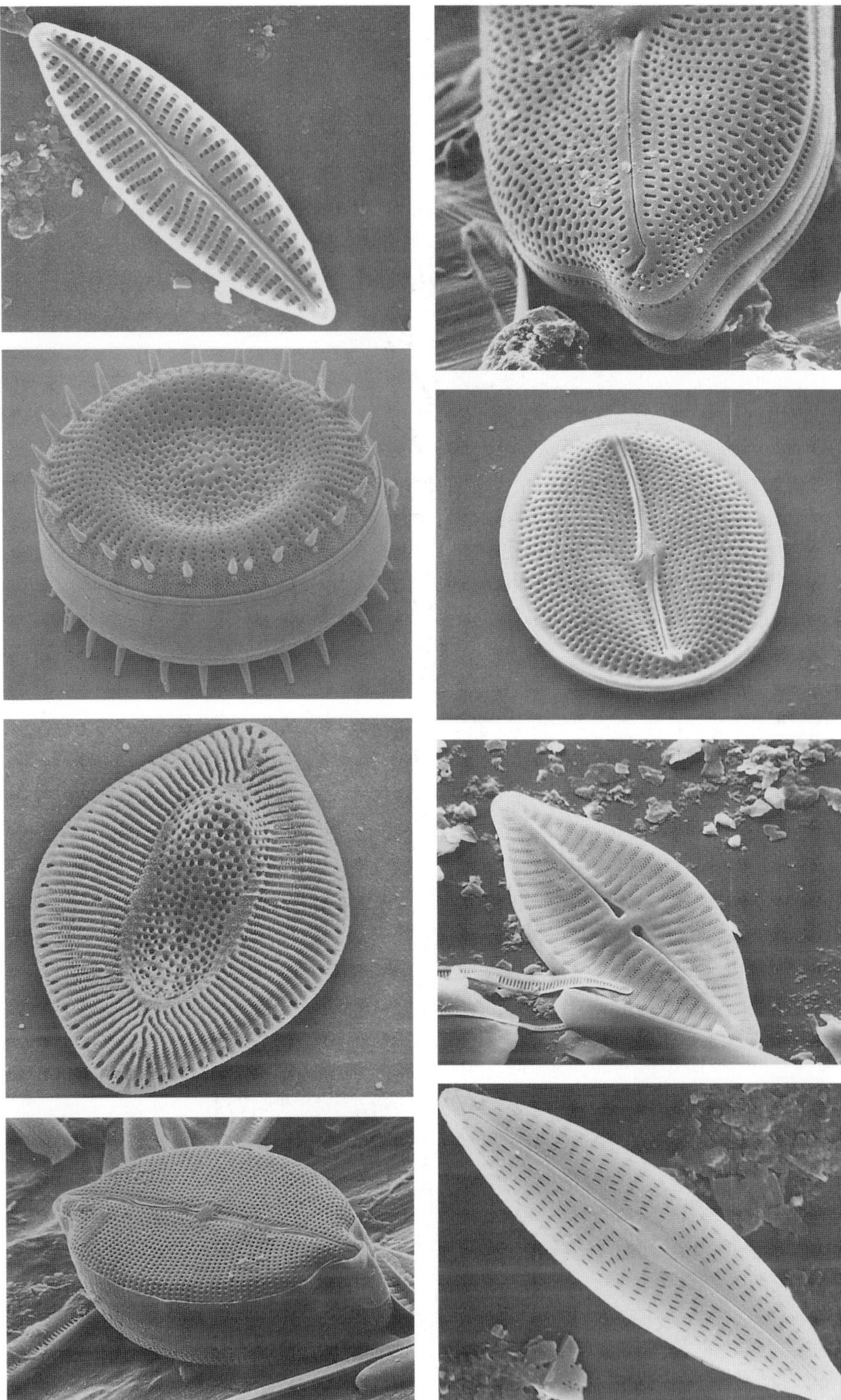

FIGURE 19.7B: ELECTRON MICROSCOPE IMAGES OF DIATOMS. COURTESY OF PROFESSOR F.E. ROUND.

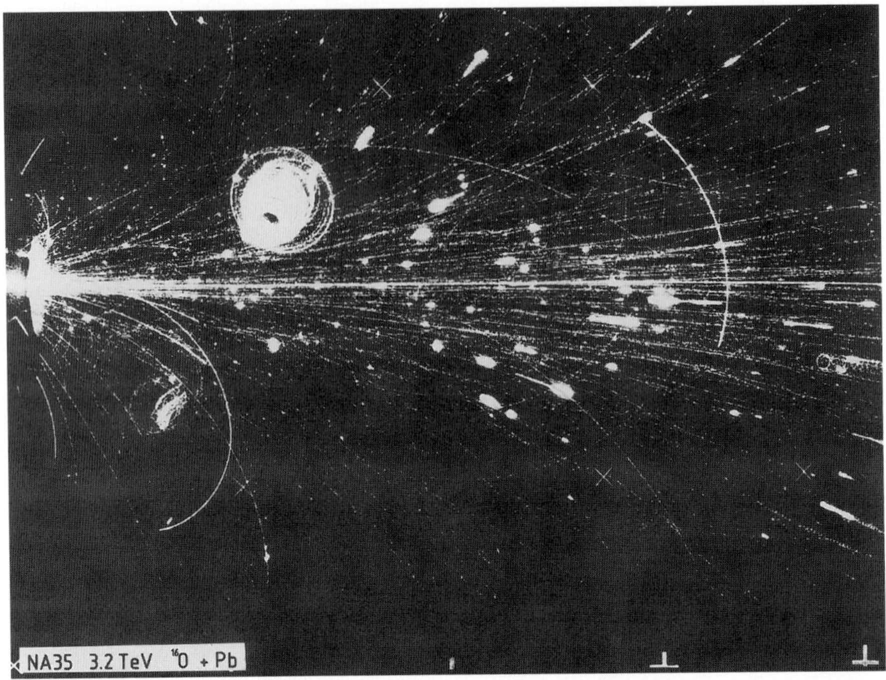

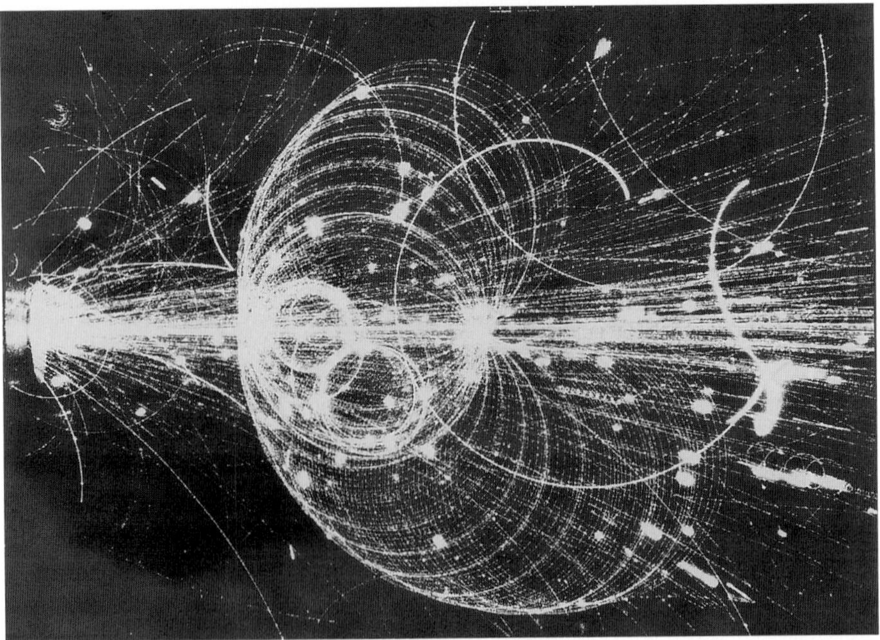

Figure 19.8a: Bubble Chamber Images of the decay of a neutron. © CERN.

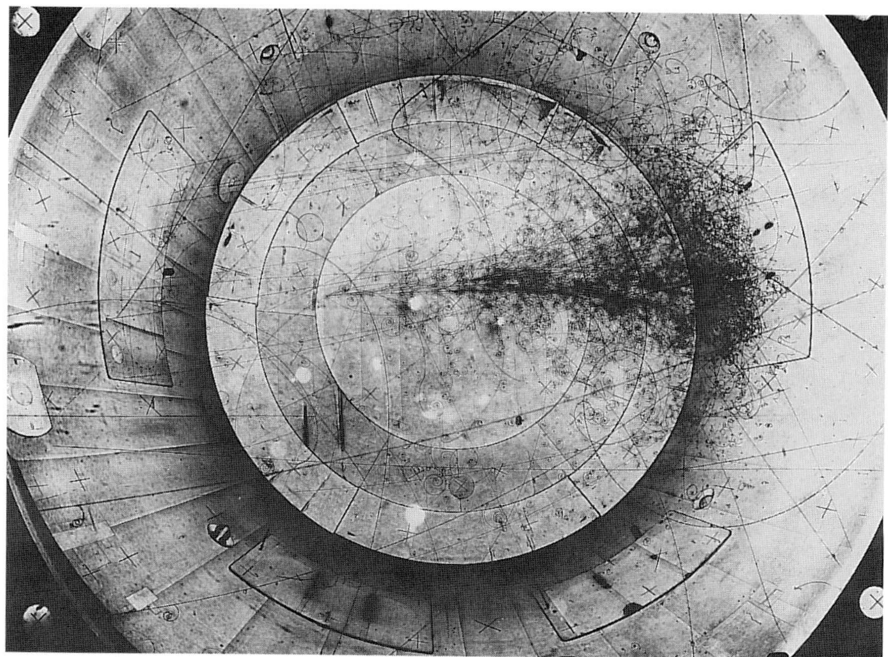

FIGURE 19.8B: BUBBLE CHAMBER IMAGES OF THE DECAY OF A NEUTRON. © CERN.

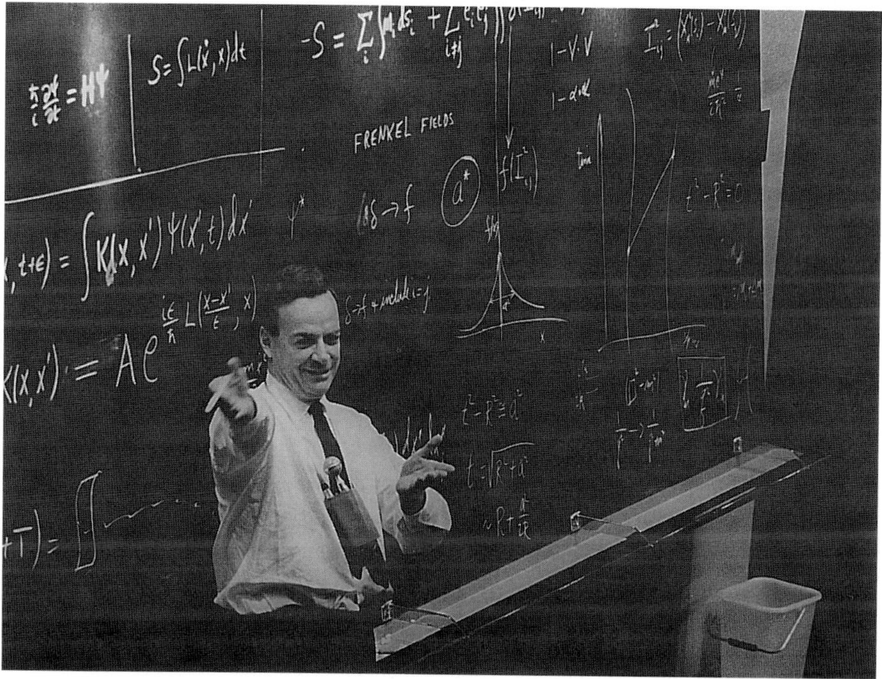

FIGURE 19.9: RICHARD FEYNMAN LECTURING AT CERN. © CERN.

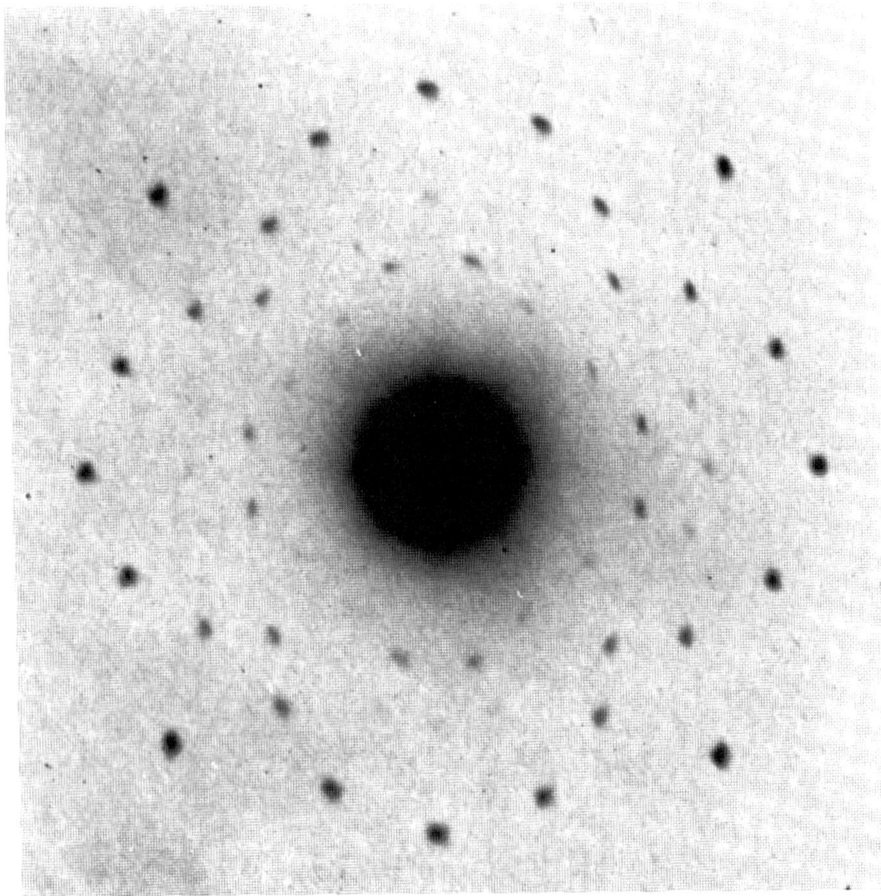

FIGURE 19.10: LAUE PHOTOGRAPH OF ROCK SALT.

features of the object or phenomenon being modeled. Our eyes are what we have to use, and ultimately we have to work within their parameters if we wish to demonstrate things convincingly in visual form.

When we turn from the model to the map, we find they share much in common. In a sense, a molecular model is a kind of map, and the notion of mapping is itself much wider in modern usage than the older concept of a plotting of the topography of the surface of the earth. Topographical maps have been invented to convey a wide range of information — climatic, economic, political, etc., — generally by color coding and sometimes by the erection of three dimensional features on the surface of the land as a form of concrete graph of the data. Maps have been produced of underground strata, the core of the earth, the sea bed and so on, and many of the maps of more inaccessible features have inevitably relied on filling the gaps between patches of incomplete data, much like the earliest maps of America added coastlines on the basis of reasonable supposition about those regions that had not been surveyed. Maps have been produced of the topography of past times, and maps which chart prehistoric climates with models extrapolated from the very incomplete fossil record. Geographical maps, like other models, show a wide range of variability in the closeness of their match to the particular properties under investigation. The properties under investigation are themselves directed by a variety of personal and wider interests, and even the most soberly objective plotting of information stands within a nexus of choices which originate in complex human situations.

Especially important forms of chemical and biochemical mapping in the later part of the century have involved the realization of data on a visual plot which graphically reveals the composition of a substance under investigation or the actual distribution of its constituents in a linear sequence. An example of the former is the technique of gas chromatography, in which a small sample of the subtance to be analyzed is vaporized in a steam of pre-heated inert gas (such as helium or argon) and the constituent parts are separated according to the varied rate at which they migrate through an absorptive column. A detector produces a graphical plot or chromatogram in which peaks indicate the presence of the components in the compound. By contrast, genetic mapping produces models for the actual distribution of genes on the same or different chromosomes, but using statistical rather than physical methods. Exploiting the observation that some genetic peculiarities seemed to be inherited in pairs, it proved possible to infer the relative proximity of two genes many years before the clarification of the role and structure of DNA and the devising of more direct markers. Linkage maps for individual chromosomes are mathematically generated models of physical locations which can potentially be used in a predictive fashion in genetics, for the purposes of identification of individuals in forensic science — and even as tools in processes of genetic engineering.

Maps can also be made of theories. A molecular or genetic model is a visualization of a kind of theory, as is a Feynman diagram — and the expression of an idea in visual form can often give it a particular kind of potency, both in terms of endowing it with concrete form and as a memory image. One of the most compelling of the graphically realized theories is that of the evolutionary tree or 'bush,' in which the Darwinian principles of natural selection are envisaged in terms of branching systems. Variable lengths are used to represent the origins and fates of species over time, with time indicated by the vertical ascent of the tree. The patchy nature of the fossil record, and recent challenges to the pure version of natural selection, mean that the configuration of the model is, even on its own terms, extremely hypothetical, and some of the trees do not aspire to present a joined-up version of all the various branches. However, judged in terms of the kinds of presuppositions and values it displays, the evolutionary tree becomes a much more problematic artefact. The adoption of the model of the tree is hardly surprising, given our inevitable propensity to envisage things by analogy to objects of which we have a clear mental picture. But the adoption of the tree then places evolutionary succession on a ascending scale, since the 'more evolved' organisms appear towards the top of the system. Although the system has not been overtly designed to say that the later forms, most notably the recently appeared species of *homo sapiens,* are superior or 'higher,' it does visually reinforce the kind of values that are embodied in such phrases as 'higher forms of life' and 'primitive organisms.' The familiar graphic representations of the succession of stages of man from animal and ape-like prototypes, which have so insinuated themselves into the popular imagination and encouraged the notion of the 'missing link,' carry almost unavoidable implications of the 'rise' or 'ascent' of 'man,' as 'he' (and they are generally male) assumes his upright gait with proud airborne head. We should be alert to the way that a diagram of a theory may implicitly convey (or may be read as conveying) values beyond those the theory is directly addressing.

From Mathematics to Modeling

Some branches of mathematics, most notably geometry, have always involved visual exposition, both in the procedures of speculation and at the stage of demonstration. Mathematicians still fill pages and blackboards (or flip-charts) with diagrams in much the same way as they have always conducted thought experiments in drawn form. I suspect they always will. More recently, computer-driven techniques have been pushing into new areas of mathematical modeling, using the powers of the computer to handle multiple iterations with a speed and complexity impossible for the human mind. In particular, major advances have been made in the modeling of chaotic systems, often on the basis of relatively simple formulae or physical set-ups. One of the simple mathematical formulas has been that of the Julia set investigated and developed in visual form by Benoit Mandlebrot, which has resulted in models of extraordinary complexity, and sometimes of peculiar beauty, especially with the artificial addition of color. One particular form is that of fractals, in which apparently asymmetrical configurations of extravagantly complex shapes exhibit a self-repeating quality as the parts are looked at on an increasingly small scale (or, effectively, under greater magnification). In a similar way models have been developed of actual physical mechanisms, such as that of a double-jointed pendulum turning around a horizontal axis, which moves in an apparently unpredictable manner. What has been discovered, by predominantly visual means as the patterns have unfolded themselves on the computer screen, is that chaotic systems exhibit compelling patterns of underlying orderliness which prevents them from being truly random. The plotting of the phenomena on graphs has revealed the presence of 'strange attractors,' as visual maps of the probabilities which arise from principles of self-organization in the chaotic system. They are chaotic but not truly random.

Initially regarded as mathematical curiosities, and even occasionally hailed as solving the problems of aesthetics, the modeling of chaotic systems and the discerning of fractal orders has begun to exhibit wide applicability in the analysis of a whole range of natural systems, as it is realized that both physical features and dynamic systems in the world may present such systems in realization and action, not in a purely mathematical manner but in the rougher terms of terms of material reality. Such natural phenomena as the configuration of coastlines and the spirals of phyllotaxis in plant growth find suggestive analogues in the mathematical models that can be generated by powerful computers. However, the bridges between the models and the physical behavior of actual materials often remain elusive. Whether or not fractals and the like solve the problems of the definition of beauty in art — which they almost certainly will not — they have resulted in a particular visual style of science, whose brightly dynamic patterns and general sense of electronic visual noise speak very much of the values of a high tech enterprise in a public context.

Humans, Machines and Contexts

The predominant images of science as an activity center upon the persons of the scientists themselves. The production of portraits of scientists arises when two main factors come into play: the conviction that the *individual* scientist is important, as a kind of hero of his or her enterprise; and the collective social desire to establish scientists as persons of note, as members of a group deemed worthy of portrayal and ultimately of recording for posterity. During the eighteenth and nineteenth centuries it had become accepted that great scientists, particularly those who rose to high position in scientific institutions, would be suitable subjects for portraiture, normally in oil paintings. Often, there would be nothing or little to distinguish the portrait of a scientist from that of another important person — perhaps a few attributes, a specimen or piece of apparatus, but not much more. As the means of recording became less expensive and exclusive, above all through photography, so portraits of practitioners proliferated in a series of contexts — personal, institutional, national and international — as a result of private initiatives, official activities and media interest. A famous scientist might, typically, appear in a family album, in a University's documentation of a research team, in the publication of an official body, in press reports and more recently on television. Such is our human interest in what people look like that it is hard for a biographer to refrain from illustrating a portrait of his or her subject, though there is no reason why the physiognomies of great scientists should provide any clue as to why they differ from average members of the profession.

Just as we see images of scientists as heroes of their own enterprise, so we find 'portraits' of machines, instruments and devices as the technological 'heroes' of scientific advance. Photographic images of such devices as an electron microscope or ultra-sound scanner are familiar in a range of professional and public contexts, though their external appearance generally tells us little about how they are used and virtually nothing about how they actually function. In a related way images of sites for science have proliferated, showing the latest laboratory, exterior or interior, in the favored architectural mode. Such images stand as visual certifiers as the style of the enterprise, speaking the most advanced language of what a high-tech device and its site of operation is expected to look like. A Victorian laboratory, in a Gothic style and decorated with mosaics of great inventors, speaks of a different set of values from a 1960s 'lab' in the international modernist style. When images of scientists, devices and settings come together in a single representation — frequently with the particpants posed in 'action portraits' in order to look un-posed — they act as reinforcing icons of the business of science as then constructed. It may seem obvious to us that an 'action portrait' is a good way to show a scientist or a scientific team, but in earlier eras it was more important to be shown as a 'gentleman' rather than as a mere operative. The hands-on representation is predicated upon changes in how the scientist wishes to be seen and is viewed by others.

All such representations, to lesser or great degrees, are loaded with social significance, and were intended, overtly or covertly, to make a point. Sometimes the propaganda becomes very overt, as when a politician is depicted visiting a laboratory to communicate his or her commitment to advances in science — particularly national advances — even when the same politician might actually be responsible for a contracting budget devoted to scientific research. Less overt, but no less purposeful, is the use of the images of scientists, machines or buildings for the promotion of science within the literature generated by the profession itself. Thus a brochure advocating the latest form of cancer treatment might well show the most advanced kind of radiological device to attack the growth as a way of giving the patient confidence that the most up-to-date and expensive technologies are to be deployed. In another visual context, the same machine could be regarded as threatening rather than benign, reminding us that a visual image is a slippery thing, and highly dependent on the framework for viewing and the range of expectations that are at work.

Complementing those representations which are produced in close conjunction with the professional business of science, are those which are generated in the more popular arena. These range from controlled popularization, as in magazines of popular science and the more serious branches of the media, to the most fanciful inventions of science fiction. The more wholly fictional the image, the less it might seem to reflect the actual activity of professional science. However, the fictional images, from Mary Shelley's *Frankenstein* in 1818 to Steven Spielberg's *Jurassic Park* in 1993, may be richly informative in two significant respects. They can provide a clear sense of dominant fashions in science, whether the vitalism of Shelley's day or the chaos theory of the later twentieth century. However much they may seem to depend upon exaggeration and caricature, such portrayals may highlight aspects of contemporary practice that would otherwise be taken for granted. They also provide an excellent mirror of the public perceptions of science and the scientist. Scientists may tend to rail against the crudity of the characterizations, but they are in part responsible for the stereotypes, and such stereotypes radically affect the political (and thus the financial) climate within which science comes to be pursued. Looking back at science fiction of the past, sometimes from the date beyond the future the authors were imagining, it is striking how much the visual paraphernalia of the 'prophetic' fictions tend to be locked into the parameters of the science of their own time. The visual images in books, magazines and films rarely escape variants of the styles of contemporary technology or work extravagant variations on the 'Gothic' style of earlier melodramas. Thus the image a craft for landing on the moon in a 1950s comic is more likely to look like a German rocket of World War II than the spidery lunar module which actually landed in 1969.

The present book, like most publications concerned with the history of twentieth. century science, is richly endowed with images which relate to the practice of science in institutional and political contexts. As historians of science have become increasingly aware of the way in which science can be characterized as a social activity, so the images used as evidence have tended to shift away from the visual material of science as a technical activity, upon which I have chiefly concentrated, towards the kind of images I have been surveying in this final section. Like science itself, the way that visual images are used in the telling of its history are subject to their own fashions.

In conclusion, it is worth reminding ourselves that the historian of the visual should learn never to take any image as a straightforward piece of 'evidence,' simply showing us what something was 'really' like. Every representation works within a specific system of communication and is richly endowed with intentional and inadvertent significance. Looking at every image in this book, we can ask why was it made, what meaning it has conveyed, and why has it been chosen.

REFERENCES

Dr. Fiona Gray of the Department of Chemistry at the University of St. Andrews acted as my research assistant in the preparation of this chapter and provided vital help in searching the range of techniques used in the various sciences.

FURTHER READING

(The following selection from the vast territory of imagery in the twentieth century is intended to provide further reading in some representative areas and issues.)

Ars Medica. Art. Medicine and the Human Condition, exhibition catalogue by D. Karp, Philadelphia Museum of Art, 1985.

Horace Barlow, Colin Blakemore and Miranda Weston-Smith, *Images and Understanding.* (Cambridge: Cambridge University Press, 1990).

Brian Baigrie (Ed.), *Picturing Knowledge. Historical and Philosophical Problems Concerning the Use of Art in Science.* (Toronto: University of Toronto Press, 1996).

Roger R. Bruce (Ed.), *Seeing the Unseen: Dr. Harold E. Edgerton and the Wonders of Strobe Alley.* (Rochester, NY: George Eastman House, 1995).

Stephen A. Drury, *Image Interpretation in Geology.* (London: Allen and Unwin, 1987).

Stephen S. Hall, *Mapping the Next Millenium.* (New York: Random House, 1992).

Roslynn D. Haynes, *From Faust to Strangelove. Representations of the Scientists in Western Literature.* (Baltimore: Johns Hopkins University Press, 1994).

D. Fox and C. Lawrence, *Photographing Medicine. Images and Power in Britain and America since 1840.* (New York, 1988).

Brian Ford, *Images of Science.* (The British Library, London, 1990).

Bass van Frassen, *The Scientific Image.* (Oxford: Oxford University Press, 1980).

James Gleick, *Chaos: Making a New Science.* (New York: Viking Penguin, 1987).

Roald Hoffmann, 'Art and Science?,' *Q. A Journal of Art.* (New York: Cornell University, May, 1990), pp. 62–5.

Michael Lynch and Stephen Woolgar, *Representation in Scientific Practice.* (Cambridge, Mass., 1990).

Pamela E. Mack, *Viewing the Earth: the Social Construction of the Landsat System.* (Cambridge, Mass.: MIT Press, 1990).

Arthur I. Miller, *Imagery in Scientific Thought: Creating Twentieth Century Physics.* (Cambridge, Mass.: MIT Press, 1986).

Arthur I. Miller, *Insights of Genius. Imagery and Creativity in Science and Art.* (New York: Springer-Verlag, 1996).

Harry Robin, *The Scientific Images from Cave to Computer.* (Harry Abrams Inc., 1992).

Lee Sider (Ed.), *Introduction to Diagnostic Imaging.* (New York: Churchill Livingstone, 1986).

Edward E. Tufte, *The Visual Display of Quantitative Information.* (Cheshire, Conn.: Graphics Press, 1983).

Edward E. Tufte, *Envisioning Information.* (Cheshire, Conn.: Graphics Press, 1990).

CHAPTER 20

The Earth Sciences and Geophysics

RONALD E. DOEL

A late nineteenth-century geologist would have been perplexed if asked to define the 'earth sciences.' No institutional structure or body of knowledge corresponded to this term. Yet the component fields of what became the earth sciences already existed by this time. Research programs in meteorology, oceanography, solid earth geophysics and seismology, terrestrial magnetism, hydrology, tectonics, and related fields flourished at various institutions. However, what was not yet shared was a conviction that these fields, as well as related research problems in geochemistry, planetary astronomy, and high-pressure physics, were interrelated parts of a conceptual *unity*, able to address such issues as the nature of earthquakes, the substance of terrestrial-solar relationships, and the future evolution of Earth's atmosphere.

The formation of the earth sciences was one of the most important intellectual and institutional transformations of twentieth-century science, comparable to the unification of physiology, bacteriology, and medical chemistry within a new field of 'biology' in the late nineteenth century. This chapter will trace two sets of developments. First, it examines major breakthroughs and conceptual developments in the earth sciences, focusing on the plate tectonics revolution and the recognition of catastrophic processes, including celestial impact, as geological forces. Second, it addresses significant social, professional, and disciplinary developments, such as military funding for geophysics after 1945 and growing concern with human impact on Earth's climate and related environmental issues, which helped to create a distinct entity called the earth sciences. Because geophysics expanded rapidly in the second half of the twentieth century, and several key concepts were primarily developed after 1950, this period receives particular emphasis.

GLOBAL PHYSICS: FORGING THE MODERN SYNTHESIS

Scientific views of the Earth, and the physical processes which governed it, changed considerably during the twentieth century. Although many new ideas were debated by earth scientists in this time, two fundamental conceptual shifts came to overshadow all others. The first was the rejection of the idea that the present-day continents and ocean basins are fixed, immutable features of the Earth in favor of a much-modified theory of continental drift called plate tectonics. Codified in the late 1960s after years of debate, this framework offered an explanation for mountain-building, earthquake zones and ocean bottom rifts in terms of a global theory of tectonic forces. The second development involved growing recognition by earth scientists, particularly after the early 1960s, that catastrophic events have deeply influenced Earth's evolutionary history, a sharp contrast to the late nineteenth-century scientific view that gradual or 'uniformitarian' processes provided sufficient explanation for geological phenomena. The collision of solid bodies is the most important geologic process to be recognized during this century. Since the late 1970s many geophysicists have come to regard comet and asteroid impacts on Earth as a proximate cause of mass extinctions, including the Cretaceous-Tertiary extinction 65 million years ago in which the dinosaurs perished. But uniformitarian interpretations of geological history have also lost ground in other ways, as new evidence for massive volcanic outpourings in Earth's past and numerous, severe shifts in Earth's climate gained acceptance among geophysicists during the 1980s.

It would be unfair, however, to present the late nineteenth-century view of Earth's past as simply one of gradualist processes played out over the mists of endless time, or to characterize the twentieth-century viewpoint as one where momentous, sometimes violent forces alone are significant. In practice many earth scientists embraced aspects of both viewpoints in their theories. A better way to understand the evolution of thought in the earth sciences is to appreciate that this intellectual field encompassed a wide range of scientific disciplines, including astronomy (age and origin of the earth), seismology (internal structure of the earth), geochemistry (planetary evolution), oceanography (origin and evolution of ocean floors and the oceans themselves), and meteorology (evolution and behavior of Earth's atmosphere), as well as traditional field geology (the structure and evolution of particular regions of Earth's land surface). Members of these disciplines devised new instruments, launched new expeditions to acquire data on terrestrial conditions, and offered new theories that encompassed ever larger realms of terrestrial phenomena. The history of the earth sciences in the twentieth century above all reflects the jostling *between* members of these disciplines to develop new, synthetic theories about Earth's physical processes.

The major question early twentieth-century geologists faced was to explain the large-scale features of our planet, particularly the formation of mountain ranges and the existence of continents and oceans. Through the end of the nineteenth century, many accepted a concept, articulated by the eminent Austrian geologist Eduard Suess, that general contraction of the Earth since its formation had caused these features, which are analogous to the skin of a wrinkled apple. A central assumption of Suess' theory was that Earth was slowly cooling, but the discovery of radioactivity in 1896 soon cast doubt on Suess' model. One result of radioactivity's discovery was to remove a long-standing objection to deep geological time favored by most geologists (the physicist Lord Kelvin had sharply limited estimates of Earth's age if its internal heat came strictly from gravitational contraction, which the existence of radioactivity negated). But the discovery also cost geology the intellectual coherence it had achieved under the theoretical framework of contraction. By the 1930s many earth scientists retained key assumptions from contraction models — particularly the permanency of continents and ocean basins — but sought new explanations for mountain formation, such as convection currents in Earth's mantle.

Many deep geological puzzles remained, however, including similarities in the stratigraphic sequences on the western shoreline of Africa and the eastern coast of South America, and equally curious similarities in paleontological data from Africa, Madagascar, India, and South America. One explanation of these puzzles voiced by turn-of-the-century geologists was that Earth's crust had flexed vertically, creating temporary land bridges between continents. But to others — a distinct minority — these relationships indicated that Earth's continents were not permanently fixed, but mobile. Both interpretations were reactions to the general failure of Suess' contraction theory. The most well-known mobilist theory came from the German meteorologist and geophysicist Alfred Wegener. In a series of articles and books published from 1912 until shortly before his death in 1930, Wegener proposed that current-day continents had drifted apart from one another following the breakup of an ancient mega-continent, which he termed Pangaea.

Given the similarity of Wegener's concept of continental drift to the modern theory of plate tectonics, which became geology's reigning paradigm a half century later, it is important to place the contributions of Wegener in perspective. Wegener's ideas inspired mobilist geologists through the mid-twentieth century, and for this reason Wegener is often celebrated as the intellectual father of plate tectonics theory. Yet his intellectual contributions, while important, should not be exaggerated. Wegener believed drift accounted for the observed geological and biological continuities between South America and Africa, and furthermore explained the supposed jigsaw fit between these continents' boundaries as well as the existence of mountains along continental margins. Yet his theory articulated none of the central features of plate tectonics theory, including the existence of subduction zones, earthquake zones, mantle hot spots, or rift valleys. He could not address the extraordinary complexity of the ocean floor, although ocean floor structure, not continental features, would eventually provide key evidence for plate tectonics. Moreover, Wegener provided little insight into the physical forces needed to give continents a horizonal motion (although two of Wegener's early twentieth-century supporters, John Joly and Arthur Holmes, did offer improved mechanisms for drift in the 1920s and 1930s). At the same time, it must be noted that Wegener's ideas, bold and original, provoked vigorous debate about the *possibility* of continental drift, and the idea gained limited acceptance among German and British geologists (particularly those familiar with exposed Mesozoic strata in Africa and South America). The concept fared least well among earth scientists in the United States, who largely rejected his exuberant embrace of grand theoretical models. For detractors like Bailey Willis, Wegener's reliance on homologues (patterns of similarities between different continents) was insufficiently rigorous for a geological community increasingly at home with petrographic analyses and other forms of physical data.

The evidence that caused plate tectonics to become the defining concept of modern geological science surfaced in several stages, beginning in the late 1940s. Although critical contributions ultimately emerged from several fields of earth science, the most important stimulus came from post-World War II studies of the ocean floor, until then little studied and marginally surveyed. By the 1950s, W. Maurice Ewing, Bruce Heezen, and Marie Tharp, all of Columbia University's Lamont Geological Observatory, found that ocean bottom sediments were younger than most geologists expected. They also discovered that ocean basins were bisected by an extensive mountainous ridge, soon recognized as the longest continuous geologic feature on Earth. Seeking to explain the surprisingly young ages of rocks sampled on the mid-Atlantic Ridge, the Princeton geologist Harry Hess, in a now-famous 1960 essay, proposed that convection currents were responsible for bringing new mantle material to the surface along these ridges. Hess' concept, which he termed sea-floor spreading, actually embodied two related ideas: new crust spread out from mid-oceanic ridges while old crust was simultaneously forced down to depths at the major oceanic trenches, typically found along the continental margins and large island arcs.

Earth scientists were not immediately drawn to Hess' idea of sea-floor spreading. What convinced many of its validity was a series of additional discoveries and observations that emerged between 1960 and 1967. Important evidence came from the relatively young field of paleomagnetism, the study of remnant magnetic fields in rocks. In the early 1960s Keith Runcorn presented paleomagnetic data indicating that the magnetic north pole had wandered in recent geologic time. While this result was not unexpected, Runcorn's paleomagnetic measurements demonstrated that North America and Great Britain had similar but *divergent* paths, a finding most easily explained if these landmasses had shifted relative to one another over time. Another surprising finding, also drawn from paleomagnetic studies, was that rocks in certain geological strata seemed *reversely* magnetized, suggesting that Earth's entire magnetic field had flipped over, such that compass needles would have pointed south at many times in the past. In 1963 Fred Vine and Drummond Matthews (and independently Lawrence W. Morley) used this finding to predict that, if sea-floor spreading was real, the sea-floor adjoining the mid-oceanic ridges ought to show stripes of alternating polarity caused by Earth's alternating magnetic field. Just such a pattern was recognized two years later in magnetic data recorded by the research ship *Eltanin.* Simultaneous efforts to establish an *absolute* time scale for these magnetic reversals, corrected with geological history, also proved successful. One key reversal, called the Jaramillo event, was dated at 0.9 million years before the present by the US researchers Allan Cox, Richard Doell, and Brent Darymple. Their historical chronicle of reversals, matched to records of magnetic stripes along mid-ocean ridges, not only confirmed the Vine-Matthews-Morley hypothesis, but revealed sea-floor spreading to be a steady, continuous process.

This was not yet a theory of plate tectonics. Two contributions from other fields of geophysics were required to transform Hess' concept of sea-floor spreading into a comprehensive theory of crustal movement and deformation on a global scale. One was the confirmed existence of a new class of geologic faults, called transform faults, which the Canadian geophysicist J. Tuzo Wilson had predicted in 1963; such faults provided a physical explanation of the movement of crustal blocks past one another along plate margins. Equally important was the determination, through studies of continental and oceanic earthquake data, of the outlines of the major 'plates,' seven in all, each built around continental landmasses. By late 1967 the core concepts of modern plate tectonics theory coalesced within the North American geological community. The authority of plate tectonics as the reigning interpretation of Earth's structure is not simply its theoretical elegance but its predictive ability. For example, in addition to resolving the geological and biological puzzles earlier addressed by Wegener, plate tectonics explains the existence of the Himalayas by the inexorable push of the Indian plate into its northern neighbor, the Eurasian plate. Under this theory, two prevailing assumptions of late nineteenth-century geology have been turned upside down: continents, rather than the ocean basins, are now viewed as permanent features, and the horizontal motions of continents are no less critical than their vertical adjustments.

A second major shift in twentieth-century geologic thought involved a substantive challenge to Lyell's definition of uniformitarianism: the realization by earth scientists that celestial impact is a fundamental geological force. The idea that collisions between solid bodies in the early solar system had created the Earth and other planets was not new, for it had been debated since Forest Ray Moulton and Thomas C. Chamberlin articulated their planetesimal cosmogony at the turn of the century. But the possibility that such impacts might continue into geologic and contemporary time received little credence. One reason was that virtually no impact sites on Earth were recognized until the early 1930s. Yet a more significant factor was that Lyell's uniformitarianism, with its emphasis on orderly processes and natural laws, sought to draw clear boundaries between the science of geology and catastrophic episodes described in Judaeo-Christian accounts of Earth's history, such as Noah's flood. By the 1950s, however, astronomers in the United States and Europe became increasingly convinced by the work of Ralph B. Baldwin, which highlighted quantitative correlations between chemical explosion craters, terrestrial impact craters, and lunar craters. The Apollo lunar program, coupled with the accelerated discovery of major impact features on Earth, gradually convinced most scientists by the 1970s that Earth had been subject to cataclysmic strikes in recent history as well as in its remote past.

Not all modern earth scientists would accept the claim, voiced by the eminent geologist-astronomer Eugene M. Shoemaker, that "the impact of solid bodies is the most fundamental of all process that have taken place on the terrestrial planets."[1] Yet the debate has now shifted to assessing the *frequency* of large collisions and the influence these impacts have on Earth, including the evolution of life. In 1980 Luis Alvarez, joined by his son Walter, called attention to a layer of iridium at the border between the Cretaceous and Tertiary era 65 million years ago, a crucial interval in paleontological history when over sixty-five percent of all living species became extinct. Since iridium is relatively rare in terrestrial rocks, but common in meteorites, Alvarez's group argued that a 10-kilometer body had struck the Earth at that time, throwing massive quantities of dust into the atmosphere and causing the dinosaurs and other species to perish. Debate over this issue became particularly heated in the 1980s, as many paleontologists argued that wholly terrestrial forces, such as unusually violent volcanic eruptions or climate shifts, were more likely to have led to this episode of mass extinction. But the late 1980s discovery of a 160-kilometer-wide buried crater under the Yucatan Peninsula, dated as 65 million years old, has provided what many astronomers and geophysicists regard as the 'smoking gun' predicted by this theory. Earth scientists now see impact events causing a broader range of environmental changes. The impact of Comet Shoemaker-Levy into Jupiter in the summer of 1994, a tremendous collision observed by ground-based telescopes and spacecraft, has further invigorated the study of impact processes in Earth's history.[2]

Solid earth geophysics was not the only branch of the earth sciences to advance rapidly in the early and mid-twentieth century. The increased application of physical methods to meteorology and climate, already subjects of theoretical inquiry in the nineteenth century, led to several critical developments. One was the polar front concept, articulated by the German-born Vilhelm Bjerknes in Bergen, Norway in the 1920s, which gave physical interpretation to the behavior of storm systems and the interactions of air masses. By the mid-twentieth century these studies were extended to the global circulation of the atmosphere by Carl-Gustav Rossby and others trained in Bjerknes' informal Bergen school. Bjerknes' hope that the future behavior of weather systems could be calculated, much like one could compute the future orbital motion of planets, was addressed by the British mathematician Lewis Fry Richardson, who in the 1920s and 1930s sought to make numerical forecasts using partial differential equations. Following the rapid advance of computers in World War II, attempts were made by John von Neumann and other researchers to realize Richardson's aim. By the 1970s, as computing power and the sheer volume of meteorological data mushroomed, meteorologists and climatologists used computer models to describe the general circulation of the atmosphere, including the globe-circling jet streams. Sophisticated models were developed by the 1980s to forecast the behavior of complex oceanographic-atmospheric systems such as El Niño, the warm, episodic Pacific Ocean current responsible for dramatic weather fluctuations in North America, Africa, and Europe. Predictions by planetary scientists Carl Sagan and Richard Turco in 1984 of a global 'nuclear winter' following a massive nuclear exchange were based on contemporary computer models, as are modern assessments of global warming. While such models yield testable predictions about large-scale changes, Edward Lorenz in 1960 found that very small effects, such as massive local thunderstorms or blizzards, can have an extremely large influence on hemispheric and global weather patterns. His fundamental discovery ended the hopes of Richardson's followers to calculate the weather with Newtonian certainty but led directly to chaos theory, a major frontier of late twentieth-century science.[3]

The history of Earth's past climate, including the onset of ice ages, became a major topic of inquiry during the twentieth century as well. Following the turn-of-the-century work of Albrecht Penck and Eduard Bruckner, the prevailing view among geologists through the 1930s was that Earth had experienced just four distinct episodes of glaciation. But geochemical studies of ocean-floor sediments and microscopic fossils in the 1950s challenged this interpretation. Using precise radiocarbon dating techniques, the nuclear chemist Cesare Emiliani argued that many more ice ages had occurred, some beginning and ending with remarkable swiftness. Emiliani's own reconstruction of glacial episodes seemed to support an astronomical theory proposed by the European mathematician and engineer Milutin Milankovitch in 1924, who assigned their cause to small variations in Earth's orbit around the Sun over time scales of 26,000 to 105,000 years. More recent studies of deep ice cores in Greenland and in Antarctica have confirmed the volatility of Earth's climate, although the precise mechanisms responsible for global temperature shifts remains controversial. Taken together, these theories support a view of Earth's evolutionary history marked by greater vicissitudes and abrupt upheavals than those anticipated in the syntheses of Lyell, Suess and other leaders of nineteenth-century geology.[4]

Other fields of geophysics have also benefitted greatly from major advances in description and theory. By studying the Gulf Stream and more recently discovered deep-ocean currents, the physical oceanographers Henry Stommel and Walter Munk mapped oceanic circulation patterns. By the late twentieth century such motions, like those in atmospheric circulation, were reproducible in computer models. Studies of large-scale oceanic structures, such as warm core rings spun from the Gulf Stream, alerted scientists to the role these immense eddies played in momentum transport.[5] In solid earth geophysics, seismological observations of earthquake waves (and later artificial explosion waves, such as from nuclear blasts) also revealed with increasing precision the structure of Earth's interior. The diameter of Earth's core was first determined by the German-born geophysicist Beno Gutenberg in 1912, roughly the same time that the Croatian geophysicist Andrijn Mohorovičić first discovered the shallow layer now regarded as the boundary between the Earth's outer crust and underlying mantle. Later observations allowed seismologists to interpret physical structures along the mid-oceanic ridges, which helped define the boundaries of the continental plates, and provided evidence for hot mantle 'plumes' that geophysicists believe are responsible for isolated volcanic chains, most notably the Hawaiian islands.[6]

In recent years, earth scientists have increasingly focused on phenomena that cross disciplinary lines, such as the capacity of Earth's oceans to absorb carbon dioxide, and large-scale shifts in the atmosphere's mean temperature. They have also studied the interrelations between geophysical phenomena at vastly different scales, such as observed links between massive thunderstorms and ionospheric emissions.

Institutions and Patronage

The body of ideas that came to constitute the modern earth sciences was shaped and informed by a wide range of professional, disciplinary, and social factors including the universities, government bureaus, and institutes where geophysicists worked, and the professional societies to which they belonged. The earth sciences were also influenced by factors outside geophysics proper, including demands from petroleum and mining industries, the needs of military planners, and more recently the concerns of government leaders over such environmental issues as stratospheric ozone depletion and global warming. Modern geophysics research must be seen within this social context; indeed, the late twentieth-century perception that the environmental sciences form a conceptual unity largely derives from our modern perception of an 'environment' placed at risk by human activity.

At the turn of the twentieth century, neither 'geophysics' nor the 'earth sciences' connoted a distinct field of research nor a common set of instrumental techniques. Geophysical research in this period lends support to the concept of 'national styles' in science, for institutional structures and conceptual emphasis varied widely from one intellectual center to another. In the late 1800s — and well into the following century — geophysical research was largely conducted within government-funded research centers. Although fundamental research was undertaken in these facilities, their general orientation remained towards applied problems in surveying and mining. Only at Pottsdam, Berlin, Munich, and other major German university centers did such fields as terrestrial magnetism, atmospheric electricity, hydrology, seismology, oceanography, and meteorology tend to share common intellectual as well as brick-and-mortar homes. Emil Wiechert's famed *Geophysikalisches Institut*, established by Wiechert at Göttingen in 1901, was the first distinct academic center for geophysics in the modern sense of the term. Its inclusive structure was reinforced by the synthesizing tradition of Alexander von Humboldt, the geologist,

TABLE 20.1: GEOPHYSICS IN NORTH AMERICA: MAJOR INSTITUTIONAL DEVELOPMENTS.

1903	Carnegie Institution of Washington (Geophysical Laboratory) founded (Department of Terrestrial Magnetism established 1904)
1925	Scripps Institution of *Oceanography* (formerly the Marine Biological Association of San Diego [1903–1912] and Scripps Institution for Biological Research [1912–1925]) established
1927	Colorado School of Mines (geophysics curricula established) Department of Geophysics, St. Louis University (established; focus on solid earth geophysics and seismology)
1928–35	Peak of Rockefeller Foundation funding for earth sciences; Earth Sciences Program from circa 1932–35
1930	Woods Hole Oceanographic Institute (founded)
1931	Harvard University establishes Committee on Experimental Geology and Geophysics; name varies through 1950s
1946	University of Alaska (Geophysical Institute) approved by Congress University of California at Los Angeles (Institute of Geophysics founded) Office of Naval Research (ONR) geophysics funding begins.
1947	Research and Development Board (RDB) Geophysics Committee established Graduate training begins at SCRIPPS (ONR funding)
1949	Air Force Cambridge Research Laboratories (later AFCRC; formerly Watson Laboratories Cambridge Field Station) Columbia University established Lamont Geological Observatory (later Lamont-Doherty Earth Observatory)
1950–51	US Navy/ONR helps establish departments of oceanography at University of Washington, Brown, and Texas A&M
1953	Earth Sciences Program established at the National Science Foundation
1955	Smithsonian Astrophysical Observatory (relocated to Harvard University; emphasizes upper atmospheric physics) Planning begins for what becomes University of California at San Diego (closely linked to Scripps Institution of Oceanography)
1956	National Science Foundation Conference held to discuss need for Institute of Theoretical Geophysics
1957	University of British Columbia develops Geophysics Division
1957–58	International Geophysical Year
1959	Massachusetts Institute of Technology (Earth Sciences Center) (consolidated through philanthropy of Cecil H. Green) Rice Institute establishes center for earth sciences (later Lunar Science Institute). University of Hawaii Geophysical Institute approved by Congress
1961	National Center for Atmospheric Research (NCAR) established
1962	School of Earth Sciences (Stanford University) established
1960s	Six additional schools of 'earth sciences' established at North American universities, including Toronto and John Hopkins
1966	US Environmental Science Services Administration established
1973–74	US Bureau of Oceans and International Environmental Affairs (with US State Department) established

explorer and naturalist whose insistence on systematic study of Earth's physical and organic phenomena became a major influence on scientific thought by the early nineteenth century. It also benefitted from the director-dominated institute structure of German universities, which bore little resemblance to the departmental organization of American universities. In England, by contrast, mathematical geophysics emerged as the dominant field, reflecting the emphasis of Cambridge University physical scientists on physico-mathematical approaches. Here the analytical traditions of George Darwin and Harold Jeffreys, which stressed the applicability of mathematics to geological phenomena, were pursued. In the United States, the 'earth sciences' — represented by the US Geological Survey and the Coast & Geodetic Survey, and by such leading intellectual lights as physically trained geologists Samuel Becker and Grove Karl Gilbert — became preeminent during the era of the great Western surveys of the 1870s. But by 1900 these fields were in decline, owing to shifts in political backing and the surge of industrialization and factories, which favored the exact physical sciences over those relating to earth processes. This trend was reflected in the academic structure of US and Canadian universities through the 1920s: various component fields of geophysics, including seismology, terrestrial magnetism, and meteorology, were occasionally represented as research *specialties* in American universities (typically within departments of physics). No inclusive centers of *geophysics* existed at any of them.

By the 1960s, however, the earth sciences were major fields at many universities and institutions around the globe, notably represented by such facilities as the Lamont Geological Observatory of Columbia University and the Shmidt Institute of Earth Physics of the USSR Academy of Sciences. The challenge thus becomes: what stimulated the growth of academic geophysics? National needs — that is to say, society's common, shared demands — provides a partial answer. By the early twentieth century, most advanced industrial nations already possessed variants of the US Coast & Geodetic Survey and Canada's broadly conceived Dominion Observatory, whose research missions included fundamental surveys of coastlines and magnetic studies which were critical for navigation. These facilities nurtured geophysical work and practice, and became informal training centers for young geophysical workers. Later in the twentieth century, new technological developments created further markets for geophysical information and expertise. The rise of the aviation industry in the 1920s was a particularly important example, for it brought about a tremendous expansion in meteorological observations and forecasts, improved government support for meteorological research centers, new professional organizations, and new career opportunities for scientists. National needs had still broader meaning in major imperial powers like Holland and Germany, where geophysical research stations in colony-states in the Far East and South America served both scientific and diplomatic needs. Such service roles continued to shape the earth sciences in significant ways throughout the twentieth century, including the Cold War and the post-1960 modern 'environmental' era.

The expansion of geophysics was also driven by its utility in finding natural resources, particularly petroleum. Shortly after World War I, physicists in Germany and the United States sought to apply wartime advances in sound-ranging as a way to discover oil deposits, inspired by predictions of a pending catastrophic oil shortage by geologists. By the early 1920s, petroleum companies began employing various geophysical methods to locate oil deposits, including magnetic, electrical, and gravitational, the last approach made possible by the double-beam torsion balance (devised by the Hungarian experimental physicist Baron Roland von Eötvös). However, the most successful approach — pioneered and later refined in the United States, principally by Everette de Golyer — was seismic reflection, which involved recording the echoes from carefully placed dynamite charges to trace underground geological structures. By the 1930s, seismic reflection teams became a common sight in oil-producing regions of North America, causing the petroleum industry to boom. Exploration geophysics became a central component of geophysics programs at the Colorado School of Mines, Houston's Rice Institute (later University), and other major universities. Large petroleum firms became a principal source of employment for geophysicists (by 1938 Humble Oil's Geophysics Department grew to over 300), but equally important were the independent geophysical consulting firms stimulated by this new market, including the European-based Schlumberger.

Yet another factor which stimulated the growth of geophysics in the years before World War II was private patronage, including the major philanthropic foundations. The Carnegie Institution of Washington, founded in 1902 with a ten million dollar gift from the steel magnate and financier Andrew Carnegie, created what quickly became one of the world's most advanced centers for geophysical research. Its Department of Terrestrial Magnetism and the somewhat misnamed Geophysical Laboratory (which actually focused on geochemistry) were the envy of geophysicists at Europe's premiere but less well-funded research laboratories. By the 1930s leaders of the Rockefeller Foundation, having poured unprecedented sums into the established scientific disciplines of chemistry, physics, and biology, became interested in the earth sciences. Rockefeller grants accelerated Harvard University's nascent entry into geophysics, sustaining path-breaking research in high-pressure geophysics, earth tides, the structure of the continental shelf, and the population of small bodies in the solar system. Rockefeller Foundation funding also aided the growth of geophysical centers overseas, including the Leipzig Geophysical Institute. Such philanthropic support, unlike that normally available to traditional academic departments, enabled geophysicists to gain a niche within universities, as well as additional faculty appointments and graduate students.

One sign of the burgeoning growth of twentieth century geophysics was the expansion of professional societies and journals in this field. Few geophysics societies existed prior to World War I, but they proliferated rapidly thereafter. The International Union of Geodesy and Geophysics (IUGG) was founded in 1919, the same year as the American Geophysical Union (AGU). This trend was worldwide, although with local variants. In Great Britain, for example, geophysics was made a new, distinct section within the older Royal Astronomical Society, founded in 1820. Professional societies and journals were themselves important in bringing increased coherence to the field of geophysics. The emphasis on petroleum geophysics within the United States gave rise to a number of specialized organizations, including the Society of Exploration Geophysics, but it was the AGU which most influenced research programs across the earth sciences. The AGU's ambitious interdisciplinary commissions encouraged researchers to take up difficult but fundamental problems. Interest shown by AGU leaders Richard M. Field and John A. Fleming in the mid-1930s in the structure of the continental shelf and ocean floor stimulated energetic research programs, not least of which were the pioneering studies by the young W. Maurice Ewing, who later founded the geophysics-centered Lamont Geological Observatory. New journals were created to accommodate this increasing flow of results: *Terrestrial Magnetism*, founded in 1896, was reconstituted as the *Journal of Geophysical Research* in 1949, and within two decades *Earth and Planetary Science Letters, Tectonophysics, Physics of the Earth and Planetary Interiors* and others appeared to absorb the overflow of material.[7]

As was the case in many fields of science, World War II marked an important watershed for the development of geophysics in the United States, the Soviet Union, and other leading scientific nations. Wartime advances in aviation, missile construction, submarine warfare, and military communications focused attention on the contributions of geophysics to warfare. By the early Cold War, military patrons declared that military applications existed for each component branch of the earth sciences: missile ranging and guidance problems in geodesy, weather forecasting and possibly control in meteorology, guided missile design in upper atmospheric physics, underseas warfare in oceanography. Military and federal funding for the earth sciences soared to unprecedented levels. In the US, new governmental and military centers of geophysics were constructed, including the Geophysics Directorate of the Air Force Cambridge Research Center, but universities were a primary beneficiary of Cold War support for the earth sciences. Four American universities, among them Columbia and the University of California at Los Angeles, created new institutes of geophysics by 1949, and numerous others added graduate programs and substantial research projects. Not all scientists liked the rapid rise of military funding for geophysics. Harvard astronomer Harlow Shapley, disgruntled by new contracts for meteor studies early in the Cold War, tartly noted that "shooting stars perform in that same level of the earth's atmosphere where the shooting rockets and rocket ships of the future are planning to operate."[8] Even so, military attention to the earth sciences stimulated this discipline more than any other single factor, making possible sustained, systematic observations of Earth's atmosphere, unprecedented studies of the ocean floor, and extended research into earthquakes and Earth's deep interior.

COMMITTEE ON GEOPHYSICAL SCIENCES

MEMORANDUM:

FROM: Secretariat, Committee on Geophysical Sciences

TO: Addressees

SUBJECT: Guided Missiles Problems and Geophysical Sciences

1. Many panels of the Committee on Geophysical Sciences have discussed the contributions these sciences can make to the field of guided missiles. Some reports pertaining to guided missile do not utilize the present knowledge available to the geophysical sciences applicable to the guided missiles field. There are abstracted below some of the statements that have been made in this regard for the purpose of having them verified by the experts on various panels.

2. The geophysical sciences are intimately concerned with the phenomena involved in the navigation of guided missiles. One of the most discussed systems of navigation is the use of the earth's magnetic field for midcourse guidance. The conclusion is that the system of magnetic navigation would have much to commend it if it were possible to develop it satisfactorily. The greatest difficulties are the unpredictable variations in the magnetic field from time to time. These are assumed to be too great to offer hope of solution in the immediate future. It is essential, in such a system that the magnetic field of the earth be well understood and charted and that instruments be developed that could be depended upon to follow lines of the magnetic field to a sufficient degree to accuracy. In this connection, it is open to question whether our knowledge of magnetic isolines over possible enemy areas is sufficiently accurate for the purposes of precise navigation. It has been stated that "even where this information is correct at a given time it is subject to change with the passage of time." Resumes on this point have stated that the rates of change are different in various parts of the world and insufficiently known. Other drawbacks quoted are the diurnal variation of the magnetic field and sudden magnetic storms. Some believe that the secular and

This document contains information affecting the national defense of the United States within the meaning of the Espionage Act, 50 U.S.C., 31 and 32. The transmission or the revelation of its contents in any manner to an unauthorized person is prohibited by law.

FIGURE 20.1: THE IMPORTANCE OF THE EARTH SCIENCES TO MILITARY AUTHORITIES WAS CLEARLY RECOGNIZED WHEN THE COLD WAR BEGAN. THIS RECENTLY DECLASSIFIED DOCUMENT FROM THE JOINT RESEARCH AND DEVELOPMENT BOARD (THE POST-WORLD WAR II SUCCESSOR TO THE OFFICE OF SCIENTIFIC RESEARCH AND DEVELOPMENT) ILLUMINATES THE LINKS THAT CIVILIAN SCIENTISTS AND MILITARY OFFICIALS PERCEIVED BETWEEN GUIDED MISSILE DEVELOPMENT AND VARIOUS FIELDS OF GEOPHYSICS.
© BOX 227, RESEARCH AND DEVELOPMENT BOARD RECORDS, NATIONAL ARCHIVES AND RECORDS ADMINISTRATION, WASHINGTON, D.C.

Another significant boost to the earth sciences came from the International Geophysical Year (IGY). Modeled after the First (1882–3) and Second (1932–3) Polar Years, designed to gather ionospheric, magnetic, and meteorological data worldwide, the IGY (July 1957–December 1958) became a dramatically larger undertaking, involving twelve fields of geophysics, several thousand scientists from sixty-seven participating countries, and expenditures of several billion dollars. Because of its tremendous scale and intellectual range, the IGY was a singularly important benchmark in the growth of geophysical research. US support for IGY activities must be understood within the Cold War context. As recent scholarship has shown, federal and military planners saw considerable advantage in using the IGY to collect synoptic, global data with potential military applications, and US geopolitical strategy involving Antarctica was furthered by allowing this continent to be 'constituted for science.' But these mixed motives should not diminish the fact that the IGY was the largest internationally cooperative scientific activity until this time, one which contributed enormous volumes of information about the physical properties of the planet, and began an era of international observation and regulation involving the Earth's environment.

To be sure, the growth of geophysics after 1945 was not entirely due to federal and military sponsorship. Petroleum and natural resource firms, including Shell Development and Standard Oil of New Jersey, remained significant centers for geophysical research and employment, and continued to sponsor geological research at universities in North America and Europe. Private philanthropy also remained an important factor. The co-founder of Geophysical Services Inc. and later of Texas Instruments, Cecil H. Green, endowed earth science research programs at the University of Toronto, the Colorado School of Mines, and the Massachusetts Institute of Technology. Green's philanthropic gifts to geophysics buildings, endowed chairs, and fellowship programs at these schools, in excess of 100 million dollars, helped to accelerate the progressive integration of earth sciences departments already occurring in these institutions.[9]

Space Exploration and Global Research

These developments alone do not give a full picture of the earth sciences in this century. An important branch of geophysics emerged once direct explorations began of Earth's upper atmosphere and ionosphere, the Moon, and other planets in the solar system. Like expeditions to Antarctica and balloon flights into the lower stratosphere in the first third of the twentieth century, these explorations revealed abundant new geophysical phenomena. More importantly, the era of space exploration (which dates from the launch of Sputnik in 1957) has led to new geophysical theories that consider Earth as one member in a class of objects with common evolutionary traits. Investigations of Earth from space, coupled with continued ground-based studies, have also yielded new information about long-term changes in the behavior of ocean currents and the climate, stimulating interest in global warming and other planet-wide concerns.

Finding a suitable historical framework for evaluating the space age is not an easy task. Stephen J. Pyne has characterized the post-Sputnik period as a "Third

Great Age of Exploration," comparable to the era of ocean-circling voyages in the seventeenth and eighteenth centuries and the exploration of the continental interiors of the Americas, Africa and Asia in the nineteenth century.[10] Certainly space exploration could not have come about without fundamental technological advances during the twentieth century, including rocket propulsion, computers, advanced telecommunications, and the Big Science support structures which they required. The modern space age was also propelled and strongly shaped by Cold War rivalries between the United States and the Soviet Union, heightened by the emblematic value of space studies as symbols of national prestige and technological competence. That this is so should not detract from the importance of geophysical results obtained from planetary science research, particularly those involving perceptions of planetary evolution and the interrelation of geological and biological processes in Earth's history.

Many fields of geophysics benefitted from space research. One such field was terrestrial magnetism. In 1957 Sydney Chapman, following suggestive research by Ludwig F. Biermann, proposed that the Sun's outermost atmosphere extended past the orbit of Earth and effectively filled the entire solar system. His theory suggested causative links between high-energy particles and the behavior of Earth's ionosphere. In 1959 the existence of the solar wind, composed of hydrogen nucleii streaming outward from the Sun, was confirmed by the Soviet spacecraft *Luna 2* and *Luna 3.* Meanwhile investigations of near-earth space by the physicist James Van Allen, using instruments onboard the US satellite *Explorer 1,* revealed the general structure of Earth's magnetic field. His instruments disclosed a large torus of charged particles encircling the Earth at 2000 to 12000 miles above its surface, which were named the Van Allen Radiation Belts. Subsequent planetary exploration revealed active magnetic fields surrounding Mercury, Jupiter, Saturn, Uranus and Neptune, and the virtual lack of magnetic fields for Venus and Mars, the two planets most resembling Earth in overall composition and size. These observations appear to support theories pointing to the origins of magnetic fields in rotating fluids, particularly iron, deep in planetary interiors, following the theoretical foundations laid by the German-American researcher Walter Elsasser. The precise nature of this mechanism nevertheless has remained a subject of debate among planetary physicists.[11]

One of the most significant conceptual developments to emerge from planetary exploration was *comparative planetology* — the study of planetary evolution, in which the Earth is considered one member among a family of related objects. Comparative planetology was not entirely born of the space age; astronomers had already collaborated with geophysicists to understand Earth's internal structure. But data from planetary exploration by spacecraft have led to important extensions of the earth sciences. Investigations of Venus, Earth's closest planetary neighbor in terms of size and distance, were achieved by the US *Mariner 2* (1962), *Pioneer Venus* (1978–79), and *Magellan* (1990–94) missions, the latter employing a high resolution orbiting radar system to view its surface. Soviet spacecraft have also visited Venus, including the *Venera* (1975 and 1982) missions which successfully placed camera-equipped landers on its surface. These explorations established that Venus evolved quite differently than Earth, revealing a planet wrapped in a thick, super-heated atmosphere of carbon dioxide and a surface lacking evidence of plate tectonic movements. Photographic reconnaissance of immense ancient flood plains on the surface of Mars has reinforced non-uniformitarian explanations for certain puzzling geologic landscapes on Earth, most notably perhaps the catastrophic flood hypothesis offered by the University of Chicago geologist J Harlen Bretz in the 1930s to account for the channeled scablands region of eastern Washington state. The US *Voyager* missions, which visited the outer gaseous planets Jupiter, Saturn, Uranus and Neptune (as well as their solid moons) between 1979 and 1989, yielded unprecedented volumes of information on these largely unexplored worlds, including the discovery of active volcanism on Io, a moon of Jupiter, and Triton, a moon of Neptune. *Voyager* instruments recorded the solar wind, planetary radio emissions, and cosmic ray events, and yielded particularly large volumes of information, including time-lapse movies of meteorological phenomena and photographic maps of over a dozen worlds.[12]

Assessing the significance of planetary exploration on the earth sciences is difficult. Conceptual development is clearly evident in virtually all fields. The American geologist-geophysicist Eugene Shoemaker successfully defined a system of geologic ages on the Moon in the early 1960s using impact characteristics, establishing that basic geologic concepts could be extended from Earth to other planets. In the mid-1970s photographic reconnaissance of the heavily cratered surface of Mercury by *Mariner 10* led Gerald Wasserburg, George Wetherill, and Bruce Murray to define the "Great Bombardment" period early in the solar system's history, reinforcing the importance of catastrophic collisions between solid bodies as a fundamental geologic force. The concept of massive impact has been brought to bear on previously difficult problems. In 1974 the US planetary scientist William K. Hartmann proposed that the Moon originated from a violent collision between the infant Earth and a Mars-sized body, an idea that gained the backing of most planetary scientists the following decade.[13] But the importance of remote planetary explorations on the earth sciences ought not be exaggerated. They have not overturned key concepts in the modern earth sciences, such as plate tectonics, the structure of Earth's interior, or the origin of planetary magnetic fields.

Surveys of Earth's general characteristics during the space age have also provided new insights into earth science phenomena. Satellites never intended to leave Earth's orbit, particularly the *Tiros, Nimbus,* and succeeding generations of meteorological satellites, joining natural resource mapping satellites such as the US *Landsat* and the French SPOT system, have yielded sufficiently fine-grained data to inspire theorists to analyze large-scale phenomena. Meteorological measurements from satellites allowed climatologists to test models of global circulation;

FIGURE 20.2: EARTH SCIENCE IN THE LABORATORY: HIGH PRESSURE RESEARCH AT THE CARNEGIE INSTITUTION OF WASHINGTON, CIRCA 1906.

ARTHUR L. DAY (LEFT), FIRST DIRECTOR OF THE GEOPHYSICAL LABORATORY OF THE CARNEGIE INSTITUTION OF WASHINGTON, AND AN UNIDENTIFIED COLLEAGUE, WORKING WITH A CARBON ARC AND RESISTANCE FURNACE. ONE OF THE MOST IMPORTANT SCIENTIFIC INSTITUTIONS ON THE LANDSCAPE OF EARLY TWENTIETH-CENTURY AMERICAN SCIENCE, THE PRIVATELY FUNDED CIW LAUNCHED INNOVATIVE RESEARCH PROGRAMS IN GEOCHEMISTRY AND GEOPHYSICS AFTER ITS FOUNDING IN 1906.

FIGURE 20.3: THE RESEARCH SHIP VEMA, OPERATED BY COLUMBIA UNIVERSITY'S LAMONT GEOLOGICAL OBSERVATORY, LEAVING NEW YORK HARBOR. FROM 1953 UNTIL HER RETIREMENT IN 1981, THE VEMA LOGGED OVER A MILLION MILES AT SEA, SETTING A NEW RECORD. ACCESS TO VESSELS WAS VITAL FOR EARTH SCIENTISTS INTERESTED IN STUDYING THE STRUCTURE AND PROPERTIES OF THE OCEAN FLOOR.

their results improved both short-term forecasts (including the paths and intensities of hurricanes) and predictions of climatological variations. In this same period, remote satellite mapping of Earth's biological and mineral resources has extended and in some instances replaced aerial and ground-based surveys, dramatically expanding the application of geophysical studies to human concerns.

The most important earth science programs to benefit from global satellite research, in a social sense, were those designed to find evidence of significant environmental change, particularly ozone depletion and global climate shifts. Here space research cannot be usefully segregated from studies conducted in more traditional laboratory and field settings. Depletion of stratospheric ozone — a thin blanket roughly 15 kilometers above Earth's surface which absorbs incoming solar ultraviolet (uv) radiation harmful to agriculture, wildlife, and exposed skin — was first recognized as a possible outcome of the proposed supersonic transport (SST) plane. Subsequent work on chemical compounds implicated in ozone destruction, particularly chlorofluorocarbons (CFCs), was carried out by among others the atmospheric chemists Mario J. Molina and F. Sherwood Rowland. Their studies indicated that CFCs, then widely used as aerosol propellants and in refrigerators, were capable of rapidly eroding the ozone layer through catalytic processes. Aircraft and balloon studies in the late 1970s indicated that ground-level CFCs indeed diffused upward to the stratosphere. Orbiting satellites have provided further significant data about ozone loss and its link to human-produced CFCs.

Much the same may be said about global climate change. The concept that human industry, by releasing carbon dioxide and other greenhouse gases into the atmosphere, could cause global warming was hardly new at the dawn of space exploration. Svante Arrhenius, the Swedish geochemist, had suggested just such a process in 1896, and in 1938 G.C. Callendar, a British engineer, made an early calculation about the pace of Earth's expected warming. Ground-based measurements by Charles D. Keeling, who first placed a carbon dioxide detector in the pristine mountain air atop Mauna Loa in Hawaii in 1958, were also the first to yield clear evidence that atmospheric CO_2 concentrations are rising, and investigations of the limited capacity of Earth's oceans to absorb excess atmospheric carbon dioxide led Hans Suess and Roger Revelle to declare in 1959, with regard to global warming, that "Human beings are now carrying out a large-scale geophysical experiment of a kind that could not have happened in the past."[14] Models developed by the early 1990s predicted that if global temperatures were to increase 1 to 3 degrees Celsius or more over the next century, sea levels will rise as Greenland and Antarctic glaciers melt, inundating coastal regions and threatening cities. Moreover, productive agricultural regions on the planet will shift as some deserts receive rain, and presently fertile areas experience drought.

The Earth Sciences as a Field of Knowledge

In the 1960s, an important reconstitution of the earth sciences took place, first in the United States and then worldwide [Table 20.1]. The result of this transfor-

TABLE 20.2: THE GROWTH OF GEOPHYSICS IN NORTH AMERICA, 1920–1990

Period	Patrons	Professional Structures	Research Problems	International Programs/ Treaty Obligations
Inter-war era (1920–c. 1945)	Carnegie Institution of Washington, geophysical prospecting firms (incl. Geophysical Services, Inc.), foundations, government (incl. US Coast & Geodetic Survey)	academic depts. of geology, meteorology, or physics; laboratories or depts. of geophysics	continental/oceanic structure; isostasy; high pressure rock deformation; polar front (air mass analysis); applied geophysics (petroleum explor.)	(Jesuit) seismic network, world magnetism surveys (CIW/Division of Terrestrial Magnetism) International Union of Geology and Geophysics (IUGG) (member-initiated research programs)
Cold War era (1945–1960s)	government (esp. military agencies, Office of Naval Research, Air Force Cambridge Research Center, Advanced Research Projects Agency), Atomic Energy Commission, National Science Foundation, geophysical prospecting firms, philanthropists	institutes of geophysics, depts. of geophysics, depts. of 'geological sciences,' laboratories or branches of geophysics	upper atmospheric phenomena; global circulation (computer modeling); deep earth structure; continental/ocean floor tectonics (plate tectonics)	International Geophysical Year (IGY) Nuclear Test Ban Treaty (seismology, atmospheric sciences, oceanography)
Environmental era (late 1960s–present)	government (NSF, National Oceanic and Atmospheric Administration, NASA, defense), petroleum firms, foundations, private philanthropists	institutes of geophysics or earth sciences; depts. of earth sciences, interdisciplinary groups	global climatic change, cross-coupled phenomena, comparative planetology, impact collision research (neo-catastrophism interpretations)	Global Atmospheric Research Program (GARP), Scientific Committee for Antarctic Research (SCAR), World Climate Research Programme (WCRP), International Geosphere-Biosphere Program (IGBP).

mation was to give the earth sciences a new place within the intellectual landscape of ideas and their institutional embodiments. Departments of geological science replaced long-standing departments of geology at many universities in America and abroad. Earth sciences curricula and research programs came to integrate regional and historical studies of the solid earth with investigations of the atmosphere and oceans, the Earth's chemical and physical properties, and the global effects of its biological inhabitants. This recasting of disciplinary boundaries also influenced professional societies and government agencies. In the US, the newly created Environmental Science Services Administration (1965), which embraced research in meteorology, geodesy, and ionospheric physics, was joined by the Environmental Studies Board of the National Academy of Sciences (1967) and the Bureau of Oceans and International Environmental and Scientific Affairs of the State Department (1973–4). Similar shifts followed worldwide, encouraging such undertakings as the International Council of Scientific Unions' International Geosphere-Biosphere Program (1986). This socio-intellectual transformation was of fundamental importance. The post-1960 period can be termed an 'environmental' era.

What caused this transformation? One critical factor was certainly the intellectual fervor that accompanied the plate tectonics revolution of the late 1960s. The success of a unified physical theory in providing explanations for such long-standing geological puzzles as mountain-formation, island arcs, and fossil continuities between remote continents caused a massive shift in authority away from traditional field geology toward the physics-dominated wing of the discipline. At many universities, geophysicists used their heightened intellectual status and access to Big Science resources to steer research towards problems suitable to physics-based approaches, in turn training new students in these techniques. The geographical tradition within many geology departments receded, and while many geologists objected to what one termed "geologic thought not founded deeply in careful observations of the Earth itself," earth science departments clearly placed greater emphasis on links to physics and chemistry than at any previous time.[15]

But this is hardly the only factor. Departments of geophysics and earth sciences in the US began their explosive proliferation in the early 1960s, well before the conceptual revolutions of sea-floor spreading and continental drift began to gain widespread support. Already by 1960, for instance, the Massachusetts Institute of Technology announced an ambitious effort to consolidate its geophysical research programs, and two years later Stanford created its new Department of "Earth Sciences." Moreover, many of these schools — joined by research agencies such as the Environmental Science Services Administration — emphasized studies in fields remote from plate tectonics, including climate studies and ionospheric research.

Several trends were responsible for this fundamental reconceptualization of the earth sciences. All were closely interrelated, and cannot be understood apart from

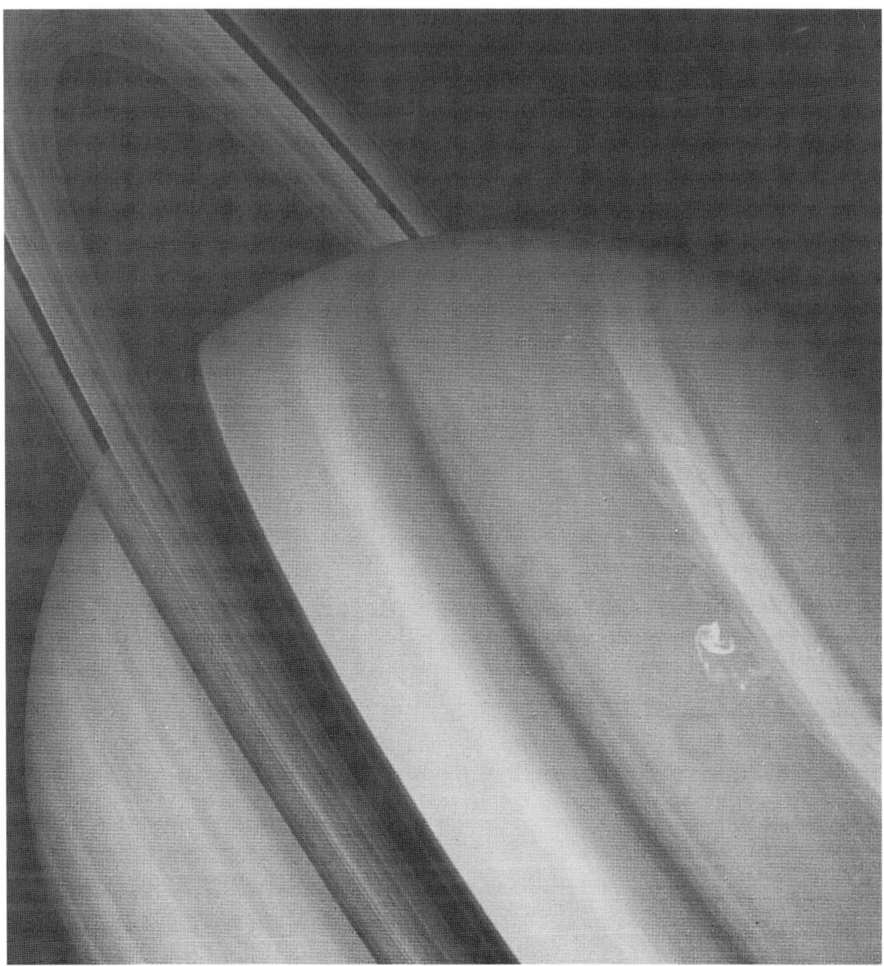

FIGURE 20.4: PLANETARY SCIENCE IN THE SPACE AGE: THE ADVENT OF SPACE EXPLORATION IN THE LATE 1950S LITERALLY PERMITTED THE DIRECT STUDY OF NEW WORLDS, CHALLENGING LONG-HELD PERCEPTIONS ABOUT PLANETARY PROCESSES WHILE INSPIRING NEW INTEREST IN ENVIRONMENTAL CONCERNS. THE *VOYAGER 2* SPACECRAFT PHOTOGRAPHED SATURN AND ITS VAST RING SYSTEM IN AUGUST 1981. COURTESY OF JET PROPULSION LABORATORY/NASA.

the others, although four separate strands can be distinguished. One was the rapid development of new instruments in physics and chemistry applicable to geological science, spurred by military demands imposed during World War II and the subsequent Cold War. Air- and shipborn magnetometers, advanced seismometers, computers, remote sensing devices, and improved sonar, gravimetric, and seismic reflection techniques dramatically expanded the range of geophysical data and encouraged new theoretical investigations. Mass spectrometry is a revealing example. A technology important for refining radioactive materials for atomic weapons programs, mass spectrometers provided a means to determine the ages of Precambrian materials older than fossil-bearing rocks and yielded extremely precise abundances for terrestrial and meteoritic materials. One of its more important results obtained through mass spectrometry in the mid-1950s was the determination of the modern age of Earth of circa 4.5 billion years, a central datum for the earth sciences and for cosmology. By the mid-1950s mass spectrometers were widely incorporated in academic, industrial and government centers of geological research, and nuclear chemists (or geochemists) became active participants of the earth sciences.[16]

A second important factor is one already noted: the military's growing interest in the earth sciences during the Cold War. Maintaining modern weapon systems, including submarine surveillance programs, bio-chemical weapons, and long-range communications required military planners to treat oceanographic, meteorological, and other geophysical phenomena as part of an integrated, comprehensive structure. Successful deployment of ballistic missiles, central to the defense strategies of the United States and the Soviet Union, demanded accurate geodetic and gravity surveys (for trajectory computation), an understanding of ionospheric and magnetic variations (to maintain communications and guidance) and detailed understanding of atmospheric-ocean interactions, atmospheric circulation, and upper atmospheric conditions (to maintain the integrity of flight paths and mechanical components). Such needs encouraged military planners to define the 'environment' as a theater of operations, and encouraged research programs to ascertain if local environmental conditions (such as weather and climate) could be modified for military advantage. The military was not alone in stressing operational definitions of the environment: farmers and ranchers who sought weather modification to bring rain to parched ranchlands also called for a unified theoretical perspective on environmental phenomena, at least indirectly.[17] But intense interest by military officials in the environmental sciences gave impetus to interdisciplinary research programs, and encouraged grand-scale undertakings such as the International Geophysical Year.

A third factor was the growing recognition, particularly after 1960, that threats to the natural environment were sufficient to require national and especially international regulation. One of the first efforts of this kind involved restricting mining and military operations in Antarctica, preserving it as a 'continent for

TABLE 20.3: ENVIRONMENTAL TREATIES RELATING TO GEOPHYSICS (SELECTED)[1]

Year	Treaty	Details
1959	Antarctic Treaty	Twelve national signatories, including US; at issue: environmental damage from military activities, nuclear waste, radioactive waste dumping. Subsequently amended and expanded as Antarctic Treaty System (ATS) *Disciplines*: hydrology, seismology, volcanology, atmospheric sciences
1963	Limited Nuclear Test Ban Treaty	117 national signatories, including US and USSR, bans above-ground testing of nuclear weapons. *Disciplines*: seismology, atmospheric sciences, hydrology, volcanology, oceanography, ionospheric physics
1972	Convention on the Prevention of Marine Pollution	31 national signatories; addresses prevention of oceanic pollution. *Disciplines*: oceanography, atmospheric sciences
1979	Convention on Long Range Transboundary Air Pollution	34 national signatories; addresses acid rain. *Disciplines*: atmospheric sciences, oceanography, soil sciences, hydrology
1987	Montreal Protocol on Substances that Deplete the Ozone Layer	75 national signatories; increased to 93 after 1990 amendment. Establishes phase-out schedule for CFCs and additional ozone-depleting chemicals. *Disciplines*: atmospheric sciences, oceanography, hydrology
1991	Protocol on Environmental Protection	Forty national signatories, including (with reluctance) US; at issue: 50-year ban on mineral exploration and exploitation in Antarctica. Has stimulated further research into possible consequences of economic activity in Antarctic environments. *Disciplines*: atmospheric sciences, hydrology, oceanography, seismology
1992	The Convention on Climate Change (Rio Conference)	153 national signatories; sets informal guidelines to return emissions of 'warming' gases to 'earlier' levels by 2000. *Disciplines*: atmospheric sciences (including geochemistry), oceanography, volcanology; broadly interdisciplinary

[1] Compiled from Aant Elzinga, "Antarctica: The Construction of a Continent By and For Science," in Elisabeth Crawford *et al.*, (Eds.), *Denationalizing Science.* (Dordrecht: Kluwer, 1993), pp. 73–106; Eugene B. Skolnikoff, *The Elusive Transformation: Science, Technology, and the Evolution of International Politics.* (Princeton: Princeton University Press, 1993); and Lawrence E. Susskind, *Environmental Diplomacy: Negotiating More Effective Global Agreements.* (New York: Oxford University Press, 1994).

science.' Another was the attempt to eliminate the testing of nuclear weapons, which resulted in the Limited Nuclear Test Ban Treaty of 1963, signed by the United States, the Soviet Union, and 115 other nations. Both of these undertakings relied on the input of geophysicists from many fields [Table 20.3], who influenced and in turn were influenced by their inherently interdisciplinary nature. Subsequent international treaties, which have addressed stratospheric ozone depletion and the global climate change, have demanded even more rigorously an interdisciplinary understanding of the interactions between geophysical and biological systems. Treaty systems like the Montreal Protocol on ozone depletion and the Rio Conference on global climate change have created important new markets for research in the earth sciences, but even more importantly — although less often recognized — they have served to highlight problems and issues that transcend narrow professional and disciplinary interests.[18]

Finally, in the post-1945 period, scientists have shown revived interest in Humboldtian approaches to the earth sciences, stressing integrative and holistic interrelations rather than the deeply reductionist approaches that have typified major research programs of twentieth century physics. Not all of these approaches can be traced to the period following World War II. For example, the concept of the 'biosphere' as the most chemically active part of Earth's crust and lower atmosphere was defined in the 1920s by the Russian chemist Vladimir Ivanovich Vernadsky.[19] Only after the 1950s however did this idea gain widespread respect, stimulated by as popular anxieties about nuclear fallout and pesticides, the budding environmental movement, and the singular influence of Rachel Carson's *Silent Spring* (1962). Increased numbers of physical scientists embraced a definition of environmental research, voiced by the American geophysicist M. King Hubbert in 1967, that stressed "integration of the traditional areal and historical aspects of the solid earth and its biological inhabitants with the earth's physical and chemical aspects."[20] One of the boldest expressions of this philosophical orientation is the Gaia hypothesis. Proposed by the British researcher James Lovelock in the 1970s, the Gaia hypothesis argues that Earth must be understood as a self-regulating system which maintains conditions suitable for life. Many biologists and earth scientists reject Lovelock's theory — some of them disturbed by its teleological overtones — but they have shown sympathy for the kind of interdisciplinary approaches that his theory relies upon. Their support shows in numerous ways, including contemporary research on interlinked terrestrial phenomena operating on vastly different scales (such as lightning flashes and immense ionospheric features which accompany them).

What is especially striking about geophysics and the earth sciences in the twentieth century is that the problems deemed worthy of investigation — and the scientific methods used to explore them — cannot be understood independently of the major social forces which have shaped the period. If the results of geophysicists are important products of modern science, we do well to remember that such factors as the utility of geophysical results for Cold War missile development, and public concern over exploitation of the natural environment, stimulated and supported this research. The earth sciences clearly demonstrate that science and its social context are intimately intertwined.

REFERENCES

1. Quoted in Kathleen L. Mark, *Meteorite Craters: How Scientists Solved the Riddle of these Mysterious Landforms.* (Tucson: University of Arizona Press, 1987), p. 235.
2. William Glen, "What Killed the Dinosaurs?" *American Scientist* (1990) **78, 3**: 354–370; and David M. Raup, *Extinction: Bad Genes or Bad Luck?* (New York: Norton, 1991).
3. James Gleick, *Chaos: Making a New Science.* (New York: Viking, 1987), pp. 12–20; Edward Lorenz, "The Problem of Deducing the Climate from the Governing Equations," *Tellus* (1964), **16**: 1–11, and Frederick Nebeker, *Calculating the Weather: Meteorology in the Twentieth Century.* (San Diego: Academic Press, 1995) on nuclear winter research see especially Carl Sagan, "Nuclear War and Climatic Catastrophe: Some Policy Implications," *Foreign Affairs* (Winter 1983–84), **62, 2**: 257–292.
4. Peter J. Bowler, *Norton History of the Earth Sciences.* (New York: Norton, 1993), pp. 229, 398 and David Fisher, *Fire and Ice: The Greenhouse Effect, Ozone Depletion, and Nuclear Winter.* (New York: Harper Row, 1990), pp. 23–29.
5. Henry Stommel, *A View of the Sea: A Discussion between a Chief Engineer and an Oceanographer about the Machinery of the Ocean Circulation.* (Princeton: Princeton University Press, 1987) and Mary Sears and Daniel Merriam (Eds.), *Oceanography: The Past.* (New York: Springer Verlag, 1980).
6. Benjamin F. Howell, *An Introduction to Seismological Research: History and Development.* (New York: Cambridge University Press, 1990) and Stephen G. Brush and C.S. Gillmor, "Geophysics," in Abraham Pais, Brian Pippard, and Laurie Brown (Eds.), *Twentieth Century Physics.* (College Park, Maryland: American Institute of Physics, 1995): 1943–2016, esp. p. 1954.
7. R.J. Tayler (Ed.), *History of the Royal Astronomical Society, Volume 2: 1920–1970.* (Oxford: Blackwell Scientific, 1987), pp. 24–25 and International Union of Geodesy and Geophysics, *Advanced Report: Commission on Continental and Oceanic Structure.* (Washington, D.C.: IUGG, 1939), pp. 4–6.
8. Shapley quoted in David H. DeVorkin, *Science with a Vengeance: How the Military Created the US Space Sciences after World War II.* (New York: Springer Verlag, 1992), on p. 96. On the military's influence on geophysics, see DeVorkin as well as Chandra Mukerji, *A Fragile Power: Scientists and the State.* (Princeton: Princeton University Press, 1989).
9. Robert Shrock, *Cecil and Ida Green: Philanthropists Extraordinary.* (Cambridge, MA.: M.I.T. Press, 1989).
10. Stephen Pyne, "Space: The Third Great Age of Discovery," in Martin J. Collins and Sylvia K. Kraemer (Eds.), *Space: Discovery and Exploration.* (Washington, D.C.: Hugh Lauter Levin Associates, 1993), pp. 14–65.
11. James A. Van Allen, *Origins of Magnetospheric Physics.* (Washington, D.C.: Smithsonian Institution Press, 1983).
12. Stephen Jay Gould, "The Great Scablands Debate," in Gould, *The Panda's Thumb.* (New York: Norton, 1982), pp. 194–203. No historical study of the Voyager mission yet exists; for a semi-technical review see chapters by James Head and Noel Hinners in J. Kelly Beatty and Andrew Chaikin (Eds.), *The New Solar System*, 3rd ed. (New York: Cambridge University Press, 1990).
13. Eugene M. Shoemaker and Robert J. Hackman, "Stratigraphic Basis for a Lunar Time Scale." *Geological Society of America Bulletin* (1961), **71, 12**: 2112 and Eugene M. Shoemaker, "Lunar Geology," in Paul A. Hanle and Von Del Chamberlain (Eds.), *Space Science Comes of Age.* (Washington, D.C.: Smithsonian Institution, 1981), pp. 51–57. The Great Bombardment concept is briefly treated in Stephen G. Brush, "Theories of the Origin of the Solar System 1956–1985," *Reviews of Modern Physics* (1990), **62, 1**: 43–112.
14. John Firor, *The Changing Atmosphere: A Global Challenge.* (New Haven: Yale University Press, 1990), p. 48.

15. Robert E. Wallace to Lee DuBridge, June 12, 1963, Box 12.2, DuBridge papers, California Institute of Technology. The place of geography within the ecology of knowledge is explored in David N. Livingstone, *The Geographical Tradition. Episodes in the History of a Contested Enterprise.* (Oxford: Blackwell, 1992).
16. Ronald E. Doel, *Solar System Astronomy in America: Communities, Patronage, and Interdisciplinary Research.* (New York: Cambridge University Press, 1996), pp. 78–114; Homer LeGrand, *Drifting Continents and Shifting Theories.* (New York: Cambridge University Press, 1988), pp. 170–177; and Francis J. Pettijohn, *Memoirs of an Unrepentant Field Geologist: A Candid Profile of Some Geologists and their Science, 1921–1981.* (Chicago: University of Chicago Press, 1984).
17. Office of the Director of Defense Research and Engineering, "International Science Activities," Nov. 27, 1961, Box 27, Frank Press files, M.I.T. archives. I thank Press for permission to use his collection.
18. Lawrence E. Susskind, *Environmental Diplomacy: Negotiating More Effective Global Agreements.* (New York: Oxford University Press, 1994).
19. Kendall Bailes, *Science and Russian Culture in an Age of Revolution: Vernadsky and his Scientific School, 1863–1945.* (Bloomington: Indiana University Press, 1989), pp. 187–190.
20. M. King Hubbert to Geology Department, Stanford University, Mar. 20, 1967, Folder 5, C. Hulett Dix papers, California Institute of Technology.

Further Reading

Bowler, Peter J. *The Norton History of the Environmental Sciences.* (New York: W.W. Norton, 1993).

Brush, Stephen G. and C.S. Gillmor. "Geophysics." In Abraham Pais, Brian Pippard, and Laurie Brown (Eds.), *Twentieth Century Physics.* (College Park, Maryland: American Institute of Physics, 1995), pp. 1943–2016.

Brush, Stephen G. *A History of Modern Planetary Physics.* (New York: Cambridge University Press, 1996), [in three volumes].

Burchfield, Joe. *Lord Kelvin and the Age of the Earth.* (New York: Science History Publications, 1975).

Doel, Ronald E. *Solar System Astronomy in America: Communities, Patronage, and Interdisciplinary Research, 1920–1960.* (New York: Cambridge University Press, 1996).

Elzinga, Aant. "Antarctica: The Construction of a Continent By and For Science," in Elisabeth Crawford, Terry Shinn, and Sverker Sörlin (Eds.), *Denationalizing Science: The Contexts of International Scientific Practice.* (Dordrecht: Kluwer, 1993), pp. 73–106.

Friedman, Robert Marc. *Appropriating the Weather: Vilhelm Bjerknes and the Construction of a Modern Meteorology.* (Ithaca, New York: Cornell University Press, 1989).

Glen, William. *The Mass Extinction Debate: How Science Works in a Crisis.* (Stanford: Stanford University Press, 1994).

Good, Gregory (Ed.) *Garland History of Geophysics.* (New York: Garland, 1997).

Greene, Mott T. *Geology in the Nineteenth Century: Changing Views of a Changing World.* (Ithaca: Cornell University Press, 1982).

Hallam, Anthony. *Great Geological Controversies.* (Oxford: Oxford University Press, 1989), 2nd ed.

Koppes, Clayton R. *JPL and the American Space Program: A History of the Jet Propulsion Laboratory.* (New Haven: Yale University Press, 1982).

LeGrand, Homer E. *Drifting Continents and Shifting Theories: The Modern Revolution in Geology and Social Change.* (New York: Cambridge University Press, 1988).

Menard, H.W. *The Ocean of Truth: A Personal History of Global Tectonics.* (Princeton, N.J.: Princeton University Press, 1986).

Mukerji, Chandra. *A Fragile Power: Scientists and the State.* (Princeton: Princeton University Press, 1989).

Nebeker, Frederick. *Calculating the Weather: Meteorology in the Twentieth Century.* (San Diego: Academic Press, 1995).

Oreskes, Naomi. *The Rejection of Continental Drift.* (New York: Oxford University Press, forthcoming).

Pyne, Stephen J. *The Ice: A Journey to Antarctica.* (Iowa City: University of Iowa Press, 1986).

Tatarewicz, Joseph N. *Space Technology and Planetary Astronomy.* (Bloomington: Indiana University Press, 1990).

Chapter 21

Neo-Darwinism and Natural History

Sharon Kingsland

'Darwinism' refers to the scientific argument that evolution can be explained by natural processes and that Darwin's theory of natural selection is the main mechanism of adaptive change. After decades of controversy about the role of natural selection as a creative force, modern Darwinism was consolidated in a group of treatises published from the mid-1930s to the 1950s. These established what is called the 'modern synthesis' in evolutionary theory, also referred to as 'neo-Darwinism.'

Although the biological mechanisms that underlie evolution have been debated throughout the twentieth century, the central idea of Darwinism has not been superseded by any new theoretical framework or worldview. While the theory has proven to be remarkably resilient, the study of the mechanisms of evolution has been profoundly changed by the development of genetics, microbiology and molecular biology. Natural history has also changed radically since the early twentieth century, when naturalists confronted the challenges of experimental biology. Systematists and paleontologists have embraced the causal analysis of the experimental sciences, and other branches of natural history have evolved into two new sciences, ecology and ethology. These changes are evident in the shift away from descriptive or story-telling narratives to a more analytical, hypothesis-testing style, complete with attempts to mathematize the biological world.

The technical development of evolutionary science and the emergence of new scientific specialties is only one side of the coin. The other is the philosophical and social debate about the nature of progress, our place in nature, and our duty toward nature. The history of Darwinism as a biological enterprise is tied to efforts to discern the larger philosophical significance of Darwin's work, that is, to find meaning in the fact of our evolution and to apply insights from evolutionary biology to ethics. The emergence of human consciousness seemed to biologists in the mid-century to reveal a genuine progressive trend in evolution. Human awareness of our own evolution was seen to provide us with a mandate to seek control over the evolutionary process. At the end of the century, concern over our destructive impact on nature, coupled with a sense of our duty to preserve it, has given evolutionary biologists a new sense of mission.

Darwinism in the Early Twentieth Century

Darwinian writing in the late nineteenth century developed the claim, implied in Darwin's work, that the theory of evolution would provide a way not only of unifying the life sciences but also of serving as the basis for a unified worldview. Many of the young naturalists who had extended Darwinian theory and research after the publication of the *Origin of Species* (1859) continued their roles as champions of Darwinism into the early twentieth century. Of these, Alfred Russel Wallace in Britain was an influential proponent of natural selection's power to produce adaptation, although other Darwinian disciples disagreed with Wallace's extreme views. Ernst Haeckel and August Weismann in Germany constructed competing versions of neo-Darwinism which stimulated many of the controversies that arose around 1900.

One of the legacies of nineteenth-century Darwinism that persisted into the first two decades of the twentieth century was the idea that there was a parallel between embryological development (ontogeny) and the evolutionary history of the species (phylogeny), such that the embryo passed through certain 'lower' evolutionary stages as it developed. This idea was expressed by Haeckel in the formula that "ontogeny recapitulates phylogeny." Haeckel's embryological approach to Darwinism combined elements of Darwin's theory with what was referred to as 'Lamarckian' inheritance, the idea that traits acquired in an organism's lifetime could be inherited and become the basis for progressive evolutionary change.

Weismann, in contrast, raised objections to the Lamarckian argument because it lacked firm experimental proof. His defense of natural selection as a creative power left no room for any synthesis of Darwinian and Lamarckian theses. Weismann regarded Lamarckism as a hindrance to deeper inquiry into the processes inside the cell that might explain how variations arose. He developed an imaginative scenario of what might be happening at the sub-microscopical level within the chromosomes to produce such variations. His critics pointed out that such unobservable speculations discouraged experimental inquiry. But Weismann's challenge prompted more systematic experimental work that continued into the first quarter of the twentieth century. This research was suggestive enough to keep scientists interested in the possibility of the inheritance of acquired characteristics. Some, particularly in Europe, remained open to the idea that further experimental work would yield valuable information on the relationship between organisms and environment. Others flatly rejected Lamarckian inheritance as unproven and unprovable.

The main lines of research at the start of the century centered on the statistical analysis of variations (biometry), studies of chromosomes during cell division and fertilization (cytology), analysis of embryological development and regeneration, and the analysis of inheritance by means of controlled breeding experiments

(Mendelism and mutation theory). These led to fresh debates about the meaning of Darwin's work and how his ideas might be revised. Partly because of the proliferation of new research, the idea that Darwin's theory could produce a unified world picture was increasingly challenged.

Many eminent botanists, zoologists, and paleontologists expressed severe anti-Darwinian sentiment at this time. Of particular concern was whether natural selection was a creative force in evolution and whether it could account for the long-term trends that appeared in the fossil record. Dominating this debate was the demand that the evolutionary hypotheses be subjected to systematic experimental or statistical investigations, getting away from the methods of indirect inference and speculative constructions of evolutionary lineages that had characterized earlier writing (Figure 21.1).

Hugo de Vries revised Darwin's ideas in a major study of discontinuities in heredity, *The Mutation Theory* (1901–1903). He argued that slow, cumulative selection of minute variations was not the way new species originated. He thought that discrete discontinuous changes, which he called "mutations," were required to produce novelty in populations. These mutations could produce in one step what he considered to be genuinely new species (Figure 21.2). Natural selection, acting on these small populations or "elementary species," then determined which ones would survive. De Vries' theory drew on his knowledge of agricultural breeding practices and on an experimental program designed to trace the appearance of mutations and their inheritance.

De Vries' theory generated enthusiasm in spite of lack of evidence to support it. This was largely because the experimental method that he employed was considered a powerful means of achieving control over the evolutionary process. It became so appealing for American biologists, for instance, that 'discoveries' of new mutations were regularly proclaimed. Hence a British naturalist wryly noted the contrast between the "volcanic energy" of American plants, with their apparently high levels of mutability, compared to the more lethargic Dutch plants that de Vries studied.[1] Research on the genetic mechanisms underlying these observations eventually showed that these were not genuine instances of speciation.

Growing interest in the causes of variation, especially variations of a discrete or abrupt nature (known as discontinuous variations), contributed to the rediscovery and belated recognition of Mendel's work around 1900. Research stimulated by Mendel's work followed much the same experimental path as that taken by followers of de Vries. (De Vries, although credited as one of the discoverers of Mendel's work, was not himself a follower of Mendel). Mendelism was also attractive to scientists engaged in breeding experiments. As with the mutation theory, the emergence of an exact approach to evolution was linked to the goal of controlling the evolutionary process through experimental intervention. The founder of the first effective school of Mendelism was William Bateson at the University of Cambridge, who launched the term 'genetics' at a scientific conference in 1906. Although he was trained in natural history and embryology, Bateson turned his back on embryology in favor of the experimental study of variation, which appeared to him a more promising field. Genetics was conceived by him as the study of the physiological processes influencing heredity.

The centenary of Darwin's birth in 1909 regenerated much support for Darwin's importance in the natural and social sciences as well as the humanities. There remained enthusiasm for Darwin's work, continued close study of his writings, and admiration for his scientific achievements, which were seen to be the basis for modern research.[2] Naturalists, while accepting the importance of experimental work, perceived experimentalists as dogmatic and too readily dismissive of natural history. They emphasized the high degree of continuity between species and accused the mutation theorists, the 'Batesonians' and other Mendelians of exaggerating the prevalence of discontinuous variability in nature. In their critique of mutation theorists, they emphasized the abundant evidence from the geographical distribution of species, from which one could infer the continuous course of evolution in time. As British naturalist E.B. Poulton put it, such evidence of continuity "rather aggressively impresses the great majority of those whose lives are devoted to the study of species."[3]

Naturalists stressed the importance of natural history to evolutionary inquiry, pointing out that Darwin's achievements were only possible because of the breadth of his knowledge of natural history. The ability to achieve such mastery of natural history was threatened by the specialization of modern science.

Bateson's advocacy of Mendelism drew criticism from quite another quarter as well. Karl Pearson, English mathematician and protégé of Francis Galton, was unmoved by the new revelations. He adhered to what he considered a truer Darwinian position, that natural selection acting gradually on small variations could give rise to permanent changes in a population. Pearson was interested in deriving mathematical definitions of heredity and variation.

Bateson and Pearson locked horns in a celebrated controversy lasting from 1902 to about 1906. The competing positions represented different strategies for developing evolutionary science in positivist directions. Bateson took natural selection for granted, but distanced himself from Darwinism, as pursued by natural historians, to focus more narrowly on the experimental study of heredity. Pearson embraced Darwin's work because of its scientific reasoning. An avid eugenicist, he agreed that natural selection was a true scientific law, but it needed more rigorous statement if it was to be the basis for social improvement. For Pearson the key to progress was not experimental science but mathematics. Expressing the laws of heredity in mathematical form represented the highest assurance of objectivity and therefore the surest ground for the development of a scientific approach to society. Pearson's biometrical school for a time seemed to be on the verge of reforming the science of heredity, but biometry was effectively eclipsed by experimental genetics.

By the 1920s developments in experimental genetics suggested the need for reassessments of Darwinism in the light of modern experimental science. Oxford mathematician Ronald A. Fisher, in a seminal treatise on Darwinism, *The Genetical Theory of Natural Selection* (1930), argued for a synthesis of Mendelism and Darwinism. Fisher tried to resolve a controversial point that was a consequence of the theory of inheritance that prevailed in Darwin's time and that was still held in the early twentieth century. When two individuals were crossed, their characteristics were 'blended' so that the offspring showed some intermediate form of the parental traits. This belief appeared fatal to the original form of the Darwinian argument, because when an individual arose possessing a 'superior' trait, that trait would be diluted by breeding with individuals lacking that characteristic. Hence the population would not evolve.

Both Mendelism and the mutation theory had focused on traits that did not appear to blend: discrete characteristics that retained their appearance through the generations. Scientists therefore divided variations into two categories: small variations that were subject to blending and more obvious and discrete variations that remained in the population as long as the individuals possessing them continued to reproduce. But by the 1920s advances in research pointed to the conclusion that all variations, even the most subtle kind, could be considered 'mutations' if they had a genetic basis. What had been taken as evidence of blending inheritance was thought to be the product of very complex processes and interactions among the genes, processes which were only barely understood.

As the modern concept of mutation evolved from genetic research, the logic supporting the argument for two categories of variation was weakened. By eliminating this distinction it became possible to justify a synthesis of Darwinism and Mendelism and to reassert the primacy of natural selection as the only directing force in evolution. This in a nutshell was Fisher's argument. It defined neo-Darwinism in the mid-century. The Lamarckian alternative, which had been brought to the fore partly to compensate for the presumed effect of blending, was discarded by Fisher as irrelevant. Many geneticists had already abandoned Lamarckian research, and it dwindled to a minor and somewhat suspect field.

The experimental program combining Mendelism and Darwinism also provided impetus for further exploration of eugenics. Indeed, one-third of Fisher's book was devoted to this subject. The interplay of Darwinism, Mendelism, and eugenics can be seen in the work of some of the most prominent Darwinians of the mid-twentieth century.

The Modern Synthesis in Evolutionary Biology

The 1930s marked a turn toward more active interest in natural selection and a return to a strongly selectionist interpretation of evolution. At the same time the distance between genetics and natural history began to narrow. In the 1920s and 1930s experimental geneticists were developing a more detailed view of the gene complex, paying more attention to interactions between genes. Population geneticists, led by R.A. Fisher and J.B.S. Haldane in Britain, and Sewall Wright in the

United States, explored the mechanisms of evolution through theoretical mathematical arguments. Certain collaborations promoted both theory and field research. The relationship between Fisher and emtomologist E.B. Ford, for example, benefited both sides, providing Fisher with the evidence he needed to analyze the power of selective forces, and prompting Ford to develop ecological genetics, the study of genetic changes in wild populations.

Developments in theoretical and experimental genetics, as well as increased interest within natural history in problems of adaptation and selection, set the stage for what is now called the "modern synthesis" in evolutionary thought. The term "modern synthesis" comes from the title of a book published by Julian Huxley in 1942, *Evolution, the Modern Synthesis,* but it refers more broadly to the collective contributions of several biologists from the mid-1930s to 1950 that set out the theoretical and observational basis of modern Darwinism. These scientists, all notable for their exceptional breadth of knowledge, agreed on basic issues concerning natural selection as a creative process. They accepted the validity of natural selection and believed that genetics needed to be integrated with the natural history disciplines. The unusual aura of cooperation that permeated certain parts of evolutionary biology has prompted different explanations which range from underlying philosophical motives, such as a general desire to unify biology, to disciplinary motives that caused scientists to create new societies and journals that would strengthen evolutionary biology and the authority of the associated fields.

The first of these texts was *Genetics and the Origin of Species* by Theodosius Dobzhansky (1937). Dobzhansky was trained in Russia as a naturalist, a specialist in entomology, and became interested in the style of experimental genetics promoted by Thomas Hunt Morgan's research school. He went to New York in 1927 to work with Morgan and remained in America for the rest of his life. Believing firmly that nothing in biology made sense except in the light of evolution, Dobzhansky synthesized the field naturalist tradition and the laboratory genetics that he learned from American sources. His collaboration with experimental geneticist Alfred H. Sturtevant in the 1930s was especially important in helping him devise a research plan, focusing on wild populations of fruit flies, for the study of evolution in nature. He analyzed how natural selection acted on these populations, changing their genetic make-up. He was also influenced by some of Sewall Wright's ideas and, although not himself a mathematician, incorporated certain of Wright's arguments into his work, thereby helping to make the esoteric mathematics of population genetics accessible to other biologists.

Mathematicians, seeking to persuade biological colleagues who were less versed in mathematics, introduced visual metaphors that helped to depict the evolutionary process. One of the most influential of these metaphors was Wright's concept of the "adaptive landscape," which was a diagrammatic representation of how fitness changed within populations (Figure 21.3). Populations were represented as peaks separated by valleys. Variation in the height of the peaks represented differences in the levels of adaptation. Natural selection was said to drive populations up the nearest adaptive peak. As Wright expressed it, the problem of evolution was to find a mechanism by which the species found its way from lower to higher peaks. Dobzhansky picked up this image, changing its interpretation somewhat so that the hills became ecological niches. This way of visualizing evolution as a dynamic process of change across a shifting adaptive landscape became a standard textbook tool for explaining modern Darwinian theory.

A major tenet of the modern synthesis was that the genetic processes that produced small evolutionary changes in the short term (microevolution) could, if extended over geological time, explain the broader patterns observable in the fossil record (macroevolution). As Dobzhansky argued, microevolution and macroevolution were parts of a single continuum, such that studies of the former helped to elucidate the latter. All evolutionary patterns could be explained by the ordering of small genetic changes by natural selection.

Dobzhansky's book was followed by a number of others, the most important of which included C.D. Darlington's *The Evolution of Genetic Systems* (1939), Ernst Mayr's *Systematics and the Origin of Species* (1942), George Gaylord Simpson's *Tempo and Mode in Evolution* (1944), and G. Ledyard Stebbins' *Variation and Evolution in Plants* (1950). The arguments in these treatises refocused attention on the detailed study of evolution as a process, with an emphasis on how natural selection operated and what its effects were. The modern aspect of evolutionary biology was its focus on causal analysis, the how and the why of evolution, rather than on simple description of the evolutionary record. Because of its emphasis on causal mechanisms, modern Darwinism is sometimes called the 'biological' interpretation of evolution, a term that Dobzhansky preferred to "neo-Darwinism," which he thought was better restricted to late-nineteenth century thought.

The emergence of the interdisciplinary approach characteristic of the modern synthesis followed a somewhat different course in different national contexts. These national differences were largely the result of differences in the institutional context of science, the economic resources available to evolutionary biologists, and the different developmental paths followed by biological disciplines in different countries. The synthesis of course has no conceptual essence, but represents an evolving dialogue played out over several decades across several disciplines, with individual views changing over time.

Agreement on certain large issues did not mean consensus on all problems. Different disciplinary perspectives produced different ideas about what was most important. It was also clear that certain branches of biology had been given short shrift in the synthesis. In part, consensus was achieved by excluding or marginalizing scientific work, for example embryology, that did not accept modern genetics or the assumption that microevolution was sufficient to explain macroevolution.

Neither were ecologists major contributors to the seminal texts of the modern synthesis, although Huxley's book, a broad multi-disciplinary survey, included a review of relevant ecological research. Most ecologists, however, tended to focus on short-term changes and wrote about evolutionary processes in a strictly token way. It was not until the 1950s that they began to contribute more prominently to the neo-Darwinian literature. David Lack, ornithologist and ecologist at Oxford, developed the ecological core of Darwin's ideas, especially in relation to the regulation of animal populations. His study of the evolution of the finches of the Galápagos Islands was a paradigm of Darwinian reasoning (Figure 21.4).[4] Lack focused especially on the role of competition in species diversification, a highly controversial subject because of the many different connotations of the term 'competition' and the difficulty of measuring its strength in natural populations.

The decision to retain the term 'neo-Darwinism' and Darwinian language in general was not just a passive continuation of tradition, but was a deliberate move designed to emphasize the continuity between modern developments and Darwin's work. As genetics developed, some biologists had raised the issue of whether Darwinism was finally to be superseded by a modern outlook and whether this change should be signaled by revisions of terminology. T.H. Morgan, although instrumental in creating experimental genetics, argued for the retention of the central language of Darwinian theory, especially the metaphor of natural selection, even though some arguments had to be reformulated in the light of modern genetics.[5] Julian Huxley also endorsed this idea, while Ernst Mayr carried this emphasis on tradition further through analyses of Darwin's scientific originality.

The retention of Darwinian language and emphasis on Darwin's importance helped to underscore the continuing relevance of natural history in modern evolutionary biology. Naturalists insisted on the importance of natural history in contributing scientific data and insights that were missing in other approaches to evolution, especially in the mathematical style of reasoning common in population genetics. Both Huxley and Mayr, who were themselves field naturalists with special interests in ornithology, stressed the origins of the modern synthesis in systematics and natural history. In Mayr's view, the crucial intellectual transitions that made it possible to apprehend and develop Darwin's argument were achieved by naturalists and by animal and plant breeders. Where Huxley had given some credit to population geneticists for their role in establishing the modern synthesis, Mayr criticized the approach as an example of "beanbag genetics," because evolution was seen in terms of input or output of genes, which he regarded as too simplified to be a source of novel contributions.[6] He credited the mathematicians for their rigor, but he located the source of key insights firmly in the naturalists' camp. He did not see the synthesis as the result of collusion or deliberate design. Individual scientists followed their own research directions in their own ways, with many disagreements over details. Mayr's views on the history of evolution reflect his own central role in the synthesis. Indeed, his historical work helped to vindicate the naturalist's perspective.[7]

A striking feature of the literature stemming from the modern synthesis is the degree to which biologists were concerned about the larger meaning of evolution

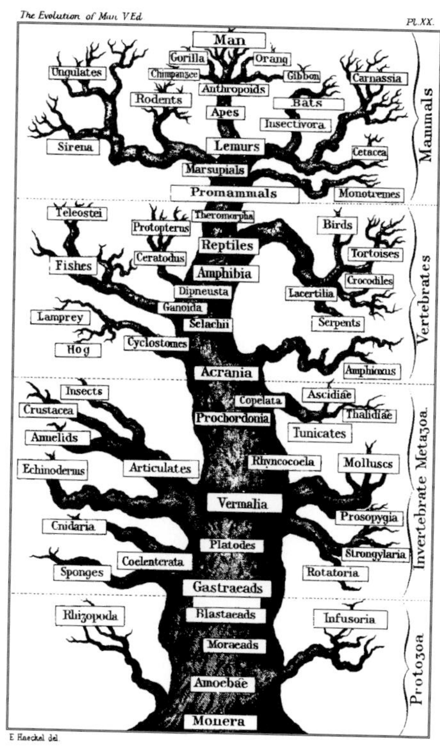

FIGURE 21.1: ERNST HAECKEL'S EVOLUTIONARY TREES WERE MEANT TO ILLUSTRATE THE UNITY OF LIFE FOR THE GENERAL AUDIENCE, BUT HIS POPULAR STYLE WAS CONSIDERED TOO SPECULATIVE BY MODERN EXPERIMENTAL BIOLOGISTS.
(SOURCE: ERNST HAECKEL, *THE EVOLUTION OF MAN: A POPULAR SCIENTIFIC STUDY*, 5TH ED., V.2, LONDON: WATTS, 1910).

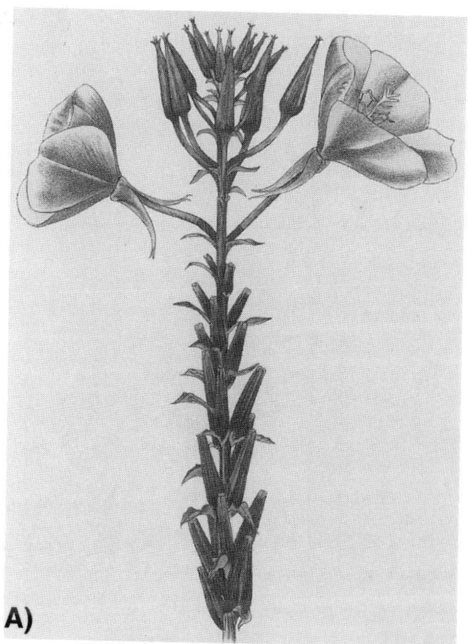

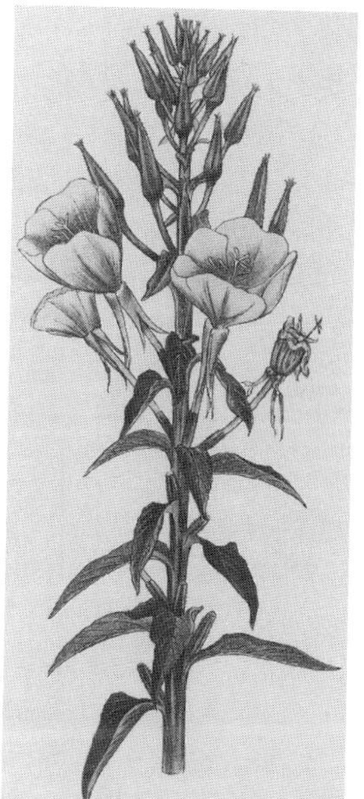

FIGURE 21.2: BOTANIST HUGO DE VRIES BELIEVED HE HAD DISCOVERED A NEW SPECIES ARISING BY MUTATION IN THE EVENING PRIMROSE. FIGURE 2A SHOWS THE PARENT SPECIES, *OENOTHERA LAMARCKIANA*, NAMED FOR THE EVOLUTIONARY THEORIST JEAN BAPTISTE LAMARCK. FIGURE 2B SHOWS A MUTANT FORM, *O. ALBIDA*, THOUGHT TO HAVE BEEN PRODUCED YEARLY BY THE PARENT SPECIES.
(SOURCE: HUGO DE VRIES, *THE MUTATION THEORY*, TRANS. J.B. FARMER AND A.D. DARBISHIRE, V. 1, CHICAGO: OPEN COURT TRADE AND ACADEMIC BOOKS, CARUS PUBLISHING, PERU. 1909.)

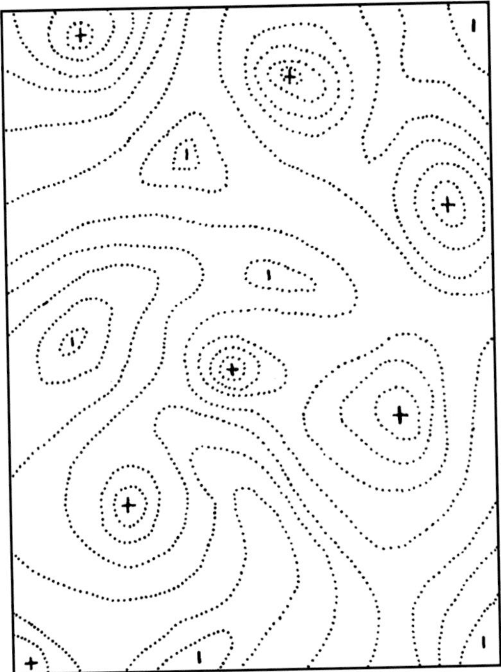

B)

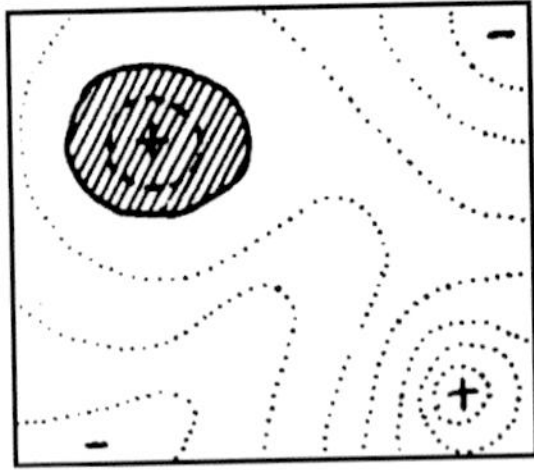

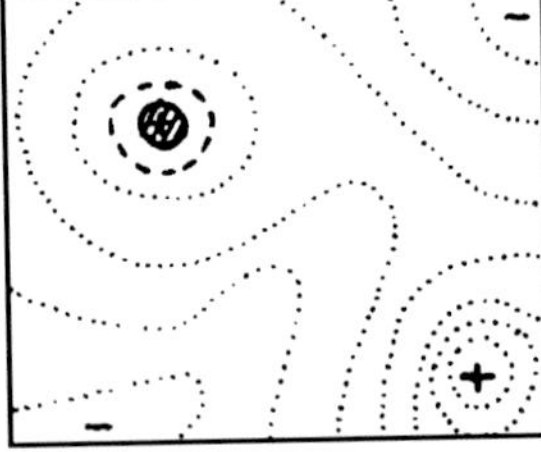

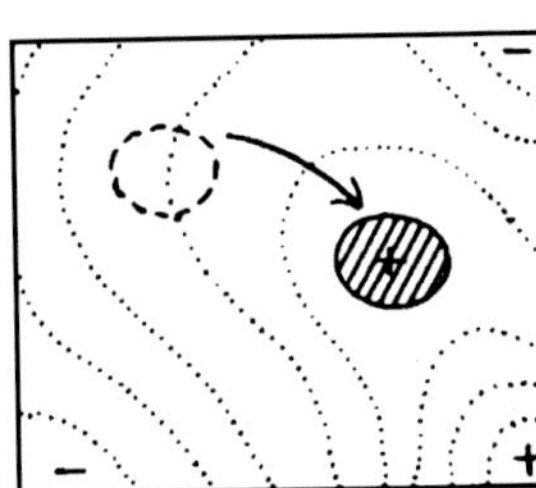

A. Increased Mutation or reduced Selection 4NU, 4NS very large

B. Increased Selection or reduced Mutation 4NU, 4NS very large

C. Qualitative Change of Environment 4NU, 4NS very large

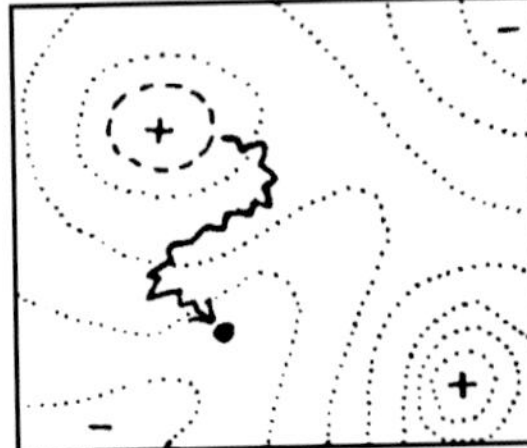

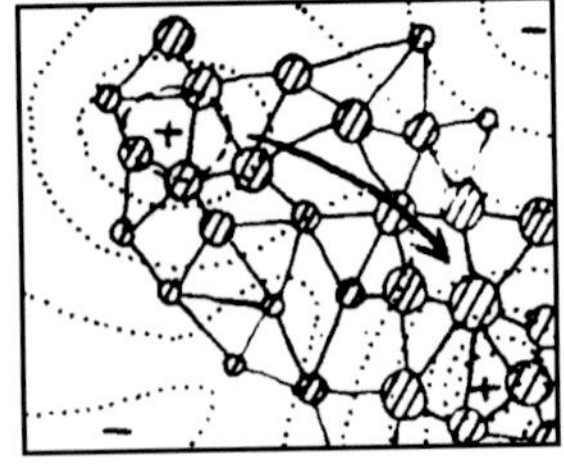

D. Close Inbreeding 4NU, 4NS very small

E. Slight Inbreeding 4NU, 4NS medium

F. Division into local Races 4nm medium

FIGURE 21.3: SEWALL WRIGHT DEPICTED THE EVOLUTIONARY PROCESS USING THE METAPHOR OF THE ADAPTIVE LANDSCAPE. A REPRESENTS A FIELD OF GENE COMBINATIONS IN TWO DIMENSIONS, WHERE SELECTION SUPPOSEDLY CARRIED THE SPECIES TO THE NEAREST PEAK. B SHOWS HOW WRIGHT THOUGHT POPULATIONS MIGHT CHANGE UNDER DIFFERENT GENETIC AND ENVIRONMENTAL CONDITIONS. (SOURCE: SEWALL WRIGHT, *EVOLUTION, SELECTED PAPERS*, EDITED BY WILLIAM B. PROVINE, CHICAGO AND LONDON: UNIVERSITY OF CHICAGO PRESS, 1986).

FIGURE 21.4: THE FINCHES OF THE GALÁPAGOS ISLANDS PRESENTED A PROBLEM OF CLASSIFICATION EVER SINCE DARWIN OBSERVED THEM, FOR SOME SPECIES DIFFER ONLY SLIGHTLY.
(SOURCE: DAVID LACK, *DARWIN'S FINCHES*, CAMBRIDGE: CAMBRIDGE UNIVERSITY PRESS, 1983).

FIGURE 21.5: EXAMPLES OF INDUSTRIAL MELANIC SPECIES OF MOTHS STUDIED BY BERNARD KETTLEWELL.
(SOURCE: BERNARD KETTLEWELL, *THE EVOLUTION OF MELANISM: THE STUDY OF A RECURRING NECESSITY*, OXFORD: CLARENDON PRESS, 1973.)

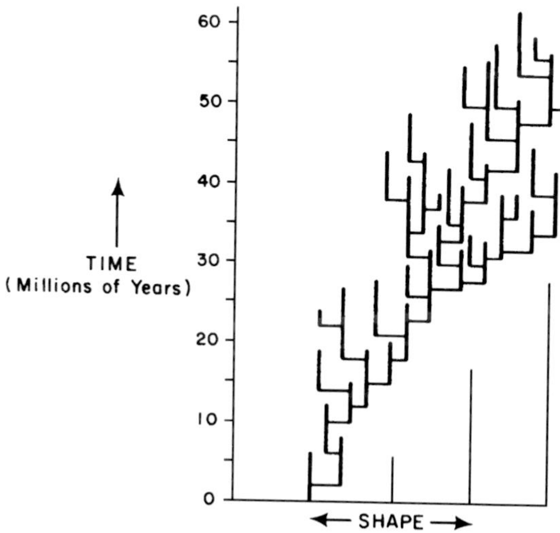

FIGURE 21.6: A HYPOTHETICAL MODEL SHOWING HOW SPECIES CHANGED OVER GEOLOGICAL TIME. SPECIES ABRUPTLY CHANGE IN SOME FEATURE IN A RELATIVELY SHORT TIME PERIOD (HORIZONTAL LINES), THEN REMAIN STABLE FOR LONG PERIODS OF TIME (VERTICAL LINES).
(SOURCE: STEVEN M. STANLEY, *THE NEW EVOLUTIONARY TIMETABLE: FOSSILS, GENES, AND THE ORIGIN OF SPECIES*, NEW YORK: BASIC BOOKS, 1981, P. 183).

for human society. In this respect they continued the debate about evolution and ethics that Darwin had stimulated. Events in the twentieth century, notably the two world wars, fueled continued discussion of the relationship between evolutionary biology, the evolution of humans as ethical beings, and the future progress of human society. In some cases biologists were seeking to regain the sense of a unified worldview that had earlier been lost. This was true of Dobzhansky for instance, who was deeply influenced by Christianity. He explored the significance of the ability to learn and to be self-reflective that had evolved as human adaptations. As humans were the only products of evolution who knew they had evolved, it was "up to man to supply the program for his evolutionary developments which nature has failed to provide."[8]

Julian Huxley, echoing the role taken in the original Darwinian debates by his grandfather Thomas Henry Huxley, stressed the need to develop a new ideology to meet the demands of the modern age, an ideology which was evolutionary, scientific, progressive, and global, yet was also concerned with the quality of life of the individual. Huxley was distressed by the pressures toward uniformity and the effects of mass culture and materialism that appeared in the 1950s. His views combined elements of nostalgia for a past age and optimism for a new future where ethics would be based on a firm scientific foundation. Developing the logic of evolutionary debates that originated in the nineteenth century, and developing broad philosophical themes that had always concerned him, he sought in effect to make modern Darwinism into a new religion.

Huxley was particularly outspoken in promoting the ideological implications of Darwinism, but he was not alone in seeking its philosophical and ultimately political meaning. Social and political issues surfaced repeatedly in the discussions that accompanied the centennial celebrations of the publication of *Origin of Species* held at the University of Chicago in November 1959.[9] This was the first time the three branches of science involved in evolutionary studies — the inorganic sciences, the life sciences, and the human sciences — had been brought together for discussion and joint criticism. The conference received wide coverage on radio and television. Concern with Cold War issues was evident and particular attention was paid to the appeal of Communism to poor, overpopulated nations and the threat of atomic war.

The relevance of modern evolutionary biology to ethics, including discussion of whether it was even possible to attain a total synthetic knowledge of the universe, concerned leading evolutionary biologists into the 1960s and many books and articles addressed this problem. These ranged from general reflections on human nature and human progress to discussions of specific issues, such as eugenics, the concept of race, the nature of aggression, the link between criminal behavior and genetics, the evolutionary differences between the sexes, and the use of IQ tests.[10]

A related trend of the 1950s was the closer relationship between biology and anthropology. With very few exceptions, cultural anthropologists in the 1930s had defined their discipline in opposition to biological approaches. By the 1950s anthropologists and biologists were finding more common ground, with biologists becoming interested in the study of social behavior and anthropologists becoming interested in evolutionary questions. However the idea that human evolution operated on a cultural plane, that is, the degree to which humans had become independent of biological evolution, was still emphasized even by biologists. This move toward a biologized anthropology would continue in the later part of the twentieth century as biologists turned their attention to the genetic mechanisms underlying social behavior.

Recent Developments and Controversies

Genetic and experimental analysis grew more sophisticated in the 1950s. Biologists introduced a genetic 'theory of relativity' with the idea that a gene did not have one absolute selective value, but a wide range of potential values that might extend from lethality to high selective superiority, depending on environmental factors and the genetic background. Experimental field research yielded observations that confounded the predictions of both geneticists and naturalists. One example is the series of studies on industrial melanism in moths (the preponderance of dark-colored moths in industrial areas polluted by soot) performed by British biologist Bernard Kettlewell in the 1950s (Figure 21.5). Geneticists were skeptical of his findings because his data showed that predation was a more effective selective agent than had been predicted theoretically. Naturalists were equally surprised at his results because they had not performed the kind of exact study needed to analyze selective pressures.[11]

In the post-war decades, an interest in mathematical approaches once again emerged. Population geneticists and population ecologists, eager to explore the wealth of new mathematical techniques that were available by the mid-1960s, disparaged naturalists as 'stamp collectors.'[12] At the same time, rapid advances in molecular biology were having a major impact on universities, squeezing out 'whole organism' biology and making any form of science reminiscent of natural history seem hopelessly out of date. Mathematical models, despite their simplifications and their deterministic nature, offered the possibility for a more analytical, hypothesis-testing style of science. Almost any suitable mathematical approach, including demography, game theory, and information theory, found its way into the evolutionary literature from the 1950s to the 1970s.

Theoretical approaches did sometimes illuminate longstanding puzzles, such as those involving social behavior. The evolution of sterile worker castes among social insects, for instance, was an outstanding problem for modern genetics. W.D. Hamilton, observing the unusual genetic relations between individuals in such species, developed an evolutionary explanation based on the concept of "inclusive fitness": the idea that fitness applies not just to individuals but to close relatives as well. John Maynard Smith developed Hamilton's arguments, using the term "kin-selection" to refer to inclusive fitness, and further developed evolutionary game theory to explain ritualized forms of aggressive behavior. He later extended game theory to explanations of peculiar sex ratios, plant growth, animal migration, male mating behavior, and the evolution of viruses.

The experimental analysis of behavior was meanwhile developing as a result of the leadership of three European naturalists, Karl von Frisch, Konrad Lorenz, and Niko Tinbergen, who shared the Nobel Prize in 1973 for their work in ethology. Von Frisch achieved renown for his study of the 'dance language' of bees, revealing the behavioral complexity of lower organisms. Lorenz's pioneering work on birds, including his discovery of imprinting, demonstrated that instinctive behavior could be subjected to concrete experiments. Tinbergen, broadly interested in the behavior of insects and birds, devised innovative ways of studying animals in the wild. The growth of ethology stimulated the field of behavioral ecology and the study of genetics in relation to behavior, opening up new research in evolutionary biology.

Connected to studies of instinct and learning, the study of social behavior from an evolutionary standpoint raised questions about the level at which selection acted. Did it act only on the individual, the closely related family group, or the larger social group? At the center of debate was the theory of "group selection" proposed in 1962 by V.C. Wynne-Edwards, who argued that selection might act on the social group as a whole and thereby promote certain forms of altruistic behavior that benefited it. His theory set off a debate about group selection and individual selection, which raged with some vehemence through the 1960s.[13]

The theory of group selection was eventually discarded, but not before it had entered into one of the most politically charged disputes of the post-war years, the debate over sociobiology. This controversy was started by the publication of *Sociobiology: The New Synthesis* (1975) by Harvard biologist Edward O. Wilson. Wilson's title echoed Huxley's book of 1942 and was intended to inaugurate a new Darwinian science of social behavior. Wilson defined sociobiology as a new discipline devoted to "the systematic study of the biological basis of all forms of social behavior, including sexual and parental behavior, in all kinds of organisms, including man."[14] It brought together a large amount of research from ethology, population genetics and population ecology, and the new ideas about group selection, kin-selection, and the evolution of altruistic behavior. In this interdisciplinary mix, Darwinian logic and the mathematical language of population genetics dominated. Social behaviors were seen as adaptive traits molded by natural selection. Sociobiology was advanced as a logical extension of Darwin's theory, but claimed to be different from past research in being methodologically more rigorous and based on modern advances in genetics and population biology.

Wilson's aggressive promotion of sociobiology and his assertion that reductionist reasoning was heuristically valuable in evolutionary biology, was also a response to molecular biology's growing dominance in the university. Wilson recalled that when James Watson came to Harvard, riding the wave of his discovery, with Francis

Crick, of the double-helical structure of DNA, the intellectual climate became very hostile to natural history.[15] These changes affected universities more widely, as molecular biology came to dominate biology departments and naturalists were made to feel third-rate. Wilson responded by dropping the word 'ecology' and proposing a reinvigorated program in evolutionary biology.

Sociobiology consisted mostly of zoology, with about ten percent of the literature in the 1970s being about humans. While Wilson's approach simply carried the trends of the modern synthesis to the next logical stage, a largely unproblematic move within zoology, it proved highly controversial when directed at human behavior. Part of the problem was that Wilson emphasized the biological basis of behavior over cultural and historical causes and tended to interpret human behavior somewhat speculatively from the point of view of the gene. Oxford biologist Richard Dawkins popularized a related version of this gene-centered approach in his best-selling book *The Selfish Gene* (1976), which was also criticized for its genetic reductionism. The ensuing controversy revolved around the question of whether a reductionist approach to behavior also implied a belief in genetic determinism. Critics sought to discredit sociobiology by attacking weaknesses in its methodology and in the logic by which explanations were constructed. Scientific critiques involved such questions as how adaptations could be identified and interpreted, the level of selection, and the appropriateness of interpreting human behavior by analogy with other animals.[16]

Despite continued resistance to sociobiology in some quarters, the evolutionary and genetic study of behavior expanded not just within biology, but also influenced the social sciences, in particular anthropology, psychology, and to a lesser extent political science. A related area of growth has been primatology, a field that had developed along several paths from the 1920s, involving both laboratory experiments and field observations on primates. The use of primates to tackle questions about parental behavior, sex differences, and aggression in humans received a boost from sociobiology. Primatology has also been the testing ground for explicitly feminist approaches to science. Interest in sexual selection, which Darwin had explored in some detail in *The Descent of Man* (1871) but which had subsequently languished as a research area, was reinvigorated in this climate. Another off-shoot of sociobiology has been renewed interest in the application of evolutionary ideas to medicine.

Overlapping with the sociobiology debate, a series of controversies raised major challenges to the modern synthesis. Systematists had been arguing from the 1960s about whether and how to depict evolutionary relations in a classification system. At issue was the reputation of taxonomy as a vibrant evolutionary science, rather than a backwater of biology. These taxonomic controversies fed into a debate sparked by a new model of evolutionary change known as the 'punctuational model' or the theory of 'punctuated equilibrium.' American paleontologist Niles Eldredge proposed the model and then developed and promoted it in collaboration with Stephen Jay Gould in the early 1970s. Other paleontologists, especially in the United States, have extended the model and strengthened its empirical basis.

The new model challenged a bias that had purportedly developed in evolutionary biology and was particularly evident in paleontology, which stressed gradual change within a lineage (phyletic evolution) rather than the relatively sudden diversification of species over a shorter period of time (speciation). Punctuationists wanted to shift the balance by emphasizing speciation over phyletic evolution. What made the new model seem especially challenging, and helped to flame the criticisms, was that it was depicted diagrammatically in such a way that the possibility of gradual change seemed to be denied (Figure 21.6). The model seemed to present itself as an alternative hypothesis that excluded views of gradual evolution, although in fact it was not so exclusionary. It did, however, contradict some cherished assumptions about gradual, linear evolutionary change. In the lively debate that ensued scientists argued over whether the new model would require a revision of the modern Darwinian synthesis, or whether it ultimately reinforced arguments that architects of the modern synthesis had already made. Mayr, for instance, had discussed years earlier how evolution might proceed rapidly in small, isolated populations, an idea fully compatible with a punctuational view of the fossil record. Eldredge, assessing the debate in the light of the history of evolutionary biology, depicted it as a battle between self-styled 'naturalists' (the punctuationists) and 'ultra-Darwinians.' Theoretical geneticists and sociobiologists came in for special criticism, while Dobzhansky's metaphor of the adaptive landscape was debunked as a pernicious "myth."[17]

One particularly controversial claim was the suggestion that human evolution was a punctuational event. The discovery of the australopithecines (members of the family Hominidae that apparently walked upright) in Africa after the First World War had raised controversies about human origins, the nature of the 'missing link' between humans and earlier primate forms, and the location of the primary center of hominid evolution. Anthropologists debated how close these hominid forms were to modern humans. The scientific literature on human evolution since the early twentieth century had emphasized the broadly progressive trends of morphological change, despite recognition of 'minor fluctuations' that caused evolution to deviate from what would otherwise have been a perfectly straight line. This general trend went from the ancestral forms of modern tree shrews, lemurs, tarsiers, New World and Old World monkeys, to anthropoid apes and finally humans. Until the 1950s fossil finds were assumed to confirm the hypothesis of a linear progression from Australopithecus to Pithecanthropus (now known as *Homo erectus*) to Homo, although Louis B. Leakey challenged the idea that the australopithecines were as close to humans in stance and behavior as had been assumed.

A series of new discoveries in Africa in the 1960s and 1970s cast more doubt on earlier assumptions of a graded morphological series. Dating of the African fossils pushed the age of hominids back to between three and four million years and raised fresh controversies about the relations of the fossil forms to each other and to modern humans. The punctuational model focused on the abruptness with which the new forms appeared, including the 'sudden' appearance of modern humans. In general it suggested an episodic sequence of events and not the gradual directional change that was assumed by scientists through most of the century.

Almost at the same time, physicist Luis Alvarez and a group of scientific colleagues in California revived a controversial explanation of extinction when they suggested that the mass extinctions of dinosaurs had been the result of an asteroid impact. Previous impact hypotheses had been rejected because they went against prevailing scientific beliefs. Alvarez's theory generated a tide of scientific publications on various aspects of the controversy and received a high level of popular attention. The idea of cataclysmic events producing mass extinctions on earth gained credibility as more research accumulated. Gould remarked that although this controversy highlighted the importance of paying attention to unique historicial events that might have a major effect on the course of evolution, it also exerted a negative impact on historical disciplines and on natural history because Alvarez disparaged paleontologists for not doing "real science."[18] Echoing similar defenses in other evolutionary controversies, Gould reiterated the importance of the natural history disciplines in advancing understanding of the evolutionary process.

In addition to these controversies, new evidence of the importance of evolutionary processes on the short-term ecological time scale came to light as a result of human interference with nature. Examples of human impact include industrial melanism, the evolution of pesticide resistance in pests, or antibiotic-resistant strains of pathogens, the effect of pesticides on the reproduction of non-target populations, the susceptibility of newly developed crops to rapidly evolving pathogens and parasites, and heavy metal tolerance in plants. These examples illustrated the need to consider the historical effects of human activity on the evolution, distribution, and abundance of species in the wild. One of the trends of late-twentieth century biology has been to try to bring humans into the picture as part of nature, a task made difficult by the way scientific method requires one to objectify the subject matter.

In ecology, scientists emerged from a period of enthusiasm with mathematical modeling and in the 1980s turned toward a more descriptive methodology, paying particular attention to the problem of historical contingency. Ecological and evolutionary writing began to take into account the full complexity of nature. Attention was drawn to the 'new natural history,' a reference to a more subtle, detailed, and interdisciplinary approach to natural history. These ecologists concerned themselves largely with problems of species diversity, asking questions about patterns of distribution and abundance and looking for evidence of the impact of specific local events, including the impact of human activity, on animal and plant populations.

In addition to drawing on population biology and evolutionary ecology, the new natural history intersected with ecosystem ecology. The term 'ecosystem' was meant to draw attention to the complex cycling of matter and energy between living and non-living parts of nature. American ecologist Eugene Odum proposed using the ecosystem concept in the 1950s as a way of organizing and unifying ecological research, and potentially giving the subject more credibility. Broad, interdisciplinary studies directed at the analysis of key ecosystems in the 1960s and 1970s had mixed success, but in general all led to greater appreciation of the complexity of ecological relationships. Some of the more focused studies revealed new threats to species, such as the discovery of the impact of acid rain on organisms. After the first major round of ecosystem studies had ended, calls resumed in the 1990s for increased funding of large-scale interdisciplinary projects in ecology and allied earth sciences. These were tied directly to the same concerns about extinction and the loss of biological diversity that had stimulated the 'new natural history,' but now scientists were challenging traditional assumptions about the 'balance of nature' and drawing attention to the role of historical accident in the evolution of ecosystems.

The emphasis on the preservation of biological diversity that emerged in the 1970s encouraged greater efforts to find and catalogue new species and to devise effective strategies for preserving regions that served as sources of diversity. Conservation biology arose as a new discipline aiming to stem the loss of biological diversity. Public awareness of the extent of this loss increased and scientists argued that such problems must be confronted through continued scientific monitoring of biological diversity. In this climate evolutionary biology was invigorated as a discipline that could serve not only to order and explain the natural world, but that would perhaps have a crucial role in saving it.

This argument had an aesthetic side as well, which appears more in the popular scientific literature. E.O. Wilson, drawing on the logic of sociobiology, argued that the evolutionary process has given us an in-built love of nature and a deep need for certain kinds of natural surroundings.[19] Therefore in preserving nature we fulfil the evolutionary imperative within us; a fulfillment that has reached its highest expression in the scientific study of nature. The aesthetic satisfaction that comes from studying nature closely continues to motivate the pursuit of natural history, irrespective of any practical considerations, adding a spiritual dimension to modern Darwinism in which we can discern faint but clear echoes of the religious sensibility that often infused and motivated the study of natural history in Darwin's time.

Conclusion

Since the beginning of the century scientists studying evolution have had several overlapping goals: to test and validate Darwin's insight into the evolutionary process; to eludicate the mechanisms by which evolution occurs in different species; to extend evolutionary reasoning to the human sciences; and to control the evolutionary process by improving human health and welfare and mitigating the destructive effects of humans on other species and ecosystems. The history of Darwinian biology has also been an evolving dialogue between three scientific approaches, those of natural history, experimental biology, and theoretical (or mathematical) biology. The controversies to which these dialogues have given rise reflect the multiple perspectives that can be brought to the study of evolution.

Running through most of these controversies is the problem of how to apply scientific methods of analysis to historical processes. The major intellectual challenge has been to find a way to be rigorous in analysis, yet sensitive to the element of contingency and unpredictability that characterizes any historical process. Meeting these challenges has involved constant examination of many of the basic concepts of evolutionary thought (such as 'adaptation,' 'competition,' and 'fitness'), close scrutiny of the metaphors in which evolutionary discussions are expressed, and frequent coining of new terms to accommodate subtle distinctions in the way evolution operates in different species. It has also involved active participation by scientists in writing the history of their disciplines. This reconstruction of history is not simply a celebratory exercise, but serves as an important rhetorical tool that justifies a scientist's perspective, method, or argument.

Natural historians have fought a constant battle to justify the value of their work in the face of the reductionist and deterministic approaches to science that have characterized the development of biology in the twentieth century. At times they have argued that a reductionist approach, such as the genetic interpretation of social behavior, is an important heuristic tool that can increase the rigor of the naturalist disciplines. At other times they have defended the importance of historical inference in biology and argued that such historical approaches simply cannot be displaced without leading to serious blunders.

One of the most profound results of modern research has been to show that evolution is not goal-directed, and that humans are not the inevitable products of a progressive trend, but are, like all other species, the accidental products of a blind historical process. Ironically, even as the evidence of neo-Darwinism mounted toward this conclusion, scientists resisted its implications and continued to seek signs of progress and ethical significance in the fact of evolution. At the end of the twentieth century, as scientists achieve more insights into the contingent nature of this historical process, they continue to grapple with the implications of these insights, especially in view of the potentially catastrophic effect of environmental changes on the survival of species.

Reference

Research for this essay was partly supported by a grant from the National Science Foundation, Program in History and Philosophy of Science (SBR-9320310).

1. E.B. Poulton, *Essays on Evolution, 1889–1907.* (Oxford: Clarendon Press, 1908), p. xx.
2. Vernon L. Kellogg, *Darwinism Today.* (New York: Henry Holt, 1907); A.C. Seward (Ed.), *Darwin and Modern Science: Essays in Commemoration of the Birth of Charles Darwin and of the Fiftieth Anniversary of the Publication of the Origin of Species.* (Cambridge: Cambridge University Press, 1909).
3. Poulton, cit. n. 1., p. xv.

4. David Lack, *Darwin's Finches.* (Cambridge: Cambridge University Press, 1983). (Reissue of original edition of 1947, with a new introduction and notes.)
5. Thomas H. Morgan, *What is Darwinism?* (New York: W.W. Norton, 1929).
6. Ernst Mayr, "Where Are We?" *Cold Spring Harbor Symposia in Quantitative Biology* (1959), **24**: 1–14. Quoted on p. 2.
7. Ernst Mayr, *The Growth of Biological Thought: Diversity, Evolution, and Inheritance.* (Cambridge, Mass.: Belknap Press, 1982); Ernst Mayr and William B. Provine (Eds.), *The Evolutionary Synthesis: Perspectives on the Unification of Biology.* (Cambridge, Mass.: Harvard University Press, 1980).
8. Theodosius Dobzhansky, "Evolution at Work," *Science* (1958), **127**: 1091–1098, quoted p. 1098.
9. Sol Tax and Charles Callender (Eds.) *Evolution after Darwin: Issues in Evolution,* 3 vols. (Chicago: University of Chicago Press, 1960).
10. Theodosius Dobzhansky, *Mankind Evolving: The Evolution of the Human Species.* (New Haven and London: Yale University Press, 1962); idem, *Heredity and the Nature of Man.* (New York: Harcourt, Brace & World, 1964); C.H. Waddington, *The Ethical Animal.* (London: George Allen and Unwin, 1960); George G. Simpson, *The Meaning of Evolution: A Study of the History of Life and of its Significance for Man.* (New Haven: Yale University Press, 1949); Gregg Mitman, *The State of Nature: Ecology, Community, and American Social Thought, 1900–1950.* (Chicago and London: University of Chicago Press, 1992).
11. Bernard Kettlewell, *The Evolution of Melanism: The Study of A Recurring Necessity.* (Oxford: Clarendon Press, 1973), p. 121.
12. Richard C. Lewontin (Ed.), *Population Biology and Evolution.* (Syracuse, New York: Syracuse University Press, 1966), p. 2.
13. George C. Williams, *Adaptation and Natural Selection: A Critique of Some Current Evolutionary Thought.* (Princeton, New Jersey: Princeton University Press, 1966).
14. Edward O. Wilson, "What is Sociobiology?" *Society* (1978), **15**: 10–14, quoted p. 10.
15. Edward O. Wilson, *Naturalist.* (Washington, D.C.: Island Press, 1994), Chapter 12.
16. Arthur L. Capland (Ed.), *The Sociobiology Debate: Readings on the Ethical and Scientific Issues Concerning Sociobiology.* (New York: Harper & Row, 1978); Howard L. Kaye, *The Social Meaning of Modern Biology, from Social Darwinism to Sociobiology.* (New Haven and London: Yale University Press, 1986); Philip Kitcher, *Vaulting Ambition: Sociobiology and the Quest for Human Nature.* (Cambridge, Mass.: MIT Press, 1985).
17. Niles Eldredge, *Reinventing Darwin: The Great Debate at the High Table of Evolutionary Theory.* (New York: John Wiley & Sons, 1995); Niles Eldredge and Ian Tattersall, *The Myths of Human Evolution.* (New York: Columbia University Press, 1982); Steven M. Stanley, *The New Evolutionary Timetable: Fossils, Genes, and the Origin of Species.* (New York: Basic Books, 1981); Albert Somit and Steven A. Peterson (Eds.), *The Dynamics of Evolution: The Punctuated Equilibrium Debate in the Natural and Social Sciences.* (Ithaca and London: Cornell University Press, 1989).
18. Stephen Jay Gould, *Wonderful Life: The Burgess Shale and the Nature of History.* (New York: W.W. Norton, 1989), pp. 280–281.
19. Edward O. Wilson, *The Diversity of Life.* (Cambridge, Mass.: Harvard University Press, 1992).

Further Reading

Adams, Mark B., *The Evolution of Theodosius Dobzhansky: Essays on His Life and Thought in Russia and America.* (Princeton, N.J.: Princeton University Press, 1994).

Bakker, Robert T., *The Dinosaur Heresies: New Theories Unlocking the Mystery of the Dinosaurs and Their Extinction.* (Kensington Publishing Corp., 1986).

Bowler, Peter J., *The Eclipse of Darwinism: Anti-Darwinian Evolution Theories in the Decades around 1900.* (Baltimore and London: Johns Hopkins University Press, 1983).

Dawkins, Richard, *The Blind Watchmaker: Why the Evidence of Evolution Reveals a Universe without Design.* (New York: W.W. Norton, 1986).

Diamond, Jared, *The Third Chimpanzee: The Evolution and Future of the Human Animal.* (New York: Harper Perennial, 1992).

Glen, William (Ed.), *The Mass-Extinction Debates: How Science Works in a Crisis.* (Stanford: Stanford University Press, 1994).

Gould, Stephen Jay, *Ontogeny and Phylogeny.* (Cambridge, Mass.: Harvard University Press, 1977).

Haraway, Donna, *Primate Visions: Gender, Race, and Nature in the World of Modern Science.* (New York and London: Routledge, Chapman, & Hall, 1989).

Kevles, Daniel J., *In the Name of Eugenics: Genetics and the Uses of Human Heredity.* (Berkeley and Los Angeles: University of California Press, 1985).

Lewin, Roger, *Bones of Contention: Controversies in the Search for Human Origins.* (New York: Simon and Schuster, 1987).

Maynard Smith, John, *The Theory of Evolution.* (Penguin, 1958).

Nesse, Randolph M. and Williams, George C. *Evolution and Healing: The New Science of Darwinian Medicine.* (London: Weidenfeld and Nicolson, 1995).

Provine, William B. *Sewall Wright and Evolutionary Biology.* (Chicago and London: University of Chicago Press, 1986).

Shipman, Pat, *The Evolution of Racism: Human Differences and the Use and Abuse of Science.* (New York: Simon and Schuster, 1994).

Thorpe, W. H., *The Origins and Rise of Ethology: The Science of the Natural Behaviour of Animals.* (London: Heinemann, 1979).

Weiner, Jonathan, *The Beak of the Finch: A Story of Evolution in Our Time.* (New York: Vintage Books, 1995).

Waters, Kenneth C., and Albert van Helden (Eds.), *Julian Huxley: Biologist and Statesman of Science.* (Houston: Rice University Press, 1992).

CHAPTER 22

Clinical Research

CHRISTOPHER LAWRENCE

Clinical research in the sense of investigating the patient at the bedside with a view to recording findings of interest to other healers could be said to be as old as medical writings. The Enlightenment in particular witnessed the widespread reporting by doctors of clinical observations with a view to promoting medical knowledge. Clinical research, however, in the sense of a bedside experimental enterprise, often relying on laboratory work, undertaken by clinicians as a significant dimension of a medical career can be traced only to the mid-nineteenth century, notably to Germany. Here doctors could aspire to paid, university teaching positions and have hospital beds at their command and laboratory facilities at their disposal. This was the teaching and research model that reformers endeavored to introduce wherever orthodox medicine was practised. It is usefully called academic medicine. In the late twentieth century, in the West, clinical research is now overwhelmingly carried out by full-time or part-time funded clinicians in university hospitals usually in association with scientists and technicians and nearly always involving laboratory work. Funded clinical research was also taking place at the end of the nineteenth century in so far as pharmaceutical companies in Britain, America and Germany established their own laboratories and tested the drugs developed within them.

A significant agent in the growth of medical research in general and clinical research in particular in the twentieth century has been funding by the state and philanthropic foundations. Such funding was the material dimension of an ideology which proclaimed that research in the basic sciences, the training of doctors in these subjects and the grounding of clinical medicine in the laboratory were the most effective means for improving the health of the population. In this respect the growth of medical research in the twentieth century has not been a simple quantitative increase of pre-existing practices. Rather it has been part of a wholescale reorganization of medicine in which university academic departments and their associated hospitals have been established as the apex of a hierarchy of laboratory-based medical services. Such restructuring has involved a fundamental reorganization of medical work: an increased integration and division of medical labor. This restructuring which began at the turn of the century was not peculiar to medicine. It was carried out in many institutions, for example those of industry and education, in conformity with the dictates of 'scientific management.' In Europe and North America the goal of scientific management was the deployment of science in the cause of national efficiency. In the case of medicine the clinical encounter was perceived as the most suitable point for science to exert its beneficent action. By the early years of the twentieth century this encounter was perceived not simply as an occasion for healing sickness but as a significant moment through which social progress could be generated. The significance of the clinical encounter in the West at this time can be measured in various terms: politically, by the increasing use of a language of 'rights' to medical treatment and morally, in a new language of duties and obligations to provide medical care. Its ongoing significance can also be measured economically by looking at expenditure on health — actually disease — in Western countries. Such expenditure, much of which goes on research, not only highlights the significance of the encounter but also indicates that particular dimensions of the encounter have themselves been privileged. Acute diseases of children or wage-earners have been the focus of much clinical research. The main (although not sole) beneficiaries of investment in clinical medicine have been disease processes amenable to specific therapeutic intervention, by surgery or drugs, rather than chronically debilitating and diffuse disorders or disturbances to which environmental manipulation might be a solution. Clinical research in the twentieth century has thus had built into it a series of assumptions about what sort of thing diseases are and what sort of diagnostic and therapeutic technology might be the appropriate outcome of any investigation.

Clinical research is also based on other explicit assumptions which have been contested. Throughout the century anti-vivisectionists have condemned the use of animals in the furtherance of medical knowledge. Two major arguments have been used. First that humans have no right to use animals for such purposes, but have a duty of custodianship towards them. Second that experiments conducted on animals do not illuminate human pathology. This last argument has sometimes been used by factions within the medical community itself. In the second half of the century the claim that animals have 'rights' has increasingly been made. Clinical research also necessarily involves experiments on human beings. In a number of instances experimenters have exposed themselves to real dangers. For instance at the beginning of the century members of the American Yellow Fever Board in Cuba allowed themselves to be bitten by mosquitoes carrying the disease. Much clinical research in the century has been conducted on patients after their fully informed consent was obtained. Sadly there is also a significant history of researchers taking advantage of vulnerable groups to conduct clinical experiments, the most infamous case of which occurred in a town in Alabama. In the "Tuskegee Syphilis Study" the course of untreated syphilis was observed in uninformed Negro men from 1932 until 1972, many years after the introduction of penicillin.

In the following I will concentrate on clinical medicine in Britain and North America but the generalizations apply to all areas where Western medicine is dominant. As a rough and ready guide it can be said that before the First World War there was little coordinated clinical research and individual initiatives predominated. In both Britain and North America experimental physiology, bacteriology and morbid anatomy were the basic sciences to which clinicians most often turned in order to elucidate bedside problems. In the inter-war years the university hospital began to be established as a significant site of clinical research and the funding of clinical appointments began on a substantial scale. Biochemistry was the science to which clinicians looked for insights, particularly in America. Most research still remained largely individual, however, and was carried out with relatively simple apparatus. The years after the Second World War saw an explosion of clinical research and a massive growth of academic medical departments. Funded clinical appointments grew rapidly. Drug company investment in research boomed. Specialization, incipient in the inter-war years, flourished. Teamwork became a significant feature of clinical investigation. Molecular biology, immunology and genetics were the sciences which clinicians saw as holding the solutions to bedside problems. The last two decades of the century have seen some concern that academic medicine and clinical research are in relative decline.

North America 1900–1914

Before the Great War most clinical research was carried out by private practitioners, usually with hospital appointments. For example, in 1910 the physician James Herrick in Chicago, working alone or with one or two other investigators on a casual basis, reported microscopic investigation of blood cells and, two years later, on electrocardiographic studies of heart disease. Herrick's work is regarded as having a significant place in the establishment of sickle cell anaemia and acute myocardial infarction as distinct diseases.

Philanthropy slowly changed this approach to the study of sickness and has played an important part in the funding of clinical research in America. In one sense, however, the creation of clinical research as a significant area of medical work was simply one aspect of the wholescale scientific reorientation of American higher education which capitalist finance procured at the turn of the nineteenth century. During this period, the so-called Progressive Era, new engineering and technical schools were established and the wealthiest men and women in the country founded universities, for example, Johns Hopkins, Tulane, Clark, Vanderbilt, Stanford and Cornell. Reformers who targeted medicine aimed at making medical schools and hospitals into university-based centers of research and seats of scientific education. The flagship of the new medicine was the Johns Hopkins medical school in Baltimore. The school and its hospital were opened for MD degree candidates in the autumn of 1893. Here an array of talented figures created an institution which was widely admired and adopted as a model for reform elsewhere. Its first professors included the physiologist, H. Newell Martin, a disciple of Michael Foster, the creator of the famous Cambridge school of physiology in England. The widely respected William Welch was professor of pathology and William Osler professor of the theory and practice of medicine. Welch, Osler, the anatomy professor, Franklin Mall, and Osler's successor Lewellys Barker had all experienced German medical education at first hand and were keen to introduce German practices into American medicine. W.S. Halsted, the German trained surgeon-in-chief and later professor of surgery, established Hopkins as a site of pilgrimage for ambitious American and foreign surgeons. At Hopkins' Hunterian Laboratory for Experimental Surgery and Medicine, Halsted's assistant, Harvey Cushing, carried out studies on the pituitary gland which were soon acknowledged worldwide.

Admired though the Hopkins School was, there was severe internal dissent over the best way to promote clinical research. The school's teachers were bitterly divided over the 'full-time' issue. At its opening Hopkins was unique. All its basic science teachers were full-time professionals devoted only to research and teaching. There was a full-time German-trained physiologist at Harvard, Henry Pickering Bowditch. Bowditch, however, had private means to supplement the small payment he received. Elsewhere in America and in many British institutions the basic sciences were taught by clinicians usually otherwise engaged in private medical practice. On the hospital wards this was universally the case. Bedside medicine was taught by men who spent most of their time in practice outside hospitals which were, in the main, charitable institutions devoted to providing medical care for the poor. At Hopkins, however, Welch and Franklin Mall were strongly in favor of a university system of clinical departments in which the professors would be full-time, not permitted to practise privately, spending their days caring for hospital patients, teaching and carrying out clinical research. Most of the clinical professors, notably Osler, were firmly opposed to this plan. Osler's perceptions were much like those of doctors in Britain where a medicine of genteel charitable service and lucrative individual private practice were much admired. In 1905 Osler became Regius professor of medicine at Oxford, England.

Philanthropic money, notably Rockefeller money, eventually ensured the establishment of the full-time system. The philanthropic affairs of the massively rich John D. Rockefeller, Sr. were managed from 1891 by a Baptist minister, Frederick T. Gates. It was Gates who had interested Rockefeller in funding educational ventures. His first significant incursion into this sphere being a gift of $600,000 to help create the University of Chicago. Gates had certainly been interested in medical matters since the early 1890s, and, by his own account, endeavored to learn more by reading Osler's *Principles and Practice of Medicine.* Quite what impression Osler's book made on Gates is difficult to discern, yet in all his later accounts he credited it with firing him with enthusiasm for scientific medicine and the founding of an institute for medical research. Gates got Rockefeller Jr. (Sr. was a committed homeopath) interested in such a project and Welch, the Hopkins pathology professor, was drawn in as an advisor. In 1901 the Rockefeller Institute for Medical Research was incorporated. At first the Institute was simply a funding body dispersing grants to researchers in various institutions. It was also the home of the *Journal of Experimental Medicine.* Gates was dissatisfied with the Institute's status as simply a funding body and largely through his efforts the Rockefeller Institute Laboratories were dedicated in New York in 1906. The laboratories were presided over by Simon Flexner formerly professor of pathology at the University of Pennsylvania. A small hospital devoted to clinical research was opened alongside the Institute in 1910. Staff at the Institute and its hospital were full-time and forbidden to engage in private practice. Rufus Cole, former director of the biological laboratory at Johns Hopkins, developed the Rockefeller hospital as a significant center for the training of clinical researchers. Under Cole's auspices a concerted clinical and experimental research program into the immune responses in pneumonia was inaugurated, the aim being to develop a curative serum. Serum and vaccine therapy at this time were regarded as two of medicine's most promising avenues of therapeutic intervention. Pneumonia was a source of concern since, not only was it a killer of the elderly, it was prevalent among young, able-bodied males, the core of the workforce. Likewise, syphilis, also perceived as a major threat to the nation's vitality, was the object of experimental enquiry. Rockefeller did not limit his gifts to his own institution. In 1901 Harvard received $1,000,000 for new research buildings at the medical school and Johns Hopkins got $500,000 after the Baltimore fire of 1905.

By the early years of the new century a small number of American physicians had developed a strong enough sense of the importance of clinical research to form an organization. The first annual meeting of the American Society of Clinical Investigation was held in 1909. The inspiration behind the society was Samuel James Meltzer: Russian born, German trained and head of the Rockefeller Institute department of experimental physiology and pharmacology. Twenty-two men were invited to the first meeting.

By the outbreak of the First World War there were a number of centers of successful clinical research in America. This followed from educational reform, philanthropic funding and the adoption of the German model in modified form. With the adoption of this model came not only a particular account of the role of science in medicine but an element of competition. Those American medical institutions which perceived themselves as forward looking sought out the most distinguished basic scientists and clinicians and rewarded them, in the first case with salaries, in the second with facilities. This strategy brought fame and high quality students and encouraged other schools to take this path.

Not all American medical schools were forward looking however. Nor was the Rockefeller Foundation the only philanthropic organization to have a significant effect on the shape of twentieth-century medicine. The Carnegie Foundation for the Advancement of Teaching had an important input too. In 1908 the Foundation funded Abraham Flexner (the younger brother of Simon), a man with great experience of education in America and Europe (although not himself medically qualified), to conduct an investigation into American medical education. The famous Flexner report appeared in June 1910.[1]

The fame of the Flexner report lies in its exposure of the appalling standards of medical education in some American medical schools, when measured against Hopkins which was Flexner's ideal. Flexner came out firmly in favor of rigorous training of medical students in the laboratory sciences, an option many distinguished physicians thought unnecessary or unhelpful, viewing the hospital and the post-mortem room as all that were needed for medical training. Even among physicians who were very keen on the basic sciences there were some who considered Flexner had over-emphasized the place of science in medicine. Flexner also argued that all medical schools should be university departments and that clinical professors should be rigorously trained in science.

Great Britain 1900–1914

In 1900 nothing quite like the Hopkins model of medical education and research existed in Britain, although the University of Edinburgh and University College London with their associated hospitals could probably most justifiably have claimed to be closest to the pattern. Nor was there yet money on the Rockefeller scale for medical research. There were similarities between Britain and North America, however. Both had full-time basic scientists teaching and researching in some of their medical schools. In neither place were full-time professors of medicine in charge of departments. To a great extent clinical research in Britain before the First World War was conducted by enthusiastic clinicians in their spare time, for example, the work of James Mackenzie on the heart or Archibald Garrod on inborn disorders of metabolism. There were few research scholarships that would enable graduates to devote time to clinical investigations. In 1909, the Beit Memorial Fellowship was established by the philanthropist Otto Beit as a memorial to his brother Alfred. Thomas Lewis was the first fellow to be elected. Lewis' electrocardiographic studies of the heart's conduction pathways, which were carried out on the wards of University College Hospital in conjunction with experiments on laboratory animals, made him world famous. Lewis was an exception among medical researchers for although basic research was being organized and funded by charitable organizations at this time (notably in the Imperial Cancer Research Fund, established in 1902) clinical research remained at the level of individual initiative. One other exception to this was work carried out at St. Mary's hospital where Almroth Wright had a large-scale, integrally organized laboratory and clinic devoted to vaccine production and testing.

The second decade of the twentieth century saw the initiation of changes in British medicine comparable to those ushered in by the Flexner Report in America. British medical reformers wanted more science in medicine and stronger university connections. Two events were particularly significant in making for change. First the appointment in 1909 of a Royal Commission on University Education in London under the chairmanship of Lord Haldane. The Commission sat for two years from 1910 to 1912 and the Haldane Reports were published between 1910–1913. Second, the founding of the Medical Research Committee in 1911.

London University was an umbrella institution to which the medical schools of the great old London hospitals, such as St Bartholomew's, Guy's and St Thomas's all belonged. Nonetheless, these hospital medical schools acted to a large extent as private institutions having little involvement with the university. Science in these hospital schools was largely equated with the basic sciences (which in many instances were still taught by part-time clinicians) or, on the ward, with pathology, the morbid anatomy of the parts studied with the naked eye or with the microscope. Clinical teaching was in the hands of clinicians who made their living in private practice. Clinical investigation coupled with laboratory-based experiment was not common in these institutions and was the prerogative of those clinicians who had an interest and sufficient private funding to support it.

The Haldane Commission besides hearing local opinion also heard evidence from Abraham Flexner who, by this time, had visited Germany and was much impressed by the organization of medicine there. In his report *Medical Education in Europe* of 1912 Flexner approved of the "practical" bias of British medical education but condemned its resistance to the "scientific attitude" and "modern methods of investigation."[2] The Haldane Commission was strongly in favor of changing things in the sorts of directions Flexner had recommended for America, that is making medical education university-based and establishing academic departments in clinical subjects where teaching and research could be carried out on patients under professorial supervision.

Specialization

One of the most striking features of the growth of clinical research during this period and central to it in the inter-war years and after was the increase in medical specialization. Specialization is often construed as a natural process, a passive response to a quantitative growth of medical knowledge. The reverse, however, is much nearer the case. In both Britain and America during these years a small number of clinicians deliberately created medical specialities by transforming the shape of medical knowledge. Clinical investigations did not simply pile up facts about disease at random, quite the opposite. New ideas about disease shaped the salient facts about what constituted a disorder. Take the case of cardiology. This speciality was created by a small number of clinicians who framed new concepts of heart disease. Around 1900 heart disease was usually thought of in morbid anatomical terms, that is in terms of changed structures, such as deformed valves. Indeed chest diseases of all sorts were thought of primarily in structural terms. Thus the heart and chest in general were investigated by identical clinical methods:

auscultation (listening with a stethoscope), percussion (tapping the chest) and, increasingly, radiography, which techniques could all reveal changes in the physical conformation of the parts. The physicians who dealt with heart diseases were not specialists but generalists who dealt with all disorders which came their way. Many of them despised specialization as quackery. These physicians regarded the tools and concepts of general medicine as sufficient to address cardiac problems. In conception and practice there was nothing *special* about the heart and its diseases.

Gradually, however, during the first decades of the twentieth century, a small number of clinicians, usually with academic connections, redefined the heart and its diseases. In doing so they utilized concepts and assumptions which had been formulated in experimental physiological laboratories not in the hospital's post-mortem room. The heart's muscle, they taught, had its own peculiar properties: irritability, rhythmicity and tonicity. These properties, they said, were the basis of normal cardiac function and in disturbances of them lay the origins of cardiac disease. These clinicians, of whom Thomas Lewis was the best known, investigated and demonstrated these properties in the normal and the diseased heart at the bedside. To do this they used devices, notably the polygraph and later the electrocardiogram, which were modifications of technologies employed in physiology laboratories. Thus the growth of cardiology as a speciality was predicated on the creation of the heart as a unique object defined in terms ultimately given meaning by the experimental laboratory sciences. Such were the ways in which researchers rooted the clinic in experimental medicine. The day-to-day demonstration of the heart's normal functioning or of its disorders was increasingly embedded in a set of specialized practices many of which were dependent on specialized technologies often only available in hospitals. Thus, just as cardiology was made into a speciality, so too was the hospital privileged in the emerging medical hierarchy. The appearance of cardiology thus involved a transformation of knowledge and a reorganization of medical work. By creating themselves as specialists, cardiologists limited the role of general practitioner to a source of referral. Similar arguments could be made for the development of other specialities, such as gastroenterology.

North America 1914–1945

These years saw philanthropic money pour into medical education and medical research. Research and education were regarded by reformers as inextricably related. Philanthropic foundations stressed the importance of the laboratory sciences in medical education and regarded researchers as the best teachers. Research itself was seen as an appropriate part of medical training. Medicine and philanthropy in the United States in this period largely meant Rockefeller money. One author has observed that between 1910–1935 "Rockefeller philanthropies were responsible for 90–95 per cent of all foundation money going into medicine."[3] One significant feature of clinical research during the inter-war years was an increasing division of medical labor. Paradoxically this division was typified by teamwork; the attempt by clinicians to coordinate their work to produce large-scale studies. This was particularly significant in the field of clinical trials which were being developed in this period. As may be imagined, however, clinicians who largely regarded bedside medicine as a field for individualism to flourish often found collaboration difficult.[4]

Following the publication of his Report, Flexner was commissioned by Gates to visit and report on the Johns Hopkins medical school. Flexner recommended reorganization of the clinical faculty on a full-time basis. Gates was enthusiastic and against quite strong opposition from some of the Hopkins clinicians was able to introduce it by pouring Rockefeller money into the school. In 1914 Theodore Janeway, who had undertaken clinical research on heart disease, became the first full-time professor of medicine at Hopkins. In the three years he served at Hopkins he was assisted by a string of resident physicians who went on to distinguished research careers. Over the next twenty years dollars from Rockefeller and other foundations were used to create full-time departments at such places as Yale, Rochester, Chicago, Harvard, Cornell, Columbia and McGill in Montreal. The move to full-time was one of the most significant innovations in clinical research. It created a profession of clinical scientists comparable to that of the basic scientists. Although generalism remained an important ideal for American physicians, specialists tightened their grip. A certification system created in 1936 recognized the discipline of internal medicine and in 1940 four of its sub-specialities inaugurated their own examinations.

The inter-war years saw the United States become the site of pilgrimage for clinical researchers from all over the world. A number of schools including Chicago, Hopkins, Cornell, Yale, Harvard, Rochester, Minnesota, and Western Reserve were acknowledged as pre-eminent. Central to this growth of American dominance was the hospital of the Rockefeller Institute under Cole and the second director, Thomas Rivers, appointed in 1937. Rivers' own work was widely acknowledged as having contributed substantially to the understanding of viral diseases. Many of the men (and later women) who would later dominate clinical research were trained at the hospital. Cole organized the hospital so that researchers had the opportunity to select a specific disease and have both clinical and laboratory facilities at their disposal. His objective was to train men and women to become heads of university-based departments of medicine. Thus at Vanderbilt University in the mid-South one of Cole's former resident physicians, George Canby Robinson, was selected as Dean in 1920. Here Rockefeller and Carnegie money was poured into a scheme to create a closely integrated hospital and medical school so that, for example, the wards, the department of medicine, the hospital clinical laboratories and the preclinical department of bacteriology were all within short walking distance. At Vanderbilt the surgeon, Alfred Blalock availed himself of these sorts of facilities to combine his observations of patients on the wards and in the operating theater with laboratory studies of dogs with experimentally-induced surgical shock.

FIGURE 22.1: STAFF AT WORK IN A BACTERIOLOGY LABORATORY, C. 1939. © ST. BARTHOLOMEW'S HOSPITAL ARCHIVES DEPARTMENT.

FIGURE 22.2: STAFF AT WORK IN A BACTERIOLOGY LABORATORY, C. 1939. © ST. BARTHOLOMEW'S HOSPITAL ARCHIVES DEPARTMENT.

FIGURE 22.3: STAFF AT WORK IN A BACTERIOLOGY LABORATORY, C. 1939. © ST. BARTHOLOMEW'S HOSPITAL ARCHIVES DEPARTMENT.

FIGURE 22.4: WELLCOME TROPICAL RESEARCH LABORATORIES IN KHARTOUM.

FIGURE 22.5: WELLCOME TROPICAL RESEARCH LABORATORIES, KHARTOUM: MAIN LABORATORY. PHOTOGRAPH, C. 1920.

FIGURE 22.6: SCIENTISTS AT WORK AT AN UNIDENTIFIED LABORATORY. (POSSIBLY A WELLCOME RESEARCH LABORATORY) EARLY TWENTIETH CENTURY.

Besides being the home of the *Journal of Experimental Medicine* the Rockefeller Institute also funded the *Journal of Clinical Investigation* which was launched in 1924 as the official journal of the Society for Clinical Investigation. As membership of this society expanded during these years other organizations were formed such as the American Federation for Clinical Research. Specialists also formed their own associations such as the American Heart Association and the American Gastroenterological Association.[5] It is noteworthy that clinical researchers in these years drew increasingly on the prestigious science of biochemistry. The isolation of insulin by Frederick Banting and Charles Best in 1921 relied on blood testing as well as animal and clinical experiments. Banting who received the Nobel Prize for this work was practising in the individualist tradition. He was an unsalaried solo researcher who availed himself of the facilities offered to him by J.J.R. Macleod, professor of physiology at Toronto (Best was a science student assigned to do the chemical testing).

Rockefeller money did not only fund clinical research narrowly conceived. It was poured into colonial medicine and public health. From 1916 onwards something over $25 million were given to found public health schools in the United States and abroad. The Foundation also targeted London for medical reform because it was the capital of the British Empire. Rockefeller money was used to create the London School of Hygiene and Tropical Medicine which opened in 1929 as a postgraduate school within the University of London. Both teaching and research were carried out at the School. The foundation also funded a massive hookworm and yellow fever eradication program. Rockefeller money was behind clinical research in Africa, Asia and Latin America. The economic benefits of this cash flow were not lost on the distributors. The improved health of indigenous populations was accompanied by economic and political control. As the Foundation president disarmingly expressed it in about 1917, "Dispensaries and physicians have of late been peacefully penetrating areas of the Philippine Islands and demonstrating the fact that for purposes of placating primitive and suspicious peoples medicine has some advantages over machine guns."[6]

Great Britain 1914–1945

Seen on a large canvas the transformations of British medicine and the organization and direction of clinical research in these years look very like those occurring in North America, with the increasing prominence of university medical departments and their associated hospitals. Indeed, Rockefeller policy and money were to a great extent instrumental in bringing this about at University College London and the Universities of Wales, Cambridge and Edinburgh. Research funding in Britain also came from charities and the Medical Research Council (MRC). A great deal of clinical research was still carried out by unfunded, highly motivated individuals. As in America teamwork did not always come easily to medical men and was more a feature of work conducted after the Second World War.

The recommendations of the Haldane commission were gradually implemented after the Great War. Clinical units with professors were established with funding from the Board of Education, a government department. On October 1, 1919 full-time directors of Medical and Surgical Units started work at St Bartholomew's Hospital. By 1925 there were five chairs of medicine at the twelve London medical schools.[7] Both Sir Arthur Ellis, first professor at the London Hospital Medical School, and Sir Francis Fraser, appointed professor at St Bartholomew's in 1920, had studied at the Rockefeller Institute hospital. By 1944 there were 13 units in Britain.[8] In 1935 a Department of Postgraduate Medicine was opened at the Hammersmith Hospital in West London, the staff having university appointments funded by the University of London. The school became world famous and, like the Rockefeller Institute hospital, was to be the site where a generation of academic researchers were trained. These men and women in turn occupied top positions in university departments.

Medical schools and universities directly funded little clinical research. Money for this came from trusts, charities and the MRC. Charities were of various sorts. There were those based on modest donations from a variety of sources. The most successful of these were the Imperial Cancer Research Fund, the British Empire Cancer Campaign (1923) and the Asthma Research Council (1936). Philanthropies and charities based on a single source of income also funded medical research. The Rockefeller Foundation put a great deal of money (in excess of $6 million) into British medical education and research. The Wellcome Trust, a medical research charity established in 1936, disbursed the profits of the Wellcome Foundation drug company. At Oxford University the motorcar manufacturer William Morris funneled profits through his Nuffield Foundation to create a postgraduate school for clinical research. Using these funds the school established chairs in clinical subjects.

The Medical Research Committee (which became the Council in 1920) was formed to administer funds, raised under the 1911 National Insurance Act, for the study of tuberculosis. It was soon steered by its managing committee towards funding various medical research projects. The MRC funded both individual researchers and departments. The first of these was established at University College Hospital where Thomas Lewis was to have beds specifically designated for "research work and higher teaching."[9]

Lewis was probably the figure most frequently consulted about the organization of clinical research in inter-war Britain. He campaigned throughout his working life for the funding and practice of what he called clinical science. He founded the Medical Research Society in 1930 and in 1932 changed the name of the journal, *Heart*, which he had edited since 1907, to *Clinical Science*. Lewis trained a great number of American and British clinical scientists. To a great extent his own work was conducted in the individualist tradition.

Clinical research was not an uncontested area in the inter-war years. While it was increasingly agreed that national health, wealth and happiness might be promoted through clinical medicine, quite how clinical medicine itself was to be promoted was the subject of marked disagreement. Debate centered on the extent to which the laboratory sciences were likely to prove the most fruitful sources of knowledge about disease. Experimental physiologists remained a very powerful interest group in British medicine. Britain led the world in experimental physiology, the departments at University College and Cambridge being particularly famous. Doctors having connections with either or both of these institutions were significant actors in the shaping of inter-war British research. In many ways experimental physiology was venerated by its practitioners as the purest of the biological sciences. It was regarded by them as intellectually the most demanding of the life sciences and the most likely source of beneficial scientific knowledge. The most prominent and influential figure connected to the experimental physiologists was Walter Morley Fletcher, secretary of the MRC. Fletcher, a Cambridge-trained physiologist and a very able framer of policies, endeavored to coordinate and control all the funding of medical research in Britain. He took the view that research in the basic sciences laid the foundations for the understanding of disease. At the other extreme from Fletcher were clinicians (usually associated with the great Royal Colleges) and public health officials who saw too much energy being wasted on the laboratory sciences and considered that investment in bedside observation or epidemiology would be more productive.

This distinction is nicely exemplified in attitudes to nutrition which was the object of a great deal of joint and independent clinical and laboratory investigation in the inter-war years in both Britain and North America. It was widely agreed in Britain that in the first half of the century the nutrition of the nation fell below all that might be desired. The attitude of Fletcher and thus the MRC was that fundamental research into the biochemistry of food was the solution. Fletcher recollected that in 1914 he met a government minister who observed "well doctor I don't hold with research. If we want to stop disease we must give the people better grub and less dirt." Fletcher said he agreed and added that he wished the minister "could tell me what better grub was and what less dirt was — for I knew no way of finding out those two things except by persistent scientific research work."[10] On the other hand, Lord Horder, one of Britain's most distinguished clinicians, took a different view. Horder was not opposed to science in medicine but he regarded clinical skills as paramount and the clinic as the most suitable site for medical research. In 1936, speaking of nutrition, he observed, "Look after the accessibility of food and nutrition will look after itself."[11] These positions did not exhaust all attitudes to the laboratory and clinical medicine. In between were figures like Thomas Lewis. Lewis was extremely well connected among the physiologists and within the MRC. But, although he was an apostle of science and experimentation, in some contexts he was hostile to the amount of attention given to the basic sciences. Lewis was drawn into open dispute on this issue with Sir Frederick Gowland Hopkins in the 1930s. Hopkins was one of Cambridge's most distinguished physiologists, having done fundamental laboratory work on vitamins. Hopkins used his Presidential Address to the Royal Society in 1934 to proclaim that the future of medicine lay with basic science. He objected to Lewis' view that clinical science was the royal road to medical progress. He was worried, he said, about a trend towards endowing research "in the Clinic on a scale which might endanger the future of research in fundamental biological science. The tenor of my remarks has been due to the conviction that in the long run such a policy would sterilize advance."[12] Others, however, saw neither the laboratory nor clinical science as the best way forward. For example, John Ryle, a man with impeccable scientific credentials who became Regius professor of physic at Cambridge in 1936. In the thirties Ryle increasingly called for research into the causes of disease by observing life styles and the environment. Such work was to be done by individual clinical study and by social survey. In 1942 Ryle moved to Oxford to become the first professor of social medicine in Britain.

It is important to see that the expansion of scientific medicine in these years was not simply an expansion of knowledge. Nor was the knowledge created by scientific medicine simply 'for its own sake' or even just to add more power to the clinical encounter. Its creation was predicated on a new set of social relations. It was integral to the running of a modern industrial society. This has been nicely exemplified in a study of Sheffield, England. Here, at the city's university and hospital, a new elite of experimentally inclined clinicians gradually displaced an older core of individualist practitioners. In doing so they integrated themselves into Sheffield's civic affairs, making the university "an institutional focus for the managerial and administrative interests that became increasingly influential in the city with the rise of large scale industry and the growth of local government."[13] They did this by creating for themselves the role of expert advisors on public health matters and by making the university's laboratories vital to the city's functioning. In turn they created themselves as specialist consulting physicians marginalizing formerly prominent general practitioners. This was partly accomplished by bringing the medicine practised at the local hospital under the control of the university. To do this full-time clinical chairs were created and the local part-timers eased out. The first incumbent of a new chair of experimental pharmacology was Edward Mellanby, a protégé of Fletcher, who had received a great deal of MRC funding for experimental work on nutrition.

The Post-War Years

The years immediately following the Second World War saw a boom in academic medicine and the flourishing of clinical research. One reason for this seems to have been the optimism which medical professionals, politicians and the public invested in medicine. Such things as insulin therapy, penicillin and the decline of the major infectious diseases were all used as evidence that, through medicine, social progress could most rapidly be facilitated. In the National Health Service (NHS) many Britons considered they had an organization for solving the nation's woes. In creating the National Institutes of Health (NIH – the equivalent of the MRC) the United States federal government committed itself to supporting medical research, especially laboratory related research. A National Institute of Health was established in 1930 and became the Institutes in 1948 but only after the Second World War was there serious federal commitment to finance medical research.[14] Foundations as well as governments poured money into medicine. Pharmaceutical companies invested unprecedented amounts in drug development. Academic clinical departments expanded and proliferated. Increasingly, research became expensive and multidisciplinary, being carried out by teams of clinicians, basic scientists and technicians. Frictions over leadership and salaries developed. Inter-war antagonisms over funding priorities — basic science or clinical medicine — persisted.

Molecular biology, immunology and genetics were installed as the basic sciences which shaped much clinical activity. From the late 1950s the structure of human proteins in health and disease was widely studied and, for example, hundreds of variants of the blood protein hemoglobin were described, some differing from the normal by only a single amino acid. Some of these abnormal hemoglobins were associated with manifest diseases which were investigated clinically and experimentally as genetic disorders. Similar approaches were used to study other diseases, for instance, hemophilia. The genes considered at fault in such disorders were mapped on chromosomes in the laboratory using techniques such as cell culture. The genes implicated in a variety of relatively uncommon diseases, such as cystic fibrosis, were identified. In the mid-1980s the Human Genome Project was originated to sequence the genes on human chromosomes. Cancer research was also the object of massive post-war investment and was also transformed by the use of genetic and molecular-biological theories and techniques. Genetics, perhaps, currently ranks as queen of the medical sciences. The growth of the highly coordinated, expensive multidisciplinary projects, has slowly killed the tradition of the individual researcher.

Everyday clinical work has itself also been transformed by the combination of laboratory studies and technological research and development. Chronic disease such as degenerative heart disease can now be treated by expensive invasive therapy, in the extreme instance by heart transplant. Whether this is the best approach to such conditions and whether there might be a disproportionate distribution of resources into such therapies are now common source of concern. Many technological innovations, such as cardiac pacemakers or resuscitation equipment, have often been conceived at the bedside and prototypes made locally. Industrial involvement, however, has usually lead to the creation of products far more sophisticated than anything that could be made in a hospital workshop. Coupled with laboratory techniques such as biochemical analysis such developments have been the basis of the management of many life threatening conditions by intensive care, resuscitation and surgery. Bedside research itself has increasingly employed numerous workers and complex technologies such as fibreoptic endoscopes, computers and imaging devices, many of which might be combined in a single research project.

North America

Nowhere was the post-war medical boom more obvious than in North America, particularly in the United States. During the war many university clinical departments received government grants for research. Funding (through the NIH) rapidly increased after the war. One observer has estimated that between 1940–1949, $763 million were spent on medical research in the United States. Two decades later, between 1970–1979, the sum was 18.4 billion.[15] In 1990 alone the federal support for health-related research was $11.3 billion. The increase in personnel is equally striking. In 1950 there were no more than 1,000 salaried academic clinicians in departments of medicine, but forty years later there were about 12,000.[16] The total number of faculty members in *all* clinical departments in 1991 was nearly 41,000. Outside of the university system a clinical center with 500 beds was set up at the NIH's massive research complex at Bethesda, Maryland. From 1960 onwards government funded general clinical research centers (GCRCs) were set up at prominent medical schools. American researchers were quick to take advantage of the growth of molecular biology and other new disciplines. Although the first randomized clinical trial — testing streptomycin for tuberculosis — was published by British workers it was in the United States that large scale cooperative clinical research groups were formed. In 1958 a group working on a national scale — the Acute Leukaemia Group — published the results of a controlled trial of leukaemia therapy.

Great Britain

After the war successive British governments saw that the MRC was relatively well provided for. In the 1960s the MRC established a clinical research center (CRC) in association with Northwick Park Hospital in North London. The British veneration of experimental physiology continued to shape the direction of research. Although British laboratory scientists had an important role in effecting transformations in biochemistry, immunology and molecular biology, British clinicians were less appreciative of these changes than their North American counterparts and much spectacular work was done on traditional lines.[17] At the Post Graduate Medical School at Hammersmith under the supervision of John McMichael invasive techniques such as liver biopsy and cardiac catheterization were developed. Catheterization remained a central research tool at the expense, says one expert, of molecular medicine and genetics.[18] Nevertheless original work in new fields was done in Britain, notably at University College Hospital where isotopes were developed as research tools. Here too, the research laboratory studying metabolic disorders became world famous.

Fin de Siecle

In the last decades of the century traditional clinical research has fallen victim to the scepticism with which medicine at large is sometimes viewed. Medicine, or at least acute care as it is currently organized, it is argued, has failed as the agent of social progress which it promised to be. There has been, it is said, a disproportionate appropriation of resources by acute medicine while the massive social problems caused by chronic degenerative disorders, let alone poverty, accrue. A question is now being asked which would have been almost inconceivable forty years ago; what is the relative place of medical research when ranking the means for achieving a better society? If this question is answered by reference to the government funding of clinical research the answer is that there has been a fall from grace. There has been a gradual decline in clinical academic staff in British medical schools since 1980. In 1990 the CRC at Northwick Park was closed. The MRC faces financial constraints.[19] In the United States, as in Britain, clinical research is acknowledged by some to be in "crisis."[20] The chance of a new US investigator receiving funding in 1965 or 1975 was 40 percent. In 1990 it was 14.2 percent.[21] The perceived "crisis" in clinical research, however, is seen by its diagnosticians to have more tangible causes. Laboratory-based studies, without reference to the hospital ward, it is claimed, are increasingly regarded as the best means to promote medical knowledge. Whether or not this is so, it is certainly true there has been a shifting in medical research funding from clinic to laboratory. There are very obvious material reasons for this. Laboratory studies are simpler to frame and easier to complete. Given the competitive scientific and medical career structure of the late twentieth century such studies preponderate in the competition for grants.

However, it may be that it is not medical research which has fallen from grace, merely the best perceived means of promoting it. The last decades of the twentieth century have witnessed, in the name of free enterprise and opposition to the 'nanny state,' a general decline and dismantling of public bodies and general reductions in government funding. This change has been accompanied by appeals to private enterprise to be the source of initiative in the arts, sciences, education, welfare, etc. In Britain it has been Conservative government policy to encourage industry to take over research from the public sector.[22] This is also true of clinical research. In Britain there has been relative decline in the fortunes of the MRC and the various higher education funding committees whereas many research charities are flourishing. The Wellcome Trust currently invests almost as much in research as the MRC. There has also been a massive growth of drug company funded research. In the United States, one observer has noted, "Adopting the entrepreneurial spirit has become acceptable in the last decade."[23] Financial capital to fund research now comes from industry or from venture capital groups backing new investigator-directed companies. Equally important in the disappearance of relatively small-scale clinical studies, however, has been the growth of so-called evidence-based medicine. Put simply this has meant that clinical knowledge is not regarded as having validity unless it is derived from massive randomized control trials, be they of therapeutic agents or diagnostic techniques. It is arguable that the meanings of clinical research now differ radically from those given to it at the start of the century.

References

I am most grateful to Bill Bynum and Steve Sturdy for their comments on an earlier draft.

1. Abraham Flexner, *Medical Education in the United States and Canada.* (New York: Carnegie Foundation for the Advancement of Teaching, 1910).
2. Abraham Flexner, *Medical Education in Europe.* (New York: Carnegie Foundation for the Advancement of Teaching, 1912), p. 66.
3. Howard S. Berliner, *A System of Scientific Medicine: Philanthropic Foundations in the Flexner Era.* (New York and London: Tavistock Publications, 1985), p. 129.
4. Harry M. Marks, "Notes from the Underground: The Social Organization of Therapeutic Research", in Russel C. Maulitz and Diana E. Long (Eds.), *Grand Rounds: One Hundred Years of Internal Medicine.* (Philadelphia, University of Pennsylvania Press, 1988), pp. 297–336.
5. A. McGehee Harvey, *Science at the Bedside: Clinical Research in American Medicine.* (Baltimore and London: The Johns Hopkins University Press, 1981), pp. 105–152.
6. E. Richard Brown, "Public Health in Imperialism: Early Rockefeller Programs at Home and Abroad", in John Ehrenreich (Ed.), *The Cultural Crisis of Modern Medicine.* (London: Monthly Review Press, 1978), pp. 252–270.
7. George Graham, "The Formation of the Medical and Surgical Professorial Units in the London Teaching Hospitals", *Ann. Sci.* (1970), **26**: 1–21.
8. Donald Fisher, "The Rockefeller Foundation and the Development of Scientific Medicine in Great Britain", *Minerva* (1978), **4**: 20–41.
9. Christopher C. Booth, "Clinical Research", in W.F. Bynum and Roy Porter (Eds.), *Companion Encyclopedia of the History of Medicine.* (London and New York: Routledge, 1993), 2 vols., **1**: pp. 205–27, quote p. 221.
10. Maisie Fletcher, *The Bright Countenance: A Personal Biography of Walter Morley Fletcher.* (London: Hodder and Stoughton, 1957), p. 179.
11. Thomas Horder, *Health and a Day: Addresses by Lord Horder.* (London: J.M. Dent and Sons Ltd., 1937), p. 152.
12. Frederick Gowland Hopkins, "Address of the President", *Pro. Roy. Soc.*. (1935), **116b**: 403–427, quote p. 427.
13. Steve Sturdy, "The Political Economy of Scientific Medicine: Science, Education and the Transformation of Medical Practice in Sheffield, 18–1922", *Med. Hist.* (1992), **36**: pp. 125–159, quote p. 133.
14. See Victoria A. Harden, *Inventing the NIH: Federal Biomedical Research Policy, 1887–1937.* (Baltimore: The Johns Hopkins Press, 1986).
15. Paul B. Beeson and Russel C. Maulitz, "The Inner History of Internal Medicine", in Maulitz and Long, op. cit., note 5, pp. 15–54.
16. Robert G. Petersdorf, "The Evolution of Departments of Medicine", *N. Engl. J. Med.* (1980), **303**: 489–96.
17. Christopher C. Booth, "Clinical Research since 1945", in Ghislaine Lawrence (Ed.), *Technologies of Modern Medicine.* (London: Science Museum, 1994), pp. 148–150.
18. M.F. Oliver, "Crisis in Cardiovascular Research in Britain", *Br. Heart J.* (1989), **6**: 325–7.
19. Christopher C. Booth, "The National Health Service, the Universities, and the Research Councils: The future of Academic Medicine", *B.M.J.* (1988), **296**: 1382–5.

20. Oliver, op. cit., note 19 and Edwin C. Calman, "The Academic Physician — Investigator: A Crisis not to be ignored", *Ann. Int. Med.* (1994), **120**: 401–10.
21. Calman, op. cit., note 21.
22. R. Smith, "Is Research to be Privatised?", *B.M.J.* (1988), **296**: 185–8.
23. Calman, op. cit., note 21.

Further Reading

Edward H. Ahrens, Jr., *The Crisis in Clinical Research.* (New York: Oxford University Press, 1992).

Joan Austoker and Linda Bryder (Eds.), *Historical Perspectives on the Role of the MRC.* (Oxford: Oxford University Press, 1989).

Michael Bliss, *The Discovery of Insulin.* (Basingstoke: Macmillan, 1987).

E. Richard Brown, *Rockefeller Medicine Men: Medicine and Capitalism in America.* (Berkeley, L.A.: University of California Press, 1979).

G. Corner, *A History of the Rockefeller Institute.* (New York: Rockefeller Institute, Press, 1965).

R.B. Fosdick, *The Story of the Rockefeller Foundation.* (New York: Harper & Bros., 1952).

A. Rupert Hall and B.A. Bembridge, *Physic and Philanthropy: A History of the Wellcome Trust, 1936–1986.* (Cambridge: Cambridge University Press, 1986).

James H. Jones, *Bad Blood: The Tuskegee Syphilis Experiment.* (New York: Free Press, 1981).

Christopher Lawrence, *Medicine in the Making of Modern Britain 1700–1920.* (London: Routledge, 1994).

Susan E. Lederer, *Subjected to Science. Human Experimentation in America before the Second World War.* (Baltimore: The Johns Hopkins University Press, 1995).

David Weatherall, *Science and the Quiet Art: Medical Research and Patient Care.* (Oxford: Oxford University Press, 1995).

CHAPTER 23

Cancer

The Century of the Transformed Cell

ILANA LÖWY

SCIENTISTS AND THE TRANSFORMED CELL

For lay persons, and for many professionals, the word 'cancer' evokes above all the 'dread disease' of Western society, while its association with the word 'science' refers to the hope that science will bring a solution to the 'cancer problem.' From the 1920s on, the official discourse of cancer experts stressed the key role of fundamental scientific research in reducing the threat of malignant disease. Experts explained that recent advances in the understanding of biochemical, biological, and genetic mechanisms of cell multiplication would be rapidly translated into the control of 'deviant cells,' that is, into the prevention and the cure of malignant growths, and into the alleviation of the plight of cancer patients. The official optimism of the 'cancer establishment' was, however, moderated in the 1980s and '90s by statistics which reflected the stagnation (and, in some areas, increase) of morbidity and mortality from malignancies. The important investments in research in the post-World War II era notwithstanding, there is still no cure or prevention in sight for the frequent cancers of the adult. Even so, the expansion of cancer studies has deeply affected biomedical research. The medical problem of cancer was translated into answerable biological questions. Such translation has often advanced biological rather than medical knowledge. Thanks to their central place in biomedical research, cancer studies have contributed to the development of new areas of biomedical investigation.

A recent article (by David Cantor) which summed up the history of cancer, stated that historical studies in fact trace two distinct histories: the internalist history of ideas about cancer since antiquity and the externalist history of growing concerns about cancer since the nineteenth century. Regrettably, these two histories are not yet reconcilable. This chapter concentrates on a few key topics within the vast domain of interactions between 'science' and 'cancer,' in an attempt to link these two histories and observe how the concepts and practices of scientists and physicians are shaped by, and act upon, their material, social, economic and political environment. Numerous important issues such as the institutionalization of cancer care, the political role of cancer charities, the subjective experience of cancer patients and their families, cancer as chronic illness, psychosocial aspects of cancer, popular representations of cancer, or attitudes to terminal disease and death are thus beyond the scope of this study.

This chapter follows the relations between the 'interior' and the 'exterior' of cancer studies, and investigates interactions between techniques and laboratory practices, theoretical tools, professional strategies, jurisdictions, institutions, and policies. It is focused on the contribution of research materials, experimental models and instruments to cancer studies, clinical practices, and the perception of the pathological entity 'cancer.' Thus the success of radiation therapy contributed to the definition of cancer as a disease of rapidly dividing cells. The high cost of radium stimulated the centralization of cancer therapy in specialized institutions. The need to test the putative anti-tumor activity of numerous natural and synthetic substances led to the elaboration of simplified (for some, over-simplified) animal models of human cancer. Finally, the observation that malignant cells (unlike normal cells) can be easily grown in the test tube and are therefore excellent research material, enhanced the attractiveness of cancer research for biologists, biochemists, genetists and molecular biologists.

Radiotherapy: Physicists and the Elimination of Rapidly Multiplying Cells

Cancer was first seen as 'tumor' (tumefaction, linked to inflammatory states) or a non-healing wound of unknown origin. In the second half of the nineteenth century pathological manifestations of cancer were linked to uncontrolled proliferation of cells. Investigators (Virchow, Remak, Cohenheim) agreed that tumors grew from the normal cells of the body, were nourished by the host's organism blood and moved to distant sites through the blood and lymph systems. Each of these statements was a potential starting point for a search of a cure. It was possible to investigate (and indeed, some researchers did so) ways to prevent vascularization of tumors, the mechanisms of dissemination of malignant cells, or the effects of generalized cancer on the organism. It was also possible to develop physiology-centered approaches to malignant disease and to view cancer as systemic, and not localized, pathology. These approaches were, however, marginal. Twentieth century cancer studies were dominated by a cell-centered perception of cancer which equated the prevention of cancer with the prevention of malignant transformation of cells, and the cure of cancer with the elimination of the totality of transformed cells.

The cell-centered view of cancer has it roots in the nineteenth century histological definition of this disease and was sustained by the (unchallenged) key role of cytology in the diagnosis of malignancies. Pathologists were mainly interested in uncovering the origin of cancer cells which, they supposed, were either displaced in space (a 'wrong' location for a given category of cells) or time (embryonic cells which subsisted in the adult). The idea that the 'essence' of cancer was the result of change in a single cell (or in a few cells) was related to developments in biochemistry (studies of metabolism on the cellular level) and genetics (chromosome studies which placed heredity on the level of the single cell). Theodor Boveri (1862–1915), one of the pioneers of chromosome studies and the first to show that embryos deficient in chromosomes developed abnormally, proposed in 1914 that cancer arises as the result of a somatic mutation, and that all the consequent pathological developments are the result of this mutation.

The perception of cancer as a disease which started as an event taking place in a single cell or a small group of cells ('malignant transformation') coincided with the pathology-based approach first advanced by Henry Dran in the late eighteenth century. Dran proposed, on the basis of his observations of breast cancer, that cancers always started as a local, and surgically curable lesions which only later became generalized and incurable. In the late nineteenth and early twentieth centuries, with development of histology and histopathology, the content of Dran's theory was translated into cellular terms, and the generalization of cancer was redefined as a migration of tumor cells. At the same time physicians began to equate cancer therapy with efforts to eliminate all the malignant cells. Cell-centered views of cancer stressed the unity of this pathology. Cancers may differ greatly in their morphology, morbid manifestations, or virulence, but they all represent variants of the same phenomenon. This unified, cell-based view of cancer was reinforced by the advent of radiotherapy: a treatment based on the principle of a systematic elimination of rapidly proliferating cells.

The observation that X-rays induced skin burns led, around 1896, to attempts to apply this technology to cancer treatment. The new method was first successfully employed in the therapy of skin cancers. By 1904, special X-ray tubes were used to treat internal tumors as well. In 1905 Drs. Jean Bergonié and Louis Tribondeau demonstrated that X-rays preferentially kill rapidly dividing cells, providing a scientific rationale for the selective use of these rays in cancer treatment. It was hoped that the new therapy would improve the (then, very low) long-term success rate of surgical treatments of cancers without the disfiguring scars of operations and without the danger that a surgery would disseminate malignant cells. The use of X-rays was, however, dangerous. The rays provoked burns and in addition it became clear that high doses of X-rays could induce cancer. At first physicians did not know how to measure and standardize X-ray radiation and they lacked precise information on the biological effects of that radiation. The therapy improved slowly, mainly through trial and error. Between 1910 and 1921 important technical advances such as the introduction of Coolidge's hot cathodic tube (1913) and later, the development by Case of a 200 kilovolt apparatus (1921), allowed better control of radiation doses and favored the diffusion of X-ray equipment among hospitals specialized in the treatment of cancer. The new equipment made radiotherapy safer and greatly increased its range of application. The observation that radium produced burns similar to the ones induced by X-rays attracted the attention of doctors to the new source of radiation. Radiation emitted by radium was directly proportional to the amount of pure compound present. It was thus easier to control than X-rays. Moreover, the use of tubes and needles containing radium (an approach developed at the Radium Institute, Paris, by Claudius Regaud and his collaborators) made possible the direct delivery of radiation to a tumor (curitherapy). Radium was rapidly introduced into the treatment of skin cancers, and its use was gradually extended to other tumors as well. The diffusion of X-ray therapy and radium therapy led to the development of a new professional group of radiotherapists: physicians specialized in the administration or clinical research of radiotherapy.

In the 1920s radium therapy was perfected through the development of radium collars and radium bombs, and, in the 1930s, of external beam machines used in the therapy of gynecological and head and neck tumors. However, the high price of radium limited the diffusion of the new therapeutic approach. The solution was the centralization of cancer treatment and the private collection of funds for the purchase of radium by cancer charities such as the American Society for the Control of Cancer and the Ligue Franco-Américaine Contre le Cancer. Cancer charities strongly encouraged the development of specialized institutions which combined pre-clinical research with cancer therapy. Cancer patients treated in these institutions were able to benefit from the latest technological advances: specialized surgery, new X-ray machines and more efficient ways of delivering radium radiation. The existence of these centers favored the diffusion of complex and expensive technologies such as radium beam machines. The shift toward 'big medicine' introduced by radiotherapy, occasionally modified the organization of health care, and its perception by physicians, cancer charities, health administrators and politicians. Therapeutic efficacy became increasingly identified with big multidisciplinary centers and with complicated and expensive instruments. The cancer treatment centers developed between the two World Wars therefore became the forerunners of the 'big medicine' era in industrialized countries. They also played an important role in the establishment of the belief, central to the cancer research in the twentieth century, that a cure for cancer would be associated with advances in fundamental research (especially in cell-biology) and 'high tech' medicine.

The Elusive Difference: The Search for Differences Between Normal and Malignant Cells

In the early twentieth century scientists did not know how to induce artificially malignancies in laboratory animals. In order to study cancer they surgically transferred a naturally occurring tumor to another animal of the same species. The highly variable rate of success of transplantation became a distinct topic of inquiry. The rejection of transplanted tumors was attributed to the body's 'resistance' to malignant growth. The observation of this phenomenon led to two lines of enquiry: the investigation of ways of intesifying this 'resistance' as a possible way of preventing or curing cancer and studies on the rules which govern transplantation of tumors. The 'resistance' studies, were an important branch of experimental cancer studies in the 1910s. They were directly linked to research on 'resistance to infectious diseases' (later immunology), seen as an expression of more general physiological mechanisms such as the inability of the grafted cells to absorb nutritional elements from the host postulated by Paul Ehrlich's theory of 'atrepsia.'

In the 1920s and '30s, leading specialists in this field arrived at the conclusion that the 'resistance' to tumors was in fact 'resistance' to transplanted foreign tissue. Thereafter, the 'resistance' studies lost much of their interest. The concomitant observation that tumors (and normal tissues) are not rejected if the donor and the recipient of a graft are genetically homogenous, led to the development of inbred ('pure') lines of laboratory animals. The inbred animals were employed to study an inherited susceptibility to tumors, but also for studies of mammalian genetics. The latter development was strongly promoted by Clarence C. Little, geneticist and cancer expert, and the founder, in 1929, of the first institution which developed and then commercialized inbred mice, the Jackson Memorial Laboratories (Bar Harbor, Maine). Genetically uniform animals (either carrying standardized transplantable tumors, or endowed with a property to develop natural tumors) gradually became an important tool for the experimental oncologist. One of the main problems of experimental studies of cancer in the early twentieth century was the great variability of naturally occurring tumors. This variability prevented the comparison of results obtained in different laboratories, and hampered the development of a unified domain of experimental cancer studies. Faced with contradictory experimental results, researchers adopted the principle of extensive exchange of research materials (tumors, tumor-carrying animals), and, in parallel, developed uniform animal models of cancer. The circulation of standardized animals and tumors and, from the 1930s on, of tumor cell-lines (that is, malignant cells maintained indefinitely in culture), restrained the variability of experimental systems. It made possible the repetition of an experiment conducted in one laboratory in numerous other laboratories, contributing thus to the stabilization of networks of scientists who studied cancer.

At the same time, scientists attempted to artificially induce tumors in order to study carcinogenesis (the formation of malignant tumors) under controlled conditions. The observation of links between specific occupations and cancer, which started with the description of a high frequency of scrotal cancer among chimney sweeps (Percival Pott, 1775), indicated that chemicals or prolonged irritation may provoke cancer. The observation, by Yamigawa and Ichigawa in 1915, in that it is possible to induce cancer in rabbits by painting their ears with coal tar, was the starting point of studies of chemical carcinogenesis. These investigations, focused, from the 1930s, on cyclic hydrocarbons, and were related to the investigation of structurally similar steroid hormones. In the 1920s radiation was employed to induce tumors in the laboratory. Scientists who studied carcinogenesis saw the external agent (be it a chemical compound, mechanical irritation, or in some cases, a virus) as an event which started a chain of independent changes in the cell, that is, a malignant disease. The scientific committee appointed in 1938 by the United States Surgeon General to set research goals for the newly founded National Cancer Institute (NCI) summed up this view: "whatever may be the contributing cause, malignancy, once acquired, becomes a fixed character of the cell (..) which is passed unchanged to the descendants."[1]

The inter-war period was dominated by the development of biochemistry and cell biology. Leading cancer experts were convinced that they should join forces with their colleagues who study the structure and the function of normal cells, because normal and malignant growth are two sides of the same coin. The biochemists hoped that their growing ability to investigate metabolic pathways, to detect minute differences in metabolism and to study functions of enzymes and of other cell proteins, would lead to the uncovering of physiological differences between normal and transformed cells. These hopes were not fulfilled. In 1938 the cancer experts stated that all the functional differences between normal and malignant cells could be attributed to a single cause: the rapid proliferation rate of cancer cells. The search for structural differences between normal and malignant cells did not fare any better. In the 1940s and '50s the perfection of highly sensitive biochemical and immunological techniques, (ultracentrifugation, electrophoresis, radiolabeling and fluorescein labeling of antibodies) led to renewed efforts to uncover differences in the constitution of normal and transformed cells. These efforts were driven by a hope that the description of tumor-specific structures (or 'tumor antigens') would lead to the development of specific immune responses against tumors. Intensive investigations failed, however, to uncover important structural differences between normal and malignant cells.

In the 1960s and '70s the question of differences between normal and malignant cells was redefined in different terms. Immunologists exchanged their focus of interest from structure to function, and replaced their (mostly elusive) search for specific 'tumor antigens' by a search for ways to stimulate immune mechanisms able to distinguish between normal and malignant cells. The body, they proposed, might be able to recognize, then to destroy, transformed cells, even if scientists were not able (yet) to identify the structural basis of such recognition. Immunotherapies of cancer developed in the 1970s were based on the principle of non-specific stimulation of cells of the immune system (usually by bacterial products). It was hoped that the activated cells would in turn selectively eliminate tumor cells. Cell biologists gradually abandoned research on differences between normal and tumor cells and returned to inquiries on causes and mechanisms of malignant transformation. In the meantime, clinical practices and the organization of cancer research were modified by the development of cancer chemotherapy: a therapeutic approach based (mainly) on the 'old' principle of elimination of rapidly multiplying cells (the only exception was hormone therapy for hormone-sensitive cancers, based on a different principle).

CHEMOTHERAPY: SCIENCE, INDUSTRY AND TRANSFORMED CELLS

The development of the chemotherapy of cancer in the aftermath of World War II illustrates the interplay between material constraints (the need to develop drugs which kill cancer cells, with as little harm as possible to normal cells) and institutional and political developments. Belief in scientific medicine was boosted by the discovery of antibiotics, while the steadily rising level of cancer deaths compared to the decreasing death rate from infectious diseases caused disquiet amongst the public and consequently politicians. The result was the important increase in credits allocated for cancer research and the development of big collaborative projects. Such projects were stimulated by successful wartime research which led to the spectacular achievements of the atomic bomb and penicillin. The development of drug treatments for cancer was directly related to studies funded during the Second World War by the US Office for Scientific Research and Development (OSRD). The first descriptions of anti-tumor activity of chemical compounds were a by-product of ORSD-funded research on poisonous gases (alkylating agents), antibiotics (actinomycin D), and nutrition (analogues of folic acid). Researchers who started studying anti-cancer drugs (mainly in the US) were therefore familiar with goal-oriented research, large-scale collaborative projects and a complex division of labor.

Between 1945 and 1954 two programs (the first at Sloan Kettering Institute in New York, directed by Cornelius Rhoads, former director of the US Army Chemical Warfare Medical Division, and the second at the National Cancer Institute (NCI) (Bethesda, MD)), screened thousands of natural and synthetic compounds in the search for molecules with anti-tumor proprieties. These programs were modeled on industrial programs which looked for new antibiotics. However, malignant cells, unlike bacteria, are very similar to normal cells. Developing substances which selectively kill tumor cells was a much more difficult task than developing new antibacterial drugs. Researchers found that the effective and toxic doses of all the potential anti-cancer drugs were very close. Extensive clinical experimentation was necessary to adapt these substances for use in patients. The effort to develop drug treatment of cancer was therefore indissolubly linked with the complicated problem of coordinating large-scale clinical trials. By the 1950s, when this problem was successfully solved, the need to link the search for anti-cancer drugs with big clinical trials had become, however, one of the strong points of the domain of cancer chemotherapy. This domain became one of the preferential sites for the interaction between basic, pre-clinical, and clinical research, and it attracted scientists and physicians interested in interdisciplinary research, the solution of practical problems, and the control of a new area of medical intervention.

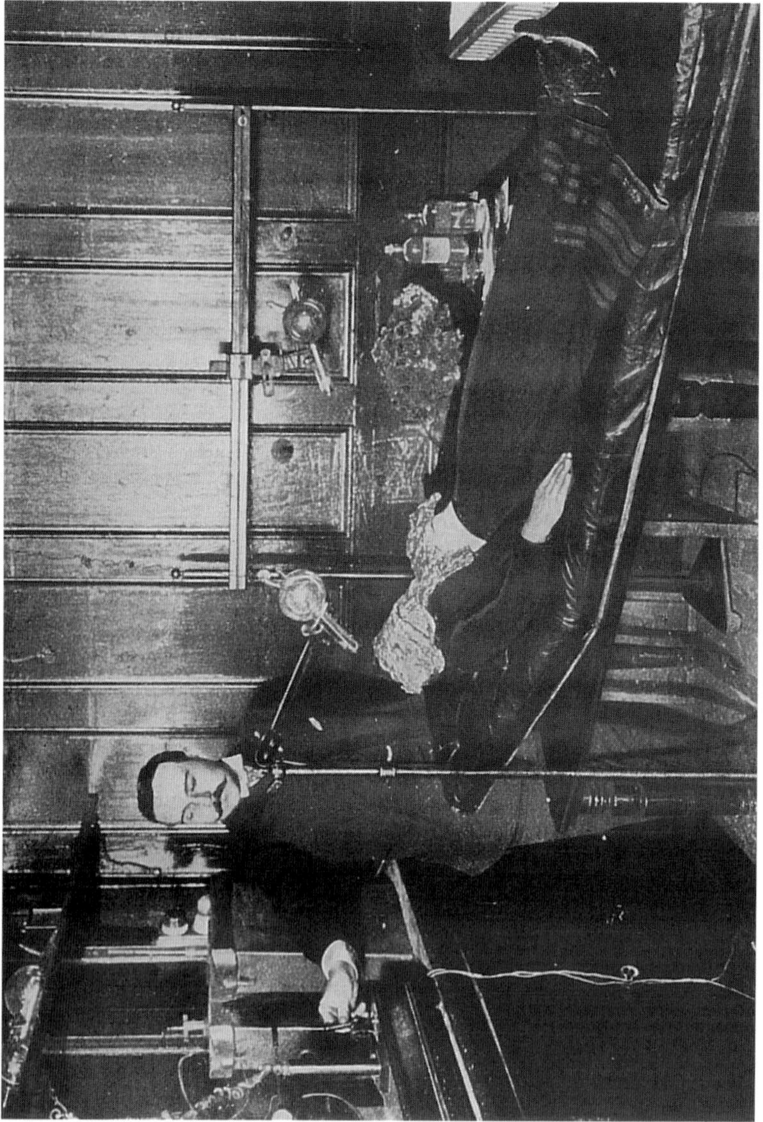

Figure 23.1: Treatment of Patient with Roentgen Rays, c. 1900. © St. Bartholomew's Hospital Archives Department.

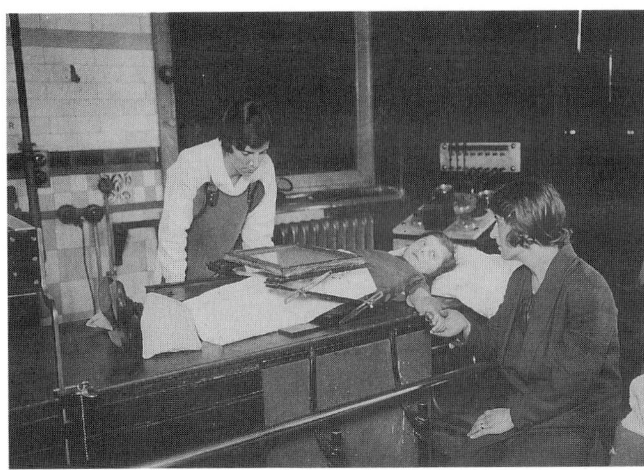

FIGURE 23.2: CHILD PATIENT IN DIAGNOSTIC X-RAY DEPARTMENT, C. 1929.
© ST. BARTHOLOMEW'S HOSPITAL ARCHIVES DEPARTMENT.

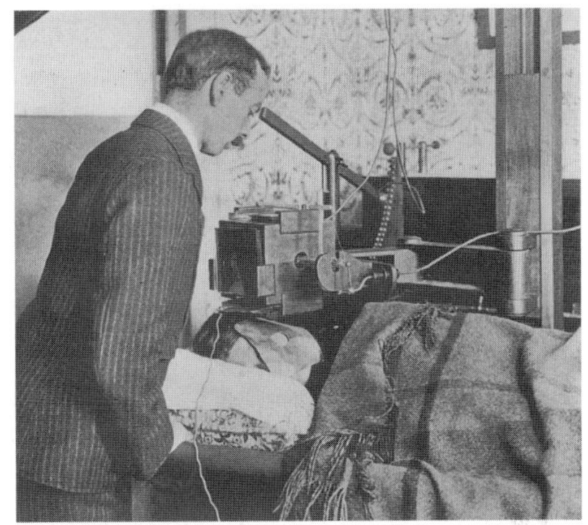

FIGURE 23.3: X-RAY THERAPY IN 1902. PATIENT IS SHOWN ON A COUCH. A SHIELD SURROUNDS THE TUBE. F.H. WILLIAMS, *THE ROENTGEN RAYS IN MEDICINE AND SURGERY*, 2ND EDITION, NEW YORK: 1902. WELLCOME INSTITUTE LIBRARY, LONDON.

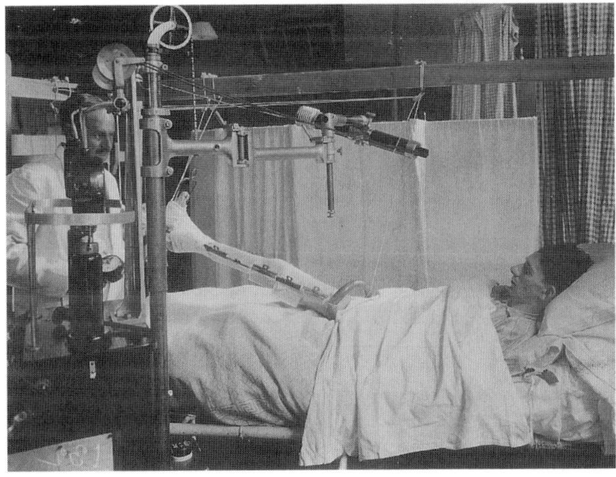

FIGURE 23.4: MALE PATIENT IN BED AWAITING LEG X-RAY, C. 1929.
© ST. BARTHOLOMEW'S HOSPITAL ARCHIVES DEPARTMENT.

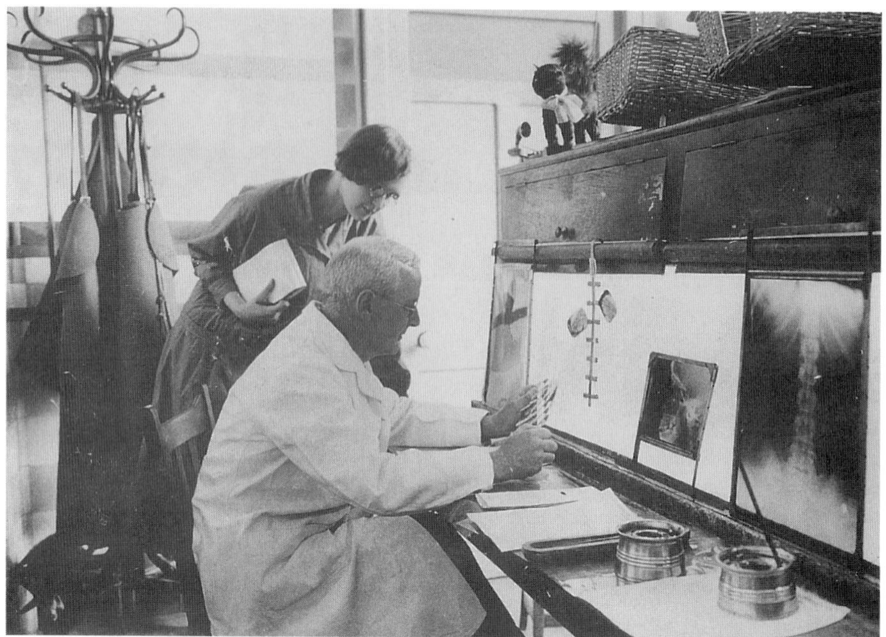

FIGURE 23.5: DOCTOR EXAMINING X-RAYS, C. 1929. © ST. BARTHOLOMEW'S HOSPITAL ARCHIVES DEPARTMENT.

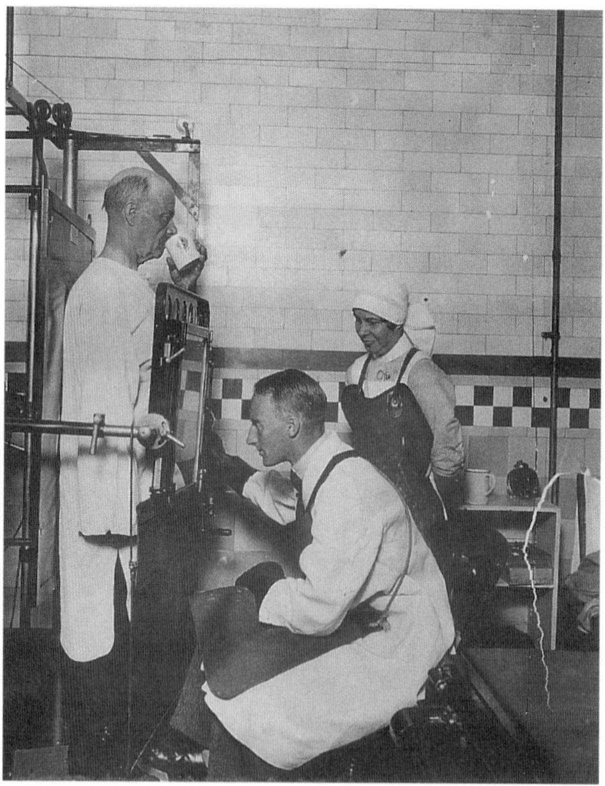

FIGURE 23.6: DIAGNOSTIC RADIOLOGY: BARIUM MEAL EXAMINATION, C. 1929. © ST. BARTHOLOMEW'S HOSPITAL ARCHIVES DEPARTMENT.

In the 1950s the activity of several US lobbies (leading cancer specialists, the American Cancer Society, the pharmaceutical industry) transformed chemotherapy into a political issue. Direct pressure of the US Congress led in April 1955 to the establishment of the Cancer Chemotherapy National Service Center (CCNSC). The CCNSC received 5.6 million dollars in 1956, 20 million in 1957 and 28 million in 1958. The aim of the new Center was to organize and coordinate clinical trials of new chemotherapies for cancer, and "to set up all the functions of a pharmaceutical house run by the NCI."[2] In order to better approximate the ideal of industrial-type research it was necessary to reduce the variability of two elements: the tumor-bearing mice employed in screenings of chemical compounds, and the patients who participated in clinical trials of new chemotherapies. Mice and tumors were controlled through the CCNSC inbred-mouse production program, conducted in collaboration with the Jackson Laboratories. The NCI developed also a complex system of sub-letting of production of mice and tumors to semi-private and private organisms. The CCNSC experts coordinated this production, fixed the norms for transplantable tumors and for screening tests, and established a 'quality control' over these tests. To facilitate mass-screening, the CCNSC employed between 1955 and 1961 only three mouse tumors to screen anti-tumor drugs. The assumption that these tumors adequately represented all human malignancies was challenged by numerous cancer experts. Nevertheless there was a (near) general agreement that cancer research should be conducted with highly uniform animals and tumors.

Control of patients was achieved through centralization and the standardization of large-scale clinical trials of new chemotherapies. The Clinical Studies Panel of the CCNSC stimulated the development of cooperative clinical study groups. Members of a cooperative group elaborated a system of mutual surveillance of the quality of laboratory analyses, randomization procedures, and the strict application of therapeutic protocols in diverse institutional settings. They also collectively defined the criteria of therapeutic success, distinct from the patients' subjective feelings and 'soft' parameters, such as the improvement of the quality of life. They developed objective (that is, quantifiable) indicators of the response to drugs, such as a measurable reduction of the tumor's size. Researchers were organized in 'task forces.' These were goal-oriented structures which promoted intensive cooperation between laboratory scientists, clinical investigators, pharmacologists, statisticians, industrial researchers and members of the CCNSC staff. In the mid-1960s, a task group dedicated to the cure of acute lymphatic leukemia (ALL) in children was able to annouce the development of the cure: the first success of large-scale collaborative research in clinical oncology.

The success of the ALL task force led to the enthusiastic adoption, in the US, of the principle of big collaborative studies of drug therapies of cancer. Leading oncologists, health administrators and politicians hoped that the extension of methods elaborated by this task force would lead to the discovery of efficient drug therapies for common malignancies. These hopes were not fulfilled in the following three decades. Cures were found for several kinds of relatively rare tumors (the most important being those found in children) but not for the most frequent cancers of adulthood. In the meantime, chemotherapy played an important role in the organization of cancer therapy in the US. The diffusion of drug therapies led to the development of a distinct medical specialty: medical oncology. In the US, this specialty became a distinct professional body, organized by groups involved testing new anti-cancer therapies. The adoption of the National Cancer Act of 1971 by the US Congress led to the expansion of budgets for pre-clinical and clinical cancer research, and stimulated the extension of clinical trials. Chemotherapy became an accepted treatment for cancer in other Western countries too. These countries adapted the US model to their specific conditions, and often differed substantially in the extent of the application of chemotherapy to specific cancers, the number and the tasks of medical oncologists and the diffusion of clinical trials. These differences notwithstanding, from the 1960s on, drug therapies (which often aim at alleviation of suffering and prolongation of life) occupied an increasingly important place in cancer treatment in the West. They were (and are), however, perceived as a partial and crude solution to the 'cancer problem.' Fundamental research in oncology aimed to replace the indiscriminate killing of rapidly dividing cells by radiation and drugs by specific methods of prevention and cure which would stem from a better understanding of mechanisms of the malignant transformation of cells.

Viruses and Oncogenes: From Transformed Cells to Modified Nucleic Acids

After World War II there was a revival of interest in hypotheses about the infectious nature of carcinogenesis. The viral hypothesis of cancer causation advanced by the French scientist Amadé Borrel in 1907, was sustained by the transfer of chicken leukemia (Ellermann and Bang, 1908) then chicken sarcoma (Rous, 1911) by tissue extracts. The transmission of malignant diseases in chickens by 'filterable agents' was, however, usually perceived as a bizarre phenomenon, found in birds only, and irrelevant for the understanding of human cancer. In 1933 the description of a transmissible tumor of the rabbit (Shope papilloma), and then, the possibility of the infectious transmission of breast tumors in mice, attracted once again attention to the viral hypothesis of tumor causation. The rabbit skin tumor was not universally recognized in the 1930s as a 'true' cancer, and did not seriously challenge the notion that viruses do not induce cancer in mammals. The story of transmissible breast cancer in mice is more complicated. In 1933 John Bittner from the Jackson Laboratory observed that if female mice from a 'low cancer' strain were nursed by mothers from a 'high cancer' strain they developed breast cancer, just like the females of the 'high cancer' strain. Bittner at first believed that the 'milk factor' was an auxiliary, 'facilitating agent,' possibly a hormone.

However he changed his definition of the 'milk factor' between 1942 and 1948. The factor became a 'colloid,' then a 'protein,' and finally, in 1948, a (likely) virus. This change might have be related to increased use of ultracentrifuge in biochemistry and virology, and to Bittner's growing realization that the 'factor' he was studying behaved when ultracentrifugated like a 'classical' virus.

In the 1950s scientists first described a virus-induced leukemia in mice, then several other cancer-inducing viruses in laboratory animals. The study of these viruses became an important research subject in the 1950s and '60s. The revival of interest in cancer viruses can be viewed as a result of a combination of several events: the introduction of new techniques such as ultracentrifugation, electron microscopy and cell culture into the virology laboratory, the growing interest of fundamental biologists in viruses (mainly in bacterial viruses, the bacteriophages), the visibility of practical achievements of virologists exemplified by the production of polio vaccine, and finally the increasing importance of mammalian cell lines (often malignant cells) as research material for the biologist. The widespread diffusion of transformed cells as a standard research material/tool in the biology laboratory facilitated a definition of numerous fundamental investigations of cell growth and differentiation as a potential contribution to the solution of the 'cancer problem.' In the 1960s virologists described numerous cancer-inducing viruses in laboratory animals, and then affirmed that standard mammalian cell-lines were often contaminated with tumor viruses. They identified numerous cancer-inducing viruses in laboratory animals and observed that mamalian cell-lines were often contaminated with tumor viruses. The conviction that tumor viruses were 'everywhere' (for example, researchers had found in the 1960s that anti-polio vaccine, prepared in monkey cells, was contaminated with a monkey tumor virus which, after an initial scare, was declared harmless for humans) had, in turn heightened the biologists' interest in these viruses. The new visibility of tumor viruses stimulated the establishment of the Virus-Cancer Program of the NCI in 1968. This generously funded program (described by some as 'moonshot-style enterprise') aimed at the ultimate discovery of 'anti-cancer vaccines.' It involved intensive exchanges of cells, viruses and anti-sera between participating laboratories. It also involved a complex division of labor between these laboratories and commercial firms.

In the 1970s two theories attempted to explain relationships between viruses and cancer. Temin's 'protovirus' theory stipulated that cancer is directly induced by the recent integration of viral genes into the cell's DNA. Todaro and Huebner's 'oncogene' theory proposed that normal cells permanently harbored cancer-inducing genes (possibly of remote viral origin). Both theories exemplified a reductionist shift from the study of transformed cells to the study of macromolecules, and both looked for a direct or indirect vertical transmission of hidden, tumor inducing factors. The description, in 1976, by Bishop, Varmus and their colleagues, of the presence of a gene homologous to the *ras* gene of the Rous sarcoma virus in normal avian cells, and then the description of similar genes in mammalian cells, were received as strong arguments in favor of the oncogene theory, and stimulated studies of the mammalian genome. At the same time, genetic engineering techniques provided tools for studying mammalian genes. The biotechnology industry became increasingly involved in cancer studies. Rapid growth of oncogene studies in the 1980s was directly related to industrial developments and the perceived commercial interest of oncogenes for the diagnosis of cancer. The description of 'cellular oncogens,' present in all normal cells put an end to search for vertically-transmitted tumor inducing elements. In the 1970s it was hoped that the search for 'tumor inducing infectious particles,' would open up ways to prevent cancer. In the 1980s this goal was replaced by interest in oncogene-based diagnostic tests. Oncogenes were initially the research materials which circulated within the network of scientists linked to the NCI's Tumor Virus Program. In the second stage of their career they became industrial products. This development is exemplified by the careers of some of the leading scientists of the Virus Cancer Program (Todaro, Scolnick) who left NCI in the 1980s for industry.

Oncogenes have reached the cancer clinics, but they have mainly become auxiliary tools for the diagnosis of selected (and usually infrequent) tumors, such as neuroblastoma. Their principal use has been in the research laboratory. Oncogenes (for some biologists, a misleading name) were redefined as a group of genes which play a central role in cell differentiation, and which, consequently, are of great interest for biochemists, embryologists and cell biologists. In the mid-1990s biotechnology firms which produce tools for the detection of oncogenes (DNA probes, monoclonal antibodies) work mainly for the fundamental research market. The interest of physicians shifted in the late 1980s to a new group of genes with an especially high mutation rate. These genes were described in the famework of the human genome project, and were related to inherited 'susceptibility to cancer.' The pathway to the description of the 'susceptibility genes' was different from the one which led to description of oncogenes. While oncogenes were found through the deliberate search for viral structures in malignant cells, 'susceptibility genes' were described following a 'molecularized' form of a classical genetic investigation: linkage studies, mapping and statistical evaluation. The 'susceptibility genes' are believed to be important in the transition from normal to malignant cells. Such a process, molecular biologists proposed in the early 1990s, involves six distinct stages, each regulated by numerous factors. The complexity of this proposed mechanism casts some doubt on earlier proposals that the shift from the study of cells to the study of nucleic acids would facilitate the understanding of the genesis of naturally-occurring tumors and makes improbable, some experts have argued, a rapid development of cures for cancer based on the understanding of the functions of 'susceptibility genes.' In the mid-1990s these genes started to be employed in diagnostic tests which spot individuals who have high chances of developing hereditary malignancies (such as retinoblastoma or hereditary forms of breast cancer): a complicated ethical problem in the absence of efficient therapies for cancer. These tests, and even more the possibility of their use in a search for an imprecise 'susceptibility' to the more frequent non-hereditary cancers, has provoked interrogations about the possible influence of professional and economic interests — of cancer specialists, industrialists, employers, insurance companies, health administrators, and politicians — on the development of such tests.

CONCLUSIONS: TOWARDS THE DISAPPEARANCE OF THE TRANSFORMED CELL?

The practice of cancer studies has changed dramatically in the last twenty years. Tumor carrying animals and histological preparations has been replaced by DNA probes, gene maps and computerized data banks. This change is restricted, however, to the research laboratory and the biotechnology plant. Outside this confined space cancer is still the terrifying 'enemy within' which treacherously destroys its victims. The enemy has been redefined in the late nineteenth and the early twentieth century: the imprecise 'crab' became a 'transformed,' 'deviant,' 'mad' cell, characterized by its non-controlled proliferation. This anarchistic proliferation, the essence of the disease cancer, was also the key to the development of therapies based on selective elimination of rapidly dividing cells: radiotherapy and then chemotherapy. The important professional, institutional, industrial, and in some cases governmental investment in the development of radiotherapy and chemotherapy, and the large diffusion of these therapies, strengthened the perception of cancer as a cellular disease.

The cell-centered view of cancer, combined with the facility to grow malignant cells in the test-tube and the redefinition of the 'cancer riddle' as the problem of cell growth and multiplication, allowed investigators who were interested in a wide range of biological problems to benefit from public and private resources earmarked for cancer research. The growth properties of cancer cells contributed to the stabilization of experimental systems which used these cells and the proliferation of experts who studied them. The proliferation of cancer experts has been, this chapter argues, linked to the material properties (low cost, robustness, transportability) of experimental systems which employed "transformed cells." The immunologist Niels Jerne linked the "proliferation of immunologists" in the 1960s to the development of a new theoretical framework, the clonal theory of antibody formation. (in his words, "Sir Macfarlane Burnet must have been pleased (...) to see how his stimulating ideas have led to a great proliferation of immunologists").[3] This proliferation was greatly accelerated after the Second World War. The conjunction of growing visibility of cancer for the lay public and politicians, the central role of the study of cells for biologists, industrialists' interests in diagnosis and therapy of cancer, and the expansion of biomedical research after World War II favored he development of a cluster of distinct professional groups (or, to use a term borrowed from interactionist sociology, an 'arena') aggregated through a common interest in human cancer, and linked through uses of a 'boundary object,' the 'transformed cell.' The groups within this cluster has only maintained loose relationships, but occasionally formed alliances and collaborative ventures focused on specific goals and held together through circulation of research materials and techniques.

As the century closes there are signs of a weakening of the perception of cancer as a cell disease. Molecular biologists are trying to redefine the 'problem of cancer' as a problem of genetic information which need to be solved on the DNA, not on the cellular level. This proposal has a solid material base. Molecular biology techniques (for example transfer of mammalian genes into lower organisms, amplification of minute quantities of DNA by the PCR technique), reduce the importance of the intact cancer cell as a research material for the biologist. On the other hand, scientists who want to maintain their identity as biomedical, not biological, investigators, and to benefit from the resources allocated to biomedical research and from professional rewards attached to this activity, need to legitimate their activity by its contribution to the solution of concrete medical problems. The persistant failure to develop efficient therapies for common cancers, in spite of the impressive accumulation of studies of normal and malignant cells, questions the validity of the translation of the medical problem of malignant disease into the biological one of the 'transformed cell.' The possible demise of this particular practice may lead to an attempt at a different translation of the 'cancer problem' into biological terms, and/or to a return to a clinic-centered perception of this disease.

References

The writing of this chapter was greatly facilitated by David Cantor's excellent summing-up of developments in cancer research and therapy, and his critical review of literature.

1. 'Fundamental Cancer Research: Report of a Committee appointed by the Surgeon General', *Public Health Reports* (1938), **53**: 2112–2130.
2. Kenneth M. Endicott, 'The chemotherapy program', *Journal of the National Cancer Institute* (1957), **19**: 275–293.
3. N.K. Jerne, 'Summary: Waiting for the end,' *Cold Spring Harbour Symposia in Quantitative Biology* (1967), **32**, 591–603, on p. 601.

Further Reading

Joan Austoker, *A History of the Imperial Cancer Fund, 1902–1986.* (Oxford, New York, Tokyo: Oxford University Press, 1988).

Theodor Boveri, *The Origin of Malignant Tumors* (transl. Marcella Boveri) (Baltimore: Williams & Wilkins, 1929 (1914)).

David Cantor, 'Cancer,' in W.F. Bynum and Roy Porter (Ed.), *Companion Encyclopedia of the History of Medicine.* (London & New York: Routledge, 1993).

Barrie R. Cassileth, 'The evolution of oncology as a sociomedical phenomenon,' in *The Cancer Patient: Social and Medical Aspects of Care,* B.R. Cassileth (Ed.). (Philadelphia: Lea & Febiger, 1979), pp. 3–15.

Jean Paul Gaudillière, 'Oncogens as metaphors for human cancer: Articulating laboratory practices and medical demands,' in *Medicine and Change: Innovation , Continuity and Recurrence,* I. Löwy (Ed.). (Paris & London: John Libbey, 1993).

Jean Paul Gaudillière, 'The molecularization of cancer etiology: Molecular biology, health policy, and biotechnology in the cold war United States,' in, *The Molecularization of Biology and Medicine*, H. Kamminga, S. de Chadarevian, (Eds.). (Harwood Academic Publishers, forthcoming).

Michael Gold, *The Conspiracy of Cells.* (Albany, N.Y., State University of New York Press, 1986).

Clarence C. Little, 'Genetics and the cancer problem,' in *Genetics in the XXth Century*, L. Dunn (Ed.). (New York: The Macmillan Company, 1958), pp. 431–472.

Dorothy Nelkin and Lawrence Tancrady, *Dangerous Diagnostic: The Social Power of Biological Information.* (New York: Basic Books, 1989).

James T. Patterson, *The Dread Disease: Cancer and Modern American Culture.* (Cambridge, Mass.: Harvard University Press, 1897).

Patrice Pinell, *Naissance d'un fléau: La lutte contre le cancer en France, 1890–1940.* (Paris: Editions Métailié, 1992).

L. J. Rather, *The Genesis of Cancer: A Study in the History of Ideas.* (Baltimore and London: The Johns Hopkins University Press, 1978).

Stephen P. Strickland, *Politics, Science and the Dread Disease: A Short Story of the Unites States Medical Research Policy.* (Cambridge, Mass.: Harvard University Press, 1972).

Robert Teitelman, *Gene Dreams: Wall Street, the Academia and the Rise of Biotechnology.* (New York, Basic Books, 1989).

Jacob Wolff, *The Science of Cancerous Disease from the Earliest Times to the Present.* (Science History Publications, 1989 (1906)).

CHAPTER 24

A Science 'Dans Le Siecle'

Immunology or the Science of Boundaries

ANNE MARIE MOULIN

Unlike most science discussed in this volume, immunology is the scientific progeny of the twentieth century, and its history spans hardly more than one hundred years. The name 'Immunology' appeared in most Western languages by the turn of the century. It only emerged after August Comte had completed his famous classification of sciences in which an important place was given to the new biology and sociology. Though it developed in the wake of triumphant positivism, it still had to find a place in the encyclopedic order of knowledge and by turns oscillated between expansion and recession, variously solicited by its two poles, biology and medicine.[1] More medical than other biological disciplines, yet holding the key to many biological phenomena, how can this late-born science be described?

Conflicting views of the history of immunology have been offered. Some commentators have emphasized its modest beginnings as a spin-off from late nineteenth century microbiology.[2] Others have shown that while drawing strength from cognate sciences, immunology has impressively contributed to them.[3] The late Niels Jerne has recalled its continuous "private line to medicine," while contemporary immunologists try to cut off immunology from its applications, for example vaccination. Most explanatory accounts of the history of immunology[4] however propose the following phases:

1. a preliminary period mapping the field of immunity and constructing the first data as 'facts.' This period was marked by the shared Nobel Prize awarded in 1908 to the discoverers of the cellular and humoral components of immunity, the German Paul Ehrlich (1840–1915) in Frankfurt and the Russian-born Elie Metchnikoff (1845–1916) in Paris;
2. a phase of generalization of immunological phenomena with the demonstration by Landsteiner that almost any molecule, if appropriately used, could be antigenic and trigger an immune response;
3. after an explosion of studies of cellular immunity in the 1950s, all components, humoral and cellular were united into a vast system, called the immune system,

taking place in the body beside the other better known apparatuses such as the nervous or the endocrine systems.

Far from being a truism, this definition, still admitted today, recognizes the existence of a new function of the body: the defense of the organism in its integrity. From this vantage point, immunologists can reconstruct a tradition of research which leads to the happy marriage of biology and medicine and confirms the great expectations afforded this alliance. They subscribe to the vision of a natural object, missed by preceding generations, which they are required to describe in detail in order to understand and manipulate it.

This definition linked immunology with the prestigious science of systems and with computer science. It opened the way to molecularization and description in the fashionable style of modern times.

Early Days, Immunity and Infection

Early immunology focused on the relationship between organisms and germs, according to a defensive paradigm. In short, it represented an appendix to microbiology, itself the latest catalogue of the living world.

Immunology followed the development of bacteriology, with its description, both at the clinical and experimental level, of resistance against infection, either innate or acquired. However, preventive immunization had preoccupied physicians for more than two centuries. Let us briefly recall variolization (whose precise dating is unknown), used in China and the Ottoman empire to protect children against the terrible scourge.[5] Launched at the beginning of the nineteenth century, the Jennerian vaccine progressively replaced variolization and after decades of legal constraints and technical improvements of the product led to the decline of smallpox. When the World Health Organization (WHO) celebrated this unique event of disease eradication on 1979, its officers paid tribute to the efforts of generations of doctors enforcing immunization without scientific knowledge of how immunity worked.

At the end of the nineteenth century, the germ theory of disease provided a general framework for the understanding of cancer and chronic diseases as well as acute infections. The rapid development of programs for attenuation of all germs in the laboratory and generalized vaccination were among the many arguments to support the idea of a radical revolution in medicine. Serotherapy, initiated in 1894, consisted of transferring protective serum from one immunized being to another, while vaccination challenged the body with pathogens artificially modified in their virulence. For doctors the immunological approach became the first choice for prevention as well as treatment of many infections. Vaccinotherapy in various guises, undertaken even after the onset of the disease, encapsulated the modern management of disease by appropriate stimulations of the body. In this way, immunology could be identified entirely with modern medicine, as claimed by the British physician Almroth Wright, the father of antityphoid vaccine: "The physician of tomorrow will be an immunizator."

In these years, the main *raison d'être* of immunology seemed to lie in its impact on medical practice. In the following years, although it became clear that the Pastorian model of germ attenuation *in vitro* would not provide the universal solution to the diseases, the active quest for vaccines developed on a previously unknown scale. The field trials of Salk and Sabin vaccines against polio involved millions of children, respectively in the United States in 1954 and in the Soviet Union in 1961: the grand enterprise of vaccination was at best described as an attempt at "patenting the sun"[6] or tapping previously unexploited energy from the living beings.[7] Purification in the laboratory helped to make the vaccines safer, without yet totally suppressing the "hazards of immunization."[8]

Immunity and Evolutionary Issues

But immunity appeared simultaneously as an important survival factor in the natural history of species, to be taken into account in all evolutionary thinking. After Metchnikoff had observed in 1883 that cells of starfish larvae could overwhelm foreign particles, he described the process of phagocytosis or the way that amoeba-like cells engulf bacteria in an organism, in distant species such as vertebrates (including of course humans) and invertebrates. This was an attempt at a synthetic presentation of the body in immunological terms and a preliminary sketch of our modern view of the immune system. He believed that living phagocytes were the exclusive agents of immunity and dismissed the humoral (or molecular) components of immunity such as antitoxins, later more generally called antibodies, which were just being described in the blood, as of secondary importance.

Metchnikoff's work made it clear that, while immunity is crucial for the survival of the individual, it also reflects the coevolutionary balance reached between the "system of phagocytes" and the threatening viruses in the environment. Later, the discovery of the transfer of maternal antibodies to the offspring helped to understand how protection of the young operates before its immune system becomes mature, but also contributed to figure the transmission of immunity patterns inside populations and anticipate the epidemiological changes in the same populations.

Beyond Infection. From 'haptens' to Molecular Biology

Immunology had stemmed from a reflection on the organism's resistance to infectious diseases. However, one important event was the experimental discovery, before the First World War, that almost any kind of molecule could trigger an immune response and induce immunological memory or biologically altered reactivity of the organism. This move is exemplified by Jules Bordet's observation (in 1898) that not only aggressive pathogens but apparently harmless red blood cells can induce a response. Following this general trend, the chemically-oriented Karl Landsteiner demonstrated in a series of studies that immune response succeeded the inoculation of chemical molecules deprived of any easily recognizable function, provided that they were appropriately presented. Experimental sera could discriminate between molecules which differed only slightly in composition or structure. They displayed an exquisite specificity which has been considered as the hallmark of immune responses, and demonstrated the analytic power of immunochemistry.

This discovery allowed a range of molecules of known molecular differences to be used as antigens and to lay the physico-chemical foundations for immunity between the two World Wars. Scientists focused on the chemical aspects, marked by the precise identification of antibodies as gammagobulins and the elucidation of their structure, an enterprise anticipating the mapping of macromolecules and genes, so typical of the entrance of biology into the era of Big Science. By stimulating work on the production of antibodies and their patterns of formation as a paradigmatic study for the understanding of proteins, immunology has played an unnoticed role in the emergence of molecular biology in the 1950s.[9]

Antibodies, whose precise nature was elusive for a long time, were involved in reactions widely used by doctors for the sake of diagnosis, or serological tests. These tests relied on phenomena such as germ agglutination triggered by patients' sera, thence labeled as 'positive.' The first tests, in 1896, were intended for typhoid (Gruber-Widal reaction) and Malta relapsing fever (Wright reaction).

A Case Study from Immunology

Serology developed as an indirect way of diagnosing disease without necessarily identifying the germ responsible for it. Drawing attention to this indirect character of the proof, a pioneering essay putting serology into sharp focus shed a new light on science studies.

This study was published in 1935 by the philosophically-oriented Polish physician Ludwik Fleck. Wassermann's serological reaction, a complex procedure pivoting around immune hemolysis or destruction of red blood cells by antibodies, introduced in 1907, had for many years been used to screen syphilitic patients. In spite of the evidence that neither the antigens nor the antibodies involved displayed the specificity required by the theory. Doctors in the nineteenth century of course had a vested interest in detecting syphilis that was considered a social scourge of the mutual shaping of biomedical and social interests and eagerly adopted the test in spite of its many flaws and obscurities.

Fleck concluded that such a medical 'fact' was actually a construct or selection of significant features in accordance with a 'style of thought' typical of a specific community[10] and reflecting their professional agenda and intellectual *a priori* choices: in other words, a scientific concept revealed some of the mutual shaping of biomedical and social interests. This approach to scientific thought has proven immensely fruitful for social studies of science in the last two decades.[11] Clearly, the climate of uncertainty generated by the controversies on the nature of antigens and antibodies as well made immunology attractive for a social researcher, more than other disciplines more ensconced in the cloak of the quest for 'pure' scientific truth.

A Science Made up of Controversies

While more attention has recently been paid to controversies as the very substance of scientific community life,[12] the history of immunology seems to have involved an unusually high level of controversy concerning all of its main issues. The history of immunology may even be summarized as a sequence of controversies more than sweeping scientific revolutions. This remark suggests that immunologists work with a unit which is larger and more complex than the received view understood as a theory. It is difficult to follow either a general theory or even a central paradigm in the loose and versatile Kuhnian sense.

The first controversy had to do with the very nature of immunity or resistance to bacteria: was it cellular, the work of phagocytes, the large devouring cells which are today called macrophages or was it rather humoral, (an antiquated word borrowed from the Hippocratic tradition), and related to mysterious substances initially called antitoxins than more generally antibodies and directed to bacterial components called antigens? Paul Ehrlich, on the German side and Elie Metchnikoff on the French, each championed one of the two theories.

Before the First World War, other heated debates focused on the mechanism of the antigen-antibody reaction. Was it a physical or a chemical reaction? Was it specific or largely non-specific? Did it follow an all-or-nothing law, or did it admit a whole range of intermediate responses? Were antibodies one and the same molecule, or multiple entities, different according to the reactions for which they were responsible in the test-tube (hemolysins, agglutinins, precipitins . . .)? 'One or two?' 'One or many?' These questions were repeated for all cells or molecules hypothetically involved in immunity phenomena, such as complement, a blood substance which seemed to be necessary for the destruction of obnoxious bacteria as well as harmless red blood cells, once sensitized by the fixation of specific antibodies. In each case, the debate followed the pattern familiar to the readers of Plato's *Sophist.* But in the *Sophist,* Socrates makes the point that the pattern of dichotomies or dialectical alternatives does not lead to the positive essence of the truth, since binary division is an evergoing process.

From the 1920s, as chemists focused on antibodies, these controversies were replaced with the same logical pattern by the debate on antibody formation. Did the antigen directly command the folding and shaping of a molecule, a good-for-everything flexible template ('instructive' theory and its many variants)?[13] Or did the antigen select a preexistent antibody[14] or cell,[15] according to the 'selective' theory, leaving it for genetics to understand the underlying mechanism?

The question of an adaptive response was presented in Darwinian selective terms only in the 1960s when the dogmas of molecular biology forbade all direct action of the environment on the structure of antibodies, which were now recognized by the tools of analytical biochemistry as proteins genetically engineered. Thus immunology officially joined the Darwinian mainstream of biology, at the Congress of Cold Spring Harbor in 1967 which marked the rally of the community to the selective view.

In the 1950s, other debates kindled the interests of the scientific community. On the identity of the antibody-producing cell: lymphocyte (a cell hitherto mysteriously deprived of any known function) or plasmocyte, a cell obviously well equipped to manufacture molecules? On the nature of the elements responsible for graft rejection: (again) antibody or cell . . .? Then on the role of the thymus, a small vestigial organ located at the anterior part of the neck, which emerged in the 1960s as a candidate for the role of maestro[16] of the immune orchestra and where originated cells responsible for cellular immune response before migrating through the body.

Later, after two kinds of cells, called B cells and T cells (respectively for bone-marrow and thymus), had been identified as the main immunocompetent cells, a debate flared up on the existence and modes of their cooperation, on the nature of the receptors located at their surface: antibodies or antibody-like structures or unknown patterns, still to be discovered. On the role of newly described sites on cells called major histocompatibility molecules (analogous to the substances indicating distinct groups on blood red cells), in turn divided into different classes and finally understood as 'antigen-presenting' structures to the effectors of the immune response.

But the violence of polemics has not ultimately involved any major disruption in the scientific development of immunological science. On the contrary, with the passing of time, the alternatives, whatever the dichotomy on which they were grounded, were eventually dissolved. A mellowed synthesis of what seemed to be once fierce oppositions emerged quietly, putting an end to radical choices, so that the history of immunology follows a 'soft' version of the hegelian pattern of thesis-antithesis-synthesis.

Contemporary knowledge seems to echo the former questions and bring tempered responses to historical debates. Contrary to the ideology of great scientific revolutions, contemporary solutions look like scholarly compromises currently renegotiated among the scientific community. Where periodization typically saw succeeding eras of theoretical hegemony as separated as imperial dynasties, immunological science appears more as a step-by-step process or a series of chemical reactions, permanently up and downregulated according to the input of fresh data.

For example, it is acknowledged today that macrophages (or phagocytes) and antibodies play an equally important role in immunity, and that the immunocompetent cell can turn into a lymphocyte or a plasmocyte. Even the triumph of the clonal theory of antibody selection at the Cold Spring Harbor Symposium (1967), which asserted immunology's true Darwinian colors did not sweep away all doubts and objections. There was no crucial experiment but only stepwise falsification of many different hypotheses loosely linked to the instructive view.[17] The present consensus does not actually exclude in the future the return of neoinstructionist views, provided they remain compatible with the current data of molecular biology. The dogma of the unidirectional flow of information from genetic material in the nucleus to cytoplasma[18] has already been challenged by the discovery of the reverse transcriptase which acts in the opposite way, from cytoplasmic nucleic acids to nucleic chromosomal acids.

The problem of the generation of antibody diversity has been solved without locating in the nuclear genes the sole source of diversity. The clonal selection theory required a great number of ready-to-use antibodies with molecular structures fitting any presenting antigen. But it seemed hardly possible that a majority of the genome be used only for the sake of providing such diversity. Exploiting the new recombinant technology, Susumo Tonegawa showed in 1979 that antibody diversity resulted from a combination of random association between different V (for variable) genes and one single C (for constant) gene located on the chromosome, assorted with mechanisms of single point mutation along the path of cellular differentiation.[19] The selection by the antigen of the appropriate antibody, then to the fittest antibody (with maximum affinity), can thus be described as an adaptive response, occurring through development and lifelong expansion of the immune system, a combination of random and teleological processes.

Alternative theories, infact, do not succeed each other with a clear dividing-line. Previous hypotheses, once officially evicted, remain available in researchers' armamentum. This is true for the interpretation of immunization procedures. While vaccination efficacy seems to depend on the biological memory, it has remained unclear what maintains this memory. Is it the antigen, hidden in a corner of the body which boosts the organism from time to time, or is memory dependent on a property of the organization maintained by regulatory networks independent of any external source of stimulation?[20] While immunologists were rather prone toward the latter solution, in keeping with the selective paradigm, the study of infectious diseases has recently led scientists to reconsider the possibility that parasites like viruses or even bacteria may remain hidden and unnoticed inside the host where they continuously activate fresh immunity. The question is still pending as to whether protective immunological memory is or is not antigen-dependent, and it may be difficult to distinguish true immunological memory from low-grade responses.

Another example can be drawn from the history of allergy,[21] observed and analyzed since the beginning of immunological studies. While anaphylactic shock or the many unpleasant reactions collected under the name allergy, struck the pioneers of immunity as a perverse deviation of an otherwise beneficial function, current immunology tends to integrate allergic reactions as normal components of a complex reaction to be judged as a whole, and prone to permanent self-control and limitation. The immunoglobulin called Ig E (e for erythema), once hailed as the ideal marker for asthma, hay fever or parasitic diseases, is now viewed as a more elusive entity, not necessarily linked to a pathological context.

Stabilizing Immunology. The Immune System in the Era of Molecular Biology

As the number of scientists calling themselves immunologists passed from a harmful to several thousands between the beginning and the middle of the century,[22] a somewhat truistic definition of immunology as the science of the immune system, cells and molecules articulated together by the means of receptors,[23] was finally articulated.

Working on antidiphteria serum in 1896, Paul Ehrlich had considered its efficacy a case of chemical neutralization of the toxin. Immunity, in short, had to do with molecules. In his correspondence, he drew two shapes joined together and commented in pedantic Latin *"Corpora nisi fixata non agunt."* "Bodies do not have any effect unless they are bound." According to Ehrlich, cells were equipped with 'side-chains' or molecules ready to bind foreign bodies present in the environment fluids. The concept of side-chain, alias antibody, otherwise **receptor**, included the idea of a molecule fixation on the cell membrane and the liberation of a soluble factor into the medium. In other words, chemical fixation and biological effects were together encapsulated in his idea of receptor.

The idea of receptor, once suggested by Ehrlich as the Ariadne's thread of immunity theories, is still the guide for contemporary biology where everything occurs through receptors and their ligands. Most biological reactions require a primary phase of recognition: this recognition develops through a phase of interaction, involving weak or strong bonds depending on the distance and obeying the laws of thermodynamics as stated by the physicists' laws of mass action.

Paul Ehrlich illustrated his thinking during his seminal conference on antibodies in 1900 with drawings representing side chains or receptors on cells as arms protruding from the protoplasma and serving to capture foodstuff. These designs, once proscribed by those who doubted the existence of antibodies strengthened confidence among those people who believed in antibodies. They still adorn most textbooks of immunology for students. Illustrating immunity as a special case of a more general reaction, they suggested to biologists the possible pathways of biological reactivity.

To what extent did they express Ehrlich's unconscious, or witness the most simple way of accommodating data in the early days? Can we consider them as the fantastic anticipation of contemporary computerized models of the interaction between antigen and antibody? Is there a link between Ehrlich's fantastic imagery and diagrams currently drawn from cristallographic data which display a three-dimensional groove on the molecule surface acting as a lure for the antigen, ready to be hidden in the antibody prehistoric cave? The issue of representing immunity is one of the most fascinating questions confronting together historians and immunologists to the sources of scientific creativity.

Immunology as the study of the immune system has both stimulated and benefited from the molecular revolution. Since the 1960s, immunologists have continued to describe an extraordinary array of molecules and cells. This has been made possible by a succession of powerful instruments that allow to isolate, clone, filter out, sequence cells and molecules.[24] Each technological innovation has generated new entities, with a constant interaction between structure and function, biology and chemistry. The immune system is embodied by multiple lattices: enzyme cascades along metabolic pathways, receptors displayed on the membrane surfaces and interacting together, networks of soluble factors. Receptors and ligands are increasingly interpreted in terms of signals, a metaphor borrowed from communication theory.

A double nomenclature is used to designate member cells and molecules, either referred to their putative function or to the technique employed. For example, 'CD_4 cells,' famous as a parameter of clinical surveillance of AIDS seropositive patients, refers to the monoclonal antibody targeted to a specific receptor and used to characterize a family of cells. But the name is also more or less synonymous for 'helper cells,' supposed to operate as a link between the presentation of the antigen to the immune system and the actual response (cell proliferation, antibody or soluble factors production . . .) Factors were named lymphokines to refer to regulating the lymphocyte activity, or more generally cytokines, a name proposed for the family of secreted proteins involved in inflammatory reactions. The list of these molecules, duly sequenced and produced through recombinant technology grows everyday and the description of their functions grows increasingly intricate, suggesting some redundancy in the system and reviving the debate between specific and non-specific reactivity.

The Uniqueness and Diversity of Living Beings

Trying, far from controversies, to make efforts converge, concerned with the multiplication of entities with an elusive function and opposed to a restrictive chemical view of immunity, the Australian biologist Frank Mcfarlane Burnet tried to assign a global sense to immunology when he referred to it in 1969 as the science of *Self and Notself*, a title for one of his books. He returned thus to the phenomenon previously described by Paul Ehrlich as *horror autotoxicus* which stated that the organism cannot defend itself against itself. Burnet suggested that at an early stage, the organism in fact actively differentiates between 'self' and 'notself.' This ability is not only based on genetic makeup but may also be influenced by the subject's environment. Even if genetics is to acquire a growing importance and a commanding position in immunology, it should still be balanced by what occurs during development.

The concept of the "uniqueness of the individual"[25] has been elaborated for centuries by theological tradition. It was related to the personal responsibility and dignity imparted to each human creature, linked in a unique way to divine transcendence. The experience of individuality was thus rooted in Western culture, inemical to metempsychosis and to the idea of soul migration through different bodies. The modern idea of the immunological self can thence be related to the philosophical experience of the subject, the discovery of an 'Ego' irreducible to the other. But, in an innovative way, the immunological 'self' pointed, beyond the moral and metaphysical meaning traditionally attached to the notion of personal freedom, to biologically-grounded individuality, originating in genes and undergoing selective pressure throughout evolution, constantly modulated and oriented by historical developments. The individual is thus made twice unique, by heritage and by history.

Immunity became the main agent for maintaining the "integrity of the body"[26] and finally the cornerstone of biological individuality. Yet the reference to self and notself retained the vague and metaphysical character of a metaphor facilitating informal interactions of immunology with other fields of research. Burnet's metaphor of the immunological self seated in the body went further by suggesting that immunology was an avenue leading into the enigmas of the living world. As a fundamental human science, immunology was called upon to solve with its own concepts the great problems of ontogenesis and phylogenesis, homeostasis and biological adaptation, reproduction and aging, infection and carcinogenesis.

The idea of the diversity of creatures has been a major concept interwoven in the general theme of the magnificence of Creation. It has been said that the immune system is a microcosm,[27] which means that it expresses and mirrors the diversity of antigens, and may then present an easy target for selective pressure. Burnet's pseudo-definition, by referring to the impact of infectious diseases and the evolutionary value of immune reactions, creates a theoretical synthesis that reflected the rich network of analogies[28] that ultimately made immunology so fascinating and multidimensional.

By choosing his formulation, Burnet also emphasized his personal program of experimental tolerance, shared by the British zoologist Peter Medawar. Animal experimentation proved that the ability to differentiate between self and notself could be manipulated by presenting a foreign antigen at an early stage of development, in other words the frontiers of the environment could be modified by biological engineering or changes in the environment. This flexibility fitted the modern medical agenda well. Therefore the biological tentative definition of immunology as the science of the self/notself distinction coincided with the medical demand for the fluidity of such borders.

Organ Transplantation

The trespassing of the immunological barriers in organ transplantation, has been celebrated as a fine example of biology and medicine walking hand in hand. Among all manipulations inspired by immunological thinking, this one has undoubtedly been the most popularly recognized.

At the beginning of the century, surgeons had proved the feasibility of such an operation, but had, in Alexis Carrel's own terms,[29] vindicated the "biological nature" of the formidable obstacle and delegated to fundamentalists the task of solving it. Between the two world wars, work focused on the nature of chemical determinants responsible for the graft failure, these determinants being considered as the hallmark of biological individuality. Using human skin grafts for treating extensive burns during World War II, the zoologist Peter Medawar asserted that foreign skin could not be used for a permanent graft. Immunology offered an interpretative framework to the transplantation problem, formulated in terms of rejection, after Medawar in 1943 analyzed the accelerated rejection of an iterated graft from the same donor ('second set') as an indication of immunological memory.[30] Firstly, there was no mechanism for immediate recognition of identities. Secondly, the rapidity of the rejection in the second set of grafts coming from the same donor displayed the characteristics of a learned immune reaction, either mediated by cells or antibodies.

Yet the conviction remained that transplantation was possible against the evidence of immune rejection in individuals grafted with organs given by another individual. The observation that in rare cases, the graft of an organ from another species survived at least for a few weeks was in the 1960s considered as a tentative proof that somehow the natural barrier between individuals could be circumvented.

In the following years, two means of circumventing rejection were invented. On the one hand, organ grafting was planned with donors selected from their pattern of leucocyte groupings (called the HLA system), on a basis analogous to blood red cell groups as used for transfusion. On the other hand, the use of corticoids and other immunosuppressive drugs, initiated in the 1960s with the discovery of azathioprin became the usual method of avoiding the deleterious effects of immune response on the graft. The advent in 1980 of cyclosporin, a powerful immunosuppressive drug, allowed organ transplantation to be performed more easily. Nowadays liver, heart and lung, intestine transplantation are currently performed, their indications are curtailed more by the scarcity of available human organs than by immunological difficulties. It has been predicted that in the twenty-first century one surgical act out of four might be some kind of grafting.

Transplantation has received a strong social support which has helped to establish the institutions and networks necessary for organizing organ transfer all over the world such as Eurotransplant, initially planned to improve immunological compatibility by careful selection of donors out of a population as numerous as possible. Transplantation has even been celebrated as an opportunity for reconciling the technical progress with the necessity for modern societies to invent new forms of solidarity and exchange between the individuals and nations and even between the living and the dead.[31]

Immunology and AIDS

Beyond the special case of transplantation, the immune system has been increasingly adopted by doctors as an interpretative framework. After the 1940s, autoimmunity had been evoked for a limited set of affections among which figured prominently anaemias and endocrine diseases. Doctors came to understand more and more pathological disorders in terms of excessive, or deficient, and in all cases inappropriate responses of the immune system. Mapping diseases onto the immune system has now resulted in attributing a variety of diseases to systemic autoimmunity,[32] including diabetes, cancer and poorly understood neurological afflictions such as multiple sclerosis and rheumatoid arthritis.

With the ongoing epidemic of AIDS, an unpredicted synchronization has occurred. AIDS has been recognized as a general disorder of the immune response before its precise case was identified. Science confronts a tailor-made adversary. HIV has been primarily described as a virus that targets and disorganizes the whole immune system[33] and generally paralyzes and destroys its master cells, which even the layman knows as 'CD4.'

A full-grown science faces a challenge. The application of immunology to the AIDS enigma has already generated an enormous amount of knowledge, without, if truth to be told, any pathbreaking discoveries in the domain of immunological therapy. This situation has generated doubts in the scientific community and fund givers, and anxiety among the lay audience.[34] Only the future will reveal the outcome of this mortal challenge and whether immunology will fulfil expectations, or whether both AIDS and Immunology will be reconceptualized in a totally different way, now hidden from our modern and even post-modern views.

Some pathogen agents are called opportunistic in the sense that their virulence is conditioned by the host's immune status. AIDS by favoring the emergence of 'new virus,' once harmless in the human and aggressive only in distant species, illustrates the contextual nature of the pathogenicity. The emphasis on a wide range of parasites of various pathogenic power once again suggests the potential of immunology to challenge clearcut classifications and received definitions. This science once considered a paragon of specificity[35] contributes to a global interrogation of the limits of knowledge and the borders between biological entities.[36]

The Science of Boundaries

Faithful to its biological and medical double nature, contemporary immunology has followed a two-pronged pathway: on the one side, the identification of cellular and subcellular events of infectious disease and resistance to infection, and the fundamental basis of cellular life or the description of molecular recognition patterns and signalization, on the other. This circumstance fully reconciles those two aspects of immunology that had mostly developed independently: the story of successful immunization and the contribution to mainstream biology, conveniently referred to as the defensive paradigm and the cognitive paradigm. In the past, immunology has oscillated between an imperialist and ancillary position. But if immunology was in search of a place in the official classification of sciences, no doubt it was a very active and fruitful mode of thinking in biology and medicine, 'une science dans le siècle.' Only it met with some difficulty with self-limitation. As the search for new molecules and cells intensifies, the frontier between the immune system and the endocrine and nervous systems becomes more difficult to establish as they share ligands and receptors.[37] While immunity was initially defined as a new sensory system, the distinction between the major physiological systems tends to be blurred. The immune system suggests a more comprehensive view or ecumenic view of the body, still to be elaborated.

The definition of immunology that made so much for its success was its definition by Burnet as the science of "Self and Notself."[38] This metaphor that kindled the imagination of many scientists and still lingers in their unconscious, refers to the constant move by which organisms interact and continuously trespass what is never in fact a fixed barrier. So, rather than plagiarizing the definition of immunology as the Science of "Self and Notself," I will suggest a description of immunology as a contribution to the understanding of biological frontiers, playing on the double meaning of boundaries: the natural limits between living bodies and the frontier between subject and object in the process of knowledge acquisition. This definition has the merit of avoiding to delineate too strict a space for immunology in the architecture of science and preserving its oscillating and dynamic character.[39]

Immunology deals with unstable frontiers such as the ones which link/divide the mother and her fetus, the host and its parasite, the tumor and carrier, the body and its internal milieu, the graftee and the graft. The sense of immunology is in the trespassing of the borders which biologists considered discrete objects. It is perhaps this instability which explains that, in the midst of its prosperity, the generation of the 1960s lamented the near extinction of its discipline.[40] Of course, the uncertainty of its epistemological status demands philosophical attention. Perhaps immunology is not a modern science at all, but a "science en attente" for a new status, as Georges Canguilhem termed it. Immunology tends to cease focusing on the human immune system, and turn to stochastic and ecological views of the biological world where man is but one accident. The emergence of 'new viruses' helps to voice these new concerns. Immunology might finally jump from being a premodern discipline to the position of post-modern science, deprived of central focus and reflecting a new epistemological agenda. As a huge number of investigators have replaced the few pioneers of the beginnings, it becomes somewhat difficult to perceive the core hypotheses and the directions of the future. The collection of disparate data has already provoked Melvin Cohn's derisive remark about "small theories" or the "one fact one theory" approach.

Yet with AIDS, immunology faces a clearcut assignment, a unique occasion to measure up to its set goals, assess its past achievements and determine its future scope. The historical perspective is the best way for making sense of one of the most exciting ventures of twentieth-century biology. The coming years will tell how immunology will respond to the challenge of emerging pathologies including the Aids epidemics. Will immunology focus more narrowly its object or will it still compete for a global sense of the body, in medicine as well as in biology?

REFERENCES

1. Frank Macfarlane Burnet, Immunology as scholarly discipline, *Perspectives in Biology and Medicine* (Autumn 1972), p. 1–10.
2. Melvin Cohn, "On thinking and wisdom", *Annual Review of Immunology* (1994), **12**: p. 3.
3. Gustav Nossal, "Choices following antigen entry", *Annual Review of Immunology* (1995), **13**: p. 1.
4. *Journal of the History of Biology*, Special issue "Immunology as a historical object" (1994), **27**.
5. Frank Fenner and Coll (Ed.), *Smallpox and its Eradication.* (Geneva: W.H.O., 1987). Derrick Baxby, *Jenner's smallpox vaccine. The riddle of Vaccinia Virus and its origin.* (London: Heinemann, 1981). Anne Marie Moulin, "La métaphore vaccine, De l'inoculation à la vaccinologie", *History and Philosophy of Life Sciences* (1993), p. 271–297.
6. Jane S. Smith, *Patenting the Sun. Polio and the Salk Vaccine.* (New York: Anchor Books, 1990).
7. *L'aventure de la vaccination,* A.M. Moulin (Ed.). (Paris: Fayard, 1996).
8. G.S. Wilson, *The Hazards of Immunization.* (London: Oxford University Press, 1967).
9. Lily Kay, "Molecular biology and Pauling's immunochemistry: a neglected dimension", *History and Philosophy of Life Sciences* (1989), **11**: p. 29–43.
10. Ludwik Fleck, *Entstehung und Entwicklung einer wissenschaftlichen Tatsache, Einführung in die Lehre vom Denkstil und Denkkollektiv.* (Basel: B. Schwabe, 1935) (English translation *Genesis and development of a scientific fact.* (Chicago: University of Chicago Press, 1979); Anne Marie Moulin, "Fleck's Style", *Cognitive and Fact. Materials on Ludwik Fleck,* Thomas Schnelle and Robert S. Cohen (Eds.). (Dordrecht, Reidel, 1986), p. 407–419; I. Löwy, *The Polish School of Philosophy of Medicine; from Tytus Chalubinski (1896–1961) to Ludwick Fleck (896–1961).* (Kluwer: Dordrecht, 1990).
11. Peter Keating and Alberto Cambrosio, "Ours is an engineering approach": Flow cytometry and the constitution of human T-cell subsets, *Journal of the History of Biology* (1994), **27**: p. 449–479.
12. H. Tristram Engelhardt and Arthur L. Caplan, *Scientific Controversies. Case Studies in the Resolution and Closure of Disputes in Science and Technology.* (Berkeley: California University Press, 1987).
13. F. Brein und F. Haurowitz, Chemische Untersuchungen, *Hoppe-Seyler Zeitungschrift* (1930), **192**, p. 45–57; L. Pauling, A theory of the structure and process of formation of antibodies, *Journal of the Chemical Society* (1940), **62**: p. 2643–2657.
14. N.K. Jerne, "The natural selection theory of antibody formation", *Proceedings of the National Academy of Sciences* (1954, 1955), **41**: p.
15. F.M. Burnet, A modification of Jerne's theory, *Australian Journal of Science* (1957), **20**: p. 67–69.
16. Jacques F. Miller, "The thymus, maestro of the immune system", *BioEssays* (1994), **16**: p. 509.
17. Kenneth F. Schaffner, "Theory changes in immunology", *Theoretical Medicine* (1992), **13**: p. 175–189.
18. Jan Sapp, *Beyond the Gene.* (Oxford: Oxford University Press, 1988).
19. Susumu Tonegawa, "Somatic generation of antibody diversity, *Nature* (1983), **302**: p. 5009–5015.
20. Niels Kaj Jerne, "Towards a network theory of the immune system", *Annales d'lmmunologie* (1974), **125C**: p. 373–389.
21. Clemens Von Pirquet, "Allergy", *Archives of Internal Medicine* (1911), **7**: p. 274.
22. Arthur M. Silverstein and Thomas Söderqvist, "History of Immunology, The Structure and Dynamics of Immunology, 1951–1972: a prosopographical study of international meetings", *Cellular Immunology* (1994), **158**: p. 1–28.
23. Anne Marie Moulin, "The Immune System; a Key Concept for the History of Immunology", *History and Philosophy of Life Sciences* (1989), **11**: 13–28.
24. Adele E. Clarke and Joan H. Fujimura (Eds.), *The Right Tools for the Job: At Work in Twentieth Century Sciences.* (Chicago: Chicago University Press, 1994).
25. Peter B. Medawar, *The Uniqueness of the Individual.* (London: Methuen, 1957).
26. Frank Macfarlane Burnet, *The Integrity of the Body. A discussion of modern immunological ideas.* (Cambridge: Harvard University Press, 1962).

27. I.R. Cohen, "Microbial immunity and the immunological homunculus", *Immunology Today* (1991), **12**: p. 105–109.
28. A. David Napier, *Foreign Bodies*, California University Press, Berkeley 1992; Une sélection qui n'est pas naturelle: les modèles sociaux du monde microbien, *L'aventure de la vaccination*, A.M. Moulin (Ed.). (Paris: Fayard, 1996), p. 409–422.
29. Alexis Carrel, "Remote results", *Journal of Experimental Medicine* (1910), **12**: p. 146.
30. Gibson and P.B. Medawar, "The fate of skin homografts in man", *Journal of Anatomy* (1943), 299–309.
31. A.M. Moulin, "The ethical crisis of organ transplants. In search of a cultural 'compatibility'", *Diogenes* (1995), **172**: 43, p. 73–92.
32. Moncef Zouali, "Development of human antibody variable genes in systemic autoimmunity", *Immunological Reviews* (1992), **128**: p. 73–99.
33. A.A. Hamburg and A.S. Fauci, "AIDS, the challenge to biomedical research", *Daedalus.* (Winter, 1989), p. 19–39; B. Seytre, *Historie de la recherche sur le Sida.* (Paris: Presses universitaires de France, 1995).
34. *AIDS. The Failure of Contemporary Science.* (London: Fourth Estate, 1996).
35. Pauline M. Mazumdar, *Species and Specificity.* (Cambridge: Cambridge University Press, 1995).
36. Dorion Sagan and Lynn Margulis, The uncut self, *Organism and the Origins of Self*, Al Tauber (Ed.). (Boston: Kluwer, 1991), p. 361–372; E. Fox-Keller, *Secrets of Life. Essays on Language, Gender and Science.* (New York: Routledge, 1992).
37. J. Edwin Blalock, "The syntax of immune-neuroendocrine communication", *Immunology Today* (1994), **15**: p. 504–510.
38. Frank Macfarlane Burnet, *Cellular Immunology. Self and Not-Self.* (Cambridge: Cambridge University Press, 1969).
39. Marc Daëron, *Le système immunitaire ou l'immunité, cent ans après Pasteur.* (Paris: Nathan, 1996), Introduction.
40. Niels Kaj Jerne, "Waiting for the End", *Cold Spring Harbor Symposia for Quantitative Biology* (1967), **32**: p. 569–575.

Further Reading

D.J. Bibel (Ed.), *Milestones in Immunology. A historical exploration.* (Madison: Science Tech Pub., 1988).

G. Corbellini (Ed.), *L'evoluzione del pensiero immunologico.* (Torino: Boringhieri, 1990).

Marc Daëron (Ed.), *Le système immunitaire ou l'immunité cent ans après Pasteur.* (Paris: Nathan, 1996).

Hermann W. Fridman, *Le cerveau mobile.* (Paris: Hermann, 1992).

Richard Gallagher, Jean Gilder, Gustav J.V. Nossal and Gaetano Salvatore (Eds.) *Immunology. The Making of a Modern Science.* (New York: Academic Press, 1995).

Edward S. Golub and Douglas R. Green, *Immunology, A Synthesis.* (Sinauer, 1991).

Jan Klein, *Immunology. The Science of Self and Not-Self Discrimination.* (New York: Wiley 1992).

Emily Martin, *Flexible Bodies, Tracking Immunity in American Culture.* (Boston: Beacon Press, 1994).

Anne Marie Moulin, *Le dernier langage de la médecine. Histoire de l'immunologie de Pasteur au SIDA.* (Paris: Presses Universitaires de France, 1991).

Pauline H. Mazumdar, *Species and Specificity. An Interpretation of the History of Immunology.* (New York: Cambridge University Press, 1995).

Arthur M. Silverstein, *A History of Immunology* (New York: Academic Press, 1990).

Alfred I. Tauber, *The Immune Self: Theory of Metaphor.* (New York: Cambridge University Press, 1994).

CHAPTER 25

The Molecular Transformation of Twentieth-Century Biology

PNINA G. ABIR-AM

The molecular transformation of biology in the twentieth century, can be seen as an ongoing historical process of 'progressive colonization' by the so-called exact sciences (i.e., chemistry, physics, mathematics, and engineering, or rather combinations of their leading sub-disciplines, such as organic chemistry or atomic physics).[1] It has undergone three distinct phases of transdisciplinary stabilization, resulting in the hegemony of post-World War I biochemistry, post-World War II molecular biology, and post-Cold War biotechnology. The transdisciplinary configuration of each phase, though invariably originating in the pre-war periods, was drastically reinforced by the international settlements that reconfigured the world order in 1918, 1945, and 1989, and which concluded World War I, World War II and the Cold War, respectively.[2] Moreover, those ever changing international spaces enabled and constrained distinct patterns of 'transnational objectivity' or scientific outcomes that transcended the control of national research traditions while invariably entailing new transdisciplinary concepts, most notably the double helix, the messenger-RNA, the operon, the genetic code, allostery.[3]

The initially metaphorical, but subsequently literal designations that each phase acquired at the time of its stabilization or successful institutionalization in academia reveal, the 'transforming' disciplines acting upon biology to produce new, hybrid fields which came to include chemistry, physics, mathematics, and technology. Thus, in the aftermath of World War I, 'biochemistry' was institutionalized as a new hybrid field while integrating biological problems with chemical techniques at the same time that experimental (as opposed to observational) biology acquired a new legitimacy beyond the overshadowing presence of evolutionary biology. The transforming discipline of chemistry emerged out of World War I endowed with a most powerful duality with regard to societal reconstruction: on the one hand it was the science behind industrial recovery including reparations, while on the other hand its performance in the then new gas warfare of the last days of World War I made it famous as a discipline capable of inducing unconventional destruction.[4]

In a similar manner, in the aftermath of World War II, 'molecular biology' acquired scientific hegemony in part as a result of its consistent deployment of a variety of physical methods, most notably X-ray diffraction and electron microscopy, to solve the structure of macromolecules, especially proteins and nucleic acids, as well as that of biological assemblies, especially viruses. Though the sub-cellular and supra-molecular domain of 'neglected dimensions' became accessible for investigating biological order in the molecular world just prior to World War II, molecular biology received an unusual boost from its capacity to 'redeem' physics, which emerged out of World War II as the most powerful science, once again in the dual sense of embodying both promises of life in the form of civilian uses and control of atomic energy, and fear of unconventional death through nuclear annihilation.[5]

In a parallel manner, 'biotechnology' quickly rose to social preeminence in the post-Cold War era, that began *de facto* in the mid-1970s and *de jure* after 1989, while including an explicit link to the world of business as the unprecedented skyrocketing of its stock in 1981 clearly demonstrated. In this case, the transforming sciences were chiefly 'high' technologies, such as automated sequencing, and computer technologies that enabled the storing and dissemination of huge databases of molecular codes. Once again, the transforming science in the post-Cold War era was computer science,[6] a hybrid field at the interface of mathematics and engineering that developed rapidly during the Cold War, in part because of its crucial applications in military intelligence. It further emerged as a most powerful 'technoscience,' in the post-Cold War era once some of its new information technologies, most notably the internet, turned out to resonate extremely well with the new socio-economic world order of increasingly permeable national borders. In this manner, 'biotechnology' became a key to policy in the twenty-first century, especially since its legal and ethical implications have increasingly come to the fore in emerging biosocial policy, further challenging for the first time the stability of evolutionary species, or seeking to improve upon 'inborn errors of metabolism' with transgenic strategies.

Since the term 'molecular biology' has now acquired a wider sociocultural currency than either biochemistry (not sufficiently 'trendy') or biotechnology (too recent and too 'applied'; this may still change due to biotechnology's ongoing prominence in business);[7] I chose to situate the rise of molecular biology in the middle third of the twentieth century in the context of other key movements of molecularization in biology, while drawing attention to structural similarities between the above three, interdependent movements.

In attempting to clarify this transformative process impinging upon biology via three molecular stages of intervention by exact sciences, rendered politically powerful and scientifically hegemonic by each of the three major wars that shaped the historical landscape of the twentieth century, the first question is whether these three phases of historicity imply a structural isomorphism that reflects recurrent causes. Was the transformation of biology by chemistry in the aftermath of World

War I causally related to its transformation by physics in the aftermath of World War II, and by engineering in the aftermath of the Cold War? Were these three successive transformations structurally equivalent in terms of their conceptual, social, institutional, and political impact?

A related question is whether these successive phases were cumulative or incommensurable, and whether each successive phase derived from scientific challenges that the previous phase failed to address in either a contingent or necessary manner. This chapter argues that each successive phase did not substitute for a previous one, which continued to grow, but rather shifted the scientific hegemony of a given strategy of molecular transformation, from 'biochemistry' to 'molecular biology' in the post-World War II period, and from 'molecular biology' to 'biotechnology' in the post-Cold War era.

In exploring each historical phase of post-war transdisciplinary stability, I look at the specificity of the research programs, scientific institutions, and science policy actions of each phase. Second, I explore the transition from phase to phase, so to clarify scientifico-historical change throughout the century. I conclude with a reflection on the careers of science statesmen in twentieth century molecular life sciences as the human nexus that linked the macro-level of international politics and the micro-level of scientific discovery.

Phase I: The Age of (Metabolic) Biochemistry, 1900–1937

The first phase in the molecular transformation of biology, stretching over the first third of the twentieth century, consisted of the world wide institutionalization of 'biochemistry,' but especially in Great Britain and Germany, the inter-war 'meccas' for biochemists. This hybrid discipline, combining techniques from organic chemistry with biological problems, drew its initial quest for a new disciplinary identity from the enzymatic uniqueness of chemical processes in the living cell. From a formerly dual marginality as a neglected province on the borderline or 'no man's land' between the antagonistic disciplinary empires of chemistry and physiology in the several decades preceding the First World War (known as 'physiological chemistry' following the establishment of the first journal in 1877); biochemistry catapulted itself in the aftermath of 1918 into a success story. This culminated in the elucidation, by the mid-1930s, of "intermediary metabolism" or the pathways of enzymatic reactions that sustain key life processes, (e.g., fermentation, photosynthesis, respiration, and digestion) while producing convertible energy and building blocks.[8]

Throughout this period the molecular transformation of biology took place at the level of small and medium size molecules, focusing on the enzymatic breakdown of carbohydrates, lipids, and proteins. The main methodological source of transforming biology was the discipline of chemistry in general, and the sub-field of organic chemistry in particular. The conceptual and experimental core of the new discipline of biochemistry revolved around the purification, isolation, and characterization of enzymes, the kinetics of the reactions they catalyzed, and their integration, into physiological effects, such as muscle contraction, or tissue growth.

The direct impetus to change the status and image of biochemistry from a twice marginal and often ridiculed province on the fringes of physiology (which treated it as a service domain remote from clinical responsibility) and [organic] chemistry (which treated it as incapable of achieving the same level of analytical accuracy), to a high profile subject of student preference and state, industry, and philanthropic patronage, was given by the recognition of biochemistry's multi-dimensional patriotic service in World War I. This process was particularly evident in Great Britain and Germany though after the Nazis' rise to power in 1933, numerous scientists, including a large contingent of Nobelist, and would-be Nobelist, biochemists were either dismissed or chose to leave,[9] to the effect that the rise of biochemistry as a discipline of growing societal prestige was particularly consistent in Great Britain. There, the school of biochemistry at Cambridge University became the subject of world wide emulation, first and foremost throughout the British Empire, but also in the US, and other countries from China to Scandinavia.

With the Biochemical Club, initially a dining club founded in 1911 becoming the Biochemical Society in 1913, and only one autonomous department in existence prior to World War I, (at Liverpool University) it is obvious that biochemistry's subsequent success story derived from its performance, as well as the perception of its performance, of diverse national services in the Great Patriotic War. First, it managed to become an authority on the home front, by providing advice on the crucial topic of adequate nutrition for both civilians and the armed forces, under conditions of scarcity, while further capitalizing upon biochemistry's spectacular association with the discovery of vitamins. Those 'miraculous' compounds captured the popular imagination, since only tiny quantities were needed to avert horrible deficiency diseases, most notably scurvy and beri-beri, afflictions that directly endangered the predominant colonialist preoccupations of the would be Allies as well as Axis empires, both before and during World War I.[10]

Second, biochemistry proved its value on the supply line to the front where the biochemists' ingenuity filled in the gap created by the naval blockade and the resulting inability to import certain raw materials from the real or 'virtual' colonies. The invention of a biochemical process for the production of aceton, a key ingredient in munition production (formerly obtained from nitrates, imported to Great Britain from Chilean mines) is well known as an intertwinned saga of science and politics, or a saga in which scientific achievement crucial to the British war effort was rewarded in (potential) political currency in the international arena.

This intimate connection of biochemistry with both the war effort and the post-World War I settlement (referring in this case specifically to the British mandate of areas transferred to its 'care' after the dismantling of the Ottoman Empire, as part of the international treaty at Versailles) was further reflected in the fact that the same person, Lord Balfour, inaugurated in the mid-1920s, both the new Dunn

Institute of Biochemistry at Cambridge, the scientific center of the British Empire, and the Hebrew University of Jerusalem, the cultural center of the Jewish homeland that the Balfour declaration enunciated in 1917 on behalf of H.M. government. The Balfour declaration had tremendous political consequences in the international arena, some of them persisting to this very day. Yet, it originated in recognition for a biochemical process that played a key role in the war effort. This process was perfected by the Manchester University biochemist Chaim Weizmann (1872–1954), a Zionist leader, founder of the internationally renowned Weizmann Institute for Science, and the first President of the State of Israel.[11]

Along similar but more domestic lines, Sir Frederic Gowland Hopkins (1860–1947), who was the first holder of a Chair in Biochemistry at Cambridge University, the 1929 co-recipient of the Nobel Prize for his discoveries in the socially charged field of vitamins, and President of the Royal Society in the early 1930s, argued publicly that biochemistry was the key to social stability and progress. At that time, social stability was threatened by the Great Depression, dramatically epitomized by the Hunger marches of the unemployed, the collapse of the Gold Standard, and the demise of the Second Labour Government. Hopkins' Address at the Centennial Meeting of the British Association for the Advancement of Science, (1931) in which he contrasted biochemistry as the analytical science of life with physics, as a science of destruction and death, played a key role in reorienting the Rockefeller Foundation's patronage from the physical to the experimental and molecular life sciences.[12]

The post-World War I international settlement impacted upon science policy for biochemistry (or for this matter upon any other scientific discipline, especially a rising one) at two levels, the national and the international. At the international level, it produced resolutions by international scientific bodies such as the meeting of Academies of Science of Allied countries that called upon governments to increase the budgets of pure science, and lay foundations for international exchanges in the spirit of the League of Nations (though a scientific Locarno came later than the political one). However, in view of the relative retreat from the international arena of the main post-World War I creditor, the US government, the majority of the international scientific exchanges remained in the hands of philanthropic foundations, most notably the Rockefeller Foundation, which began a pattern of considerable investment in science, both in the US and Europe.[13]

In the UK, however, biomedical policy was shaped by the Medical Research Council (hereafter MRC), whose first Secretary, Sir Walter Fletcher, exercised expansive leadership in general (i.e., vis-a-vis other governmental or public bodies) while specifically supporting biochemistry. The MRC's centralized and interventionist role in biomedical policy, at least during Fletcher's tenure (1919–1933) also included the administration of some of the Rockefeller Foundation's own funds, as well as the funds of some British charities in biomedical fields.[14] The MRC was one of the National Research Councils that were established in many countries, most notably the UK, Canada, and the US, at the end of World War I, while investing wartime committees with top national authority (e.g., being attached to the Prime Minister's Office as opposed to being part of the Ministry of Health).

British biomedical policy after the First World War supported the autonomous rise of basic biochemistry because of the complex relationship between two Cambridge physiological scientists who collaborated before the war on muscle physiology and biochemistry: Fletcher, an interventionist civil servant, and Hopkins, an innovative but enigmatic scientist. Fletcher's championship of an Institute of Biochemistry for Hopkins at Cambridge derived from his loyalty to his *alma mater* as well as from a hope, that was not to be fulfilled, that such an Institute might serve as a source of scientific innovation in matters of nutrition, or applied biochemistry, a subject with implications for public policy that was to erupt into a national scandal in the mid-1930s.[15] For this purpose, he secured various research grants that the MRC gave to biochemists at Cambridge and elsewhere, but more especially he was instrumental in negotiating the large private Dunn bequest as an endowment at Cambridge University that provided both a building, and various salaried positions for lecturers and other academic staff. Biochemistry at Cambridge was incorporated into the Tripos in 1934–5, becoming an 'obligatory passage' for the international traffic in biochemists.

However, by the mid- but especially the late-1930s, an ongoing transition from biochemistry to molecular biology could be observed at all of the above mentioned levels of science policy and patronage, formal institutional and informal social networks, and research programs. Sir Edward Mellanby, who succeeded Fletcher upon his death in 1933, developed an almost opposite approach to biomedical policy, one which was neither interventionist nor in favor of basic research. Lacking Fletcher's sympathy for Cambridge, Mellanby managed not only to prevent some of the Cambridge biochemists from receiving stable support from the MRC but adversely affected their prospects with the Rockefeller Foundation.[16] Even when he belatedly agreed to support some biomolecular work at Cambridge, in 1947, on the structure of haemoglobin, the 'molecular lung,' Mellanby did so in response to a plea from the Professor of Physics, not that of Biochemistry.[17] Mellanby's reluctance is reflected in the reaction of the foremost British clinical researcher Sir Thomas Lewis who appealed in vain for some MRC initiatives in retraining clinical researchers at the end of World War II: "What are you waiting for? The German surrender? The Japanese defeat? What is the plan? Am I right in thinking that there is no plan?"[18]

At the formal institutional level, Hopkins, though at the peak of his fame and formal power, did not manage to secure university commitment for expansion of his department, because the chairs of other departments, who were not as close to retirement, prevailed in the university committees. Nor did he seem able, in the aftermath of a unique relationship with Fletcher, lasting over two decades, to establish a similar rapport with the new MRC Secretary, a former research student of his. In the same vein, Hopkins did not understand or care about the Rockefeller Foundation's requirement that targeted grantees such as himself must solicit its projected investments. Hopkins was also constrained by his personal affection for Joseph Needham, his second in command whose forays into the diverse fields of biochemical embryology, philosophy, Marxism, Anglo-catholic religiosity, and socialist activism, which he further sought to unify, were not shared by other members of the department. Though they idolized Hopkins, admiring his biochemical manifestos for balancing the different biological and chemical agendas, and his liberal socialist politics, the disciples became chiefly concerned with solidifying the scientific status of biochemistry, which at that time meant the pursuit of the chemical method at the expense of new biological problems.[19]

At the level of informal social networks, Needham, who succeeded J.B.S. Haldane (yet another brilliant biochemist who transferred to population genetics) as Reader in Biochemistry in 1933, had already begun participating as a founding member in the Biotheoretical Gathering. This was a scientific Bloomsbury of theoretically minded scientists from different disciplines. They explored new relationships between the physical and the biological sciences, in light of the then novel theoretical positions afforded by the Marxist interpretation of science as well as the revolution in physics. This group, with which Needham was involved for the entire decade, not only drifted away from biochemistry but came under the influence of J. Desmond Bernal and Dorothy M. Wrinch, a physicist and a mathematician, respectively, who had little interest in metabolic biochemistry, though they would appropriate proteins for their wider structural agendas in what became known as 'molecular biology.'[20]

Last but not least, at the level of research programs, the problem of protein structure emerged as the 'secret of life' in the mid- and late-1930s. But this problem only further highlighted biochemistry's self-imposed disciplinary limitation in dealing with a topic that acquired multi-disciplinary scientific relevance. Indeed, the fact that after 1935 the Biotheoretical Gathering embraced Bernal and Wrinch's pioneering approaches to protein structure meant that even the second in command of Cambridge biochemistry, Needham, joined an agenda for which biochemistry was at best marginal. Grasping this awkward situation, Needham declined to become formal successor quoting his administrative unsuitability, and eventually left for a war mission to China from which he returned with dominant interests other than biochemistry.[21]

Cambridge University's refusal to appoint Hans Krebs, in Britain since 1933 as a refugee from Germany, as Hopkins' successor, (Krebs was later appointed to a Chair in Oxford shortly after he won the Nobel Prize in 1953) eventually led to an institutional and intellectual limbo that ensured both the decline of biochemistry and the rise of molecular biology at Cambridge in the period after World War II. A clear symbol of this change of disciplinary emphases is Fred Sanger, a twice Nobelist scientist (in 1958 and 1980, for innovations in sequencing methods for proteins and nucleic acids, respectively) who left the Cambridge department of biochemistry to align himself with the new MRC Laboratory of molecular biology.[22] As the official statements of the Biochemical Society in the late 1960s, incidentally chaired by Krebs, made clear (in a spirited response to a science policy governmental document produced by molecular biologists), the disciplinary rivalry between biochemistry and molecular biology had moved from Cambridge to the national arena, where biochemists felt compelled to challenge the ever encroaching policy initiatives of the molecular biologists.[23]

PHASE II: THE AGE OF MOLECULAR BIOLOGY, 1938–1973

Much as biochemistry in the inter-war period owed its scientific hegemony to its success in building upon its social missions during World War I, whether in nutrition or ammunition; the rise of molecular biology to scientific hegemony by the 1960s is largely, if not entirely, a result of its resonance with post-World War II preoccupations, whether in the domains of military intelligence (e.g., the concern with coding and decoding in an era of superpower confrontation) or civilian health. The viral epidemic of polio in the late 1940s and early 1950s provided large private and governmental appropriations for research on viruses, some of which becoming part and parcel of the new field of molecular biology.

Science policy after World War II and the new scientific institutions, informal collaborative networks, and new research programs that it triggered, eventually sustained a shift, by the 1960s, from the hegemony of biochemistry to that of molecular biology. This occurred in part because would-be molecular biologists projected themselves as strategic disciples, via institutional alliances and rhetorical manifestoes of spiritual kinship, with the scientific discipline that reaped most of the glory resulting from building the atomic bomb and its subsequent use to shorten the war, i.e., atomic physics.[24]

Furthermore, through its kinship with biophysics, yet another previously marginal field on the outskirts of physics and biomedical science, which received a great boost after World War II as a potential scientific and cultural redeemer of the physicists' conscience and tainted skills; molecular biology appealed to many as a hopeful outlet for scientific and societal energies. But it also cultivated a strong connection to the science policy apparatus that emerged in the context of the Cold War, a connection that had both political and scientific payoff. While budgets increased dramatically especially after the Sputnik triggered panic of the late 1950s, new concepts were imported from war related physical sciences, especially operations research, cybernetics, computer science. Though their actual scientific payoff remains debatable, the imported terms not only proved heuristically useful but gave the new field both a new, trendy allure and a claim to distinctiveness from classical biochemistry. This adaptability paved the way to the relatively rapid acquisition of autonomous status by molecular biology in the counter-cultural decade of the 1960s, when change *per se* was deemed desirable.[25]

Figure 25.1: An Autoclave from the early twentieth century. © The Pasteur Institute.

Figure 25.2: An Autoclave in a late twentieth-century laboratory.

FIGURE 25.3: THE LEGROUX LABORATORY, PARIS, 1916–1917.

FIGURE 25.4: GENE THERAPY TRIALS FOR CYSTIC FIBROSIS IN PROGRESS IN PROFESSOR BOB WILLIAMSON'S LABORATORY AT ST MARY'S HOSPITAL MEDICAL SCHOOL (WELLCOME CENTRE MEDICAL PHOTOGRAPHIC LIBRARY).

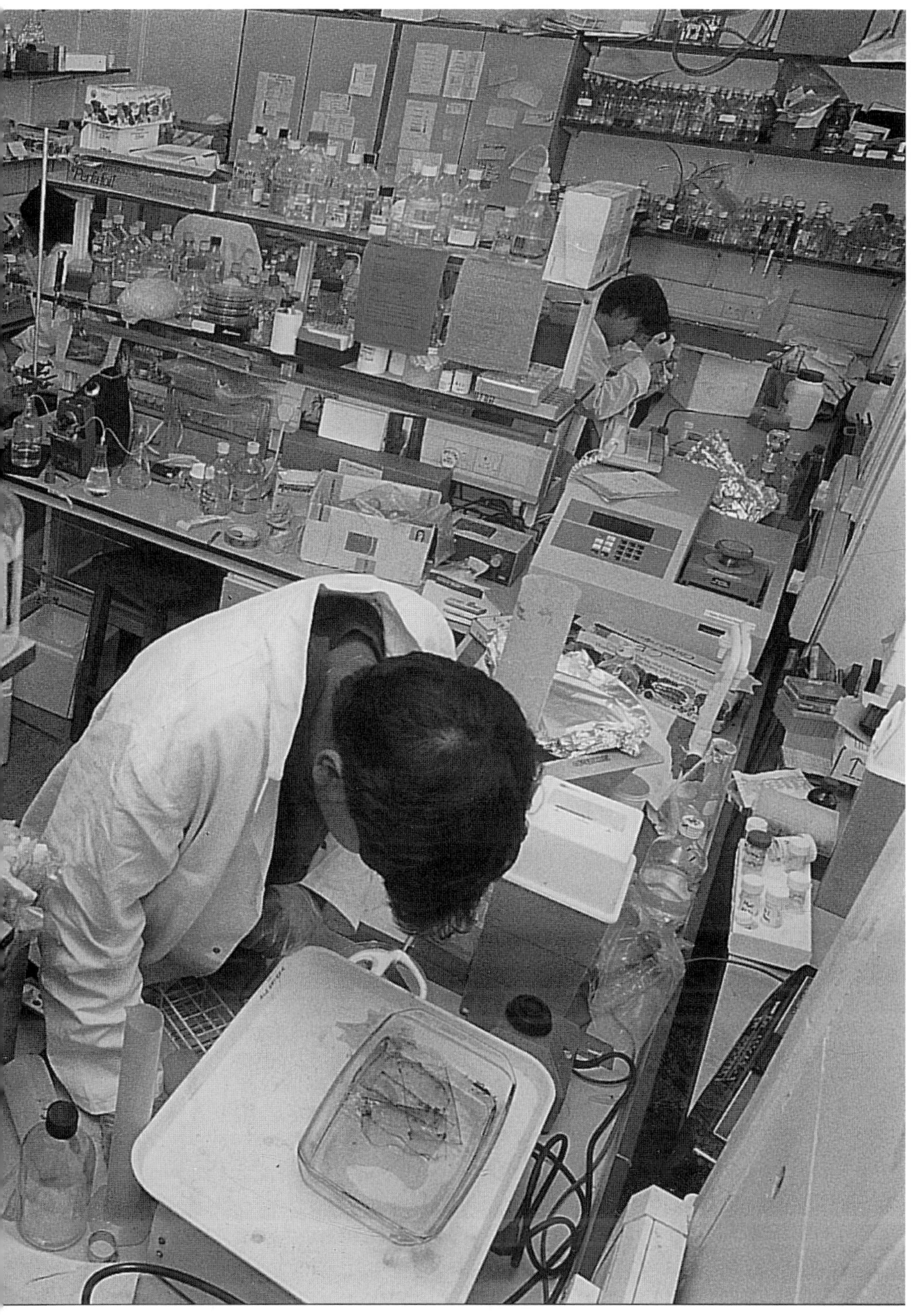

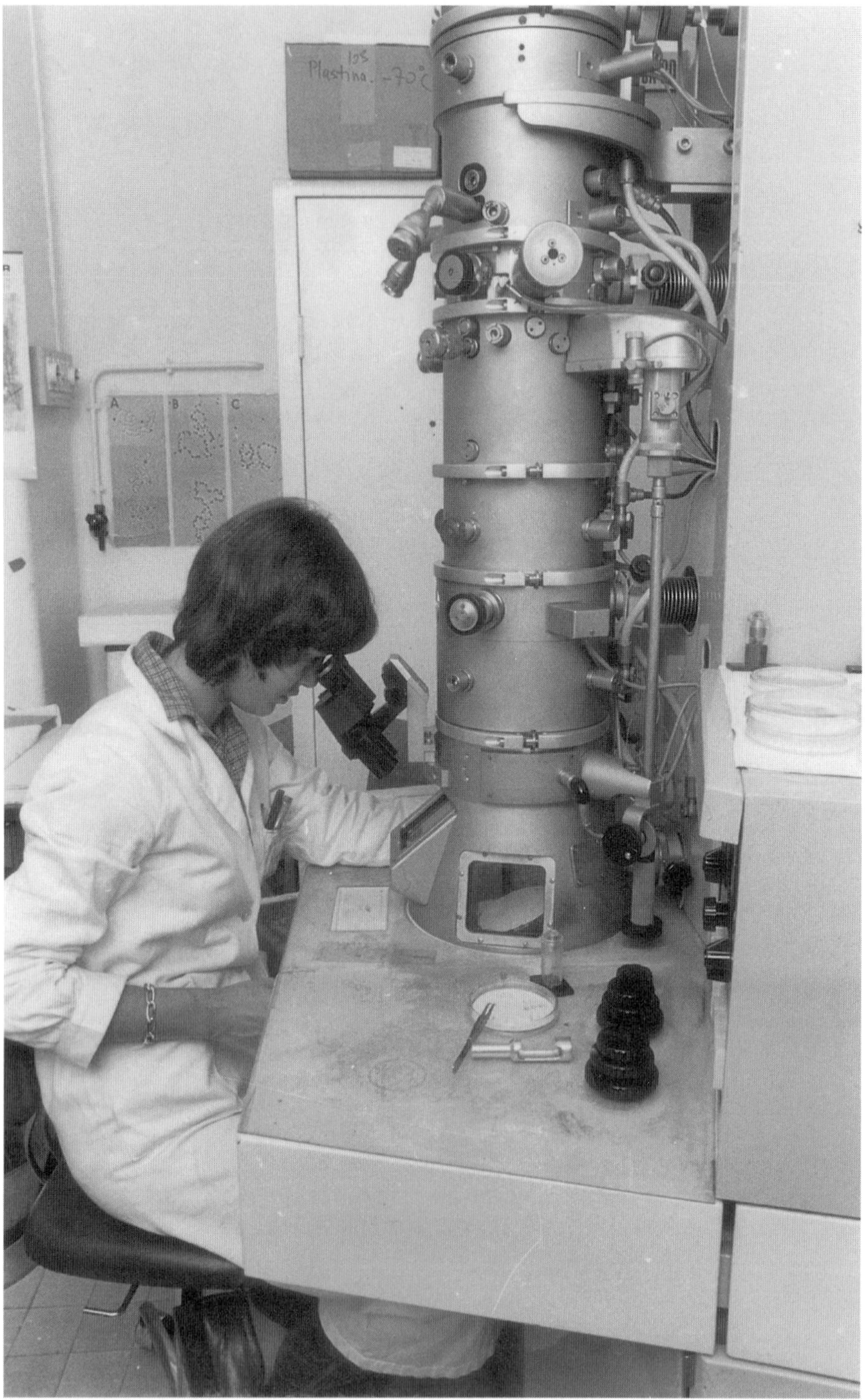

FIGURE 25.5: USE OF AN ELECTRON MICROSCOPE 1970. © THE PASTEUR INSTITUTE.

Molecular biology began to develop around empirical and theoretical innovations, such as the first protein X-ray photo (1934) and the first theory of protein structure (1936), as well as the overall discourse on protein structure, or the 'secret of life' that captivated the best scientific minds from numerous disciplines and countries in the mid- and late-1930s. Protein structure emerged as a scientific utopia that best resonated with the social utopia of the 1930s, i.e., the communist response to the largest ever, other than global war, political, economic, military, and social polarization. Indeed, by the end of the decade this massive cumulative polarization erupted into World War II. The efflorescence of transdisciplinary and international meetings on protein structure held in Paris, Klampenborg (Denmark), Cold Spring Harbor (US), London, and Edinburgh, between 1937–1939, not to mention many other specialized sessions at disciplinary meetings of biological, chemical, and physical societies, culminated with the appearance of this metaphor, in science policy documents, in 1938.[26]

The problem of protein structure had posed a transdisciplinary challenge since the mid-1930s, namely, how to solve a biologically significant problem deriving from the proteins' universal and versatile involvement in life processes, including respiration, digestion and reproduction, in terms of a complete resolution of their chemical and spatial structures, structures that because of their complexity required a battery of physical and physico-chemical techniques, ranging from X-ray crystallography to the ultracentrifuge, electrophoresis, chromatography, and electron microscopy, among others.[27]

As such, the problem of protein structure not only rendered biochemistry marginal but opened the way to the entry of physicists, physical chemists, and mathematicians, among other well meaning 'colonizers' who competed over a new promising territory in the name of their mastery of new technologies. Despite its 'natural' affinity with proteins, which included enzymes — biochemistry's quintessential molecular catalysts of life processes — the conceptual framework of biochemistry became constraining. It remained too closely linked to organic chemical methods, at a time when chemists were slowly recovering from their previous rejection of the concept of the macromolecule. Furthermore, biochemists had no tools to solve the spatial structure of macromolecules, such as proteins, since the hegemony of chemists also meant a resistance or delayed acceptance to methods such as X-ray diffraction. Indeed, in the generation between the mid-1930s and the mid-1950s, the concept of the macromolecule and the technique of X-ray diffraction became pillars of molecular biology, via their contribution to the solution of protein structure, the problem which was long believed to hold the ultimate explanation of biological order (hence, its flashy designation 'the secret of life').

The fact that the problem of protein structure also resonated with the great socialist utopia of the 1930s gave it a hard-to-match scientific urgency. There was an unmistakable resonance between the collective effort required to build a new socialist society and the effort required to solve protein structure with its tens of thousands of reflections, of different intensities. The enormity of the task was in both cases only matched by the great promise of a utopia. Similarly, international exchanges around the revolutionary problem of protein structure paralleled socialist hopes at the time for an international embrace of the new revolutionary social order. International cooperation was as much a socialist as a scientific slogan of the era in which the master organizations for intense international operations were both the Comintern and the Rockefeller Foundation.

The irony of the early days of molecular biology should not be lost if we recall that protein structure was mostly bankrolled by the Rockefeller Foundation whose own international and interdisciplinary agenda, though geared to salvage capitalism from the 'social unrest' of the 1930s, also construed protein structure as a multiple rationale for social progress and managerial efficiency. Protein structure meant anything and everything to anybody who was somebody in the late 1930s. But how did the utopian discourse on protein structure lead the molecular transformation of biology from biochemistry to molecular biology in the post-World War II period?

By the outbreak of World War II, it was already clear that traditional disciplinary methods were no longer sufficient for carving meaning in a transdisciplinary universe. At the end of the war, as a result of the drastic use of atomic bombs that signified the synergy of both transdisciplinary and international collaboration, the valorification of such collaboration on a wide social scale became a tenet of the post-World War II science policy that flourished in response to the strategical implications of the bomb. A new 'poetics and politics of movement' across disciplinary boundaries became not only possible, but a constitutive routine. Grants specifically designed to enable physical scientists to redeem themselves after their contribution to large scale destruction by retooling their skills in more peaceful domains boosted molecular biology as an outlet that enabled those cohorts to retain their pretenses to basic science without the disgrace of entering naturalist 'stamp collecting.' Furthermore, acquiring the respectable aura of working on the secret of life, and thereby becoming 'politically correct' and morally righteous was irresistable.

In this context of rapid increases in the research population, largely though not entirely as a result of legislation entitling World War II recruits to university education, a broader and looser transdisciplinary identity which did not insist upon strict credentials in any single discipline was necessary to accomodate a larger number of transient and interdisciplinary recruits. Prolonged apprenticeship in fields such as meticulous chemistry, the methodical basis of biochemistry became outdated and impractical in an era of rapid turnover of both career moves and short lived compounds. Leading biochemists, whose formative years coincided with the golden age of biochemistry in the 1920s, the era of bourgeois hegemony in the aftermath of the collapse of four aristocracy-based empires, were slow to grasp the significance of the social professional climate in the post-World War II era. That new milieu, in the shadow of the technoscientific feat of the atomic bomb, increasingly intermingled biochemists, biophysicists, microbiologists, geneticists, but especially their hybrid combinations, under conditions of frequent moves across institutions and countries.

By the time some biochemical leaders began complaining that molecular biology was biochemistry without a license, the old system of strict licensing was rendered outmoded by science policy imperatives of the super-powers race, especially in space, as the Sputnik triggered a new scale and speed of scientific growth. Professional identity was no longer to be acquired through prolonged apprenticeship in a single discipline to the legacy of which one was expected to remain loyal in the long run, but from rapid, opportunistic moves across transdisciplinary, transinstitutional and transnational band wagons, as the story of messenger-RNA so amply demonstrates.[28] The era of 'DNA tycoons' and those who 'made a killing in RNA' had arrived; proteins, with their complex structures and biochemical origins soon became outmoded, even derided, as not leading to fast results in an era where scientific competition mimicked strategic rivalry between the superpowers, thus providing an advantage in molecular biology to those working on fast-reproducing organisms, especially bacterial viruses. A preoccupation with information transfer and nucleic acid thus came to substitute for the previously predominant concern with conformation and proteins.[29]

Nowhere is the transition of hegemony from protein structure and the era of biochemistry to that of DNA structure and the era of molecular biology more dramatically exhibited than in the double helix story. There the convergence of transdisciplinary, transinstitutional, and international movements, more political than poetic (for almost any scientist of distinction was prevented from entering or leaving the US, the new center of action, by the raging McCarthyism), resulted in a discovery that assembled formerly isolated scientific problems and results through the contingent agency of the only scientist who was young enough (less than 25) not to merit a dossier with the FBI during the Cold War, or to be hindered by pre-Cold War standards of social or scientific grace.[30]

While all the empirical work relevant to solving the structure of DNA was done by scientists operating within the framework of classical biochemistry, microbiology or crystallography (the teams of Avery, Chargaff, Wilkins, Franklin, Pauling); the structure was solved by those who bypassed those classical approaches grounded in experimental skills and disciplinary loyalties in favor of transdisciplinary opportunism, theoretical adventurism, linguistic flexibility, deceptive rhetoric, abuse of cross-national and cross-disciplinary rivalries, and clever patronage from physicists *qua* policy advisers, thus pioneering or anticipating the new structural conditions that prevailed in the period after World War II.

After the British fiasco of protein structure of 1950, and the make believe American solution of the alpha-helix of 1951, the shift toward DNA structure and discourse (or later the structure and functions of various RNA types, and eventually the relationship between DNA, RNA, and proteins as in the biochemical cracking of the genetic code, by the mid-1960s) took place in the 1950s for both scientific and political reasons. On the scientific level, technological spin-offs from the war effort produced new techniques, and expanded the scope, range, and affordability of instruments developed prior to or during World War II (e.g., electron microscopy, isotopic tracers). All experimental biology, whether molecularly minded or not, was affected by this technological trend coupled with an equally powerful demographic influx.

On the political level, the moral and social fallout from using the atomic bomb, and its successor arsenal, especially the hydrogen bomb, and the sheer economic growth, driven by super-power competition supported science lavishly. Big Science had increasingly become an asset throughout the total Cold War and its scientists had become the only Presidential arbitrators of resource allocation between the rival armed services. This post-World War II climate of generous support for research also led to increased support for biomedical purposes. Virus research which together with protein, DNA and RNA structure came to constitute the main staple of distinct research objects of molecular biology, benefitted from the multiplying agencies (such as NIH, ONR, AEC, and NSF's in the US) and foundations so eager to support biomedical research, that basic research, without any remote medical benefit, could be pursued both in the US and its allied European countries. The redeeming value of biology for counteracting the social reverberations of atomic death further reinforced the fortunes of molecular biology; research on microorganisms promised basic insights into life, possible payoffs ranging from an edge in military warfare to disease free society. In 1962, it became obvious from the relegation of the Nobel Prize for protein structure to the chemistry section, that protein structure, once considered the 'secret of life' was displaced from the core of biology, by DNA structure, which received the Nobel Prize for Physiology.

What was, therefore, the significance of DNA displacing proteins from the core of molecular biology, being accompanied by heredity displacing biochemistry and nutrition, in the 1950s and the 1960s, from the hegemonic place they held in science and society, in the post-World War I era? Ironically, the Lysenko scandal in 1948 heralded the growing role of heredity during the Cold War, though at that time heredity in plants and microorganisms rather than human heredity took the center stage, as the latter remained discredited by the eugenicist abuses of the totalitarian regimes, especially the Nazi concentration camps. Nevertheless the principles of molecular genetics that were enunciated with the Central Dogma in the late 1950s, (unidirectionality of the transfer of biological information from DNA to RNA to proteins, and its irreversibility from proteins to nucleic acids) were rapidly claimed to have universal validity, thus enhancing the status of molecular biology as a science of unifying principles.[31] The pattern was to be repeated in the early 1960s with the model of the operon that first explained cellular regulation in terms of molecular feedback mechanisms between the genome and its environment; and again, with the genetic code in the early and mid-1960s, that established the co-linearity of nucleic acids and proteins on the basis of triple base codons, anti-codons, and a comaless, redundant, code.

Are there any similarities between the above described transition from biochemistry to molecular biology and the subsequent transition from molecular biology, which reached a peak in the 1960s with its academic stabilization and institutionalization, independently and apart from either biochemistry and organismic or classical biology, to biotechnology, a phase that began in the mid-1970s but rapidly intensified in the 1980s. The discovery of recombinant-DNA, after several years of relative silence converged with the international oil crisis that initiated the end of the Cold War (the October 1973 war in the Middle East was the last time the US put its forces on nuclear alert) and the rise of multi-polar global economic competition that has become much more obvious since 1989.

The radical shift in the 1970s triggering the decline of the Cold War and the rise of biotechnology included the women's liberation movement that legalized many forms of contraception and popularized societal interest in controlling one's health; the war on cancer that encouraged scientific short-cuts as well as social activism around biomedical issues; and the investment climate that peaked in the early 1980s as a result of electoral changes in both US and UK toward regimes bent on deregulation. All these factors contributed to the rise of biotechnology as a new force not only in science but also in the corporate world of international economic competitiveness, and its grounding in high tech innovation.

Phase III: The Era of Biotechnology, 1974–Present

The third phase in the molecular transformation of biology began in the mid-1970s, when the Asimolar conference, summoned in 1974 at the initiative of leading academic molecular biologists, alerted the public, including the government and the media, to the hazardous implications of a new biotechnological breakthrough that enabled direct manipulation of the genetic material and the formation of 'new forms of life' — the phenomenon of artificially controlled recombinant-DNA.[32] Initially, the main concern was that biohazards arising from the 'new forms of life' could escape from the laboratory into the environment spreading unknown and possibly dangerous combinations of genes, viruses, or bacteria (bacteria, especially E. coli, which is common in the human digestive tract, served as the cloning hosts for new combinations of genes and viruses, including those involved in cancer.)

The most spectacular aspect of the early part of the biotechnological phase (1974–1977) was a highly publicized voluntary moratorium on research with recombinant-DNA until the National Institutes of Health (NIH) in the US (and comparable governmental bodies elsewhere) could establish guidelines to monitor the potential hazards. However, once guidelines were established, it became obvious that they could not be implemented outside governmental and academic laboratories dependent on public funds. Moreover, the rising market and moral value of some of the initial products of biotechnology (hormones to correct life threatening deficiencies, most notably insulin and human growth hormone, or anti-viral and anti-cancer drugs such as interferon) rapidly shifted the lucrative activity to the new biotech firms, usually in partnership between universities, industry, and the government, which split the revenues, real or projected, from patenting the new processes and organisms.[33]

Eventually, in the early 1980s, especially after Genentech skyrocketed on the Stock Exchange (1981), and governments favoring deregulation were elected in both the US and the UK, governmental regulations to contain the biohazards of biotechnology were dismantled. This occurred largely as a result of the growing concern of governments in the 1980s with economic competitiveness in the international arena, an arena dominated by multi-nationals which eventually began to absorb, from the mid-1980s, the small up-starts that flourished in the preceding decade.[34]

The increasing importance of economic competitiveness in the international arena, which transformed biotechnology into such a hot asset despite the fact that very few innovations became marketing or medical successes, derived from the gradual decline of the Cold War climate after the mid-1970s, when the retreat of both superpowers from unrealistic adventures in Asia (most notably, following the loss of wars in Vietnam and Afghanistan, but also the loss of face in post-revolutionary Iran), the oil crisis, the environmental danger resulting from the depletion of the ozone layer, the Geneva agreements for nuclear disarmament (for which Alva Myrdal received the 1982 Nobel Peace Prize), the rise of Asian centers (chiefly Hong Kong, Japan, South Korea, Taiwan) of industry and capital based on international exchanges, and related global concerns, shifted the dominant policy from national economies driven by strategic confrontation and its ideological substratum, to multi-national interdependent economies driven by commercial advantage, an advantage which came increasingly to depend upon high tech innovation.

However, if the biotechnology industry had increased its scale from the regional and the national to the international, while diversifying from health related products to agricultural ones, its most promising products remained various proteins that perform crucial bodily functions or suppress pathological conditions (e.g., insulin, interferon, blood clotting factors) and modified organisms, both plants and animals. Of course, the biomanufacturing of such products requires 'cutting' and 'stitching' of functional segments of nucleic acids, or genes, and their transfer across species, but the aspiration is placed upon marketable medically and agriculturally relevant products that are either proteins or modified organisms.

In contrast to mainstream biotechnology's emphasis upon proteins as products and protein-nucleic acids as crucial complexes required to artificially 'engineer' new products, products that in their turn may benefit society and individuals in a positive manner by increasing their respective welfare; the more recent and perhaps even more spectacular rise of the Human Genome project, or the mapping of all genes on the human chromosomes and sequencing of all DNA segments in all the genes, carries social, legal, and ethical implications, that on the whole are more divisive and disruptive for both society and individuals. The once benign concern with the hazards of 'new forms of life' gave way to reasonable fear of generating genetic underclasses, resulting from discrimination in employment, medical insurance, or social intercourse, on the basis of being identified as carriers of incurable genetic disorders, even though such disorders may take a long period of time to manifest themselves clinically.[35]

Though the effort to diagnose and eventually cure rare genetic diseases that destroy the quality of life for those afflicted is a noble goal, the accompanying preoccupation with genetic determinism reflects the post-Cold war era's increasingly visible failure to sustain collective institutions based on ideological aspirations for a utopian future, and the resulting legitimation of a narcisstic individualism which blames social ills on genetic flaws rather than on socially deficient institutions, outdated concepts of social order, and poorly implemented policies.

Despite a plethora of metatexts on the Human Genome project, all seeking to grapple with the numerous problems posed by a project that elevated the combined impact of computer science and technology upon biology to the new heights of playing directly with the blueprint of life itself;[36] it is still premature to guess the future course of the third phase of the molecular transformation of biology. Whether biotechnology's endless promise for new biological diversity will sustain an increasingly healthy and better fed international society, or whether the Human Genome project will diagnose a few rare genetic disorders while exacerbating social disorder in the name of a return to genetic differences, remains to be seen. Conceivably, the 'turn' of the Human Genome project in the early 1990s to recording and preserving the world's cultural diversity, may transform molecular biology into the basis for social stability in the post-ethnic twenty-first century of multi-culturalism without the need for three post-war settlements.[37]

Conclusions

The discovery of recombinant-DNA in the early 1970s opened a new, third major phase of transdisciplinary stabilization in the molecular transformation of biology in the twentieth century. This was a phase in which neither proteins, (as in the first historical phase following World War I) nor nucleic acids (as in the second historical phase following World War II), and their respective structures and functions, but interactive protein-DNA complexes (and the processes of selective transcriptions that they regulate) became the center of technical and conceptual effort.

To recapitulate, in the aftermath of the First World War and its Versailles international settlement, the 'exact science' exerting hegemony upon the transformation of biology was chemistry, namely the science most implicated in World War I military strategy, including both production of conventional munitions and tactical use of nerve and mustard gas on the front line, was chemistry. The outcome of that transformation was the world wide institutionalization of biochemistry, with Great Britain and its Colonial Empire, the greatest beneficiary of the post-World War I settlement, also becoming the leading country in the autonomous rise of biochemistry.

In the same vein, in the aftermath of World War II and its Yalta and Potsdam international settlements, the exact science that exerted hegemony upon the transformation of biology was [atomic] physics. Once again, this was the science most implicated in military strategy including the earlier and unconditional surrender of the enemy as a result of dropping the atomic bombs on Japan while the US, the greatest beneficiary of World War II, also became the leading country for the rising fortunes and eventual stabilization and autonomous institutionalization of molecular biology.

Similarly, in the aftermath of the Cold War (whether counting from the formal change of international borders in 1989 or from the more informal 'symptoms' of the incipient demise of the superpowers as sole determinants of international strategic welfare and warfare, since the mid-1970s); the exact science that exerted hegemony upon the transformation of biology was a combination of mathematics, especially computer science, and engineering, especially automated sequencing. Once again, those were precisely the disciplines that played key roles in the Cold War, and its dependence upon both coded military intelligence and an arsenal of ever innovative, high technologically intensive weaponry.

Much as biochemistry emerged after World War I as the hegemonic transdisciplinary outcome of chemistry's impact upon biology, and much as molecular biology emerged after World War II as the hegemonic transdisciplinary outcome of physics's impact upon biology; biotechnology emerged after the mid-1970s, as the hegemonic disciplinary outcome of the combined impact upon biology of technology, especially computerized automation.

However, if the geopolitical center of biochemistry's rise to power was Great Britain and the British Empire in the aftermath of World War I, while after World War II the US became the center of molecular biology's rise to power; in the post-Cold War era the center has become more fragmented and decentralized, as befitting the post-modernist transition to the 'global village.' Represented in political economy by continent-wide supra-nation-state alliances such as EEC or NAFTA, multi-national corporations, supra-national scientific organizations such as EMBO or CERN, information networks such as internet, and satellite communication, such as CNN; world international competitiveness in the increasingly interdependent commercial arenas of high technology came to define the context in which biotechnology rose to prominence, while emerging as a favorite object of technoscientific innovation, large scale financial investment and economic policy.

Two questions remain to be addressed in order to complete this twentieth century grid of post-War stability eras superimposed upon transdisciplinary phases of molecular transformation of biology. The first question pertains to the precise nature of the science policy, institutions, concepts, and techniques that mediate between the global political economy and a given type of hegemonic transdisciplinary stability within each era, i.e., to what extent the rise of biochemistry after World War I, of molecular biology after World War II, and of biotechnology after the Cold War were scientific outcomes predicated upon the historical specificity and structural coherence of each such post-War period. The second question pertains to the continuity or lack of it between types of transdisciplinary stability that were triggered by historical transition across successive eras, namely by the transition from biochemistry to molecular biology to biotechnology, across the twentieth century.

The rise of biochemistry in the inter-war period, as manifested through progress in clarifying pathways of intermediary metabolism as its conceptual mainstream; through growth of world leading new academic departments and research institutes, most notably those at Cambridge University, Columbia University, and the Kaiser Wilhelm Research Institutes in Berlin; through patronage by philanthropic foundations, most notably the Carlsberg Foundation in Copenhagen, the Rotschild Foundation in Paris, and the Rockefeller Foundation in New York; was both an outcome of chemistry's rise to the status of a strategic science in World War I, and of post-World War I (temporary) stabilization.

The economic processes of stabilization whether including war reparations, new control of national resources, or access to new markets including new colonies and dominions, made biochemistry — the basic science underlying biological combustion or the production of energy through enzymatic degradation of food — a key component of the bourgeois society. Then recovering its political muscle in a favorable social space (World War I was more devastating for both the aristocracy and the working class than the bourgeoisie which often profiteered from it), post-World War I liberal society's new energy sources ranged from colonial oil fields and the chemical industry at home to biochemistry, or the universal science of life sustaining bioenergetics, as well as the fermentation processes underlying the beer and wine industries that remain so central to the bourgeois life style.

The rise of molecular biology after World War II, as manifested through progress in solving the structure of proteins and nucleic acids while also decoding their functions in the expression and transmission of biological specificity, as its conceptual mainstream; was both an outcome of physics' rise to the status of a strategic science in World War II, under most possible dramatic circumstances afforded by the atomic bombs, and of post-World War II stabilization via the Cold War. This change was manifested through establishing many new laboratories, academic programs, research institutes, journals, most notably the MRC Laboratory at Cambridge University, the Services de Biochimie and Genetique Cellulaire at the

Pasteur Institute in Paris, the Virus Laboratory at Berkeley, or expansion of previous institutions such as the Cold Spring Harbor Laboratory in the State of New York, and the divisions of Biology and Chemistry at Caltech, in southern California, among others. Molecular biology was further sustained by a vast network of science policy agencies, both new and old, most notably the Medical and the Agricultural Research Councils (MRC and ARC) in Great Britain, the Office of Naval Research (ONR), the National Science Foundation (NSF), the National Institutes of Health (NIH), the Atomic Energy Commission (AEC) in the US, and the Centre National pour la Recherche Scientifique (CNRS) and la Delegation Generale pour la Recherche Scientifique et Technique (DGRST) in France.

The economic process of stabilization, most notably the Marshall plan for Western Europe and COMECON for Eastern Europe, created regions of militarily and ideologically protected international markets for American and Soviet products respectively, while making molecular biology — the basic science underlying the transmission of biological specificity via the so-called 'genetic code' — a central discipline in a world maintaining its equilibrium through superpower spying on each other's intelligence codes. Post-World War II welfare and socialist states, recovering from unprecedented losses of (civilian) population during World War II, found in molecular biology a science able to restore symbolically their loss of generations by uncovering the ultimate and universal mechanisms for the reproduction of life, while also promising to redeem the survivor generation's guilt over the use of the atomic bomb, as well as the more 'conventional' blitz-krieg.

The rise of biotechnology after the Cold War, as manifested through progress in artificially engineering functional proteins, segments of nucleic acids, genes, and transgenic organisms, (at the level of research programs); and through establishing new up-start companies as well as national and international regulatory science policy bodies; was both an outcome of technology (especially materials science, computer science, and genetic engineering), and of the reconfiguration of the world in the post-Cold War era as an increasingly integrated global village. The economic process of stabilization, most notably the decline of military or cost-inefficient production and the rise of sharp competition in consumer goods, especially from Asian manufacturers combining large reservoirs of labor and rapid adaptability to world market, made biotechnology — the technoscience underlying the modification of nature's own errors — whether defective proteins, genes, or organisms — the high tech of the future, in a world seeking to transcend forever its political, cultural, even biological, boundaries. With biotechnology, the ability to the perfect human if not quite social evolution may just be around the corner, much as biochemistry held the key to improving nutrition of the liberated and liberal masses after World War I or molecular biology held the key to genetically coded biopower in the post-World War II era of polarized ideological identities.

As to the transition from the disciplinary hegemony of biochemistry in the post-World War I era to that of molecular biology in the post-World War II era, and eventually to biotechnology, in the post-Cold War era, being continuous or discontinuous, cumulative or incommensurable, the answer differs according to the level of social action that is examined. At the conceptual level, the transition from small to large macromolecules, or even to biological assemblies, entails both cumulative elucidation of enzymatic biosynthesis and new surplus value of meaning in the sense that macromolecules also embody a coded order or information; while the transgenic organisms engineered by biotechnology operate beyond the species barriers. Though cumulative, these transitions entailed new universes of biological meaning, thus expanding, via manipulation at the molecular meaning, the concepts of 'classical' biology.

At the complementing level of technique and method, there is turn taking by the exact sciences. Though chemistry, physics, and engineering began to have an impact on biology from the start of the twentieth century, it so happened that chemical techniques in biology enjoyed precedence after the First World War I, physical techniques after the Second World War, and engineering techniques after the Cold War. At the same time, biochemistry continued to grow even though it lost cultural hegemony to molecular biology especially in the 1960s, while molecular biology continues to grow even though it lost, in the post-Cold War era, the cultural hegemony to biotechnology.

These structural changes can be followed with particular sharpness in the careers of 'biomolecular' science statesmen, that combined basic discoveries with policy making while acquiring their dual social role under specific historical circumstances. Notable examples include, Chaim Weitzmann and Frederic Hopkins during World War I and the inter-war period; Ernst Chain, John Kendrew and Jacques Monod during the post-World War II period, especially in the 1960s and early 1970s; Paul Berg, Maxine Singer, during the early days of recombinant-DNA in the mid-1970s; and Sydney Brenner, Francois Gros, Robert Sinsheimer, James Watson, Walter Gilbert, Leroy Hood, among others, during the heydays of the Human Genome project since the mid-1980s.

These historical transitions can be further observed at the level of individual careers. While the majority of scientists in a given era, post-World War I, post-World War II, or post-Cold War, follow the hegemonic discipline of that era, a few prove to be pioneers; others cannot make the transition to the next phase, while still others prove adept at jumping on the band-wagon of the next generation. Thus, among leading biochemists active both before and after World War II, Hans Krebs, Arthur Kornberg, and Erwin Chargaff, are among those who never quite accepted molecular biology as a distinct domain. In contrast, Fritz Lipmann, Severo Ochoa, Marshall Nirenberg, Robert Holley, and Francois Gros, shifted from biochemistry to molecular biology through the golden opportunities of 'macromolecular biochemistry,' especially messenger-RNA, transfer-RNA, and their role in decoding the genetic code.

In a similar manner, many leading molecular biologists and biochemists showed no interest in biotechnology, but Berg, Brenner, Dulbecco, Gilbert, Sinsheimer, Watson, Zinder were among those who became leaders of large enterprises in the era of biotechnology and especially its most spectacular offshoot, the Human Genome project. Still, each phase produced new leaders that were not part of the previous phase and were even antagonistic to it. Hence, only a collective biography of these three generations of 'biomolecular' science statesmen could clarify the social and political history of these three distinct periods of molecularization.

It is further remarkable that while in the inter-war era of biochemistry's hegemony there are no women biochemists involved in science policy, since the rise of the women's liberation movement in the 1970s, women biomolecular scientists began to play leading roles in science policy. For example, molecular biologist Maxine Singer, now President of the Carnegie Foundation, has emerged as one of the Asilomar leaders in the mid-1970s; marine biologist turned biotechnologist Rita Colwell of the University of Maryland, and biochemist and molecular biologist Marianne Grunberg-Manago of the Institut de Biologie Physico-Chimique in Paris, (since 1994 the first woman president of the French Academy of Sciences) have also emerged as science policy leaders since the 1980s, or during the third phase of biotechnology as a hegemonic phase in the molecular transformation of biology. However, biotechnology's apparently greater responsiveness to gender issues than that of either biochemistry or molecular biology, may reflect socio-historical change at large and fluidity of all types of boundaries late in the century, rather than a greater compatibility with gender on the part of biotechnology.

Similarly, at the level of institutions, the transition from biochemistry to molecular biology followed various options. In some institutions they co-existed (Cambridge, Harvard, Caltech); in others, one discipline transformed into the other (UC-Berkeley, Pasteur Institute); in others biochemistry prevailed (Columbia, Yale), while elsewhere it was molecular biology (University of Oregon, University of California in San Diego). At the fortieth anniversary of the double helix held in April 1993 at UNESCO headquarters in Paris under the auspices of its Secretary General, Federico Mayor, a biochemist, the participants included all three disciplinary species (namely, biochemists, molecular biologists, and biotechnologists) though Nobelist molecular biologists the leaders of the second phase, were the most prominent, as befits the anniversary of the most symbolic discovery in molecular biology.

A historical ethnographer of scientific commemorations can only speculate as to the social composition of the participants at the half-centennial, let alone the centennial of the double helix. It remains to be seen whether the current phase of biotechnology, dominated by business and industry, (for example, the May 1996 program of the Mass Biotech Council's annual meeting reflects a prevalence of titles such as Chief Executive Officers, senior partners, vice-presidents, and presidents) will also lead to a changing of the guard among the totems of molecular biology. Only the future can tell whether DNA's cultural hegemony, well enshrined ever since the Dali declaration, during the counter-culture movement of the 1960s, that DNA structure was proof of God's existence, will also continue into the twenty-first century, thus preserving the wider social currency established by molecular biology in the post-World War II era.

References

1. Pnina G. Abir-Am, "The discourse on physical power and biological knowledge in the 1930s: A reappraisal of the Rockefeller Foundation's 'policy' in molecular biology," *Social Studies of Science* (Aug. 1982), **12**: 341–382; idem, "Reply to four responses," ibid., (May 1984), **14**: 252–263; idem, "'New' trends in the history of molecular biology," *History Studies in the Physical and Bilogical Science* (1995), **26**: 167–196. See also Donald Fleming, "Emigre physicists and the biological revolution," *Perspectives in American History* (1968), **2**: 153–189; Nicolas Rasmussen, "The midcentury biophysics bubble: Hiroshima and the biological revolution in America, revisisted," *History of Science*, vol. 34, (in press).
2. Charles S. Maier, "Empires or nations? 1918, 1945, 1989" in Mark Rosemont (Ed.) *Three post-war eras*, forthcoming. See also Eric Hobsbawm, *The Age of Extremes, A History of the World, 1914–1991.* (New York: Pantheon, 1994); Mikulas Teich, 'The scientific-technical revolution: An historical event in the twentieth century? in Roy Porter and Mikulas Teich (Eds.), *Revolution in History.* (Cambridge University Press, 1986), 317–330.
3. Pnina G. Abir-Am, "From multi-disciplinary collaboration to transnational objectivity: International space as constitutive of molecular biology, 1930–1970," in *Denationalizing Science: The International Context of Scientific Practice.* (Dordrecht: Kluwer, 1993; Sociology of Science Yearbook, vol. 16–1992), 153–186. Horace F. Judson, *The Eighth Day of Creation, The Makers of the Revolution in Biology.* (New York: Simon and Schuster, 1979); and Michel Morange, *Histoire de la Biologie Moleculaire.* (Paris: La Decouverte, 1994).
4. Joseph S. Fruton, *Molecules and Life.* (New York: Wiley, 1972); Marcel Florkin, *A History of Biochemistry.* (Amsterdam: Elsevier, 1972); P.R. Srinivasan, J.S. Fruton, J.T. Edsall, (Eds.), *The Origins of Biochemistry, A Retrospect on Proteins.* (New York: N.Y. Academy of Sciences, 1979); Robert E. Kohler, *From Medical Chemistry to Biochemistry, The Making of a Biomedical Discipline.* (Cambridge University Press, 1982); Mikulas Teich with Dorothy M. Needham, *A documentary history of biochemistry, 1770–1940.* (Farleigh Dickinson University Press, 1992).
5. See Rasmussen 1996, op.cit. note 1.
6. Robert Bud, *The Uses of Life, A History of Biotechnology* (Cambridge University Press, 1993); Vivian Walsh, "Demand, public markets and innovation in biotechnology," *Science and Public Policy.* (June, 1993), 138–156; Susan Wright, *Molecular Politics: Developing American and British Regulatory Policy for Genetic Engineering, 1972–1982.* (University of Chicago Press, 1994).
7. See Robert Teitelbaum, *Gene Dreams: Wall Street, Academia, and the Rise of Biotechnology.* (New York: Basic Books, 1989); idem, *Profits of Science: The American Marriage of Business and Technology.* (New York: Basic Books, 1994); Barry Werth, *The Billion-Dollar Molecule.* (New York: Simon and Schuster, 1994), Walsh, 1993, op.cit. note 6.
8. See note 4; Fruton, *Contrasts in Scientific Style, Research Groups in the Chemical and Biochemical Sciences.* (Philadelphia: American Philosophical Society, 1990); Frederic L. Holmes, *Hans Krebs, The Formation of a Scientific Life, 1900–1933, vol. 1; Architect of Intermediary Metabolism, 1993–1937, vol. 2.* (Oxford University Press, 1991, 1993); Harmke Kamminga and Mark Weatherall, "The Making of a Biochemist, I: Frederick Gowland Hopkins' Construction of Dynamic Biochemistry," *Medical History* (1996), **40**: 269–292.
9. See David Nachmansohn, *German-Jewish Scientists, 1900–1933.* (New York: Springer Verlag, 1979; Arthur Kornberg and others (Eds.), *Reflections on Biochemistry, In Honour of Severo Ochoa.* (New York: Pergamon Press, 1974); Erwin Chargaff, *Heraclitean Fire, Sketches of a Life before Nature.* (New York: Rockefeller University Press, 1978);

10. See Jennifer Beinart, "The inner world of imperial sickness: the MRC and research in tropical medicine" in Joan Austoker and Linda Bryder (Eds.), *Historical Perspectives on the Role of the MRC.* (Oxford University Press, 1991), 109–136; Celia Petty, "Primary research and public health: the priorization of nutrition research in inter-war Britain," ibid., 83–108. See also Robert E. Kohler, "Walter Fletcher, F.G. Hopkins and the Dunn Institute of Biochemistry: A case-study in the patronage of science," *Isis* (1978), **69**: 331–355; Mark Weatherall and Harmke Kamminga, *Dynamic Science, Biochemistry in Cambridge, 1898–1949.* (Cambridge Wellcome Unit Publications, 1992; companion to an exhibition);
11. See Yehuda Reinhartz, *Chaim Weitzmann, A Biography.* (Boston, 1985); Bud, 1993, op.cit. ch. 37–43.
12. For details see Abir-Am, "The recasting of the relationship between physics and biology at two International Congresses in 1931," *Humanity and Society* (1985), **9**: 388–427; idem, 1982, op.cit., note 1.
13. See Brigitte Schroeder-Gudehus, *Les Scientifiques et la Paix.* (Montreal: University of Montreal Press, 1978); Alice Teichova & P.L. Cottrell (Eds.), *International Business and Central Europe, 1918–1930.* (Leicester University Press, 1983); Robert E. Kohler, *Partners in Science, Foundations and Natural Scientists, 1900–1945.* (Chicago: University of Chicago Press, 1991), Part II; Lily E. Kay, *The Molecular Vision of Life: Caltech, The Rockefeller Foundation and the New Biology.* (New York: Oxford University Press, 1993); Paul Weindling, "Public health and political stabilisation: The Rockefeller Foundation in Central and Eastern Europe between the Two World Wars," *Minerva* (1993), **31**: 253–267; Abir-Am, "The assessment of interdisciplinary research in the 1930s: The Rockefeller Foundation and physico-chemical morphology," *Minerva* (1988), **26**: 152–175.
14. See Donald Fisher, "The Rockefeller Foundation and the development of scientific medicine in Great Britain," *Minerva* (1978), **16**: 20–41; Christopher Booth, "Clinical research" in Austoker & Bryder (Eds.) 1989, op.cit., 205–242; Abir-Am 1988, op.cit.
15. See Petty, 1989, op.cit.; Weatherall & Kamminga, 1992, op.cit.; B.S. Platt, "Sir Edward Mellanby," *Annual Review of Biochemistry* (1956), **25**: 1–28.
16. Abir-Am 1988, op.cit., note 13.
17. Max F. Perutz, "Origins of molecular biology," *New Scientist* (January 1981), 28–31.
18. Booth, 1989, op.cit., 229.
19. On F.G. Hopkins see the collective tributes by students and disciples, Needham and D. Green (Eds.), *Perspectives in Biochemistry.* (Cambridge University Press, 1938); Needham and E. Baldwin (Eds.), *Hopkins and Biochemistry.* (Cambridge University Press, 1949); Kohler 1978, op.cit.; Kamminga and Weatherall, 1996, op.cit.
20. See Abir-Am 1987, op.cit.
21. See Gary Werskey, *The Visible College.* (London: Lane, 1978); Needham, "The making of an honorary Taoist" in Mikulas Teich & Robert Young (Eds.), *Changing Patterns in the History of Science.* (London: Heinemann, 1973), 20–40; Abir-Am 1987, 1988, op.cit.
22. See Soraya de Chadarevian, "Protein sequencing, biochemists and molecular biologists in Cambridge, in the 1950s," *Journal of the History of Biology* (1996), **28**: xxx–zzz.
23. See Holmes, 1993, op.cit.; see also T.W. Goodwin, *The History of the Biochemical Society, 1911–1986.* (London: The Biochemical Society, 1987); Abir-Am 1992, op.cit.
24. See Fleming 1968, op.cit.; Edward J. Yoxen, "The role of Schroedinger in the rise of molecular biology," *History of Science* (1979), **17**: 17–52; Abir-Am, 1982, 1985, 1988, op.cit.; Robert Olby, "From physics to biophysics," *History & Philosophy of the Life Sciences* (1989), **11**: 305–309; Evelyn Fox Keller, "Physics and the emergence of molecular biology: A history of cognitive and political synergy," *Journal of the History of Biology* (1990), **13**: 389–409; William Lanouette, *In the Shadow of Genius: A Biography of Leo Szilard.* (New York: Scribner's, 1992); Rasmussen 1996, op.cit.
25. See Sahotra Sarkar, "Decoding 'coding': Text, context, and DNA," forthcoming in S. Sarkar (Ed.), *The Philosophy and History of Molecular Biology: New Perspectives.* (Dordrecht: Kluwer, 1996); idem, "Biological information: A skeptical look at some Central Dogmas of molecular biology" (Boston Colloquium for the Philosophy of Science, April 1993); Evelyn Fox Keller, *Reconfiguring Life: Metaphors in Twentieth Century Biology.* (Columbia University Press, 1995); See also Abir-Am 1992, 1993, op.cit.; Morange 1994, op.cit.

26. See Abir-Am 1993, op.cit.
27. Abir-Am, 1982, op.cit.; idem, "A historical ethnography of a scientific anniversary in molecular biology: The first protein X-ray photo (1984, 1934)," *Social Epistemology*, vol. 6 (1992), no. 4, 323–354; 361–2; 372–3; 380–7; See also Boelem Elzen, "Two ultracentrifuges: A comparative study of the social construction of artifacts", *Social Studies of Science* (1986), **16**: 621–662; Lily E. Kay, "Laboratory technology and biological knowledge: Tiselius and electrophoresis," *History & Philosophy of the Life Sciences* (1988), **10**: 51–72; Doris T. Zallen, 'The Rockefeller Foundation and spectroscopy research: The Programs at Chicago and Utrecht,' *Journal of the History of Biology* (1992), **25**: 67–89; Nicolas Rasmussen, "Making a machine instrumental: RCA and the war time beginning of the electron miscorscope," *Studies in History and Philosophy of Science* (1996), **27**: 311–349.
28. See Judson 1979, Part II, op.cit.; Morange 1994, op.cit., ch. 13; Abir-Am 1992, 1993, op.cit.
29. See John Kendrew, "Conformation and Information in Biology" in Alexander Rich and Norman Davidson (Eds.), *Structural Chemistry and Molecular Biology*. (San Francisco: Freeman, 1968), 270–279.
30. See Abir-Am 1993, op.cit. For general background on the double helix see Judson 1979, op.cit. Part I; Olby 1974, op.cit. (reprinted in 1994); Anne Sayre, *Rosalind Franklin and DNA*. (New York: Norton, 1975); Chargaff, 1978, op.cit.; BBC, "The race for the Double Helix" (London, 1984); Donald A. Chambers (Ed.), *The Double Helix: Perspective and Prospective at Forty Years*. (New York: N.Y. Academy of Sciences, 1995).
31. Sahotra Sarkar, 1993, 1995, op.cit.; Morange 1994, op.cit.; Chambers 1995, op.cit.
32. Sheldon Krimski, *Genetic Alchemy, The Social History of the Recombinat DNA Controvery*. (Cambridge: MIT Press, 1982); Wright 1994, op.cit.
33. Charles Weiner, "Anticipating the consequences of genetic engineering: Past, present, future," in Carl F. Cranor (Ed.), *Are Genes Us? The Social Consequences of the New Genetics*. (Rutgers University Press, 1994), 31–51; Walsh 1993, op.cit.; Wright, 1994, op.cit.
34. See Harvey Brooks, "The relationship between science and technology," *Research Policy* (1994), **23**: 477–486; Skolnikoff, 1993, op.cit., chapter 4; Lewis M. Branscomb, "America's emerging technology policy," *Minerva* (1992), **30**: 317–136. See also Teitelbaum 1989, 1994, op.cit. Wright, 1994, op.cit.
35. See Richard C. Lewontin, *Biology as Ideology: The Doctrine of DNA*. (New York: Harper Perenial, 1992); Ruth Hubbard and Elijah Wald, *Exploding the Gene Myth, How Genetic Information is Produced and Manipulated by Scientists, Physicians, Employers, Insurance Companies, Educators, and Law Enforcers*. (Boston: Beacon Press, 1993); Dorothy Nelkin and Lawrence Tancredi, *Dangerous Diagnostics: The Social Power of Biological Information*. (University of Chicago Press, 1994, 2nd. edition); Daniel Kevles and Leroy Hood (Eds.), *The Code of Codes, Scientific and Social Issues in the Human Genome Project*. (Harvard University Press, 1992); Derek Chadwick and others (Eds.), *Human Genetic Information: Science, Law, and Ethics*. (New York: Wiley, 1990).
36. Richard Doyle, "Vital language" in Cranor (Ed.), 1994, op.cit., 52–68;
37. See George Marcus (Ed.), *Technoscientific Imaginaries, Cultural Studies for the End of the Century, vol. 2*. (University of Chicago Press, 1995); David A. Hollinger, *Postethnic America, Beyond Multiculturalism*. (New York: Basic Books, 1996); Daedalus [AAAS Journal], *The Quest for World Order*. (Cambridge/Mass.: AAAS, 1995, summer issue); Abir-Am, "The apolitical poetics of emigre biochemists: From scientific innovation to aesthetic alienation," in Jeffrey S. Timon (Ed.), *Vagabondage: The poetics and politics of movement*. (Berkeley: University of California Press, 1997).

Further Readings

Abir-Am, Pnina G., "From multi-disciplinary collaboration to transnational objectivity: International space as constitutive of molecular biology, 1930–1970," in *Denationalizing Science: The International Context of Scientific Practice*. (Dordrecht: Kluwer, 1993; Sociology of Science Yearbook, vol. 16–1992), 153–186.

Bud, Robert, *The Uses of Life, A History of Biotechnology*. (Cambridge University Press, 1993).

Chadarevian, Soraya, "The architecture of proteins: Building the laboratory of molecular biology at Cambridge/UK" in M. Hagner, H-J. Rheinberger and B. Wahrig-Schmidt (Eds.), *Objekte, Differenzen und Konjunkturen, Experimental systemeim Historischen Kontext*. (Berlin: Academie Verlag, 1994), 181–200.

Fruton, Joseph, *Contrasts in Scientific Style, Research Groups in the Chemical and Biochemical Sciences.* (Philadelphia: American Philosophical Society, 1990).

Holmes, Frederic L. *Hans Krebs, The Formation of a Scientific Life, 1900–1933, vol. 1; Architect of Intermediary Metabolism, 1993–1937, vol. 2.* (Oxford University Press, 1991, 1993).

Gaudillere, Jean-Paul "Chimie biologique ou biologie moleculaire? La biochimie au CNRS dans les annees soixante," *Cahiers pour l'Histoire du CNRS, no. 7.* (1990), 91–147.

Judson, Horace F., *The Eighth Day of Creation, The Makers of the Revolution in Biology.* (New York: Simon and Schuster, 1979).

Krimski, Sheldon, *Genetic Alchemy, The Social History of the Recombinat DNA Controvery.* (Cambridge: MIT Press, 1982).

Lewontin, Richard C., *Biology as Ideology: The Doctrine of DNA.* (New York: Harper Perenial, 1992).

Morange, Michel, *Histoire de la Biologie Moleculaire.* (Paris: La Decouverte, 1994).

Nelkin Dorothy, and M. Susan Lindee, *The DNA Mystique, The Gene as a Cultural Icon.* (San Francisco: Freeman, 1995).

Rabinow, Paul, *Making PCR, A Story of Biotechnology.* (University of Chicago Press, 1996).

Sarkar, Sahotra, "Decoding 'coding': Text, context, and DNA," forthcoming in S. Sarkar (Ed.), *The Philosophy and History of Molecular Biology.* (Dordrecht: Kluwer, 1996).

Sayre, Anne, *Rosalind Franklin and DNA.* (New York: Norton, 1975).

Teich, Mikulas, with Dorothy M. Needham, *A documentary history of biochemistry, 1770–1940.* (Farleigh Dickinson University Press, 1992).

Teitelbaum, Robert, *Gene Dreams: Wall Street, Academia, and the Rise of Biotechnology.* (New York: Basic Books, 1989).

Weiner, Charles, "Anticipating the consequences of genetic engineering: Past, present, future," in Carl F. Cranor (Ed.), *Are Genes Us? The Social Consequences of the New Genetics.* (Rutgers University Press, 1994).

Wright, Susan, *Molecular Politics: Developing American and British Regulatory Policy for Genetic Engineering, 1972–1982.* (University of Chicago Press, 1994).

CHAPTER 26

Biochemistry, Molecules and Macromolecules

HARMKE KAMMINGA

From modest beginnings around the turn of the century, the discipline of biochemistry began to flourish in the inter-war period and has expanded vastly since then. By the 1960s, most universities worldwide offered courses in biochemistry, and university-trained biochemists were working in virtually all areas of biology and medical science, in clinical laboratories, in state institutions for biomedical research, testing and standardization, and in the research laboratories of industrial companies producing pharmaceuticals, fertilizers, foods and food supplements, beer, detergents, synthetic fibers, and so on.

The massive expansion of biochemistry in the twentieth century was linked to new scientific outlooks, research objectives and technical resources, new forms of science funding, organization and training, as well as the enlistment of biological science in new social, economic and political agendas.

Biochemistry deals with the chemistry of life, that is, the chemical investigation and explanation of biological processes. It encompasses the identification of molecules taking part in chemical reactions in living organisms, the investigation of these chemical reactions and their relation to physiological phenomena, and the study of the regulation, coordination and organization of the chemical reactions within the organism and its constituent cells. Biochemists have aimed to provide new chemical understanding of traditionally defined biological functions such as growth, respiration, nourishment, digestion, movement, sensation and reproduction, across the living world.

In its alliance with medicine, biochemistry has provided new chemical explanations of the human body's workings in health and disease, and devised new diagnostic procedures and means of intervening in disease processes at the chemical level. Over time, this intervention has taken on very precise forms of administering specific molecules to adjust specific chemical reactions in the body, or to block vital chemical reactions in pathogenic microbes so as to kill or neutralize them. Hormones (such as insulin) and vitamins belong to the former category, antibiotics (such as penicillin) to the latter.

Biochemists have not been the only scientists engaged in the study of biological molecules and their reactions in the living organism. Nor were biochemists the major players in the study of all biological molecules at all times. Because the discipline and the research objects do not overlap neatly, I shall first address some questions about perspectives on biochemistry as a discipline and then outline its initial focus on the study of metabolic pathways of molecules. Finally, I shall discuss some studies of proteins, which came to be classified as very large molecules, or 'macromolecules,' using hemoglobin as an exemplar. I shall draw attention to the multidisciplinary character of this research, and sketch its links with a number of specific social objectives.

Disciplinary Issues

Biochemistry has been an exceptionally complex field, in terms of the relationships between its scientific content, its institutional settings, and its links with neighboring domains. Historically, the pursuit of research concerned with the chemistry of life did not coincide with the formation of biochemistry as a scientific discipline. By taking this historical disjunction seriously, we can bring into view the innovatory nature of the efforts of scientists who actively pursued the formation and consolidation of the new discipline.

The chemistry of life was investigated well before biochemistry was established as a scientific discipline. The rise of organic chemistry and experimental physiology in the nineteenth century generated new research practices that were applied to questions about the chemical composition of living organisms. The chemical constituents of food, of body fluids, tissues and excretory products of organisms were actively investigated, as were some of the chemical changes implicated in respiration, digestion, nutrition and fermentation. New specialized fields of research were created under names such as 'agricultural chemistry,' 'animal chemistry,' 'physiological chemistry,' 'medical chemistry' and 'chemical pathology.' Much of this research was undertaken in the medical faculties of universities, but later also in independent research institutes for medical research, agricultural research stations, and laboratories established by the brewing industry.

In the first decade of the twentieth century concerted attempts began to be made, in several countries, to bring together the study of the chemistry of physiological processes under the name of biochemistry. Indicative of this trend is the foundation at that time of numerous societies and journals with 'biochemistry' (or 'biological chemistry') in the title. With respect to journal titles, the term was first used in 1902 in the subtitle of *Beiträge zur chemische Physiologie und Pathologie: Zeitschrift für die gesamte Biochemie,* followed in 1903 by an abstract journal, *Biochemisches Centralblatt* (renamed *Zentralblatt für Biochemie und Biophysik* in 1911). In the United States, the *Journal of Biological Chemistry* began publication in 1905, and in 1906 the *Biochemical Journal* and the *Biochemische Zeitschrift* were founded in Britain and Germany, respectively. The very first departments of biochemistry were also created in this period, such as that established in Liverpool in 1902.

We do not have a definitive answer to the question of what, in general, motivated these moves towards an independent biochemistry. In stressing the novelty of their endeavors, a number of prominent advocates of the new biochemistry claimed that advances in chemistry and biology had reached a point in the late nineteenth century which allowed a new fusion of the two subjects, centered on a new notion of the cell as an organized, chemically dynamic unit. This perspective of biochemistry as a novel, twentieth century pursuit was promoted persistently, for instance, by British biochemist Frederick Gowland Hopkins:

> As a progressive discipline [biochemistry] belongs to the present century. From the experimental physiologists of the last century it obtained a charter, and, from a few pioneers of its own, a promise of success; but for the furtherance of its essential aim that century left it but a small inheritance of facts and methods. By its essential or ultimate aim I myself mean an adequate and acceptable description of molecular dynamics in living cells and tissues.[1]

Similarly, historian of biochemistry Robert Kohler suggested in 1973 that "the enzyme theory of life," which became prominent around the turn of the century, generated the intellectual motivation for the proliferation of biochemical research in the early twentieth century.[2] The notion that all chemical reactions in living organisms are regulated by biological catalysts, called enzymes, not only became the spur for multifarious investigations of the chemical changes involved in all living processes; it also gave the investigators themselves a sense of communal identity, a sense of belonging to the new and distinctive discipline of 'biological chemistry' or 'biochemistry.'

Kohler later argued that conceptual innovation, while important, is insufficient by itself to explain the rise of a new discipline,[3] and eventually denied that "particular theories have, in general, a causal role in the creation of disciplinary institutions":[4] instead of viewing scientific disciplines as research and teaching programs centered on a set of communal theoretical and experimental commitments, disciplines should be considered as institutional creations within the (competitive) political economy of science, with special emphasis on the actions of "scientific entrepreneurs."

While attention to institutional politics has enriched the history of biochemistry, this move has not yet resulted in a more coherent picture of biochemistry as a discipline. It is true that biochemistry, in any given period, is not easily captured in terms of a set of theoretical and experimental commitments. As the papers published in biochemical journals show, these have been extraordinarily heterogeneous from the start. The situation is further complicated by the multiple connections of biochemistry, in terms of its subject matter, with an enormous range of neighboring disciplines: organic, physical and colloid chemistry, physiology, pharmacology, medicine, agriculture, botany, zoology, microbiology, embryology, immunology, endocrinology, nutrition science, cell biology, genetics, biophysics, molecular biology, and so on. These manifold connections have kept the disciplinary boundaries of biochemistry permeable and mobile. Furthermore, the strength of each of these connections has varied significantly over time and within different institutions.

If we look at biochemistry principally as an institutional creation, we find that the institutional base of biochemistry has also been remarkably diverse. In German universities, biochemical research and teaching (usually under the heading of 'physiological chemistry') long remained tied to physiology and medicine on the one hand or to organic chemistry on the other hand. In Britain, a number of separate departments of biochemistry were established early on, but in most of them research and teaching remained closely allied to medicine and pathology. In the United States, the Medical Reform Movement created opportunities for biochemists to gain a strong academic foothold in the preclinical medical curriculum, though not generally in separate departments of biochemistry.

However, this characterization of the institutional heterogeneity of biochemistry rests on certain assumptions about the aims and content of biochemical research and teaching. Without such assumptions, there would be no justification for categorizing the research done by people working in departments of physiology, pharmacology, organic chemistry, etc., as being biochemical. At the level of research questions, most histories of biochemistry, whether written by biochemists or historians, assume strong continuities between nineteenth century chemical physiology and twentieth century biochemistry. Such continuities are demonstrable in specific cases of certain individuals, certain experimental practices and certain lines of research, but they cannot be assumed to hold generally if we are to avoid circularity in our characterization of biochemistry as a distinctive discipline. In particular, continuity assumptions obscure the question why a critical mass of scientists began to call themselves biochemists, and to introduce the new term in the titles of journals, societies, university syllabuses and departments.

Historians seem to be faced with a dilemma: if we count as 'biochemical' only that which went on in laboratories or departments of biochemistry, we exclude too much. Conversely, if we define biochemistry as a set of questions about the chemistry of life, and then project these questions back into the past, we are liable to include too much. One way out is to consider in detail the stated intentions of individual scientists who self-consciously promoted and practised the new biochemistry, under that name, and to examine any similarities in overall scientific, institutional and social objectives among them, as well as any major changes in objectives over time. Detailed analyses of this kind are not yet available for a large enough number of early biochemists to draw any firm comparative conclusions. This limitation should be borne in mind in the following outline of early priorities in biochemical research.

Biochemistry and the Dynamics of Molecules

One of the distinctive features of the new biochemistry was an emphasis on the dynamics of chemical reactions associated with biological processes. A number of influential advocates stressed not only that investigations of metabolism should move beyond crude input-output studies, and be aimed at unravelling the serial transformations undergone by molecules in intermediary reactions; but they should also incorporate new physiochemical approaches to the study of reaction kinetics in order to elucidate the ways in which these reactions are controlled and coordinated in the living cell. Perspectives on the cell as a chemically dynamic unit, in which myriad reactions are controlled by biological catalysts (enzymes), were promoted successfully in Germany, by Franz Hofmeister and Albrecht Kossel for example, and in Britain, especially by Frederick Gowland Hopkins.[5] This general outlook was inspired by, and in turn generated, manifold studies of intermediary metabolism.

For decades, these studies were focused on the reactions of small molecules, especially the breakdown products of foodstuffs such as carbohydrates, fats and proteins; and on the role of enzymes, vitamins and hormones in these reaction pathways. Techniques were developed to follow metabolites through their conversions in fine chemical detail; to isolate specific enzymes and study their functions in catalyzing specific steps in reaction chains; and to elucidate the ways in which oxidation reactions yield energy that is utilized in energy-consuming processes such as the biosynthesis of tissue components or muscular work.

An early spur to these investigations was Eduard Buchner's demonstration, in 1897, that a cell-free extract from yeast can bring about the fermentation of glucose to alcohol — a process that had previously been thought to require the integrity of the living yeast cell. Buchner called his preparation zymase, which he believed to be a single enzyme. Others soon showed that this fermentation does not take place in a single step, but in a series of reactions, each requiring its own enzyme. An important intermediate reaction product, isolated by Arthur Harden and William Young at the Lister Institute in London, was shown by Otto Meyerhof in Berlin to be involved also in glucose metabolism in muscle. This and other points of similarity between metabolic processes in cells from different organisms gradually began to be taken as suggestive of an underlying unity in patterns of metabolism, across the living world.

A central innovation in perceptions of the ways in which different metabolic reactions in the cell are integrated came with the exploration of metabolic cycles, culminating in Hans Krebs' formulation of the citric acid cycle in 1937. The famous 'Krebs cycle' has come to represent the major pathway of cellular respiration: a common breakdown product of carbohydrates, fats and proteins is oxidized in a cyclic sequence of chemical conversions, linked to the so-called 'respiratory chain,' which eventually releases carbon dioxide. Not only does this finely tuned pathway generate energy, it is interlinked, through intermediates in the cycle, with the major biosynthetic processes of the cell. The fine details of the cycle and its interlinked processes were worked out by an enormous number of biochemists worldwide. The intermediates, enzymes and co-enzymes of the respiratory chain were eventually located very precisely in the mitochondria, subcellular organelles now looked upon as the 'powerhouses' of the cell.

Research on intermediary metabolism, a biochemical subject *par excellence*, had a very heterogeneous institutional base. Some of the recognized authorities in this field worked in biochemistry laboratories, but the majority worked in medical institutions, departments of physiology, pharmacology, chemistry, and so on. Nor was the institutional autonomy of biochemistry pursued with uniform enthusiasm. Several leading investigators in the field, especially in Germany, were positively hostile to attempts to separate biochemistry from physiology. Nobel Laureate Meyerhof, for one, maintained persistently that such separation would undermine the unity of physiology. When he moved from Berlin to the new Kaiser Wilhelm Institute for Medical Research in Heidelberg in 1929, he insisted on calling his own institute 'Institut für Physiologie.'

The institutional diversity of biochemical research shows up clearly in individual careers. Here is just one telling example. After his medical training, Krebs worked from 1926 until 1930 in the laboratory of Otto Warburg, a leading authority in the biochemistry of cell respiration, at the Kaiser Wilhelm Institute for Biology in Berlin. Here Krebs began his studies of metabolic reactions associated with biological oxidation. After a year in the Department of Medicine at Altona, he moved in 1931 to the Department of Medicine at Freiburg, where he worked out the 'ornithine cycle' implicated in the synthesis of urea in the liver. When the Nazis came to power in 1933, Krebs left Germany and joined Hopkins' Department of Biochemistry in Cambridge. In 1935 he was appointed Lecturer of Pharmacology in Sheffield, where he formulated the citric acid cycle. Throughout these moves, Krebs was pursuing research normally categorized as biochemical. Eventually he became the first incumbent of the new Chair of Biochemistry in Sheffield in 1945 and was appointed Professor of Biochemistry at Oxford in 1954.

Krebs had felt very much at home in the Cambridge Department. Not only did Hopkins, who admired his work greatly, offer him refuge in a turbulent time, but Krebs there found a community that shared his passion for correlating the fine detail of chemical conversions in the cell with basic biological processes. Hopkins had publicly promoted 'dynamic biochemistry' as a fundamental science of life since 1913, and he continued to do so until the end of his very long career. His Department of Biochemistry, created in 1914, merits attention for the uniquely wide-ranging program of research and teaching that was put into place there in the inter-war period. There were many other places where basic biochemical research was pursued successfully, especially in Germany, but the particular combination of academic autonomy and the wide scope of Hopkins' program was highly unusual at the time.

With Hopkins' encouragement, research in the department focused on fundamental biological questions, separate from medical concerns. It encompassed reactions of intermediary metabolism and their control, mechanisms of biological oxidation and reduction, biological catalysis and enzyme kinetics, biochemical genetics, biochemical embryology and comparative biochemistry. The research objects were not only humans and vertebrate animals, but also invertebrates, plants, and microorganisms. The conviction that the great variety of life rests on certain common biochemical patterns grew with this research.

Besides constituting a thriving research community, Hopkins' school established separate teaching programs in biochemistry, so that it became possible for students to obtain a degree in the subject, and for Hopkins and his colleagues to pass on to new generations the style of biochemistry practised in Cambridge. Cambridge-trained biochemists obtained posts elsewhere and took with them a particular approach to the subject. Similarly, the many overseas visitors who worked in the department were, according to autobiographical accounts, inspired by their experience of Cambridge biochemistry. Hopkins' school came to be seen as a model internationally and Hopkins himself as the father of British biochemistry.

With the active support of his colleagues, Hopkins worked extremely hard to establish and defend the institutional autonomy of his department's research and teaching. He also benefited from conditions that were generally favorable to scientific development in Britain in the 1910s and 1920s. His close links with Walter Morley Fletcher, the first Secretary of the Medical Research Committee (renamed Medical Research Council in 1919), were particularly consequential. Hopkins and Fletcher had collaborated very fruitfully in the Cambridge Physiological Laboratory in the first decade of this century. They enjoyed a relationship of mutual respect and admiration, and, in 1913, Fletcher was appointed Secretary of the new Medical Research Committee on Hopkins' recommendation.

This committee was one of the products of the ideology of 'National Efficiency', stemming from anxieties that Britain was severely lagging behind its major competitors, especially Germany, in terms of social organization and planning. In line with this ideology, and with explicit inspiration from the Bismarckian model, the government stimulated and supported scientific research pursued for social and economic ends. On the same model, social welfare legislation was introduced, including the 1911 National Insurance Act, with the explicit aim of warding off socialist revolution. The Medical Research Committee was established under the National Insurance Act to support research into tuberculosis, a disease that threatened a serious drain on the sickness benefit provided for under this Act.

When war broke out in 1914, the committee was soon enlisted in organizing war-related medical research on subjects such as nutrition, trench fever and gas warfare. The traumas of the war massively amplified pre-existing anxieties about Britain's backwardness in organizing science in the nation's interest, and in the post-war reconstruction the government made a concerted effort to promote science with state support. As part of this endeavor, the Medical Research Committee, under Fletcher's Stewardship, was transformed into the Medical Research

Council (MRC), the prime governmental funding body for medical research. Building on his prior contacts among the British scientific establishment, Fletcher created a powerful network which both advised on and benefited from the MRC's funding policies. He took a long-term view as regards clinical applications and strongly promoted fundamental research in physiology and biochemistry. Fletcher's political power and influence in directing and funding research were instrumental in creating the climate and resources that allowed Hopkins' department to flourish. Not only did the MRC provide the department with research funds, Fletcher also persuaded the Dunn Trustees to grant Hopkins a very generous bequest to build, staff, and equip a large institute, which was opened in 1924. It was in the setting of the the Dunn Institute of Biochemistry that Hopkins' school acquired international renown.

In Hopkins' department and in biochemical laboratories elsewhere, research was not centered exclusively on the identification of molecules taking part in the reaction pathways of intermediary metabolism. There was a great deal of investigation of the catalysis of these reactions by enzymes under various conditions. The crucial role played by enzymes in regulating and coordinating metabolic reactions generated enormous interest among biochemists in the structure of these substances. Unlike the relatively simple molecules constituting metabolic reaction chains, however, enzymes long remained elusive as chemical entities.

Enzymes were widely thought to be either proteins or small organic molecules bound to proteins. Protein matter had been viewed as the basis of life from the nineteenth century, but proteins were problematic entities from the chemical point of view. Around the turn of the century, the renowned German organic chemist Emil Fischer identified amino acids as the characteristic breakdown products of proteins, and presented evidence that these small nitrogen-containing molecules are linked together through peptide bonds. Intact proteins seemed to be heterogeneous aggregates of polypeptide chains, forming large colloidal particles of variable composition. Many chemists, including Fischer, thought that proteins were too large to have a stable molecular identity. Eventually, however, proteins came to be viewed as very large individual molecules, each with a constant structure.

The structural study of proteins was pursued widely, as holding a key to the distinctive processes of life. Biochemists played their part in this endeavor, especially in the determination of amino acid sequences, but it also involved, and was at times dominated by, scientists who would not have viewed themselves as biochemists: organic chemists, physical chemists, colloid chemists, X-ray crystallographers, and others. Because of its multidisciplinary nature, work on protein structure has a very complicated history and the successes scored in this history cannot be attributed to biochemistry alone. In the following section, I shall illustrate this point by sketching some episodes in the multidisciplinary research on one protein, hemoglobin.

FIGURE 26.1: RESEARCH ROOM OF THE DEPARTMENT OF BIOCHEMISTRY IN THE CORN EXCHANGE BUILDING, CAMBRIDGE, BEFORE THE DEPARTMENT'S MOVE TO THE DUNN INSTITUTE IN THE 1920S. (FROM THE COLLECTION OF THE DEPARTMENT OF BIOCHEMISTRY, UNIVERSITY OF CAMBRIDGE. REPRODUCED WITH PERMISSION.)

FIGURE 26.2: HANS WEIL (LEFT) AND HANS KREBS (RIGHT), BOTH REFUGEES FROM NAZI GERMANY, IN THE DUNN INSTITUTE OF BIOCHEMISTRY, CAMBRIDGE, *C.* 1934. (FROM THE COLLECTION OF THE DEPARTMENT OF BIOCHEMISTRY, UNIVERSITY OF CAMBRIDGE. REPRODUCED WITH PERMISSION.)

FIGURE 26.3: UNDERGRADUATE PRACTICAL CLASS IN BIOCHEMISTRY, CAMBRIDGE, 1937. THE DEMONSTRATORS IN THE FOREGROUND ARE BARBARA HOLMES (LEFT) AND RICHARD SYNGE (RIGHT). BARBARA HOLMES, MARRIED TO FELLOW BIOCHEMIST ERIC HOLMES, WAS F.G. HOPKINS' DAUGHTER. RICHARD SYNGE WAS TO SHARE THE 1952 NOBEL PRIZE IN CHEMISTRY WITH ARCHER MARTIN, FOR THEIR DEVELOPMENT OF PARTITION CHROMATOGRAPHY, WHICH OPENED THE WAY TO AMINO ACID SEQUENCING OF PROTEINS. (FROM THE COLLECTION OF THE DEPARTMENT OF BIOCHEMISTRY, UNIVERSITY OF CAMBRIDGE. REPRODUCED WITH PERMISSION.)

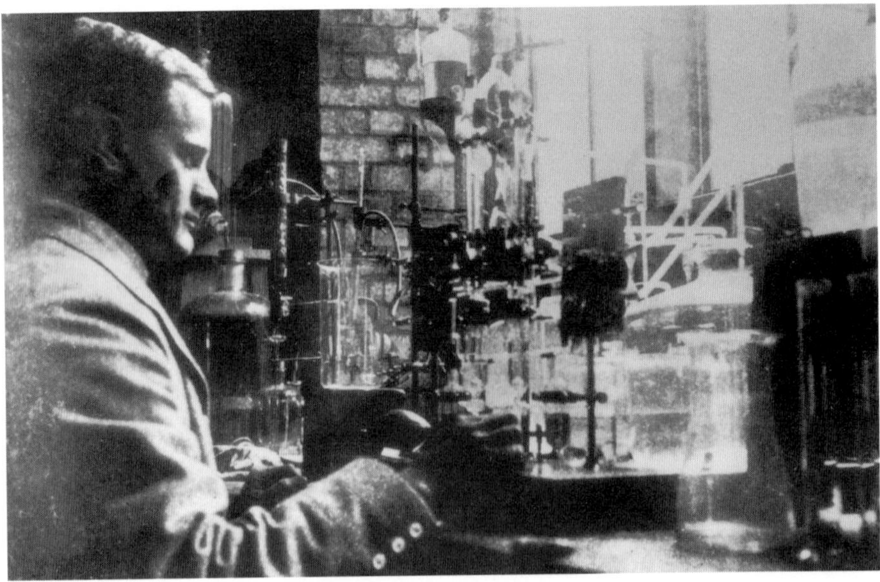

FIGURE 26.4: WALTER MORLEY FLETCHER WITH MUSCLE RESPIRATION APPARATUS IN THE CAMBRIDGE PHYSIOLOGICAL LABORATORY, *C. 1900*. FLETCHER WAS THE FIRST SECRETARY OF THE MEDICAL RESEARCH COMMITTEE (LATER MEDICAL RESEARCH COUNCIL) FROM 1913 UNTIL HIS DEATH IN 1933. (FROM THE COLLECTION OF THE DEPARTMENT OF BIOCHEMISTRY, UNIVERSITY OF CAMBRIDGE. REPRODUCED WITH PERMISSION.)

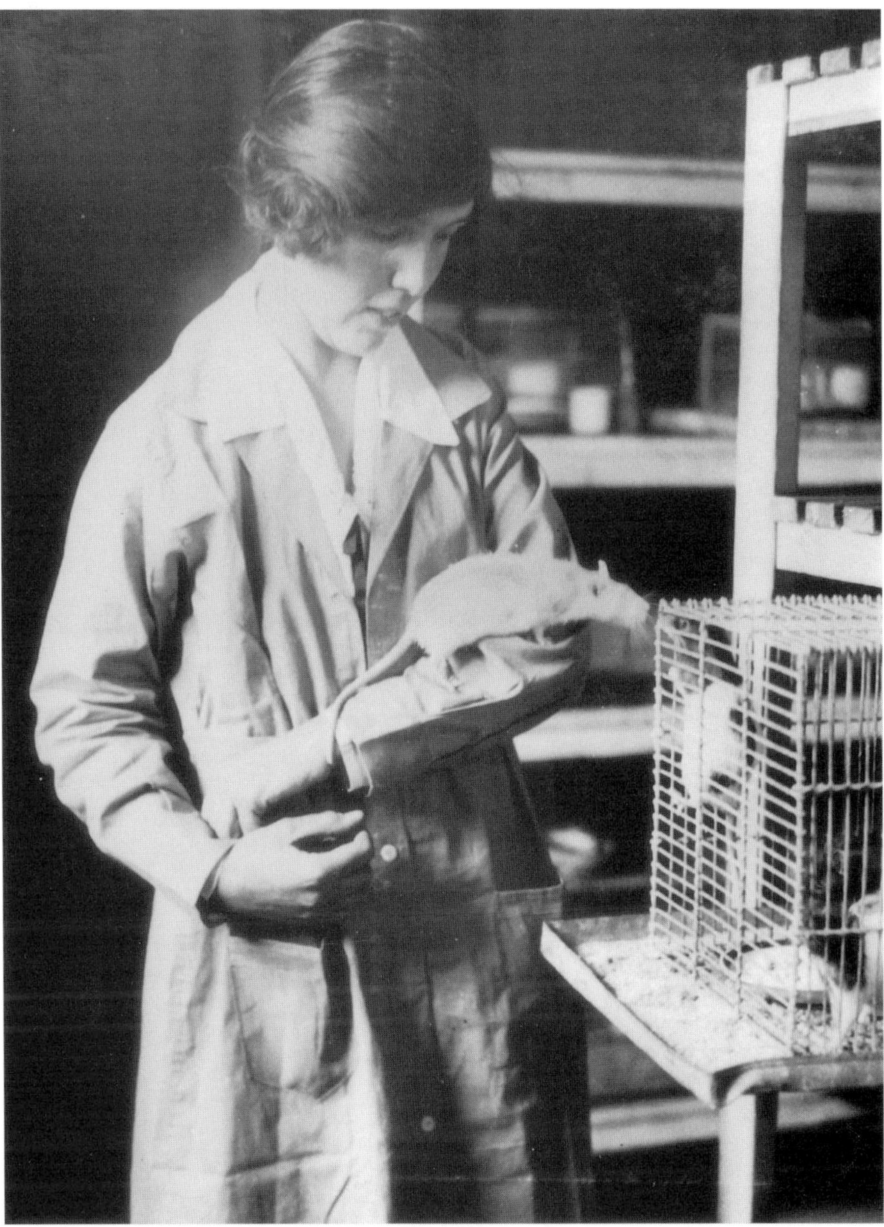

FIGURE 26.5: RUBY LEADER, TECHNICIAN IN CHARGE OF THE ANIMAL ROOM OF THE DEPARTMENT OF BIOCHEMISTRY, CAMBRIDGE, FOR NEARLY 30 YEARS, FROM 1917. (FROM THE COLLECTION OF THE DEPARTMENT OF BIOCHEMISTRY, UNIVERSITY OF CAMBRIDGE. REPRODUCED WITH PERMISSION.)

FIGURE 26.6: FIRST WORLD WAR SOLDIER IN THE TRENCHES, WEARING APPARATUS WITH ANTIDOTE TO POISON GAS. (FROM THE COLLECTION OF THE WELLCOME INSTITUTE LIBRARY, LONDON. REPRODUCED WITH PERMISSION.)

FIGURE 26.7A&B: LATE TWENTIETH-CENTURY LABORATORY. BELOW ON THE RIGHT A BACTERIOLOGICAL STERILE HOOD; ON THE LEFT VERTICAL ACRYLAMIDE GEL FOR THE ANALYSIS OF SMALL DNA FRAGMENTS.

FIGURE 26.8: ULTRACENTRIFUGES FOR SPINNING CELLULAR EXTRACTS.

FIGURE 26.9: ULTRAVIOLET TRANSILLUMINATOR FOR VISUALIZING DNA.

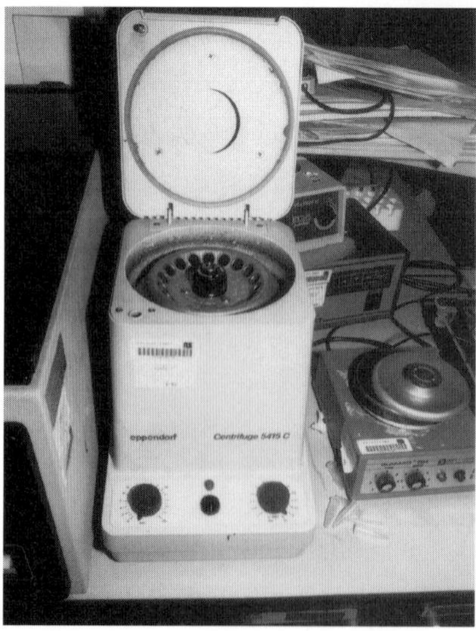

FIGURE 26.10: MICROFUGE ON A LATE TWENTIETH-CENTURY LABORATORY DESK TOP.

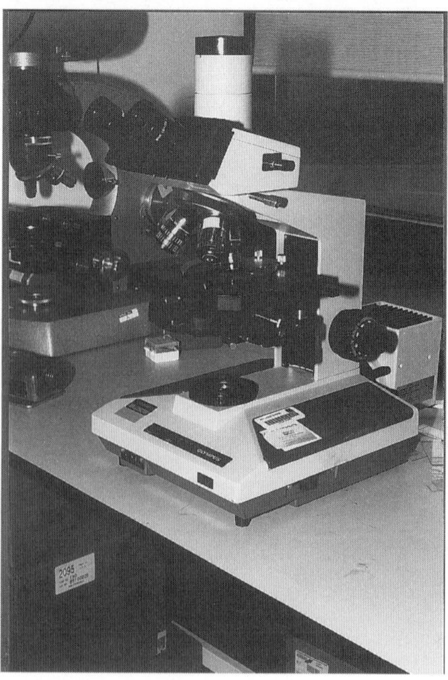

FIGURE 26.11: BINOCULAR MICROSCOPE WITH OBJECTIVE LENSES OF DIFFERING MAGNIFICATION.

Biochemistry and Macromolecules: The Case of Hemoglobin

Every biochemistry textbook devotes attention to hemoglobin: its role in respiration, its mechanisms of binding gases in the blood, its structure, its clinically significant variants, and so on. These different aspects of hemoglobin are united in an internally consistent picture of the molecule. This coherent textbook concept of hemoglobin was built up from a multiplicity of scientific approaches, many of them far beyond the confines of biochemical laboratories.

Hemoglobin was identified in the nineteenth century as the iron-containing proteinaceous substance which gives blood its characteristic red color. Blood was understood to play a vital role in respiration: it takes oxygen from the lungs to the body cells, where the oxygen takes part in oxidation reactions releasing carbon dioxide, which the blood then carries back to the lungs. By the late nineteenth century, hemoglobin was understood to be the oxygen carrier of the blood and the oxidation properties of its iron atoms were thought to be implicated in the respiratory process.

Early in the twentieth century, physiological investigations of the uptake of gases by blood shifted in focus from whole blood to solutions of hemoglobin. The Cambridge physiologists Joseph Barcroft and A. V. Hill were at the forefront of this research. Their studies of hemoglobin-oxygen binding and dissociation at different pressures of oxygen and carbon dioxide provided insights into mechanisms of control of the respiratory functions of the blood. Physiologists also investigated the responses of the respiratory system to environmental perturbations. For instance, much of Oxford physiologist John Scott Haldane's research was aimed at understanding physiological responses to high carbon monoxide levels in the blood in relation to occupational respiratory diseases, such as those affecting miners. He used the colorimetric properties of hemoglobin as indicators of states of the respiratory system as a whole, notably its level of oxygenation.

Respiratory physiology became a subject of central concern during the Great War, with the introduction of gas warfare. The British government, with the help of bodies such as the Royal Society and the Medical Research Committee, actively promoted wartime projects to improve understanding of the effects of poison gases, to devise protective measures, and to find effective ways of treating poison gas victims. To this end, leading respiratory physiologists were enlisted in the war effort, including Haldane and Barcroft. This boost to research in respiratory physiology had important repercussions after the war.

When the MRC began to support medical research after the war, respiratory research was one of the privileged areas. The war had very forcefully given this subject prominence and there was a body of specialists whose expertise could be built upon. More specifically, in 1919, the MRC founded the Haemoglobin Committee, under the chairmanship of Barcroft. For about ten years, this Committee sponsored hemoglobin research in different centers in Britain. At this time biochemists intensified their efforts to elucidate the chemical basis of the respiratory properties of hemoglobin, paying special attention to chemical structure of the heme group and the fine detail of oxidation, reduction and oxygenation of this group.

Basic research of this kind was compatible with the MRC's funding policies, as shaped by Fletcher and his circle. More generally, the increasing involvement of funding agencies, whether public or private, has often steered research in particular directions, in accordance with objectives defined by administrators of research funds and the peer reviewers they select. Another powerful example directly relevant to hemoglobin research is the Rockefeller Foundation. This private body injected enormous funds into biochemical and biophysical research in the interwar period, in Europe as well as the United States. Its funding policies have been examined by several historians, most recently by Lily Kay.[6]

Kay draws attention especially to the Rockefeller Foundation's promotion of a molecular vision of life as part of its "Science of Man" agenda, which aimed, in the long term, at social control. Social theory was to be modeled on natural science, in order to provide a rational basis for social intervention, grounded in the rigorous methods of the natural sciences. These far-reaching objectives had important implications for the kind of scientific research that was supported by the Rockfeller Foundation, focusing on the physical and chemical investigation of potentially controllable and manipulable systems. With biological systems being seen as part of a continuum ranging from atoms to nations, a great deal of funding was targeted on biochemistry and biophysics. Structural studies featured strongly in this program, including investigations of proteins such as hemoglobin.

Until the 1930s, proteins were widely seen as colloidal particles of unstable and variable structure, although the 'macromolecule' concept was vigorously debated in the 1920s. This notion of very large molecules with a well-defined and stable atomic configuration was proposed in 1922 by Staudinger in Zürich, on the basis of his chemical studies of rubber and other organic polymers. Initially meeting vigorous opposition, especially when applied to proteins, it gradually gained support during the 1930s.

In 1939, the Royal Society organized a meeting entitled 'A Discussion on the Protein Molecule,' which brought together leading investigators of proteins from several countries and different disciplines. In his opening address, Theodor Svedberg, Professor of Physical Chemistry at Uppsala University, drew attention to the radial shift in perspective implied by the title of the meeting:

> The proposal of the subject of this discussion is in itself a remarkable thing and a symbol of the spirit of this meeting. A few years ago the proposal would have looked preposterous. Proteins were known as a mysterious sort of colloids, the molecules of which eluded our search. What is it then that has happened in these years? Why is the most distinguished scientific society of this country inviting a discussion on the protein molecule?[7]

Svedberg's own answer to these questions and the other contributions to the meeting bear witness to a key element in the eventual acceptance of the idea that proteins are discrete (macro)molecules: the evidence produced by new physical and physiochemical techniques and instruments, often applied and developed with the active support of the Rockefeller Foundation. In the 1920s and 1930s, scientists studied the mass, shape, structure and physiochemical properties of proteins in novel ways, for instance by means of osmotic diffusion studies, ultracentrifugal sedimentation, electrophoresis, and X-ray diffraction analysis of protein fibers and crystals. From these different directions, increasing support was obtained for the view that proteins are discrete molecules of stable configuration, and that they are of very high molecular weight; in other words, that they are macromolecules.

Few of these new techniques were designed specifically for protein research at the outset. For instance, X-ray diffraction by crystals was first studied, in 1912, as a means of determining whether X-rays have a wave-like or particulate character. That the diffraction patterns provide information about the three-dimensional arrangement of atoms within crystals was shown in the same year by the physicists William Bragg and his son Lawrence. First applied to very simple inorganic crystals, X-ray crystallography began to be used for the structural analysis of more complex minerals and simple organic molecules in the 1920s, of organic polymers and protein fibers in the late 1920s, and of crystals of globular proteins in the 1930s. In the process, novel methods were devised to extract structural information from diffraction patterns. Early X-ray crystallographic work on proteins provided independent support for the polypeptide chain theory of protein fibers, for the constancy of structure of crystalline proteins, and for their very large size.

Nor was the ultracentrifuge 'purpose-built' for protein research. Svedberg initially built this instrument for the study of colloidal dispersions of metals, work for which he was awarded a Nobel Prize in chemistry. He later used it to investigate proteins, especially hemoglobin. Finding that the sedimentation patterns of this protein indicated uniform particle size, suggestive of stable molecular identity, Svedberg shifted his research to protein chemistry and pioneered the ultracentrifugal determination of the molecular weights of proteins. In the course of his investigations, Svedberg redesigned the ultracentrifuge at several stages, adapting it to new purposes. The export of this massive instrument to other centers, and making it work, involved a complex process of modification and standardization.

Although the macromolecule notion of proteins was consolidated by the use of these techniques, the detailed structure of proteins did not become immediately transparent. For a long time, it was hoped that there were basic structural patterns common to all proteins. For example, ultracentrifugal and crystallographic investigations suggested that most globular proteins are composed of a number of subunits, four in the case of hemoglobin. Svedberg himself thought that there was a basic unit of molecular weight 17,600, common to all globular proteins, but varying in number for individual proteins. Biochemists attempted to find regularities in the amino acid sequences of proteins, looking for repeat patterns. In retrospect, the structural complexity and individuality of proteins turned out to have been vastly underestimated.

Hemoglobin was widely used in structural investigations using the new techniques. It was easy to obtain, from blood; its color offered the advantage that its sedimentation in the ultracentrifuge could be followed visually; it crystallized comparatively easily; and, of course, it was considered to be of great physiological importance in terms of its respiratory functions. Novel insights into the mechanisms mediating these functions came from the X-ray crystallographic studies of Max Perutz and his colleagues at the Cavendish Laboratory, and later the Laboratory of Molecular Biology, in Cambridge.

It took Perutz over 20 years of intensive structural investigation to produce a three-dimensional model of the hemoglobin molecule. This structure, presented in 1960, showed the arrangement of the molecule's four subunits, the general course of its polypeptide chains and the positions of its heme groups. Perutz and his group went on to refine the model and, in 1968, determined the positions of individual atoms within the structure. The route towards this landmark achievement was by no means linear. It involved several steps that were subsequently seen as wrong turns or set-backs, and its ramifications were myriad. Not only did the research motivate, and rely on, technical and theoretical innovations in X-ray crystallography, but it also involved amino acid sequencing methods developed by biochemists, Linus Pauling's proposal of the 'alpha-helix' as the basic configuration of the protein polypeptide chain, consequential detours into studies of structural variants of the hemoglobin molecule, and advances in computing.

The full structural analysis of hemoglobin had deeper consequences. The precise three-dimensional configuration of the heme groups in the hemoglobin subunits provided clues about the structural basis of the distinct characteristics of oxygen binding by hemoglobin. It was known that the four heme groups of hemoglobin influence each other's affinity for oxygen: once one oxygen molecule is bound to the heme group of one of hemoglobin's subunits, then the successive binding of oxygen molecules to the heme groups of the other three subunits takes place more rapidly. Similarly, when one oxygen molecule has been released, then those bound to the other subunits are released more readily. Perutz presented evidence that oxygenation of hemoglobin is accompanied by a configurational change in the subunits which is triggered by a shift in orientation of the iron atom to which oxygen is bound. The change in one subunit induces a change in configuration of the other subunits, such that oxygen molecules have easier access to them. With this insight, coherence was produced between explanations of hemoglobin at the levels of structure, physiological function, and binding mechanisms. It is this strong explanatory coherence which has consolidated the meaning of hemoglobin for different scientific and medical communities: a patchwork of different perspectives has stabilized into a robust, shared concept.

This example illustrates the fruitfulness of looking at the history of biochemistry via explorations of research objects rather than general subject areas or institutions alone. It highlights the fluidity of the disciplinary boundaries. Moving in and out of various institutional settings, hemoglobin has remained a biochemical object, but it has not been confined to biochemistry. By following the research object itself, we see that the construction of a biochemically powerful concept cannot be attributed to biochemistry — or any other discipline — alone.

There is a further dimension to the current concept of hemoglobin, a genetic one, which came from a medical direction and which moved out into society in controversial ways. The medical, genetic, and structural aspects of hemoglobin came together for the first time in relation to sickle cell anemia. This debilitating disease is prevalent in certain areas of Africa and among Afroamericans. It is characterized by sickle-shaped red blood cells with seriously reduced oxygen-binding capacity.

In 1949, the geneticist James Neel, working in Ann Arbor, demonstrated with rigor that sickle cell anemia is inherited in recessive Mendelian fashion. This means that individuals who inherit a sickle cell gene from one parent only are not seriously affected. However, if they inherit sickle cell genes from both parents, they will develop sickle cell anemia. Almost concurrently, Linus Pauling at the California Institute of Technology arrived at the notion of sickle cell anemia as a "molecular disease" from a structural direction. Pauling and colleagues demonstrated differences in the electric charges of hemoglobins from healthy people, from sickle cell anemia patients, and from people displaying the milder form, sickle trait. This study indicated sharp structural differences in the hemoglobin molecules of the three groups. Pauling argued that these structural differences were the basis of the different oxygen-binding capacities of the red blood cells among the three groups, and that they resulted from differences in the genes.

In Cambridge in 1956, Vernon Ingram used the innovative amino acid sequencing technique developed by biochemist Fred Sanger to compare normal and sickle cell hemoglobin. He located the structural difference in just one amino acid of the hemoglobin molecule, which was then attributed to a point mutation in DNA. Since the 1950s, clinical biochemists have studied a great number of other variants of hemoglobin, some of them with important implications for the understanding of pathological conditions and their patterns of inheritance.

Of great relevance in this context is the expansion of research in human genetics after World War II, especially in the United States. This expansion can be attributed at least in part to war-related concerns with the mutagenic effects of radiation, in relation to the atomic bomb project. It was the US Atomic Energy Commission which initiated the large-scale funding of research in human genetics. Neel himself was appointed to an official survey team for mutation studies in Hiroshima and Nagasaki in 1946. On his return to the US in 1948, he began to investigate the heredity of blood disorders, including sickle cell anemia. His interests and expertise could be catered for within the favorable funding situation for work in human genetics.

With the characterization of disorders such as sickle cell anemia as hereditary diseases with a specific molecular basis, the possibilities of intervention began to be explored, in the form of genetic counselling and genetic screening. One of the first genetic advisory services in the US was the Heredity Clinic at the University of Michigan in Ann Arbor, opened in 1940, and headed by Neel from 1946. Screening programs targeted at high-risk groups came later. By 1971, sickle cell screening laws were enacted in 17 states and in 1972 Congress passed the National Sickle Cell Anemia Act, which provided for research, screening, counselling and education about the disease. This Act, the first of its kind, was soon followed by similar legislation for thalassemia, another inherited hemoglobin disorder, prevalent among immigrant families from Eastern Mediterranean countries. There was no cure for these hereditary diseases, but high-risk individuals identified by screening were advised not to have children. If women in high-risk families nevertheless became pregnant, intrauterine diagnosis and abortion legislation made it possible to prevent the birth of affected babies.

Shortly after the introduction of the National Sickle Cell Anemia Act, black activists in the US began to speak out against sickle cell screening, expressing fears that the program could be used as a tool for anti-black eugenics. This reaction illustrates that the translation of knowledge acquired in the laboratory into health policy in the social domain can have unanticipated repercussions. If insights into hereditary disorders are seen strictly as milestones in the march of scientific progress which open up unprecedented opportunities for the relief of suffering, then opposition to genetic screening programs is liable to be judged irrational. However, black opposition to sickle cell screening in the early 1970s takes on completely different meanings when it is viewed in the light of a political climate of racial tension and a long history of eugenic legislation in the United States, much of it targeted at groups defined in terms of race.

In the late 1960s and 1970s, racism was being challenged vigorously by black power movements more militant than the Civil Rights Movement which had initiated collective resistance to discriminatory laws and practices. In the same period, there was a revival of a much publicized and profoundly controversial hereditarianism focused on race, notably with respect to the purported inheritance of intelligence as measured by IQ tests. The new generation of black activists (and fellow white radicals) was sharply alert to attempts at supposedly scientific justifications for racial discrimination: it was pseudoscientific to ascribe what were alleged to be black inferiorities to inborn genetic differences. Instead, the explanation should be sought in quite obvious social inequalities, which ought to be eradicated forthwith.

In this polarized climate, sickle cell screening programs, especially where participation was made compulsory for blacks, could all too easily be interpreted as schemes promoting differential reproduction, at the expense of the black population. By suggesting that we can make sense of this response, I do not mean that intervention in the spread of hereditary disorders is sinister as a matter of principle, but that great weight should be attached to the sociopolitical dimensions of such interventions. However impressive the history of insights into hemoglobin-based disorders is from the biomedical point of view, the genetic screening programs are part of that history. The problems surrounding the implementation of these programs are to be understood in the light of sociopolitical priorities and conflicts prevailing at the time.

CONCLUSIONS

The boundaries of biochemistry have been extraordinarily fluid. If defined loosely as the chemistry of life, biochemistry can be taken to encompass an enormous range of investigations and institutional settings over the past two hundred years. If interpreted as an intentionally promoted endeavor in the early years of this century, we can identify certain new approaches, with a heterogeneous institutional base. Once we focus on particular research objects of interest to biochemists, the massive traffic across institutional disciplinary boundaries becomes obvious.

Taking hemoglobin as an example, I have tried to illustrate that research in one area has invariably been informed, influenced and transformed by research in other areas. A full history of hemoglobin research would require detailed attention to the many complex interactions involving the transfer, transformation and stabilization of knowledge, instruments, skills and techniques which drove the research. All these interactions were mediated by people, pursuing different agendas: not only scientists and clinicians, but also administrators of research funds, science and health policy makers, legislators, technologists, and so on. It is an enormous set of such interactions that, together, has shaped hemoglobin research in the twentieth century, research which has produced a robust biochemical concept of hemoglobin.

Furthermore, I have indicated that hemoglobin research did not stand in isolation from the cultures within which it was embedded: the research was shaped, not only by scientists' norms of what constitutes sound scientific procedure, nor just by its institutional organization and settings, but also by values and priorities decided upon by society at large. These have helped to define the objectives of the manifold research programs involving hemoglobin, and to set parameters for judging the significance of their outcomes. In that respect, hemoglobin research is typical of most scientific endeavor, within and beyond biochemistry. In general, the feedback processes between scientific and other cultural values have been subtle and intricate. By bringing them into view, we can understand better how scientific meanings are made, and by whom.

REFERENCES

I want to thank Andrew Cunningham (Cambridge) and Bert Theunissen (Utrecht) for their encouragement with this chapter and their fruitful comments on an earlier draft. I am also grateful to Tilli Tansey and Sir Christopher Booth for inviting me to participate in the History of Haemoglobin Summer School in London, July 1993, where I presented the hemoglobin material included here.

1. F.G. Hopkins, "Some Chemical Aspects of Life," *Report of the Proceedings of the BAAS* (1933), 1–24 on p. 3.
2. R.E. Kohler, "The Enzyme Theory and the Origin of Biochemistry," *Isis* (1973), **64**: 181–196.
3. R.E. Kohler, "The History of Biochemistry: A Survey," *Journal of the History of Biology* (1975), **8**: 275–318.
4. R.E. Kohler, *From Medical Chemistry to Biochemistry: The Making of a Biomedical Discipline.* (Cambridge University Press, 1982), p. 3.
5. M. Teich with D.M. Needham, *A Documentary History of Biochemistry, 1770–1940.* (Leicester University Press, 1992), pp. 504–520.
6. L. E. Kay, *The Molecular Vision of Life.* (Oxford University Press, 1993).
7. T. Svedberg, "Opening Address," *Proceedings of the Royal Society of London Series B* (1939), **127**: 1–17.

FURTHER READING

D.J. Kevles, *In the Name of Eugenics.* (Cambridge, MA: Harvard University Press, 2nd edn, 1995).

P.G. Abir-Am, "The Politics of Macromolecules: Molecular Biologists, Biochemists and Rhetoric" *Osiris* (1992), **7**: 164–191.

J. Austoker and L. Bryder (Eds), *Historical Perspectives on the MRC: Essays in the History of the Medical Research Council of the United Kingdom and its Predecessor, the Medical Research Committee, 1913–1953.* (Oxford University Press, 1989).

J.T. Edsall, "Blood and Hemoglobin: The Evolution of Functional Adaptation to a Biochemical System," *Journal of the History of Biology* (1972), **5**: 205–257.

J.S. Fruton, *Molecules and Life: Historical Essays on the Interplay of Chemistry and Biology.* (New York: Wiley Interscience, 1972).

F.L. Holmes, *Between Biology and Medicine: The Formation of Intermediary Metabolism.* (Berkeley: Office for History of Science and Technology, 1992).

F.L. Holmes, *Hans Krebs: The Formation of a Scientific Life, 1900–1933* (Oxford University Press, 1991); and *Hans Krebs: Architect of Intermediary Metabolism, 1933–1937.* (Oxford University Press, 1993).

H.F. Judson, *The Eighth Day of Creation: The Makers of the Revolution in Biology.* (London: Cape, 1979), Part III: Protein.

H. Kamminga and M.W. Weatherall, "The Making of a Biochemist, I: Frederick Gowland Hopkins' Construction of Dynamic Biochemistry," *Medical History* (1996), **40**: 269–292.

L.E. Kay, *The Molecular Vision of Life.* (Oxford University Press, 1993).

R.E. Kohler, *From Medical Chemistry to Biochemistry: The Making of a Biomedical Discipline.* (Cambridge University Press, 1982).

R.E. Kohler, *Partners in Science: Foundations and Natural Scientists.* (University of Chicago Press, 1991).

R.C. Olby, *The Path to the Double Helix.* (Seattle: University of Washington Press, 1974), Sections I and IV.

S. Sturdy, "From the Trenches to the Hospitals at Home: Physiologists, Clinicians and Oxygen Therapy," In J.V. Pickstone (Ed.) *Medical Innovations in Historical Perspective.* (London: Macmillan, 1992), pp. 104–123.

M. Teich with D.M. Needham, *A Documentary History of Biochemistry, 1770–1940.* (Leicester University Press, 1992).

M.W. Weatherall and H. Kamminga, "The Making of a Biochemist II: The Construction of Frederick Gowland Hopkins' Reputation," *Medical History* (1996), **40**: 415–436.

CHAPTER 27

Polymer Chemistry

YASU FURUKAWA

The birth of polymer chemistry, or the chemistry of macromolecules, marks a new epoch in the history of modern science. The concepts put forward in this field not only expanded the theoretical and methodological outlook of chemistry but also provided a base for the growth of other new sciences, notably molecular biology and molecular physics. The practical aspects of polymer chemistry are more familiar to us: the synthetic fibers, synthetic rubbers, and a wide variety of plastics on which our modern culture has come to depend, all stem from its applications.

Polymer chemistry is the field of science which deals with a class of substances that have special properties: their colloidal nature in solution and fibrousness or elasticity in the solid state, as exemplified by rubber, cellulose, proteins, starch, resins, and numerous synthetic polymers. Historically, this field emerged from a fundamental reinterpretation of existing objects of inquiry in which the German organic chemist, Hermann Staudinger, played a central role during and after the First World War. His theory of the nature and properties of macromolecules stimulated a new wave of polymer studies. By the end of the Second World War polymer chemistry had won wide recognition as a new and expanding scientific discipline by interacting with industry and the biological sciences. Germany, the United States, and Japan were the leading nations in this field in its early stages.

COLLOID CHEMISTRY AND THE AGGREGATE THEORY

The words 'polymer' and 'colloid' were coined in the nineteenth century. In 1832 the Swedish chemist, Jöns Jacob Berzelius, recognized the existence of compounds with the same proportionate composition but different numbers of atoms. He called such cases 'polymeric' from the Greek words *polys*, many and *meros*, part. Thus, a series of olefines C_nH_{2n} were said to be polymeric, since they had the same percentage composition but different molecular weights. Berzelius' polymer concept was soon adopted by other chemists.

In 1861 the Scottish chemist, Thomas Graham, reported that certain polymeric substances when in solution showed an extremely slow rate of diffusion through membranes such as parchment. He named such substances 'colloids' from the Greek word meaning glue, as distinguished from 'crystalloids' that crystallize and possess a high diffusibility. By the end of the century, a class of colloidal substances, such as rubber, cellulose, starch, and synthetic resins, were recognized as high polymers, although they were not then thought to have large molecules.

The great success of classical organic chemistry in the second half of the nineteenth century lay in the field of crystalloids in Graham's classification. Synthetic dyes, medical drugs, and numerous organic chemicals on which chemists successfully worked were all crystalloids. Organic structural chemistry, founded by August Kekulé and others in the mid-nineteenth century, formed the framework that led to the elucidation of the relationship between molecular structures and properties of classical organic substances.

Colloids were at odds with the organic chemists' research objectives. Unlike ordinary compounds, colloidal substances were gelatinous and could neither be crystallized from solution nor distilled without decomposition. Dubbed 'grease chemistry' (*Schmierenchemie*), the study of these substances did not attract practitioners of organic chemistry, since such materials did not respond to established methods for isolation, purification, and analysis.

However, a powerful stimulus to the study of polymers came from industry. In fact, the polymer industries had sprung up just before the science of polymers began to take shape. In the 1900s the production of synthetic plastics such as Bakelite (a phenolic resin) was already under way. Cellulose (the material that constitutes the cell walls of plants) and its derivatives, for example, cellulose nitrate and cellulose acetate, were being converted into such useful products as celluloid, viscose rayon, lacquers, films, cellophane, and explosives. The rubber industry was also growing rapidly, due largely to the rising demand for automobile tires.

Early advances in these new industries relied mostly on experience or trial and error rather than scientific understanding. In short, practice was more advanced than theory. But industry in turn stimulated scientific investigations into polymers. For example, an increasing demand for rubber around 1910, resulting in the exhaustion of wild rubber from the Amazon basin and rising prices, led a number of chemists to work on synthetic equivalents. Important studies of rubber by academic organic chemists such as Carl D. Harries, Samuel S. Pickles, and Hermann Staudinger, initially arose from their interest in synthetic rubber.

The rapid growth of colloid chemistry in the early decades of the twentieth century took place concurrently with industrial development. Indeed, enthusiasts for this science boasted of its industrial importance. This area was most actively cultivated and popularized by the son of the great physical chemist, Wilhelm Ostwald, at Leipzig. Wolfgang Ostwald, founder of the Colloid Society, and editor of two leading German journals in this field, was responsible for the foundation of colloid chemistry as an independent division of the physico-chemical sciences. Expanding upon Graham's concept, he defined colloids as dispersed systems consisting of particles ranging in size from 1/1,000,000 to 1/10,000 millimeters, which were too small to be seen and too large to be regarded as molecules. A magnitude he called "the World of Neglected Dimensions."[1] These particles, he claimed, were not themselves molecules but their aggregates. Ostwald and his followers believed in the unity of matter. Rejecting organic chemistry's traditional molecular-structural approach, they claimed that a colloid was a physical state of matter into which any substance might be brought. Under appropriate conditions any compound could form a colloidal solution. Thus, the properties of colloids should be determined not by their peculiar molecular structure but by the degree of dispersion of any material.

In many ways, organic chemists were under the influence of the colloidalist views during the 1910s and the 1920s. A number of leading German organic chemists, including Emil Abderhalden, Carl D. Harries, Max Bergmann, Paul Karrer, Rudolf Pummerer, Hans Pringsheim, and Kurt Hess, elaborated the so-called 'aggregate theory' that explained the sub-particulate structure of colloid polymers, the task which colloid chemists had left aside. According to this theory, polymers, such as rubber, cellulose, starch, and proteins, were the aggregates of small molecules held together by certain physical forces, other than Kekulé's normal chemical valence bonds.

A brief examination of Harries' view of rubber structure illustrates the general grounds of the aggregate theory. By the time he began his study, it had been found that natural rubber, an elastic solid obtained from a milk-like fluid (latex) of certain tropical trees, was made up of only two elements, carbon and hydrogen, the proportions of which were respectively five to eight. Harries proposed a structural formula for this 'isoprene' unit (or the C_5H_8 unit) which might be a constituent of the polymeric rubber molecule. The apparent total absence of end groups in his chemical analysis seemed to preclude the idea of any linear-chain structure but to indicate a ring structure instead. His explication appeared in 1905 when he presented the formula of an eight-membered cyclic molecule, consisting of two isoprene units. Colloid particles in a rubber solution were, in his opinion, the aggregates or physical molecules of the cyclic 'chemical molecules' held together by 'partial valences.' Furthermore, the partial valence forces were derived from the carbon-carbon double bonds in the 'chemical molecule.'

The rise of the aggregate theory affected the terms chemists used. The word 'polymerization' was used as a synonym for molecular aggregation. Likewise, the term 'molecular weight' referred to the weight of the physical aggregate or a colloidal particle. Thus, the apparent high molecular weights of colloid polymers, reported by a number of chemists who employed the customary methods of boiling point elevation, freezing point depression, or osmotic pressure, were not generally taken literally as the weights of the real chemical molecules. Bergmann called colloidal substances 'pseudo-high molecular substances.'[2] He insisted that the classical structural theory was ill-suited for the study of these pseudo-high molecular substances because the structural theory was based on Avogadro's molecule in gaseous phase. Hence, the structural formula would provide little information about variations that the molecule underwent in solidification, liquefaction, and solution processes. The cause of colloidal properties lay largely in the magnitude of the aggregating forces, a factor to which the classical theory could not be applied.

The aggregate theory gained further support in the 1920s from X-ray crystallography when X-ray diffraction was employed to examine the structure of polymers. This type of research was carried out intensively at the Kaiser Wilhelm Institute for Fiber Chemistry in Berlin-Dahlem, where Reginald O. Herzog directed a number of physicists and physical chemists, including Michael Polanyi, Karl Weissenberg, and Herman F. Mark. Rubber (when stretched), and a part of cellulose, were then known to exhibit a crystalline form to which X-ray analysis was applicable. Their study of these polymers showed that the unit cell, the recurring atomic group in the crystalline lattice, was as small as the ordinary molecule. During this period, many crystallographers assumed that the molecule could not be larger than the unit cell. From this, scientists such as Herzog concluded that the molecular size of polymers was likewise small. To exponents of the aggregate theory, this conclusion appeared as clear-cut empirical evidence supporting their view of the aggregate structure of polymeric substances. By the mid-1920s the aggregate theory had gained overwhelming support and offered the first unitary theory of structures for diverse polymeric substances before the macromolecular theory emerged.

Staudinger and Macromolecules

Hermann Staudinger, Professor at the Federal Institute of Technology in Zürich, first expressed his new theory of polymer structure at the 1917 meeting of the Swiss Society of Industrial Chemistry, and in 1920 he began publishing a series of papers in German journals.[3] He had developed an interest in this field during the early 1910s while investigating polymerized ketenes that decomposed into isoprene, the basic unit of natural rubber. Consequently, he attempted to polymerize isoprene into synthetic rubber, which in turn stimulated his interest in the polymer structure.

Trained in traditional organic chemistry, Staudinger aimed to reinterpret Ostwald's 'neglected dimensions' in terms of organic structural theory rather than colloidalist views or the aggregate theory. He firmly maintained that the molecule was the entity from which all physical and chemical properties stemmed: colloid polymers formed no exception to this rule. He thought that colloidal particles were in many cases 'macromolecules' (a term he coined in 1922) consisting of between 10^3 and 10^9 atoms linked together by normal bonds. Hence there would be no need to assume physical forces whose origins remained obscure. Macromolecular compounds, such as rubber, cellulose, and synthetic polymers, were thus structured on the same principles as those of organic chemistry, namely Kekulé's. The difference between his structuralist view and the previous aggregationist view was most apparent in the interpretation of the relation of properties of polymers with their chemical structures. The former attributed properties of a polymer to its large molecular structure, and the latter to external forces outside the molecule. In opposition to the ring structure, Staudinger proposed long-chain molecular formulas for a number of polymers. He suggested (as it happens, incorrectly) that unsaturated free valences at the ends of the chain molecule surely existed, but could remain non-reactive due to the great size of the polymer molecule (1,000–100,000 times the size of low molecules).

Some scientists had assumed the existence of large molecules before Staudinger. For example, the organic chemist, Emil Fischer, considered proteins to be giant molecules consisting of 'polypeptides' in which various amino acids were linked. However, he implicitly placed an upper limit for the molecular weight of organic molecules (about 5000). In 1906 the English rubber chemist, Samuel S. Pickles, proposed a large molecular structure for natural rubber, but maintained a cyclic shape for it. Although inspired by aspects of these earlier works, Staudinger developed coherent general principles of polymers with a new concept of matter and also created new methods to demonstrate the existence of macromolecules.

Staudinger's evidence was based on classical techniques of organic chemistry. For instance, the aggregate theory predicted that hydrogenation of rubber should yield a normal low molecular substance, because saturation of the double bonds would occur and destroy the alleged partial valences between the molecules (which were claimed to be the source of colloidal properties of polymers). In his experiment, however, Staudinger obtained a contradictory result. The properties of the saturated hydro-rubber resembled those of natural rubber; it did not crystallize but formed a colloidal solution like rubber. Thus, he concluded that colloidal particles of rubber were not aggregates of small molecules held together by partial valences, but were instead giant molecules.

Staudinger used synthetic polymers as model substances that he boldly thought would serve to explain the structure and behavior of the more complex natural polymers. Thus, polyoxymethylene was used as a model for cellulose, and polystyrene for rubber. He prepared various lengths of polymer chains from simple compounds in order to demonstrate that physical properties, such as the viscosity of their solutions, correlated with the degree of polymerization, a characteristic shared with the homologous series of paraffin hydrocarbons. He argued that this result strongly supported his view of the macromolecular structure of polymer substances. He also subjected these polymerization products to chemical reactions, such as hydrogenation, methylation, nitration, and hydrolysis, but found that the degree of polymerization was not affected. In other words, polymers could be converted into their derivatives without their sizes being changed. The method whereby such conversions were made, later called 'polymer analogous reaction,' demonstrated to him the concept of macromolecularity. In 1929 Staudinger developed the well-known 'Staudinger law of viscosity' to express a relationship between the viscosity and the molecular weight of polymers in dilute solutions. His law paved the way for a new, cheap method to determine the molecular weight of polymers.

Despite Staudinger's confidence, the evidence he marshaled did not convince his contemporaries who were suspicious of such huge molecules. The controversy was to continue into the early 1930s, but the debate climaxed earlier at the Düsseldorf symposium of the Society of German Scientists and Physicians in the fall of 1926, shortly after Staudinger moved to the University of Freiburg. At the meeting, Staudinger found himself isolated in the face of strong criticism from the supporters of the aggregate theory, including Max Bergmann and Hans Pringsheim. The symposium ended with no agreement on the issue of macromolecules. Even Staudinger's admiring colleague, Heinrich Wieland, a 1927 Nobel laureate, advised him:

> Dear colleague, drop the idea of large molecules; organic molecules with a molecular weight higher than 5000 do not exist. Purify your products, for example, your rubber, then it will crystallize and prove to be a low molecular compound![4]

Herman F. Mark was among the earliest X-ray experts who disputed the preconception that the molecules of polymers must be smaller than an X-ray elementary cell. In 1928, he and Kurt H. Meyer at I.G. Farbenindustrie developed a new theory that appeared to be a compromise between Staudinger's macromolecular theory and the aggregate theory, on the basis of their sophisticated X-ray investigations. According to them, colloidal particles in a solution were not themselves macromolecules but 'micelles,' a term that had originally been used in the previous century by the Swiss botanist, Carl W. von Nägeli, for crystalline building blocks of starch. Because Mark and Meyer assumed that micelles were aggregates of primary valence chains or long-chain molecules held together by 'special micellar forces,' they believed that the molecular concept could not be applied to the micelle, that is, the colloidal particle. They argued that the weights of these particles, as determined by physical methods (e.g., osmotic pressure measurements), represented not molecular weights but 'micellar weights.' Furthermore, micelles were stable in solution because of the considerable cohesive power of the Van der Waals-type micellar forces. The colloidal properties were therefore dependent on the micellar structure of the colloidal particles rather than on the structure of the long-chain molecules. Mark and Meyer estimated the size of micelles from the widths of X-ray diffraction spots. They suggested, for example, that a cellulose micelle was formed by 40–60 primary valence chains, each of them composed of 30–50 glucose units. A micellar size of 30–50 glucose units was taken to be the molecular length of cellulose. A micelle could be compared to a match box. The match box (micelle) would be comprised of 40–60 matches (long-chain molecules), each match being the same length as the match box. Apart from the size of primary valence chains, the basic structure they proposed would prove to be a substantial contribution to modern cellulose chemistry.

Although Staudinger and Mark and Meyer were by and large in agreement on the long-chain structures of polymers, the former strongly criticized what he called the 'new micelle theory.' Extremely sensitive about priority, Staudinger made clear the difference between his theory and their new micelle theory in published papers and personal correspondence. He argued that their estimate of the main chains fell short and that the 40–60 chains should be, according to the match-box model, connected through primary valences to form a very long albeit thin chain-molecule. In particular, he attacked their physicalist bent which stressed molecular cohesion and physical molecules in explaining colloidal properties of polymers: a trait shared with the aggregate theory.

By 1930 the climate of opinion in German academic research was shifting from the aggregate to the macromolecular theory. Although the Staudinger-Mark-Meyer conflict continued well into the 1930s, the impact of the latter's theory, which acted as a bridge between the macromolecular view and the aggregate theory, facilitated this shift. Proponents of the aggregate theory, such as Pummerer, Bergmann, Pringsheim, and Hess, began to cite Meyer and Mark's work, if not Staudinger's. Leading organic chemists such as Richard Willstätter considered Mark and Meyer's work powerful evidence for Staudinger's macromolecular theory.

Staudinger's theory now found support from a few scientists outside Germany. Among significant new developments were molecular weight measurements with the ultracentrifuge, introduced by the Swedish scientist The (Theodor) Svedberg in 1926. This tool estimated the molecular weight of certain proteins to be several millions. Perhaps more important, the young American organic chemist, Wallace H. Carothers, established the macromolecular view through a series of investigations inaugurated in 1928. Taking up a subject which the Staudinger school had neglected, he worked on the mechanism of polymerization, especially polycondensation. Carothers' intensive study restated and extended the ideas of Staudinger, Mark, and Meyer, providing indisputable evidence for macromolecules. By the mid-1930s, many of the arguments about polymers concerned details of the long-chain structures rather than questioning the existence of very large molecules. This shift was vividly illustrated by the Faraday Society's 1935 meeting in Cambridge, the first international conference devoted exclusively to polymers and polymerization. The principal speakers included such macromolecularists as Staudinger, Carothers, Meyer, Mark, Johann K. Katz, Karl Freudenberg, Eric K. Rideal, and Harry W. Melville.[5]

The Shaping of a New Discipline

From the outset, Staudinger had considered it his mission to return the study of polymers from the hands of colloidalists to the respectable science of organic chemistry. As its title suggests, his textbook, *Organische Kolloidchemie* (1940), was intended to direct chemists' attention from traditional colloid doctrines to the new principles of macromolecules in dealing with organic colloids. The triumph of the macromolecular theory in the 1930s forced colloid chemists to narrow their definition of colloids. Although colloid chemistry would continue to exist, it no longer enjoyed the strong claim made by colloidalists such as Wolfgang Ostwald in the previous decades.

While stressing organic chemistry, Staudinger saw his field not simply as an extension of classical organic principles but as a new branch of organic chemistry, since in many respects polymers showed characteristics different from ordinary organic compounds. The molecular weights of polymers can only be expressed by average values rather than by precise numbers. Owing to their large molecular sizes, the properties of polymers come not only from the atomic arrangements within the molecules but also from the shapes of these macromolecules. In this respect, his macromolecular conception differed from the traditional notion of organic compounds. As a result, a separate treatment was in order. Anticipating the vast possibilities in this new branch of organic chemistry, he speculated that the field of organic chemistry was far from exhausted as many classical organic chemists thought. In 1926 he announced, "we are only standing at the beginning of the chemistry of true organic compounds and have not reached anywhere near a conclusion."[6]

As teacher and proselytizer, Staudinger wasted no time in spreading his science. He used all possible means at his disposal (students, associates, lectures, and publications) to spread his views and spur those interested in the field into action. However, many of Staudinger's contemporaries regarded these methods as somewhat aggressive. Nevertheless, the cities of Zürich and Freiburg, where he had been advocating his macromolecular theory since the early 1920s, eventually earned the name of the 'high boroughs of high polymers' for providing a generation of German chemists who shared the disciplinary identity of polymer chemistry. Between 1920 and 1927, Staudinger trained 17 doctoral students in the new field at Zürich. Between 1928 and his retirement in 1954, Staudinger directed 57 doctoral students at the University of Freiburg. The total sum was 74 doctorates in thirty-five years of his academic career, including such brilliant students as Rudolph Signer and Werner Kern. Günther V. Schulz and Elfriede Husemann, though they did not earn doctorates from Staudinger, served as his chief collaborators at Freiburg. Staudinger was also a prolific writer, publishing over five hundred papers on macromolecular chemistry in various German scientific periodicals as well as several textbooks. He published these treatises with approximately 100 co-workers.

Although unwilling to be directly involved with the industrial applications of his science, Staudinger maintained good connections with the industry. After graduation, most of his students went to German chemical firms, undoubtedly resulting in the dissemination of his views in industry. The political upheaval in Germany had a profound impact on its academic circles. When Adolf Hitler came to power, Staudinger's opponents who had Jewish antecedents left Germany (including Bergmann, Herzog, Mark, Meyer, and Pringsheim). Staudinger, who was not Jewish, stayed in Germany but suffered from political oppression under the Nazi regime, due in part to his pacifist activities during World War I. Despite his growing fear that German polymer chemistry would be persecuted by the Nazis, Staudinger was able to continue his research. Some of Germany's industrial leaders protected him from dismissal and arrest. From its early stages, they had recognized the practical significance of Staudinger's work, incorporating it into their development of synthetic rubber and plastics.

Staudinger preferred to call his field of activity "macromolecular chemistry" (*makromolekulare Chemie*) rather than polymer chemistry. He sought to avoid terminological confusion surrounding the name 'polymer,' which had once been used by his opponents to designate the small molecular compound. When he took over the editorship of the *Journal für praktische Chemie* in 1940, Staudinger transformed this oldest surviving German chemical periodical by adding the sub-title *Unter Berücksichtigung der makromolekularen Chemie* (In Consideration of Macromolecular Chemistry). The revised issue started in the same year. Four years later, Staudinger ventured to change the whole title into *Journal für makromolekulare Chemie.* As a result of the wartime conditions in Germany and poor distribution, the journal ended after only two volumes. Since publication in Leipzig became virtually impossible after the war, he founded a new journal, titled *Die makromolekulare Chemie,* in Basel in 1947. This was effectively the first German periodical devoted exclusively to macromolecular chemistry. Staudinger was editor for the next two decades until his death at the age of eighty-four.

Carothers and the Rise of American Polymer Chemistry

In the United States, the institutionalization of polymer chemistry took place as early as in Germany. However, the cradle of American polymer chemistry was not the university but industry, and specifically the Du Pont Company. At Du Pont, Wallace H. Carothers initiated his study of polymers shortly after a fundamental research program was established in 1927. Carothers was a product of American pragmatic education, trained as an organic chemist and lacking the background of German chemical training from which earlier generations of American chemists had often benefited. He earned his Ph.D. from the University of Illinois shortly after World War I, having studied under the master of organic synthesis, Roger Adams. In 1927, while Carothers was an instructor at Harvard University, Du Pont offered him the position of group leader in its program. He became interested in the academic debate about polymers from reading German journals.

After moving to Du Pont in 1928, he chose to attack the problem of polymers from the synthetic side in a scientific style reflecting that of Adams. By 1930 his group had successfully synthesized artificial giant molecules from small molecules, using step-by-step organic reactions to demonstrate the existence of macromolecules.[7] Within an industrial framework, his basic research on macromolecular synthesis was transformed in a remarkably short period into highly successful commercial products, such as neoprene rubber and nylon. Nylon, in particular, turned out to be a large scale confirmation of his theory of condensation polymerization. Rapid industrialization doubtless helped legitimate polymer chemistry in a country where there had been hardly any academic debate about macromolecules. Carothers' work would become a prototype for science-based industrial research on polymers, ushering in an epoch of synthetic polymers in the post-war period.

Unlike the academic Staudinger, the industrial researcher Carothers did not train students in the field of polymer chemistry. Yet, under Carothers' influence, a new generation of polymer chemists was to emerge in American universities in the 1930s. After his suicide in 1937, Paul Flory and Carl S. Marvel were among those who continued Carothers' tradition of research. Flory was introduced to the field of macromolecules when he joined the fundamental research group at Du Pont in 1934. Flory, who was trained as a physical chemist at Ohio State University, provided support for Carothers' study from a mathematical perspective, a skill that Carothers admittedly lacked. At Du Pont, Flory investigated molecular size distribution in linear condensation polymers on the basis of statistical methods. His research on the kinetics of addition reactions introduced the important concept of chain transfer, whereby a growing chain molecule could be saturated with an atom from another molecule that might be a monomer, a polymer, or a solvent molecule. In 1938 Flory left Du Pont for the University of Cincinnati and later Cornell University, where he continued theoretical studies in the physical chemistry of macromolecules, the work that was to win him the Nobel Prize in chemistry in 1974.

Marvel's interest in polymers was also stimulated through close contact with Carothers, both as a close friend and as a consultant for Du Pont in 1928. Heavily influenced by Carothers' work, Marvel devoted most of his research at the University of Illinois from 1933 onwards to the organic chemistry of polymers. His areas of investigation ranged from sulfur dioxide addition polymers and the mechanism of vinyl polymerizations, to the development of synthetic rubber. While at Illinois, Marvel trained a large group of doctoral students and published over four hundred papers.

Thus, Carothers' successors brought the study of macromolecules into the academic setting as early as the 1930s. American polymer chemistry, which first arose from a basic research program in industry, gradually spread as an academic discipline by the late 1940s. Du Pont's venture into fundamental research paid off with nylon that contributed to the doubling of the Company's size. Industrial research thus played a crucial role in the emergence of a new science in America. In this respect, Carothers' example reversed the traditional relationship between science and industry, in which the latter only followed university science. His scientific activity at Du Pont demonstrated to his contemporaries that industry might now provide the initiatives traditionally expected of pure science.

In 1940 the American chemical community welcomed a powerful organizer of the newly established field from Europe. When Hitler's troops invaded Austria in 1938, Herman Mark fled Vienna to Canada, where he worked for the Canadian International Paper Company for two years. After losing Carothers, Du Pont eagerly sought out the best polymer chemists. They showed a growing interest in Mark, who had corresponded with Carothers since 1934. Mark moved to the United States in 1940, where he was appointed to a joint position as Du Pont consultant and Adjunct Professor at the Polytechnic Institute of Brooklyn, New York, with the financial backing of Du Pont. In the same year he started an influential monograph series titled *High Polymers* published by Interscience, New York.

At Brooklyn, Mark taught his students an introductory course in polymer chemistry and established a series of weekly symposia and intensive summer courses, involving outside scholars as well as industrial researchers. By mid-decade the 'Brooklyn Poly' had established a graduate program leading to M.Sc. and Ph.D. degrees with a major in polymer chemistry. A number of newly educated polymer chemists, such as Turner Alfrey, Frederick R. Eirich, and Charles Overberger, worked and helped to teach this field at the Institute under Mark's directorship. In 1946, as the new, independent 'Polymer Research Institute' Mark's Brooklyn school became a mecca for advanced students and polymer researchers, along with Marvel's in Illinois.

World War II accelerated polymer research in American universities and industries. The government's synthetic rubber research program, begun in 1942 in the face of Japan's occupation of the Pacific area, involved many leading academic and industrial scientists, including Marvel at Illinois, Morris Kharasch at Chicago, Izaak M. Koltoff at Minnesota, Flory and Peter Debye at Cornell, and Frank Mayo at U.S. Rubber. This program greatly advanced the science of polymers. In the same year, Du Pont's entire production of nylon was allocated for vital military uses such as parachutes, flak vests, and military tires by the War Production Board.

Shortly after the war, as the polymer industry grew to be one of the major industries in the United States, publications on polymer chemistry reached the point that the *Journal of the American Chemical Society* could hardly accept the deluge of manuscripts. The first English-language periodical devoted to the field of macromolecules, the *Journal of Polymer Science,* was inaugurated under Mark's editorship in 1946. Its first volume carried fifty-seven papers, and the journal was soon to become the leading vehicle for the growing number of polymer scientists in post-war America. Building on foundations firmly laid by Carothers and his successors, Mark thus served as an important organizer and teacher in establishing the discipline of polymer chemistry in the United States.

FIGURE 27.1: THE BAKELIZER, USED BY LEO HENDRIK BAEKELAND AROUND 1907, TO PRODUCE BAKELITE UNDER PRESSURE AT HIGH TEMPERATURE. COURTESY SMITHSONIAN INSTITUTION.

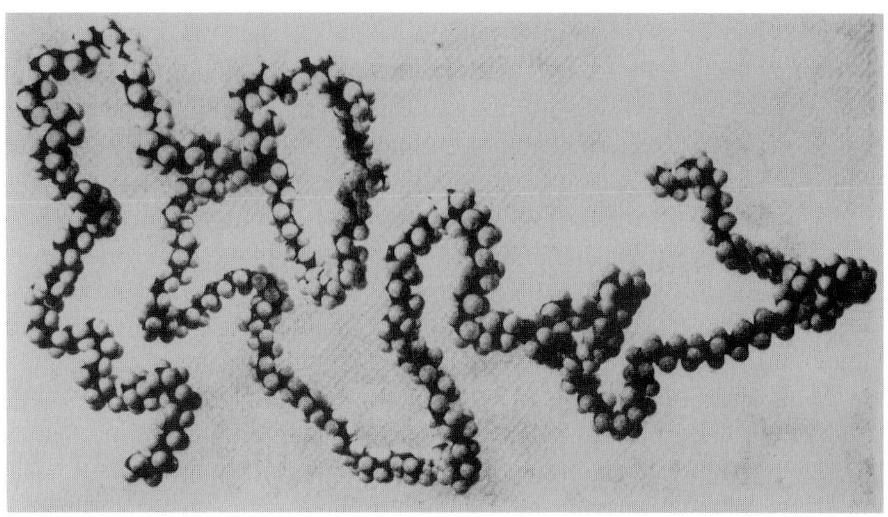

FIGURE 27.2: WERNER KUHN'S MODEL OF A MACROMOLECULAR CHAIN IN SOLUTION. COURTESY HANS KUHN.

FIGURE 27.3: HERMANN STAUDINGER HOLDING A MODEL OF A MACROMOLECULE. COURTESY DEUTSCHES MUSEUM.

Polymer Chemistry in the Japanese Context

Japan had a long tradition in the fiber and textile industry. Around the mid-1930s the country was a major exporter of silk to the United States and the world's biggest producer of rayon. Correspondingly, early Japanese studies of polymers centered on fibers.

The macromolecular theory was introduced to Japan by those individuals who had studied in Germany in the 1920s and 1930s. While Staudinger had two doctoral students from Japan, more Japanese students studied under his chemical opponents. Ironically, it was not Staudinger's pupils but the latter group who played a major role in the formation of macromolecular chemistry in Japan. Among them was Ichiro Sakurada, a student of Kurt Hess at the Kaiser Wilhelm Institute for Chemistry in Berlin-Dahlem. While in Germany, Sakurada was a harsh critic of Staudinger's work, but after assuming a position at Kyoto Imperial University in 1931, he changed his mind and became the most active propagator of the macromolecular view of polymers in Japan's chemical community. By that time he and other Japanese students sensed that the macromolecular theory was clearly gaining significant acceptance in Europe and America.

Japan's boycott of Australian wool in 1936, which signaled the beginnings of her self-sufficiency program, resulted in the establishment of the Japanese Institute for Chemical Fibers Research at Kyoto Imperial University. Here, Sakurada and his students, including Seizo Okamura, began to study the synthesis of 'artificial wool' from cellulose acetate. Du Pont's announcement about the invention of nylon in 1938 was a serious shock to Japan, as many industrialists were concerned that this new synthetic fiber would deal a fatal blow to the Japanese silk industry. Three years later, the government, industry, and universities together created the Research Association of Synthetic Fibers to promote research and development. Under its umbrella, Sakurada's group at Kyoto studied polyvinyl alcohol leading to the invention of the 'vinylon' fiber, and a group at Tokyo Institute of Technology, including Toshio Hoshino (a former student of Wieland) and Yoshio Iwakura, worked on nylon and polyurethane fibers. In 1940 Osaka Imperial University set up the first program on polymer chemistry and the Institute for Fiber Science under the directorship of Yukichi Go, who had intensively studied cellulose structure under Karl Freudenberg, R. O. Herzog and Kurt H. Meyer between 1929 and 1934.

In early 1943, in the midst of World War II, the Research Association of Synthetic Fibers was transformed into the Association of Polymer Chemistry, the world's first independent scholarly society in this field and the forerunner of today's Society of Polymer Science. It published the first issue of its journal, *Kobunshi Kagaku* (Polymer Chemistry) in 1944, two years before the appearance of Mark's *Journal of Polymer Science,* and three years before Staudinger's *Die makromolekulare Chemie.* The Society played an indispensable role in the phenomenal growth of the polymer community in post-war Japan. The rapid growth of polymer chemistry as a discipline can therefore be seen as a representative example of characteristic interactions between science, industry, and society.

Conclusion

In retrospect, Staudinger succeeded in creating polymer chemistry on the basis of organic chemistry. Missing from Staudinger's intellectual scope however, was a physico-chemical insight into polymers. For example, he devoted himself primarily to elucidating the static state of macromolecules like their molecular structures and shapes. Because macromolecules move, the properties of polymers should therefore be affected. Ironically, physical chemistry, for which Staudinger showed a distaste, would become a vital part of polymer research and greatly expand the scope and understanding of the dynamic behavior of macromolecules. Important studies in the physical chemistry of polymers, which employed statistics, kinetics, thermodynamics, and hydrodynamics, emerged after the mid-1930s, as represented by the work of a new generation of physical chemists, such as Werner Kuhn, Paul Flory, and Maurice Huggins. These new studies disproved some of Staudinger's initial concepts, such as those dealing with the rigidity of macromolecules and the non-reactivity of end groups of macromolecules.

Later in his career, Staudinger's interest turned more and more to biological aspects of macromolecules. As early as 1926 he had argued for the biological implications of such macromolecules as proteins. Such thoughts were further encouraged by his wife, Magda Staudinger, a plant physiologist who helped to develop new concepts relating macromolecules to physiological and philosophical questions. In 1946, Hermann Staudinger published a monograph, *Makromolekulare Chemie und Biologie,* in which he attempted to explain the life processes of the living cell from the point of view of macromolecular chemistry. Although his own work on bio-macromolecules only touched the fringes of the subject, the theory itself opened a door for a new approach for biologists, chemists, and physicists in attacking the enigma of life. Symbolically, Staudinger received his 'belated' but most coveted Nobel Prize in chemistry as the founder of polymer chemistry at the age of 72 in 1953, the same year in which the Watson-Crick double helix theory of the giant molecule DNA emerged as a landmark of the new science of molecular biology.

The year 1953 symbolized the maturity of polymer chemistry in two other respects. Paul Flory published a landmark textbook entitled, *Principles of Polymer Chemistry,* which was to play a definitive pedagogical role in this growing field. It served as the bible for a new generation of polymer scientists throughout the world. In the same year Karl Ziegler, the former director of the Kaiser Wilhelm Institute for Coal Research at Mülheim, discovered catalysts for producing linear crystalline polyethylene, which was soon followed by the Italian Guigio Natta's preparation of crystalline isotactic polypropylene with a Ziegler catalyst. This work on the syntheses of stereoregular polymers, which for the first time enabled chemists to control the shape of synthetic polymers (even to synthesize natural rubber possessing a specific three-dimensional configuration), was a major breakthrough in the polymer industry, and it made the two scientists the co-winners of the 1963 Nobel Prize in chemistry. As illustrated by these developments, the second half of the twentieth century saw an explosive growth of the science-based polymer industry, developing a flood of new and better synthetic fibers, rubbers, and plastics that brought about a material revolution. Due to its industrial and biological importance, and the pure intellectual pleasure it provided, the study of macromolecules became a major new field in twentieth-century science.

References

I am grateful to John Krige, Peter Morris, and Elizabeth Sandager for their helpful suggestions.

1. Wolfgang Ostwald, *An Introduction to Theoretical and Applied Colloid Chemistry: The World of Neglected Dimensions*, trans. Martin H. Fischer. (New York: John Wiley and Sons, London: Chapman and Hall, 1917).
2. Max Bergmann, "Allgemeine Strukturchemie der komplexen Kohlenhydrate und der Proteine," *Berichte der deutschen chemischen Gesellschaft* (1926), **59**: 2973–2981, on p. 2973.
3. *Das wissenschaftliche Werk von Hermann Staudinger*, Magda Staudinger, Heinrich Hopff, and Werner Kern (Eds.), 7 vols. (Basel and Heidelberg: Hüthig & Wepf Verlag, 1969–1976).
4. Hermann Staudinger, *From Organic Chemistry to Macromolecules: A Scientific Autobiography Based on My Original Papers*, trans. Jerome Fock and Michael Fried. (New York, London, Sydney, and Toronto: Wiley-Interscience, 1970), p. 79.
5. See *Transactions of the Faraday Society* (1936), **32**: Pt. 1.
6. Hermann Staudinger, "Die Chemie der hochmolukularen organischen Stoffe im Sinne der Kekuléschen Strukturlehre," *Berichte der deutschen chemischen Gesellschaft* (1926), **59**: 3019–3043, on p. 3043.
7. Wallace H. Carothers, *Collected Papers of Wallace Hume Carothers on High Polymeric Substances*, Herman F. Mark and G. Stafford Whitby (Eds.). (New York: Interscience Publishers, 1940).

Further Reading

Furukawa, Yasu. *Staudinger, Carothers, and the Emergence of Macromolecular Chemistry.* (Philadelphia: University of Pennsylvania Press, forthcoming.)

Hounshell, David A. and Smith, John Kelly. *Science and Corporate Strategy: Du Pont R&D, 1902–1980.* (New York, New Rochelle, Melbourne, and Sydney: Cambridge University Press, 1988).

Mark, Herman F. *From Small Organic Molecules to Large: A Century of Progress.* (Washington, D.C.: American Chemical Society, 1993).

McMillan, Frank M. *The Chain Straighteners: Fruitful Innovation: The Discovery of Linear and Stereoregular Synthetic Polymers.* (London and Basingstoke: The Macmillan Press, 1979).

Morawetz, Herbert. *Polymers: The Origins and Growth of a Science.* (New York, Chichester, Brisbane, Toronto, and Singapore: Wiley-Interscience, 1985).

Morris, Peter J.T. *Polymer Pioneers: A Popular History of the Science and Technology of Large Molecules.* (Philadelphia: Center for History of Chemistry, 1986).

Morris, Peter J.T. *The American Synthetic Rubber Research Program.* (Philadelphia: University of Pennsylvania Press, 1989).

Mossman, S.T.I. and Morris, Peter J.T. (Eds.). *The Development of Plastics.* (London: Royal Society of Chemistry, 1994).

Priesner, Claus. *H. Staudinger, H. Mark und K.H. Meyer: Thesen zur Grösse und Struktur der Makromoleküle.* (Weinheim; Deerfield Beach, Florida; and Basel: Verlag Chemie, 1980).

Seymour, Raymond B. (Ed.) *History of Polymer Science and Technology.* (New York and Basel: Marcel Dekker, 1982).

Seymour, Raymond B. (Ed.) *Pioneers in Polymer Science.* (Dordrecht, Boston, and London: Kluwer Academic Publishers, 1989).

Staudinger, Hermann. *From Organic Chemistry to Macromolecules: A Scientific Autobiography Based on My Original Papers.* Translated by Jerome Fock and Michael Fried. (New York, London, Sydney, and Toronto: Wiley-Interscience, 1970).

CHAPTER 28

Atomic and Molecular Science 1900–1960

MARY JO NYE

ARE ATOMS AND MOLECULES REAL?

The contemporary 'atom' and 'molecule' belong to both chemistry and physics. For much of the nineteenth century, the words 'atom' and 'molecule' were used interchangeably to describe any invisible and massy particle that was capable of expressing motion and force. The term 'molecular physics' was common throughout the century for describing the investigation of the mechanical properties of these particles. After the time of John Dalton (1766–1844), the word 'atom' increasingly came to be identified with the chemical atom as an element's experimentally-determined combining weight relative to the standard of oxygen or hydrogen. Chemists frequently used the term 'compound atom' rather than 'molecule' for combinations of atoms into complex chemical substances.

After discussions at an international chemical conference in Karlsruhe in 1860, more and more chemists distinguished the elementary chemical 'atom' from the composite 'molecule' made up of atoms. Stanislao Cannizarro (1826–1910) was among those chemists who argued that the 'chemical atom' and 'physical atom' were the same real and elementary simple particle, while August Kekulé (1829–1896) was among those who resisted the presumption that the chemical atom was simple in the sense of indivisible. Many chemists, like Marcellin Berthelot (1827–1907), conjectured that there must be an inner structure for the atom in order to account for its energy, while natural philosophers and physicists like James Clerk Maxwell (1831–1879) and Ernst Mach (1838–1916) proposed that the atom was a structure in the ether and that spectral lines emitted by excited atoms must correspond to vibrations of something inside atoms.

Arguing the impossibility of establishing the existence of the atom or molecule through direct experimental evidence, the physical chemists Wilhelm Ostwald (1853–1932) and Pierre Duhem (1861–1916) strongly spoke out around 1900 in favor of reconceptualizing the fundamental principles of chemistry by eliminating the atomic hypothesis in favor of the principle of energy and the laws of thermodynamics. These anti-atomist arguments not only assailed the century-long Daltonian atomist tradition, but also censured the newer molecular-kinetic theory of gases that was closely identified with the work of Rudolf Clausius (1822–1888), Maxwell, and the theoretical physicist Ludwig Boltzmann (1844–1906).

In 1857 Clausius had revived the assumption of a mechanical theory of heat in which gas particles are presumed to be elastic spheres moving about randomly with velocities proportional to their energy. He defined the energy of a gas in terms of the velocities of these molecular spheres, using the general formula:

$$pV = 1/3nm\bar{v}^2$$

where $\bar{v}^2$ is the molecule's mean square velocity, p is the pressure on the walls of a container, V is the volume of the container, n is the number of particles, and m is the mass of each particle.

Knowing the density at a given pressure, Clausius deduced that the average speed of gas molecules must be several hundred meters per second. The mean free path of a molecule, the Gaussian distribution of the velocities of molecules at a given temperature, and the probabilistic character of the irreversibility inherent in the second law of thermodynamics were among the many theoretical derivations that followed in the work of Maxwell and Boltzmann.

Still, Ostwald was not alone in objecting to the explanation of energy and thermodynamics by means of the atomic-molecular hypothesis, both on epistemological grounds that it was unprovable and on empirical grounds that certain assumptions, like the equipartition of energy among a few degrees of freedom within a molecule, were not consistently corroborated by experimental results. Despite Boltzmann's probabilistic interpretation of the law of entropy, there also remained objections to deriving the thermodynamic law of irreversibility from reversible mechanical processes.

So were atoms and molecules real? Did theoretical calculations of their mean free paths or velocities or masses correspond to real objects? Among the best theoretical estimates of molecular size were those of Boltzmann's Viennese colleague Josef Loschmidt (1821–1895) who had deduced an upper limit for the volume of a molecule from considerations of packing, as well as a minimum value for Avogadro's number N and an upper limit for molecular diameter. Loschmidt's and others' calculations placed the diameter of molecules between 10^{-7} and 10^{-8} centimeters (between 0.001 and 0.0001 micron, where 1 micron = 10^{-3} mm). Experimental work by John William Strutt, the third Lord Rayleigh (1842–1919), estimated the thickness of oil films on water at no more than 0.002 micron or between two and twenty molecules.

Among those who became thoroughly, even passionately, committed to proving the validity as well as the usefulness of atomic and molecular theories in the early decades of the twentieth century were the French physicist and physical chemist Jean Perrin (1870–1942) and the Swedish physical chemist Theodor Svedberg (1884–1971).

Perrin's early experimental work focused on X-rays and cathode rays. In 1903 he began studying colloids and interested himself in their electrical properties. He soon calculated the kinetic energy of colloid particles in motion by using the law of Johannes Dederik van der Waals (1837–1923) that kinetic theory is equally valid for liquids and gases.

In 1905 Perrin turned to the phenomenon of Brownian motion, first noted by the English botanist Robert Brown (1773–1858) as the spontaneous motions of pollen granules in water under a microscope. By the time Perrin concerned himself with this topic, the many explanations of the phenomenon included surface tension, capillarity, temperature effects, electrical action, and, more recently, the kinetic motion of the visible microscopic particles as they are bombarded by the invisible molecules of the liquid in which they are suspended.

With the physicist Paul Langevin (1872–1946) advising him, Perrin estimated the mean energy of yellow vegetable latex granules from observations of their motions, but he was disappointed when his value came out some 100,000 times smaller than that required by the kinetic theory. However, the new ultramicroscope, invented by Henry F.W. Siedentopf (1872–1940) and Richard Zsigmondy (1865–1929) in 1903, made possible observations at the limits of 0.005 microns.

Perrin soon recognized that the apparent mean velocities of observed particles varied in size and direction, depending upon the length of the time of observation, without tending toward a limit. Thus, the observed motions were not velocities at all, but rather, displacements which were the visible results of the many collisions and changes of direction of the microscopic particles moving at faster than visible speed.

From 1905 through 1912 Perrin organized a laboratory program of investigations on colloidal suspensions of latex, mastic, and resin particles. He studied three principal phenomena. These included the vertical distribution of colloid particles after they reach equilibrium (this corresponds to the Laplacian distribution of particles in air under the influence of gravity); the translational displacement of particles, and the rotation of particles. Perrin's experimental procedures were complicated and precise, requiring separation (by centrifugation) of particles with identical radius; establishing their density; marking them with small faults or occlusions in order to study their rotation; and taking thousands and thousands of photographs of particles to be counted under well-defined conditions.

His very first series of experimental investigations, focusing on the distribution of particles under the influence of gravity, turned out to be an effect predicted by Albert Einstein (1879–1955) in a theoretical paper on Brownian motion. In this paper, "On the Movement of Particles Suspended in Fluids at Rest, with the Aid of the Molecular-Kinetic Theory of Heat" (1905), Einstein used considerations based on entropy, free energy, and osmotic pressure to calculate the diffusion and the displacements of small suspended particles. The displacements, he predicted, would be proportional to the square root of the time and inversely proportional to the square root of Avogadro's number *N*. Perrin's data confirmed the predicted displacements and allowed him to calculate an improved value for *N*.

In 1907 Theodor Svedberg completed a thesis on colloids at the University of Uppsala. He, mistakenly, as it turned out, represented particle motion by sinusoidal curves corresponding to oscillating motions and he successfully demonstrated that Brownian motion is in no way an electrical effect. Svedberg worked with finely dispersed inorganic colloids, for example gold, platinum, and silver hydrosols, and he used the ultramicroscope for making particles visible down to five microns. He related his results to Einstein's theoretical predictions, as well as to those of the physicist Maryan Smoluchowski.

Ostwald reviewed Svedberg's thesis for a journal, noted its support for Einstein's theoretical work, and conceded that Svedberg's was the experimental proof that Ostwald himself had been demanding. Ostwald wrote later that the investigations of Perrin "entitle even the cautious scientist to speak of an experimental proof for the atomistic constitution of space-filled matter."[1] By 1913 the matter was settled. John D. Bernal, the physicist and X-ray crystallographer, recalled his impression of Perrin book's *Atoms*, published in 1913. Perrin, said Bernal, had finally "demonstrated the existence of atoms and the possibility of counting them."[2]

An aspect of both Perrin's and Svedberg's work that persuaded many skeptics to belief in molecular reality was the work's visual character. Svedberg, for example, aimed to refine the principles of the ultramicroscope so that it would be possible to see a particle the size of "one-tenth of a normal atom."[3] He also wanted to be able to determine the distribution of particle sizes, not just average particle size, and it was this quest that led him and his co-workers to devise the centrifuge that he called the "ultracentrifuge" by analogy to the ultramicroscope (discussed in the next section).

There were other important visual images associated with the atom and the molecule in the early twentieth century. One was the result of the project in the Manchester laboratory of Ernest Rutherford (1871–1937) to test hypotheses about the distribution of electrical charge within the atom by studying the paths of alpha particles from naturally radioactive sources as they were directed against metal targets.

The method of following the alpha-particles was extraordinarily tedious. In a dark room one or two observers used a movable low-powered microscope to count flashes of light on a zinc sulfide screen as charged particles hit it one-by-one. At each flash the observer pressed a key which recorded his observation. Then correlations were made among the differences in counts in order to come up with a reliable number.

What the method did was to give objective existence to discrete individual particles in the laboratory. When not all investigators could duplicate the results from Rutherford's laboratory, demands were made for a counting method that was both less tedious and less subjective. The method and its instrument were the invention of Hans Geiger (1882–1945), one of Rutherford's co-workers who perfected the "Geiger counter" with Walther Müller in 1928 after returning to Germany. As a single charged particle crossed the central tube of the counter, it caused an electric discharge, which was converted into an audible and recordable 'click.'

Another instrument for seeing and counting particles was designed initially for a very different purpose. Charles T.R. Wilson (1869–1959) conceived the idea of reproducing cloud effects in a laboratory after reveling in the magnificent early-morning vistas seen from the summit of Ben Nevis in the Scottish Highlands. He found that in the complete absence of dust particles, condensation of moist air occurred if the expansion container had been exposed to X-rays. By 1911 the cloud chamber was reliably displaying the paths of single charged particles as trails of minute water droplets.

In 1924 Patrick M.S. Blackett (1897–1974) obtained dramatic photographs in a cloud chamber of the track of a hydrogen nucleus as well as the track of the recoil atom during the disintegration of a nitrogen atom bombarded with alpha-particles. Eight years later, again in the Cavendish Laboratory, Blackett and the Italian physicist G.P.S. Occhialini combined two Geiger counters with a cloud chamber so that expansion of the gas in the chamber was triggered by charged particles lying exactly in the plane of the chamber. They published photographs the next year of the track of a new particle first observed by Carl Anderson (1905–1991) at the California Institute of Technology that became known as the positron.

It may seem ironic that atoms and molecules became objectively real at the same time that atoms were becoming decomposable. Yet there is considerable logic in this historical development, since much resistance to the reality of atoms and molecules was rooted in the inconceivability of inconsistent properties (like inner vibrations and impenetrability) that were proposed to coexist in a chemical atom or a physical molecule.

How Large Can Molecules Be?

For many physicists and physical chemists during the course of the twentieth century, the molecule was the object of mathematical derivation in thermodynamics, chemical kinetics, low-temperature physics, and solid-state physics. For these scientists, their visualized molecule was hardly different from the simple sphere imagined by Rudolf Clausius in the mid-nineteenth century. For many chemists, however, especially the large and growing numbers of organic chemists and biochemists, the molecule was an exceedingly complicated bit of work.

During the course of the nineteeth century, from roughly the 1830s through the 1860s, the fundamental principles of organic structural chemistry were established, including the linking together of carbon atoms in straight, branching, or cyclic hydrocarbon structures, the tetravalency of the carbon atom, and chemical addition and substitution on single, double, triple, and benzene-type bonds. In 1874 Jacobus H. van't Hoff (1852–1911) and Joseph Achille Le Bel (1847–1930) each proposed the distribution of carbon valences in the three-dimensional space of a tetrahedron, firmly establishing the causal principles of the optical isomerism discovered earlier by Louis Pasteur (1822–1895).

One of the greatest of organic chemists in the early twentieth century was the Berlin professor Emil Fischer (1852–1919), whose research interests moved from the production and characterization of relatively simple aromatic compounds in the 1870s and 1880s to the study of increasingly complex molecules of biological significance, including the sugars, the purines, and proteins.

In 1902 Fischer and Franz Hofmeister of Strassburg each proposed that proteins are constituted by amino acids joined together by the condensation of the amino group (NH_2) of one amino acid with the carboxyl group (COOH) of another, forming amide bonds (–CONH–). Using a method pioneered by Theodor Curtius for the formation of the dimer 2,5 diketopiperazine, Fischer formed a glycylglycine through a –CONH– link.

$$H_2N{-}CH_2.CO.NH.CH_2COOH$$

Fischer and his group went on to combine more than two amino acids, proceeding from what Fischer called 'dipeptides' to 'polypeptides,' with the triumph in 1907 of an 18-amino acid polypeptide: leucyl-triglycyl-leucyl-triglycyl-leucyl-octaglycyl-lycine. These polypeptides behaved like intermediate products of protein hydrolysis.

How big, then, would a completely synthesized protein be? The method of François Raoult (1830–1901) of inferring molecular weight from the depression of the freezing point of a solution had been used to estimate the molecular weight of natural 'soluble starch' at 32,400. However, Fischer explicitly rejected a reported estimate of molecular weight for natural proteins (including hemoglobin), of 12,000 to 15,000 although he accepted as valid the figure of 4,021 for a starch derivative that he synthesized with Karl Freudenburg. Fischer thought it probable that natural proteins had smaller weights than artificial ones.

In 1915, Wolfgang Ostwald (1883–1943), the son of Wilhelm Ostwald, published a popular work entitled *The World of Neglected Dimensions (Die Welt der vernachlässigten Dimensionen)*. He lowered Fischer's cap on molecular weight, laying out principles of what came to be called the aggregate theory of colloids. Others, too, including Carl Harries, Rudolf Pummerer, Kurt Hess, Paul Karrer, and Max Bergmann subscribed to a physico-chemical theory that colloidal substances such as cellulose, rubber, starch, proteins, and resins are the physical aggregates of small molecules held together by some intermolecular forces, perhaps similar to secondary valence or partial valence.

Theodor Svedberg was among those who had a very different point of view from the Ostwald group. During the 1920s Svedberg developed an "ultracentrifuge" that had the same purpose as the ultramicroscope: to see and to record colloid particles of different sizes using their rates of sedimentation. In 1924 he and Robin Fahraeus discovered that particles of the protein hemoglobin settled at the same speed and were of the same size with a molecular weight of about 64,800, a result reconfirmed in 1926 with an improved machine. Svedberg shortly afterwards found that a class of proteins called hemocyanins appeared to have molecular weights in the millions.

A new method of analysis was soon to complement Svedberg's. In 1920, a year after Fischer's death, his colleague Freudenberg offered experimental data from cellulose degradation that cellulose is a long-chain structure with very high molecular weight. The next year Michael Polanyi (1891–1976), a Hungarian physical chemist working at the Kaiser Wilhelm Institute for Fiber Chemistry, presented a lecture commenting on recent X-ray analysis of cellulose samples.

Polanyi concluded that the measured X-ray diffraction spots were consistent with either of two structures: a long glucosidic chain or an aggregate made up of units of a small number of glucose anhydrides. However, Polanyi declined to rule one way or the other on the basis of the physical evidence alone. Chemists at Berlin strongly favored lower molecular weights for natural biological molecules and regarded Polanyi as an interloper in matters having to do with organic chemistry.

Such debate about the structure and weight of natural substances like cellulose and protein demonstrates the rivalry and misunderstanding which was not unusual among organic chemists, physical chemists and physicists in the late nineteenth and early twentieth centuries. The discovery of X-rays in 1895 gave rise within twenty years to the application of the methods of X-ray crystallography to the problem of molecular structure. But just as some chemists in the late nineteenth century had resisted the identification of new elements by spectroscopy or distrusted the estimation of molecular weight from freezing-point depression, many chemists initially were unenthusiastic about X-ray data.

William Henry Bragg (1862–1942) and his son William Lawrence Bragg (1890–1971) were among the founders of the field of X-ray crystallography. One of their first collaborative studies was work on diamond that proved the correctness of the tetrahedral model of the directed valences of the carbon atom. Together at Leeds the Braggs, who shared the Nobel Prize in Physics in 1915, worked out the exact positions of atoms in diamond, copper, potassium chloride, and other crystals.

By the mid-1930s the methods of X-ray crystallography were leading to notable breakthroughs for establishing structures and not simply for reconfirming them. One such study was carried out at Cambridge by Dorothy Crowfoot (later Hodgkin) (1910–1994) with J.D. Bernal. They placed a crystal of the stomach enzyme pepsin in a glass capillary, in front of an X-ray beam, and found that the photographic film behind the crystal displayed a sharp pattern of spots. This result confirmed that an enzyme has a specific atomic structure and that X-ray analysis potentially could resolve the structure of a protein molecule. In 1944 Hodgkin and her research group at Oxford completed the determination of the structure of penicillin (molecular weight 334) by X-ray diffraction methods before organic chemists themselves had agreed on a structure.

When Polanyi in 1921 had failed to convince organic chemists in Berlin of the long-chain and high-molecular weight structure of cellulose, purely chemical arguments along these lines had just been published by Hermann Staudinger (1881–1965) at the Zurich Polytechnic Institute. In 1922 Staudinger provided strong chemical evidence for very long chains by refuting predictions from Harries' aggregate theory of rubber.

According to Harries, partial valences generated from unsaturated double bonds hold together relatively small molecules in rubber; thus hydrogenation of rubber should yield a low-weight molecular substance. Staudinger and his associate J. Fritschi found, in contrast, that properties of hydrorubber were similar to those of natural rubber, not of a distinct, low-weight molecular substance. They used the term 'macromolecule' for the first time.

Hermann Mark (1895–1992) was another pioneer in the application of X-rays in natural product chemistry. Working with Kurt Meyer in Berlin, Mark developed a theory of long-chain molecules held together in "banks" or "micelles," using the widths of X-ray diffraction spots to estimate the size of the micelles at 50 primary chains wide with each chain composed of some 50 glucose units. By the mid-1930s he had estimates of macromolecular weights as high as a million, supported by both physical and chemical investigations.

Of the interplay between physical and chemical methods, Mark said in retrospect:

> We [Staudinger and Mark] both favored the concept of long-chained molecules. He did, on the basis of organic chemistry; and I did, on the basis of X-ray diffraction. He only trusted organic chemistry. I said trust both (techniques); we have two methods which do not contradict. My God, they could have contradicted![4]

The American industrial chemist Wallace Hume Carothers (1896–1937) brought macromolecules into ordinary industrial application as an outcome of pursuing theoretical questions. Working in the research department of the E.I. du Pont de Nemours firm in Wilmington, Delaware, Carothers set out to make synthetic compounds of very high molecular weight. His first success was the rubber analogue neoprene (polychloroprene), the first important synthetic rubber. Carothers then prepared a series of polyesters that seemed not to have promising characteristics.

By 1930, Carothers and his team were synthesizing "superpolymers" with molecular weight of 10,000 or more, and Carothers' co-worker Julian Werner Hill (b. 1904) found that the superpolymers could be mechanically drawn out or dry-spun from solution into fibers and threads. These fibers had properties similar to natural fibers of cellulose or silk, with a similar X-ray pattern. After these discoveries, interest shifted from theoretical research on macromolecules to the production of a useful fiber. "Nylon 66," made by condensation polymerization from adipic acid and hexamethylene-diamine, went into commercial production in 1938.

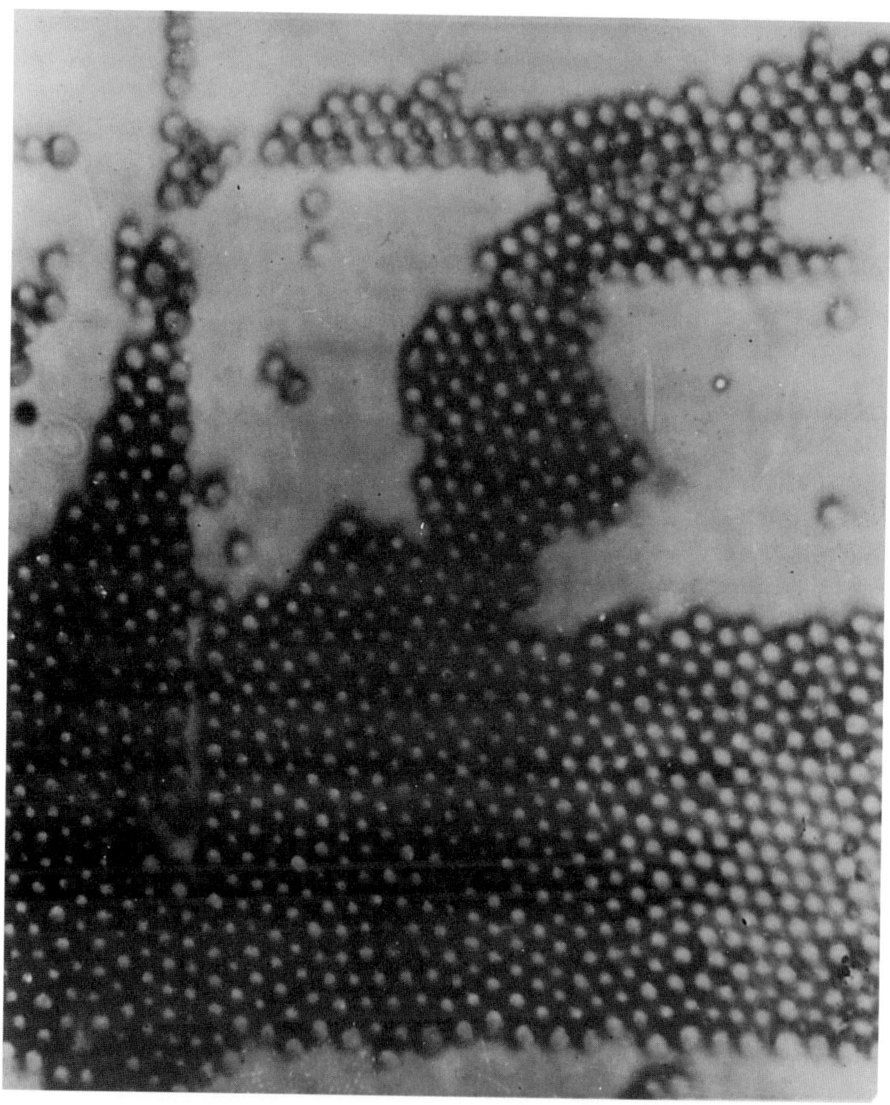

FIGURE 28.1: A PHOTOGRAPH OF GRANULES OF THE RESIN GAMBOGE USED BY JEAN PERRIN IN DETERMINING THE RADIUS OF PARTICLES UNDER AN ULTRAMICROSCOPE. COURTESY OF THE LATE FRANCIS PERRIN.

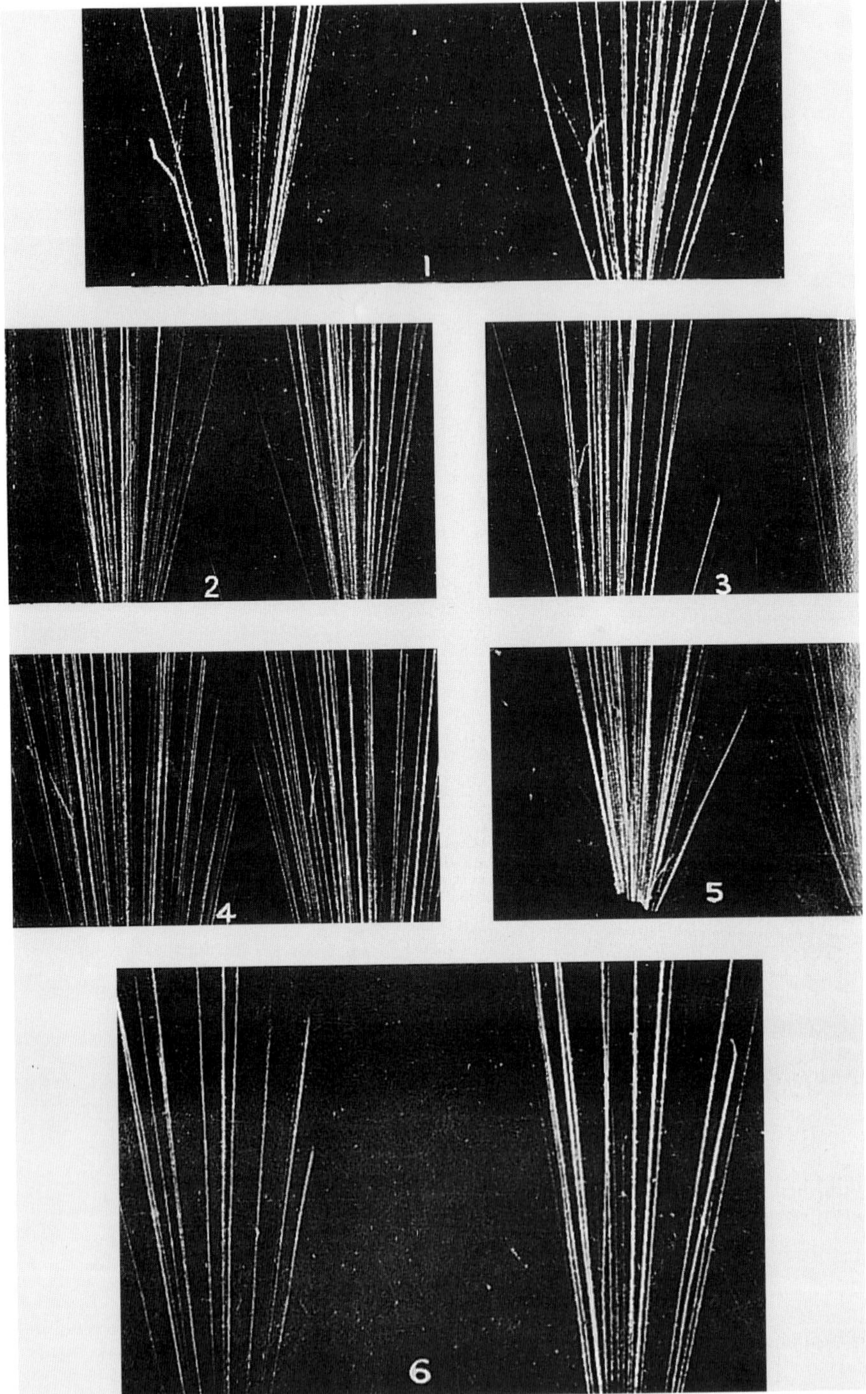

FIGURE 28.2: PHOTOGRAPHS OF THE EJECTION OF PROTONS FROM NITROGEN NUCLEI BY FAST ALPHA-PARTICLES. REPRODUCED IN P.M.S. BLACKETT, "THE EJECTION OF PROTONS FROM NITROGEN NUCLEI, PHOTOGRAPHED BY THE WILSON METHOD," *PROCEEDINGS OF THE ROYAL SOCIETY OF LONDON*, A107 (1925), 349–360, PLATE 7 (FACING P. 361).

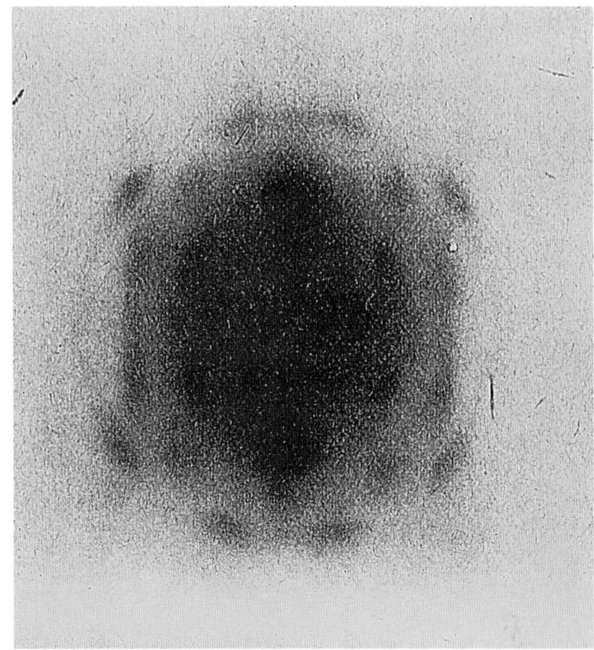

FIGURE 28.3: X-RAY DIFFRACTION PHOTOGRAPH OF CELLULOSE FIBER. REPRODUCED IN MICHAEL POLANYI, "FASERSTRUKTUR IM RÖNTGENLICHTE," *DIE NATURWISSENSCHAFTEN* (1921), 9: 337–340, FIGURE 2 ON P. 339.

FIGURE 28.4: WALLACE HUME CAROTHERS IN 1931 DRAWING OUT SYNTHETIC RUBBER AT E.I. DUPONT DE NEMOURS RESEARCH LABORATORY. PERMISSION OF HAGLEY MUSEUM AND LIBRARY, WILMINGTON, DELAWARE.

FIGURE 28.5: LINUS C. PAULING WITH A VACUUM PUMP AT HIS OFFICE LABORATORY AT THE CALIFORNIA INSTITUTE OF TECHNOLOGY AROUND 1940. COURTESY OF THE ARCHIVES AT THE CALIFORNIA INSTITUTE OF TECHNOLOGY AND THE AVA HELEN AND LINUS PAULING PAPERS, SPECIAL COLLECTIONS, KERR LIBRARY, OREGON STATE UNIVERSITY.

FIGURE 28.6: JAMES D. WATSON AND FRANCIS F.C. CRICK WITH THEIR MODEL OF DNA, CAVENDISH LABORATORY, CAMBRIDGE, MAY 1953. PERMISSION OF ANTONY BARRINGTON BROWN.

If the sciences of big molecules were transforming biochemistry and industrial chemistry, they were also helping to create the new field of molecular biology. In 1930 Linus Pauling (1901–1994) visited Hermann Mark in Germany. Mark's use of the methods of X-ray diffraction and electron diffraction intrigued Pauling, as did Mark's theories of protein structure and flexibility of polypeptide chains.

In 1932 Pauling, already well-known for his application of quantum wave mechanics to the chemical electron bond, sought funding from the Rockefeller Foundation in support of research on structural chemistry. This work would include investigations of the structures of protein, hemoglobin and other complex organic substances. When the Foundation shifted its interest more squarely from chemistry and physics to biology and medicine, Pauling targeted hemoglobin for study in a grant application of 1934.

Using techniques of crystallization of amino acids and X-ray investigations of their dimensions, Pauling began building up pictures of the way the protein must look. From 1937 until 1951 he and his co-workers at Caltech discovered the role of hydrogen bonding and assigned positions to amino acids on a coiled chain, using measurements of interatomic distances and bond angles, as well as three-dimensional molecular models, to develop a representation of what became known as the alpha helix. In 1958 John Cowdery Kendrew (b. 1917) fully determined the structure of the protein myoglobin, and in 1960 Max F. Perutz (b. 1914) succeeded with hemoglobin. These were the first complete X-ray determinations of protein structure, making use of computers available after the Second World War.

James D. Watson (b. 1928), who had been a postdoctoral fellow for four months at Caltech in 1950, knew Pauling's methods. Watson enlisted Francis Crick (b. 1916) at the Cavendish Laboratory to find the molecule of the gene. They employed precisely Pauling's method of model-building, concentrating on the structure of deoxyribonucleic acid, the molecule that the bacteriologists in Oswald Avery's laboratory at the Rockefeller Institute had identified as the transforming principle or genetic material for bacterial types.

When he was refused a passport to travel to England for a Royal Society symposium on protein structure in April 1952 because of his political activism against the testing of nuclear weapons, Pauling may have missed an opportunity to see the X-ray diffraction photographs of deoxyribose nucleic acid made by Rosalind Franklin (1920–1958). Basing their work on photographs by W.T. Astbury of a mixture of two forms of DNA, Pauling and Robert B. Corey finished a paper at Caltech in February of 1953 modeling DNA with three intertwined chains having phosphates near the fiber axis and bases on the outside.

In April, Watson and Crick proposed an alternative structure of two helical chains each coiled round the same axis with bases on the inside of the helix and phosphates on the outside. Their structure was based on the superior photographs taken by Franklin of the so-called "B" form of DNA, as well as Erwin Chargaff's requirement of equality between the number of thymine and adenine groups and the number of cytosine and guanine groups in nucleic acids. This work confirmed and popularized the notion of the molecular basis of life.

What Makes Atoms and Molecules Do What They Do?

Attempts to explain material properties in solids, liquids, and gases, as well as endeavors to account for chemical combination, stability, and instability have been at the heart of theories about atoms and molecules since ancient times. Two very different approaches were represented in force theories (attractions, repulsions, elective affinities) and kinetic theories (collisions, reaction rates, equilibria). Chemical thermodynamics and chemical kinetics became thriving areas of research in the late nineteenth and early twentieth centuries. Among leaders in the first decades of the twentieth century were Walther Nernst, Max Bodenstein, Michael Polanyi, Frederick Donnan, Irving Langmuir, Cyril Hinshelwood, N.N. Semenov, and Henry Eyring.

Within force theories, the hypothesis that electrostatics or electrodynamics might account for chemical affinities was a speculation largely eclipsed during the mid-nineteenth century by the rise of the structure theory in organic chemistry. However, the ionic theory of Svante Arrhenius (1859–1927), claiming that molecules of electrolytes exist in dilute solution as charged ions, revived considerable interest in the supposition that matter might be electrical in nature.

More dramatic was the demonstration at the end of the nineteenth century by J.J. Thomson (1856–1940) that cathode-rays are negatively-charged sub-particles of the hydrogen atom. And Thomson, who had always been interested in chemistry, proposed models of the hydrogen atom and the ethane molecule using the electron as a tube of force binding one atom to another in a molecule. The arrangement of electrons within an atom, he proposed, might confer specific chemical properties upon the different chemical elements, with a group of eight electrons corresponding to the special stability of the inert gases in the periodic classification of the elements.

After studies at Leipzig, Berlin, and MIT the American physical chemist Gilbert N. Lewis (1875–1957) offered in 1916 a model of eight electrons, or four duplets of electrons, arranged at the corners of a cube as the fundamental unit of the atom. The sharing of a pair of electrons between two atoms, he proposed, constitutes the fundamental bond of the chemical molecule, although this bond may vary from strongly polar to nonpolar depending on the gradations of electron-pair distribution between atoms. In 1919 Irving Langmuir (1881–1957) gave the names "covalent" and "electrovalent" to two types of chemical bonds, consciously extending Lewis' theory. "Electron rearrangement," said Langmuir, "is the fundamental cause of chemical action."[5]

By 1919, as scientists returned to investigations they had left behind during the First World War, work in atomic and molecular science was often taken up where it had been left in 1914. The laboratory group of Ernest Rutherford (1871–1937) at Manchester had established during 1908 to 1911 that the atoms in a metal target contain positively charged nuclei that are large and massive in comparison to the size and weight of an electron. Henry G.F. Moseley (1887–1915), who died in the war, had proven a connection between the nuclear charge and what he called the atomic number based on X-ray studies.

Like Moseley, Niels Bohr (1885–1962) had worked in Rutherford's laboratory before the war and he sought to come up with a model of the atom that would serve both chemistry and physics. In 1913 Bohr modeled a simple diatomic molecule as a girdle of electrons rotating in a circle at a right angle to the axis connecting two atoms. He attributed the chemical properties of the atom to electrons well outside the center of the atom and any radioactive properties to the interior nucleus. He also substituted the language of electron 'shells' and 'subshells' for the prevalent terminology of 'orbits' or 'spheres.' His most revolutionary idea was the application of Max Planck's hypothesis of the quantum of energy to the molecule's electrons by way of explaining the emission of discrete wavelengths of light in the line spectra of hydrogen.

Bohr's work came to the attention of the theoretical physicist Arnold Sommerfeld (1868–1951) who incorporated the relativistic effect of velocity on mass in orbital electron motion. He also calculated elliptical electron orbits, for which Bohr's circular orbit was a special case, and defined additional quantum numbers to supplement Bohr's 'first' quantum number for the radius of the principal axis of orbit (n_r or n). Sommerfeld's second quantum number specified the orbit's eccentricity as an effect of its quantized angular momentum. A third quantum number specified quantized orientation in space of the plane of the orbit in a magnetic field. This property of orientation was related empirically to the splitting of spectral lines in a magnetic field (the "normal Zeeman effect").

Following on Sommerfeld's work, Bohr outlined an improved general model of the structure of atoms in 1922 using what he called the "Aufbauprinzip" (building-up principle) for feeding electrons into an atom's subshells, a principle he had already proposed in 1912. Correlating the electrons in these subshells with spectral-line data for the so-called "strong," "principal," "diffuse," and "fundamental" lines, he designated the subshells *s, p, d,* and *f.*

However, the spectral lines attributed to the so-called "anomalous Zeeman effect" seemed to require half-quanta formulas. To meet this problem, Wolfgang Pauli (1900–1958) proposed the use of a fourth quantum number which can have two values, along with the rule that no two electrons in an atom can have the same four quantum numbers. This principle of "exclusion," which had no logical or physical basis, was given meaning in the late fall of 1925 by the Dutch physicists George Uhlenbeck (1900–1988) and Samuel Goudsmit (1902–1978), who proposed that the electron may spin in one of two directions, each direction designated by one of two values for the fourth quantum number.

While Bohr and many European scientists focused on understanding the structure and dynamics of the atom in the early 1920s, many American chemists and physicists who were interested in spectroscopy centered their work on the molecule, making use of the traditional notion of the diatomic molecule as a vibrating rotator capable of translational motion and using infrared band spectra from excited diatomic gases to study the energy of the molecule. The Danish physical chemist Niels Bjerrum was an early leader in developing a quantum theory for these kinds of molecules and this general approach was taken up by Edwin C. Kemble at Harvard, Raymond T. Birge at Berkeley, and Harrison M. Randall at the University of Michigan.

Both Robert S. Mulliken and John Clarke Slater worked with Kemble at Harvard University. Along with Walter Heitler, Fritz London, Friedrich Hund and Linus Pauling, they were among the founders of a quantum chemistry that equally came to be called theoretical chemistry or chemical physics. The approaches of these theoreticians are usually distinguished as the Heitler-London-Pauling-Slater method of valence bonds (building up a molecule from discrete atoms) and the Hund-Mulliken method of molecular orbitals (feeding electrons into orbitals around the nuclei of the molecule). Heitler and London appeared committed to the reduction of chemistry to first principles, while Pauling, Slater, and Mulliken were more pragmatic and semi-empirical in their methods.

The new quantum molecular science made use of a quantum mechanics that had been transformed by confirmation of the theory of Louis de Broglie (1892–1987) that a wavefront is associated with the electron. In 1926 Erwin Schrödinger (1887–1961) developed a very useful equation defining electron intensity or density from the presupposition that the electron is a concentration of waves in a small packet in space. Max Born (1882–1970), influenced by experimental data on atomic and molecular collisions, reinterpreted Schrödinger's function (in fact, the function squared) as the probability density for an electron in configuration space.

In 1927 Walter Heitler (1904–1981) and Fritz London (1900–1954) successfully worked out a mathematical description of the diatomic hydrogen molecule. They assumed that the wave function of each of two electrons is centered on one of the hydrogen nuclei, i.e., that the H_2 molecule consists of two hydrogen atoms. In doing this, they used the resonance theory for electron waves developed in 1926 by Werner Heisenberg (1901–1976). In subsequent papers they specifically applied their results to chemical valence theory, deducing well-known chemical facts, including the combination of hydrogen atoms to form a hydrogen molecule and the non-existence of a diatomic helium molecule or ion.

In contrast, Friedrich Hund published an entirely different approach. Hund assumed that an individual electron moves in a potential field that results from all the nuclei and from the other electrons present in the molecule. As Robert Mulliken, who met Hund in Göttingen where he was Born's assistant, later described it, this approach "regards each molecule as a self-sufficient unit and *not* as a mere composite of atoms."[6]

Robert Mulliken (1896–1986) and Erich Hückel (1896–1980) refined the "molecular-orbital" approach, a method which favors some particular region of space and disfavors others. In contrast to the Heitler-London method, which was perfected by Linus Pauling (1901–1994), the MO method overemphasizes, rather than underemphasizes, the ionic character of a molecule.

John C. Slater (1900–1976), who had studied molecular spectrosopy with Edwin Kemble at Harvard University and returned to teach there, developed an approach (the "determinantal method") which offers a way of choosing among linear combinations (essentially sums and differences) of the polar and non-polar terms in the Hund-Mulliken equations, in order to bring their method into better harmony with the non-polar emphasis characteristic of the Heitler-London-Pauling approach where polar terms do not figure in the wave equation.

However, it was Pauling's approach that won over most chemists, at least before 1940. Pauling was concerned to explain chemical bonding in hydrocarbons and in unsaturated molecules like the nitrogen oxides and benzene. The carbon atom has six electrons, which should be distributed on the basis of quantum principles into energy states of $1s^2$, $2s^2$, $2p^2$. However, carbon has four valence electrons, suggesting that one of the $2s$ electrons has been promoted to the higher p energy state so that the electron configuration in carbon is $1s^2$, $2s$, $2p_x$, $2p_y$, $2p_z$. If this were the case, one of the CH bonds (the $2s$ one) in CH_4 would be different from the others. It is not.

By 1931 Pauling had the answer, as did Slater. The basic idea was the mixing or 'hybridization' of the s and p energy levels, so that a new valence-bond wave-function has a lower energy value intermediary in character between the energy values associated with either a s (spherical) or a p (elliptical) wave function or orbital. Pauling and Slater demonstrated that certain types of wave functions project out in characteristic directions, p waves, for example, represented by three dumbbell-shaped distributions or contour-lines at right angles to one another, and the s wave a distribution which is spherically shaped.

Hybridization of these wave functions, or orbitals, produces electron distributions identical in kind and oriented toward the corners of a tetrahedron, rather than at right angles, i.e., with C–H–C angles of 109.5° rather than 90°. If a bond angle is expected to be 90° and departs from that figure (109.5° in methane, 107° in ammonia, 104.5° in water), hybridization could be suspected. There was now a quantum mechanical justification for the chemist's tetrahedral atom.

Pauling turned in earnest to unsaturated molecules and to conjugated molecules with alternating single and double bonds. He extended the notion of mechanical resonance to molecules like carbon dioxide and benzene, reasoning that a wave function might be set up to represent each of the possible classical valence, or electron-pair, bonds in these compounds. Each equation corresponds to a combination of ionic and covalent character for a bond, as well as to its energy content. The actual electronic structure is a resonance hybrid which is none of these, just as it is none of the Kekulé structures ordinarily drawn for benzene. Rather the resonance hybrid structure is something with elements from each equation and a lower energy value than any of them.

With Pauling's theory, the chemical and the physical molecule were in harmony with one another. While many chemists continued to use graphic representations of molecular structure and electron transfers in their everyday work, an increasingly sophisticated mathematical analysis of the cause of atomic and molecular activity became available in the following decades.

The hydrogen 'nucleus' that figured in many of these calculations had been called a "proton" by Rutherford as early as 1919. In 1932 James Chadwick (1891–1974), reasoning from experiments by W.G. Bothe (1891–1957) and Irène (1897–1956) and Frédéric Joliot-Curie (1900–1958) recognized the existence of the neutron which also had been predicted by Rutherford. The discovery of the neutron was decisive in creating the high-energy sciences of nuclear and particle physics that emerged from the events of the Second World War.

As noted above in connection with the role of X-ray diffraction studies in atomic and molecular science, the incorporation of new generations of computers into physical and chemical laboratories after the Second World War was to transform the speed with which atomic and molecular structures and reaction pathways could be computed. New generations of spectroscopic instruments similarly revolutionized the study of atoms and molecules in the 1950s and 1960s, with many of these instruments aggressively marketed by entreprenurial firms to industrial and academic laboratories.

Optical spectroscopy had made possible the identification of individual elements and compounds after the mid-nineteenth century. By the 1920s, infrared and Raman spectroscopy were beginning to provide information about molecular structure, including the presence and location of functional groups or unsaturated bonds.

Nuclear magnetic resonance (NMR) became available in the 1950s, making use of the discovery that an atomic nucleus subjected to a powerful magnetic field selectively absorbs high-frequency radio waves. In 1946 Felix Bloch measured the absorption for protons in water and in 1947 Rex Richards first reported the use of NMR to distinguish between three types of hydrogen atom in ethanol, each with its own characteristic electron environment. Mass spectroscopy, which was used initially to determine the masses and abundances of isotopes, was also modifed in the 1940s and 1950s into a powerful tool for determining molecular structure from the fragmentation of molecules.

The reality of atoms and molecules, the extraordinary range in size and structure of the molecule, and the seat of atomic and molecular activity in mechanics, electrodynamics, and three-dimensional structure all were established during the course of the twentieth century. As recently as the turn of this century, the study of atoms and molecules was thought by some scientists to be a speculative pursuit of little practical value to anyone. Few scientific fields demonstrate more clearly the transformations of esoteric knowledge into practical knowledge.

REFERENCES

1. Wilhelm Ostwald, Vorbericht to *Grundriss der physikalischen Chemie.* (Grossbothen, November 1908).
2. Quoted in Mary Jo Nye, *Molecular Reality. A Perspective on the Scientific Work of Jean Perrin* (1972), pp. 161–162.
3. Quoted in Anders Lungren. "The Ideological Use of Instrumentation: The Svedberg, Atoms, and the Structure of Matter." In Svante Lindqvist (Ed.), *Center on the Periphery: Historical Aspects of Twentieth Century Swedish Physics.* (Canton, MA: Science History Publications, 1993), p. 333.
4. Herman F. Mark, "Interview with Herman F. Mark," *Journal of Chemical Education* (1979), **56**: 83–86, on p. 84.
5. Irving Langmuir, "The Structure of Molecules" *BAAS Reports. Edinburgh, 1921* (1922), pp. 468–469.
6. From Robert Mulliken, "Spectroscopy, Molecular Orbitals, and Chemical Bonding," pp. 131–160 in *Nobel Lectures: Chemistry, 1963–1970.* (New York: American Elsevier, 1972), on p. 137.

FURTHER READING

Assmus, Alexi Josephine. "Molecular Structure and the Genesis of the American Quantum Physics Community, 1916–1926," *Historical Studies in the Physical and Biological Sciences* (1992), **22**: 209–231.

Elzen, Boelie. "The Failure of a Successful Artifact. The Svedberg Ultracentrifuge." In Svante Lindqvist (Ed.), *Center on the Periphery: Historical Aspects of Twentieth Century Swedish Physics.* (Canton, MA: Science History Publications, 1993), pp. 347–377.

Furukawa, Yasu. "Staudinger, Carothers, and the Emergence of Macromolecular Chemistry." University of Oklahoma Ph.D. thesis, 1983.

Gavroglu, Kostas and Ana Simoes. "The Americans, the Germans and the Beginnings of Quantum Chemistry," *Historical Studies in the Physical and Biological Sciences* (1994), **25**: 47–110.

Hudson, John. *The History of Chemistry.* (New York: Chapman and Hall, 1992).

Laidler, Keith J. *The World of Physical Chemistry.* (Oxford: Oxford University Press, 1993).

Lenoir, Timothy and Christophe Lécuyer. "Instrument Makers and Discipline Builders: The Case of Nuclear Magnetic Resonance," *Perspectives on Science: Historical, Philosophical, Social* (1995), **3**: 276–345.

Lundgren, Anders. "The Ideological Use of Instrumentation: The Svedberg, Atoms, and the Structure of Matter." In Svante Lindqvist (Ed.), *Center on the Periphery: Historical Aspects of Twentieth Century Swedish Physics.* (Canton, MA: Science History Publications, 1993), pp. 327–346.

Nye, Mary Jo. *Molecular Reality. A Perspective on the Scientific Work of Jean Perrin.* (London: Macdonald and New York: American Elsevier, 1972).

__________. *Before Big Science: The Pursuit of Modern Chemistry and Physics, 1800–1940.* (New York: Twayne-Macmillan, 1996).

Perutz, Max F. "Molecular Biology in Cambridge." In Richard Mason (Ed.), *Cambridge Minds.* (Cambridge: Cambridge University Press, 1994), pp. 193–203.

Servos, John W. *Physical Chemistry from Ostwald to Pauling: The Making of a Science in America.* (Princeton: Princeton University Press, 1990).

CHAPTER 29

Solid State Science

LILLIAN HODDESON

The ancient picture of a universe filled with matter composed of but a few fundamental elements faded after the scientific revolution. By the nineteenth century, it had been replaced by a far more lively and complex view in which the materials in nature were seen as a rich array of elements in countless combinations. Three of the four Aristotelian elements — earth, water, and air — survived as categories for describing the ordinary states of matter: solid, liquid, and gas. The fourth element — fire — eventually became associated with ionized gases (plasmas).

The emerging fields of physical science could then proceed in their studies of matter on several levels: the cosmic (explored in astronomy and astrophysics); the microscopic (atomic, and much later nuclear and particle physics); and the level we focus on here, of collections of atoms and molecules, as found in materials. These studies of collections included some time-honored scientific themes, such as elasticity and mineralogy, that had been of interest to practical people ever since men and women were interested in using real materials for useful purposes, such as making swords, tiles, bowls, tools, or jewelry. Over the centuries, artists, crafts persons, builders, and other groups outside what became the scientific establishment created a vast fund of empirical knowledge about the materials. This fund was eventually subsumed by the 'new' scientific subfield which by the mid-1940s came to be known as 'solid state science,' or more commonly, 'solid state physics' (and more recently 'condensed matter physics').

The new subfield formed an umbrella over many areas of study previously thought of as independent — old areas, such as crystallography, strength of materials, electrical conduction, ferroelectricity, and magnetism, as well as new ones, such as superconductivity, which developed out of more recent discoveries. The enormous number of existing materials and the multitude of their properties (e.g., material structure, cohesion, diffusion, plasticity, electrical conduction, and magnetism) gives the subfield enormous scope. The field's usefulness in virtually every sphere of human activity caused the field to be heavily populated. That the term 'solid-state' was widely popularized in the 1960s by advertisers' tags on transistorized devices is one expression of the field's usefulness.

Yet in comparison with other areas of twentieth-century science, solid state physics has been little examined by historians. In one heroic attempt during the 1980s, The International Project on the History of Solid State Physics, a group of historians and physicists put a massive effort into opening up the field for historical study by identifying and studying the spines of the solid state umbrella. In this project's major publication, Spencer Weart pointed out three principal spines (using a different metaphor, that of an edifice, he called them "pillars"): X-ray crystallography, which provided the first look at the atomic lattice inside solids; quantum mechanics, which provided the theoretical framework for the modern theory of solids; and structure sensitive studies, which focused attention on the dependence of matter's properties on the detailed nature of its structure, including deviations of that structure from perfection.[1] Several other historical efforts added more detail and further dimensions. For example, Dominique Pestre, studying a particular post-war solid state group (that of Louis Néel in Grenoble), noted the multidisciplinary and collaborative nature of the field, as well as its strong ties with industry, government and the military. He also pointed out that certain scientific leaders in solid state were successful because they were also entrepreneurs with the "soul of an engineer."[2]

Despite such historical work, the community of historians who have worked on solid state physics has remained relatively small and there are many aspects of the huge field still to be explored by science scholars. For example, no one has yet adequately examined the extent to which solid state physics reflects larger cultural trends. It is this aspect that I comment on here, following in the footsteps of historian Paul Forman, one of the few who have noted the postmodern transition in the area of physics.

The argument — which I summarize in this introduction and illustrate in the sections below — starts with the observation that before the twentieth century, many workers in the areas that later were subsumed by solid state physics developed a variety of particular theoretical schemes to help them explain the observations they made about the properties and behavior of solids. Yet certain nagging fundamental questions remained unanswered, for example: Why do metals, insulators and semiconductors have such different behaviors? Further difficulties included the fact that although particular theories could be constructed to fit certain observations, they frequently did not fit others, and the different theories often contradicted each other. As more studies were made, and as the fund of good data grew larger, fitting theories to the data required making an increasing number of *ad hoc* assumptions (most of which eventually proved untenable).

Paradoxically, to deal adequately with real materials, a small number of the best theoretical physicists in the world had to leave the domain of real materials for about three decades and develop a theory for unreal solids. While most workers (e.g., the magnetics research physicist employed at Bell Laboratories, Richard Bozorth) continued their studies of real materials using empirical models, the elite group who diverted into unreality focused between 1900 and 1930 on developing by 1933 the esoteric 'quantum theory of solids,' a theory that worked rather well in describing ideal solids (model materials having far fewer parameters than the simplest real materials).

Only after this modern theory of ideal solids was in place could the community contend with the far more complex problems of explaining the behavior of real materials within a unified conceptual framework — or in the words of the late historian and metallurgist Cyril Stanley Smith, with "far more complex things than have been allowed in the domain of respectable physics in the past." The return to real materials occurred between 1933 and 1945. It was a critical transition in solid state physics, for it marked the end of relegating real structures to "philosophically inferior positions."[3] In the following 'postmodern' period, it became possible for solid state physicists to address a wide range of ancient and modern questions about real materials — old questions, such as why certain materials are fragile, brittle, or ductile, why others conduct heat or electricity, what causes electrical resistance or thermoelectricity, and why some materials are magnetic; and new questions, such as why some materials exhibit superconductivity.

The answers proved useful in communications, travel, commerce, manufacture, art, and defense. In consequence, solid-state physics was generously supported by institutions of industry, government, and academia, and the field flourished. Not only did the character of solid state research become heavily multidisciplinary (chacterized by collaborations among workers in different fields), but the field absorbed many different styles of research (from purely theoretical to purely empirical). Quantum mechanical specialists saw themselves as working in the same area as physical chemists, industrial engineers, metallurgists, and experimental physicists. It was no accident that the first comprehensive textbook designed to bring together the entire solid-state field — *The Modern Theory of Solids*, by Frederick Seitz, published in 1940 — was aimed not only at physicists in academia but at engineers and others interested in practical application. The name 'solid-state physics,' which had been almost unknown before 1940, became a familiar term, applied not only to an intellectually and socially connected field, but to university chairs, journals, schools, conferences, and even buildings, as the physicists studying the various states of matter formed what Spencer Weart called a "solid community."[4]

1900–1926 — The Quantum and the Crystal Lattice

Among the important events that culminated in the invention of quantum mechanics in 1925–26, two stand out in the prehistory of the modern theory of solids: Albert Einstein's 1907 theory of the specific heats of solids (the heat needed to raise the temperature of a unit of mass by one degree) and the experiments conducted in Munich in 1912 by Max von Laue, Walther Friederich, and Paul Knipping demonstrating that metallic crystals diffract X-rays. Once Einstein successfully applied the new theory of the quantum to a problem of solids, many other physicists were encouraged to get on the band wagon and refine the calculations, for example Peter Debye and Theodore von Kármán, who worked in Göttingen in 1912. These works pointed to the need to create a full quantum-mechanical framework for explaining solid-state phenomena.

Similarly, once the remarkable Munich X-ray diffraction experiments gave physicists their first look at the crystal lattice, the way was open for many other diffraction studies that explored the arrangements of atoms in many simple crystals — for example, in the extensive series of X-ray studies conducted by the British father-and-son team of Sir Lawrence and William Henry Bragg, and in the pioneering electron diffraction experiments that Clinton Davisson and Lester Germer carried out in the mid-1920s at the newly incorporated Bell Telephone Laboratories in New York City. The Davisson and Germer experiments were also the first experimental demonstration of the French physicist Louis de Broglie's controversial theory that particles can behave like waves, a key step in the development of quantum mechanics (by Werner Heisenberg, Erwin Schrödinger, and others).

1926–1932 — The Quantum Mechanical Theory of Ideal Solids — Modernity

An essential ingredient for extending the quantum theory to solids was the development in 1926 by Enrico Fermi and Paul Dirac of a quantum statistics for treating the particles in solids. With that ammunition, Wolfgang Pauli took the first step toward developing the quantum theory of solids in 1926. Pauli was not particularly interested in solids, but rather in using the observed weak paramagnetism of metals as a test problem to help him address a fundamental question: which statistics — those developed by Fermi and Dirac or those statistics developed by Einstein and the Indian theorist Satyendranath Bose for treating light photons — apply in the case of metals? In working this problem out Pauli developed a useful free-electron quantum theoretical approach for treating metals. In the process he unintentionally opened the door to the modern theory of solids, an area he soon realized would become objectionably messy and applied. He eventually became famous for his pointed criticisms of the new subfield as a physics of dirt "that one shouldn't wallow in."[5]

Pauli left the job of extending his free-electron theory of metals to his mentor, Arnold Sommerfeld, who in 1927 used Pauli's methodology to structure a more comprehensive 'semiclassical' free-electron theory that succeeded in addressing a wide range of long-standing problems of metals (e.g., specific heats, electrical conductivity, and thermal conductivity). But it remained a puzzle why any theory that was based on the assumption that electrons are *free* should work at all for solids, in which electrons are obviously not free.

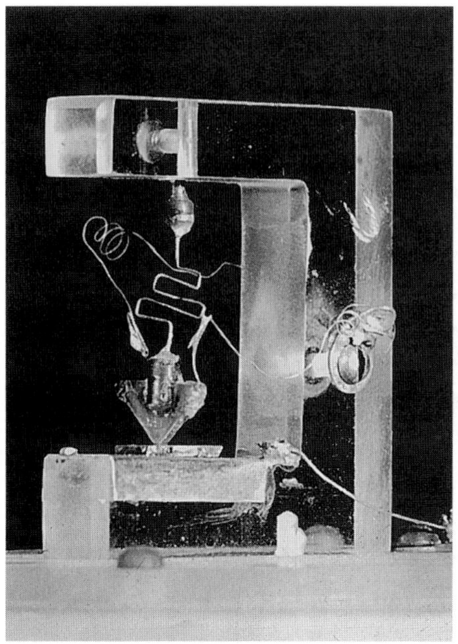

FIGURE 29.1: FIRST TRANSISTOR EVER ASSEMBLED. 1947. COURTESY OF AT&T.

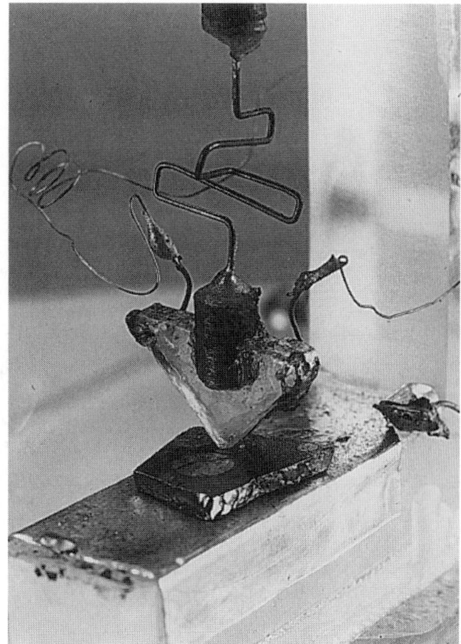

FIGURE 29.2: THE ORIGINAL POINT CONTROL TRANSISTOR. COURTESY OF AT&T.

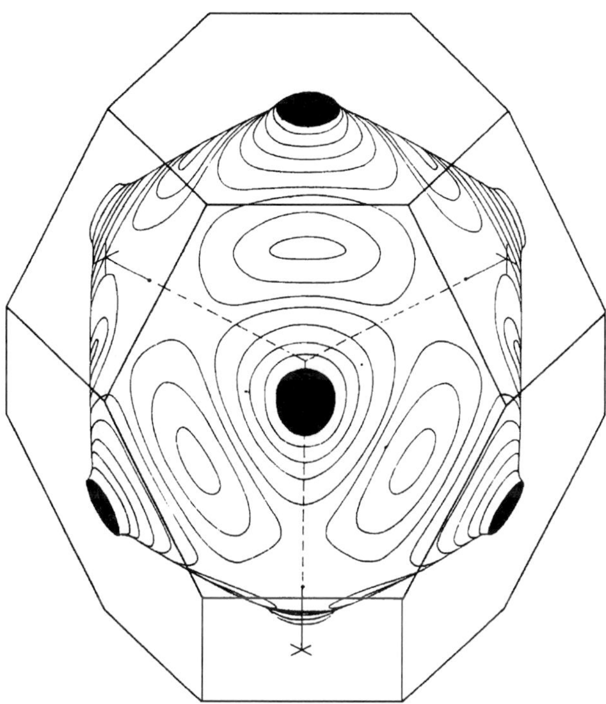

FIGURE 29.3: THE FERMI SURFACE OF COPPER DETERMINED EXPERIMENTALLY BY A.B. PIPPARD IN 1957. COURTESY OF THE ROYAL SOCIETY.

FIGURE 29.4: X-RAY SPECTROMETER.

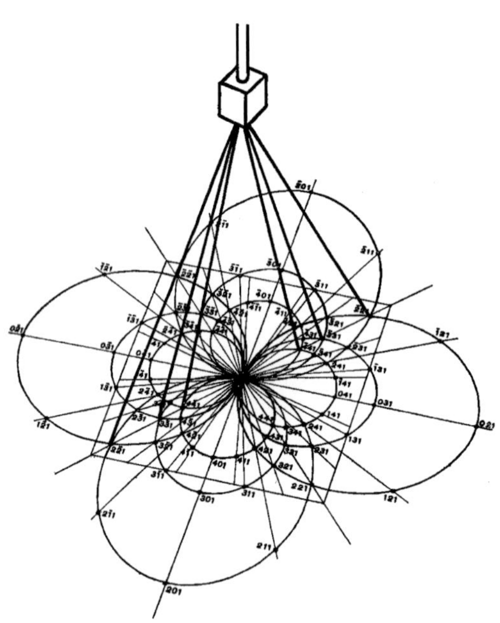

FIGURE 29.5: SIDE DRAWING FROM A MODEL SHOWING THE PROCESS OF THE LAUE PHOTOGRAPH. FROM *AN INTRODUCTION TO CRYSTAL ANALYSIS* BY SIR WILLIAM BRAGG. G. BELL AND SONS, LTD. 1928.

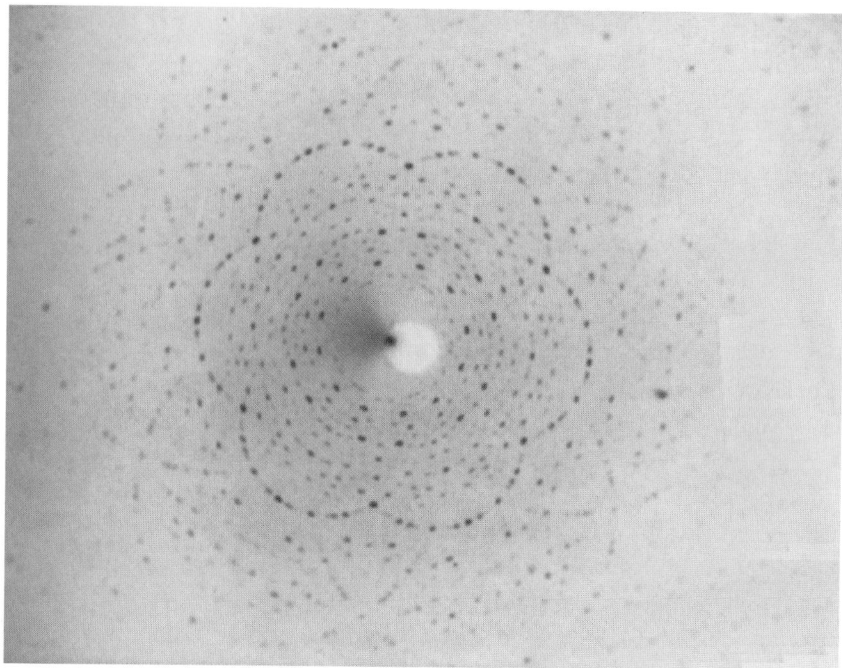

FIGURE 29.6: LAUE PHOTOGRAPHS OF RONTGEN RAYS PASSING THROUGH A BERYL CRYSTAL. (WM. LEHMANN).

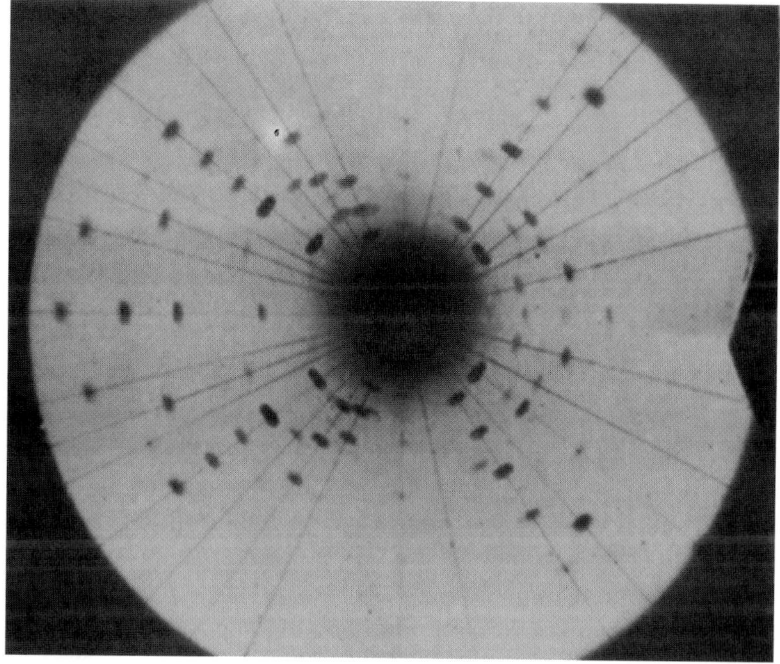

FIGURE 29.7: STEARIC ACID CRYSTAL (MÜLLER). NOTE THE DIFFERENCE IN SYMMETRY BETWEEN THIS AND THE ABOVE.

Heisenberg's first doctoral student, Felix Bloch, offered part of the answer in 1928. Drawing on several reasonable approximations, Bloch reduced the complicated problem of treating the many electrons in a crystal to a one-body model problem. Then using a theorem he discovered, Bloch proved that electrons move through a periodic crystal lattice *as though* they were free particles. Buried in this work, was the fertile concept that electrons in solids occupy quantum-mechanical energy bands; within the framework of what became known as 'band theory,' physicists would in time be able to explain most of the electronic properties of solids.

Several more pieces of the quantum theory of solids were then filled in over the next four years by a handful of workers, most of them students centering about Pauli in Zurich or Heisenberg in Leipzig. For example, Rudolf Peierls put forth the useful concept of the "hole," the idea that empty electron states near the top of an otherwise filled band behave exactly like positively charged particles. Alan Wilson, on a visit in 1931 from Cambridge to Heisenberg's institute, assembled the available pieces of band theory so as to explain the long-standing puzzle of the difference between metals and insulators. Seizing upon the clue that a material with a filled band carries no current, he recognized that insulators must have completely filled bands while metals must have partially filled bands. He portrayed semiconductors as insulators having a gap between their filled and partially filled bands; conduction would occur when electrons crossed the gap. Wilson also used his model later that year to forge a pioneering theory of rectification at metal-to-semiconductor junctions, thus forming the basis for subsequent rectification theories developed in the late 1930s and early 1940s by Walter Schottky, Nevill Mott, Boris Davydov, and others.

1933 — The Turning Point

The modern period of development of the quantum theory of ideal solids reached a turning point in 1933. By then some of the most dramatic properties of ideal solids had been explained qualitatively. The pace of fundamental work was slowing down in part because making further progress was becoming harder, and because available materials contained too many impurities and imperfections to be approximated by the assumption that they were ideal perfect solids.

Meanwhile, the new 'quantum theory of solids' was being written up in at least half a dozen major review articles, including the most influential one, the massive 1933 *Handbuch der Physik* by Sommerfeld and his student Hans Bethe (who did 90 percent of the writing).[6] These texts were used in training the new generation of physicists who would become the first to view themselves as "solid-state physicists." Three of the first graduate programs to offer training in the modern theory of solids were those at Princeton (around Eugene Wigner), MIT (around John Slater), and Bristol (around Nevill Mott and Harry Jones).

The social structure of the field began to shift. While the new generation was in graduate school in the early 1930s, most of the theorists who had been prominent in the creation of the quantum theory of ideal solids were turning to other areas, such as nuclear physics and quantum electrodynamics, which were then offering exciting fundamental problems. This change of interest was especially pronounced among the physicists (including Peierls, Bethe, and Bloch) who emigrated to the United States or Great Britain to escape fascist domination.

A major breakthrough at Princeton in 1933 marked the start of a new phase. Wigner and his first graduate student Frederick Seitz developed a novel approximation procedure for computing the band structure of the simplest real metal, sodium. Until this time quantum-mechanical band structure calculations had been done only for ideal solids. The pioneering work by Wigner and Seitz then led to an industry of calculations of real band structures at many institutions throughout the world.[7] The modern theory of solids, now no longer led by a handful of leading physicists studying ideal solids, advanced beyond its modern program and joined the efforts of the many industrial and academic scientists who had been devoting themselves to studies of the real materials in the world. The intrinsic imperfections ('dirtiness') of real solids were embraced and conquered using new methods of approximation and numerical analysis tailored to fit specific problems. Although the new solid state program appeared crude and inelegant to Pauli and others in his generation, the new generation considered it beautiful to be able to solve for the first time many problems that had been insoluble within the theory of ideal solids.

Impact of World War II — The Transition to the Postmodern Period

Events during World War II reinforced the new direction solid state physics was taking. What was valued for military and industrial applications (e.g., the building of atomic bombs and radar systems) was reliable performance of devices made of real imperfect materials. Postmodern solid-state physics helped meet these objectives, and the industrial and military establishments offered copious funding to the growing field both during and after the war.

In wartime research and development, overarching theories were valued only to the degree they could be tailored to solve particular practical problems. And it was often not possible to take the time to develop a full understanding of phenomena. The pressures to make functioning devices quickly encouraged solid state physics (and other scientific areas) to assume aspects of engineering practice; e.g., limited trail and error methodologies, iteration procedures, and numerical analysis.[8]

The war also contributed to the field's growth by offering new tools, new organizations, and new materials, and by encouraging cross-fertilization of fields. For example, neutrons, intensely studied by the atomic bomb project, proved to be effective probes of solids in diffraction experiments. Thermal neutron scattering proved useful in studying helium-4 and eventually helium-3. New digital computing methods developed during wartime allowed solid state theorists to achieve far greater accuracy in solving problems of quantum mechanics. New resonance techniques deriving from the wartime radar program proved useful in studying solids because the microwave radiation could be tuned to coincide with natural vibrational or rotational frequencies of atoms and molecules in applied magnetic fields. (Thus in nuclear magnetic resonance, energy transfer from a radiofrequency circuit to a system of nuclear spins can yield sharp absorption lines.) Wartime studies also made many new research materials available. For example, the helium liquefier which Samuel Collins of MIT developed, using technology he developed for portable oxygen generators to be used on airplane missions, enabled any laboratory to obtain bulk liquid helium at reasonable cost and hence engage in low temperature physics research. Wartime research also helped to develop theoretical techniques, including quantum electrodynamics and many-body theory.

On a social level, many younger scientists, stimulated by the experience of working closely with leading scientists in the large defense projects, were encouraged to enter the solid state field. And the wartime experience of working within highly organized military institutions characterized by large-scale, goal-oriented, multidisciplinary research influenced the post-war organization of solid state research in many industrial and academic institutions.

The war created so many connections and influences on the emerging solid state field that it is impossible to discuss them in any single account. We are forced to limit our scope to case histories — e.g., of particular devices, such as the transistor, the laser,[9] or radar;[10] particular schools or groups, like the one surrounding Néel at Grenoble;[11] or particular discoveries like the microscopic theory of superconductivity.[12]

In the case of the transistor, the cornerstone of our modern microchip technology, the war's influence derived from the fact that semiconductors are a critical component of crystal rectifiers used in radar detectors and frequency converters. Enormous programs of research, for example, at the MIT Radiation Laboratory in the United States, were directed toward every aspect of crystal rectifiers. In addition, by contracting research out to numerous industrial and academic groups (e.g., Westinghouse, Bell Laboratories, General Electric, Sylvania, Du Pont, the University of Pennsylvania, Purdue University, British Thompson-Houston, the British Telecommunications Research Establishment, and Oxford University) the wartime program caused many groups to be informed about the cutting edge of semiconductor research.

That interchange accelerated the steps toward the invention of the transistor. Even before the war ended, Mervin Kelly, Bell Labs Director of Physical Research, recognized that the free technical exchange enjoyed during wartime would create a fierce post-war competition. In anticipation of this competition, he established in January 1945 a large multidisciplinary solid-state division modeled on the large wartime defense projects. Multidisciplinary teams within the division included first-rate theorists, experimental physicists, chemists, circuit experts, and other specialists. The research head of the new division, William Shockley, one of Slater's graduate students during the 1930s, encouraged Kelly to hire John Bardeen, one of Wigner's graduate students in the early 1930s who was then working (with less than full enthusiasm) on geophysical problems at the Naval Ordnance Laboratory in Washington D.C.

At Bell Laboratories, Bardeen formed a close collaboration with experimental physicist Walter Brattain. Studying recent developments in semiconductors, they drew on several research developments emerging from the wartime crystal rectifier program: research on the physics and technology of point-contacts, on the physics of silicon and germanium, on the technology for producing pure and perfect crystals, and on the theory of rectification at semiconductor interfaces. With his thorough grounding in the quantum theory of solids, Bardeen recognized that transistor action is a consequence of holes being injected into the germanium structure which he and Brattain were exploring. They demonstrated the historic first point-contact transistor in December 1947.

The transistor then spawned a massive industrial effort, which included Silicon Valley in California and the industrial efforts in Japan that eventually brought that country to lead the world market in transistorized consumer electronics. More generally, the transistor acted as a world-wide cultural transmitter, a 'crystal fire' encouraging new modes of cultural transmission (e.g., via FAXes or the Internet) that have been revolutionizing society.[13]

Further Studies of Real Materials

The rapid growth of solid state physics during the first two post-war decades produced a quantitative understanding of the behavior of many specific materials. This understanding drew on a variety of new microscopic considerations, for example, of the internal structure of crystals, the role of impurities and imperfections, the energies associated with interfaces, magnetic and electrical behaviors, and phase transformations. During the 1950s, much interest centered on determinations of the 'Fermi surface,' a quantum-mechanical feature so characteristic of particular solids that it often has been referred to as a solid's 'face.'

Two areas — 'point defects' and 'dislocations' — dealt explicitly with imperfections of crystal structures. Pre-war studies (especially by members of the Göttingen school of Robert Wichard Pohl) underscored the practical implications of studying localized positions of atomic dimension where the lattice structure is imperfect. Studies of the mechanical properties of crystals were immediately useful in metallurgical processes, such as the conversion of pig iron into ductile steel through heat treatment. In the post-war period, plastic deformations of solids came to be understood in terms of a distorted or defective lattice structure, a study that in time gave way to the subfield of 'linear crystal defects,' or 'dislocations.'

Magnetism, one of the centuries-old areas of natural philosophy, was transformed between 1930 and 1960 into a modern solid-state area.[14] The first crude quantum mechanical theories of magnetism (by Niels Bohr, Pauli, Heisenberg, and others) acted as a bridge between the earlier empirically-based studies (e.g., by Pierre Curie, Pierre Weiss, Paul Langevin, Hendrik Antoon Lorentz, and Kataro Honda) and later microscopic treatments. The concept of electron spin by Uhlenbeck and Goudsmit (1925) and Heisenberg's pioneering quantum theory of ferromagnetism (1928) opened the way toward explaining ferromagnetism as a consequence of a quantum-mechanical coupling between neighboring atoms. But despite major contributions to the theory of magnetism before and after the war by many researchers, ferromagnetism remained relatively intractable using the one-electron quantum theory of solids.

Collective Phenomena, including Superconductivity

The problems solid-state physicists encountered in their studies of ferromagnetism during the 1930s proved to be far more basic than they realized at that time. Only a few probing thinkers, among them Felix Bloch, suspected already then that the cluster of problems untreatable by the one-electron quantum theory of solids — including ferromagnetism, superconductivity, superfluid helium, ferroelectricity, the cohesive energy of metals, and phase transitions — might have a common root. Bloch speculated that the basic difficulty in all these cases was lack of a proper theory for the interactions of the particles in solids. Treating what came to be called 'collective phenomena' would require a proper 'many-body' theory.[15]

Various pieces emerged in the context of different problems. For example, Wigner in the United States constructed a piece while working in 1934 and in 1938 on the cohesive energy of metals. He developed a perturbation approximation for coping with the interacting particles inside metals. Lev Landau in the Soviet Union introduced the concept of an 'order parameter' in 1935, while studying phase transitions. Several years later, while studying superfluid helium, he realized that a microscopic theory for that system required postulating weakly interacting 'elementary excitations' above the ground state, for example quantized sound waves ('phonons'). He also suggested that superconductivity could be explained in terms of an energy gap in the elementary excitation spectrum. Later still, in studying phase transitions and ferromagnetism, Lars Onsager published his theory of the two-dimensional Ising model (1944), the first rigorous statistical evaluation demonstrating the existence of a phase transition and associated critical behavior in the thermodynamic limit.

The events of World War II influenced the evolving theory of collective phenomena heavily, as it did so many post-war developments in science. Consider for example the development by John Bardeen, Leon Cooper, and J. Robert Schrieffer of the microscopic theory of superconductivity. One reason so many had failed to explain the discovery made in 1911 by the Dutch physicist Heike Kamerlingh

Onnes, that certain materials lose all traces of electrical resistance below a critical temperature, was that several major pieces of the solution were missing. In particular, the factors still not available in 1950 were: (1) experimental verification of Fritz and Heinz London's speculative phenomenological theory developed in the mid-1930s, which included the important ideas later used by Bardeen as heuristics that superconductors exhibit "long-range order" and are characterized by an energy gap between the ground state and low-lying exited states; (2) recognition that the fundamental interaction that underlies superconductivity is the electron-phonon interaction; and (3) a proper theoretical machinery for dealing with the many-body theory interactions of the electrons and phonons.

These pieces emerged in the decade following the war out of wartime developments unrelated to the theory of superconductivity. For example, A. Brian Pippard working in England and others working in the United States used their wartime radar experience to establish the London theory experimentally. Similarly, the availability of pure isotopes arising from the wartime neutron bombardment program made it possible for Emanuel Maxwell at MIT and Bernard Serin at Rutgers to discover in 1950 the 'isotope effect' (relating the mass of mercury isotopes with their superconducting transition temperature). This finding gave Bardeen the pivotal clue that the important interaction in superconductivity is the one between the electrons and the phonons. And wartime study of electron plasmas conducted in the context of the uranium isotope separation program brought David Bohm and his students to develop the many-body formalism that Bardeen, Cooper and Schrieffer used in their development in 1957 of the theory of superconductivity.

When the various pieces of the theory of collective phenomena were finally joined in the 1950s, the new edifice confirmed Bloch's earlier intuition that those problems that had been insoluble in the context of the one-electron quantum theory of solids were indeed related by common theoretical themes. The themes include elementary excitations, macroscopic wave functions, order parameters, and broken symmetry. Like different patches of mushrooms in the same forest, the different sections of the theory were recognised as parts of a common network whose different branches were attached to different areas (e.g., superconductivity, helium-3, critical phenomena). And when Landau presented his field theory of interacting Fermi liquids in 1957, he provided the long sought-after basis for understanding why the one-electron picture that Sommerfeld and others developed in the 1920s and 1930s had been so effective.

This pattern of growth, illustrated by the area of collective phenomena — a pattern in which different subareas, some old and some new, pull together because they are found to share common roots — has proved to be the essential development process in solid state physics. Unlike other subfields of physics that emerged in the 1930s, e.g., atomic, nuclear, and particle physics which evolved by branching out from new discoveries, solid state physics evolved almost entirely by convergence.[16]

REFERENCES

1. Spencer Weart, in L. Hoddeson, E. Braun, J. Teichmann, and S. Weart, *Out of the Crystal Maze: Chapters from the History of Solid State Physics.* (New York and Oxford: Oxford University Press, 1992), p. 622.
2. Dominique Pestre, *Louis Néel, Le Magnétisme et Grenoble: Récit de la création d'un empire physicien dans la province francaise, 1940-1965.* (Paris: Centre National de la Recherche Scientifique, 1990), pp. 155-170.
3. C. S. Smith, "The Prehistory of Solid-State Physics," *Physics Today* (1965), **18**: no. 12, pp. 18-30, on pp. 19,30.
4. Hoddeson, in *Crystal Maze,* pp. 616-662.
5. Hoddeson *et al.*, in *Crystal Maze,* pp. 88–160; p. 181, note 458.
6. A. Sommerfeld and H. Bethe, "Elektronentheorie der Metalle." in H. Geiger and K. Sheel (Eds.), *Handbuch der Physik,* ser. 2, (Berlin: Springer, 1933), **24**: pp. 333–622.
7. F. Seitz, *The Modern Theory of Solids.* (New York: McGraw-Hill, 1940); N. Mott and H. Jones, *The Theory of the Properties of Metals and Alloys.* (New York: Dover Publications, 1936).
8. Hoddeson in L. Hoddeson, P.W. Henriksen, R. Meade, and C. Westfall, *Critical Assembly: A History of Los Alamos During the Oppenheimer Years, 1943–1945.* (New York: Cambridge University Press, 1993), pp. 4–11.
9. J. Bromberg, *The Laser in America: 1950–1979.* (Cambridge, Mass: The MIT Press, 1991).
10. P. Forman, "'Swords into Plowshares': Breaking New Ground in Physical Research with Radar Hardware and Technique after Wold War II," *Reviews of Modern Physics Vol. 67, No. 2.* (April 1995), 397-455.
11. Pestre, *Néel.*
12. Hoddeson *et al.*, *Crystal Maze,* Chapter 8, pp. 486–616.
13. M. Riordan and L. Hoddeson, *Crystal Fire: The Birth of the Information Age.* (New York: W.W. Norton & Co., 1997).
14. Pestre, *Néel.*
15. Hoddeson *et al.*, *Crystal Maze,* pp. 487-89.
16. Weart, in *Crystal Maze.*

FURTHER READING

E. Braun and S. MacDonald, *Revolution in Miniature: The history and impact of semiconductor electronics.* (Cambridge, London and New York: Cambridge University Press, 1978).

J. Bromberg, *The Laser in America: 1950–1979.* (Cambridge, Mass: The MIT Press, 1991).

R. Buderi, *The Invention that Changed the World: How a Small Group of Radar Pioneer Won the Second World War and Launched a Technological Revolution.* (New York: Simon and Shuster, 1996).

M. Eckert and H. Schubert, *Crystals, Electrons, Transistors: From Scholar's Study to Industrial Research.* (New York: American Institute of Physics, 1990) translated by Thomas Hughes, originally published as *Kristalle, Elektronen, Transistoren.* (Reinbek bei Hamburg: Rowohlt Pocket Edition Publishers, 1986).

L. Hoddeson, E. Braun, J. Teichmann, and S. Weart, *Out of the Hoddeson et al., 1992: Chapters from the History of Solid State Physics.* (New York and Oxford: Oxford University Press, 1992).

N. Mott, *The Beginnings of Solid State Physics: A Symposium Organized by Sir Nevill Mott, F.R.S. Held 30 April–2 May 1979.* (London: The Royal Society, 1980).

D. Pestre, *Louis Néel, Le Magnétisme et Grenoble: Récit de la création d'un empire physicien dans la province francaise, 1940-1965.* (Paris: Centre National de la Recherche Scientifique, 1990).

H. Quiesser, *The Conquest of the Microchip: Science and Business in the Silicon Age.* (Cambridge, Mass. and London: Harvard University Press, 1988).

M. Riordan and L. Hoddeson, *Crystal Fire: The Birth of the Information Age.* (New York: W.W. Norton & Co., 1997).

CHAPTER 30

From 'Elementary' to 'Fundamental' Particles

S.S. SCHWEBER

In 1867, in his inaugural lecture as the first Cavendish Professor, Maxwell remarked that "Two theories of the constitution of bodies have struggled for victory with various fortunes since the earliest ages of speculation: one is the theory of a universal plenum, the other is that of atom and void." Atomism had gained ascendancy ever since Newton and by the beginning of the nineteenth century John Dalton could base his *New System of Chemical Philosophy* on the belief that there existed a considerable number of what he called "*elementary* principles," which can never "be metamorphosed, one into another, by any power we can control" and that " all atoms of the same kind, whether simple or compound, must necessarily be conceived to be alike in shape, weight, and every other particular."

By the late 1860s some 70 chemical elements were known, and in his Cavendish lecture Maxwell, echoing Dalton, noted that "the molecule . . . is a very different body from any of those with which experience has hitherto made us acquainted . . . [I]ts the mass, and the other constants which define its properties, are absolutely invariable . . . [Furthermore] there are innumerable other molecules, whose constants are not approximately, but absolutely identical with those of the first molecule, and this whether they are found on earth, in the sun, or in the fixed stars." But Maxwell concluded his address with the question: "But what if these molecules, indestructible as they are, turn out not to be substances themselves, but mere affections of some other substance . . . a uniformly dense *plenum*?" The tension between what kind of substance, whether material particles or a continuous plenum has persisted, as has the question of what is primary, substance or law. [If an underlying plenum is assumed, 'particles' are then considered localized, stable concentrations of energy, spin, charge and other constant attributes.]

By the end of the nineteenth century most physicists and chemists believed in the reality of atoms and, like Maxwell, thought that they were indestructible and indivisible, though not necessarily structureless, as some internal constitution was required to explain their observed spectra. That atoms indeed have an internal constitution became the accepted view when in 1899, two years after his discovery of the electron, Joseph John Thomson announced 'the splitting of the atom,' for he had established that in the process of ionization part of the mass of the atom got free and became detached from the original atom. Similarly, soon after its discovery in 1896, it became apparent that the explanation of radioactivity in terms of the expulsion of subatomic particles implied a 'divisible atom.' Thus Marie Curie noted in 1900, that the atoms of radioactive elements are 'indivisible' from the chemical point of view but 'divisible' when undergoing a radioactive decay. The picture which emerged in the first decade of the twentieth century, was the following: when the energy available in the laboratory setting is insufficient for ionization, (non-radioactive) atoms could be considered 'elementary' constituents of matter. But in other environments (such as ones where high voltages or high temperature exist) they may split into their constituent parts.

The search for 'the' ultimate constituents of matter has had a cyclic history since its inception at the beginning of the nineteenth century. Each stage was initially characterized by incoherence. But the confusion gave way to a measure of clarity through classification, and with the help of the latter the empirical data reduced to some measure of order. Once that order was ascertained a new level of substance and structure was discovered, and became charted with the help of new instruments and technologies. Again incoherence and confusion reigned until regularities operating at that level were discerned, classified and modeled. Four such cycles are readily identified: the unraveling of the level of atoms; then that of their nuclei [and their associated radioactivity]; then that of the latter's constituents and associated entities [i.e., the hadrons — the strongly interacting particles like the neutron, the proton, and the mesons] and most recently that of their constituents: quarks and gluons.

The case of the chemical elements is illustrative of the process. With Dimitri Mendeleev's periodic table of 1869 some measure of clarity was brought into the classification of the chemical elements. Moreover, by virtue of the gaps that existed in the classificatory scheme, Mendeleev predicted that further elements should exist with chemical properties that he could specify — and indeed, these were later discovered. However, the interpretation of the patterning had to await the discovery of the electron and the modeling of the nuclear atom. A phenomenological explication of the table, in terms of the quantized orbits that electrons could occupy in their motions around the positively charged core nucleus, was attempted by Walther Kossel in 1916. He interpreted the pronounced stability of the noble gases to mean that the electronic structures of these atoms consist of closed shells, with a sizable expenditure of energy necessary to remove an electron from such a configuration. Expanding on these insights, Niels Bohr in 1920 gave a phenomenological explanation of the Mendeleev table, including the placement of the rare earths in it, based on his *Aubauprinzip*, according to which an atom could be considered to be built up by the successive addition and binding of electrons into orbits described by the old quantum theory. A more 'fundamental,' dynamical explanation had to await the invention of quantum mechanics and the formulation within that framework of the Pauli exclusion principle. Quantum mechanics indicated that structured systems can only have discrete states, each having a characteristic energy. The smaller the size of the system the greater is the scale of characteristic energies. Thus the excitation energies of the valence electrons of atoms, whose sizes are of the order of 10^{-8} cm, are of the order of electron-volts (eV). The vibrational and rotational spectrum of molecules, composite systems made up of atoms, have characteristic energies of a tenth and a hundredths of an eV respectively. Nuclei, which have radii of the order of 10^{-13}cm, have characteristic excitation energies ranging from thousands (KeV) to millions of electron-volts (MeV). In all composite systems made up of electrically charged particles, the excited states tend to have short lifetimes. They decay to the stable ground state by emitting electromagnetic radiation.

In the early 1960s a situation similar to the one that had prevailed for atomic structure in the early 1920s, existed for the then known 'elementary particles,' the various hadrons. A pattern, classification and taxonomy had been discerned by Murray Gell-Mann and by Yuval Ne'eman, and as had been the case with Mendeleev's table, gaps were noted, and new particles were predicted and then found. And as with the periodic table, the further elucidation by Gell-Mann and by Zweig of the patterning for the baryons and the mesons — the 'eightfold way' — was based on the properties of assumed constituents of the objects being classified. These constituents were called quarks by Gell-Mann. Quarks could interact with one another — and the resulting bound structures were identified with the observed hadrons. This was the analogue of the Bohr *Aufbauprinzip* of the 1920s. And as in the case of atomic structure and chemical binding, the more 'fundamental' explanation had to await the formulation of quantum chromodynamics. Quarks and leptons, together with the *W*s, the *Z*, and the photon (the spin 1 gauge bosons that mediate the forces between the spin 1/2 quarks and leptons) are the 'fundamental' particles of the Glashow-Weinberg-Salam electroweak theory, the quantum field theory (QFT) that has successfully unified the weak and the electromagnetic interactions. Leptons is the generic name for electrons, muons and tau particles, and their associated neutrinos. All leptons are spin 1/2 particles; they partake in the electromagnetic and the weak interactions but do not interact strongly. [The 'standard' representation of the electroweak interactions also includes scalar Higgs particles, that are responsible for the dynamical symmetry breaking that gives quarks their masses.] Quarks and gluons, the latter being the spin 1 gauge particles that mediate the strong interactions between quarks, are the 'fundamental' entities of quantum chromodynamics, the QFT that correctly describes the strong interactions up to energies of 1 TeV, or equivalently down to distances of 10^{-15}cm. [10^{-15}cm is about 1/100th the proton radius; 1 TeV, or tera-electron volt is 10^{12} eV].

The standard model that describes the strong and electroweak interactions observed in nature in terms of quarks and leptons and of the gauge bosons that mediate the forces between them, is one of the great achievements of the human intellect. It will be remembered — together with general relativity and quantum mechanics — as one of the outstanding advances in the physical sciences in the twentieth century. But much more so than general relativity and quantum mechanics it is the product of a communal effort. It required experimenters and theorists to work closely together. It was accomplished by virtue of the skills of the experimenters and engineers who invented the necessary technologies to build the accelerators, detectors and computers, who designed the experiments and who analyzed the data. Moreover, it required that the experimenters, who directed the large teams that carried out the experiments establishing the empirical basis of the model and who confirmed its adequacy, had to possess exceptional organizational skills and unusual leadership qualities in addition to their great technical competence. But analyzing the technical dimensions of the community only highlights the necessary components that made success possible. They were not sufficient. High energy physics is an example of Big Science, and one of the characteristic feature of big science is that among all the factors that enter into the enterprise — scientific, technological, economic, sociological, political — the political, by virtue of scale and cost, is the most critical and consequential. However, in the present chapter I shall only be concerned with the so called internal factors relating to the growth of knowledge in nuclear and particle physics.

Although there is clearly an intimate relation between experimental and theoretical advances in nuclear and particle physics it is fair to say that for the most part experiments have driven theory. There have been some notable exceptions: Yukawa and the postulation of the meson, Lee and Yang and parity non conservation, and Bjorken and scaling in deep inelastic scattering of leptons off nucleons. Sheldon Glashow has stated the matter as follows:

> To a large extent the success is the result of experimental discovery. Generally, the essential basis of our theory is the result of patient, even plodding endeavor. It is the accumulated knowledge due to very many scientists. It has been a history, though, which is punctuated by dramatic and surprising events.[1]

The list of unanticipated experimental discoveries is indeed long: X-rays (Roentgen 1895); radioactivity (Becquerel 1896, Pierre and Marie Curie, Rutherford, Soddy, Laborde 1896–1905, Chadwick 1914); the nucleus (Geiger and Marden, Rutherford 1911); nuclear transmutation (Rutherford 1919); the neutron (Chadwick 1932); the μ-meson (Anderson 1935); the π meson (Powell 1947); the 'strange' particles (1947–19??); neutral kaon behavior (1955); parity non-conservation (Lee and Yang 1955); violation of time reversal symmetry (Cronin and Fitch 1957); the μ neutrino (Schwartz and Steinberger 1958); charm (Ting 1973); the tau and its associated neutrino; the upsilon . . . And the list could easily be extended.[2] One characteristic of the enterprise seems to have been constant from its beginning. Already in 1900 Rutherford wrote his mother: 'I have to keep going, as there are always people on my track. I have to publish my present work as rapidly as possible in order to keep in the race. The best sprinters in this road of investigation are Becquerel and the Curies.' The race has continued to the present day.

For the most part, new technologies and new or improved instrumentation have been responsible for the detection of the new particles or new phenomena, for access to the new domains and then for charting them. Thus a plot of the known number of chemical elements from the time of Lavoisier to the present reveals that the growth is essentially linear over the last two hundred and fifty years, except for bursts of discovery occurring as a result of newly introduced technologies and instrumentation, (such as large voltaic batteries in the 1820s, spectroscopy in the 1860s, and the large accelerators from the 1950 to the 1980s that made possible the creation of transuranic elements). The same is true for the subnuclear particles. The use of nuclear emulsions were responsible for the discovery of the pions. Nuclear emulsions lofted in high altitude balloons, a technology perfected during World War II, and cloud chambers triggered by fast electronic circuitry, likewise developed during the war, made possible the discovery of the so called 'strange' particles in the late 1940s. Synchrocyclotrons, electron synchrotons, linear accelerators of ever greater intensity and energy, storage rings, evermore complex and sensitive detectors have played a crucial role in the discovery of new particles and new phenomena. A corollary of the above is that as Galison has stressed the history of the instrumental practice is often decoupled from that of the theoretical one, and the periodization of the experimental and instrumental component of high energy physics does not necessarily match that of the theory.[3]

I will be concerned primarily with the conceptual dimensions of nuclear and particle physics and the present chapter consists of an overview of nuclear physics and of the standard model. Until very recently, nuclear physics has been primarily a phenomenological theory, and one aspect of its history is its relation to more 'fundamental' theories from which could be derived the potentials that enter into the phenomenological theory. Relativistic quantum field theories (RQFT) have been the principal candidates for such 'fundamental' representations and they have played a central role in the history of high energy physics. In a RQFT both the particles and the forces that act between them are represented by quantized fields.

Nuclear Physics

The history of nuclear physics during the twentieth century presents an interesting study not only of the interaction between theory and experiment, but also of the transplantation and cross-fertilization of theories and models from one field of physics to another. As might be expected, one can delineate fairly distinct periods during which novel instrumentation and experiments propel the field, with theory being essentially non-existent. The first thirty years of the field — from the discovery of radioactivity in 1896 to Chadwick's discovery of the neutron in 1932 — was one such period. Radioactivity gave the first indication of an ontic nature of chance and probability in the physical world. André Debierne, a pupil and collaborator of Marie Sklodowska Curie and Henri Poincaré, noted that the 'atoms' of radioactive bodies 'disintegrate at random.' The model of the nuclear atom that Rutherford advanced in 1911, based on his derivation of the Coulomb scattering cross-section by a point charge and Gamov's and Condon and Guerney's quantum mechanical explanation of α-decay discipline in 1928 — stands out as the major theoretical insights of that period. Conversely, there are periods after 1932, when particular theoretical advances or breakthroughs set the agenda, and experiments are attempting to corroborate the theoretical perceptions and predictions.

One overarching question can be said to have driven the field since the early 1930s: to describe the forces that hold the nucleus together and once these forces are known, to predict the properties of stable nuclei. High energy physics is an outgrowth of this central question. Ironically, for all the great achievements in the unravelling of the subnuclear domain and in formulating a microscopic theory of quarks and gluons, only limited insight have been gleaned into the explanation of the nuclear forces. Yet nuclear physicists have resolved many of the problems that are encountered in treating quantum mechanically systems of neutrons and protons interacting via short range spin and iso-spin dependent forces and have been able — using computers — to calculate properties of stable nuclei starting with *empirically* determined nuclear forces.

The history of the conceptual developments in nuclear physics can be periodized[4] into phases that can be distinguished by the rather different character of the central questions being asked during each period:

1896–1932. During this phase the variety of nuclear phenomena was discovered (radioactivity, α and β rays, nuclear transmutation), the validity of the nuclear model of the atom established and the structural properties of nuclei (such as their mass, electric charge, size, magnetic moment) became accurately measured. The discovery of induced nuclear reactions by Rutherford in 1919 directly exhibited the possibility of changing and exchanging the elementary building blocks of nuclei. This first phase was initiated by Antoine Henri Becquerel's discovery of 'uranic rays,' the phenomenon that opened up the field of radioactivity. There followed the work of Marie and Pierre Curie, of Rutherford and Soddy, and of others unravelling the structure of the uranium and thorium decay series. The outstanding achievement of the period is that of Rutherford establishing the existence of a positively charged nuclear core of the atom, that is responsible for most of its mass.[5] Furthermore, Rutherford with his α-particle scattering experiment and his derivation of the Rutherford scattering formula established a new style of reasoning: all physical measurements and interactions can be considered as scattering processes. As defined by Hacking, 'A style of reasoning makes it possible to reason toward certain kinds of propositions, but does not of itself determine their truth value.' A style of reasoning determines what may be true or false. Similarly, it indicates what has the status of evidence. As Hacking has noted, styles of reasoning tend to be slow in evolution, are vastly more widespread than paradigms, and are not the exclusive property of a single disciplinary matrix.[6]

These early endeavors established that nuclei formed a large, and discernably ordered family. Furthermore, there accumulated compelling evidence for the view that nuclei are composite entities, made up of more elementary constituents . Moseley's (1913) experiments established the quantization of nuclear charge — that is, that it is a multiple of the proton's charge — and together with the work of Aston (1920) gave credence to Prout's hypothesis (1815) that nuclei are composed of a discrete number of elementary constituents with a mass approximately equal to that of a proton. In fact, a nucleus could be specified by two numbers: its integer mass number, A, and its electric charge, Z.

By 1915, it was accepted that the nucleus is the site of all radioactive processes, including beta-radioactivity in which a nucleus ejects an electron. Furthermore, convincing evidence had been given in 1914 by James Chadwick that the energy spectrum of the emitted electron was continuous. It was therefore natural to assume that electrons existed in the nucleus. Already in 1914 Rutherford had assumed the hydrogen nucleus is the positive electron — he called it the *H*-particle — and he conjectured that nuclei are made of *H*-particles and electrons. During the 1920s the generally accepted model of a nucleus was that it consisted of the two elementary particles then known: protons and electrons. Rutherford in his Bakerian Lecture of 1920 had suggested that a proton and electron could bind and create a neutral particle, which he believed was necessary for the building up of the heavy elements.

The advent of quantum mechanics led George Gamov, and independently Edward Condon and Ronald Gurney, in 1928 to explain α-particle decay as a quantum mechanical tunneling of an α-particle (assumed to exist as a subnuclear constituent) through the nuclear potential barrier. Their explanation obtained as a derived byproduct the previously established empirical Geiger-Nuttall law connecting the energy and lifetime of the emitted α-particle. However if nuclei were assumed to be composed of protons and electrons it was difficult to understand the spin of N^{14} which was known to be one, whereas the proton-electron model required it to be composed of an odd number of spin 1/2 particles. Similarly, should there be electrons in the nucleus their magnetic moment — as determined by the hyperfine structure of atoms — ought to much larger than determined experimentally. Confusion reigned before the discovery of the neutron!

1932–1935. During this phase the nuclear problem was formulated. 1932 is surely the annus mirabilis of nuclear physics. The discovery of the neutron by Chadwick led quickly to the view that nuclei are composite systems built out of protons and neutrons with Z = number of protons and A = number of protons + number of neutrons. The neutron — which was assumed to be a electrically neutral, spin 1/2 particle with a mass roughly equal to that of the proton — made possible the application of quantum mechanics to the elucidation of the structure of the nucleus. Heisenberg in a series of papers formulated a quantum mechanical model of nuclear structure based on phenomenological (static) two-body forces between protons and neutrons and neutrons and neutrons. These nuclear forces had to be of short range (only acting over distances of the order of 10^{-13}cm), strong (since nuclear binding energies are of the order of MeVs as compared to eVs for atomic systems), and had to account for saturation (i.e., that the binding energy of nuclei is proportional to A).

The two outstanding achievements of theory during the period were:

1. the characterization — by Heisenberg, by Majorana, and by Wigner in 1932–3 — of the nuclear forces that are needed to understand the binding energies of the light nuclei and the general features of nuclei such as saturation and radii proportional to A; and
2. Fermi's theory of β-decay of 1934 which freed nuclear physics from having to assume the existence of electrons in the nucleus.

In 1932, J.D. Cockroft and E.T.S. Walton accelerated protons to 500 kiloelectron volts (KeV) of energy by means of a high voltage generator they had built at the Cavendish Laboratory of Cambridge University. They then used these protons to disintegrate lithium nuclei, thereby ushering in a new era in physics wherein experimenters no longer depended on naturally occurring α-particles as their projectiles in scattering experiments. Cockroft and Walton's machine was the first of a series of instruments designed to generate ever higher dc voltages with which to accelerate protons, deuterons, and α-particles to be used as projectiles in nuclear physics experiments.[7] The double tandem van der Graaf generators built in the 1950s, capable of accelerating protons to about 20 million electron volts (MeV), mark the endpoint of these developments. The attainable voltages in such direct high voltage machines is limited by insulation breakdown. These difficulties were bypassed in the cyclotron, the accelerator that Ernest O. Lawrence conceived in 1929 after reading a paper by R. Wideröe. In the cyclotron, a magnetic field bends the charged particles in circular paths and are accelerated as they cross a gap between electrodes. Since non-relativistic charged particles moving in a constant magnetic field have a constant period of revolution (irrespective of speed), the particles in a cyclotron can be accelerated on each traversal of the gap if their motion is in resonance with the applied radiofrequency electric field across the gap. Thus high-energy particles are produced without the need of high voltages. The first cyclotron — with pole pieces 10" in diameter — was built by Lawrence and Livingston at the University of California in Berkeley in 1932. It was able to accelerate protons to 1.2 MeV. During the thirties ever larger cyclotrons were built, primarily in the United States and in Japan. Lawrence's Radiation Laboratory at the University of California in Berkeley was the site where the principal developments took place. Among the many students Lawrence trained during the '30s were Edwin McMillan, Luis Alvarez, Wolfgang Panofsky, and Robert Wilson. They were the experimental physicists who shaped high energy physics after World War II.

1935–1952. The developments during this period defined the nuclear paradigm. The seminal work was Bohr's liquid drop model of the nucleus developed in 1936 to explain Fermi's findings in neutron induced nuclear reactions. Bohr recognized that the densely packed nuclear system being analyzed in the neutron reactions required one to consider the collective many-body features of the nuclear dynamics. His compound nucleus model was elaborated by Breit and Wigner and by Bethe, by Peierls and others in the late 1930s. There are two other important landmarks during this period. The first is the discovery of nuclear fission by Hahn and Stassman in 1938, and the explanation of the phenomenon by Lise Meitner and Otto Frisch, and by Bohr and Wheeler in 1939. The second is the explication by Maria Goeppert Mayer and Hans Jensen in 1948 of nuclear structure in terms of an independent particle model wherein the individual particles move in the average potential due to the other particles. To account for the particular stability of certain nuclei they indicated that this potential had to have a strong spin-orbit component.

Some of the most spectacular successes of nuclear physics in the period from 1935 to 1952 — and some of its most consequential impact — have been in areas that might be called applied nuclear physics. The successes are in astrophysics where nuclear physics has supplied the solution to the problem of energy generation in stars, has explained their life cycles and the concomitant process of nucleosynthesis. It has allowed calculations of the internal constitution of stars, and in particular of our sun, to a remarkable degree of precision. In more recent years nuclear physics has provided important theoretical insights to explain the mechanism of supernova explosions and the internal constitution of neutron stars and other stellar remnants. Its most consequential impact is of course in the realm of nuclear weapons. There is no better indication of how far the theoretical understanding of nuclear structure and nuclear reactions had progressed by the early 1940s than to state that the first atomic bomb — the gun assembly to obtain a critical mass of U^{235} — was never tested prior to its use over Hiroshima.

The period from 1948 to the present can be characterized as trying to discover 'the feel of nuclear matter.' The past decade has seen concerted efforts to derive the nuclear forces from quantum chromodynamics.

From Pions to the Standard Model: The Conceptual Development in Particle Physics

Modern particle physics can be said to have begun with the end of World War II.

Peace and the Cold War ushered in an era of new accelerators of ever increasing energy and intensity that were able to artificially produce the particles that populate the subnuclear world. Simultaneously, there developed the expertise to construct particle detectors of ever increasing complexity and sensitivity that allowed recording the imprints of high energy subnuclear collisions. Challenges, opportunities, and resources attracted practitioners: the number of 'high energy' physicists worldwide grew from few hundred after World War II, to some 8,000 in the early 1990s.

The state of affairs in 'elementary particle' physics after the war was summarized by John Archibald Wheeler in an important address delivered at a joint meeting of the National Academy of Sciences and the American Philosophical Society in the fall of 1945.[8] Wheeler observed that the experimental and theoretical researches of the 1930s had made it possible to identify four fundamental interactions: a) gravitation, b) electromagnetism, c) nuclear (strong) forces, d) weak-decay interactions. Wheeler believed that the interesting and exciting areas of research were the investigations of the electromagnetic, the strong, and the weak interactions, and indeed, these became the traditional domain of high energy physics.

At the time electromagnetic forces were conceived as arising through the exchange of photons, integer spin massless bosons. It was known that in a quantum field theoretical description of the interaction between particles, the range , R, of the force between them is related to the mass, m, of the boson exchanged between them by the relation $R = h/ mc$. Massless bosons such as photons thus give rise to forces of infinite range.

Although such a mechanism had been advanced by Yukawa to account for the strong nuclear forces between nucleons, namely the exchange of a *massive* spin 0 boson, the meson that had been discovered in the showers produced by cosmic rays did not seem to have the property required for it to be the strongly interacting Yukawa particle. On the other hand, the weak β-decay interactions had been described by Fermi in 1934 by a field theory that assumed that the process $n \rightarrow p + e^- + \nu$ was a fundamental direct process. Fermi's theory of beta-decay and Yukawa's theory of nuclear forces established the model upon which all subsequent developments were based.[9] The model postulated new 'impermanent' particles to account for interactions and led to a description of nature in terms of a sequence of families of elementary constituent of matter with fewer and fewer members; and the model assumed that relativistic quantum field theory was the natural framework in which to attempt the representation of high energy phenomena. However, all relativistic QFTs are beset by divergence difficulties that manifest themselves in perturbative calculations beyond the lowest order: they yield infinite results.[10] These difficulties impeded progress and throughout the 1930s most of the workers in the field doubted the correctness of QFT in view of the divergence difficulties that plagued all formulations of relativistic quantum field theories.

Two important developments in 1947 shaped the further evolution of particle physics. Both were the result of intense discussions following experimental results presented to the Shelter Island conference. At that conference the curious results that Conversi, Pancini and Piccioni had obtained regarding the decay of mesons observed at sea-level led Marshak to formulate the 'two-meson' hypothesis. He suggested that there existed two kinds of mesons. The heavier one, the π meson which was identified with the Yukawa meson responsible for the nuclear forces, is produced copiously in the upper atmosphere in nuclear collisions of cosmic ray particles with atmospheric atoms. The lighter one, the μ meson observed at sea-level is the decay product of a π meson and interacts but weakly with matter. Within a year, Powell (1948) using nuclear emulsions sent aloft in high altitude balloons corroborated the two meson hypothesis by exhibiting $\pi \rightarrow \mu$ decays. During the early 1950s the data pouring out of the plethora of pion producing accelerators led to the rapid determination of the characteristic properties of the three varieties of pions, the positively charged one, the negatively charged one, and the neutral such as their mass, their spin and their lifetime for decay. The two meson hypothesis also suggested that there were two distinct kind of matter. There are particles like the electron, the muon and the neutrino that do not experience the strong nuclear forces; these are the leptons. Then there are the hadrons, particles like the neutron, proton and mesons that do interact strongly with one another. Subsequently, it proved to be useful to further split the hadrons into two separate families called baryons and mesons. Baryons are the 'heavy' strongly interacting particles, with the proton and neutron the lightest members of the family. They all have spin 1/2, and except for the proton are unstable and decay, one of the decay product always being a proton. Mesons have integer spin, are unstable and when free ultimately decay into leptons or photons.

In January 1949, Jack Steinberger gave evidence that the μ-meson decays into three light particles

$$\mu^+ \rightarrow e^+ + \nu_e + \nu_\mu$$

and Tiomno and Wheeler, and Lee, Rosenbluth and Yang shortly thereafter indicated that the process can be described by a Fermi-like interaction. Moreover, they found that the coupling constant describing this interaction is of the same magnitude as that occurring in nuclear β-decay. Thus the pre-1947 period can be characterized as that of *classical beta-decay*. The post-war period initiated the modern period of *universal Fermi interactions.*

The second important development in the immediate post-World War II period was a theoretical advance. It stemmed from the attempt to quantitatively explain the discrepancies between the predictions of the relativistic Dirac equation for the level structure of the hydrogen atom and the value it ascribed to the magnetic moment of the electron and the empirical data. It resulted in a procedure that has become known as mass and charge renormalization. The ideas of mass and charge renormalization, made it possible to formulate and to give physical justifications for algorithmic rules to eliminate all the ultraviolet divergences that had plagued QED and to secure unique finite answers. The success of renormalized QED in accounting for the Lamb shift, the anomalous magnetic moment of the electron and the muon, the radiative corrections to Compton scattering, pair production and bremsstrahlung, were spectacular.[11] This success secured a firm justification for believing that local QFT was the framework best suited for the unification of quantum theory and special relativity, and that QFT also provided a malleable framework into which to incorporate symmetries — both space-time and internal symmetries.

High energy physics during most of the fifties was dominated by pion physics, although cosmic rays experiments indicated the presence of new 'strange' particles. Mesons were being produced with ever greater energy and intensity in the score of synchrocyclotrons and electron synchrotrons being built around the world. In 1953 the 3 GeV Cosmotron went into operation at Brookhaven National Laboratory and in 1956 the 6 GeV Bevatron at the Berkeley Radiation Laboratory came on line. During the first half of the decade theoretical attempts to explain pion-nucleon scattering and the nuclear forces were based on field theoretical models emulating QED. The success of QED rested on the validity of perturbative expansions in powers of the coupling constant, e^2/hc, which is small, $\approx 1/137$. However, for the pseudoscalar meson theory of the pion-nucleon interaction, which in the early 1950s was believed to be the theory of the strong interactions, the coupling constant had to be large — of the order of 15 — for the theory to yield nuclear potentials that would bind the deuteron. No valid method was found to deal with such strong couplings. Furthermore it was clear that mesons theories were woefully inadequate to account for the properties of all the new hadrons being discovered.

Thus at the end of the 1950s QFT faced a crisis because of its inability to describe the strong interactions and the impossibility of solving any of the realistic models that had been proposed to explain the dynamics of hadrons. Efforts to develop a theory of the strong interactions along the model of QED were generally abandoned, although a local gauge theory of isotopic spin symmetry advanced by Yang and Mills in 1955 was to prove influential later on. There were two responses to the crisis. One was Chew's *S*-matrix program, which rejected QFT and attempted to formulate a theory that made use only of observables embodied in the *S*-matrix, the matrix elements of which were the amplitudes for the possible scattering processes. The program attempted to extract physical consequences using only such general properties as analyticity, crossing symmetry, unitarity, etc., without recourse to any dynamical field equations. The other response to the crisis was to make symmetry considerations central. The progress in classifying and understanding the phenomenology of the ever increasing number of hadrons during the 1950s and 1960s was made not on the basis of a fundamental theory but by making use of symmetry principles and their associated group theoretical methods. This approach tried to make use only of kinematical principles embodying what were considered essential features of a relativistic quantum mechanical description, and eschewed dynamical assumptions.

Symmetry is one of the fundamental concept of modern particle physics. It is used not only as a classificatory and organizing tool but also as a foundational principle to describe dynamics. The notion of symmetry was enriched by two developments in the 1950s. The first was the realization by Lee and Yang (1956) that parity is not conserved in the weak interactions and that this could be explained by assuming that only left-handed neutrinos and right handed anti neutrinos exist in nature. The second fundamental advance in symmetry consideration was made by Yang and Mills in 1955. In analogy with QED, they extended the global isotopic spin symmetry of nucleons to a local symmetry, thereby introducing a gauge theory of the strong interactions:

> The electromagnetic field is described by a (quantized) vector field, $A_\mu(\mathrm{x})$, the (quantized) electromagnetic potential. To this vector potential can be added the gradient of an arbitrary (space-time) dependent scalar function, $\mathrm{X}(x)$, and this addition leaves QED invariant provided that the phase of all (charged) matter fields are changed by an amount proportional to $\mathrm{X}(x)$. More precisely, QED is invariant under the gauge transformation:
>
> $$A_\mu(x) \rightarrow A_\mu(x) + \partial_\mu \mathrm{X}(x)$$
>
> $$\Psi(x) \rightarrow exp(ie\, \mathrm{X}(x))\, \Psi(x)$$
>
> This symmetry is called local gauge invariance. One speaks of a global symmetry when $\mathrm{X}(x)$ is constant over all space-time. The latter is sufficient to insure the existence of a conserved current operator of the charged particle fields, $j_\mu(x)$. Local gauge invariance demands that the photon be massless. The requirement of relativistic invariance, gauge invariance and that the coupling constant gauging the strength of the interaction be dimensionless, determines that the interaction term in the Lagrangian describing the interaction between charged particle fields and the em field be of the form: $e\, j_\mu(x)\, A_\mu(x)$.

The local gauge invariance however implies that the gauge bosons are massless, that is have zero mass. This is not the case for the pion and thus Yang and Mills's theory was considered an interesting model but not relevant for understanding the strong interactions.

Interest in gauge theories was revived after the discovery of spontaneous symmetry breaking (SSB), the mechanism which indicates how a 'foundational' theory can have a much richer underlying symmetry than is observed. The microscopic explanation of the phenomena of superconductivity that Bardeen, Cooper and Schrieffer (BCS) advanced in 1957 exhibited a spontaneously broken symmetry. SSB was introduced into particle physics by Nambu who had also clarified symmetry breaking in the BCS theory. The formulation, by Goldstone, Nambu, Higgs and others, of explicit mechanisms for breaking symmetries through the introduction of scalar bosons, provided ways to generate masses for the gauge bosons. Gauge theory, the mathematical framework for generating dynamics incorporating local symmetries into a QFT has played a crucial role in the development of QFT. *It can rightly be said symmetry, gauge theories and spontaneous symmetry breaking have been the three pegs upon which modern particle physics rests.*

QUARKS

The phenomenological theorizing of the 1950s led to the view that the 'elementary' constituents of hadrons are quarks. Initially, to account for the observed spectrum of hadrons, Gell-Mann and Zweig assumed that were three kinds of quarks (generically indicated by q), called up (u), down (d) and strange (s), that had spin 1/2, isotopic spin 1/2 for the u, d, and the s, and strangeness 0 for the u and d, and –1 for the s quark. The three quarks were to carry baryonic charge of 1/3 and an electrical charge that is 2/3 (for the u) and –1/3 (for the d and s) that of the proton's charge. This was a rather startling assumption since there is no experimental evidence for any macroscopic object carrying a positive charge smaller than that of a proton or a negative charge smaller that of an electron. Since a relativistic quantum mechanical description implies that for every charged particle there exists an 'anti-particle' with the opposite charge, it was assumed that there are likewise anti-quarks (generically denoted by $\bar{q}$) having the opposite electric charge. Quarks were assumed to interact with one another and to form bound states giving rise to the observed hadrons. Thus a meson was assumed to be a bound up and anti-down quark. Similarly a proton was 'made up' of two up quarks (with electrical charge 2/3 e) and a down quark (with electrical charge –1/3e), giving rise to a structure with an electrical charge of +1 e. However in order to satisfy the Pauli principle in structures like the omega minus that are presumably constituted of three identical strange quarks all in s-states, quarks had to be given a new attribute, a new form of charge, called color. Color is a 'three dimensional' analogue of electric charge

An entire phenomenology grew out of this classifacatory scheme. In the early 1960s the flavor *SU(3)* quark model, in which the u, d, and s quarks are considered the building blocks, could classify all the then known hadrons into three families: an octet of spin 0 mesons (that included the π and K mesons); an octet of spin 1/2 baryons (that included the neutron and the proton, the Λ and the Σ's) and a decuplet of spin 3/2 baryons.

With the discovery of the J/Ψ and the charmed meson, and subsequently of the upsilon, the tau and the bottom meson, six different 'flavors' of quarks are needed to account for the observed hadron spectroscopy. They are all spin 1/2 particles that partake in the strong, electromagnetic and the weak interactions and they come in pairs: up and down (u,d), charm and strange (c,s) and top and bottom

(t, b). The first member of each pair has electric charge 2/3 and the second −1/3. Each flavor comes in three colors.

From the time they were introduced as 'hypothetical' particles, there was an important problem connected with quarks: if all hadrons are indeed made up of fractionally charged quarks, why is it that one does not eventually reach an energy high enough to liberate the constituent quarks in a collision process and thus allow a fractionally charged hadron to be observed? This is the so-called confinement problem. And even were one able to provide a mechanism that accounts for the confinement of quarks what meaning is to be attached to the reality of quarks as constituents of hadrons if they can never be observed empirically?

QCD, the gauge theory of color interactions, only emerged a decade after the introduction of quarks,[12] and it took another five years or so for it to be generally accepted. The discovery that gauge theories are asymptotically free was crucial in this development.

In QFT the vacuum is a dynamical entity. Within any small volume of space-time the root mean square values of the field strengths (electric and magnetic in QED, color gluon field in QCD) averaged over the volume, do not vanish. Similarly for the charge and current density. Virtual particle-antiparticle pairs are constantly spontaneously being created and as demanded by the energy-time uncertainty relations, particle and anti-particle annihilate themselves shortly thereafter without travelling very far. These virtual pairs can be polarized in much the same way as molecules in a dielectric solid. Thus in QED the presence of an electric charge e_0 polarizes the 'vacuum' , and the charge that is observed at a large distance differs from e_0 and is given by $e = e_0/\varepsilon$, with ε the dielectric constant of the vacuum. The dielectric constant depends on the distance (or equivalently, in a relativistic setting on energy) and in this way the notion of a 'running charge' varying with the distance being probed, or equivalently varying with the energy scale, is introduced. Electric dielectric screening tends to make the effective charge smaller at large distances. Similarly virtual quarks and leptons tend to screen the color charge they carry.

It turns out however that non-Abelian gauge theories like QCD have the property that virtual gluons 'antiscreen' any color charge placed in the vacuum (and in fact overcome the screening due to the quarks). This means that a color charge that is observed to be big at large distances originates in a charge that is weaker at short distances, and in fact vanishes as $r \to 0$. Thus the force between two quarks is vanishingly small when they are close together but increases as the separation between them increases. This phenomenon has been called asymptotic freedom. The discovery of the anti screening in spin 1 non Abelian gauge theories was made independently by Politzer and by Gross and Wilczek in 1973, and immediately suggested to Gell-Mann, Fritzch and Wess, to Weinberg, and to Gross that QCD was an attractive candidate for be the theory of the strong interactions.

The past two decades has seen a large number of successful explanations of high energy phenomena using QCD. The empirical data that can be accounted for quantitatively in such diverse phenomena as lepton and photon deep inelastic scattering, jets, . . . is impressive. At a conference in 1992 devoted to an assessment of QCD since its initial formulations Altarelli, in his review of 'QCD and Experiment' remarked that

> [Since the late '80s] [M]any relevant calculations, often of unprecedented complexity, have been performed. As a result of 3 years of really remarkable progress , our confidence in QCD has been further consolidated, . . . and a lot of additional checks from many different processes have become possible.[13]

At that same conference Frank Wilczek, one of the important contributor to the field, could assert in his opening remarks that

> QCD is now a mature theory, and it is now possible to begin to view its place in the conceptual universe with appropriate perspective.[14]

Although QCD has had spectacular successes in explaining high energy phenomena. its relation of nuclear physics is more ambiguous. As mentioned earlier, nuclear physics is essentially a phenomenological theory. It is a many body theory that tries to explain the structure of systems composed of fermions interacting through strong, short range forces — forces that are determined empirically. The basic problem of relating nuclear physics to QCD is the same as relating chemistry to atomic QED: there is a mismatch of energy scales. The basic energy scale in QED is .5 MeV as given by the mass of the electron. The energy scale of molecules is on the other hand a few eV or less. Similarly the fundamental scale of QCD is of the order of 200 MeV, whereas nuclear physics is concerned with energies of 1/10th to 1/100th of this. Perhaps, the most fundamental contribution that QCD has made to nuclear physics is to inform physicists which phenomena are simple and fundamental, and which are intrinsically complex and secondary. Thus, by virtue of its success of being able to calculate fairly accurately hadron masses and to make quark confinement inside hadrons plausible, QCD has indicated that physicists need not seek new fundamental laws in that domain. The challenge for nuclear physicists is to relate to QCD the important qualitative phenomena upon which nuclear physics rests such as the hard core feature of the nuclear forces and their saturation.

The standard model is of course not a final theory. Although the model was corroborated with the discovery of the *W* and *Z* particles at CERN in 1983, and accepted by the community as the framework for describing the interactions of leptons, quarks and gluons below 1 TeV, there are too many parameters that enter the theory (e.g., the masses of the quarks, the various coupling constants, the Weinberg angle in the electroweak theory) for it to be considered a final theory. A further important feature of the standard model is that it invokes the existence of particles known as Higgs bosons, which are responsible for the generation of the masses of the *W* and *Z* particles. Whether such particles exist is a crucial test for the theory.

In the early 1980s it was hoped that the features of a more foundational theory would be discerned by exploring physics in the TeV region. In 1982 the Reagan administration, for reasons of national prestige and international economic competitiveness, encouraged the American high energy community to devise a challenging national accelerator project. The administration regarded such an undertaking as a way to bolster the United States' sagging sense of its scientific and technological prowess. High energy physicists responded with plans for a Superconducting Super Collider (SSC), the most ambitious particle accelerator ever attempted. It was to be a proton collider an order of magnitude more energetic than any earlier accelerator which was to make use of the superconducting magnet technology that had recently been developed at the Fermi National Laboratory. The search for the Higgs bosons — and the mapping of the new high energy landscape in the TeV range — became the primary justification for the building of the machine. The best theoretical estimates for the mass of the Higgs boson indicated that it was above 300 GeV which required a multi-TeV collider for its discovery. The contemplated machine was designed to produce 40 Tev protons, which in turn determined the size — a 54 mile ring — and the initial cost of the machine — $4.4 billion — both far greater than of any previous machine. The project was begun in 1983 but abruptly terminated by Congress in October 1993 after $2 billion had been spent on planning and construction.[15]

Although a new collider at Cern facility will eventually make possible experimentation in the TeV range the future of high energy physics based on ever higher energy accelerators will clearly come to an end. Cosmology and the study of the early universe are already becoming the new laboratories for investigating high energy physics.

References

1. Glashow "Does elementary particle physics have a future ?", in der Boer, J. *et al.*, *The Lesson of Quantum Theory.* (New York: Elsevier Scientific Publishers, 1986).
2. Pickering, A. *Constructing Quarks.* (Edinburgh: University of Edinburgh Press, 1984) and Pais, A., *Inward Bound: Of Matter and Forces in the Physical World.* (Oxford: Clarendon Press, 1986).
3. Galison, P. *How Experiments Find.* (Chicago: University of Chicago Press, 1987).
4. Mottelson, B. "The study of the nucleus as a theme in contemporary physics", in der Boer, J. *et al.*, *The Lesson of Quantum Theory.* (New York: Elsevier Scientific Publishers, 1986).
5. Pais, A. *Inward Bound: Of Matter and Forces in the Physical World.* (Oxford: Clarendon Press, 1986).
6. Hacking, I. "Styles of scientific reasoning" in Raichman, J. and West, C., (eds) *Post-analytic Philosophy.* (New York: Columbia University Press, 1986), pp. 145-165.
7. Livingstone, M.S. *Particle Accelerators: A Brief History.* (Cambridge: Harvard University Press, 1969).
8. Wheeler, J.A. "Problems and prospects in elementary particle research," *Proceedings of the American Physics Society* (1946), **90**: 36-52.

9. Brown, L.M., and Hoddeson, L., *The Birth of Particle Physics.* (Cambridge: Cambridge University Press,1983); Brown, L.M. *et al.*, *Pions to Quarks: Particle Physics in the 1950s.* (Cambridge: Cambridge University Press, 1989); Darrigol, O. "The quantum electrodynamical analogy in early nuclear theory," *Revue d'Histoire des Sciences* (1988), **41**: 225-97.
10. Weinberg, S., "Particles, fields and new strings," in der Boer, J. *et al.*, *The Lesson of Quantum Theory.* (New York: Elsevier Scientific Publishers, 1986); Weinberg, S., *Dreams of a Final Theory: The Search for the Fundamental Laws of Nature.* (New York: Pantheon, 1992); Pais, A., *Inward Bound: Of Matter and Forces in the Physical World.* (Oxford: Clarendon Press, 1986).
11. Schweber, S.S., *QED and the Men Who Made It.* (Princeton: Princeton University Press,1994).
12. Gross, D.J., "Asymptotic freedom," in *Physics Today* (1987), **40/1**: 39-44; Gell Mann, M., "Quarks, colours and QCD," in Zerwas, P.M., and Kastrup, H.A., *QCD 20 Years Later.* (Singapore: World Scientific, 1993); Pais, A., *Inward Bound: Of Matter and Forces in the Physical World.* (Oxford: Clarendon Press, 1986).
13. Altarelli in Zerwas, P.M., and Kastrup, H.A., *QCD 20 Years Later.* (Singapore: World Scientific, 1993).
14. Wilczek in Zerwas, P.M., and Kastrup, H.A., *QCD 20 Years Later.* (Singapore: World Scientific, 1993), p. 16.
15. Kevles, D.J., "Preface 1995. The death of the superconducting super collider in the life of American physics," in *The Physicists: The History of a Scientific Community in Modern America.* (Cambridge: Harvard University Press, 1995).

CHAPTER 31

Computer Science

The Search for a Mathematical Theory

MICHAEL S. MAHONEY

COMPUTERS, MATHEMATICS, AND COMPUTER SCIENCE

The discipline of computer science has evolved dynamically since the creation of the device that now generally goes by that name, that is, the electronic digital computer with random-access memory in which instructions and data are stored together and hence the instructions can change as the program is running. The first working computers were the creation not of scientific theory, but of electrical and electronic engineering practice applied to a longstanding effort to mechanize calculation. The first applications of the computer to scientific calculation and electronic data processing were tailored to the machines on which they ran, and in many respects they reflected earlier forms of calculating and tabulating devices. Originally designed as a tool of numerical calculation, in particular for the solution of non-linear differential equations for which no analytical solution was available, the computer led immediately to the new field of numerical analysis. There, speed and efficiency of computation determined the initial agenda, as mathematicians wrestled large calculations into machines with small memories and with basic operations measured in milliseconds.

As the computer left the laboratory in the mid-1950s and entered both the defense industry and the business world as a tool for data processing, for real-time command and control systems, and for operations research, practitioners encountered new problems of non-numerical computation posed by the need to search and sort large bodies of data, to make efficient use of limited (and expensive) computing resources by distributing tasks over several processors, and to automate the work of programmers who, despite rapid growth in numbers, were falling behind the even more quickly growing demand for systems and application software. The emergence during the 1960s of time-sharing operating systems, of computer graphics, of communications between computers, and of artificial intelligence increasingly refocused attention from the physical machine to abstract models of computation as a dynamic process.

Most practitioners viewed those models as mathematical in nature and hence computer science as a mathematical discipline. But it was mathematics with a difference. While insisting that computer science deals with the structures and transformations of information analyzed mathematically, the first Curriculum Committee on Computer Science of the Association for Computing Machinery (ACM) in 1965 emphasized the computer scientists' concern with effective procedures:

> The computer scientist is interested in discovering the pragmatic means by which information can be transformed to model and analyze the information transformations in the real world. The pragmatic aspect of this interest leads to inquiry into effective ways to accomplish these at reasonable cost.[1]

A report on the state of the field in 1980 reiterated both the comparison with mathematics and the distinction from it:

> Mathematics deals with theorems, infinite processes, and static relationships, while computer science emphasizes algorithms, finitary constructions, and dynamic relationships. If accepted, the frequently quoted mathematical aphorism, 'the system is finite, therefore trivial,' dismisses much of computer science.[2]

Computer people knew from experience that 'finite' does not mean 'feasible' and hence that the study of algorithms required its own body of principles and techniques, leading in the mid-1960s to the new field of computational complexity. Talk of costs, traditionally associated with engineering rather than science, involved more than money. The currency was time and space, as practitioners strove to identify and contain the exponential demand on both as even seemingly simple algorithms were applied to ever larger bodies of data. Yet, central as algorithms were to computer science, the report continued, they did not exhaust the field, "since there are important organizational, policy, and nondeterministic aspects of computing that do not fit the algorithmic mold."

There have been some notable exceptions to the notion of computing as essentially a mathematical science. In a widely read letter to *Science* in 1967, Allan Newell, A.J. Perlis, and Herbert Simon, each a recipient of the ACM's highest honor, the Turing Award, argued that the computers were as much phenomena as instruments and hence that the study of them constituted an empirical science.[3] Other winners have echoed that view, among them Marvin Minsky, one of the creators of artificial intelligence, and Donald Knuth, whose two-volume *Art of Computer Programming* has long been the standard reference tool for programmers.

Thus, in striving toward theoretical autonomy, computer science has always maintained contact with practical applications, blurring commonly made distinctions among science, engineering, and craft practice. That characteristic makes the field both a resource for the reexamination of these distinctions and an elusive subject to encompass in a short historical account. What follows, therefore, focuses on the core of the search for a theory of computing, namely the effort to express the computer and computation in mathematical terms adequate to practical experience and applicable to it.

In tracing the emergence of a discipline, it is useful to think in terms of its *agenda,* that is, what practitioners of the discipline agree ought to be done, a consensus concerning the problems of the field, their order of importance or priority, the means of solving them, and perhaps most importantly, what constitute solutions. Becoming a recognized practitioner of a discipline means learning the agenda and then helping to carry it out. Knowing what questions to ask is the mark of a full-fledged practitioner, as is the capacity to distinguish between trivial and profound problems. Whatever specific meaning may attach to 'profound,' generally it means moving the agenda forward. One acquires standing in the field by solving the problems with high priority, and especially by doing so in a way that extends or reshapes the agenda, or by posing profound problems. The standing of the field may be measured by its capacity to set its own agenda. New disciplines emerge by acquiring that autonomy. Conflicts within a discipline often come down to disagreements over the agenda: what are the really important problems? Irresolvable conflict may lead to new disciplines in the form of separate agendas.

As the shared Latin root indicates, agendas are about action: what is to be *done?* Since what practitioners do is all but indistinguishable from the way they go about doing it, it follows that the tools and techniques of a field embody its agenda. When those tools are employed outside the field, either by a practitioner or by an outsider borrowing them, they bring the agenda of the field with them. Using those tools to address another agenda means reshaping the latter to fit the tools, even if it may also lead to a redesign of the tools. What gets reshaped, and to what extent, depends on the relative strengths of the agendas of borrower and borrowed.

Machines That Compute

At the outset, computing had no agenda of its own from which a science might have emerged. As a physical device, the computer was not the product of a scientific theory from which it inherited an agenda. Rather, computers and computing posed a constellation of problems that intersected with the agendas of various fields. As practitioners of those fields took up the problems, applying to them the tools and techniques familiar to them, they gradually defined an agenda for computer science. Or, rather, they defined a variety of agendas, some mutually supportive, some orthogonal to one another, and some even at cross purposes. Theories are about questions, and where the nascent subject of computing could not supply the next question, the agenda of the outside field often provided it.

When the computer came into existence, there were two areas of mathematics concerned with it as a subject in itself rather than as a means for doing mathematics. One was mathematical logic and the theory of computable functions. To make the notion of 'computable' as clear and simple as possible, Alan Turing proposed in 1936 a mechanical model of what a human does when computing:

> We may compare a man in the process of computing a real number to a machine which is only capable of a finite number of conditions $q_1, q_2, \ldots, q_R$ which will be called '*m-*

> configurations.' The machine is supplied with a 'tape' (the analogue of paper) running through it, and divided into sections (called 'squares') each capable of bearing a 'symbol.'[4]

Turing imagined, then, a tape divided into cells, each containing one of a finite number of symbols. The tape passes through a machine that can read the contents of a cell, write to it, and move the tape one cell in either direction. What the machine does depends on its current state, which includes a signal to read or write, a signal to move the tape right or left, and a shift to the next state. The number of states is finite, and the set of states corresponds to the computation. Since a state may be described in terms of three symbols (read/write, shift right/left, next state), a computation may itself be expressed as a sequence of symbols, which can also be placed on the tape, thus making possible a universal machine that can read a computation and then carry it out by emulating the machine described by it.

Turing's machine, or rather his monograph, fitted into the current agenda of mathematical logic. The *Entscheidungsproblem* stemmed from David Hilbert's program of formalizing mathematics; as stated in the textbook he wrote with W. Ackermann:

> The *Entscheidungsproblem* is solved when one knows a procedure by which one can decide in a finite number of operations whether a given logical expression is generally valid or is satisfiable. The solution of the *Entscheidungsproblem* is of fundamental importance for the theory of all fields, the theorems of which are at all capable of logical development from finitely many axioms.[5]

Having written the paper for W.H.A. Newman's senior course at Cambridge, Turing turned for graduate study to Alonzo Church at Princeton, who had recently introduced an equivalent notion of 'computability,' which he called 'effective calculability.' Turing subsequently showed that his machine had the same power as Church's lambda calculus or Stephen Kleene's recursive function theory for determining the range and limitations of axiom systems for mathematics.[6]

The other area of mathematics pertinent to the computer originated in a 1938 paper by Claude Shannon of MIT and Bell Telephone Laboratories.[7] In it Shannon had shown how Boolean algebra could be used to analyze and design switching circuits. It took some time for the new technique to take hold, and it came to general attention among electrical engineers only in the late 1940s and early '50s. The literature grew rapidly in the following years, but much of it took a standard form. Addressed to an audience of electrical engineers, most articles included an introduction to Boolean algebra, emphasizing its affinity to ordinary algebra and the rules of idempotency and duality that set it apart. Either by way of motivation at the outset, or of application at the end, the articles generally included diagrams of circuits and of various forms of switches and relays, the former to establish the correspondence between the circuits and Boolean expressions, the latter to illustrate various means of embodying basic Boolean expressions in concrete electronic units. The articles gave examples of the transformation of Boolean expressions to equivalent forms, which corresponded to the analysis of circuits. Then they turned to synthesis by introducing the notion of a Boolean function and its expression in the normal or canonical form of a Boolean polynomial. The articles showed how the specification of a switching circuit in the form of a table of inputs and outputs can be translated into a polynomial, the terms of which consist of products expressing the various combinations of inputs. From the polynomial one could then calculate the number of switches required to realize the circuit and seek various transformations to reduce that number to a minimum.

As the notion of an algebraic theory of switching circuits took root, authors began shifting their attention to its inadequacies. They were not trivial, especially when applied to synthesis or design of circuits, rather than analysis of given circuits. For circuits of any complexity, representation in Boolean algebra presented large, complicated expressions for reduction to minimal or canonical form, but the algebra lacked any systematic, i.e., algorithmic, procedure for carrying out that reduction. The symbolic tool meant to make circuits more tractable itself became intractable when the number of inputs grew to any appreciable size.

Neither Turing's nor Shannon's mathematics quite fit the actual computer. Turing effectively showed what a computer could not do, even if given infinite time and space to do it. But Turing's model gave little insight into what a finite machine could do. As Michael Rabin and Dana Scott observed in their seminal paper on finite automata:

> Turing machines are widely considered to be the abstract prototype of digital computers; workers in the field, however, have felt more and more that the notion of a Turing machine is too general to serve as an accurate model of actual computers. It is well known that even for simple calculations it is impossible to give an *a priori* upper bound on the amount of tape a Turing machine will need for any given computation. It is precisely this feature that renders Turing's concept unrealistic.[8]

By contrast, Boolean algebra described how the individual steps of a calculation were carried out, but it said nothing about the nature of the computational tasks that could be accomplished by switching circuits. A theory of computation apposite to the electronic digital stored-program computer with random-access memory lay somewhere between the Turing machine and the switching circuit. Computer science arose out of the effort to fill that gap between the infinite and the finite, a region that had hitherto attracted little interest from mathematicians.

Just what and who would fill it was not self-evident. In a lecture "On a logical and general theory of automata" in 1948, John von Neumann pointed out that automata and hence computing machines had been the province of logicians rather than mathematicians, and a methodological gulf separated the two:

> There exists today a very elaborate system of formal logic, and, specifically, of logic as applied to mathematics. This is a discipline with many good sides, but also with certain

> serious weaknesses. This is not the occasion to enlarge upon the good sides, which I certainly have no intention to belittle. About the inadequacies, however, this may be said: Everybody who has worked in formal logic will confirm that it is one of the technically most refractory parts of mathematics. The reason for this is that it deals with rigid, all-or-none concepts, and has very little contact with the continuous concept of the real or of the complex number, that is, with mathematical analysis. Yet analysis is the technically most successful and best-elaborated part of mathematics. Thus formal logic is, by the nature of its approach, cut off from the best cultivated portions of mathematics, and forced onto the most difficult part of the mathematical terrain, into combinatorics.
>
> The theory of automata, of the digital, all-or-none type, as discussed up to now, is certainly a chapter in formal logic. It will have to be, from the mathematical point of view, combinatory rather than analytical.[9]

Von Neumann subsequently made it clear he wanted to pull the theory back toward the realm of analysis, and he did not expand upon the nature of the combinatory mathematics that might be applicable to it.

It was not clear what tools to use in large part because it was not clear what job to do. What should a theory of computing be about? Different answers to that question led to different problems calling for different methods. Or rather, it opened the theory of computing to different approaches, shaped as much by the taste and training of the theorist as by the parameters of the problem. As Figure 31.1 illustrates, people came to the theory of computing from a variety of disciplinary backgrounds, most of them mathematical in form yet diverse in the particular kind of mathematics they used. Some of them were centrally concerned with computing; others took problems in computing as targets of opportunity for methods devised for other purposes altogether. In the long run, it all seemed to work out so well as to lie in the nature of the subject. But computers had no nature until theorists created one for them, and different theorists had different ideas about the nature of computers.

Theory of Automata

As von Neumann's statement makes clear, automata traditionally lay in the province of logicians, who had contemplated the mechanization of reasoning at least since Leibniz expounded the resolution of disagreements through calculation. Since logical reasoning was associated with rational thought, automata formed common ground for logicians and neurophysiologists. Before Turing himself made the case for a thinking machine and proposed his famous test in 1950, his original paper provoked Warren McCulloch and Walter Pitts in 1943 to model the behavior of connected neurons in the symbolic calculus of propositions and to show that in principle such 'nervous nets' were capable of carrying out any task for which complete and unambiguous instructions could be specified.[10] The behavior of the nets did not depend on the material of the components but only on their behavior and connectivity, and hence they could equally well be realized by switching circuits.

Automata thus brought convergence to the agendas of the neurophysiologists and of the electrical engineers. Shannon's use of Boolean algebra to analyze and design switching circuits spread quickly in the late 1940s and early '50s. In its original form, the technique applied to combinational circuits, the output of which depends only on the input. Placing a secondary relay in a circuit made its behavior dependent on its internal state as well, which could be viewed as a form of memory. Such sequential circuits could also be represented by Boolean expressions. For both circuits, the basic problem remained the reduction of complex Boolean expressions to their simplest form, which in some cases (but not all) meant the least number of components, a major concern when switches were built from expensive and unreliable vacuum tubes.

The analysis of sequential circuits gave rise then to the notion of a sequential machine, described in terms of a set of states, a set of input symbols, a set of output symbols, and a pair of functions mapping input and current state to next state and output, respectively. Starting in an initial state, a sequential machine read a sequence of symbols and outputted a string of symbols in response to it. The output was not essential to the model; one could restrict one's attention to the final state of the machine after reading the input. 'Black-boxing' the machine shifted attention from the circuits to the input and output, and thus to its description in terms of the inputs it recognizes or the transformations it carries out on them. With the shift came two new sets of questions: first, what can one say about the states of a machine on the basis simply of its input-output behavior and, second, what sorts of sequences can a machine with a finite set of states recognize and what sorts of transformations can it carry out on them?

In 1951, working from the paper by McCulloch and Pitts and stimulated by the page proofs of von Neumann's lecture quoted above, Stephen C. Kleene established a fundamental property of finite-state automata, namely that the sets of tapes recognized by them form a Boolean algebra. That is, if L_1 and L_2 are sets of strings recognized by a finite automaton, then so too are their union, product, complements, and iterate or closure.[11] Still working in the context of nerve nets, Kleene called such sets 'regular events,' but they soon came to be called 'regular languages,' marking thereby a new area of convergence with coding theory and linguistics.

By the end of the 1950s a common mathematical model of the finite automaton had emerged among the convergent agendas, set forth in the seminal paper by Michael Rabin and Dana Scott mentioned above. Their formalism, aimed at retaining the flavor of the machine and its ties to Turing machines, became the standard for the field, beginning, significantly, with tapes formed by finite sequences of symbols from a finite alphabet Σ. The class of all such tapes, including the empty tape Λ:, is denoted T. A finite automaton A over an alphabet Σ is a quadruple (S, M, s_0, F), where S is a finite set of states, M is a function which for every pair consisting of a state and a symbol specifies a resulting state, s_0 is an initial state, and F is a subset of S comprising the final states. As defined, the automaton works sequentially on symbols, but by an obvious extension of M to sequences, it becomes a tape machine, so that $M(s,x)$ is the state at which the machine arrives after starting in state s and traversing the tape x. In particular, if $M(s_0,x)$ is one of the final states in F, the automaton 'accepts' x. Let $T(A)$ be the set of tapes in T accepted, or defined, by automaton A, and let T be the class of all sets of tapes defined by some automaton.

Rabin and Scott pursued two related sets of questions: first, what can be determined about the set of tapes $T(A)$ accepted by a given automaton A, and, second, what is the mathematical structure of $T(A)$ and of the larger class T? Taking a lead from Kleene's results, they established that T constitutes a Boolean algebra of sets and hence that it contains all finite sets of tapes. Two central decision problems were now immediately solvable. First, if an automaton A accepts any tapes, it accepts a tape of length less than the number of its internal states, and therefore an effective procedure exists for determining whether $T(A)$ is empty (the 'emptiness problem'). Second, if A is an automaton with r internal states, then $T(A)$ is infinite if and only if it contains a tape of length greater than or equal to r and less than or equal to $2r$; again, an effective procedure follows for determining whether the set of definable tapes is infinite.

At two points in their study, Rabin and Scott identified the set T with the semigroup, an algebraic structure that had until then had received very little attention from mathematicians. But they did not make much use of the concept, approaching their model combinatorially and in essence constructing automata to meet the conditions of the problem. That reflected the agenda of mathematical logic, a subject still largely on the periphery of mathematics, outside its increasingly algebraic agenda. Yet, as the reference to the semigroup suggests, it was making contact with algebra. The link had been forged by way of coding theory.

Coding theory arose from another path-breaking paper of Shannon's, his *Mathematical Theory of Communication,* first published in 1948. There Shannon offered a mathematical model of information in terms of sequences of signals transmitted over a line and posed the fundamental question of how to preserve the integrity of the sequence in the presence of distortion. The answer to that question took the form of sequential codes that by their very structure revealed the presence of errors and enabled their correction.

In a vaguely related way, circuit theory and coding theory verged on a common model from opposite directions. The former viewed a sequential machine as shifting from one internal state to another in response to input; the transition might, but need not involve output, and the final state could be used as a criterion of acceptance of the input. The main question was how to design the optimal circuit for given inputs, a matter of some importance when one worried about the cost and reliability of gates. Coding theory focused on the problem of mapping the set of messages to be transmitted into the set of signals available for transmission in such a way that the latter permit unambiguous retranslation when received. That means, first and foremost, that the stream of signals can be uniquely scanned into the constituents of the message. In addition, the process must take place as it occurs, and it should be immune to noise. That is, the code must be unambiguously decipherable as encountered sequentially, even if garbled in transmission.

It was in the context of coding theory that the agendas grouped around the finite automaton intersected with an important mathematical agenda centered in Paris. In "Une théorie algébrique du codage," Marcel P. Schützenberger, a member of Pierre Dubreil's seminar on algebra and number theory at the Faculté des Sciences in Paris, linked coding to the 'fundamental algebraic structure' of Bourbaki:

> It is particularly noteworthy that the fundamental concepts of semigroup theory, introduced by P. Dubreil in 1941 and studied by him and his school from an abstract point of view, should have immediate and important interpretations on the level of the concrete realization of coding machines and transducers.[12]

Schützenberger translated the problem of coding into the conceptual structure of abstract semigroup theory by viewing a sequence of symbols as a semigroup and codes as homomorphisms between semigroups. The details are too complex to follow here, but the trick lay in determining that the structural properties of semigroups and their homomorphisms gave insight into the structure of codes. Schützenberger showed that they did and extended his analysis from the specific question of codes to the more general problem of identifying the structures of sequences of symbols with the structure of devices that recognize those structures, that is, to the problem of automata.

The resulting semigroup theory of machines thus gave a mathematical account of sequential machines that pulled together a range of results and perspectives. More importantly, perhaps, it suggested mathematical structures for which there was at the time no correlate in automata theory. Within Schützenberger's mathematical agenda, one moved from basic structures to their extension by means of other structures. In "Un problème de la théorie des automates" he set out the semigroup model of the finite automaton and then looked beyond it, aiming at an algebraic characterization based on viewing a regular language as a formal sum, or formal power series, Σx of the strings it comprises and considering the series as a 'rational' function of the $x \in X$:

> We then generalize that property by showing that if, instead of being finite, S is the direct product of a finite set by **Z** (the additive group of integers) and if the mapping $(S, X) \rightarrow S'$ and the subset S_1 are defined in a suitable fashion, then the corresponding sum is in a certain sense 'algebraic.'[13]

That is, the extended sum takes the form $\Sigma<z,x>x$, where $<z,x>$ is the integer mapped to the string x. The terminology reflected the underlying model. Were the semigroup of tapes commutative, the sums would correspond to ordinary power series, to which the terms 'rational' and 'algebraic' would directly apply. The generalization was aimed at embedding automata in the larger context of the algebra of rings.

Initially, Schützenberger suggested no interpretation for the extension; that is, it corresponded to no new class of automata or of languages accepted by them. In fact, both the languages and the automata were waiting in the wings. They came on stage through the convergence of finite automata with yet another previously independent line of investigation, namely the new mathematical linguistics under development by Noam Chomsky. The convergence of agendas took the concrete form of Schützenberger's collaboration with Chomsky on context-free grammars, then just getting underway. The result was the mathematical theory of formal languages.

Formal Languages

Formal language theory similarly rested at the start on an empirical base. Chomsky's seminal article, "Three models of language" (1956), considered grammar as a mapping of finite strings of symbols into all and only the grammatical sentences of the language:

> The grammar of a language can be viewed as a theory of the structure of this language. Any scientific theory is based on a certain finite set of observations and, by establishing general laws stated in terms of certain hypothetical constructs, it attempts to account for these observations, to show how they are interrelated, and to predict an indefinite number of new phenomena. A mathematical theory has the additional property that predictions follow rigorously from the body of theory.[14]

The theory he was seeking concerned the fact of grammars, rather than grammars themselves. Native speakers of a language extract its grammar from a finite number of experienced utterances and use it to construct new sentences, all of them grammatical, while readily rejecting ungrammatical sequences. Chomsky was looking for a metatheory to explain how to accomplish this feat, or at least to reveal what it entailed. (Ultimately the physical system being explained was the human brain, the material basis of both language and thought.) In 1958 he made the connection with finite automata through a collaborative investigation with George Miller on finite state languages and subsequently joined forces with Schützenberger, then a visitor at MIT.

As powerful as the semigroup seemed to be in capturing the structure of a finite automaton, the model was too limited by its generality to handle the questions Chomsky was asking. It recognized languages produced by translations from tokens to strings of the form $\alpha \rightarrow x$, $\alpha \rightarrow x\beta$ (α,β non-terminals, x a string of terminals), by then called 'finite state,' or 'regular' languages, but not languages generated by recursive productions such as $\alpha \rightarrow u\alpha v$, $\alpha \rightarrow x$, by which sequences may be embedded in other sequences an arbitrary number of times. These were the languages determined by the phrase structure grammars introduced by Chomsky in 1959, foremost among them the context-free grammars, and modeling them mathematically required more structure than the monoid could provide. That was reflected in the need for internal memory, represented by the addition of a second (two-way) tape. It was also reflected in the fact, quickly ascertained, that the languages are not closed under complementation or intersection and hence cannot be represented by something so tidy as a Boolean algebra.

What made the context-free grammars so important was the demonstration by Seymour Ginsburg and H.G. Rice in 1961 that large portions of the grammar of the new internationally designed programming language, Algol, as specified in John Backus' new formal notation, BNF (Backus Normal Form, or later Backus-Naur Form), based on the productions of Emil Post, are context-free. When the productions are viewed as set equations, the iterated solutions form lattices, which have as fixpoints the final solutions that correspond to the portions of the language specified by the productions. Of course, were Algol as a whole context-free, then the language so determined would consist of all syntactically valid programs. Here, Schützenberger could bring his model of formal power series to bear. In a group of papers in the early 1960s, he showed that the series he had characterized as 'algebraic' were the fixpoint solutions of context-free grammars. The approach through formal power series also enabled him to show that the grammars were undecidedly ambiguous and that they could be partitioned into two distinct subfamilies, only one of which fitted Chomsky's hierarchy.

Several agendas again converged at this point, as a mix of theoretical and practical work in both the United States and Germany on parenthesis-free notation, sequential formula translation for compilers, management of recursive procedures, and syntactic analysis for machine translation more or less independently led to the notion of the 'stack' (in German, *Keller*), or list of elements accessible at one end only on a last in-first out (LIFO) basis. In the parlance of automata theory, a stack provided a finite automaton with additional memory in the form of a tape that ran both ways but moved only one cell at a time and contained only data placed there by the automaton during operation. Schützenberger and Chomsky showed that such 'pushdown automata' correspond precisely to context-free languages and embedded the correspondence in Schützenberger's algebraic model.

The development, hand-in-hand, of automata theory and formal language theory during the late 1950s and early 1960s gave an empirical aspect to the construction of a mathematical theory of computation. The matching of Chomsky's hierarchy of phrase-structure grammars with distinct classes of automata enlisted such immediate assent precisely because it encompassed a variety of independent and seemingly disparate agendas, both theoretical and practical, ranging from machine translation to the specification of Algol and from noise-tolerant codes to natural language. These it united theoretically in much the same way that Newton's mechanics united not only terrestrial and celestial phenomena but also a variety of agendas in mathematics, mechanics, and natural philosophy. By the late 1960s, the theory of automata and formal languages had assumed canonical shape around an agenda of its own. In 1969, John Hopcroft and Jeffrey Ullman began their text, *Formal Languages and Their Relation to Automata,* by highlighting Chomsky's mathematical grammar, the context-free definition of Algol, syntax-directed compilation, and the concept of the compiler-compiler:

> Since then a considerable flurry of activity has taken place, the results of which have related formal languages and automata theory to such an extent that it is impossible to treat the areas separately. By now, no serious study of computer science would be complete without a knowledge of the techniques and results from language and automata theory.

Far from merely complementing the study of computer science, the subject of automata and formal languages became the theoretical core of the curriculum during the 1970s, especially as it was embedded in such tools as lexical analyzers and parser generators. With those tools, what had once required years of effort by teams of programmers became an undergraduate term project.

While the bulk of that new agenda focused on grammars and their application to programming, part of it pulled the theory back toward the mathematics from which it had emerged. Increasingly pursued at a level of abstraction that tended to direct attention away from the physical system, the theory of automata verged at times on a machine theory of mathematics rather than a mathematical theory of the machine. "It appeared to me," wrote Samuel Eilenberg, co-creator of category theory, in the preface of his four-volume *Automata, Languages, and Machines* (1974)

> that the time is ripe to try and give the subject a coherent mathematical presentation that will bring out its intrinsic aesthetic qualities and bring to the surface many deep results which merit becoming part of mathematics, regardless of any external motivation.

Characterized by constructive algorithms rather than proofs of existence, those results constituted, he went on, nothing less than a "new algebra, ... contain[ing] methods and results that are deep and elegant" and that would someday be "regarded as a standard part of algebra." That is, its ties to the computer would disappear.

Formal Semantics

Yet, even when the computer remained at the focus, some practitioners expressed doubt that formal language theory would suffice as a theory of computation, since computing involved more than the grammar of programming languages could encompass. As noted above, the practical activity of programming created the theoretical space to be filled by a mathematical model of computation. As the size and complexity of the programs expanded, driven by a widely shared sense of the all but unlimited power of the computer as a 'thinking machine,' so too did the expectations of what would constitute a suitable theory. One concept of a "mathematical theory of computation" stems from two papers in the early 1960s by John McCarthy, the creator of LISP and a progenitor of 'artificial intelligence,' yet another agenda of computing, with links both to neurophysiology and to logic. It was McCarthy who invoked Newton as historical precedent for the enterprise:

> In this paper I shall discuss the prospects for a mathematical science of computation. In a mathematical science, it is possible to deduce from the basic assumptions, the important properties of the entities treated by the science. Thus, from Newton's law of gravitation and his laws of motion, one can deduce that the planetary orbits obey Kepler's laws.[15]

In another version of the paper, he changed the precedent only slightly:

> It is reasonable to hope that the relationship between computation and mathematical logic will be as fruitful in the next century as that between analysis and physics in the last. The development of this relationship demands a concern for both applications and for mathematical elegance.[16]

The elegance proved easier to achieve than the applications. The trick was to determine just what constituted the entities of computer science.

McCarthy made clear what he expected from a suitable theory in terms of computing: first, a universal programming language along the lines of Algol but with richer data descriptions; second, a theory of the equivalence of computational processes, by which equivalence-preserving transformations would allow a choice among various forms of an algorithm, adapted to particular circumstances; third, a form of symbolic representation of algorithms that could accommodate significant changes in behavior by simple changes in the symbolic expressions; fourth, a formal way of representing computers along with computation; and finally a quantitative theory of computation along the lines of Shannon's measure of information. What these criteria, recognizably drawn from the variety of agendas mentioned above, came down to in terms of McCarthy's precedents was something akin to traditional mathematical physics: a dynamical representation of a program such that, given the initial values of its parameters, one could determine its state at any later point. Essential to that model was the notion of converging on complexity by perturbations on a fundamental, simple solution, as in the case of a planetary orbit based initially on a two-body, central force configuration. Formal language theory shared with machine translation a similar assumption about converging on natural language through increasingly complex artificial languages.

For McCarthy, that program meant that automata theory would not suffice. Its focus on the structure of the tape belied both the architecture and the complexity of the machine processing the tape. As McCarthy put it:

> Computer science must study the various ways elements of data spaces are represented in the memory of the computer and how procedures are represented by computer programs. From this point of view, most of the current work on automata theory is beside the point.[17]

It did not suffice to know that a program was syntactically correct. One should be able to give a mathematical account of its semantics, that is, of the meaning of the symbols constituting it in terms of the values assigned to the variables and of the way in which the procedures change those values.

Such an account required a shift in the traditional mathematical view of a function as a mapping of input values to output values, that is, as a set of ordered pairs of values. To study how procedures change values, one must be able to express and transform the structure of the function as an abstract entity in itself, a sequence of operations by which the output values are generated from the input values. Seeking an appropriate tool for that task, McCarthy reached back to mathematical logic in the 1930s to revive Alonzo Church's lambda calculus, originally devised in 1929 in an effort to find a type-free route around Russell's paradox. The attempt failed, and Church had long since abandoned the approach. The notation, however, seemed to fit McCarthy's purposes and, moreover, lay quite close structurally to LISP, which he had been using to explore mechanical theorem-proving.

The developments leading up to the creation of denotational, or mathematical, semantics in 1970 are too complicated to pursue in detail here. McCarthy's own efforts faltered on the lack of a mathematical model for the lambda calculus itself. It also failed to take account of the peculiar problem posed by the storage of program and data in a common memory. Because computers store programs and data in the same memory, programming languages allow unrestricted procedures which could have unrestricted procedures as values; in particular a procedure can be applied to itself. "To date," Dana Scott claimed in his "Outline of a mathematical theory of computation" in 1970,

> no mathematical theory of functions has ever been able to supply conveniently such a free-wheeling notion of function except at the cost of being inconsistent. The main *mathematical* novelty of the present study is the creation of a proper mathematical theory of functions which accomplishes these aims (consistently!) and which can be used as the basis for the *metamathematical* project of providing the 'correct' approach to semantics.[18]

One did not need unrestricted procedures to appreciate the problems posed by the self-application facilitated by the design of the computer. Consider the structure of computer memory, representing it mathematically as a mapping of contents to locations. That is, a state σ is a function mapping each element l of the set L of locations to its value $\sigma(l)$ in V, the set of allowable values. A command effects a change of state; it is a function γ from the set of states S into S. Storing a command means that γ can take the form $\sigma(l)$, and hence $\sigma(l)(\sigma)$ should be well defined. Yet, as Scott insisted in his paper, "[t]his is just an insignificant step away from the self-application problem $p(p)$ for 'unrestricted' procedures p, and it is just as hard to justify mathematically."

Recent work on interval arithmetic suggested to Scott that one might seek justification through a partial ordering of data types and their functions based on the notion of 'approximation' or 'informational content.' With the addition of an undefined element as 'worst approximation' or 'containing no information,' the data types formed a complete lattice, and monotonic functions of them preserved the lattice. They also preserved the limits of sequences of partially ordered data types and hence were continuous. Scott showed that the least upper bound of the lattice, considered as the limit of sequences, was therefore the least fixed point of the function and was determined by the fixed point operator of the lambda calculus. Hence self-applicative functions of the sort needed for computers had a consistent mathematical model. And so too, by the way, did the lambda calculus for the first time in its history. Computer science had come around full circle to the mathematical logic in which it had originated.

The Limits of Mathematical Computer Science

Beginning in 1975 and extending over the late 1970s, the Computer Science and Engineering Research Study, chaired by Bruce Arden of Princeton University, took stock of the field and its current directions of research and published the results under the title *What Can Be Automated?* The committee on theoretical computer science argued forcefully that a process of abstraction was necessary to understand the complex systems constructed on computers and that the abstraction "must rest on a mathematical basis" for three main reasons:

(1) Computers and programs are inherently mathematical objects. They manipulate formal symbols, and their input-output behavior can be described by mathematical functions. The notations we use to represent them strongly resemble the formal notations which are used throughout mathematics and systematically studied in mathematical logic.

(2) Programs often accept arbitrarily large amounts of input data; hence, they have a potentially unbounded number of possible inputs. Thus a program embraces, in finite terms, an infinite number of possible computations; and mathematics provides powerful tools for reasoning about infinite numbers of cases.

(3) Solving complex information-processing problems requires mathematical analysis. While some of this analysis is highly problem-dependent and belongs to specific application areas, some constructions and proof methods are broadly applicable, and thus become the subject of theoretical computer science.[19]

Defining theoretical computer science as "the field concerned with fundamental mathematical questions about computers, programs, algorithms, and information processing systems in general," the committee acknowledged that those questions tended to follow developments in technology and its application, and hence to aim at a moving target — strange behavior for mathematical objects.

Nonetheless, the committee could identify several broad issues of continuing concern, which were being addressed in the areas of computational complexity, data structures and search algorithms, language and automata theory, the logic of computer programming, and mathematical semantics. In each of these areas, it could point to substantial achievements in bringing some form of mathematics to bear on the central questions of computing. Yet, the summaries at the end of each section sounded a repeating chord. For all the depth of results in computational complexity, "the complexity of most computational tasks we are familiar with — such as sorting, multiplying integers or matrices, or finding shortest paths — is still unknown." Despite the close ties between mathematics and language theory, "by and large, the more mathematical aspects of language theory have not been applied in practice. Their greatest potential service is probably pedagogic, in codifying and given clear economical form to key ideas for handling formal languages." Efforts to bring mathematical rigor to programming quickly reach a level of complexity that makes the techniques of verification subject to the very concerns that prompted their development:

> One might hope that the above ideas, suitably extended and incorporated into verification systems, would enable us to guarantee that programs are correct with absolute certainty. We are about to discuss certain theoretical and philosophical limitations that will prevent this goal from ever being reached. These limitations are inherent in the program verification process, and cannot be surmounted by any technical innovations.

Mathematical semantics could show "precisely why [a] nasty surprise can arise from a seemingly well-designed programming language," but not how to eliminate the problems from the outset. As a design tool, mathematical semantics was still far from the goal of correcting the anomalies that gave rise to errors in real programming languages.

If computers and programs were "inherently mathematical objects," the mathematics of the computers and programs of real practical concern had so far proved elusive. Although programming languages borrowed the trappings of mathematical symbolism, they did not translate readily into mathematical structures that captured the behavior of greatest concern to computing. To a large extent, theoretical computer science continued the line of inquiry that had led to the computer in the first place, namely, the establishment of the limits of what can be computed. While that had some value in directing research and development away from dead ends, it offered little help in resolving the problems of computing that lay well within those limits. For that reason, the broad field referred to as 'computer science' remains an amalgam of mathematical theory, engineering practice, and craft skill.

References

1. "An Undergraduate Program in Computer Science — Preliminary Recommendations," *Communications of the ACM* (1965), **8,9**: 543–552; at 544.
2. Bruce W. Arden (Ed.), *What Can Be Automated?: The Computer Science and Engineering Research Study (COSERS).* (Cambridge, MA: MIT Press, 1980), 9.
3. *Science* (1967), **157**: 1373–74.
4. "On Computable Numbers, with an Application to the Entscheidungsproblem," *Proceedings of the London Mathematical Socxiety* (1936), **ser. 2, vol. 42**: 230–265; at 231.

5. D. Hilbert and W. Ackermann, *Grundzüge der theoretischen Logik.* (Berlin: Springer, 1928), 73–4.
6. Stephen C. Kleene, "Origins of Recursive Function Theory," *Annals of the History of Computing* (1981), **3,1**: 52–67.
7. Claude E. Shannon, "A symbolic analysis of relay and switching circuits," *Transactions of the AIEE* (1938), **57**, 713–23.
8. Michael O. Rabin and Dana Scott, "Finite automata and their decision problems," *IBM Journal* (April, 1959), 114–25; at 114.
9. Published in *Cerebral Mechanisms in Behavior — The Hixon Symposium,* L.A. Jeffries (Ed.). (New York: Wiley, 1951), 1–31; repr. in *Papers of John von Neumann on Computing and Computer Theory,* William Aspray and Arthur Burks (Ed.). (Cambridge, MA/London: MIT Press; Los Angeles/San Francisco: Tomash Publishers, 1987), 391–431; at 406.
10. Alan M. Turing, "Computing Machinery and Intelligence," Warren S. McCulloch and Walter Pitts, "A logical calculus of the ideas immanent in nervous activity," *Bulletin of Mathematical Biophysics* (1943), **5**, 115–33.
11. "Representation of events in nerve nets and finite automata," in *Automata Studies,* Claude Shannon and John McCarthy Eds.). (Princeton: Princeton University Press, 1956), 3–41.
12. M.P. Schützenberger, "Une théorie algébrique du codage," *Séminaire P. Dubreil et C. Pisot* (Faculté des Sciences de Paris), Année 1955/56, No. 15 (dated 27 February 1956); p. 15–02. Cf. "On an application of semi groups[!] methods to some problems in coding." *IRE Transactions in Information Theory* (1956), **2,3**: 47–60.
13. M.P. Schützenberger, "Un problème de la théorie des automates," *Seminaire Dubreil-Pisot* (1959/60), **13**, no. 3 (November 23, 1959), 3–01.
14. Noam Chomsky, "Three models of language," *IRE Transactions in Information Theory* (1956), **2,3**: 113–24; at 113.
15. John McCarthy, "Towards a mathematical science of computation," *Proc. IFIP Congress 62.* (Amsterdam: North-Holland, 1963), 21–28; at 21.
16. John McCarthy, "A Basis for a Mathematical Theory of Computation," in P. Braffort and D. Hirschberg (Eds.), *Computer Programming and Formal Systems.* (Amsterdam: North-Holland Publishing Co., 1963), 33–70; at 69.
17. John McCarthy, "Towards a mathematical science of computation," *Proc. IFIP Congress 62.* (Amsterdam: North-Holland, 1963), 21–28; at 21.
18. Dana S. Scott, "Outline of a mathematical theory of computation". (Technical Monograph PRG-2, Oxford University Computing Laboratory, 1970), 4.
19. Bruce W. Arden, *What Can Be Automated?: The Computer Science and Engineering Research Study (COSERS).* (Cambridge, MA: MIT Press, 1980).

FURTHER READING

ACM Turing Award Lectures: The First Twenty Years, 1966 to 1985. (New York: Reading, Mass.: ACM Press; Addison-Wesley Pub. Co., 1987); lectures since 1986 published annually in *Communications of the ACM.*

William Aspray (Ed.), *Computing Before Computers.* (Ames: Iowa State University Press, 1990).

Bruce W. Arden (Ed.), *What Can Be Automated?: The Computer Science and Engineering Research Study (COSERS).* (Cambridge, MA: MIT Press, 1980).

Thomas J. Bergin, Jr. and Richard G. Gibson, Jr. (Eds.), *History of Programming Languages II.* (New York: ACM Press; Reading, Mass.: Addison-Wesley Pub. Co., 1996).

Martin Campbell-Kelly and William Aspray, *Computer: A History of the Information Machine.* (New York: Basic Books, 1996).

Jan van Leeuwen (Ed.), *Handbook of Theoretical Computer Science.* (2 vols., Amsterdam; New York: Elsevier; Cambridge, Mass.: MIT Press, 1990).

Herbert Mehrtens, *Moderne Sprache Mathematik: Eine Geschichte des Streits um die Grundlagen der Disziplin und des Subjekts formaler Systeme.* (Frankfurt am Main: Suhrkamp, 1990).

Richard L. Wexelblat (Ed.) *History of Programming Languages.* (New York: Academic Press, 1981).

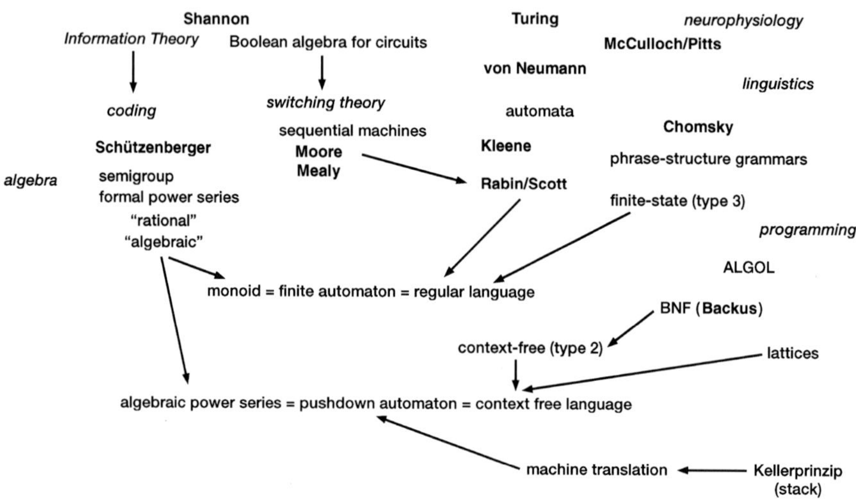

FIGURE 31.1: THE AGENDAS OF COMPUTER SCIENCE.

Chapter 32

Astronomy

Karl Hufbauer

In the *Encyclopedia Britannica*'s classic eleventh edition (1910–11), the era's preeminent historian of astronomy described the discipline's purview as "all the matter of the universe which lies outside the earth's atmosphere." Agnes Clerke went on to inform readers that the science had two main branches — the hoary field of astrometry, "which treats of the motions, mutual relations and dimensions of the heavenly bodies," and the neoteric field of astrophysics, "which treats of their physical constitution." Believing that astronomy was "among the more progressive sciences," she was optimistic that its "new and improved methods of research [would] ultimately lead to a comprehension of the universe." Today, as the twentieth century is closing, Clerke's general description of astronomy still rings true. Indeed, in portraying the acquisition of "a comprehension of the universe" as the discipline's ultimate goal, she seems almost to be anticipating the emergence of its third main branch — cosmology.

Astronomy was, as suggested by the adequacy of Clerke's characterization, already a mature science by 1900. This chapter on the discipline's continuing evolution during the twentieth century first considers the motives for and manifestations of social support for astronomy. It goes on to examine how astronomers, and physicists working on astronomical problems, have used this support to improve the discipline's organization and toolkits. Then it describes some of the key steps taken by such scientists in advancing our species' understanding of the stars, the galaxies, the origin of the elements, and the origin of the universe. By way of concluding, it considers astronomy's prospects in the decades ahead.

Social Support for Astronomy

Social support for astronomy has increased substantially during the twentieth century. This support has been forthcoming mainly because of deep-seated cultural beliefs and values. Many, adhering to the traditional belief that the majesty of the heavens attests to divine providence, have viewed astronomy as spiritually uplifting. Many have also taken vicarious pride in the imposing, and very subtle, technologies required to observe ever more remote celestial phenomena and, more recently, to send instrument-laden spacecraft on ever more challenging missions. And growing numbers, thirsting for credible natural accounts of humanity's place in the cosmos, have derived emotional satisfaction from astronomy's inquiries into origins. Although not so important as these basic cultural motives, practical concerns have also inspired increasing support for astronomical research. Time and again, instrument-making firms have contributed to the development of telescopes, auxiliaries, and sensors as a strategy for bettering corporate proficiency and markets. And chiefly since the beginning of World War II, the armed forces have supported some lines of research in the hope that this patronage would improve their operational capabilities or, more generally, enhance their appeal to the scientific-technical labor force.

Support for astronomy has taken two forms — moral and material. The public and elites alike have expressed their moral support by showing increasing interest in the science. Multitudes have gone beyond simple wonderment at starry heavens to observe comets and eclipses, visit planetariums and observatories, watch coverage of lunar and planetary missions, and view television series about the investigation of celestial phenomena. Within this burgeoning public, fair numbers have acquired semi-popular books about astronomical researches and subscribed to astronomical monthlies. Many have also attended general lectures by astronomers. Both responding to and nurturing this interest in astronomy, the news media have gradually paid more attention to the field and sci-fi authors and film-makers have gone increasingly out of their way to provide astronomically realistic backdrops for their adventures. Meanwhile, at the other end of the popular-elite spectrum, the committee for the Nobel Prize in Physics has displayed an appreciation for astronomy from time to time ever since 1967 when, in nominating Hans Bethe, it highlighted his solution of the stellar-energy problem.

Following up on the expanding moral support for astronomy, those with moderate-to-great wealth or power have provided the discipline with increasing material support since 1900. Astronomy's benefactors have devoted a sizable fraction of their patronage for the science to research. Like Andrew Carnegie who endowed Mt. Wilson Observatory in 1904, philanthropists, foundations, military branches, and scientific agencies have funded an ever larger number of major observatories. Since the launch of *Sputnik* in 1957, military branches and scientific agencies have also funded — at a much greater cost — increasingly sophisticated astronomical missions. Not only the patrons of these emblematic disciplinary resources but also universities have backed up initial capital costs with essential budgets for auxiliary instrumentation, sensors, staff, data analysis, interpretation, and scientific publication. Complementing all this support for research, astronomy's patrons have also funded training in and recruitment of needed experts for the discipline. In particular, they have made it possible for universities to greatly expand the number and size of astronomy graduate programs and fellowships and for observatories and space-science missions to multiply the number of physicists working on astronomical problems.

Astronomy's Organization

As social support for astronomy has grown, professional astronomers — i.e., scientists holding jobs in which they were expected to pursue astronomical research — have become increasingly numerous and widely distributed. Although good counts are not available, the general pattern of growth and dispersion is clear. Around 1900 a few hundred scientists occupied such positions. Although some worked in the Southern Hemisphere and Asia, the vast majority lived in Europe and North America. By the eve of World War II, the global population of professional astronomers had approximately doubled. Now, near the end of the century, the numbers might exceed 15,000. Growth in the Southern Hemisphere and Asia, especially in Australia and Japan, has been sufficient to reduce — but not nearly enough to challenge — the Euro-American preponderance in the global astronomical community. The discipline's expansion has been accompanied by numerous changes in the ways in which professional astronomers have organized their community's research, communications, and training and recruitment. Having moderate effects before 1940, such changes have transformed their work lives since World War II.

Enriching a disciplinary research tradition based mainly on specialized, long-term observing programs, professional astronomers extended the international cooperation that had characterized a few nineteenth-century projects to the whole of astronomy between 1900 and 1940. The first step here was taken in 1904–05 when solar physicists, upon the urging of George Ellery Hale (then Director of the Yerkes Observatory and Editor of the *Astrophysical Journal*), established the International Union for Cooperation in Solar Research. The primary task of the Solar Union's working committees was to promote cooperation in each of the field's main specialties. A victim of World War I's hostilities, the Solar Union was succeeded in 1919 by the International Astronomical Union (IAU). The IAU, through its more than thirty committees, encouraged astronomers to regularly assess ongoing research in their respective fields and, where desirable, organize cooperative endeavors. Meanwhile, theoretical astrophysicists — a new breed — were developing an opportunistic approach to research. Instead of working steadily in one problem area as did most observers, they eagerly sought out fresh opportunities for using their interpretive toolkits. Their strategy was so successful that one senior observer derisively referred to their strategy of exploiting hard-won observational data as cream-skimming.

Astronomers also augmented disciplinary channels of communication in the decades before World War II. Throughout most of the period, the Solar Union and then the IAU served as triennial forums for announcing recent results, evaluating trends, and promoting consensus. Meanwhile, Harvard College Observatory, which was directed by Harlow Shapley, began an announcement service in the mid-1920s, giving astronomers a handy way to report their latest findings to an extensive set of subscribers. German astronomers founded the *Zeitschrift für Astrophysik* (1930), which soon joined the *Monthly Notices of the Royal Astronomical Society* and the *Astrophysical Journal* as a leading disciplinary forum. Lastly, following the physicists' example, astronomers began holding small research conferences in which groups of experts shared their most recent results, speculations, and strategies.

Besides diversifying their research strategies and communication channels, astronomers were improving their training of Ph.Ds and their recruitment of outside experts. The strengthening of graduate programs was usually the result of many incremental improvements in faculty and instruments. However during the 1930s, under the leadership of Shapley and Otto Struve, the programs at Harvard and the University of Chicago moved well ahead of most competitors by incorporating a fair amount of theoretical astrophysics into their instruction. Meanwhile, astronomers sought in various ways to elicit contributions from physicists with desirable instrumental and interpretive skills. Sometimes they arranged for the employment of such physicists, the most successful such appointment being in 1906 when the recruitment to Greenwich of young Arthur Stanley Eddington launched what became a brilliant career in theoretical astrophysics. Astronomers also encouraged talented physicists to immigrate or, more often, to make short-term excursions from physics into astronomy by hosting them as postdoctoral fellows, by seeking them out for collaborations on borderland problems, and in the case of Harvard's Shapley by inviting them to participate in high-level summer sessions.

During the half-century since World War II, astronomers have simultaneously become far more cooperative and far more opportunistic in their approaches to research. Their cooperativeness has been engendered by the immense growth in the patronage for and the scale of their undertakings. Increasingly they have been obliged to collaborate at national and/or international levels to get priority status for a major new observing system, to shepherd it through complex approval processes, to supervise its construction and operation, to use it for observing, and to analyze and elucidate its results. A necessary feature of scores upon scores of big-astronomy projects since 1945, this kind of cooperation may have peaked with what became known as the *Hubble Space Telescope.* All the while, an opportunistic approach has appealed to an ever larger fraction of astronomers for at least part of their research. The main reason has been that the astronomical community's growth has given rise to intense competitive pressures for positions and grants. These pressures have, in turn, stimulated desires for quick returns. Such desires have often led observers who are bogged down in major projects to seek out fresh opportunities. They have also reinforced the opportunistic orientation of those in astronomy's rapidly swelling subcommunities of theorists.

Astronomers have greatly amplified the complexity of their communication system since 1945. Here their mushrooming numbers and budgets have been the main driving forces. At first, these forces swelled the discipline's main journals and societies. However, by the mid-1960s many astronomers were so eager to have closer contacts with fellow specialists that they began urging that the disciplinary channels of communication be supplemented with specialty-oriented journals and societal divisions. During the next decade, despite disciplinary leaders who fretted about astronomy's disintegration, this campaign was increasingly successful — e.g., the journal *Solar Physics* (1967) and the American Astronomical Society's Solar Physics Division (1970). Moreover, thanks not only to generous budgets but also to jet travel, astronomers were holding more and more problem-oriented conferences with published proceedings. As publication venues have multiplied, growing numbers of astronomers have begun making it a practice to publish several variants of the same piece so that their work will have a better chance of being seen. More recently, having become increasingly concerned about publication delays, growing numbers have availed themselves of new technologies for electronic mail and posting so as to speed up communications.

In the meantime astronomers have significantly expanded both professional training and outside recruitment. This expansion, which occurred mainly between 1945 and 1980 because of a robust demand for astronomical personnel, has been followed by consolidation during the last decade and a half. Astronomers have developed Ph.D programs in all prosperous and several not-so-prosperous nations around the globe. In doing so, they have sought to build up strong departments and institutes with ample access to opportunities for student and faculty research. They have also incorporated ever more physics into Ph.D curricula in recognition of the ever greater role of physics-based tools in the discipline's advance. Concurrently astronomers have done much to recruit physicists possessing valuable tools. In addition to their pre-war recruitment strategies, they have been particularly successful at using plaudits and prizes to inspire continuing contributions from physicists who, like the post-war radio physicists, demonstrated the effectiveness of their tools for astronomical research. In recent years astronomers have also played successfully on the physicists' excitement about research prospects in astrophysics and cosmology to persuade many physics departments to start up Ph.D tracks in astronomy.

Astronomy's Tools

Astronomers and physicists, aided by growing social support, have brought ever more diverse and versatile toolkits to bear on astronomical problems during the twentieth century. The main new observational tools have been increasingly complex systems for gathering and recording emissions from celestial sources. The main new interpretive tools have been novel physical theories and, after World War II, ever more powerful computers.

Far and away the most significant advance in astronomy's observational capabilities between 1900 and 1940 was George Ellery Hale's establishment of the Mt. Wilson Observatory near Los Angeles. Founded with Carnegie money in 1904, this observatory soon had three solar telescopes and two large reflectors — including the Hooker Telescope with its 2.5-meter mirror (operational 1919) — as well as a spectroscopy laboratory, superb shops, and a large staff which was continually upgrading the auxiliary instrumentation. The remaining additions to the observational toolkit during these decades, although comparatively modest taken one by one, were numerous, and cumulatively significant. In fact, two were sufficiently novel that the Royal Astronomical Society recognized their developers with its Gold Medal (astronomy's highest award) — i.e., the physicist Albert A. Michelson's interferometer which when installed on the Hooker Telescope in 1920 enabled his collaborators to measure a distant star's diameter for the first time and the astrophysicist Bernard Lyot's coronagraph which when used high in the Pyrenees in 1930 enabled him to observe the solar corona outside eclipse for the first time.

Rushing ahead of those seeking to augment observational capabilities, theoretical astrophysicists together with theoretical physicists enriched astronomy's interpretive toolkit in four important ways before World War II. First Arthur Stanley Eddington and others set about improving upon nineteenth century efforts to harness thermodynamics as a tool for interpreting the sun and stars. Next Eddington and others started using Albert Einstein's theory of general relativity in their cosmological researches. Concurrently the theoretical astrophysicist Henry Norris Russell and others employed the increasingly puissant quantum theories of atomic and molecular processes to interpret the physical meaning of celestial spectra. Lastly in the late 1930s the theoretical physicist Hans Bethe and others began making effective use of nuclear data and theory to expand understanding of the internal constitution of the stars.

Those engaged in the extraordinary expansion of astronomy's observational capabilities since 1945 have exploited three principal strategies. The first, in which physicists have usually taken the initiative, has centered on the development of means to observe celestial emissions outside the visible spectrum. In the two decades after the war, physicists in England, Australia, and Holland led the way in following up on earlier detections of galactic and solar radio emissions by initiating development of the distinctive and surprisingly powerful toolkit of what soon came to be known as radio astronomy. Meanwhile, physicists in the United States and Soviet Union used post-war missile and space programs as a means of getting instruments above the obscuring atmosphere and thereby instituted ultraviolet and X-ray astronomy and direct observations of the solar wind and low-energy cosmic rays. As a result of all these successes, astronomers and physicists were in general agreement by the mid-1960s on the importance of opening up new windows on celestial phenomena. Accordingly several teams enthusiastically went on during the next decade to initiate observations of solar neutrinos, galactic infrared radiation, and cosmic gamma-ray emissions and to inaugurate the as-yet unsuccessful quest for gravity waves.

The second strategy has been to construct ever larger observing systems with ever greater signal-gathering power. Here optical astronomers have been particularly active. Since the completion in 1948 of Palomar Observatory's Hale Telescope with its 5-m mirror, they have supervised the construction of about fifteen single-mirror telescopes larger than Mt. Wilson's famous Hooker Telescope on mountains in both hemispheres and islands near the equator. In addition, they have begun developing multiple-mirror telescopes. The most impressive so far is the first Keck Telescope (1993), a 10-m segmented-mirror instrument on Mauna Kea, which will soon be coupled with a second Keck Telescope so as to achieve binocular vision. Finally optical astronomers played a major role in the lengthy development of the *Hubble Space Telescope*, including devising a correcting lens for the 2.4-m mirror the figuring of which was botched by a military contractor. Meanwhile those interested in developing radio, infrared, ultraviolet, X-ray, gamma-ray, neutrino, and gravity-wave observational systems have ardently pursued the strategy that bigger is better.

The third strategy has been to develop better auxiliary instrumentation and sensors so as to extract more information from those emissions that are captured. Increasingly the province of astronomical and physical instrumentalists, this line of work has involved a myriad of incremental advances that have enabled observers to secure data with higher spatial, spectral, and/or temporal resolution. The most impressive advance by far has been the deployment since the early 1980s of charged-couple devices (CCDs) as sensors. Having a sensitivity more than 30 times higher than photographic plates, installation of these solid-state detectors on existing and new instruments has resulted in an impressive improvement in telescope performance. Moreover, since the output from a CCD can be fed directly into a computer, observers can go about their work much more interactively.

Almost keeping pace with the ongoing expansion in astronomy's observational capabilities, theoretical astrophysicists working mainly with theoretical physicists have steadily increased the power of the discipline's interpretive toolkit during the last half-century. Legions have extended and refined the applications of relativity theory, quantum mechanics, and nuclear theory to astronomical problems. Between 1955 and 1970, meanwhile, Hannes Alfvén, Eugene Parker, and many others began bringing magnetohyrodynamics to bear on astrophysical problems involving the interaction of particles and fields. During the ensuing decade, Roger Ulrich, Franz Deubner, and others introduced seismological techniques into the analysis of stellar structure. And since the late 1970s, Alan Guth and many, many others have worked at integrating elementary-particle theory and cosmology. All the while, the theorists have been making increasing use of ever faster and more versatile computers to analyze data, check out alternative models, and develop increasingly persuasive interpretations.

FOUR KEY ASTRONOMICAL ADVANCES

As astronomy's social support, size, organizational complexity, and observational and interpretive toolkits have grown, astronomers and physicists have advanced humanity's knowledge of the heavens. Their researches have been devoted to discerning new classes of celestial objects and new properties of familiar classes, to ascertaining the positions, magnitudes, and spectra of as many members of each class as possible, and to theorizing about the structure, evolution, and interactions of all the known classes. Most of this work, though technically challenging and, for those doing it, intriguing, yields results that seem fairly routine when considered from a distance. Yet these everyday results constitute a crucial background for those relatively rare advances in astronomical understanding that strike experts and laymen alike as being of the first importance. Four such advances in the twentieth century have centered on the internal constitution of the stars, the nature of the galaxy and spiral nebulae, the origin of the elements, and the triumph of 'big-bang' cosmology.

Around 1900 most astronomers, taking our sun as a model, supposed that the stars are immense gaseous spheres composed of the familiar elements heated to incandescence by ongoing gravitational contraction. And thinking that they had found evolutionary clues in nebular structures and stellar spectra, they surmised that spiral nebulae collapse into blue- or white-hot young stars which, as they continue to contract and radiate, gradually traverse all the spectral types ending up as relatively cool red stars. On this view the sun is a middle-aged star that, given the limited store of its gravitational energy, would at best sustain life on the earth for a few more tens of millions of years. Notwithstanding their confidence in the soundness of this conception of the stars, astronomers were driven to doubts by several developments during the century's first decade and half — the discovery of subatomic energies, the dating of some some radioactive rocks at immense ages, the identification of giant stars much larger than the sun, and the abandonment of the view of spirals as nearby stars-in-formation.

Cambridge University's Arthur Stanley Eddington led the way between 1916 and 1926 in developing a more robust theory of the stars. His starting point was the idea that the recently discovered giant stars are so rarefied that they behave in accord with the perfect-gas laws. If so, he went on to argue, the inward gravitational pressure in giants should be counterbalanced by the outward pressure from escaping radiation. He soon set about trying to generalize his theory of the "radiative equilibrium" of giant stars so that it would also apply to ordinary stars like the sun. Eddington succeeded in 1924 when, in another brilliant insight, he realized that ionization would be so great at the central temperatures in a star that the resulting gas would consist mainly of bare nuclei and electrons. Such a gas, unlike one consisting of much larger neutral atoms, should obey the perfect-gas laws at the relatively high central densities of ordinary stars. Establishing his theory's broad applicability by showing that 44 stars, including the sun, all fell on or near its predicted mass-luminosity curve, Eddington went on to explore its implications for stellar evolution. He saw two possibilities. If helium-building nuclear reactions were the main energy source, stars would evolve along tracks of essentially constant mass over billions of years. But if mutual annihilation by electrons and protons were the source, stars would dissipate most of their mass over lifetimes of trillions of years. Eddington preferred the second option because it supported the view that the distribution of stars on a Hertzsprung-Russell diagram represents an evolutionary track with stars originating as massive red giants and, as they gradually consume their fuel reserves, moving onto and then down the main-sequence line as ordinary stars. Still, although preferring the annihilation hypothesis, he emphasized the openness of the issue in his classic *The Internal Constitution of Stars* (1926).

Three developments in the ensuing decade set the stage for the identification of the energy-generating process in the late 1930s. First, Cecilia Payne, Henry Norris Russell, and others established that hydrogen is by far the most abundant element in the stars, thereby laying the basis for improved estimates of the conditions under which energy is generated. Second, Edwin Hubble discovered the distance-redshift relation for spiral nebulae (see below), leading Eddington to infer that the universe began expanding a billion or so years ago and hence that stellar lifetimes are of the order predicted by the helium-building hypothesis. Third, while physicists consistently failed in attempts to coax electrons into annihilating protons, they found a mounting number of element-building reactions in the early and mid-1930s. At this juncture several young theoretical physicists took up the stellar-energy problem. Encouraged by Eddington and other theoretical astrophysicists, they all tried to find a match between their respective areas of expertise and the requirements of the problem. In 1938 Hans Bethe of Cornell University emerged triumphant from their lively competition. In a compelling quantitative analysis, he argued that ordinary stars are powered by helium-building reactions, with less massive, cooler stars depending on a reaction chain beginning with proton-proton collisions and more massive, hotter stars relying on a reaction cycle beginning with proton-carbon collisions. Bethe's solution to the stellar-energy problem was immediately hailed by astronomers. Closing a crucial gap in Eddington's theory, it constituted a transformative breakthrough in research on stellar structure and evolution.

In 1918, two years after Eddington began theorizing about the stars, Harlow Shapley challenged the prevailing view among astronomers about the nature of the galaxy and spiral nebulae. On this view the solar system was near the center of a smallish galaxy (diameter 10,000 light years) and the spirals were "island universes," i.e., remote star systems that might well indicate how our galaxy would appear if seen from such a distance. Then a junior observer at Mt. Wilson Observatory, Shapley based his challenge on a study of almost seventy globular clusters of stars. His basic idea was that these clusters outline our galaxy. If so, their spatial distribution — as determined by using Cepheid variable stars as standard candles — indicated that the galaxy's center is 50,000 light years from the sun in the direction of Sagittarius and that its overall diameter is 300,000 light years. Inclined by the very immensity of his galaxy to think that the spirals must be within it, Shapley was receptive to his colleague Adriaan van Maanen's recent, and as it later turned out spurious, measurements of internal motions in a spiral nebula which suggested its proximity. Over the next decade Shapley's revolutionary cosmology had a mixed reception. Most astronomers accepted his general ideas about the galaxy's immensity and the direction of its center. However all but a few remained skeptical about his and van Maanen's view of spirals as constituents of the galaxy.

The astronomer who settled the question about the nature of spiral nebulae was Edwin Hubble of Mt. Wilson Observatory. In 1923–24 he found 34 Cepheid variables on plates of the spirals M31 (Andromeda) and M33 taken with Mt. Wilson's two big reflectors. Like Shapley, who had left for Harvard three years before, Hubble used these Cepheids as standard candles for estimating the spirals' distances. The outcome for both spirals — around one million light years — was well beyond the borders of Shapley's big galaxy. Hubble, who was concerned about the conflict between his own and his colleague van Maanen's results, cautiously delayed the formal announcement of his discovery for more than a year. Once announced, however, his huge distances were rapidly accepted as clinching the case for the idea that spiral nebulae are island universes or simply, as they gradually came to be called, galaxies. Hubble followed up on his finding with major papers on the types and distances of "extra-galactic" nebulae. His most important result, announced in 1929, was a relation between spiral distances and velocities. Here he was putting together his measurements of the distances of 24 spirals with the measurements of their redshifts, many of which were first made during the preceding decade and a half by Vesto Slipher of the Lowell Observatory. His startling conclusion was that, on average, the further away the spirals are from our galaxy, the faster they are rushing away from us.

Hubble's discovery elicited a spate of interpretations from theoretically-minded astronomers and physicists between 1929 and 1933. Those who were sympathetic toward general relativity came to regard the distance-redshift relation as additional evidence for Einstein's theory. Here Eddington and Willem de Sitter, drawing on a 1927 paper of Georges Lemaître, led the way. They argued that for some unknown reason the universe had entered into the present era of relativistic expansion a billion or so years ago. Lemaître soon went beyond Eddington and other relativists to speculate that the universe began as a primeval atom, a quantum transition in which initiated its explosive expansion. Not everyone, of course, was eager to invoke general relativity as an explanation for Hubble's relation. For instance, while accepting the universe's expansion, E. Arthur Milne sought to show that this phenomenon was a consequence of what he called "kinematic relativity."

By contrast, denying expansion, Fritz Zwicky suggested that the light from the remote galaxies had lost energy during its long journey to us and hence been redshifted. Most astronomers could see no clear basis for deciding among these possibilities in the 1930s. Indeed Hubble himself soon became rather reticent about interpreting the galactic redshifts as recessional velocities.

Surprisingly the hypothesis that the universe began with an explosion received indirect support from Bethe's theory that main-sequence stars are powered by helium-building nuclear reactions. His investigation indicated that the internal temperatures and densities in ordinary stars are too low for any appreciable production of elements beyond helium. Hence all the heavier elements in such stars must have originated prior to their formation. While Bethe stopped at this point, Carl Friedrich von Weizsäcker — a theoretical physicist who had independently solved the stellar-energy problem — was quick to argue that the extreme conditions needed for building heavier elements up to their observed abundances were present in the primeval fireball that initiated the universe's expansion. After World War II the theoretical physicist George Gamow and his collaborators Ralph Alpher and Robert Herman did the most to develop this idea. In their scenario our universe began with the explosion of a very small and hot ball of energy, which during the first hour or so of rapidly expanding and cooling produced first neutrons, then protons and electrons, next deuterium and helium, and finally heavier nuclei that through radioactive decay soon yielded today's elemental-abundance curve. Their initial successes in interpreting this curve as a remnant from the 'big-bang' led many to regard this approach to the origin of the elements as promising. However, their admitted inability to suggest any plausible reactions that, during the short time available, would get beyond the gaps caused by the extreme instability of nuclei with masses 5 and 8, led in the early 1950s to a general wariness about the approach's validity.

Taking a different tack, a few astronomers and physicists thought that the heavier elements must be built up in the stars. The Cambridge theoretical astrophysicist Fred Hoyle, this idea's leading advocate, began arguing in 1946 that supernovae not only attain the high temperatures needed for the genesis of heavy elements but also are sufficiently violent to scatter the newly-formed elements throughout their celestial neighborhoods. Developments in both observational astronomy and theoretical astrophysics greatly increased the plausibility of Hoyle's speculative proposal by the mid-1950s. Observers lent credence to the idea that the heavy elements are formed in and scattered by exploding stars by finding indications that the younger a star is the higher the fraction of elements beyond helium that it contains. Particularly striking evidence that element building occurs in stars was provided by the discovery in 1952 of the short-lived element technetium in a red giant's atmosphere. Meanwhile theorists bolstered confidence in Hoyle's approach by proposing nuclear reactions that, acting at very high temperatures and densities over much longer times than allowed by the 'big-bang' hypothesis, could produce elements beyond helium. Of special importance here were proposals in 1951–52 that collisions of three helium nuclei could form carbon nuclei (mass 12), thereby laying the basis for the production of heavier elements by bridging the gaps at 5 and 8. Also important were suggestions in 1954 of specific reactions that would yield neutrons for element building.

Two years later Hoyle teamed up at Caltech with the visiting Cambridge observer E. Margaret Burbidge, her theorist husband Geoffrey Burbidge, and the nuclear physicist William Fowler to work out a comprehensive theory of stellar nucleogenesis. Their goal was to elucidate all the major and minor features in the detailed abundance curves that had just become available. Although admittedly failing to develop cogent explanations for the exceptionally low abundances of deuterium, lithium, beryllium, and boron, they proposed seven different processes occurring during the evolution of massive stars that accounted for the abundances of all the heavier elements. According to the scenario advanced in B^2FH (their paper's acronym), element building commences in a main-sequence star with hydrogen burning. Once most of the hydrogen in the hot core has been transformed into helium, the star's central region collapses and heats to such high temperatures that element building by means of helium burning and four other nuclear processes continues in a succession of stages in the resulting red giant. Finally, once these energy-generating reactions could no longer counterbalance the force of gravity, the star would collapse into a white dwarf. Such a collapse, B^2FH argued, would be accompanied by an explosion that would disperse a substantial fraction of the cooked-up elements through the surrounding space. If the star were exceptionally massive, the explosion would be a supernova that itself would synthesize many additional heavier elements (including, alas, uranium). B^2FH did such a magisterial job of bringing the latest work in astronomy and physics to bear on the abundance problem that it was recognized, even while circulating as a preprint in early 1957, as a decisive turning point in research on the origin of the elements.

In the late 1940s, almost a decade before his research on stellar nucleogenesis culminated in B^2FH, Hoyle went beyond seeking an alternative to 'big-bang' element building to challenge 'big-bang' cosmology in its entirety. He and, separately, Hermann Bondi and Thomas Gold trenchantly subverted 'big-bang' creationism by emphasizing that the age of the universe as reckoned from Hubble's distance-redshift relation was considerably smaller than the age of the earth. In their alternative 'steady-state' cosmology, the universe has never really changed. Hence it should have the same basic appearance no matter what the vantage point in space *or* time. For example, just as the universe appears to us to have been expanding for the Hubble time (then less than two billion years), so would it have appeared ten billion years ago. This appearance of expansion, the three Cambridge mavericks postulated, is the consequence of a continuous creation of hydrogen atoms throughout space at just the rate needed to compensate for the matter disappearing in the distance. Certain that the 'big-bang' theory was beyond rescue, they defied observers to come up with convincing evidence that the universe of the past differed in essential ways from the universe of our era.

The ensuing cosmological controversy was lively, and inconclusive. On the one hand, Walter Baade of Mt. Wilson and Palomar Observatories improved the prospects for 'big-bang' theory in 1952 when he initiated what became a series of upward revisions in the Hubble time by making a persuasive case that the distances to the external galaxies are at least two times greater than estimated by Hubble. But, despite various tries, no one managed to develop a robust line of evidence that, as the 'big-bang' theory implied, the universe of earlier times differed substantially from our own. On the other hand, the manifest success of B^2FH in 1957 bolstered 'steady-state' cosmology by providing an account of the origin of the elements that would be equally valid at any time in the universe's history. However, none of its partisans could advance a plausible physical explanation for continuous matter creation. The contending arguments were so evenly matched that many astronomers were reluctant to take sides. A poll of 1959 revealed, for instance, that while 33 percent favored the 'big-bang' hypothesis and 24 percent the 'steady-state' worldview, over 40 percent were dissatisfied with both theories or undecided.

During the 1960s a succession of findings gave 'big-bang' cosmology the upper hand. In 1961 and 1965 respectively, the Cambridge radio astronomer Martin Ryle and the Caltech astronomer Allan Sandage undermined the 'steady-state' theory's postulate of temporal isotropy by making good cases that the populations of radio galaxies and quasars in the universe have decreased with the passage of time. And in 1964 Hoyle conceded the insufficiency of stellar nucleogenesis so far as helium is concerned when he argued that its cosmic abundance is so great that much of it must have been generated elsewhere than in the stars. While these various results redounded to the advantage of the 'big-bang' theory, a serendipitous finding in radio astronomy was what clinched the argument. Arno Penzias and Robert Wilson of Bell Laboratories announced in 1965, after more than a year of analysis, that over half of the thermal noise emanating from a new and highly sensitive microwave antenna that they were transforming into a radio telescope was cosmic, not atmospheric or instrumental, in origin. The noise indicated that in all directions space has a temperature of a few degrees above absolute zero at the antenna's wavelength of operation. Puzzled by their finding, Penzias and Wilson were happy to learn that Robert Dicke's group at Princeton University had an interpretation ready at hand — i.e., that the thermal noise is a remnant of the 'big-bang.' Dicke and his colleagues, who had been planning to search for just such a background radiation, were disappointed to be scooped. Still they marshaled a strong case that the noise indicates that the universe's expansion and cooling since the 'big-bang' fireball has resulted in a background temperature of some 3° K. The evidence of Penzias and Wilson together with the argument of Dicke's group struck most astronomers and physicists interested in cosmology as conclusive. Indeed, the triumph of 'big-bang' cosmology was so rapid that in 1967 Gamow, Alpher, and Herman were indignantly protesting that their prediction of the background radiation's existence some two decades earlier had been lost sight of during the cosmological disputes of the 1950s.

Each of the four major advances recounted above — the development of a robust theory of main-sequence stars (1924–38), the confirmation of the galactic nature of spiral nebulae and discovery of their distance-redshift relation (1924–29), the elaboration of a convincing scenario for the stellar genesis and dispersal of the elements beyond helium (1952–57), and the triumph of 'big-bang' cosmology with the serendipitous discovery of a cosmic background radiation and its interpretation as a remnant of the primeval fireball (1964–65) — has been followed up in interesting and significant ways. The Eddington-Bethe theory of main-sequence stars has served astronomers and physicists as a firm reference point in their investigations of stellar structure and evolution which, besides the important work on nucleogenesis in post-main-sequence stars, have come to include such fascinating entities as pulsars and black holes. Hubble's researches in extragalactic astronomy have played a foundational role not only in the development of 'big-bang' cosmology but also in studies of the distribution of galaxies and quasars. B^2FH's analysis of the origin of the elements has been an important starting point in investigations of nuclear processes in massive stars, possible supernova triggering of the solar system's formation, and the creation of deuterium and helium in the big bang. And the triumph of 'big-bang' cosmology has inspired increasingly subtle studies of the bearing of elementary-particle physics on the universe's earliest history and present structure.

Astronomy's Prospects

In all likelihood astronomy's prospects are good for the next few decades. Social support for the science should, at the least, keep pace with economic growth so long as the developed nations can avoid debilitating cold and hot wars or profound environmental and resource crises. Drawing on this support, astronomers should attract sufficient new talent to sustain the current rapid pace of improvement in the discipline's observational and theoretical toolkits. So armed, astronomers together with their recruits from physics should easily continue the process of following up on the key advances of prior generations of astronomers. They could do significantly more. They might identify the dark matter responsible for preventing the dissipation of galaxies, clusters of galaxies, and perhaps the universe. And, having recently found unambiguous evidence for planetary systems around other stars, they might even succeed in their search for extraterrestrial intelligence and in so doing completely transform our species' self-conception, worldview, and ways of life.

References

My thanks to astrophysicist Virginia Trimble whose responses to my original outline and first draft were not only sharp but constructive.

Further Reading

R. Berendzen, R. Hart, and D. Seeley, *Man Discovers the Galaxies.* (New York: Science History Publications, 1976).

A. Blaauw, *History of the IAU: The Birth and First Half-Century of the International Astronomical Union.* (Dordrecht: Kluwer Academic Publishers, 1994).

A.M. Clerke, *A Popular History of Astronomy during the Nineteenth Century*, 4th ed. (London: Adam and Charles Black, 1902).

D.H. DeVorkin, *Science with a Vengeance: How the Military Created the US Space Sciences after World War II.* (New York: Springer-Verlag, 1992).

R.E. Doel, *Solar System Astronomy in America: Communities, Patronage, and Interdisciplinary Research, 1920–1960.* (Cambridge: Cambridge University Press, 1996).

D.O. Edge and M.J. Mulkay, *Astronomy Transformed: The Emergence of Radio Astronomy in Britain.* (New York: John Wiley & Sons, 1976).

O. Gingerich (Ed.), *Album of Science: The Physical Sciences in the Twentieth Century.* (New York: Charles Scribner's Sons, 1989).

O. Gingerich (Ed.), *Astrophysics and Twentieth Century Astronomy to 1950: Part A.* (Cambridge: Cambridge University Press, 1984).

N.S. Hetherington (Ed.), *Encyclopedia of Cosmology: Historical, Philosophical, and Scientific Foundations of Modern Cosmology.* (New York: Garland Publishing, Inc., 1993).

K. Hufbauer, *Exploring the Sun: Solar Science since Galileo.* (Baltimore: The Johns Hopkins University Press, 1991).

K. Krisciunas, *Astronomical Centers of the World.* (Cambridge: Cambridge University Press, 1988).

K.R. Lang and O. Gingerich (Eds.), *A Source Book in Astronomy and Astrophysics, 1900–1975.* (Cambridge, Mass.: Harvard University Press, 1979).

E.R. Paul, *The Milky Way Galaxy and Statistical Cosmology, 1890–1924.* (Cambridge: Cambridge University Press, 1993).

R.W. Smith, *The Expanding Universe: Astronomy's 'Great Debate,' 1900–1931.* (Cambridge: Cambridge University Press, 1982).

R.W. Smith, *The Space Telescope: A Study of NASA, Science, Technology, and Politics*, 2nd ed. (Cambridge: Cambridge University Press, 1993).

O. Struve and V. Zebergs, *Astronomy of the Twentieth Century.* (New York: The Macmillan Company, 1962).

CHAPTER 33

Mathematics in the Twentieth Century

AMY DAHAN DALMEDICO

One can take the second international congress of mathematicians held in Paris in 1900 as the starting point for the study of mathematics in this century. About 200 mathematicians participated, representing some ten countries, mostly from western Europe. At the end of the century more than 4000 attended the 21st international congress held in Kyoto in 1990, or the 22nd held in Zurich in 1994. By that time more than 50 countries were members of the International Mathematics Union. The number of published research papers is now close to 90,000 annually and the mathematics community — people engaged in research in this discipline — comprises several tens of thousands of members distributed throughout the world. These figures apart, the impact of mathematics on industrial societies has increased enormously, and the crucial role played by the circulation and digitalization of information offers it another important field of application whose eventual impact we cannot predict. Throughout the century mathematics has always had the lion's share of education and all countries, whatever their educational systems, have chosen to give mathematics a decisive place in the curriculum and in the selection of elites, so that there has been great demand for teachers in this discipline.[1]

FROM 1900 TO THE 1930S, THE SUPREMACY OF THE HILBERT SCHOOL

Three countries, France, Britain and Germany, were the most influential in European mathematics at the end of the nineteenth century, to which should be added an Italy which was making great strides. These four countries published more than two-thirds of the hundred or so journals circulating in 1870. National mathematics societies were founded everywhere. A second phase in the organization of mathematics activities had just been inaugurated with the meetings of international congresses.

Two mathematicians dominated the Paris congress: the Frenchman Henri Poincaré (1854–1912) who was its chairman, and the German David Hilbert (1862–1943). The latter gave a paper on "the future of mathematics," which became very famous, and in which he spelt out 23 problems which ought, in his view, to shape mathematical research in the new century, and which in fact came largely to play this role, at least during the first half. It has been said that Poincaré and Hilbert were the last 'universal mathematicians' in the sense that they worked in or at least followed the development of almost all fields of mathematics. They were also the leading figures in the two most powerful national mathematics communities in Europe. That said, their legacy to the twentieth century proved to be very different.

Poincaré's subjects covered the classical fields of nineteenth century mathematics: the theory of functions, notably elliptic, differential equations, and celestial mechanics. He revitalized them completely by introducing the concept of the group, and using analogies taken from non-euclidean geometries. By this means he created fuchsian functions, later called automorphics, a field of study related to Lie group representations which is still very active today. Poincaré also invented *qualitative methods* for the study of dynamic systems. By combining local and general points of view, these explored the relative relationships and general behavior of trajectories, their stability, their complexity . . . and allowed one to look for mathematical solutions even when they were not quantifiable. It is from methods of this kind that we have been able to develop current theories of complexity and chaos. Poincaré applied his qualitative methods to celestial mechanics — the three-body problem above all — of which he was the greatest specialist after Newton, Lagrange and Laplace. Some of his findings remain unsurpassed today. Poincaré was interested above all, to use his own words, in "problems which presented themselves, and not those which one posed oneself." By this he meant that he was particularly concerned with major questions emerging from the natural sciences, and that he had little time for problems which were artificially fabricated by mathematicians. To resolve these 'objective' problems he created the necessary tools, and laid the foundations of entirely new disciplines like topology and algebraic topology. His mode of thought was above all *intuitive* and *geometrical,* in the very contemporary sense of the domination of geometry in all of mathematics and physics today.

Poincaré was a mathematician who mostly worked alone. He did not train students, he did not teach the subjects in which his own research was particularly active and novel, and he did not found a school. Of course his fame and prestige were exceptional: his colleagues at the university or at the Academy of Sciences recognized his outstanding scientific capability and they admired him, but they barely risked exploring the new fields he opened, Hadamard apart, they concentrated their efforts on the theory of functions. After his premature death, the First World War would considerably weaken the potential of French mathematics, notably through the loss of almost half the students at the Ecole Normale Supérieure, the cradle of future scientific talent.

In general one might say that during the first decades of the twentieth century the French mathematical community had a number of outstanding members, and produced important theories and results, but that these mathematicians did not found schools and did not have an institutional weight commensurate with their mathematical talent. One thinks here of people like H. Lebesgue and integral theory, or of P. Levy and probability theory. Elie Cartan (1869–1951), another great French mathematician, was also a rather isolated savant whose work was only recognized belatedly. Yet his research (on the calculation of external differential forms, on spinors, on Lie groups, and so on) changed modern geometry and was of great importance for theoretical physics.

Along with Poincaré, Hilbert was already a mathematician of exceptional importance. His reputation was built on many and very different researches.[2] His first breakthrough occurred in 1888 with the proof of the *existence* of a finite base in all invariant systems by showing that every contrary hypothesis led to a contradiction. "It is not mathematics, it is theology," his opponents objected. However this kind of proof, which enabled one to describe abstractly the structure of the objects considered without exhibiting them was, according to Hilbert, typical of modern mathematics. Shortly thereafter he worked on number theory for a report, the *Zahlbericht* commissioned in 1897 by the German mathematical society. He synthesized, simplified and unified the results obtained during the nineteenth century developing them further and laying the foundation for theories (algebraic geometry, homological algebra . . .) which would be used half a century later. He then turned to analysis and mathematical physics: Dirichlets' problem, calculus of variations, and integral equations, notably developing the theory of quadratics having an infinite number of variables. This was the beginning of Hilbert spaces and of spectral theory which were to become basic tools in physics. Hilbert published his *Foundations of Geometry* in 1899, which sanctioned in a stunning manner the advent of the axiomatic method. He classified the axioms of geometry which now became an abstract science whose relationship to reality was broken. He spoke of points, lines and planes but one could just as well, he said, use the terms tables, chairs, etc. All that mattered were relationships of logical dependence and compatibility. What Hilbert did for geometry was soon to be an ideal for all mathematics, and he taught the great majority of mathematicians to *think axiomatically.*

During the first decades of the twentieth century Hilbert turned his attention to logic, and notably to the consequences of the paradoxes identified by E. Zermelo and B. Russell in set theory. He undertook the simultaneous reformulation of the foundations of arithmetic and logic, creating a new discipline, "*metamathematics*" or the theory of proof, based on "finitisme." Gödels' proof, in 1931, for the indetermination of the axioms of arithmetic rendered some aspects of the finitist program null and void, forcing modifications and transfers, without however triggering a deep crisis in the discipline itself. There was rather a controversy over what should be considered the foundation of the discipline and a redefinition of its boundaries with logic and the philosophy of mathematics both of which were, for Hilbert, integral parts of mathematics.

The institutional location of Hilbert in the German mathematics community was very different to that of Poincaré in France. In particular he inaugurated the collective dimension of modern mathematics which has remained in place throughout the century. Arriving in Göttingen in 1895 at the request of Felix Klein (1849–1925) who represented a mathematical tradition somewhat different to his, rooted in the geometry of the nineteenth century which emphasized the role of intution and imagination however Hilbert built an alliance with Klein based on the shared conviction regarding the universal and predominant role which mathematics should have in the sciences.[3] Together they turned Göttingen into the international fatherland for mathematics for thirty years. The number of students in the mathematical sciences grew from 90 in 1892 to almost 800 in 1914. Ten to fifteen percent of them were sufficiently competent to benefit fully from the mathematical resources at Göttingen and to participate in research seminars. At the same time several research institutes in physics, applied mathematics and mechanics, electrotechnology and geophysics, all close to the Department of Mathematics, were established and financed by a consortium of industrialists and scientists set up at Felix Klein's initiative. The arrival of C. Runge and L. Prandtl in 1904 was decisive for the development of these sectors, even if, after World War I, the efforts to integrate pure and applied mathematics came to nothing, since, under the influence of Hilbert, mathematics underwent a profound and dramatic transformation.

Between 1895 and 1914 Hilbert supervised more than 60 doctoral students. He built a stimulating intellectual environment in his courses and seminars but also through more informal and convivial contacts (afternoon teas, weekly walks, etc). About 20 of the Privatdozents whom he welcomed to Göttingen went on to become very prestigious scientists: Hermann Weyl, Sommerfeld, Caratheodory, Max Born, Courant, Von Karman, Van der Waerden, Zermelo, etc. Minkowski, C. Runge, E. Landau and Emmy Noether (despite the difficulties encountered due to the prevailing mysogeny) were amongst the outstanding teachers who also emerged from this group.

The influence of the Hilbert school was global. It attracted young talent from France, where later the Bourbaki group was formed; from Italy; from the Netherlands (Van der Waerden); from central and eastern Europe (Von Neumann, Feller . . .); from Russia (Alexandroff, Ostrowski); from the United States (Norbert Wiener); from Japan (Takagi). No center in or outside of Germany could compete with the attraction of David Hilbert's Göttingen for young mathematicians. It was the point of departure for modern algebra (E. Noether, E. Artin, van der Waerden . . .), arithmetic (Hecke, Siegel), quantum mechanics (Born, Heisenberg, von Neumann), and for a school of logic (P. Bernays, W. Ackermann, G. Gentzen), which in its turn stimulated the research of Herbrand, Gödel and Tarski.

The 23 problems posed at the conference held in 1900[4] could be divided into six categories: algebra, number theory, analysis, geometry, algebraic geometry, and foundations of mathematics. Although this did not cover all Hilbert's foci of interest — e.g., topology and functional analysis were not included — the list offered a vision and a choice of what was important and 'basic' in mathematics, and this was progressively transformed into a veritable hierarchy of mathematical disciplines. The trilogy algebra, algebraic geometry, and number theory was at the summit of this hierarchy for much of the century with a clear dominance of problems at the intersection of several subdisciplines. Stress was thus placed on the internal interfaces of mathematics. Probability and statistics were totally absent, and had a long and hard struggle to be integrated into the institutional and intellectual corpus of mathematics.

In Hilbert's tradition, strongly rooted in axiomatics, set theory and modern algebra, every effort was made to see modern mathematics in structuralist terms. In the Hilbertian approach, here directly opposed to that of Poincaré, the good mathematician creates his problems, and either solves them or shows that they do not exist. However, in principle, the unknown does not exist in mathematics. This approach became increasingly fruitful and useful for physics, as several examples reveal: Riemanian geometry for Einsteinian relativity, matrix algebra and Hilbert spaces for quantum mechanics, and group theory for theoretical physics. In this way the development and the autonomy of mathematicians were both assured: the axiomatization and the abstractionism of theories extended their domain of relevance. From this point of view, the more the mathematical tool was 'purified' and turned in on itself, the greater became its potential for application. The philosophical, indeed ideological discourses of Hilbert and his colleagues on the discipline were perfectly coherent with their conviction and their practice, and they too were opposed to the relationship favored by Poincaré between mathematical thinking and reality. Hilbert was the editor in chief of the major journal *Mathematische Annalen* until 1930, and this ensured the dominance of his ideas throughout the international community.

Nevertheless, the hegemony of the Hilbert school was not total during this period. Other national mathematics communities existed; other research programs were developed, other results were obtained. I can only mention a few of them in the framework of this chapter. The examples I give from amongst many have been chosen either because of their subsequent importance, or because they are indicative of a robust line of research in parallel to the dominant tendency:

1. A very focused school of geometry emerged in Italy around the work of S. Séveri and Enriquez.
2. In England the field of refined analysis (special functions, inequalities, complex variables) became preponderant with the work of the mathematician Hardy and this remained the case for much of the century.
3. The mathematical study of Brownian motion, whose physical theory was explored by A. Einstein (1905) and J. Perrin (1909) was taken up by the American Norbert Weiner in 1923.[5] In 1933 the pathbreaking work of Kolmogorov on

the foundations of probability further reoriented the domain. At this time a number of individual contributions (like that of the Frenchman P. Lévy) enriched a field of cardinal importance today.

4. Extremely valuable results on statistics were obtained by the British researchers Karl Pearson and Ronald Fischer, notably for biometry.
5. Finally some early research on modeling was done in the 1920s and 1930s, notably in population dynamics (Volterra, Lotha), in population genetics (Wright, Fischer, Haldane), on the diffusion of epidemics (Kermack, McKendrick), etc. Among these Van der Pols' mathematical model of the heartbeat using a non-linear differential equation merits special mention.[6] The origins of mathematical modeling in economics are also to be traced to these decades.

The Time of Dramatic Changes

Hitler's establishment in power in 1933 and the pursuit of Jews in Europe by the Nazis ruined the celebrated center in Göttingen in a few months, dismantled the German school of mathematics, and also destroyed centers in Prague, in Budapest, in Poland, etc. There was a massive migration of mathematicians to the United States,[7] with H. Weyl, J. von Neumann and K. Gödel being among the most famous to go. Nor should we omit two other giants in mathematics in the USA who arrived earlier from the Soviet Union: Solomon Lefschetz and Oscar Zariski. Finally we should add a few French mathematicians who moved across the Atlantic during the Second World War: Léon Brillouin, R. Salem and André Weil.

These immigrants encountered a mathematical school in America whose leaders were George D. Birkhoff at Harvard and Oswald Veblen at Princeton, and which was noteworthy for its emphasis on abstraction. Even though US industry was rising to prominence at the time, very few American academic mathematicians were interested in applications or in contacts with industry in the 1920s and 1930s. All the same one should mention Max Mason and Warren Weaver, coauthors of a book on the electromagnetic field, and Vannevar Bush, a brilliant expert on the theory of electronic circuits and pioneer of the application of advanced mathematical techniques to the transmission of energy, a problem on which he had already collaborated with Norbert Wiener.

Applied mechanics and the mechanics of continuous media were almost ignored in the USA at this time, while in Europe L. Prandtl (1875–1953) had begun to build a school of fluid mechanics. Levi-Civita was particularly influential in this school and indeed the congress on theoretical and applied mechanics held in 1922 brought together scientists like von Karman, von Mises, G.I. Taylor, Southwell, D.M. Burgers, S. Goldstein and so on. The arrival of von Karman at Caltech's Guggenheim Aeronautical Laboratory in 1930 heralded the education in aerodynamics of a generation of experts who were to play a crucial role during World War II and in the American aerospace industry. Von Mises, another immigrant from Berlin, developed along with Bergman, Geiriger and Prager sophisticated methods for studying the mechanics of continuous media. In New York Richard Courant, formerly of Göttingen where he was Hilbert's student and colleague before being the director of the Mathematics Institute, arrived with two collaborators Kurt Friedrichs and Hans Lewy. Courant published a book with Hilbert in 1924 which was destined to become very well-known, the *Methoden der mathematischen Physik*, and he tried to establish a center of applied mathematics.

As the international political situation deteriorated the need to educate a number of applied mathematicians assumed a new urgency.[8] Summer schools held at Brown University proved decisive for the emergence of that which would become a group of applied mathematicians. The United States' entry into the war reshaped the course of events. Vannevar Bush was able to have a federal agency set up to control the Office of Scientific Research and Development. The scientists, many of whom were Jewish emigrés who had fled Hitler, collaborated massively and without hesitation with the military in research directed towards the war effort. The 'purest' mathematicians, or those most attached to basic research (like Hermann Weyl, Solomon Lefschetz, S. MacLane, S. Ulam and Garrett Birkhoff) changed their fields, or reoriented their goals, their work habits, and their criteria of rigor. In a situation of this kind the mathematicians added a number of new weapons to their usual arsenal of vigorous proofs: specific solutions, asymptotic descriptions, simplified equations, even experiments in laboratories. The question of the legitimacy of these methods was not asked at a time of urgency, and this facilitated their later generalization and banalization among applied mathematicians. The most important sites for this collaboration with the military were the government laboratories of the different sections of the armed forces (like the Aberdeen Ballistic Laboratory and Los Alamos) and also numerous laboratories situated at universities (the Radiation Laboratory at MIT, Berkeley, Brown and New York University). In 1942 the OSRD was reorganized and a new agency, the Applied Mathematics Panel (AMP) was formed. Directed by Warren Weaver, it was conceived as an organization for mathematicians who were to assist other scientists and the military in the war effort. The AMP was a contract organization whose practice continued after the war. It changed mathematicians, bringing them in closer touch with other scientists and opening a large variety of research problems.

The first major field to emerge was at the intersection between fluid mechanics, the mechanics of deformable media, and wave theory. A major study by Hermann Weyl on waves was followed, notably in New York, by considerable research on gas dynamics (notably supersonic), on the theory of explosions in air and water, and on shock waves. The book *Supersonic Flow and Shock Waves* written by Friedrichs and Courant and published in 1948, tried to develop a systematic theory of the propagation on non linear waves in relation to gas dynamics. Other work on submarine ballistics gave interesting results on orthogonal analytical functions and on the theory of functions of many complex variables. Finally there was important work done on aeroballistics, the theory of tracking curves (important for missile guidance) the theory of the control of explosions, problems of stabilization, and radar.

The second major mathematical field which expanded enormously was that of probability and statistics. For example, statistical models were developed to study the impact of fragmentation and to improve anti-aircraft defense. Norbert Wiener used his previous results from ergodic theory to formulate a statistically based predictive theory which enabled one to improve the chances of intercepting an aeroplane on the basis of available information on its trajectory. Similarly, at the Rad. Lab, there was research on the statistical properties of noise, and the detection of a signal in a background of noise. Abraham Wald developed sequential analysis, which was later to become of major importance. At Los Alamos the statistical approach to differential equations, to integro-differentials and to other analytical equations was developed within the framework of research on nuclear explosions. Ulam and John von Neumann published results on the Monte Carlo method and on the possible application of probability theory to hydrodynamic calculations.

Finally new disciplines came into being. Operations research was one of them. During the early years of the war, the kind of questions were how to optimize the placement of radar stations? What was the best way to dispose ships in a convoy in the war against enemy submarines? How to optimize the logistic management of the available human and material resources? Initial results were obtained in Great Britain, subsequently Philip Morse's group at MIT became very active.[9] Based as it was on probabilistic models, operational research explicitly introduced chance and uncertainty as intrinsic characteristics of the phenomena to be modeled. Von Neumann brought in methods from game theory: given specified targets, given too the specifications of the weapons to destroy them, how should the effort be distributed between the targets so as to inflict maximum damage? The game aspect entered the picture when it was assumed that both the attacker and the defender could act as they chose, and that the best approach for one depended on what the other did. In 1943 John von Neumann and Oskar Morgenstern published their *Game Theory and Economic Behaviour*, which started the mathematical study of decision-making. At the end of the 1950s operations research gave way to systems analyses, which dealt with more complex and more open problems, and which assumed a central place in economics and other social sciences.

There was something of a redistribution of the field immediately after the Second World War. The United States had become the world leader in mathematics: by the size of its scientific community and the variety of domains which that covered, and the dynamism of its university and research system. Only Great Britain, France and the Soviet Union could still hope to compete. Above all though, a far greater field of research than before was now open to mathematicians, and the possibilities of cross-fertilization with other scientific and technical disciplines were immense.

FROM THE 1950S TO THE 1970S

The first post-war international congress of mathematicians was held in 1950 in Cambridge (Massachusetts). Its proceedings provide one with a rather good idea of the new and now enlarged space occupied by the discipline. Mathematicians like those whose wartime activities we have just described — von Neumann (theory of shock waves), N. Wiener (theory of the statistics of prediction), C. Shannon (information theory), A. Wald (statistics) — were invited to present their latest results alongside speakers on pure mathematics, algebraic geometry (O. Zariski, A. Weil), theory of analytic manifolds (H. Cartan), homotopy (W. Hurewicz), algebraic groups (C. Chevalley) whose work could be situated in the noble tradition of the most prestigious mathematical sub-disciplines. That said, one notices, in the successive four-yearly international congresses a clean 'take-over of power' by structural Hilbertian mathematicians. The number of sessions on, for example, algebra, algebraic geometry, and algebraic topology increased from congress to congress, while the number of sessions dedicated to mathematical physics, to statistics and to applied branches fell regularly. Large parts of classical analysis and differential equations were considered as more or less exhausted of interest. The interactions between mathematics and the social sciences disappeared from international congresses. Of the 22 Field medals awarded between 1950 and 1978 four were for number theory,[10] seven for algebraic geometry, five for differential topology and algebra, while only three were for analysis[11] and there were none for probability theory. The domain associated with partial differential equations was only given the medal in 1994!

Mathematicians generally explain their transformation by appealing to the internal dynamics of mathematics itself. In their view, abstraction and formalization were necessary steps for progress in most fields. Indeed we note that the truly *algebraic* character of mathematics — construction of algebraic tools, use of algebraic methods — rose irresistibly to prominence between the 1930s and the late 1950s. By contrast, during the subsequent period we find a progressive return to *geometrical* language and intuition, and this up to the present day, when people now speak of the universal domination of geometry, even in the fields of analysis, topology or statistics.

These changes were beautifully illustrated by Hassler Whitney, one of the main proponents of algebraic topology, in a history of his discipline written in 1962.[12] This, he said, had achieved a certain measure of success in the 1930s with the work of S. Lefschetz, P.S. Alexandroff and H. Hopf. The sudden arrival of the theory of cohomology in 1935 raised a new set of problems. The powerful and generalized algebraic machinery of the 1940s enabled one to treat problems which had seemed too complex and without solution, while the new mathematical techniques also expanded in themselves. New concepts like exact series, schemas and homological algebra were developed with enormous success, bestowing the appearance of pure algebra on the topics of topological algebra. At the end of the 1950s a number of major discoveries of a geometrical kind, like R. Thom's theory of cobordism were made. A vast new field was opened up, that of differential topology. A few years later, Whitney goes on, these tools had completely transformed the field and a new sensitivity to 'useful' problems began to develop again.

Similar changes, more or less dispersed in time, can be seen in other domains like algebraic geometry (with, notably, a long period for the fabrication of very general and abstracts tools – J. Leray, A. Weil, J-P. Serre and A. Grothendieck) functional analysis, or harmonic analysis. However was this explosion in the algebraic and formal aspect of disciplines (recognized by many mathematicians themselves) necessarily inscribed in the internal dynamics of the discipline itself? Or was it rather one consequence of a vision of the architecture of the mathematical corpus which was inherited from the Hilbertian tradition and subsequently taken up and accentuated in the middle of the twentieth century, notably by the mathematicians of the Bourbaki group?

We believe that the internal maturation of mathematics cannot alone explain what is also a consequence of a choice (political, institutional and intellectual) of hierarchy of values. The case of the French school of mathematics is particularly interesting here. At the 1950 congress its importance was recognized by the award of the Fields medal to Laurent Schwartz for his research on distributions, and by the fact that twelve of the invitees were French. The young founders of the Bourbaki school, who began to establish themselves in the Germany of the 1930s, were fully mature and productive. Henri Cartan reined from 1940 to 1966 at the Ecole Normale Supérieure in rue d'Ulm where he attracted some outstanding minds. Emulators and disciples became some of the greatest mathematicians of the period (J-P Serre, R. Thorn and A. Grothendieck all received Fields medals), and all took university posts in the 1950s. And the preferred subject of the 'Normalians,' that choice which was implicitly recommended from the middle of the 1950s to the early 1970s, was algebraic geometry. Other fields which were also legitimized were number theory, the theory of groups and automorphic forms, and algebraic and differential topology in which R. Thom in particular excelled.[13] A discipline as fundamental as probability was marginalized, albeit in higher mathematical education, in research, and the institutions of the professional community. The same can be said for branches like statistics and numerical analysis.

Jean Dieudonné (1906–1992), who was an emblematic figure of the French mathematical school, made explicit and theorized this ideology of pure, abstract and structural mathematics in what he called "*Bourbaki's choice.*"[21] The more abstract a theory was, the more it could offer to intuition since it had then eliminated all contingent, that is to say concrete, aspects. One can visualize the serious consequences of such a dogmatic position when it is implemented, for example, in reforms of the so-called modern mathematics in secondary education. The philosophy of mathematics corresponding to this conception sees the discipline as "*a reservoir of abstract forms*" into which reality, more or less miraculously, moulds itself *a posteriori*.[15] The 'profound,' 'brilliant' mathematician is thus invited to turn his back on the real (i.e., problems emerging from the natural sciences) these being more concrete branches having direct applications.

The situation in the United States was different and far more open. The diversification of the mathematical community and the decentralization of the decision-making structures were sufficient to ensure that a number of different schools could co-exist. At the Institute for Advanced Study at Princeton, or Harvard, for example, exceptionally talented pure mathematicians predominated, while the University of New York's Courant Institute was a very active bastion of applied mathematics, specializing in the domain of partial differential equations and numerical analysis. Here too outstanding progress was made on the dynamics of fluids and magneto-fluids, including the use of computers. This research, on the one hand, provided engineers with the conceptual tools needed to control the technological systems involving dynamic flow (turbines, pipe-lines, aerodynamic shapes, etc.) and, on the other, it gave to theoreticians the keys to the understanding of the possible behavior of fluids, and the study of the evolution of solutions to equations like those of Navier-Stokes, etc.

An excellent American school of probability also emerged whose three leaders were J.L. Dobb (general theory of martingales), W. Feller (theory of Markov processes), and M. Kac (spectral methods in probability). This school was at the interface with the physicists. Nor should one forget the work done on stochastic differential equations by the Japanese K. Ito, in close collaboration with the Americans. Abraham Wald developed decision theory, and so established a link between von Neumann game theory and the earlier work of Neyman and Pearson. Indeed Neyman, who originally worked in England, developed a very famous school of statistics at Berkeley.

Since the eighteenth century we have been able to distinguish two motors of development in the history of mathematics: one related to problems in mechanics and physics, the other due to the internal exigencies of the mathematics itself. We have seen though that the political and social context of World War II, the demands of the military - industrial complex which it created, and the technical needs which it engendered, served in the USA as a powerful driver generating new fields and new problems. At the same time it must be said that in the post-war period the international community of mathematicians, at least in the West, made every effort to ignore or to marginalize these topics. It undoubtedly privileged, from the 1950s to the 1970s, internal interfaces between mathematical disciplines, which led to a substantial introversion of the mathematicians themselves.

The Mathematical Strength of the Soviet Union

Mathematicians were extremely productive in Russia and the Soviet Union throughout the twentieth century and, as far as we know, the development of their research does not follow the periodization that has been adopted earlier. The collapse of the Soviet regime, the enormous economic difficulties in the country, and the massive exodus of the most important scientists to the West make one wonder whether this brillant tradition can survive. The reason why mathematics and mathematicians as a whole enjoyed a privileged position are many and various, and still not properly understood.

From the dawn of the century Russia had a tradition of mathematical and scientific schools in St Petersburg, Moscow and later even in Kiev, Gorki, etc ... Those persisted from generation to generation and produced a continual flood of diverse results of the highest quality. The oldest was the Moscow school of function theory founded by Egorov (1869–1931) and his brilliant student N.N. Lusin (1883–1950).[16] Around them an active group, called the "Lusitania" was built up, amongst whose members we should mention A.Y. Khinchin, M.Y. Suslin and, above all, Lusin's students, P.S. Alexandroff and A.N. Kolmogorov, great friends who were to become leaders of two very important mathematical schools.

P.S. Alexandroff (1896–1982) who traveled successively in the 1920s to Göttingen, to Paris with Urysohn and to Princeton, was one of the founders of topology. He then turned his attention to algebraic topology. His students in Moscow were V.V. Niemytzki, L.A. Tamarkin, A.N. Tychonov and L.S. Pontragin. His collaboration with H. Hopf resulted in a joint publication in 1935 which played a crucial role in diffusing new ideas in topology. The schools of Lusin and Suslin also made important contributions to the field of Fourier series and analysis, while the school of Guelfond and Shafarevitch excelled in number-theory.

The Russian school was particularly original and fruitful in the area of dynamic systems. A.A. Andronov was an heir of both Poincaré (rather forgotten from this point of view in the West) and of Lyapunov, and during the 1930s he transferred their results on Hamiltonian mechanics to the field of dissipative systems. He developed qualitative methods for studying non-linear oscillations of actual systems encountered by scientists and by engineers (the stability of mechanical systems, vibrations and resonance phenomena, electric circuits, etc.). This was also the case with N.M. Krylov and N.N. Bogoliubov who concentrated on quantitative and asymptotic methods for nonlinear physics. Andronovs' school worked, with Neimark and Mitropolski on the theory of control and automation in the post-war period. These results, which were important for the development of Soviet technology (stabilization of tanks, missile guidance) were avidly studied and translated in the United States by Solomon Lefschetz at Princeton. From the end of the 1940s he piloted a research program on differential equations, a domain that was then dormant, and which the US agencies hoped to reactivate.[17]

A.N. Kolmogorov (1903–1987) was one of the greatest mathematicians of the twentieth century. His output was immense, both on fundamental and applied topics. During the 1920s he worked on function theory, set theory and geometry. During the 1930s it was the turn of the axiomatization of probability, work on chance functions, Markov chains and stochastic functions. In the 1940s he was remembered for statistical theory and turbulence, while in 1954 his famous theory subsequently baptized the Kolmogorov-Arnold-Moser (KAM) theory on the localization of the orbits of perturbed Hamiltonian systems was developed. Nor should we omit information theory, the foundations of mathematical logic, research on algorithmic complexity.[18] Kolmogorov's work, as he has suggested himself, is divided into two main areas: mathematics and mechanics on the one hand, probability and information theory on the other, the overriding objective being to grasp the essence of the concepts of order and chaos. This was a vast program whose ambitious nature is comparable only to that of Hilbert's, but whose spirit is fundamentally different. The Kolmogorov school (Manin, Arnold, Novikov, Sinaï, etc.,) has also had an exceptionally high output and has done work of the first order.

It seems that the structure of Soviet society favored abstract sciences, like mathematics, which were beyond politics, which required no particular logistic or technological infrastructure, and whose internal development was not controlled by the authorities. This autonomy vis-à-vis the official ideology conferred a privileged status on mathematics and made it particularly attractive for students undertaking a scientific career. What is more there was a long-standing and popular tradition of training young mathematicians. The first olympiads were held in Leningrad in 1936, while they took place in Moscow the year after. Indeed the system functioned for some forty years without a real official character, thanks to the enthusiasm of young mathematicians. They organized themselves into more or less informal 'mathematical circles,' which were situated in most towns to attract the most gifted youth.

In the 1960s, the years of Khruschevian liberalism, mathematicians grasped the new opportunities of attracting young talents to the mathematical profession: elite schools of mathematics and physics for adolescents, tours in the provinces to recruit the most gifted on the occasion of the local olympiads, correspondence schools. No chance was lost. Courses were given by the most influential mathematicians: Kolmogorov, who was personally engaged in all these initiatives, Lavrentiev, Guelfond, etc. The physicist I. Kikoin was also heavily committed. All of the mathematicians of repute today, who did their studies in the 1960s, passed through this system.

Another factor contributing to the popularization of good quality mathematical research was the journal *Kvant,* founded in 1969, to which the most important whom we have already mentioned contributed, and which had the extraordinary circulation of 360,000 copies! This golden age was not to last for long, however. The entry of the tanks into Prague and the hardening of the Brezhnev regime were accompanied by an upsurge of anti-semitism in universities and in research, by abrupt changes in administrative directives, and by various 'clean-ups.'

THE RETURN TO THE 'CONCRETE'

From the beginning of the 1970s, the general picture in mathematics gradually began to change, mostly in response to the new scientific and technical demands of contemporary society. Fields which had lain dormant for many decades were revitalized the hierarchy between different sectors of mathematics was reorganized and new domains of research emerged.

As L. Carleson has noted,[19] for about twenty years the rumors that classical analysis was dead were widespread; and the future of this discipline, one of the oldest in mathematics, and one closely connected to applications, seemed to reside in branches which were increasingly abstract and generalized. Then suddenly, beginning in the 1970s, harmonic analysis, in conjunction with the theory of martingales and that of Brownian motion made a dramatic come back. Mathematicians set about *analyzing* increasingly complex mathematical objects, for which periodicity and invariance in translation no longer existed. This involved solutions to equations of linear or non-linear partial derivatives, functions of several complex variables, and operators derived from formal algorithms whose robustness and efficacity was to be proven. Alongside the traditional Fourier series the toolbox was extensively diversified and sophisticated with various decompositions, atomic, molecular, and wave packets, and wavelets, etc. Harmonic analysis — this "leading edge technology" — was found to be extraordinarily efficient while the "huge ideological machines", to quote Y. Meyer's criticisms of abstract and structural mathematics,[20] became obsolete, notably for the study of chance objects as demanded by contemporary physics and chemistry.

Another field of analysis, at the intersection between pure mathematics, industrial applications and informatics, made considerable progress: numerical analysis in liaison with partial derivative equations and their methods of approximation. In effect most of the phenomena encountered in applications (calculation of structures, elasticity, aerodynamics, etc.) or in macroscopic physics (new materials, liquid crystals, plasmas, etc.) were modeled by such equations, usually non-linear. Their solution called for the use of huge computers and is known today as scientific calculus. The contribution made by the French school of applied mathematics, developed since the 1970s under the direction of Jacques Louis Lions, was essential for this.[21]

In the last three decades the theory of dynamic systems has become one of the most important in mathematics, in interaction with physics, engineering sciences and biology, and is at the heart of theories of complexity and chaos. Whereas it was originally confined to studying differential equations, its scope was progressively enlarged to cover group action, and iterative applications. A number of new areas emerged in a spectacular way along with this: fractals, objects having auto-similar properties (i.e., invariant under a change of scale) deterministic or stochastic, the study of barely regular structures and of singularities. Chaos theory, born of the confrontation between the theory of dynamics systems and computer technology, covers all these objects and touches a large number of domains: celestial and ballistic mechanics, turbulence, percolation, polymers, catalysis, meteorology, and econometry.

Several currents converged to make probability one of the most important domains in contemporary mathematics. One flowed from the tools of analysis needed for the statistical investigation of complex phenomena: another from the problems posed by various modelings, theory of signals, of detection, of filtering and of control; queuing problems and operational research which represented most computer and telematic systems; quantum theory and particle systems from physics. In particular the increasing importance of statistical physics, i.e., the study of the macroscopic properties of systems comprising a very large number of interacting microscopic systems, also stimulated the development of probability theory which entered, through ergodic theory, into problems of diffusion, and percolation. Chance became omnipresent in mathematical modeling, be that in the models themselves, in the acquisition of data (size of the sample), or in measurement errors. What is more, models which introduced chance were found, paradoxically, to be very efficient, and able to deal with their own weaknesses, (supraparameterization, for example), simply by manipulating the laws of probability. Indeed today we can say that, from physics to economics, from lotteries to sociology, in all domains of the natural, human and social sciences, we turn to probability to model phenomena.

Finally, even traditionally the 'purest' and the most structural of disciplines, like algebra, number theory, or algebraic geometry have not escaped from this 'return to the concrete.' Algebra has become a basic tool for coding, compressing, and correcting errors in the transmission of information. A class of highly performant correcting codes has been built with the help of arithmetic and algebra, using rational fractions on algebraic curves. Number theory (prime numbers, p-adic numbers . . .) is crucial in cryptography and in research into the reliability of computers.

As for geometry, it has found its relationship with physics (and with mechanics), which has a long and rich history, intensified even more. On the one hand there has been a clear shift of interest to overarching problems of differential geometry which have their counterpart in physics. In the 1970s, in particular, when one regarded the 'Standard Model'[22] as the basic framework for unifying electromagnetic, weak and strong interaction in the nucleus, symmetries and geometric structures played an important role in the understanding of the physics of elementary particles. On the other hand, 'images' have become privileged objects of study both as regards the technical problems surrounding the treatment of images, and for the power for modeling in very different fields: for example the chemistry of complex molecules and alloys.

In 1966 René Thom wrote: "It must be admitted that classical problems, like Fermats' theorem or Riemanns' conjecture can wait a few more years to be solved;

but if science wants to use the differential tool to describe natural phenomena, it cannot ignore for much longer the topological structure of the attractors of a structurally stable dynamic system since every 'physical state' has a certain stability, a certain permanence and is necessarily represented by an attractor . . .'[23] Thirty years later mathematicians did not have to choose, they had achieved two things. Fermat's theorem had just been dismantled by Andrew Wiles, crowning three centuries of profound and very difficult mathematical research! And the study of attractors foreign to dynamic systems had made great strides, as had that of many other tools for understanding the natural and social sciences.

Two conclusions can be drawn from this survey of mathematics in the twentieth century. Firstly, today nothing seems likely to put a brake on the expansion of mathematics and its interface with a range of human activities: and this even though mathematics, like most sciences, can only reply to the precise, narrow, and perfectly well formulated questions asked of it. Indeed one is always tempted to take a simplified and purified model, a cut through the thickness of the real, for reality itself. And to spend all one's time trying to understanding this model better and better, breeding the illusion that one is increasingly more able to grasp the world as a whole. Secondly, mathematics, whose 'purity' is so often appealed to, has nevertheless been subject, in a particularly brutal way, to the shock of factors which were, *a priori*, external to it, and in particular the major political dramas of this century (the emergence of nazism, the Second World War and, most recently, the collapse of the Soviet Union) which hinder us from having an exclusively introspective vision of its future.

Translated by John Krige.

References

1. J.P. Bourguignon. "Un siècle de mathématiques" in *Les Polytechniciens dans le Siècle*, J. Lesourne (ed.). (Paris: Dunod, 1994).
2. H. Sinaceur and J.P. Bourguignon. "David Hilbert et les mathématiques de XXème siècle", *La Recherche*. (Paris: September 1993).
3. D. Rowe. "Klein, Hilbert, and the Göttingen Mathematical Tradition" *Osiris*, 2nd series, (1989), **5**: pp. 189–213.
4. J.M. Kantor. "Hilbert (Problemes de)" in *Encylopaedia Universalis* (1985), **vol. XI**: pp. 412–418.
5. J-P. Kahane. "Des séries de Taylor au mouvement brownien, avec un aperçu sur le retour" in *Developments of Mathematics. 1900–1950*, J-P. Pier (ed.). (Birkhauser, 1994), Basel.
6. G. Israël. "The Emergence of biomathematics and the case of population dynamics. A revival of mechanical reductionism and darwinism" *Science in Context* (1993), **6, 2**: pp. 469–509.
7. L. Bers. "The migration of European Mathematicians to America" in *A Century of Mathematics in America*, P. Durren (ed.). *American Mathematical Society, History of Mathematics*, **vol. 1**: pp. 231–243.
8. A. Dahan Dalmedico. "L'essor des mathématiques appliquées aux Etats-Unis: l'impact de la seconde guerre mondiale", in *Revue d'Histoire des Mathématiques* (1996), **2(2)**: pp. 149–213.
9. S. Schweber and M. Fortun. "Scientists and the legacy of World War II: the case of Operations Research", *Social Studies in Science* (November 1993), **vol 23(4)**: pp. 595–642,.
10. Selberg (1950); K.F. Roth (1958); A. Baker (1970); Bombieri (1974).
11. Schwartz (1950); Hörmander (1962); Fefferman (1978).
12. Speech made by J. Milnor at the prizing giving ceremony for the Fields medal. Proceedings of the international Congress of Mathematicians. (Stokholm, 1962), p. xlviii.

13. M. Andler. 'Les mathématiques à l'Ecole normale supérieure au XXème siècle: une esquisse" in *Ecole nomale superieure, le Livre du Bicentenaire.* J-F. Sirinelli (ed.) (Paris: P.U.F., 1994).
14. *Panorama des mathématiques pures.* (Paris: Gauthier-Villars, 1977).
15. N. Bourbaki. "L'Architecture des mathématiques" in F. Le Lionnais (ed.) *Les Grands courants de la pensée mathématique.* (Paris: Lib A. Blanchard, 1962), pp. 46–47.
16. S.S. Demidov. "The Moscow School of the Theory of Functions" in S. Ziravkovska et P. Durren (eds.) *Golden Years of Moscow Mathematics.* History of Mathematics, vol. 6 AMS. (London Mathematical Society, 1991).
17. A. Dahan Dalmedico. "La renaissance de systèmes dynamiques aux Etats-Unis après la deuxième guerre mondiale: l'action de Solomon Lefschetz" *Supplemento ai Rendiconti del Circolo Matematico di Palermo,* Serie II, N. 34 (1994) and A. Dahan Dalmedico, "Le difficile héritage de Henri Poincaré en systèmes dynamiques" in J.L. Greffe, G. Heinzmann, K. Lorenz (eds.) *Henri Poincaré, Science et Philosophie,* pp. 13–35.
18. S. Diner. "Les voies du chaos dans l'école russe' in A. Dahan Dalmedico, J-L. Chabert, K. Chemla (eds), *Chaos et Déterminise Le Seuil* (1992).
19. Speech made at the prize giving ceremony for the Fields medal by C. Fefferman. Proceedings of the International Congress of Mathematicians. (Helsinki, 1978), vol. 1, p 53.
20. C. Houzel (ed.) *Rapport de Prospective en Mathématiques.* (Paris: CNRS, 1985).
21. A. Dahan Dalmedico. "Polytechnique et l'Ecole française de mathématiques appliquées" in B. Belhoste, A. Dahan Dalmedico, D. Pestre and A. Picon (eds.) *La France des X.* (Paris: Economica, 1995).
22. S. Schweber's article in this volume.
23. Speech made at the prize giving ceremony for the Fields medal by S. Smale. Proceedings of the International Congress of Mathematicians. (Moscow, 1966), vol. 1, p 28.

Further Reading

The work of Hélène Gispert and especially her "La France mathématique (1870–1914), Cahiers d'Histoire et de Philosophie des Sciences. (Paris: 1992).

H. Sinaceur. "Hilbert. Du formalisme à la constructivité: le finitisme", *Revue internationale de Philosophie* (4/1993), vol 47, n. 186.

See H. Mehrtens. *Moderne-Sprache-Mathematik. Eine Geschichte des Streits um die Grundlagen der Disciplin und des Subjekts formaler Systeme.* (Berlin: Suhrkamp Verlag, 1990).

Constance Reid. *Hilbert,* Springer-Verlag, 1st edition. (New York: 1970).

H. Sinaceur. *Corps et Modèles, Essai sur l'histoire de l'algèbre réele.* (Paris: Vrin, 1991).

The documents prepared for *Notices of the American Mathematical Society* (February 1993), vol 40, N. 2, in particular the article A.B. Sossinksy: 'Russian Popular Math Traditions — Then and Now".

CHAPTER 34

Material Culture, Theoretical Culture and Delocalization

PETER GALISON

It was 1993, and superstrings, the 'theory of everything,' was the rage. Arthur Jaffe, a senior member of the physics department at Harvard and for several years chair of the mathematics department, along with Frank Quinn, a mathematician at Virginia Tech, penned the following in the pages of the *Bulletin of the American Mathematical Society:*

> Theoretical physics and mathematical physics have rather different cultures, and there is often a tension between them. Theoretical work in physics does not need to contain verification or proof, as contact with reality can be left to experiment. Thus the sociology of physics tends to denigrate proof as an unnecessary part of the theoretical process. Richard Feynman used to delight in teasing mathematicians about their reluctance to use methods that "worked" but that could not be rigorously justified.[1]

Jaffe and Quinn quickly added that mathematicians, unsurprisingly, retaliated: as far as they were concerned, physicists' proofs carried about as much weight as the person who claimed descent from William the Conqueror . . . with only two gaps. Nor has tension between cultures been restricted to the axis of theory/ mathematics. Albert Einstein and Paul Dirac famously derided putative experimental refutations of major theories, and experimentalists have never hesitated to mock what they considered to be the aimless speculation of theorists. One cartoon, widely circulated in the physics community during the 1970s, portrayed a balance scale with thousands of offprints labeled "theory" heaped on one side, outweighed by a single paper marked "experiment" on the other. Beneath these cross-currents of jibes and jests lie substantive disagreement about what constitutes an adequate demonstration, and, ultimately, a clash over whose pilings sink sufficiently deep to stabilize further construction. What vouchsafes knowledge, and for whom?

Like Jaffe, I find it useful to talk about the difference in cultures between the interacting groups that participate in physics. In fact, as we look around the national and international laboratories — now and throughout the last two centuries — the diversity of such cultures is striking: there are electronic engineers, cryogenic engineers, experimenters, computer programmers, field theorists, phenomenologists, just to name a few.

What motivates talk of 'cultures' or perhaps better, the subcultures of physics? Part of the appeal of the distinction between cultures is driven by historical concerns. To address the question 'Why does this happen there and then?' we want to identify affiliations between certain activities inside the walls of the laboratory and others outside, all the while being careful not to exaggerate the distinction between 'inside' and 'outside.' The world of the electrical engineer fashioning circuitry for the central tracking detector at a major colliding beam detector is a world apart from that of the theorist who may eventually be a consumer of its data. By contrast, the electrical engineer may well share a world with other electrical engineers also concerned with shielding their delicate printed circuits from massive pulses of X-rays — engineers, for example, preparing electronic devices to survive a nuclear battlefield. A condensed matter theorist may have more to say to a quantum field theorist than to his own condensed matter experimentalist colleagues as they struggle to lay atom-thin films of metal, or build new ceramics.

Anthropologists generally understand the cultural to embrace not only social structures *per se*, but crucially the values, meanings, and symbols associated with them. Now it is true that the term 'culture' has always and continues to reside in disputed anthropological territory. Clifford Geertz, Marshall Sahlins, Grananath Obeyesekere, for example, sharply disagree on how constraining, how overarching a 'culture' is. But the important point for the historico-philosophical characterization of the production of science is that we cannot pretend that meanings, values, and symbols are mere window dressing. When a mathematician derides a computer-based demonstration as a horrendous violation of the very idea of mathematics; when a theoretical physicist recoils from renormalization, pronouncing it a 'trick'; when an experimenter asks, in shocked tones, if future generations of experimentalists will get their data from 'archives' rather than through the concerted application of screwdrivers, soldering irons, and oscilloscopes — at these and other moments like them, *values*, always present, have surfaced. Meanings too can differ: when theorists speak about a particle, say an electron, they may, through usage, deploy a concept quite distinct from the usage of 'electron' spoken of by an experimenter.

It is useful to separate subcultures of physics on more philosophical, specifically epistemic grounds. We can ask: what is it, at a given time that is required for a new particle or effect to be accepted among theorists (in contrast to the requirements that must be satisfied among experimentalists)? This can be stated more precisely: what, at a given time and place, are the *conditions of theoreticity*? What is it that a theory must exhibit for it to count as reasonable even before it faces new experiments? From time to time these requirements change, and we can specify the circumstances of these alterations. A theory in many branches of physics is not out of the starting gate if it is not relativistically invariant, if it does not conserve charge. Such constraints may not be forever. For generations, *conservation of parity* was such a rigid requirement. (Conservation of parity is the demand that any process allowed by physical law ought also to allow a mirror-reflected image of the same process.) Parity, along with time reversibility, fell as absolute demands, leaving behind only approximate conservation laws. Renormalizability (the demand that a theory be constructed with a fixed and finite set of parameters that can then be used to predict to arbitrary precision) was another such broken constraint. Thought to be a rigid and exact stricture on theory in the early 1970s, by the 1980s renormalizability too reappeared as an approximate constraint.

In a similar way, we can speak of *conditions of experimentality*, focusing on the constraints that allow (or disallow) forms of laboratory argumentation. How are probabilistic arguments to be treated? Would a single instance of an event be considered a persuasive demonstration? Are witnesses necessary to secure experimental closure? Do experimental results without error bars count? Or, if results do include errors, how much statistical power is demanded? How much and what kind of knowledge can be deferred to other fields, literally or figuratively 'black-boxing' components parts of the experiment?

Conditions of instrumentality can also be distinguished as those constraints that delimit the allowable form and function of laboratory machines themselves. These conditions governing the material culture of physics may be of different temporal structures: there are broad, long-lasting classes of instruments (picture-producing instruments or statistic-producing instruments). Such classes provide constraints of the longue durée, as they set out the conditions under which (for example) picture-producing apparatus will be judged distortion free, or by which statistics-producing instruments will be assessed as having a certain loss rate. Then there are middle-term conditions on 'species' of instruments that may achieve legitimacy as knowledge facilitating objects (bubble chambers or optical telescopes or spark chambers). And at the short-range of temporality there are the individual tests and conditions that certify individual instruments, *this particular* bubble chamber, or even more specifically this particular bubble chamber and its production of *this particular* bubble chamber photograph. All such conditions of possible theorizing, experimentation, and instrumentation are temporal: they change with time; the dynamics of those changes constitute some of the most interesting and difficult questions facing the study of science.

Taken together, the historical, anthropological, and philosophical concerns suggest that periodization is a far more complicated business than might be suspected from older models of the philosophy of science. We ought at least look to see if the rhythms of change in one domain (experiment, for example) are the same as that of another (theory, for example). That is, instead of assuming that theory, instruments, and experiments change of a piece in one great rupture of 'conceptual scheme,' 'program,' or 'paradigm,' we would do better to see how the various practice domains change, piece by piece. Dates of conceptual breaks such as 1905 (special relativity), 1915 (general relativity), 1926 (non-relativistic quantum mechanics), and 1948 (quantum electrodynamics) may have been points of discontinuity in theory; they were not so in the development of the material culture that surrounded instrumentation and experimentation. And while there may be good reasons not to jettison widely accepted experimental practices at precisely the same moment the community is entertaining the radical reformation of theoretical practice, none of this is to say that co-periodization *cannot* occur. But it is to say that co-periodization ought be shown, not assumed to follow from the dictates of antipositivist philosophy of science in any of its forms.

An objection springs to mind. The separation of theory and experiment holds good in many branches of twentieth-century physics. Certainly this sociological division is so in atomic, cosmic-ray, nuclear, and particle physics, but also astrophysics, planetary physics, plasma physics, condensed matter physics. In each of these domains separate societies, meetings, and reprint exchange networks have long existed. But what of areas of inquiry where the separation is incomplete or nonexistent? What of the physics of the broad middle of the nineteenth century where a James Clark Maxwell, a Heinrich Hertz, or a Lord Kelvin could hardly be classified as a pure theorist or a pure experimentalist? And what of whole domains of other kinds of science, biology, to take perhaps the most powerful example? Does it make sense to speak of separate cultures of experimentation and instrumentation in such instances? The question could be rephrased: are there clusters of practices in experimental work tied to practices outside the laboratory differently from the way theory is situated? An example might be found in the introduction of nuclear magnetic resonance just after World War II. At least for Robert Pound and Edward Purcell, though they worked both theory and experiment, their theoretical efforts drew on classical electrodynamics and basic quantum mechanics — both long since established — while their instruments and procedures drew heavily on then-recent wartime radar developments. I would argue this: there is no universal answer to the question of whether it pays to speak of distinct cultures of theory and experiment in the absence of sharp sociological lines between the groups. Everything rides on how bundled together certain practices are — and that cannot be settled in advance, but only in the thick of historical inquiry.

The periodization picture sketched here might be represented in Figure 34.1, designating *intercalated* practice clusters. Looked at more finely, even 'theory' ought to be broken up in a similar arrangement, with some forms of theoretical practice lasting for the long term, while others are of shorter duration. These same considerations hold good for instruments and experiments. Such a scheme contrasts directly with two others. In one (Figure 34.2), denoted in a too-rough designation as 'positivist,' the view is that observations build aggregatively and continuously, while the level of theory breaks seriatim. Because observation builds cumulatively and intertheoretically, many followers of logical positivism and logical empiricism held it in special esteem.[2] Theory in a sense builds on the foundation of observation, and numerous metaphorical systems have designed to capture this primacy of the neutral observation language. In the other metaphorical scheme — designated (also crudely) as 'antipositivist' — the scheme of Figure 34.2 is stood on its head: now instead of viewing a neutral observation language as primary and theory as secondary, the reverse is true (see Figure 34.3). 'Theory' is everywhere, as Benjamin Whorf, Thomas Kuhn, Paul Feyerabend, N.R. Hanson and so many others taught us. To enforce the notion of a conceptual scheme, antipositivists assumed the equivalent of the co-periodized picture of Figure 34.2. When theory changed, it precipitated a break in meaning that extended 'all the way down.'

Suppose we stay with the argument presented so far, conservatively focusing on that sector of twentieth-century physics in which the separation of cultures of theory, experimentation, and instrumentation are reasonably well defined. Still, a new and more serious objection arises: if the culture of theory, for example, is really all that distinct in its contextualization, meaning, values, and argumentative structure from that of experiment, how do the domains relate at all? Reformulating the problem we could say this: to escape from the problems of noncommunication that arise from the confrontation of various block-periodized 'conceptual schemes' we moved to specify more local, intercalated subcultures of physics. Doesn't this just multiply the original problem, leaving us with hundreds of incommensurabilities, ruptures, revolutions, or epistemic breaks where before we had a few?

Behind this objection is a picture of language that is fundamentally holistic. 'Mass,' 'time,' and 'space,' are thought of as fixed terms, fully specified along with their connotations in one conceptual scheme (the 'paradigm of classical or Newtonian physics') and carrying with them a particular set of instruments and experimental procedures that are only understandable in terms of that conceptual scheme. Equally fully specified, or so goes the argument, is another, incompatible conceptual scheme (the 'paradigm of Einsteinian physics') in which 'mass,' 'time,' and 'space' have utterly different meanings. Because Einstein and Newton and their respective followers 'speak different languages' any putative communication between them amounts to little more than puns, a homophonic happenstance. Out of this picture come some of the most famous metaphorical structures used to capture the radical untranslatability of languages: Gestalt shifts and systematic visual-perceptual misconstruals on the basis of prior conceptions. If Gestalt shifts or total shifts of conceptual scheme are the model for what happens at the boundary between languages then indeed, the repartition of schemes like the positivist periodization of Figure 34.2 into the antipositivist periodization of Figure 34.3 may be of use historically, but analytically we advance not one inch.

What does happen at the boundary between cultures, where people face one another? Do people, in fact, translate with the sudden, Gestalt-like character of the duck-rabbit switch? Here we can learn from the burgeoning field of anthropological linguistics, a field that at least in one of its forms has dealt extensively with the historical and structural development of *trading languages*, highly specific linguistic structures that themselves fit between two or more extant full-blown languages. Very roughly a 'trading jargon' or 'foreigner talk' designates a few isolated terms used to facilitate inter-linguistic communication, 'pidgin' refers to a more developed language, with sufficient structure to allow more complex modes of exchange between speakers. Generally, pidgins are characterized by more regularized phonetic, syntactic, and lexical structure than the 'parent' languages that the pidgin links. For example while one of the parent languages may carry multiple consonant clusters (CCC), pidgins tend to be routinized into consonant-vowel-consonant (CVCV) form. Pidgins may — but do not necessarily — develop into full-fledged creoles, where 'creole' designates a language with sufficient structure to allow people to 'grow up' within it. Creoles, unlike pidgins, can be a *first* language. Intriguingly, it seems to be the case that among our linguistic capacities is the ability to shift the register of the language we speak: we are able to restrict vocabulary, regularize syntactic as well as phonetic form.

On this view, linguistic borders are not the thin-line mathematical idealizations that they appear to be in the Gestalt-switching picture of the antipositivists. Instead, linguistic borders appear as thick and irregular, more like the creoles one in fact finds in border regions in many areas of the world. So it is, I argue, in the *trading zones* between theories or between experiment and theory, or between a physics subculture and an engineering subculture grounded in the industrial or military traditions in contact with physics. Here I would suggest that we drop the attempt to gloss the interaction among the electron theorists Lorentz, Abraham, and Poincaré with Einstein in the early twentieth century as 'Classical' physicists 'talking past' 'Relativistic' physicists, a forced Gestalt switch between old and new notions of 'mass,' 'space', and 'time,' with language shifting as a whole. Instead, we ought to examine the ways in which each of these physicists actually went about coordinating their theories with the results of experimenters like Kaufmann and Bucherer. For despite these philosophical protestations about incommensurability, these laboratory ventures were precisely aimed at comparing the various electrodynamic theories. How, they asked, did the deflection of fast electrons as a function of velocity come into contact with (for example) Abraham's and Einstein's specific theoretical proposals? Or consider the interaction of experimenters and theorists around track images in bubble chambers. While each may have very different ideas about adequate demonstration, or even about the nature of particles, it is nonetheless often possible for both groups to come to common ground of 'interpretation' — e.g., 'this particle decays here to two lighter particles, one of which escapes and the other is absorbed.' The point is that some meaning gets stripped away in the trading zone where theory meets experiment, where engineering meets theory, where, in general, the scientific subcultures encounter one another. What thrives in the interstitial zones is neither trivial nor purely instrumental, it is a form of scientific exchange language.

In the interstitial zone I have not distinguished sharply between locally shared terms and mathematical-syntactic relations on one side, and the material objects on another. This is deliberate: we ascribe meaning to machines as surely as we do to mathematical symbols. And in new material and functional contexts, the meaning of machines can alter as well — we need only look around the laboratory to see a myriad of technologies now performing functions (and carrying meaning) a long way from their site of origin. For this reason, it may be helpful to think of the process by which a computer logic circuit, vacuum tube, or clock mechanism becomes modularized as a form of pidginization. But because this process is not, at least in the first instance, purely linguistic, it is helpful to think of the production and elaboration of such common objects as 'wordless pidgins' and 'wordless creoles.' Wordless grammar corresponds to the rules of combination allowed for these objects — circuit design rules, for example. Similarly, an element in such a wordless interlanguage (should we call it the analogue of a noun?) corresponds to the useful notion advanced by Leigh Star and James Griesemer of the 'boundary object,' an entity participating simultaneously in two or more fields of inquiry.[3] Here a cautionary note is worth sounding. To speak of wordless creoles is not to commit oneself to a position in which concepts can be extricated from language altogether; I mean to emphasize that pieces of scientific objects are often transferred without words.

Throughout this discussion, I intend the pidginization and creolization of scientific language to be treated seriously just the way one would address such issues for language more generally (that is, not as a 'model' or 'metaphor'). First, physicists themselves regularly refer to problems of language in terms of idioms, meaning compatibility, and translation. There are books such as *Computers as a Language of Physics* and physicists characteristically speak about 'putting this current algebra argument into the language of field theory' or ask 'can you express that relation in the language of effective field theory?' Second, the standard account of science in terms of 'conceptual schemes,' 'epistemic breaks,' and 'paradigm shifts' presupposes and relies on talk of meaning change between terms; the argument presented here simply says that the account of linguistic work at the boundary is oversimplified and unhelpful. For example, 'translation' as used by Kuhn to describe scientific change explicitly borrows from notions of translation from ordinary language: "in time," says Kuhn's expositor Paul Hoyningen-Huene, "the members of one group will be able to translate (in the everyday sense of 'translate') portions of their counterparts' theory."[4] Finally, 'model' and 'metaphor' talk presupposes a radical break between scientific and ordinary use of language, and philosophical attempts to enforce such a dichotomy have notoriously foundered. The point I want to make is this: the characterization of different registers (such as jargons, pidgins, and creoles) is helpful in distinguishing between different modes of scientific and non-scientific uses of language. My intention is to expand the notions of interlanguages to include both the discourse of scientific and nonscientific utterances, and both material and abstract systems. It is not to 'apply' the results of one field to another.

How literal is the notion of a 'trading zone'? At one level, I have in mind the most literal sorts of spatialized scientific practice. Laboratories, scientific campuses, often reflect architects' and scientists' expectations of intellectual proximity and meeting points in the walls, hallways, and stairwell landings. Conferences, informal meetings, visits, present other transient sites for face-to-face interactions. If one is trying to understand the development of the bubble chamber one would do well to register the circumstance that in the 1950s cryogenic engineers from the Air Force were meeting with staff from Lawrence Berkeley Laboratory. If one is trying to understand the development of computer simulations, it is essential to track the direct encounter of mathematicians, physicists, weapons designers, and statisticians with one another in a series of conferences from UCLA to Endicott, New York. In the end, however, the issue is one of coordination and regularization of different systems of values and practices; if that takes place outside of any spatial configuration, this too is of interest. In a typical colliding beam experiment, the full set of 500 or so experimenters may never be in the same place at the same time. Many will never meet at all. Webs and computer links bind interlocking laboratories and subgroups together and conventions, standards, and procedures are often established with no single point of authority or geographical center. Exchange zones are written into architectures, but the architectures may not be physical or spatial.

As such considerations suggest, there are many avenues opened up by the less restrictive image of substantive as opposed to mathematically thin boundaries between the subcultures of physics. And in the conceptual sphere, the stripped-down view of shared, specific meanings as opposed to total translation between full languages may offer a better vision of how knowledge moves in and across boundaries. Other questions arise as well. One could follow the development of a highly restrictive inter-field into a full-fledged 'creole' as in the gradual articulation of physical chemistry or biochemistry out of highly localized shared techniques.[5] Or one could examine instances in which the interlanguage more or less died off, as did eighteenth century boundary field of iatromechanics — or, in more recent times, many of the myriad of unification schemes such as those that would (à la Millikan) join cosmic ray physics to the genesis of higher elements in deep space; or Einstein and Weyl's notion that they would be able to unify *directly* electrodynamics with general relativity. As in linguistic boundaries, there is nothing to block the attempt to put languages together, but at the same time there is no guarantee that either social or intellectual coherence will follow. Some pidgins get resorbed back into one of the 'parent' languages, some stabilize as pidgins, and some grow into full-fledged creoles, and then ultimately into a language, full stop.

Here I would like to explore a different point: *delocalization.* The problem is this. Over the past years, we have again and again seen how tied specific laboratory practices are to their conditions of origin. An instrument made in one spot is often difficult to replicate in another without the bodily transport of things and people. Paper instructions are often not enough. We have learned from scholars like Harry Collins how necessary it is to attend to the site-specificity of particular materials, skills, and resources available at a given time and place: the little tricks of the trade that made one group of people able to build a certain kind of laser where others could not. Eventually, though, these objects do travel, the lasers, prisms, accelerators, detectors, tubes, and films — hence the problem. If the original production of scientific knowledge is so reflective of local conditions — whether they are craft techniques or religious views, material objects or forms of teamwork, how does *de*localization take place?

One solution, referred to by Simon Schaffer as the 'multiplication of contexts,' accounts for delocalization by a process in which the original context is imposed elsewhere. Methods are devised which "distribute instruments and values which make the world fit for science."[6] One pictures maps of imperial expansion, the dark black arrows of distributed context radiating outward like the footsteps of a conquering army from Oxford, Cambridge, London, or Paris to newly-acquired sites elsewhere. To be replicated, air-pumps required a particular set of machines, facticity, and witnessing to be in place. To be enforced values and methods of standardization stamped their mark on distant sites. Multiplying *these* contexts was a precondition for replication, and the lines of power that designate the creation of those conditions extend from center to periphery. Similarly, Bruno Latour focuses on ways in which the world is modified to make it possible for an instrument, even one as simple as a clock, to "travel very far without ever leaving home."[7]

A different solution, building on the notions of discursive and wordless pidginization within trading zones of limited exchange, would put greater emphasis on two features of the 'transfer.' First, it would stress the *activity* of interpretation that takes place on the receiving end of the objects, techniques or texts. Varying Latour's apt phrase, scientific practices can also travel to very *different* homes. However imposed the more powerful set of techniques might be, the site of their application fundamentally alters the way those technologies manifest themselves. Many creolists insist, for example, that French creoles can only be understood if one abandons the attempt to view them as merely simplified French, and instead considers the combination of the lexical French structure with a variety of syntactic elements from African languages. Just such a nuanced view is needed in the domains of material and theoretical cultures — pieces of apparatus can circulate without the whole, devices can move without their scientific contexts, functional specifications can move without a trace of their original material form. An abbreviated example: working against low-flying attack fighters, radar engineers during World War II designed a 'memory tube' that would store radar returns and cancel out the signal of any stationary object — leaving only the plane. In travelling to a different context not long afterward, the tube became a recirculating memory to store information for the early computers. Again, cut loose from its 'meaning' the device is then appropriated by particle physicists who want to use it to locate the position of a passing particle (by measuring the time it takes for a particle-induced pulse to reach the end of the tube). We have, in a sense, a hermeneutics of material culture. At every stage of such multiple transfers we need to ask *both*: How does the stripping-down process occur by which local circumstance is removed? And then how does the re-integration into a *new* context take place?

If the first feature pointed to is *activity*, the second is *locality*. Elements of the meaning of a scientific practice are pared away. Theoretical physicists drop many properties that they ascribe to an electron or quark before they bring those notions to the bubble chamber scanning table at which they meet the experimentalist. The embedding of 'electron' in a quantum field theoretical description, or a particular unified theory might be deleted altogether. Conversely, before encountering the theorists, the experimentalists bracket many of the concerns about the nature of the film, the optics and compression pattern of the liquid hydrogen. The tracks being discussed carry, in this interaction, a shared local meaning. Experimentalist and theorist can pore over the scanning table, gesturing at the tracks and arguing interpretations: "this kaon here into two unseen gammas, which then produce these two electron-positron pairs over here . . .' But such exchanges, though widely shared, are not pieces of a universal protocol language interpretable in *any* epoch of physics. Against the positivists: there is no 'neutral observation language' in the activity of track interpretation — what goes on in the process of sorting out the particle decay scheme does not in any way, shape or form constitute a 'pure' observation to be positioned against 'hypothesis.' Against the conceptual-schemers: we do not find a homogenized amalgam in which experimentation and theory become a single, undifferentiated whole in which every experimental statement is utterly fixed by the theory to which it is inseparably attached. The pidginized hybrid of the language of track interpretation has *both* elements of theory and of experiment, while recognizing the self-maintained distinct identity of each. A pidgin is neither a linguistic passe-partout nor a subset 'baby-talk' of a 'full' language; it facilitates complex border interactions. Coordination around specific problems and sites is possible even where globally shared meanings are not.

The history of physics can be profitably seen as a myriad of such productive, heterogeneous confrontations. Field theorists met radio engineers in the American radar laboratories of World War II. British cosmic ray experimentalists encountered colloidal chemists in the production of nuclear emulsions sensitive enough to 'photograph' all known elementary particles. Such heterogeneity continues. One collaboration joining mathematics and physics expertise began a 1980 *Physics Report* by recalling the halycon days in which Newtonian mathematics and Newtonian physics could develop together, only to separate under the pressures of specialization:

> a formidable language barrier has grown up between the two. It is thus remarkable that several recent developments in theoretical physics have made use of the ideas and results of modern mathematics.... The time therefore seems ripe to attempt to break down the language barriers between physics and certain branches of mathematics and to re-establish interdisciplinary communication.[8]

This exchange was bilateral: physical techniques from field theory were solving problems in algebraic geometry, and mathematical tools were at the root of the 'string revolutions' that, for many late-twentieth century physicists, promised a final unification of gravity and the short-range forces that held matter together. When algebraic geometry 'traveled' to the physicists, it was often precisely by *shedding* many of the values with which it was practised in mathematics departments. To the horror of many mathematicians, 'speculative,' 'intuitive,' and 'physical' argumentation were supplanting the hard-won rigor of mathematics proper. When these practices traveled, some of the mathematically constitutive values associated with them stayed behind.

This observation — that meanings, values and symbols often stay home or switch identities when scientific theories and instruments travel — lies at the heart of the alternative (exchange language) picture of delocalization I have in mind. Donald Glaser, inventor of the bubble chamber, desperately hoped the device would 'save' small-scale physics (and the life that went with it) from the onslaught of Luis Alvarez's factory-laboratory. But substantial portions of Glaser's original device — stripped of those material components that were tied to the small-scale — were reappropriated into the massive chambers that became the very symbol of Big Physics, Alvarez's sector of the Lawrence Berekely Laboratory. This is not to say that values played an inessential role in the constitution of particle physics at any stage, it is, instead, to say that the particular guiding values altered radically as they shifted from a defense of individual craft-style work to a form of scientific life that emerged from the massive nuclear weapons and radar projects of World War II.

What then is the relation between these two accounts of delocalization — 'multiplication of contexts' and 'exchange language'? Let us return briefly to the relation of algebraic geometry to quantum field theory in string theory. For there the practitioners of both sides were of roughly equal stature — such an exchange did not resemble (for example) the relation between technicians in Alvarez's laboratory to the Nobel Prize winning physicists who ran various groups. And perhaps here lies a clue. In examining situations in which the balance of power was maximally unequal — it might well be the case that one group could impose a fuller set of contextualized values along with specific practices on the other. In other words, we might keep in mind that in terms of power relations, there are two interesting limits to the confrontation of languages. At one extreme, the two languages enter into contact in states of roughly equal power. Linguistically, such situations typically result in pidgins in which the lexical admixture of the two languages is markedly heterogeneous. This is of great interest from the scientific standpoint, because we are often faced with situations of this type. String theory, for example puts quantum field theorists on one side and algebraic geometers on the other: a situation in which the balance of power is roughly equal. At the other extreme, in which one group is far more powerful than the other, very different linguistic structures might be expected. For example, in very unequal balance of power, it is common to find that the lexicon emerges overwhelmingly from the superordinate language and a regularized, restricted syntactic structure from the less powerful language. It is also well documented that in very unequal situations, pidgin languages can be reabsorbed back into the superordinate language. Such instances occur in the domain of science and technology. In large-scale collaborative ventures, such as the Manhattan Project, one sees sectors of almost every possible power relation, from the Du Pont (engineer run) effort in Chicago to the Los Almos (scientist run) laboratory.

We are now in a position to understand the relation between the interlanguage and context-multiplying accounts of delocalization. Context multiplication is the limit case of interlanguage coordination just when the power imbalance is so pronounced that the recipients' constitutive values and technical practices were thoroughly subordinated, or where relevant local values did not enter.

Let me end with one final thought. In a sense, the last fifteen years of studies in the history and sociology of science have left us with a powerful set of tools for understanding the local origins of scientific ideas, practices, and methodological precepts. But awkwardly we have grafted to this local description, a picture of language (broadly conceived) that remained global, rigid, and holistic. No wonder we often end with a peculiarly bad set of choices. At one extreme we anchor our notion of science in a global picture of language imagining that moving machines and ideas is automatic, completely ignoring the contextualized circumstances in which scientific work originates. At the other, we imagine that practices are so tied to local circumstances that we either descend into a radical nominalism in which no one is 'really' talking to anyone else, or we wrongly conclude that the full context of origin is packed off, kit and caboodle, to every distant site of application. If, as the anthropological linguists are trying to teach us, meanings don't travel all at once in great conceptual schemes, but rather hesistantly, partially, and nonetheless efficaciously, perhaps, for those studying the development of science, there is an exit from this impasse.

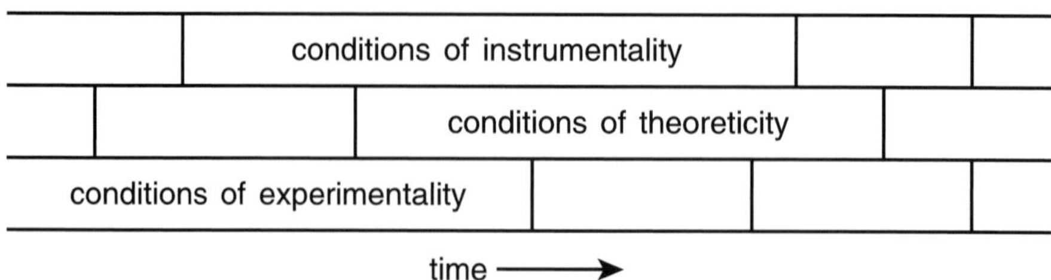

FIGURE 34.1: "INTERCALATED" PERIODIZATION.

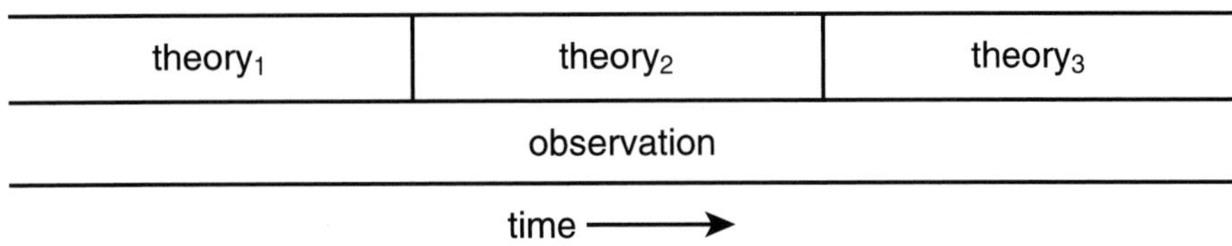

FIGURE 34.2: "POSITIVIST" PERIODIZATION.

observation$_1$	observation$_2$	observation$_3$
theory$_1$	theory$_2$	theory$_3$

time ⟶

FIGURE 34.3: "ANTIPOSITIVIST" PERIODIZATION.

REFERENCES

1. Arthur Jaffe and Frank Quinn. "Theoretical Mathematics": Toward a Cultural Synthesis of Mathematics and Theoretical Physics. *Bulletin of the American Mathematical Society* (1993), **29**: 1–13, on 5.
2. Galison. "Aufbau/Bauhaus: Logical Positivism and Architectural Modernism," *Critical Inquiry* (1990), **16**: 709–52.
3. Galison. "The Trading Zone: The Coordination of Action and Belief. Paper presented at TECH-KNOW Workshops on Places of Knowledge, their Technologies and Economies. UCLA Center for Cultural History of Science and Technology, 1989; Susan Leigh Star and James R. Griesemer, "Institutional Ecology, 'Translations' and Boundary Objects: Amateurs and Professionals in Berkeley's Museum of Verterbrate Zoology, 1907–39," *Social Studies of Science* (1989), **19**: 387–420.
4. Paul Hoyningen-Huene. *Reconstructing Scientific Revolutions.* (Chicago: University of Chicago Press, 1993), p. 257.
5. John W. Servos. *Physical Chemistry from Ostwald to Pauling.* (Princeton: Princeton University Press, 1990); Robert E. Kohler, *From Medical Chemistry to Biochemistry.* (Cambridge: Cambridge University Press, 1982).
6. Simon Schaffer. "A Manufactory of Ohms," in Robert Bud and Susan E. Cozzens, *Invisible Connections. Instruments, Institutions, and Science.* (Bellingham, Washington: SPIE Optical Engineering Press, 1991), p. 23.
7. Bruno Latour. *Science in Action.* (Cambridge: Harvard University Press. 1987), 251.
8. Tohru Eguchi, Peter B. Gilkey, and Andrew J. Hanson. "Gravitation, Gauge Theories and Differential Geometry," *Physics Reports* (1980), **66**: 213–393, on 215.

FURTHER READING

Barnes, B. *T.S. Kuhn and Social Science.* (New York: Columbia University Press, 1982).

Carnap, R. *The Logical Structure of the World and Pseudoproblems in Philosophy* translated by Rolf George. (Berkeley and Los Angeles: University of California Press, 1967); "Intellectual Autobiography," in *The Philosophy of Rudolf Carnap*, Library of Living Philosophers, Volume 11, edited by Paul Arthur Schilpp. (La Salle, III: Open Court, 1963), 3–84.

Feyerabend, P. *Problems of Empiricism: Philosophical Papers*, Volume 2. (Cambridge: Cambridge University Press, 1981); *Realism, Rationalism and Scientific Method: Philosophical Papers*, Volume 1. (Cambridge: Cambridge University Press, 1981).

Galison, P. *Image and Logic: A Material Culture of Microphysics.* (Chicago: University of Chicago Press, 1997); *How Experiments End.* (Chicago: University of Chicago Press, 1987); "Introduction: The Context of Disunity," and "Computer Simulations in the Trading Zone," in Galison and Stump, *The Disunity of Science.* (Stanford: Stanford University Press, 1996); "Context and Constraints", in Buchwald (ed.), *Scientific Practice.* (Chicago: Chicago University Press, 1995).

Galison, P. "Theory Bound and Unbound: Superstrings and Experiment" in Friedel Weinert (ed.), *The Laws of Nature.* (Berlin and New York: Walter de Gruyter, 1995), 369–408.

Hanson, N.R. *Patterns of Discovery.* (Cambridge: Cambridge University Press, 1958); *The Concept of the Positron.* (Cambridge: Cambridge University Press, 1963).

Kuhn, T.S. *Structure of Scientific Revolutions.* (Chicago: Chicago University Press, 1970).

Latour, B. *Science in Action.* (Cambridge: Harvard University Press, 1987), esp. 251ff.

Pickering, A. *Constructing Quarks: A Sociological History of Particle Physics.* (Chicago: University of Chicago Press, 1984).

Shapin, S. and Schaffer, S. *Leviathan and the Air-Pump.* (Princeton: Princeton Univ. Press, 1985).

Schaffer, S. "Glass Works: Newton's Prisms and the Uses of Experiment," in D. Gooding, *et al.*, *The Uses of Experiment.* (Cambridge: Cambridge Univ. Press, 1985), 67–104; "A Manufactory of Ohms," in Robert Bud and Susan E. Cozzens, *Invisible Connections. Instruments, Institutions, and Science.* (Bellingham, Washington: SPIE Optical Engineering Press, 1992).

CHAPTER 35

Biologists at Work

Experimental Practices in the Twentieth-Century Life Sciences

JEAN-PAUL GAUDILLIERE

The life sciences are usually defined with concepts such as cell metabolism, genes, or macromolecules that are characteristics of three specialties which have emerged in the twentieth century: biochemistry, genetics and molecular biology.[1] Notwithstanding the richness of these cognitive changes, studies of contemporary biology have recently focused on materials, techniques, and work arrangements. One motive for this interest may be that the accumulation of biological knowledge has been nurtured and constrained by the interaction between laboratories, industrial settings, hospitals, farms, and administrative bodies. The question of how biologists *work* has become an important topic and historians have turned their attention to aspects of biological research which were once viewed as mundane: the collection of research materials, the production of instruments, the purification of biological agents, the definition of standards, or the circulation of organisms.

This chapter demonstrates how the making of biochemical compounds, genes, or macromolecules have been related to changes in laboratory practices. Although the picture emerging from the case studies is one of local and conflicting patterns of work rather than a coherent vision of grand reshuffling, two general features will be emphasized. Firstly, the processes by which new biological objects have acquired stability during the twentieth century have increasingly mobilized material, social and cultural technologies which are often perceived as characteristic of the industry. Secondly, an intense traffic between biological laboratories and medical settings resulted in the translation of medical issues into workable experimental systems. Changes in the experimental life sciences therefore suggest that ours is the century of biomedical technology.

THE COLLECTION AND PURIFICATION OF RESEARCH MATERIALS

In June 1881, Louis Pasteur invited a large crowd of government officials, veterinarians, farmers, and journalists to observe the results of an anthrax vaccine trial which had been organized in the small town of Pouilly-le-Fort. The local veterinary surgeon Hippolyte Rossignol had offered fifty sheep, half of which had been injected by Pasteur's collaborators with an 'attenuated' preparation of anthrax bacilli. The other twenty-five sheep received no injections until both groups were inoculated with a culture of virulent anthrax bacilli. Pasteur's prophecy was that all the vaccinated sheep would survive, while the unvaccinated sheep would all die from anthrax.[2] The Pouilly-le-Fort trial was a triumph of experimental skills. A few weeks after the injection of virulent bacilli, all the vaccinated sheep were still alive and most of the unvaccinated sheep had already died. All the participants in the display could testify that Pasteur had mastered a method for preparing immunizing anthrax bacilli.

As soon as the results of Pouilly-le-Fort were known in Berlin, the physician and bacteriologist Robert Koch claimed that Pasteur's success was mere chance.[3] Koch's criticisms were targeted at several aspects of the procedures supposedly employed in Paris. Firstly, Pasteur did not cultivate his bacilli on a solid-state medium as Koch and his collaborators used to do: he employed liquid broth. He did not collect isolated colonies on agar plates but transferred liquid samples whose purity could not be ascertained. Secondly, Pasteur explained that attenuation originated in the cultivation of bacteria in the presence of oxygen. This physiological treatment seemed to confirm Pasteur's vision of bacteria as complex living beings that could vary and gain new properties in response to environmental changes. To Koch, the variability of pathogenic properties was suspicious. Indeed, it reinforced the notion that Pasteur was operating with heterogeneous bacteria and contaminated brew. Koch's argument was based on the belief that bacterial species could be characterized by shape, staining pattern, ability to use specific nutrients, and ability to induce particular diseases. This stability was essential to physicians like Koch who used bacteriological procedures for diagnostic purpose.

Twenty years later, the Franco-German debate about the variability and purity of bacterial species was closed. Rather than being based on theoretical assumptions this achievement originated in shared experimental practices. Microbiological diagnosis had become common place and routine procedures had been established for isolating, staining, testing the bacteria associated with the most common epidemic diseases. The research materials used by bacteriologists had been put under control through innovations including collections of reference strains, defined protocols for collecting body fluids or preparing culture media, industrially produced dyes or nutrients, standardized nomenclature, etc. These were not isolated changes, the expansion of laboratory studies with (and on) antibodies, antigens, hormones, vitamins, and enzymes in the early twentieth century shows that a purification culture of chemical origins grounded the development of many specialties.

The history of sex hormones may illustrate the complex networks biologists established in order to collect, prepare and stabilize biobiological entities of a chemical sort.[4] The concept of hormones was first introduced in 1905 by the physiologist Ernest Starling and it typified the shift from a physiology of electrical influx and nerves to a physiology of circulating chemicals. Sex hormones originated in attempts to prepare extracts of the testis and ovaries. In the 1890s, the French physiologist Brown-Séquard strongly advocated organotherapy, i.e., the use of animal organs as therapeutic agents. Brown-Séquart argued that testes and ovaries controlled the sexual development of the organism through 'internal secretions.' Sex hormones however did not become material reality until the 1930s when biochemical purification superseded physiological work. In the 1920s notion of a single female sex hormone secreted by the ovaries and a single male sex hormone secreted by the testes was blurred by the observation of male hormones in women and female hormones in men. By 1935, the diversification and disembodiment of substances controlling sex resulted in a biochemical vision of sex hormones as steroids that could be transformed one into another. Thus, different sets of chemical reactions would operate in the male and female organisms.

The making of sex hormones as chemicals was rooted in the development of biological assays and collection strategies. In the early decades of the twentieth century 'female' principle would be any substance that could restore properties of the organism which changed after the removal of the ovaries. A wide range of tests fitted this definition and the test of choice varied from laboratory to laboratory. Physiologists for example used the growth of mammary nipples in mice, the level of blood calcium and sugar, or the development of feathers in domestic fowl. The tests were not only developed to analyze the control of physiological functions by hormones, but also to evaluate the quality of commercial ovarian preparations. However, these procedures were criticized by gynecologists who relied on their special interest in pregnancy, and employed the increased weight of the uterus in rabbits. In contrast, the measuring of male sex hormones was less debated. There was no medical specialty backing the use of male hormones and the development of tests was the exclusive field of physiologists and biochemists. In the 1930s, the comb test was the assay most widely used. The practice consisted in measuring the size of the comb in castrated roosters before and after hormone treatment. Though keeping roosters was inconvenient, the animals could be used several times, and growth of the comb was easy to detect, record by photography, and quantify.

Any attempt at the isolation of the sex hormones faced the problem of gaining access to research materials.[5] Alliances between laboratories, food manuafctures, pharmaceutical companies, and hospitals were important. How did endocrinologists first collect ovaries and testes? Some physiologists obtained large ovaries from whales caught by fishermen. Others purchased or obtained gonads from pharmaceutical companies connected with meat packing plants or slaughterhouses. The Dutch physiologist Ernst Laqueur, who headed one of the three groups which reported the purification of the female hormone in 1929, founded a company with the director of a slaughterhouse. *Organon* was established to manufacture "organ preparations on a scientific basis." Laqueur was one of the directors in charge of the scientific and medical management of the company. Cow ovaries were useful, but scientists continued to look for cheaper and more easily available sources. In 1926, two German gynecologists working on pregnancy and fertility tests reported that the urine of pregnant women was more active than most ovarian extracts of female hormones. Urine was a biological liquid which was easy to process with routine biochemical tools. Moreover, once contact had been made with hospitals, urine was abundant and inexpensive. It was readily adopted by biochemists who became key-players in the field. In 1929, the isolation of the female hormone was reported. Likewise, the isolation of the male hormone from male urine collected in police barracks was announced by the German chemist Adolf Butenandt.

The final stage in the making of sex hormones under the chemical umbrella of 'steroids,' was a standardization stage. Isolation triggered substantial investments into the development of purified preparations which resulted in an increasing need for analysis and quality control. Colorimetric assays based on organic reactions emerged in the early 1930s. They were used to measure the steroid content of blood or urine, and entered hospital laboratories as a means of making endocrine imbalance visible. Biological assays, however, remained the most important devices used to control the quality of preparations for medical uses. The Health Organization of the League of Nations organized two conferences on international norms for sex hormones. Laboratory physiologists and chemists dominated these meetings. During the first one, in 1932, the vaginal smear test was chosen as the standard test for female hormones. It was based on the analysis of cyclic changes in the epithelial cells of the vagina of mice or rats. Female hormones became substances "producing in the adult female animal completely deprived of its ovaries, an accurately recognizable degree of the changes characteristic of the estrus" or substances with "specific-producing activity."[6] In other words, "oestrogenes."

The Production of Model Organisms

If molecules and chemical purification are to be viewed as particularly characteristic of twentieth-century biology the study of heredity is exceptional since no procedures for the isolation of invisible genetic factors attracted the attention of geneticists until World War II. This however does not mean that genes were less operational than viruses, differentiation factors or enzymes. On the contrary, for half the century hereditary factors were defined in terms of offspring distribution, pedigree analysis, and artificial crosses between selected strains differing by a few traits. Geneticists not only built on the traditions of seed-men and breeders, they also selected highly engineered organisms for experimental research. Their model systems embodied theoretical ideals such as the mendelian laws, the linear distribution of genes along chromosomes, or the one gene-one enzyme relationship.[7] Comparison of the laboratory histories of laboratory flies and of laboratory mice reveals the contrasting dynamics that determined the production of these systems and the gathering of their users. While the work done on *Drosophila* by the group headed by Thomas Hunt Morgan resulted in the development of chromosome maps, the makers of genetically pure strains of mice generated models of human diseases.

In 1909, the American biologist J. East complained about the wild behavior of the beastly little flies he had brought into his laboratory in order to carry out experiments on evolutionary change. Soon he gave up raising flies and focused on more well-behaved systems such as corn. Ten years later, the beastly little fly had been transformed into the genetic system of everyone's choice.

Flies of the *Drosophila* type crossed the threshold of biological laboratories after 1900.[8] Like East, the Columbia biologist Thomas H. Morgan started to breed flies in 1909 as a means of simulating evolution. Morgan was looking for periods of morphologic variations which would reveal something of the mechanism of species formation. He viewed mutations as adaptative events which could be induced by altering the physical environment. The fly project complemented studies on sex determination which were conducted on other insects, and studies of mendelien factors carried out on mice.

As Morgan and his assistants raised and inspected an increasing number of flies, a gathering flood of mutants mounted: body-color mutants, eye-color mutants, and wing-shape mutants surfaced in a few months. This 'mutational crisis' pulled Drosophila back into the Mendelian framework. Morgan's matings of white-eyed mutants and wild flies, for example, revealed intractable discrepancies with expected Mendelian ratio. Somehow Morgan came to the conclusion that the factor controlling the white eye trait was linked to the distribution of X-chromosomes, characteristic of the female flies. Many crosses with two different mutants however showed unstable association of traits. In addition, Morgan had trouble integrating the many new factors into the physiological schemes in use to explain the formation of visible traits. To account for his five eye-color mutants, Morgan postulated the existence of a color determiner gene and three modifying factors. But a few months later, a sixth and a seventh mutant appeared, and the whole series of factors had to be reworked. This happened again and again with most mutant groupings.

The mapping program of the Columbia group emerged as a way out of this messy situation. Morgan's notion of 'crossing-over' was both an explanation for statistical abnormalities and a means to classify the mutants. Postulating that the probability of rearrangement between genes linearly distributed on one single chromosome would be a linear function of the distance separating the two genes, Morgan adopted the percentage of flies showing 'recombined' traits as a measure of the distance between genetic factors.[9] By 1912, two students — Bridges and Sturtevant — were assigned a large-scale attack at mapping two Drosophila chromosomes. This 'student project' expanded so rapidly that by 1915 chromosomal maps and gene locations were the main products of Morgan's laboratory. Large-scale mapping was an 'autocatalytic' process: "experiments were scaled-up to produce more precise measurements of distance, and larger experiments generated a large number of new mutants that had to be mapped and assimilated in the mapping process."[10]

In order to use recombination percentages as a direct measurement of gene distances, new tools and new skills were required. The variability of quantitative results was reduced in three different ways: husbandry was standardized, flies were engineered, and workers were recruited and trained. Morgan's boys raised flies in hundreds of milk bottles. An early improvement was to breed only one pair of flies in each bottle. Better conditions of food, temperature and population density made crossing-over experiments easier to reproduce. A special apparatus designed to 'etherize' flies eased transfers and long microscopical examinations which were necessary to count offspring and isolate mutants. The reconstruction of *Drosophila* as a tool for mapping consisted of two strategies. One aim was to select stocks combining several mutations. A second was to rebuild the chromosomes in order to get rid of factors which modified or suppressed crossing-over. Stocks were discarded because results were too unstable, because expected classes of recombinants were absent, or because they produced inconsistencies in the map. Thus, a dozen stocks characterized by a stable rate of crossing-over a given stretch of chromosome were established. In this way, the laboratory fly was "made to conform to the ideal of Mendelian theory and chromosomal cartography."[11] Finally, mapping required a new division of labor. By 1925, the laboratory was maintaining and using 395 mutants. Scaling-up not only required technicians who washed the bottles, prepared the banana diet, or surveyed the shelves, but also a small army of college teachers — mostly experimental zoologists, who were part-time researchers analyzing a few mutants. Morgan's group thus established an efficient and well-integrated system whose development was actively supported by the Carnegie Institution of Washington and later by the Rockefeller Foundation.

In the 1920s, by virtue of the work accomplished by Morgan's team, basic genetics had become synonymous with chromosomal mapping, and studies of gene interactions. To a considerable extent, this was an American specialty which was based on a closely integrated community. Most drosophilists were trained in Morgan's laboratory and for fifteen years or so this shared experience greatly helped to integrate the group. The free circulation of mutants became a fundamental custom which accelerated the pace of the work and made it harder for those who entered the game to doubt the credibility and the value of results and conceptions which originated in the main center of production and maintenance, i.e., Morgan's laboratory. This paternalistic structure prevailed until the 1930s.

By virtue of post-war developments which linked molecular biology, heredity in bacteria and *Drosophila* genetics, Morgan's group has achieved special historical visibility. It would, however, be misleading to take it for granted that *Drosophila* was the sole or even the most employed model organism in American genetics before

1940. Within American academic circles, mouse genetics probably attracted most students of evolution, development, behavior, and pathology. During the 1910s and 1920s, Castle's laboratory at Harvard probably formed the core group of 'mousers.' Castle consciously positioned his work in mammalian genetics in opposition to Morgan's style of work and vision of chromosomal mechanics.[12] Castle argued for a long time against Morgan's postulates. He favored the existence of modifier genes and non-linear arrangements. But the most striking contrast between fly and mouse workers originated in the intimate relations which linked the studies of inheritance in mice and the fight for the control of human pathologies. While *Drosophila* was considered as a good supplier of chromosomes, *Mus musculus* was viewed as a good representative of humans.

This is probably best exemplified by the career of C.C. Little and the creation of the Jackson Memorial Laboratory. C.C. Little was a student of Castle who completed a Ph.D thesis that stressed the combination of Mendelian factors controlling coat color in mice. In 1909, expanding on Johansen's practice with plants, Little started systematic brother-sister matings in order to establish pure lines of mice. Most lines could not be continued because the animals became so weak and prone to diseases that they died before reaching a reproductive age. One single group of animals passed the 'degeneration' phase, survived the selection, and was later used in breeding experiments.

In 1913, Little who was in charge of "an homogenous stock" of brown mice, was prompted into cancer research by E.E. Tyzzer, the professor of pathology at Harvard. Tyzzer used mice to transplant tumors in order to study the changing ability of hosts to inhibit the development of cancer. Tyzzer concluded that the successful transplant of tumors in mice depended upon the method of inoculation, the nature of the tumor and the racial differences in the host. Tyzzer and Little crossed their 'inbred' stocks and successfully reduced the variability of the transplantation experiments. Thus, 'susceptibility' to cancer transplants was viewed to be determined by hereditary factors. This and similar attempts led to a new emphasis on genetic homogeneity as an important element of cancer research, but they undermined the medical value of transplantation experiments, and the medical meaning of inbred mice. Since it was agreed that 'resistance' was not specific to the transplanted tissue, the phenomenon could hardly be enhanced, and it was of little interest to clinicians.[13] Transplantation experiments then became a tool for geneticists who were assessing the purity of mice and controlling the practice of inbreeding.

In the 1910s and 1920s, the murine genetic community was characterized by two problems: the low number of workers and the correlative lack of colonies and stable stocks. The creation of the Jackson Laboratory reversed the situation. The Jackson laboratory was a private non-profit organization established by C.C. Little in 1929 with the support of automobile tycoons. Rooted in Little's instrumental vision of mouse genetics, the 'Jax' was conceived as a center devoted to the study of mammalian genetics and cancer.

At the Jackson Laboratory, husbandry did not aim at the search for mutants, but at the selection and preservation of strains that could be used as models of cancer and other human pathologies. A typical product of the laboratory was, for instance, the 'A line.' The females in A lines were affected with a high incidence of mammary tumors but virgin females were almost unaffected. This strain was therefore used as a system which exemplified the respective role of genetic factors and breeding in the causation of breast cancer. Inbreeding was standardized to avoid deviation from patterns of medical interest. An important activity at the Jax was the control of genetic homogeneity of established lines through skin grafting experiments.

The Jackson Laboratory began resemble an industrial plant in the 1930s. During the Depression, the big entrepreneurs who supported Little radically reduced funding, and Little made the decision to stop giving mice for free. In the following years, the Jax production of 'useful' strains of mice grew rapidly by virtue of major supply contracts with large biomedical research institutions such as the Rockefeller Institute, and later, the National Cancer Institute. In the aftermath of World War II, the development of chemotherapy of cancer and the organization of the large-scale screening of chemical compounds used in clinical trials resulted in a radical change of scale. Although the systems adopted in order to screen putative anti-cancer drugs were constantly discussed and rearranged, the basic commitment was to test the effects of chemicals supplied by the research departments of pharmaceutical companies on the growth of tumors in mice. The US NCI standardized the screening procedures, contracted testing to commercial firms, and coordinated the production of genetically homogenous mice. The Jackson Laboratory was the most important partner of the National Cancer Institute in this venture. NCI committees regulated the production of mice for chemotherapeutic research. They defined standard (and mandatory) rules for feeding, housing, or bedding laboratory mice. The Jackson Laboratory was to supply controlled breeding pairs to other commercial breeders, and to participate in mass production. Between 1955 and 1960, the local output increased from 200,000 to 1,000,000 mice a year. The radically increased demand for homogenous mice in turn prompted a major reorganization of the relations between research laboratories and production departments at the Jackson. Laboratory stocks once maintained by one geneticists helped by one or two technicians were complemented by 'production expansion stocks' which controlled genetic quality and provided breeding pairs for the production units and external breeders.

These 'trading' units enhanced a new regime of research by mediating between the geneticists and the developers. Systematic chromosomal mapping was triggered by mass production. In 1945 the "Special Jax-Mouse issue" of the *Journal of Heredity* included the first mouse chromosomal map of this organism. The number of genes was slightly above what it had been in the 1920s: 59 known mutants against 20. By the mid-1960s, over 300 mutant genes had been already been located and the

Jackson Laboratory was the organizing center of mammalian genetics.[14] With the support of the National Science Foundation, the Jackson workers were maintaining (and supplying) reference stocks of mouse mutants. No wonder if nomenclature and mapping results were negotiated against the Jax standard.

Specificity stemming from the medical value of mice prevailed over the alignment with genetic work of the type achieved with *Drosophila.* In contrast to the practice of the early 1900s when coat color illustrated the transmission of Mendelian hereditary factors, the mutants admired and actively sought after World War II were those representing human pathologies. Accordingly, the booming production of mice at the Jackson Laboratory resulted in the appearance, selection, and study of numerous pathological models. In the 1950s and 1960s, strains such as *obese, diabetic,* or *dystrophia muscularis* were generic devices which circulated between many biomedical research settings. The Jax biologists working with these mutants were simultaneously achieving genetic, biochemical, and physiological studies, collaborating with pathologists, improving the (re)production techniques and quality control, and organizing the production for outside users.

Biophysical Instrumentation and Macromolecules

In the 1920s, 'glass and iron' instruments used by biologists were of three types that can be traced back to the nineteenth century. Inscribing machines such as the kymograph translated physiological patterns into circulating and quantifiable graphs. Optical tools magnified invisible objects. Manipulating devices were used to cut, handle, transfer, and fix organs, tissues or organisms. After 1945, an increasing number of physico-chemical instruments were developed to visualize and to measure the new elementary units of life: genes, viruses, proteins, and other macromolecular entities.[15] The establishment of these tools was facilitated by the Rockefeller Foundation. There is hardly any large instrument developed by physicists or chemists, which came to be used in biochemistry or physiology, that the foundations' division for natural sciences had not contributed in developing in the 1930s. The record includes the ultracentrifuge, X-ray diffraction, electron diffraction, electrophoresis, spectroscopy, electron microscopy, radioactive isotopes and accelerators.[16] This rise of biophysical instrumentation has sometimes been viewed as a case for the technological origins of 'molecular biology.' Rather than exemplifying technological determinism, the history of ultracentrifugation and of electron microscopy shows how new patterns of work created new machines and new biological entities. The development of this form of instrumentation was associated with a new character in the biological sciences, namely the 'machine-developer' whose career focused on engineering and usage of one complex piece of equipment. This radically changed a few biological specialties including biochemistry and bacteriology. Proteins ceased to be colloid and became individual molecules, while viruses came to be viewed as "replicating genetic units" rather than "small, invisible bacteria causing diseases."

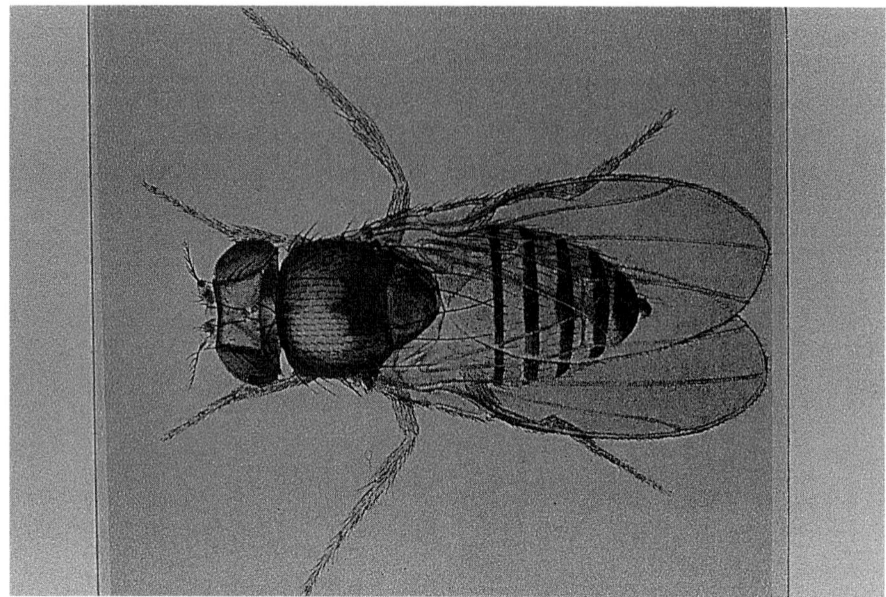

FIGURE 35.1: AN EARLY DRAWING OF THE LABORATORY FRUIT FLY. REPRODUCED FROM C.B. BRIDGES AND T.H. MORGAN, *CONTRIBUTION TO THE GENETICS OF DROSOPHILA MELANOGASTER*, CARNEGIE INSTITUTION, 1919.

FIGURE 35.2: FLY CULTURE. COURTESY OF AMERICAN PHILOSOPHICAL SOCIETY.

FIGURE 35.3: A *DROSOPHILA* LABORATORY IN THE 1930S. COURTESY OF AMERICAN PHILOSOPHICAL SOCIETY.

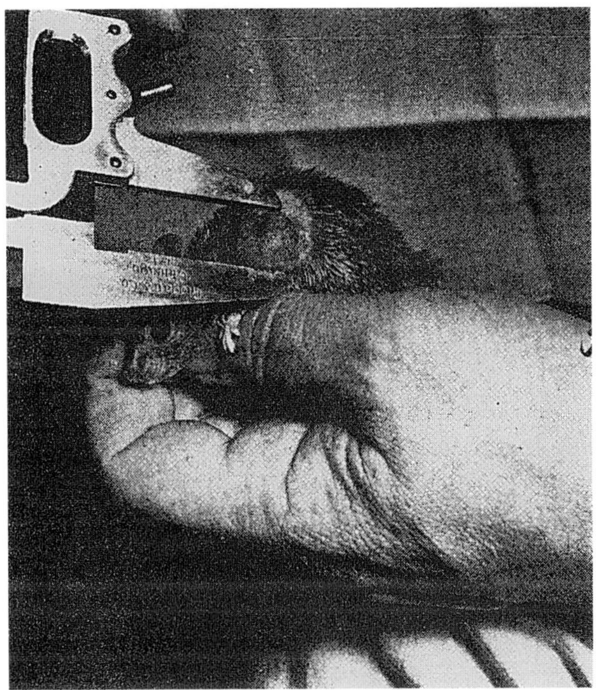

FIGURE 35.4: MEASUREMENT OF A BREAST TUMOR IN A GENETICALLY STANDARDIZED MOUSE. REPRODUCED FROM *CANCER NEWS*. COURTESY OF THE AMERICAN CANCER SOCIETY.

FIGURE 35.5: A MOUSE PRODUCTION ROOM IN THE 1960S. COURTESY OF THE JACKSON MEMORIAL LABORATORY.

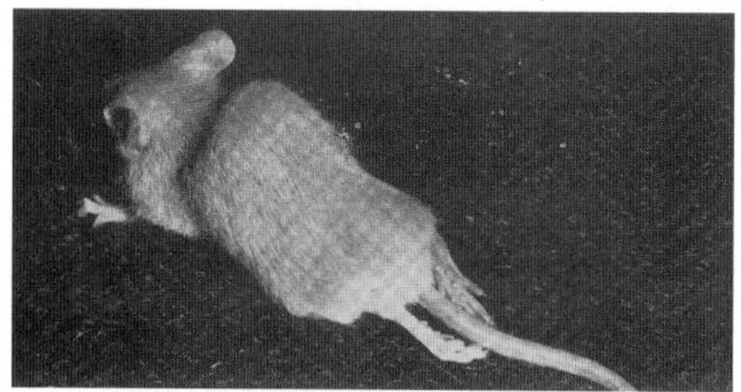

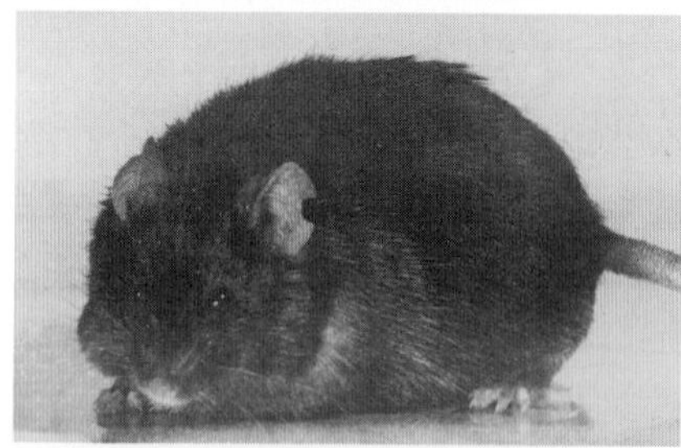

FIGURE 35.6: MUTANT MICE FROM THE JACKSON LABORATORY: DYSTROPHIA MUSCULARIS, OBESE, AND HAIRLESS. REPRODUCED FROM *HANDBOOK ON GENETICALLY STANDARDIZED JAX MICE*, 1968.

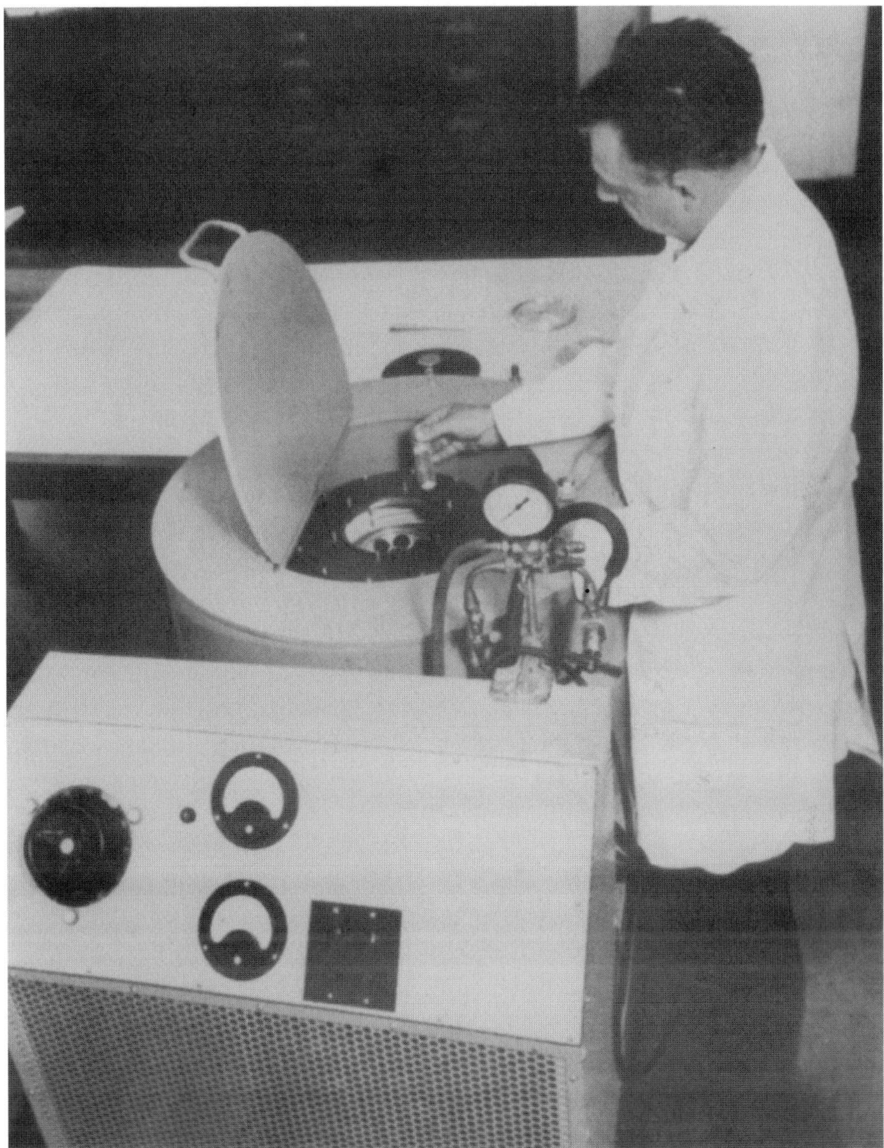

FIGURE 35.7: ULTRACENTRIFUGE AT THE CRELLEN LABORATORY (CALIFORNIA INSTITUTE OF TECHNOLOGY) CIRCA 1950. COURTESY OF THE ROCKEFELLER ARCHIVE CENTER.

The initial setting for the development of the ultracentrifuge was Theodore Svedberg's laboratory in Uppsala (Sweden). What made the apparatus special was the optical system which facilitated observation of the rotating material. Svedberg adapted the industrial milk centrifuge by including a system of lenses for viewing the sedimentation through a window. Svedberg's intention was to assess the distribution of colloids. The ultracentrifuge legend then explains that he radically changed his mind about the uses of the instrument while watching the sedimentation of an hemoglobin solution. Svedberg considered hemoglobin to be a colloid which formed aggregates of heterogenous sizes. In contrast to the expected fuzziness of the rotating solution, he observed a sharp boundary between colored and clear solution. This indicated that the particles sedimented together and that their behavior could be described with one single measurable sedimentation rate. Moving from the measurement of sedimentation rate to the computation of molecular weight, Svedberg transformed the ultracentrifuge from a tool for analyzing suspensions of non-homogenous aggregates into a device 'showing' the molecular nature of proteins through its transparent sedimentation cell and lenses. Systematic measurement and comparison of the sedimentation patterns of different preparations of proteins eventually resulted in a theory that most proteins were made up of multiples of a unit protein with a molecular weight of 17,500 that defined the Svedberg unit. By 1931, with a Nobel Prize, local patrons, and the support of the Rockefeller Foundation Svedberg had transformed his laboratory into a unique center for ultracentrifugation receiving material from many researchers. Svedberg's choice of an oil-turbine with a rotor spinning in a low pressure hydrogen atmosphere resulted in large machines which were extremely expensive to build and to maintain. Since each molecular weight determination took several days or weeks, four medium-speed machines were alternatively run in his laboratory. An alternative 'simple' ultracentrifuge was built in the United States by physicists James Beams and Edward Pickels with very different prospects.[17] They developed engines for rotating mirrors in order to measure physical changes taking place in short time intervals. In contrast to Svedberg's centrifuge, Beams' machines were bench-top air-driven apparatus, "easy to construct, that it may be widely varied to suit the need of the experimenter."

Up to 1940, half a dozen machines were operating in the United States. The Rockefeller Foundation encouraged scientists to send their material to Uppsala for analysis, and supported the construction of oil-turbine machines in a few laboratories including the Department of Physical Chemistry at Harvard Medical School (Edwin Cohn), the Princeton Laboratory of Animal and Plant Pathology of the Rockefeller Institute (Wendel Stanley), and Jack Williams' laboratory at the University of Wisconsin. Medical research during World War II triggered new uses. Ultracentrifuges were employed not only to determine molecular weights, but to prepare large quantities of homogenous fractions of blood proteins or large quantities of viruses required for vaccination experiments.[18] Biological fractions showing specific properties started to be described with numbers employed to measure the centrifugation forces.[19] Viruses, for instance, were described as '15,000 g material' or '20,000 g fraction.' In other words, entities previously described in terms of pathological effects animals (or man) were increasingly presented as macromolecules which were purified by running the ultracentrifuge. After the war, expecting a greater need for the ultracentrifuge, Pickels and Maurice Hanafin (from a glassware company) founded "Spinco," a company that would develop and sell analytical ultracentrifuges. Their first commercial model would have both analytical and preparative rotors. After two years, Spinco had orders for a few machines. Analytical use required skilled workers to operate the optical system and much interpretative work and calculations to get values of molecular weights. Spinco then changed strategy: the company tried to establish a larger market for a centrifuge which was only preparative. The new model was designed to be simple and operable by biologists and technicians untrained in the physical sciences. The preparative model secured analytical uses in two ways. Firstly, it kept Spinco financially viable. Secondly, preparative ultracentrifugation routinized the characterization of biological 'macromolecules' by sedimentation parameters. In the late 1950s, biologists were performing daily runs with ultracentrifuges in most biochemical laboratories. This way, they prepared enzymes, labeled fractions, membranes, mitochondria, chloroplasts, etc. This practice increased the demand for physicochemical investigations performed with the analytical models. Thus, a few laboratories, for instance the Virus Lab at the UC Berkeley, developed research strategies centered on the application of ultracentifugation to a wide range of biological issues translated in terms of structure of biological macromolecules.

In contrast to the ultracentrifuge, the electron microscope was never a benchtop device. Biological electron microscopy emerged as a specialty whose practice was based on the alliance of a few academic scientists and industrialists. Several electron microscope projects started in Europe in the early 1930s. By 1935, these prototypes were employed to investigate metals or conducting materials. Analysis of biological objects initially focused on materials used by the textile industry. The Radio Corporation of America, in the United States, launched a commercial electron microscope project in 1938.[20] During the war, RCA invested much time and resources in the building of the machine and, at the same time, in the development of techniques for applications in biology. Wartime research on biological objects such as bacteria or viruses was easier to advertise than classified physical applications in military research. RCA's investments resulted in a prototype operated by a postdoctoral fellow trained in physical chemistry and biology, Thomas F. Anderson. RCA established a Committee on Biological Applications of the Electron Microscope under the patronage of the National Research Council. The committee monitored Anderson's work and organized the collaboration of biologists and chemists interested in RCA's machine. The role of the committee was to evaluate the projects, assess on-going research and allocate time slots. The

RCA-NRC committee operated until 1942 when it was endorsed by the NRC as a committee of reviewers in electron microscopy.

This regulatory body included two virologists, Mudd and Stanley, contributed to stabilizing the identity of viruses. Anderson and visiting scientists had to decide whether the pictures they shot were facts or artifacts. The technique was new and every biologist was aware of the artificiality of the procedures that were employed, i.e., vacuum, desiccation, chemical treatments of specimens. One aspect of the problem was that no natural complementarity between new pictures and established practices existed. It was necessary for instance to 'triangulate' electron microscopy and ultracentrifugation.[21] Debates on the nature of 'bacteriophages' — a group of viruses infecting bacteria — illustrate the process. In November 1941, S. Luria came to RCA to discuss the possibility of using the electron microscope to measure the size of bacteriophages.[22] Most plant and animal virologists then used two different techniques to determine the size of these viruses: diffusion data and analytical ultracentrifugation. Sometimes results coincided, at other times they differed. Initially, pictures of Luria's bacteriophages were unhelpful and opposed both diffusion and irradiation measurements. Anderson's collaboration with Luria aligned irradiation and electron microscopy. They refined the pictures and stressed the fact that the images of bacteriophages revealed viruses with a complex organization which were granted with heads and tails. Alignment was made possible by taking for granted that the pictures of heads were not homogenous but displayed a genuine internal structure consisting of granules, and that these 'granules' could be genes, the elementary unit which were viewed to be measured by irradiation experiments.

By the end of the war, regulation was taken over by the Electron Microscopic Society of America. The first meeting held at the Chemical Exposition in Chicago in 1942 was a gathering of about 60 scientists representing almost all of RCA's customers. A few years later, EMSA was an enlarged professional society informally making decisions about the right tools and the right methodology.

Conclusion

The historiography of the life sciences has recently focused on the development of techniques, instruments, and research practices. This perspective has provided a richer picture of how contemporary biologists work. The case studies presented in this chapter depict arrangements of settings, tools, organisms, and people that grounded the twentieth-century interests in biochemicals, genes, and macromolecules. This exposition shows that biological knowledge depended on the mobilization of cognitive as well as industrial, social, and discursive technologies. More specifically, it reveals that the activities of twentieth-century biologists increasingly relied on the blending of biological, industrial, and medical practices. The responses of biologists to problems of access, preparation, and homogeneity of research tools illustrate this trend.

A major trend in the twentieth-century life sciences has been the rising importance of biological molecules such as metabolites, hormones, or vitamins. This pattern was based on the development of large-scale collection and sophisticated purification procedures. The 'molecularization' of the twentieth-century life sciences was not only associated with changing views on the nature of life, but also on changing forms of experimental work, and on new alliances, for instance between biological laboratories, slaughterhouses, pharmaceutical firms, and hospitals.

In addition to the purification of biologicals, twentieth-century biologists have displayed on-going investments in the selection and engineering of model organisms. Standardized husbandry, controlled breeding, mutant selection, and stock exchanges shaped experimental programs as well as genetic research collectives. In the 1920s, the laboratory fly became the ideal system for a small, paternalistic community of formal genetics interested in chromosomal cartography. In contrast, the laboratory mouse was transformed into a well-controlled biomedical machine useful for studying human diseases. During the post-war era, genetically homogenized mice became widely used research commodities. Their fate combined two frames which typify contemporary experimentation in the life sciences: the logic of technological development, and the logic of biomedical modeling. On the one hand, cancer researchers working in pharmaceutical companies, hospitals, and at the National Cancer Institute selected a few strains which were typical industrial objects mass produced in order to screen drugs. On the other hand, an expanding series of mouse mutants showing abnormalities which were compared with human pathologies contributed to mediate between the biological research laboratory and the hospital ward.

Finally, one should stress the changing role of research instruments after World War II. The emergence of macromolecules — viruses, proteins or nucleic acids — and the invention of molecular biology are linked to the formation of a new type of expertise based on the command of complex physical apparatus. Histories of the electron microscope or the ultracentrifuge show that this change did not only reinforce the control of research strategies by industrial firms such as RCA which produced these tools, but resulted in the appearance of a new character in the biological sciences: the scientist-entrepreneur who defined problems in terms of instrumentation and built vast research empires based on the quasi-monopolistic command of complex experimental systems.

References

1. G. Allen, *Life Sciences in the Twentieth Century.* (Cambridge: Cambridge University Press, 1978).
2. B. Latour, *The Pasteurization of France.* (Cambridge (MA): Harvard University Press, 1988); G. Geison, (1995).
3. K. Codell Carter (Ed.), *Essays of Robert Koch.* (New York: Greenwood Press, 1987), 96–115.
4. N. Oudshoorn, *Beyond the natural Body: The Making of Sex Hormones.* (London: Routledge, 1994).
5. A. Clarke, "Research Materials and Reproductive Science in the United States, 1850–1940." in G. Geison (Ed.), *Physiology in the American Context.* (Bethesda: American Philosophical Society, 1987), N. Oudshoorn, *op. cit.*, Chapter 4.

6. N. Oudshoom, *op. cit.*, p. 47.
7. J.R. Griesemer, "Laboratory Models, Causal Explanation and Group Selection." *Biology and Philosophy* (1988), **3**: 67–96.
8. G.F. Allen, "The Introduction of Drosophila into the Study of Heredity and Evolution: 1900–1910." In *Science in America Since 1820*, N. Reingold (Ed.). (New York: Science History Publications), 266–277.
9. G. Allen, *Thomas Hunt Morgan, The Man and his Science.* (Princeton: Princeton University Press, 1978).
10. R.E. Kohler, *Lords of the Fly: Drosophila Genetics and the Experimental Life.* (Chicago: The University of Chicago Press, 1994).
11. R.E. Kohler, *Lords of the Fly: Drosophila Genetics and the Experimental Life.* (Chicago: The University of Chicago Press, 1994).
12. G. Allen, "William Ernest Castle", in *Dictionary of Scientific Biography*, C. Gillispie (Ed.). (New York: Scribner's, 1970), vol. 3, pp. 120–124; W. Provine, *Sewall Wright and Evolutionary Biology.* (Chicago: University of Chicago Press, 1986).
13. I. Löwy, J.P. Gaudilliere, "Disciplining Cancer: Mice and the Practice of Genetic Purity" in Gaudilliere and Löwy (Eds.), (1997).
14. Staff of the Jackson Laboratory, *Biology of the Laboratory Mouse.* (New York, Dover Publications, 1966).
15. L. Kay, *The Molecular Vision of Life.* (Oxford: Oxford University Press, 1994).
16. R.E. Kohler, *Partners of Science.* (Chicago: The University of Chicago Press, 1991).
17. B. Elzen, "Two Ultracentrifuges: A Comparative Study of the Social Construction of Artefacts", *Social Studies of Science* (1986), **16**: 621–662.
18. A. Creager, "Producing Molecular Therapeutics from Human Blood: Edwin Cohn's Wartime Enterprise" in *Molecularising Biology and Medicine. 1930–1970*, H. Kamminga, S. de Chadarevian (Eds.). (London: Harwood, forthcoming).
19. H.J. Rheinberger, "From Microsomes to Ribosomes: Strategies of 'Representation'", *Journal of the History of Biology* (1995), **28**: 49–89.
20. N. Rasmussen, *Picture Control: The Electron Microscope and the Transformation of American Biology, 1940–1959.* (Stanford: Stanford University Press, 1997).
21. S. Leigh Star, "Triangulating Clinical and Basic Research: British Localizationists, 1870–1906." *History of Science* (1986), **24**: 29–48.
22. T. Anderson, "Electron Microscopy of Phages", in *Phage and the Origins of Molecular Bioloy*, Cairns *et al.* (Eds.). (Cold Spring Harbor Laboratory, 1966), pp. 63–77.

Further Readings

G. Allen, *Life Sciences in the Twentieth Century.* (Cambridge: Cambridge University Press, 1978).

A. Clarke, J. Fujimura (Eds.), *The Right Tool for the Job: At Work in Twentieth Century Life Sciences.* (Princeton: Princeton University Press, 1992).

J.P. Gaudillière, I. Löwy (Eds.), *The Invisible Industrialist: Manufactures and the Production of Scientific Knowledge.* (London: Macmillan, 1997).

G. Geison (Ed.), *Physiology in the American Context.* (Bethesda: American Philosophical Society, 1987).

G. Geison. *The Private Science of Louis Pasteur.* Princeton: Princeton University Press, 1995).

L. Kay, *The Molecular Vision of Life.* (Oxford: Oxford University Press, 1994).

R.E. Kohler, *Partners of Science.* (Chicago: The University of Chicago Press, 1991).

R.E. Kohler, *Lords of the Fly: Drosophila Genetics and the Experimental Life.* (Chicago: The University of Chicago Press, 1994).

N. Oudshoorn, *Beyond the Natural Body: The Making of Sex Hormones.* (London: Routledge, 1994).

N. Rasmussen, *Picture Control: The Electron Microscope and the Transformation of American Biology, 1940–1959.* (Stanford: Stanford University Press, 1997).

S.L. Star, *Regions of the mind: Brain research and the quest for scientific certainty.* (Stanford: Stanford University Press, 1989).

CHAPTER 36

Mastering Nature and Yeoman

Agricultural Science in the Twentieth Century

DEBORAH FITZGERALD

It is an irony of modern life that the production of one of life's necessities — food — is considered not a 'basic,' but an applied, science. The term agricultural science, in this sense, is considered not a disciplinary distinction, but a description of the sorts of problems other scientific areas are pointed towards. Agricultural chemistry is chemistry directed toward agricultural problems, agricultural economics is industrial economics turned to farm production, and so forth. But in the twentieth century the farm and agricultural laboratory have often themselves been the sites of knowledge production, as in the case of genetics, and have generated fundamental knowledge both within and without the context of production problems. Furthermore, such contexts arguably contribute to the character of scientific knowledge, serving not simply as empty receptacles but rather as participants in an ongoing discourse about nature's demands. To use a mathematical metaphor, agricultural sciences are not types or subsets of ordinary sciences, but rather venn diagrams that join ordinary scientific knowledge with circumstances, quite literally, in the field.

As a genre of knowledge and work, agricultural science subsumes a rather large set of locales and identities. It includes sciences such as genetics, plant physiology, and botany, livestock breeding, nutrition, parasitology, bacteriology, and chemistry. Agricultural research is conducted on the range, in the poultry-house, in the canning factory, among herds of cattle, standing in prairie grass, in orchards and fields, in migrant labor camps, feedlots, and of course in sparkling government and university laboratories. Virtually no area of agricultural work has escaped the rationalizing hand of science, and in the United States at least, the term 'scientific agriculture' is now a redundancy. Each of the agricultural sciences have developed at its own rate and for its own reasons. Some owe more to the luck and audacity of individuals, as with Luther Burbank and horticulture, than to scientific discoveries or institutional support. Others, like animal nutrition, were as related to medicine and biochemistry as to agriculture per se. Still others, such as agronomy, which deals with grasses, was practically created out of thin air to deal with the recently-settled arid prairies of the western United States. Thus, it is difficult and foolhardy to generalize too much in this brief consideration of agricultural science, and I will attempt to draw out the general trends rather than individual distinctions.

In the United States, scientific research in agriculture has been tied to the federal government and land-grant college system since 1862, when the federal government established agricultural (and mechanical) colleges in every state for the purpose of educating citizens, and particularly immigrants, in the ways of the mechanical and farming arts. When it was discovered by 1887 that few educated persons — 'experts' — possessed any systematic and reliable knowledge about agriculture, the Hatch Act was passed by Congress to fund the creation and dissemination of agricultural knowledge. In 1906 the Adams Act provided more funding for original scientific research, and in 1914 the legislation establishing the Extension Service was passed, which guaranteed that the results of college-based agricultural research would be carried out into the farming community. In the early years, virtually all agricultural research was conducted under the auspices of the United States Department of Agriculture (USDA) and/or the state agricultural experiment stations. Much of what passed for scientific research was in fact the collection of raw material and data. In each state, scientists tried to determine the natural boundaries of agricultural practice, from soil types and climatic patterns to reliable crops and livestock for regional needs. At the USDA, scientists collected and compared these reports, attempting to chart national trends. Local knowledge formed the backbone of all later research programs.

Research in agriculture emerged from simple questions, and tended to remain rather local in character. Hybrid corn, one of the notable examples of agricultural science, in many respects grew out of small, local concerns. In Illinois, as in other states where it was the major field crop, corn was the subject of much research because it was plentiful, easy to breed, and of great interest to farmers and legislators whose approval was necessary to keep the college running. In other states, too, experiment station scientists tended to ask small questions that could be answered with local materials rather than large questions that could not, with the result that there was a lot of repetition between different states, and a lot of idiosyncratic projects. Nonetheless, by the early twentieth century, agricultural scientists had collected a great deal of material, posed quite a few questions, and in some areas had completed the 'basic' work upon which more sophisticated projects depended.

Also in the early years, certain relationships and presumptions were established that would later cause some difficulties, particularly in terms of the agricultural scientists' persistent identity crisis vis-a-vis other scientists. First and foremost, patterns of funding had an enormous effect on the problems agricultural scientists chose to work on. Federal funds were linked to state funds in complicated ways, so that each year the dean of the agricultural college had to persuade state legislators that the college and experiment station had done work of use and interest to the state. Since in rural states particularly, prosperous farmers had a great deal of influence with legislators, research projects were frequently geared towards their interests rather than the interests of, for example, tenant farmers or more general scientific theories. As the twentieth century progressed, scientists were driven to seek funding beyond the statehouse. In a pattern familiar to all scientists, they accepted funding from businesses, many of whom actively cooperated in the research itself by means of their own in-house laboratories and staffs. For such companies, cooperative research bought more than mere knowledge; it garnered as well the imprimateur of the cooperating college, a big issue as agricultural businesses became more competitive.

The difference between scientific research conducted within the land grant system and that conducted within private companies was not as stark as one might imagine. In some ways it was more cooperative than competitive in the early years, partly because the community of agricultural scientists was fairly small. Thus, plant breeders working on corn in Illinois in the 1920s were all on a first name basis, because they knew each other well and exchanged both research materials and findings whether they worked at the college of agriculture or the seed company. Especially during the research phase of a project, interaction between the two groups was continuous, even though there was little or no money involved. Exchanging seed stocks, getting together to create an experiment to understand (and eliminate) root rot, promoting new tillage techniques among farmers, collaborating on frost studies — in many ways this worked as a scientific partnership. And all parties seemed to benefit. The agricultural college had access to vast experimental fields owned by the company instead of the college's meagre fields, and, as mentioned earlier, the company gained the scientific expertise of all the workers in the agricultural college — chemists, physicists, breeders, botanists, meteorologists — which the company could not support on its own. This relationship, which for a particular historical moment might be seen as collaborative, would soon come to be judged as 'cozy,' in the perjorative sense, as the differential effects of research were better understood in the 1960s and later.

The shifting character of agricultural science can be characterized in several ways. First, and perhaps most obviously, knowledge that in 1900 was tacit, rule of thumb, experiential, and held almost entirely by farmers themselves rather than non-farmers (would-be 'experts'), by 1960 had been captured by such experts (now called scientists) and codified, tested, straightened out, manipulated, and reformed into scientific knowledge. In some areas, such as chemistry, this happened just before the turn of the century, while in some others, such as livestock breeding, one could argue that the process is still underway. The relative importance of different agricultural actors changed profoundly and irrevocably during this period. In 1900 the actors in agricultural science were the farmers, who supplied raw materials and what was, in effect, laboratory space (the fields and barns), the few professors in agricultural colleges, and small-town businessmen who sold seeds, implements, and so forth. By the time of the New Deal agricultural businesses had become big players in the agricultural enterprise. Generally, most major transitions of this sort occurred between 1920 and 1960.

Second, the expanding role of corporate actors has led to an increasingly industrial mentality in agriculture, evidenced most notably by efforts to turn simple farming practices into production systems, that is, systems that are similar to industrial or factory approaches in that materials and processes are specialized, automated, and integrated. In some cases this came about through vertical integration; one thinks here of grain-based corporations such as Pillsbury, which contracted with grain farmers in the American midwest, or fruit and vegetable producers who controlled the process from growing to canning in the western United States. In other cases the combination of new scientific knowledge, such as Mendelism, was linked to family-based money and opportunity, in companies such as Pioneer Hybrids or Funk Brothers Seed Company. And in the case of machinery, research and development sometimes emerged as a result of a capability in building other metal things, such as automobiles. In all cases, the trend was for private companies to mimic and then to overwhelm the research function previously performed by the agricultural colleges. As the century progressed, few colleges could compete with the companies in terms of capital, equipment, or even attracting scientific talent.

The third major theme concerns the circular path of agricultural knowledge and materials. Both England and America, and other countries to a lesser extent, built their scientific preeminance and agricultural flexibility on plant materials indigenous to other countries. Cotton, sugar, wheat, rice, and many fruits spring most readily to mind. Not only plant explorers in the employ of the federal government, but also amateur naturalists, tourists, and businesspeople found amusement and profit in collecting new plant strains that very often were 'improved' through breeding or genetics. Plants were dug from foreign soil, deposited in American labs, tested in American fields, packaged in corporate labs, and then sold as American seeds. Some plants, such as rice or sugar, did not have to leave their tropical homes for this process to occur; Western scientists simply built agricultural scientific systems in these locales. This practice accounts for quite a sizeable portion of scientific research in agriculture today.

These themes and trends cut across the many discipline of agricultural science, and the historian can easily become overwhelmed by the multitude of idiosyncratic cases. What follows, then, is an effort to pull several scientific threads out of the larger picture to better illustrate the changing nature of agricultural science.

Plant Breeding and Genetics

As one of the earliest and most prominent examples of the application of science to agricultural problems, the development of hybrid corn provides an instructive lesson in the complex relationships between private, corporate research goals and those of the public sector; between science that is problem solving and that which is problem generating; between the boosterish rhetoric of scientists and salesmen alike, and the realities of farmers trying to puzzle out what agricultural science would mean for them.

What ended up as hybrid corn began as a problem-solving activity within at least two different communities, neither of which cared much about the other. One group, seedsmen, were people who made a living selling seeds of corn, wheat, rye and so forth to farmers for spring planting. Some of them, such as E.D. Funk in Illinois, had also been trying to cross-fertilize those individual plants that seemed better than the others in some way, but felt that once these 'improved' strains were sold to farmers the superiority of the strain was diluted by farmers who neglected to keep it separate from poorer strains. Such breeders longed for a way to protect such plant creations; they wanted biological protection and control to prevent the plant's random mating, and they wanted financial protection and control because they felt they had invested a lot in these new plants. While most such breeders were not trained in science, their experience with plants, soils, machinery, weather, animals, and chemistry gave them certain advantages over college-trained but inexperienced plant breeders.

The second group were the geneticists who were interested almost entirely in Mendelian theory. The rediscovery of Mendel's Laws in 1900 offered a revolutionary new way of understanding how organisms inherited or failed to inherit the characteristics of their parents, an idea that had implications for those seeking to direct and shape organismic change. While some geneticists used the rapidly-reproducing fruit fly to study patterns of inheritance, others used the easily-manipulatable corn plant. Using techniques of pure-line breeding and self-fertilization, geneticists were able to study the ways in which characters were inherited. One of the most prominent sites for this research was R.A. Emerson's maize genetics laboratory at Cornell, where Barbara Mcclintock did her pioneering work on transposition.

The third and most important group of people, plant breeders working at the state experiment stations, stood somewhere between the commercial breeders and the geneticists in terms of research interests and training. On the one hand, college breeders wanted to breed corn strains that could withstand nature's depredations — high winds, pests, diseases, droughts, and so on. Fewer losses in the field meant greater yields, which farmers believed would lead to higher profits, and this belief could, in turn, lead to a favorable consideration in legislative appropriations to the agricultural college. On the other hand, many such breeders had been trained in the new genetics, and they were interested in understanding patterns of inheritance as well as understanding field behavior of corn. For them corn provided a happy coincidence of materials and interests.

Not surprisingly, hybrids were developed by three people who had links to both the farm and the laboratory: E.M. East, George Shull, and Donal Jones. East and Shull discovered semi-independently in 1909 the single-cross method of breeding. By this method, ears of corn were self-fertilized for six or seven generations, a process that was thought to concentrate dominant characteristics. An inbred ear could then be cross-fertilized with another inbred ear, creating a single-cross. While most such crosses were unexceptional or even terrible from a farmer's point of view, every so often a cross would produce an ear with a desirable combination of characteristics. However, even good crosses were too small to be commercially viable, however, and in 1919 Jones developed the double-cross by cross-breeding two single-cross lines. From 1919 until the mid 1930s, breeders spent most of their time trying to find inbred lines that would 'nick,' that is, form an exceptional strain when crossed with another inbred. But such inbreds turned out to be fairly rare, and for many years all the commercial and agricultural college hybrids were composed of various combinations of only a dozen or so good inbreds.

By the mid-1940s most American farmers had switched to hybrid corn, and in the ensuing decades hybrids have kept pace with emerging trends in biotechnology. The creation of cytoplasmic male sterility in the 1960s shifted production away from double-crosses back to single-crosses, and in recent years recombinant techniques have created corn strains targeted for resistance to particular insects or diseases. The trend, begun in the 1920s, continues to emphasize scientific manipulation of the plant and the tacit goal of total environmental mastery, and an increase in yields but a decrease in the number of farmers producing those yields.

As it turned out, corn was somewhat anomalous as an example of the power that scientific technique could exert over agriculture. Lots of crops were much more difficult to hybridize; wheat, for instance, has such tiny features that inbreeding and crossing was both too tedious and too costly for most breeders, and efforts to improve the plant have centered on simple breeding. Corn was unusual too in that while corn hybrids were quickly followed by corollary inventions such as mechanical corn pickers and specialized fertilizers, the original hybrids were basically created as a single innovation rather than a package. Since then, however, many plant innovations have been conceived as part of larger systems of production.

In the case of tomatoes, scientific research was directed not toward increasing production directly, but toward making the tomato part of a system of production that included mechanical harvesting. The first goal was approached by means of varietal crossing, that is, by crossing strains that had desirable characteristics and tossing out strains that had undesirable characteristics. Such a determination depended upon whether the crop was bound for the fresh produce market or the processing market. But in the mid-1940s, University of California-Davis plant breeder G.C. Hanna abandoned the more traditional breeding goals in favor of creating a tomato that could withstand the rough handling of a picking machine. The characteristics Hanna was looking for included firmness of the fruit, simultaneous

Figure 36.1: Frank N. Meyer's cart in Turkestan in 1911. Taken from *After a Hundred Years: Yearbook of Agriculture*, 1962 (Washington, DC, Government Printing Office, 1962).

Figure 36.2: Controlled hand pollination of corn. Taken from *After a Hundred Years: Yearbook of Agriculture*, 1962 (Washington, DC, Government Printing Office, 1962).

Figure 36.3: A farmer in Oregon watches as a soil scientist sketches the boundaries of soils he has studied. Taken from *After a Hundred Years: Yearbook of Agriculture*, 1962 (Washington, DC, Government Printing Office, 1962).

ripening of fruit in the field, and ease of removing fruit from the stem. It also had to be no worse than preferred cultivars in terms of yield, disease resistance, flavor, and so forth. At the same time, Hanna's colleague, agricultural engineer Coby Lorenzen, was developing a machine that would pick the fruit mechanically, a goal inspired by the threatened demise of the *bracero* program that provided migrant labor to the California vegetable and fruit fields. Both Hanna and Lorenzen were successful in their goals, and the resulting 'hard tomato' gained notoriety both for its lack of flavor and its role in reducing the number of small tomato growers in California.

Breeding crops to meet particular desires has long been part of the agricultural landscape, although the limits on what plants and common sense could tolerate have continually been pushed further. In the late 1890s, Cyril Hopkins at Illinois tried to breed corn with high protein and low oil content once he realized how easily corn responded to selection of extreme characters. His idea was to market specialized feed corn to livestock farmers. What he found, however, was that characters were linked in such a way that the breeder could not collect all desirable traits simultaneously; high protein might link with small ears, or susceptibility to cinch bugs. There was no magic bullet. Later breeders have found that they must choose which characteristics to emphasize. For the tomato breeders at Davis, the most important character was sturdiness and toughness of skin: flavor, shape, color, and other characteristics were secondary to this goal. Green cauliflower, square broccoli, tiny squash — it would seem that by means of variation and hybridization, agricultural scientists can rationalize plant life to suit themselves.

Animals

In animal sciences, breeding, mechanization, and pharmaceuticals have transformed the humble barnyard menagerie into a highly industrial and systematized enterprise. Here again, what started as simple breeding for particular characteristics has become a complicated, multi-pronged effort to control all aspects of production. Research on animals has tended to resemble the input-output model of industrial production, in which livestock are black boxes into which scientists pour feed, vaccines, and hormones, and out of which come meat and milk. While early twentieth century animal research centered on problems involved in raising each type of animal, with little interaction between, for instance, poultry and dairy scientists, in the last two decades researchers have become fascinated by more systematic approaches to production that emphasize the common features of animal-raising rather than the differences. Such efforts combine breeding and genetics, food additives and nutritional supplements, vaccinations, and highly controlled housing.

This ultra-commodification of animals has resulted in dramatic images. Changes in the poultry industry, for example, have been driven by capabilities in barnyards and laboratories to breed chickens for specific industrial and consumer uses, to lower the chickens' vulnerability to pests and disease with pharmaceuticals, and to mechanize their upkeep with giant chicken houses in which food, water, 'daylight,' and temperature are managed from a control panel. Hogs have been similarly transformed by means of breeding and the creation of the 'confinement' system, by which they, like chickens, live out their days in artificially-controlled environments. Advocates of this type of agriculture claim that it avoids the need for dirty, tedious, hit-or-miss approaches to animal-raising that result in bad food and poor farming lives. But critics maintain that it leads to capital domination of markets, the demise of the family farm, and lax supervision by government. Whatever one's view, it is inescapable that the growing dominance of this type of agriculture in the United States and elsewhere is the culmination of both scientific practices and innovations and a scientistic ethos of production.

A particularly robust branch of research was centered on animal growth hormones and their regulation of metabolism, and particularly on the creation of synthetic hormones that allow scientists and farmers to artificially control such functions as milk production in dairy cows. Studies of bovine growth hormone (BGH), which scientists report they began in the 1940s as basic research into animal physiology, led in the 1980s to the creation of a synthetic version of BGH made with recombinant DNA techniques and standard fermentation processes. The research was conducted mostly at Cornell University, and funded by Monsanato and several smaller agricultural corporations. The idea behind the commercial release of BGH was that, with daily injections of the hormone, lactating cows could produce a remarkable increase of 10 percent to 40 percent more milk, which industry representatives argued would increase the dairy farmers' profit as well as efficiency. Shortly after the public announcement of the drug's release, but before approval by the Food and Drug Administration, public concern began to mount, and several states decided to call a moratorium on BGH. The most significant reason for the outcry was the public's concerns about BGH residue in milk products, and secondly about the effect of BGH on small dairy farmers, already struggling to survive. As with so many other agricultural innovations, the debate turned on regulation.

One of the most notorious creations by agricultural scientists, diethylstilbestrol or DES, was likewise an attempt to accelerate the rate of growth of beef cattle. DES was first synthesized in 1938, and quickly became the standard feature of endocrinology studies, as well as the most commonly used estrogen. At Iowa State University in the early 1950s, Wise Burroughs worked out a way for DES to be given to beef cattle in feed rations, and received Food and Drug Administration approval very quickly. While the relation between synthetic estrogenic hormones and cancer had been suspected, DES remained the standard feed additive until it was banned in the 1970s when cancer clearly caused by DES was diagnosed in American women.

CHEMICALS, PESTS AND DISEASE

Agricultural scientists have been tremendously active in attempts to eradicate pests and disease. These efforts have taken several forms. In a sense, the most straightforward approach to plant protection was simply to throw out seed that had been damaged by pests or disease. Plants that survived an infestation of insects or an attack of rot or wilt were, by definition, more resistant than their dead neighbors, and were often billed as such by seed salesmen. Not much science was required for this response. But from the beginning of Mendelian breeding programs, the idea of breeding for resistance to particular problems was a dominant concern, and both garden plants and field crops have been bred with pest and disease resistance as the primary concern.

Another line of research has been in biological control programs, in which the problem of insect pests was addressed by introducing natural enemies to keep the pest in check. The most heralded example of this approach occurred in the 1890s in California, when the 'cottony cushion scale' threatened to destroy the citrus orchards. After traditional poisons failed to help, USDA agent Albert Koebele brought back from Australia a ladybug known as *vedalia* that solved the scale problem almost immediately. As it turned out, this kind of success was unusual, and later attempts to introduce predators for the boll weevil and gypsy moth did not work very well. One reason success was somewhat elusive was that it often took more time for scientists to identify, locate, breed, and distribute predators, than it did for farmers to buy and apply poisons that, whatever their long-term difficulties, often seemed immediately effective. Many agricultural colleges struggled to maintain programs in biological control throughout the century but they were often neglected as funding sources, including the USA, came to favor chemical controls.

The use of chemicals to kill insect pests dates back into the nineteenth century, when treatments made with arsenic, lead, and Paris green were commonly applied to crops demonstrating insect or disease vulnerability. Even though scientists were aware that such treatments could pose serious health problems, regulation proved difficult. With the introduction of DDT after World War II, federal and state agencies began spraying campaigns to kill the beetles responsible for Dutch Elm disease, gypsy moths, and a host of others insects. Related chemicals were used to attack fire ants, which had been considered a minor nuisance, but this had the paradoxical effect of increasing the numbers of ants by eradicating its enemies. With the publication of Rachel Carson's *Silent Spring*, the public debate on the use of pesticides reached such proportions that DDT itself was banned. Nonetheless, agricultural scientists working for both state agricultural colleges and private corporations continue to work in a world in which the dominant paradigm emphasizes chemicals.

Conclusion

In many respects, changes in agricultural sciences have mirrored changes in other natural sciences over the last century. As life sciences have shifted from the organismic to the molecular, from morphology and physiology to biotechnology, so too have agricultural problems shifted from the field to the laboratory. Rather than selecting their own seed from their fields, or spreading manure from their own livestock, or breeding their own sows and mares, farmers now turn to college and corporate experts for help with virtually all production activities. As the unit of analysis gets smaller, the scale of operation and production seems to get bigger. Although agriculture is one of the few knowledge areas in which a group analogous to farmers exists as major actors, the knowledge produced is still as opaque as that produced in more academic scientific areas.

Despite the increasingly sophisticated scientific knowledge base, and the complex systems of production, agricultural science still relies upon and resides within the fields and feedlots of the rural landscape. There may be many fewer farmers practicising their trade, but wheat fields still roll across the plains, cattle still wander through meadows, and fruit still grows on trees. Agricultural scientists continue to visit the countryside in search of problems needing solutions, and farmers continue to expect the agricultural college and federal agricultural establishment to respond to his own particular needs and problems. Thus, the intellectual, academic, and social issues that shaped the agricultural sciences 100 years ago still obtain, and the tensions within the system continue to exist.

In the end, the most powerful and lasting innovation to emerge from twentieth-century agricultural science will prove to be the system of agricultural research and production itself. More than science *qua* science, the system of producing scientific knowledge and reconfiguring societies with that knowledge is already profoundly changing agricultural societies in a global context. As American agricultural scientists come to feel that opportunities for meaningful change in agriculture are diminishing in the North, they turn toward the South looking for new problems and materials. And increasingly what is sent to those societies, in addition to scientific and technical knowledge, is the organizational structure for deploying it. *That* is the nature of agricultural science at the close of the twentieth century, for better or worse.

Further Reading

Busch, Lawrence, William B. Lacy, Jeffrey Burkhardt, and Laura B. Lacy, *Plants, Power and Profit: Social, Economic, and Ethical Consequences of the New Biotechnologies.* (Cambridge, Mass.: Blackwell, 1991).

Fitzgerald, Deborah, *The Business of Breeding: Hybrid Corn in Illinois, 1890–1940.* (Ithaca, New York: Cornell University Press, 1990).

Hightower, Jim, *Hard Tomatoes, Hard Times.* (Cambridge, Mass.: Schenckman, 1973).

Kloppenburg, Jack Ralph, *First the Seed: The Political Economy of Plant Biotechnology.* (New York: Cambridge University Press, 1988).

Marcus, Alan I., *Agricultural Science and the Quest for Legitimacy: Farmers, Agricultural Colleges, and Experiment Stations, 1870–1920.* (Ames, Iowa: Iowa State University Press, 1985).

------, *Cancer from Beef: DES, Federal Food Regulation, and Consumer Confidence.* (Baltimore: Johns Hopkins University Press, 1994).

Molnar, Joseph and Harry Kinnucan (Eds.), *Biotechnology and the New Agricultural Revolution.* (Boulder: Westview Press, 1988).

Palladino, Paolo, "The Politics of Agricultural Research: Plant Breeding in Britain, 1910–1940," *Minerva* (1990), **28**: 446–68.

Perkins, John H., *Inspects, Experts and the Insecticide Crisis: The Quest for New Pest Management Strategies.* (New York: Plenum, 1982).

Rosenberg, Charles, *No Other Gods: On Science and American Social Thought.* (Baltimore: Johns Hopkins University Press, 1976).

Rossiter, Margaret, *The Emergence of Agricultural Science: Justus Liebig and the Americans, 1840–1880.* (New Haven: Yale University Press, 1975).

------, "The Organization of the Agricultural Sciences," in Alexandra Oleson and John Voss (Eds.), *The Organization of Knowledge in Modern America, 1860–1920.* (Baltimore: Johns Hopkins University Press, 1979), pp. 211–48.

CHAPTER 37

The Role of Physical Instrumentation in Structural Organic Chemistry

PETER J.T. MORRIS AND ANTHONY S. TRAVIS

During the twentieth century, many changes have taken place in the organic chemist's laboratory, but none greater than the introduction of electronic instrumentation. The widespread adoption of instrumental methods has led to tremendous advances in the determination of complex structures. Between 1940 and 1970, structural studies were reduced from being life-long Nobel Prize-winning activities for leading professors to a day's work for graduate students and technicians. Doubtlessly, this shift has transformed the chemist's work and the nature of different jobs within chemical laboratories. Some organic chemists found more time to work on organic synthesis, others transferred their attention to biomolecular topics, and yet others were forced to find alternative careers outside chemistry. Even a leading pioneer of the new methods, Carl Djerassi, found reason to regret the passing of the former intellectual and creative challenge of structure determination:

> But if eliminating the need for 'wet chemistry' (the laboratory equivalent of 'Twenty Questions') saves a lot of time and material, it also makes structure elucidation a more mechanical endeavor. Ironically, much of our own research into better flashlights [physical instrumentation] has made obsolete the traditional and often intellectually exciting ways of exploring dark rooms [organic chemical structures].[1]

Remarkably, the emergence of, and the present-day reliance upon, instrumental methods, has received little attention from historians, although it spans the history and sociology of science, the history of technology and even business history. The vastness, diversity and complexity of both the science and the technology of these developments explain the dearth of any significant historical literature. Furthermore, the displacement of classical methods by physical instrumentation was completed only in the last two decades. The story involves instrument manufacturers, chemists, physicists, government agencies and the chemical industry. Here, we will delineate the impact of instrumentation on the determination of the structures of organic compounds by focusing on the natural products that are the lifeblood of organic chemistry and the mainstays of the biomedical sciences.

During the second half of the nineteenth century, the structures of many important natural products had been established, with considerable degrees of accuracy, by lengthy and painstaking studies. By 1900, the chemical laboratories in which this work was carried out were characterized by reagent bottles and test tubes, the glass and porcelain apparatus employed in qualitative and quantitative analysis, and combustion furnaces necessary for routine, but extremely tedious, elemental analyses. Together, these made up the standard tools used to study the reactions and decompositions of natural products. They were often accommodated in cupboards and on shelves that overlooked long wooden, and highly polished, benches, with sinks installed at the ends. A few fume cupboards enabled the comparatively safe handling of dangerous chemicals. The laboratory techniques were what we now call 'wet and dry.' There were very few physical instruments to be found in such laboratories: a chemical balance, a simple apparatus for the determination of melting points, and perhaps a polarimeter, used to examine optically active compounds such as carbohydrates and proteins. By 1980, most of those methods had been superceded by physical techniques, which had not originally been developed for such studies.

Early Responses to the New Instrumental Technology

The most powerful, and ubiquitous, of the new tools used in structural elucidation is the nuclear magnetic resonance, or NMR, spectrometer. In a 1985 interview Laurie Hall, the newly appointed professor of medicinal chemistry at the University of Cambridge, observed that "NMR has developed in the past two and a half decades to the point where for many chemical scientists, it is now the dominant analytical and structural tool." Twenty-six years earlier, Hall had "quite by accident . . . 'discovered' NMR. What I didn't know at the time was that there were only half a dozen NMR machines in the whole of Britain; but I decided that I wanted to do NMR because even at that early date (1959), it was quite obviously going to become an important technique in organic chemistry."[2]

A decade earlier, it was far more difficult to appreciate the potential offered by NMR. MIT physical organic chemist John D. Roberts first heard about NMR from Richard Ogg of Stanford University:

> One day in late 1949 or 1950, he [Ogg] was at MIT, and I invited him to lunch. He was really wound up and proceeded to tell me about the wonderful new magnetic resonance spectroscopy, with such promise for chemistry. I wish I could say that I could understand even 5 percent of what he told me, but I had too little knowledge of magnetism and absorption of radio frequency radiation — indeed, hardly any knowledge of other radiation . . . It was clear there were applications to chemistry, even if I didn't understand what they were.[3]

These two responses to an emerging laboratory technology provide as good an insight as any into the ways in which organic chemists came to appreciate how, with the aid of instruments, physical properties could be related to structural features. Here we explore these developments during the middle of the twentieth century, requiring of the reader no special knowledge of the techniques, certainly no more than the level of initial understanding held by John D. Roberts in 1949. To place the story in perspective, however, it is necessary to review the classical, pre-instrumental methods used for determination of structure by chemical means.

Classical Methods

The modern representations of organic chemical structures originated with the concepts of the tetravalent carbon atom and the six-carbon benzene ring, as suggested by August Kekulé in 1858 and 1865, respectively. These enabled graphic representations of aliphatic and aromatic molecules. With the aid of chemical analysis that afforded empirical formulae, from which molecular formulae were derived, and prior knowledge of the atoms in functional groups, such as hydroxyl (OH) and carbonyl (C=O), good approximations of the structural formulae of relatively complex molecules could be drawn. A good early example is the structure of alizarin, a constituent of the madder root, and an important dyestuff. During 1868, using qualitative tests, Carl Graebe and Carl Liebermann showed that it was a quinone with two carbonyl groups, and that it probably contained two hydroxyl groups. However, it was the action of chemical 'brute force,' in this case through zinc dust distillation, that provided the main clue to the constitution, and a partial structure. The reduction with zinc gave anthracene. By 1874, the total structure of alizarin was available, as were the structures of related hydroxyanthraquinones. Since these compounds were colored, it was possible to identify them by spectral analysis, through their characteristic 'fingerprints.'

Other 'brute force' chemical methods included acid and alkali hydrolysis, and nitric and chromic acid oxidation. A series of specialized tests for particular functional groups was slowly built up, the most pungent of which was doubtlessly August Hofmann's carbylamine test for primary amines. These methods permitted the structural elucidation of indigo and purines by test-tube methods. The detection of molecular asymmetry through measurement of optical rotation with the polarimeter became an important tool in the investigation of the structures of carbohydrates and proteins by Emil Fischer, from 1891. Another important contribution to organic structural chemistry was ozonolysis, introduced by Carl Dietrich Harries during 1903–5. Because it revealed the position of carbon-carbon double bonds in molecules, ozonolysis was particularly important for the study of natural rubber by Harries between 1903 and 1916, and the investigation in the early 1930s of the vitamin A precursor α-carotene by Paul Karrer.

In general, parent molecules were reconstructed from known fragments, using intuitive assumptions, such as the 'isoprene rule' introduced by Leopold Ruzicka in 1922, and thinking by analogy. These 'wet and dry' methods, were, however, time-consuming. Complex structures could take two to four decades to resolve. The work on chlorophyll, started by Richard Willstätter in the early 1900s, led to

TABLE 37.1: NOBEL LAUREATES IN STRUCTURAL ORGANIC CHEMISTRY.

1902	Emil Fischer	Sugars, purines
1905	Adolf von Baeyer	Indigo
1915	Richard Willstätter	Chlorophyll
1927	Heinrich Wieland	Bile acids, steroids
1928	Adolf Windaus	Steroids, vitamins
1930	Hans Fischer	Blood pigments, chlorophyll
1937	W.N. Haworth	Sugars, vitamin C
	Paul Karrer	Vitamins A and B_2
1938	Richard Kuhn	Carotenoids and vitamins
1939	Adolf Butenandt	Sex hormones, steroids
	Leopold Ruzicka	Polyenes, higher terpenes
1947	Robert Robinson	Strychnine, etc.
1958	Frederick Sanger	Insulin
1962	Max Perutz	Myoglobin and haemoglobin
	John Kendrew	ditto
1964	Dorothy Hodgkin	Vitamin B_{12}
1965	R.B. Woodward	Mostly for his synthetic work

In later years it becomes harder to separate structural organic chemistry from molecular biology or biochemistry, for example, Alexander Todd's Nobel Prize (in 1957) for his work on nucleotides, or the medicine-physiology Nobel Prize awarded to James Watson, Francis Crick and Maurice Wilkins in 1962 for the determination of the structure of DNA.

a 'clover leaf' structure in 1912, and the full structural determination in 1939 by Hans Fischer. This was confirmed (using physical methods) by Patrick Linstead in 1955–6. The magnitude of these tasks was such that they were often rewarded with the highest accolades, including the Nobel Prize (see Table 37.1). Certainly they added greatly to the stock of chemical knowledge and laid the foundations for the introduction of physical techniques in structural organic chemistry.

The golden era of classical determinations began around 1930 and lasted almost twenty years. It resulted from a 'critical mass' of accumulated data. In addition to chlorophyll, the triumphs included most of the vitamins (A, B_1, B_2, B_6, C, D, E), many steroids, quinine, and, above all, strychnine. The empirical formula of strychnine had been established as early as 1838; it took another sixty years to identify a benzene ring attached to a nitrogen atom in the molecule. Between 1910 and 1932, William Henry Perkin, Jr., and Robert Robinson, made further contributions towards the elucidation of the structure of strychnine. In 1929, they showed that it was an indole derivative. The position of a second nitrogen atom could not be determined until 1948 when Robert Burns Woodward resolved several controversies and drew the first correct structural representation of strychnine. Arthur J. Birch summed up the chemist's approach during this time:

> Our classical natural products, typically extractable by organic solvents, made available interesting but not too complex molecular structures as exercises for chemical investigation and training. Many such substances were of interest in connection with human applications as dyestuffs, pharmaceuticals, tanning agents, psychedelic agents, and so forth. Classical structure determinations (as practised until about the 1960s) are at maximum difficulty with such initially totally unknown natural products . . . The principles of structure work then technically involved two aspects: to detect and interrelate functional groups by chemical means, and to obtain structural information on the atomic nature of the main nucleus. Until the mid-1970s* new substances could only be examined for their chemical transformations. When possible they were converted into previously known, or synthesizable, simpler compounds (e.g., by fission of unsaturation and by dehydrogenation into aromatic substances, for which synthetic methods were efficient). The interpretation of such work was facilitated by the efficient indexing of known compounds and their properties. For this and other reasons, organic chemistry is the best documented science (the order of a million substances on record).[4]

The indexing of information about the reactions and structures of organic molecules was, and remains, essential for both classical and instrumental approaches to structural determination. This necessity gave rise to the multi-edition, multi-volume and complex *Handbuch der organischen Chemie* founded by Friedrich Konrad Beilstein, who completed the first edition in 1882 after two decades' effort. The period between 1950 and 1970 saw the introduction of various 'atlases,' compendia of the infrared, ultraviolet, NMR and mass spectra of numerous organic compounds.[5]

Evolution of Chemical Instrumentation

Before 1930, developments in physics had already contributed towards structural knowledge of simple molecules. The relative positions of atoms, and information about the nature of chemical bonds, had been made possible by numerical data provided by X-ray diffraction and dielectric behavior. In particular, it was observed that certain organic functional groups gave characteristic absorption spectra. However, since the taking of measurements was an invariably lengthy process, requiring considerable skill, organic chemists were slow to incorporate these techniques in their laboratory practice. World War II and the demands of the new petrochemical industry provided the stimulus, and resources, for the development of chemical instrumentation, and its increasing application to organic chemistry. A notable example was the use of infrared spectroscopy (and to a lesser extent ultraviolet spectroscopy) in the study of the structure of synthetic rubber by Paul Flory and John White at Esso Research in 1942; White moved to the instrument manufacturers Perkin-Elmer in 1944.

---

*Birch appears to be in two minds about the date of the changeover. We would date the changeover to the mid-1960s, in agreement with Birch's first estimate.

FIGURE 37.1: TWO MODELS OF CRYSTAL STRUCTURES. A VIT B12 WET CRYSTALS AND A VIT B12 HEXACARBOCYCLIC ACID FRAGMENT. © SCIENCE MUSEUM/SCIENCE & SOCIETY PIC LIB.

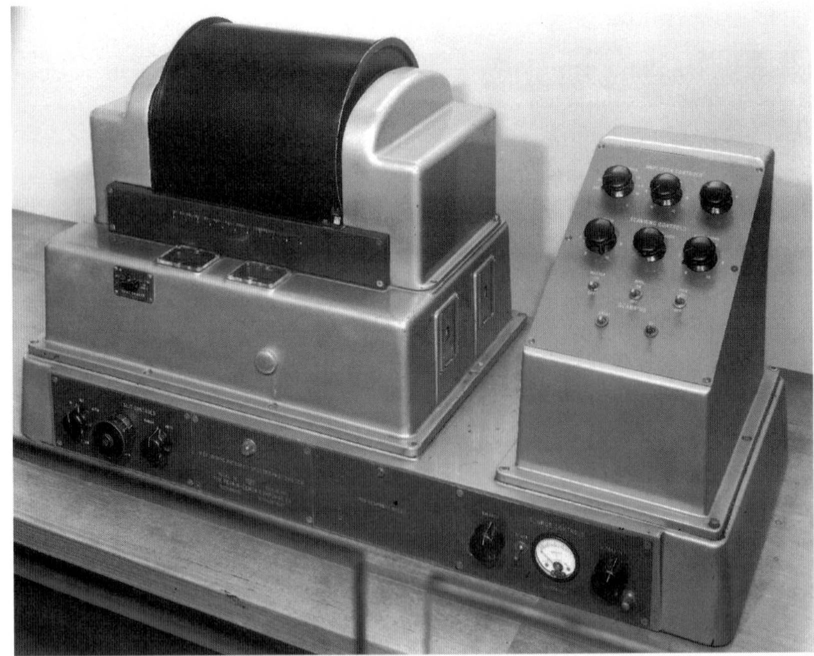

FIGURE 37.2: IR SPECTROMETER. 1960S. © SCIENCE MUSEUM.

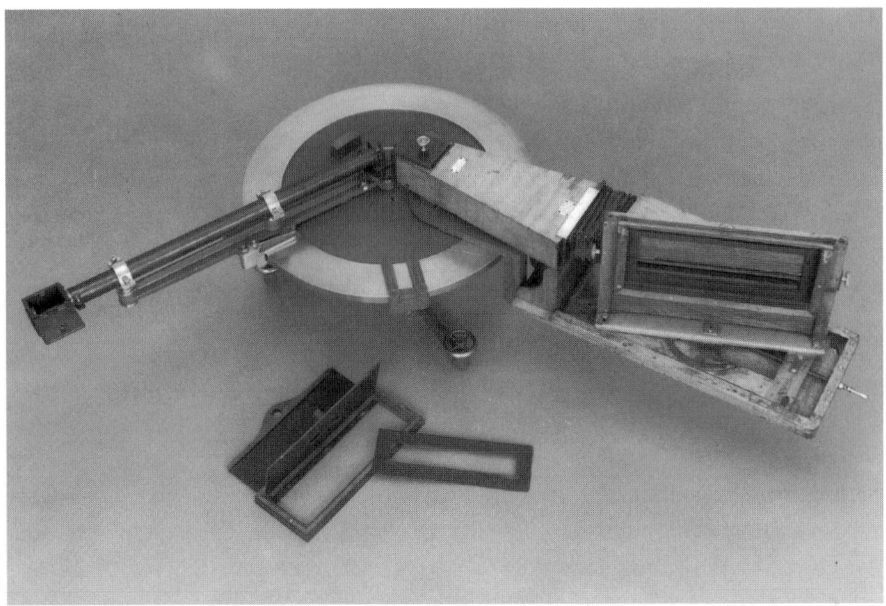

FIGURE 37.3A: HARTLEY'S UV SPECTROSCOPE (1879). © SCIENCE MUSEUM.

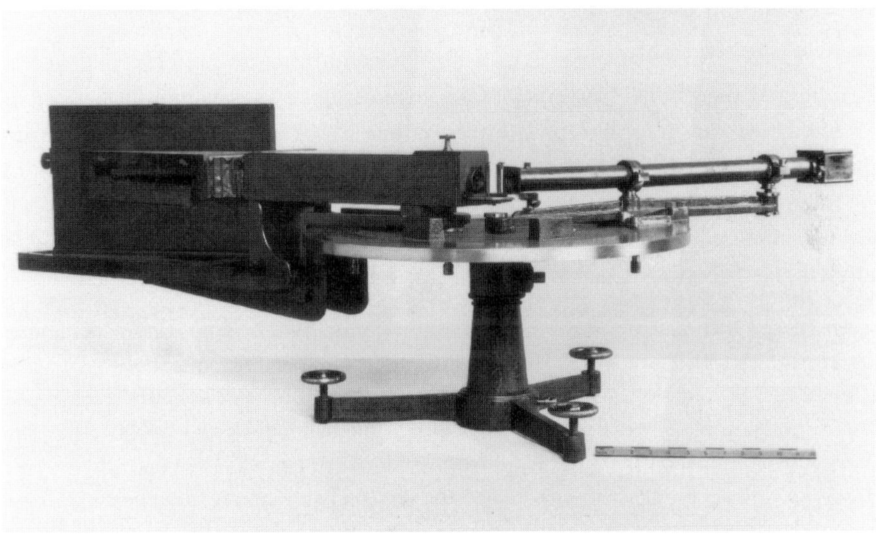

FIGURE 37.3B: HARTLEY'S UV SPECTROSCOPE (1879). © SCIENCE MUSEUM.

The mediators in this endeavor were the instrument manufacturers, notably Arnold Beckman, who often established close links with the pioneers. Initially, the instrument companies tended to work with physicists, chemical physicists, and physical chemists interested in fundamental processes. When it became clear that companies, especially petroleum and petrochemical firms, rather than universities, would be their major customers, the instrument manufacturers collaborated with industrial researchers, most notably the link-up in infrared spectroscopy between Beckman and Robert Brattain at Shell Research with the encouragement of the US government's Petroleum Administration for War. For the most part, it was only after World War II that the developers of physical instrumentation established close links with organic chemists concerned with structural problems. This collaboration quickly deepened and once it was shown that the new instruments were of value in the elucidation of complex structures, particularly of natural products such as steroid hormones, funding for academic research became available from both government agencies and the chemical and pharmaceutical industries.

Although the use of instrumentation in routine organic analysis and structural determination became widespread only during the 1960s, its prior history can be traced back to the beginning of the twentieth century, notably with X-ray diffraction, an area not usually regarded as falling within physical instrumentation, but clearly a forerunner of later developments.

Development of X-ray Methods

In 1912, Max von Laue obtained the first X-ray diffraction pattern. His success encouraged the young Lawrence Bragg to investigate the structural analysis of alkali halides, and Bragg's results were published in 1913. It soon became apparent that the method offered the possibility of identifying the location and nature of each atom in a molecule. There were of course problems, particularly the need for crystalline samples and the fact that organic molecules are much more difficult to analyze structurally than inorganic ionic lattices. Bragg later observed:

> The strong homopolar bonds between the atoms make the organic molecule a definite entity, which typically retains its individuality when the solid is melted or dissolved or even vaporized. In contrast, the forms of most inorganic structures only exist in the solid state . . . it [is] possible to investigate [inorganic] structures of quite high complexity . . . The organic molecule, on the other hand, is an entity which typically has an irregular shape and no symmetry . . . a correspondingly large number of parameters must be determined to define the structure. It is not surprising that their analysis was for long regarded as an almost impossible task. They have, however, a compensating feature which helps analysis. The atoms are linked by homopolar bonds, and the lengths of these bonds and the angles between them can be established to a high degree of accuracy by the analysis of the simpler structures.[6]

The simpler structures were those of naphthalene and anthracene, the first organic molecules investigated by Bragg in 1921. Despite the initial difficulties, the structure of hexamethylenetetramine was established by Roscoe Dickinson and Albert Raymond in 1923, of polymers by Herman Mark in the mid 1920s, of hexamethylbenzene by Kathleen Lonsdale in 1928, and of ergosterol by John Desmond (Sage) Bernal in 1932. The progress of X-ray crystallography was tied intimately to the development of computation, particularly the availability of the Patterson function (1934), Lipson-Beever strips (1936), and Patterson-Tunell strips (1942).

As in other physical techniques, X-ray structure determination depends to a considerable extent upon prior knowledge, in this case assigning likely positions of atoms in the unit cell. This led to the development of a trial and error technique that was particularly successful in the study of aromatic hydrocarbons, whose structures were already known. In cases where there was no structural information available, the incorporation of a few atoms with atomic numbers much greater than those of other atoms in the molecule provided useful information. Once the locations of these 'heavy atoms' had been established, phase constants for these atoms were used to produce an electron density distribution. From this, peaks for light atoms, particularly carbon, nitrogen, and oxygen, were then established.

This 'heavy atom' method was particularly useful in the case of centrosymmetric structures in which the heavy atom at the center of symmetry is taken as the origin. It was first used by John Robertson in 1935 on a molecule that came with the 'heavy atom' already in place, namely the first of the newly discovered synthetic dyes known as phthalocyanines, related to the natural pigments haeme and chlorophyll. Robertson and Ida Woodward (no relation to R.B. Woodward) published an electron-density map of platinum phthalocyanine in 1940. The 'heavy atom' method also enabled the full structural determination of a steroid, cholesteryl iodide, by Harry Carlisle and Dorothy Crowfoot Hodgkin, in 1945, which was in full agreement with structural information determined by chemical means in 1932 by Heinrich Wieland and Elisabeth Dane, and independently by Otto Rosenheim and Harold King. X-ray crystallography also resolved the debate over the structure of penicillin. Chemical studies during World War II had permitted two possible structures. Robinson favored two separate rings, while R.B. Woodward preferred the fused 'β-lactam' structure (which was supported by infrared studies at Shell Research). The definitive structure of penicillin and proof that it was a β-lactam was provided by the X-ray analysis of Crowfoot Hodgkin and Charles Bunn at Oxford, "working in a state of much greater ignorance of the chemical nature of the compounds we have had to study than is usual in X-ray analysis."[7] The X-ray analysis was simplified by the presence of a large sulphur atom. This study was completed in spring 1945, at about the same time that chemical work at Merck came up with the same structure. Similarly, Johannes Bijvoet determined the structure of strychnine in 1947–49, although he was narrowly beaten, as we have seen, by Woodward's classical approach.

The structure of vitamin B_{12} was an outstanding application of X-ray diffraction to organic structure determination — Bragg described it as "breaking the sound barrier . . . of telling the organic chemist something he did not already know."[8] The history of vitamin B_{12} went back to 1855 when Thomas Addison reported pernicious anemia. However, the value of liver in its treatment was not realized until 1926. Partly because of World War II, another twenty-two years were to elapse before the active principle, vitamin B_{12}, was isolated by Karl Folker's group at Merck, and, independently, by Lester Smith, at Glaxo. Subsequently, the structure of this vitamin was extensively studied by Merck, Glaxo, British Drug Houses (BDH), and Alexander Todd's group at the University of Cambridge. The chemists clarified several important features of vitamin B_{12}, including the presence of pyrrole rings, but were unable to determine completely its extremely complex structure. Smith prepared a crystalline sample that Glaxo donated to Dorothy Crowfoot Hodgkin. She published the full structure of the cobalt-containing molecule in 1957, after eight years of study. It was the first time such a complex molecule had been almost entirely elucidated by physical methods. Even the molecular formula was deduced from the X-ray work!

The determination of the structure of vitamin B_{12} was significantly assisted by the growing power and availability of computers. Hodgkin later recalled:

> And we were greatly helped by friends with computers: on a particularly happy day Kenneth Trueblood, on a casual summer visit to Oxford, walked into the laboratory and offered to carry out any additional calculations we needed on a fast computer in California, free and for nothing and with beautiful accuracy.[9]

X-ray crystallography was completely transformed by the arrival of electronic computers in the 1950s. They enabled routine determination of bond lengths, bond angles, and the spacing between non-bonded atoms. By the 1960s, three-dimensional electron-density distribution patterns, incorporating heavy atoms and isomorphous replacement, enabled the definitive solution to many structural problems. This breakthrough led to the determination of the structure of the complex biomolecules myoglobin in 1960 by John Kendrew, and haemoglobin by Max Perutz ten years later.

Ultraviolet and Infrared Spectroscopy

Spectroscopic methods based on absorption in the visible spectrum had been employed since the 1860s, mainly for 'fingerprinting.' Indeed this is how alizarin and other hydroxyanthraquinones were identified. By the early 1900s spectroscopy had found extensive application with colored compounds such as chlorophyll, haemoglobin and dyestuffs, using simple spectrometers and colorimeters. Studies into the ultraviolet region had been pioneered by Walter Noel Hartley in the latter decades of the nineteenth century. Hartley was mostly interested in the use of ultraviolet spectroscopy in metallurgical analysis, but he also studied the ultraviolet spectra of organic compounds. For instance, in 1899 he used ultraviolet spectroscopy to study the vexed issue of tautomerism, and was able to show that isatin has a keto (lactam) rather than an enol (lactim) structure. Between 1907 and 1919, the French chemist Victor Henri studied numerous natural products including strychnine, chlorophyll and cholesterol. He showed that the ultraviolet spectrum of an organic compound was characteristic of certain bonds, rather than a function of the whole molecule.

Widespread use of ultraviolet spectroscopy relied on the introduction of quartz spectrographs in which spectra were recorded photographically. An early manufacturer of these instruments was Adam Hilger; its technical director Frank Twyman collaborated closely with Hartley. In 1906, Hilger introduced the first fixed adjustment quartz spectrograph, in 1910 the sector photometer, and in 1931 the photoelectric 'Spekker' photometer. Arthur C. Hardy at MIT designed an advanced, but expensive, photoelectric spectrophotometer which was commercialized by General Electric in 1933. Ultraviolet spectroscopy assisted the determination of the structure of thiamine (vitamin B_1) by R.R. Williams in 1936. Several new photoelectric spectrophotometers were introduced during World War II. Beckman brought out the famous DU spectrophotometer in 1941, and it was followed by British-made instruments such as the Unicam SP 500 and Hilger Uvispeck. The spectra were obtained by measurement of point-by-point dial readings. Subsequently, recording spectrophotometers, with spectra recorded on paper charts, appeared on the market, although at three times the price of non-recording machines. Ultraviolet spectroscopy was particularly useful for identifying the presence of conjugated double bonds, and therefore applicable to structural work on carotenoids and steroids.

The principle behind infrared spectroscopy is similar. During 1881-82, William de Wiveleslie Abney and Edward Robert Festing were the first to relate infrared spectra to the structures of organic compounds. Much of the fundamental work was done by William Coblentz in the United States between 1903 and 1905, using a non-recording spectrometer. However, his efforts were not widely recognized. The first commercial infrared spectrometers were introduced in the early 1940s, notably by Beckman in 1942, and found widespread use in the wartime study of synthetic rubber. Infrared spectroscopy picks out functional groups, and is more useful as a backup to other methods, or for 'fingerprinting,' rather than for direct determination of structures. It is used to establish branching in long chain hydrocarbon polymers whose properties are determined by the branched chain. Most functional groups in a pure compound or mixture can be determined within minutes. The method is non-destructive, and requires only a few milligrams of sample. About 1950, Van Zandt Williams of the American firm of Perkin-Elmer gave a talk about the new Mark 21 double-beam infrared spectrophotometer to Gordon Sutherland's infrared research group at Cambridge. Everyone there was very impressed by the value of infrared spectroscopy for structure elucidation, and Alexander Todd ordered a Perkin-Elmer 21 for the new university chemical laboratory. When the bench-top Perkin-Elmer 137 came out in 1957, Ralph Elsey, a technician at the Cambridge laboratory, remarked "compared with the 21 the Model 137 is made of tin!"[10] But as it was much cheaper than earlier models, it brought infrared spectrophotometry within the reach of the ordinary chemistry laboratory and, in time, even the undergraduate teaching laboratory.

Woodward Rules

The availability of ultraviolet spectroscopic data encouraged several theoretical developments, of which the Woodward Rules for conjugated dienes and trienes in cyclic systems were outstanding. During 1941–42, the young R.B. Woodward made a significant contribution towards the establishment of structural knowledge from observed spectral data, in this case ultraviolet absorptions of steroids containing double bonds. Woodward undertook a careful numerical analysis of published spectral data for various steroidal ketones (containing the structures C=C—C=O) that gave intense absorption maxima around 230–250nm. This enabled him to develop rules that were of general applicability, involving incremental additions to the wavelength of the original absorption maximum, depending on the arrangement of the double bonds and the substituents attached to them. With characteristic self-confidence, Woodward observed in the first of four papers on the subject: "A few substances do not conform with the generalizations outlined above. The simplest and most probable explanation of this apparent anomaly is that the structures at present assigned to those compounds are incorrect."[11] And, of course, he was right. Thus he showed that the accepted structure for a methadiene derivative was incorrect.

This was the first systematic application of instrumentation, apart from the polarimeter, to a major area of natural product chemistry. There had been earlier attempts to establish numerical relationships between the ultraviolet absorption maximum in a steroid's spectrum and the number and type of conjugated double bonds and substituents that it contains, most notably by the German chemist Heinz Dannenberg in 1940, but they had little impact. By contrast, the Woodward rules, as extended by Louis Fieser in 1949, are still cited today.

Octant Rule

This was another remarkable achievement by R.B. Woodward that enabled a relationship to be drawn between structure and absorption. Again, it relied on prior studies, in particular Carl Djerassi's work with Optical Rotatory Dispersion, or ORD. This rule is based on the fact that optically active compounds are associated with the rotation of the plane of vibration of light with which they interact, thereby providing clues to the stereochemistry. The conventional polarimeter, as introduced in the 1840s, uses the yellow D-line of sodium. The principle of ORD is based on the fact that optical rotation varies with wavelength. Jean Baptiste Biot established in 1813 that the rotation of plane polarized light of an optically active substance varies with wavelength. In 1895, Aimé Cotton discovered the effect named after him, *viz.* that the sign of optical rotation changes markedly around an absorption maximum, typically in the far ultraviolet. It increases abnormally and falls off sharply close to a characteristic absorption band. However, optical rotatory dispersion languished for decades for lack of suitable instrumentation. Until 1950, nearly all measurements of optical rotation were at a single wavelength (usually the sodium D-line), which is of little value in the study of organic compounds.

In 1953, O.C. Rudolph & Sons introduced the first commercial spectropolarimeter. Carl Djerassi described it as "the prince who awoke the sleeping beauty of rotatory dispersion."[12] He measured this effect for steroids, asymmetric cyclic ketones (the same compounds behind the formulation of Woodward's rules), in the ultraviolet region: "the laborious point-by-point wavelength measurements being taken by the wives of my graduate students."[13] Woodward's involvement came six years later, when he introduced Djerassi to two theoretical chemists, William Moffit and Albert Moscowitz, and they all developed the Octant Rule during a single brainstorming session in 1958. It was published three years later. Djerassi recalled that at the 1958 meeting, "The octant rule explained virtually all of our published and unpublished results."[14]

The octant rule relates the Cotton effect (positive or negative) to the substituents on the steroidal ketone according to their arrangements in space, thus clarifying the conformation as well as the gross structure. This is expressed graphically by placing the carbon of the ketone carbonyl group at the center of an octant created out of three planes. Each segment of the octant carries a plus or minus sign. This can be demonstrated with cyclohexanone, as explained by Djerassi: "In our original version . . . the cyclohexanone model was divided into three planes corresponding to the nodal and symmetry planes of the 280nm transition of the carbonyl chromophore. These three planes create eight octants, and the presence of substituents in each octant is given a qualitative rotational contribution."[15] The sum of the differing contributions allowed the sign of the Cotton effect to be calculated. Conversely the sign of the Cotton effect could be used to choose between one of several possible structures. As with the Woodward rules, the derivation of the octant rules depended on empirical data, here the material painstakingly collected by Djerassi.

Mass Spectroscopy

Francis William Aston and Arthur Dempster independently made the first mass-spectrographs around 1919. The mass spectrometer produces positive ions from the sample, and uses a strong magnetic field to resolve them into a series of beams recorded on photographic plates or, more recently, by electronic detectors. They are then presented as a series of peaks representing mass/charge ratios. Alfred Nier developed the first high resolution mass-spectrometer at the University of Minnesota in the late 1930s. The Consolidated Engineering Company (CEC), an

American firm founded by a member of President Hoover's family, moved into this field in the late 1930s with the aim of supplying the petroleum industry with instruments that could give rapid analyses of hydrocarbon mixtures. The first CEC instrument, model 21-101, was brought out in 1942, and Metropolitan Vickers produced the first British counterpart, the MS-2, eight years later. The CEC 21-103C instrument, introduced in the early 1950s, was widely used by organic chemists until it was displaced by double-focusing instruments, namely the British MS-7, introduced in 1958, and the CEC 21-110, in 1963, which was widely used in the United States.

Until the mid-1950s, mass spectroscopy had found great value only in physics and physical chemistry. The key year for the application of mass spectroscopy to organic chemistry was 1956, when Fred McLafferty began his work on the fragmentation patterns formed when a complex organic compound breaks up in a mass spectrometer, and John Beynon used a high resolution mass spectrometer to work out the molecular formulae of organic compounds. Using fragmentation analysis, it gradually became possible to decipher the structure (or at least the partial structure) of an organic compound from the mass spectrum. An early pioneer was Ivor Reed at the University of Glasgow, who used fragmentation analysis in 1956 to determine the structure of the side-chains of various steroids. Computerized data-handling of mass spectra data also emerged in this period.

By 1959, Klaus Biemann, an Austrian chemist working at MIT, had entered this field and soon became "a renaissance man in mass spectroscopy" according to his colleagues.[16] His first paper concerned the use of mass spectroscopy to determine the amino-acid sequence of a peptide. Peptides and proteins have endured as one of his major interests. In the early 1960s, however, he mostly used mass spectroscopy to determine the structure of complex alkaloids, especially indole alkaloids. This work brought him into contact with R.B. Woodward, who used Biemann's expertise with the mass-spectrometer to identify intermediates in his famous organic syntheses. By early 1964, Biemann had acquired a CEC 21-110 that enabled the entire mass spectrum to be displayed on a single photographic plate. Using an IBM 7094, this data could be used to calculate the exact molecular mass of each fragment. Biemann presented this information in the form of an 'element map,' a table with the fragments arranged in different columns according to their heteroatom content. (A heteroatom is an atom of any element in an organic compound except carbon or hydrogen.) This was a very powerful technique and could be used to determine the structure of compounds with only microscopic samples. For instance, in 1968, he determined the structure of the marine sex hormone, antheridiol from its high-resolution mass spectrum, in conjunction with ultraviolet and infrared spectra, and the NMR spectrum. Nothing was known about its structure beforehand and the mass spectrum was obtained from a sample of "a few micrograms."[17]

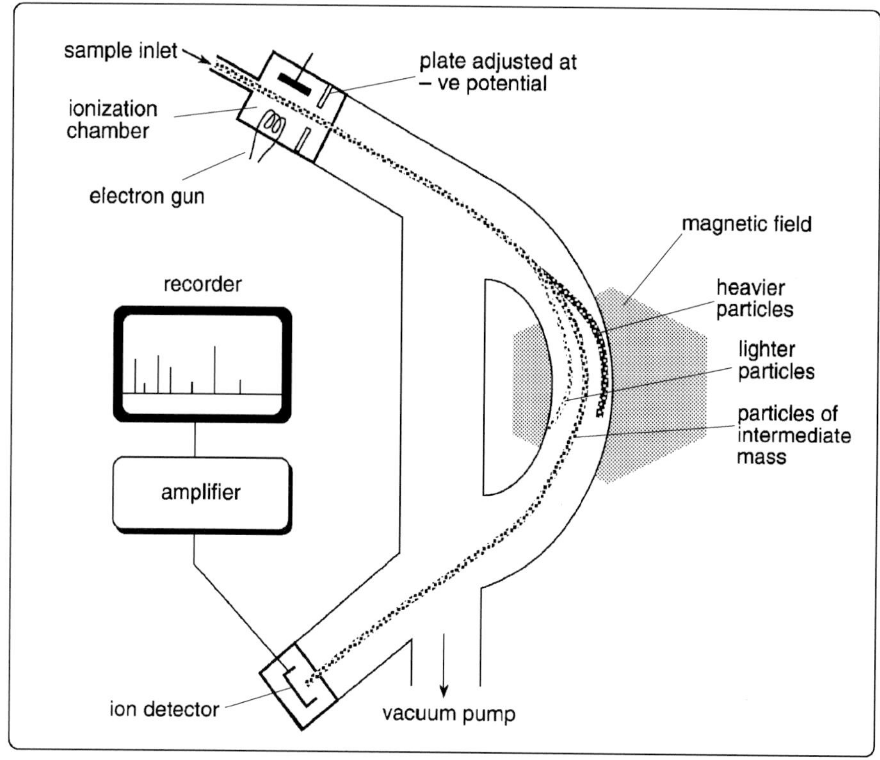

FIGURE 37.4: MASS SPECTROMETER.

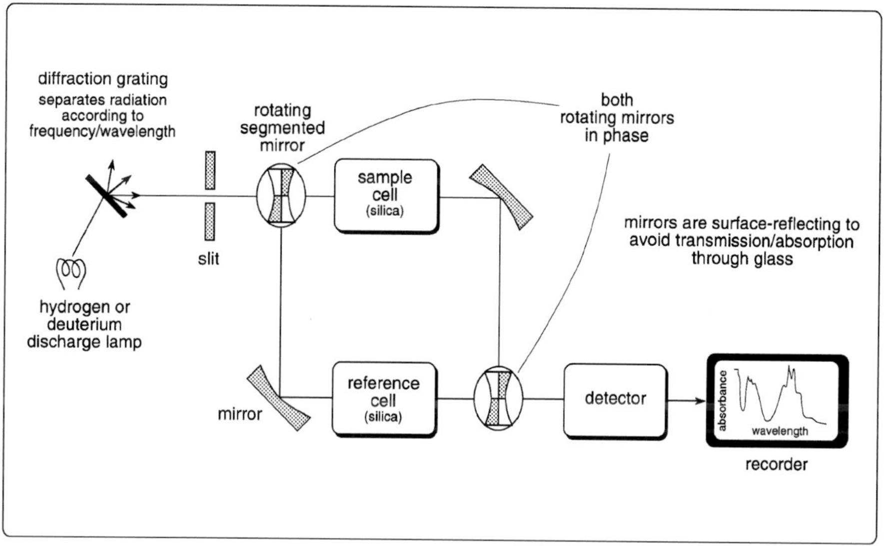

FIGURE 37.5: ULTRAVIOLET SPECTROMETER.

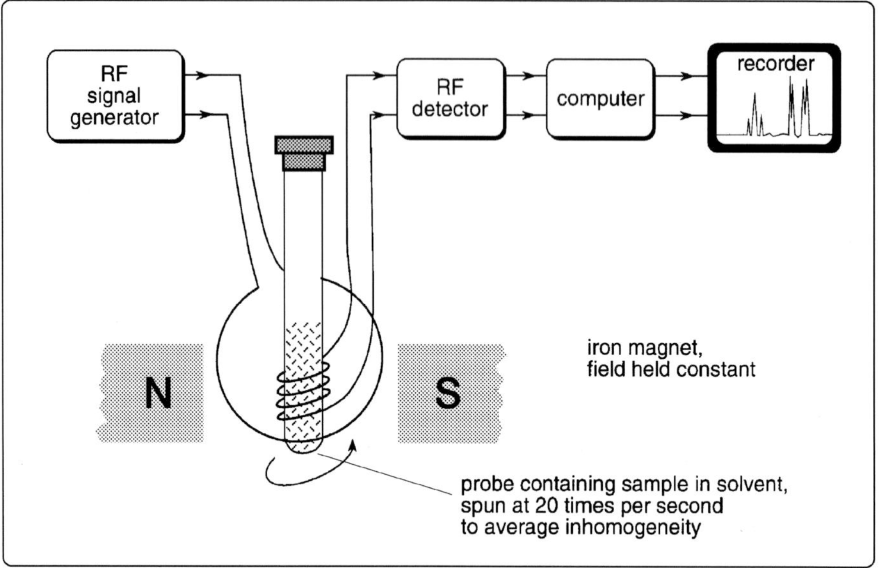

FIGURE 37.6: N.M.R. SPECTROMETER.

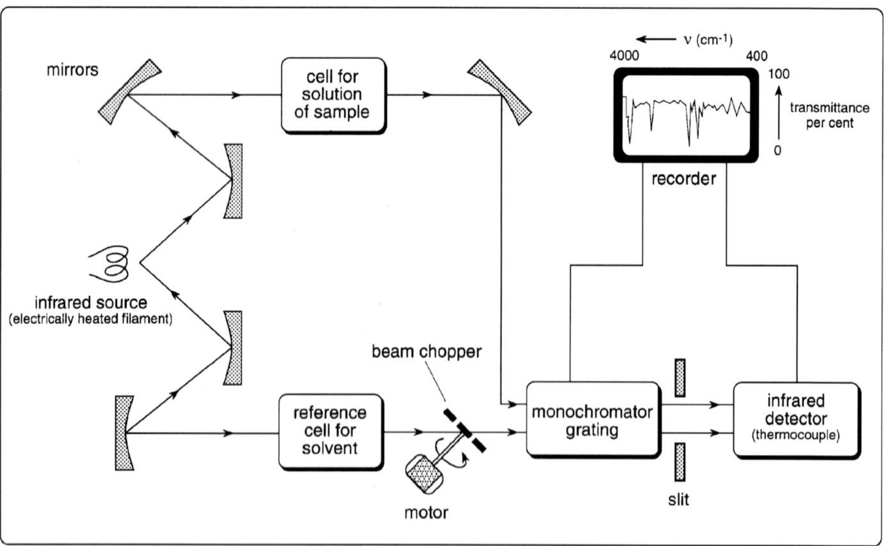

FIGURE 37.7: INFRARED SPECTROMETER.

After the mid-1960s, Biemann concentrated on the linking of mass spectroscopy with gas chromatography. The ability of mass spectroscopy to give results with a sub-milligram sample makes it of great value with gas chromatography, in which very small amounts of pure substances can be separated from previously intractable mixtures. Although primitive gas chromatographs were developed during and after World War II, especially at Innsbruck (Biemann's original university), modern gas chromatography dates from the announcement of gas-liquid partition chromatography by Archer Martin and Tony James in 1951. This technique was rapidly taken up by academic chemists and the petrochemical industry. In 1957, R.S. Gohlke achieved the first hyphenation with mass spectroscopy (GC-MS) and Biemann (with J.T. Watson) was the first to develop an effective method of removing most of the carrier gas before introducing the sample to the mass spectrometer. Biemann's work on GC-MS reached its literal high point in 1976 when, as part of NASA's Viking program, he landed a GC-MS instrument on the surface of Mars to analyze any organic compounds that might be there. The GC-MS showed that there were no organic compounds on the Martian surface with a sensitivity limit of parts per billion.[18]

It is largely through the agency of GC-MS that it has been possible to study pheromones: chemicals that control the social behavior, especially the sexual behavior, of insects. The veteran steroid chemist Adolf Butenandt carried out the first major study of an insect pheromone, the attractant of the female silkworm moth (*Bombyx mori*) in 1959. However, he did not use GC-MS which was still in its infancy. Nearly all the early work on the structure of pheromones was surprisingly traditional: mainly chemical methods such as hydrolysis, ozonolysis, permanganate oxidation and infrared spectroscopy. However, as many pheromones are relatively simple compounds containing double bonds and only one or two other functional groups, these methods were adequate if enough material was available. It appears that R.M. Silverstein, in his 1966 study of the male bark beetle (*Ips paraconfusus*), was the first to use preparative gas chromatography followed by an instrumental battery of infrared spectroscopy, mass spectroscopy and NMR. Thus he could show that a mixture of compounds was involved, thereby disproving Butenandt's idea of a single compound for a single role.

In December 1960, Carl Djerassi also entered the field of mass spectroscopy because of his interest in cacti alkaloids. His interest in these alkaloids stemmed from his early work at Syntex in Mexico City (and also gave rise to an infamous mescaline party at his home in 1954). Djerassi recalled:

> As soon as I arrived [at Stanford University] in Palo Alto, I applied to the National Institutes of Health for financial support in buying a mass spectrometer in order to conduct a systematic study of the technique, using steroids as initial model substrates before applying it to the, structurally, much more diverse group of alkaloids. We wanted to determine whether special rules of fragmention and reassembly could be developed

> . . . which would make this method of more general utility. Using steroids as substrates, we set out to 'mark' certain portions of the molecule with stable, nonradioactive isotopes of hydrogen and carbon to facilitate the reassembly of the broken pieces. Eventually we used the marking technique, which on its own involved many man-hours of synthetic effort, to establish the rules of mass spectrometric decomposition for a wide variety of molecules, such as steroids, triterpenes and alkaloids.[19]

One interesting example of Djerassi's work was the discovery of gorgosterol. In 1969, Djerassi was sent a supposedly pure sample of a steroidal marine toxin by Paul Scheuer at the University of Hawaii. Djerassi takes up the story:

> On subjecting his sample to mass spectroscopic analysis, we found it to consist of at least three sterols: two conventional ones of the cholesterol type and a third one with a seemingly unprecedented number of carbon atoms. I encouraged Scheuer to isolate more of that 'impurity,' which we then subjected at Stanford to . . . nuclear magnetic resonance and mass spectroscopy. In a joint communication with Scheuer, we published the structure of 'gorgosterol,' which has the same tetracyclic steroid nucleus as cholesterol, but an extremely unusual 'side chain' — an assembly of eleven carbon atoms [containing a cyclopropane ring] attached to position 17 of the steroid skeleton. We determined the complete structure of gorgosterol only by means of X-ray crystallography . . . and promptly got hooked on a research line [marine sterols] my group pursued for the next twenty years.[20]

In the mid-1960s, through his Stanford college Joshua Lederberg, Djerassi also became involved with the early stages of the project to find out if there was life on Mars, specifically to develop a program that could be used to analyze the data sent back by the Mars-based GC-MS. Lederberg commented that "We are trying to teach a computer how Djerassi thinks about mass spectrometry."[21]

Nuclear Magnetic Resonance

The NMR effect was first reported in 1946 by Felix Bloch at Stanford and, independently, by Edward Purcell at Harvard University. Subsequently, Stanford became a leading center for research into NMR. In the simplest version of NMR, a sample of a compound is bathed in a very strong constant magnetic field (nowadays generated by superconducting magnets) and bombarded by radio waves of a single frequency. The resonance absorption of the radio waves by protons (or other suitable atoms such ^{13}C or ^{19}F) over a narrow range of radio frequencies is recorded and studied. The dependence of the resonance frequency of a proton on its chemical environment, as measured by the 'chemical shift,' was noted by several researchers in 1949–50. Shortly thereafter, in 1951, the first NMR spectra were obtained by James Arnold, a postgraduate at Stanford. These showed separate resonances for protons located at different positions in the molecule: the field-dependent chemical shifts. The field-independent separation between peaks, spin-spin coupling, which represents the interaction between protons, was discovered by Warren Proctor and F.C. Yu at Stanford, and by Herbert Gutowsky and David

McCall at Illinois during 1952–53. Herb Gutowsky was one of the first to introduce NMR into organic chemistry.

Varian launched the first commercial NMR spectrometer in 1952, which operated at 30 MHz. It found immediate application in industry, especially at Du Pont, whose NMR expert, William Phillips, described to his visitor John Roberts the proton spectra of N,N-dimethylformamide, in which two methyl signals coalesced at elevated temperature due to increased rates of rotation about the carbon-nitrogen amide bond. That was in 1954, when at $26,000 such an instrument was beyond university budgets. With the help of Linus Pauling, Roberts convinced the newly formed National Science Foundation of NMR's value for structural studies, and received a grant equal to the purchase price. This was used to acquire a 40 MHz instrument, introduced in 1955, which, according to Roberts, "was to be the first commercial NMR spectrometer to be sited in a university. If it was not the first piece of such equipment, I'm sure it was the first to be put under the jurisdiction of an organic chemist."[22] The Varian 40 MHz instrument was also the first to be introduced into Britain, at Cambridge University. Unfortunately neither this instrument (which was transferred to the Colchester Institute in the 1960s) nor the second 40 MHz machine in Britain (formerly at Liverpool University) have survived.

Leaders in NMR structural organic chemistry in the late 1950s were: Jim Shoolery, of Varian Associates Inc. in Palo Alto, California, who worked on steroids and in 1955 introduced spin decoupling; Raymond Lemieux, University of Ottawa, noted for carbohydrate research; Basil Weedon, Imperial College, London, who worked on polyenes; and Karl Folkers of Merck who studied ubiquinones. Ray Lemieux's experience of the initial problems in 1955 was not untypical. After managing, with difficulty, to set up a suitable homogeneous magnetic field, every effort had to be made to maintain that field. However, "it often did not last through a run (about 20 minutes) because of a sudden change in line voltage, a passing truck, or simply the temperature change caused by someone opening the laboratory door. Many of the problems could be minimized by working between 1 and 5 a.m., when the city line voltage was more constant and the street traffic was relatively quiet. Under these conditions, often, the main worry was, 'Would the recorder pen work throughout the experiment?'" Some aspects of NMR were less readily apparent. Thus Lemieux observed "I well remember when we learned that our spectra would be much improved by spinning the sample tube."[23] Bloch had this bright idea in 1954 while he was stirring a cup of tea. Once the teething problems were overcome and improvements in resolution were achieved, there were rapid developments in conformational studies on six-membered ring compounds.

An early example of the use of NMR in structural work was the determination of the structure of the alkaloid aspidospermine by Harold Conroy at Brandeis University between 1957 and 1959. Initially, the NMR spectrum suggested the presence of a N-methyl group, but this was ruled out by classical chemical methods and it was then possible to arrive at the correct structure. The X-ray determination of the structure of aspidospermine was published simultaneously by John Mills and S.C. (Scan) Nyburg at University College of North Staffordshire (now Keele University) in 1959. Biemann then studied the aspidospermine series of alkaloids in the early 1960s using mass spectrometry.

The interpretation of NMR spectra was greatly improved by the growing availability of stronger magnetic fields and hence the use of higher radio frequencies. Higher field strengths produced proportionally greater chemical shift separations, which allowed chemists to distinguish between peaks created by spin-spin coupling and peaks from wholly different protons. Spin-spin decoupling could also be used to confirm such interpretations. In 1961, Varian brought out the first frequency-locked NMR machine, which operated at 60 MHz and surprisingly, Varian marketed the first commercial instrument to use a superconducting magnet (220 MHz) in the same year. Bruker introduced the first commercial Fourier transform NMR spectrometer in 1969 (90 MHz), which was followed a year later by a superconducting Fourier transform model operating at 270 MHz.

During the early 1960s, organic chemical shift data became available in tabulated form and in spectral atlases. With advances in resolution and sensitivity, chemists were able to correlate spin-spin coupling constants with physical features of molecules. The publication of the Karplus equation in 1963 provided the basis for relating the spin-spin coupling constant to the dihedral angle between bonds. The Nuclear Overhauser Effect (NOE), discovered in 1953, was introduced into chemical NMR by Frank Anet and A.J.R. Bourn in 1965. It is useful in the study of conformations, because it provides information about the positions of protons in space, rather than the formal structure. For instance, Lemieux discovered in the late 1970s that NOE could be used to study the electron-withdrawing effect of an oxygen atom in a fucose ring on a proton on another ring which was nonetheless close to it in space.[24]

As with other instrumental methods, NMR has been widely used to determine chemical structure, often requiring 'hit or miss' assumptions similar to those employed in classical structural determination. This has the danger that chemists studying the 'entrails' of a complex NMR spectrum often find the structure they want, not necessarily the correct one. Hence the widespread use of NMR (and also mass spectroscopy) has led to a degree of uncertainty creeping back into the published chemical structures.

Modern NMR spectroscopy uses a combination of ^{1}H (proton) and ^{13}C NMR for structural work and to a lesser extent, nitrogen, fluorine and phosphorus NMR. There has been an explosion of techniques since 1979, notably in one and two dimensional NMR spectral measurements by modern pulse techniques. Some of these have produced strange acronyms, such as FOCSY, NOESY, COSY, SECSY, HOHAHA, INEPT, INADEQUATE, WALTZ, DANTE, INFERNO, SIMPLE, SPOTS and DEFT.[25]

Promotion of Physical Methods Through Publications

The diffusion of instrumental methods into mainstream organic chemistry was a slow process. The new methods had to gain acceptance from chemists who had got where they were using well-established chemical techniques. Very often, organic chemists only had a broad understanding of physical chemistry and an even weaker grasp of quantum physics and electronics. To win them (and perhaps more importantly their postgraduate students) over to the new physical instrumental methods, it was necessary to use propaganda in the form of monographs and textbooks. The late 1950s was an exciting period to be a chemist working with these new techniques and it was also a golden age for publishers of chemical monographs, most notably the New York firm of McGraw-Hill. Characteristically, Djerassi was both innovative and astute in his dealings with McGraw-Hill over the publication of his *Optical Rotatory Dispersion: Applications to Organic Chemistry* in the same year:

> One aspect of my publishing contract with McGraw-Hill was unusual. When the publisher invited me to prepare this first monograph dealing with the organic chemical applications of optical rotatory dispersion, I insisted on a penalty clause, whereby my royalties would escalate by 1 percent for each week the book's appearance might be delayed beyond my requested publication date. To everyone's surprise, the McGraw-Hill lawyers accepted my proposal, provided I agreed to return the corrected page proofs from Mexico City [where Djerassi was research vice-president of Syntex] within twenty-four hours of their receipt. This was supposed to prevent a horror scenario, whereby my royalties might escalate to unprecedented heights were I simply to sit on the page proofs. As a final compromise, the publisher set each chapter in print as it was received, rather than waiting for the entire manuscript. I managed to finish the book in time by sticking to a rigid Monday-Wednesday-Friday writing schedule, and McGraw-Hill was equally diligent. The royalties from the book eventually paid for a swimming pool at my new house in California, whose steps were set in Mexican tiles reading 'built by optical rotatory dispersion.[26]

1959 and 1960 were key years for publications on the new methods and, together with what was happening in the laboratories, must be considered a major watershed. In addition to Roberts' book, L.M. Jackman, *Applications of Nuclear Magnetic Resonance to Organic Chemistry*, and J.A. Pople, W.G. Schneider and H.J. Bernstein, *High Resolution Nuclear Magnetic Spectroscopy* also helped to introduce NMR to organic chemists. J.H. Beynon, *Mass Spectrometry and its Applications to Organic Chemistry* (1960), and K. Biemann, *Mass Spectroscopy* (1962, also McGraw-Hill) played a similar role for mass spectroscopy. It must be debatable, however, how many organic chemists fully understood the very technical monograph by Pople, Schneider and Bernstein, which may explain the failure of Roberts' second foray into this field:

> I told Bill Bejamin that . . . I would write *An Introduction to the Analysis of Spin-Spin Splitting in High-Resolution Nuclear Magnetic Resonance Spectra* [published in 1961]. It was a long title

for what was to be a short book. Indeed, it was not a lot more than a fuller explanation of a relatively few pages in Pople, Schneider and Bernstein . . . Bill Benjamin wanted this book to be a showpiece to convince potential authors that he could do a very high-quality publishing job in half or less of the time normally required. The heat was on, and this was a text with a lot of mathematical equations, graphics and spectra . . . The book sold several thousand copies, but I had remarkably little feedback on its utility. Perhaps there weren't that many people who were that interested in understanding the basis of what is involved in spin-spin splitting.[27]

Surveys of the entire field also played an important role in the assimilation process. The first volume of *Determination of Organic Structures by Physical Methods* (edited by E.A. Braude, and F.C. Nachod) appeared in 1955, and had an immediate impact; the second volume (edited by F.C. Nachod and W.D. Phillips) followed in 1961. *Chemical Applications of Spectroscopy*, edited by W. West as part of the widely circulated series on *Technique in Organic Chemistry* edited by Arnold Weissberger, brought the latest developments in infrared and ultraviolet spectroscopy to a wider audience when it was published in 1956. Another important overview — *Elucidation of Structures by Physical and Chemical Methods*, edited by K.W. Bentley — was published in 1963. The task of winning organic chemists over to the new methods was assisted by the availability of undergraduate texts that emphasized the role of physical methods in organic chemistry, most notably J.D. Roberts and Marjorie Caserio's *Basic Principles of Organic Chemistry*, which was published by Bill Benjamin who had just set up on his own as W.A. Benjamin. Roberts recalled:

The first edition was, I think, a landmark book with many features which are standard today. It was also a shocker to many, especially through the very early introduction of spectroscopy; the many follow-up spectroscopic problems; the inclusion of problems, as appropriate, right within the text; and perhaps more than anything, the unusual length for an elementary book. When I asked my organic chemistry colleagues, who complained about the length, for a list of things they would cut out, I never got much help on that score. The usual response was that perhaps we might 'somewhat enlarge' the sections that covered the suggestor's own field of interest.[28]

However, Roberts and Caserio was too basic for the more specialized British degree, and Oliver and Boyd published *Physical Methods in Organic Chemistry*, edited by J.C.P. Schwarz, in the same year. McGraw-Hill countered in 1966 with *Spectroscopic Methods in Organic Chemistry* by Dudley Williams (who had worked with Djerassi) and Ian Fleming. Its reading lists helped to sell other McGraw-Hill books, and unlike the other two books, it has endured, with a fifth edition in 1995.

Conclusion

In summary, infrared and ultraviolet spectroscopy are used as 'fingerprinting' techniques for known compounds, and for identification of functional groups. Mass-spectroscopy is also a 'fingerprinting' technique; the fragmentation pattern can be used to determine both molecular mass and structure. NMR is the most powerful and widely employed technique and is particularly useful for conformational analysis. While NMR and mass spectroscopy can give excellent results for relatively small molecules (such as the pheromones), they can yield debatable structures, at least in the case of the larger more complex natural products.

While X-ray crystallography can give an unambiguous result, not every compound is studied by X-ray crystallography, which is usually performed away from the organic chemistry laboratory by different staff (who often have different priorities) and thus takes extra time and money. Furthermore, as Djerassi has emphasized, X-ray crystallography requires a crystalline sample and not all organic compounds can be crystallized in a suitable form or obtained in amounts large enough to crystallize. Taken together, however, these techniques enable us to determine chemical structures rapidly and accurately with very small amounts of material.

Thomas Kuhn has described the paradigm as a conceptual model used to provide direction to a scientific activity by "defining implicitly the legitimate problems and methods." Scientific revolutions occur when the ruling paradigm is overthrown by a new one. The introduction of instruments into the methods of structural determination, particularly in organic chemistry, is an illustration of this paradigm shift. The development of the new paradigm started in the early 1940s, began to seriously displace the old paradigm by the late 1950s, and the process was complete by the end of the 1960s. The routine of one type in chemistry was transformed into routine of another type, with major implications for the field of chemistry in general and the chemists who carry out such work. The introduction of electronic instrumentation after 1940 was nothing less than a scientific and technological revolution. It has led to the near-total displacement of classical 'wet and dry' methods in organic structure elucidation.

References

1. Carl Djerassi, *The Pill, Pygmy Chimps, and Degas' Horse: The Autobiography of Carl Djerassi.* (Basic Books, New York, 1992), p. 104.
2. "Laurie Hall: Opening up a Pandora's box". *Chemistry in Britain* (December, 1985), **21**: p. 1057.
3. John D. Roberts, *The Right Place at the Right Time.* (Washington, D.C., American Chemical Society, 1990), p. 151.
4. Arthur J. Birch, *To See the Obvious.* (American Chemical Society, Washington D.C., 1995), pp. 56–7.
5. The pioneering effort was the American Institute of Petroleum Research Project 44 which put out atlases of infrared, ultraviolet and mass spectra (mainly hydrocarbons) between 1947 and 1959. Other atlases of infrared and ultraviolet spectra were published by Stadler Research Laboratories of Philadelphia, and the *Documentation of Molecular Spectroscopy* by Butterworths in collaboration with Verlag Chemie. R. Mecke and F. Langenbucher, *Infrared Spectra of Selected Chemical Compounds,* (Heyden and Sons, London, 1965) and *The Aldrich Library of Infrared Spectra,* (1970) by C.J. Pouchert followed somewhat later. H.M. Herschenson compiled indexes (but did not reproduce the spectra) for infrared and ultraviolet spectra for the period of 1930–59 (UV) and 1945–62 (IR), which were published by Academic Press. The earliest NMR atlas was N.S. Bhacca, D.P. Hollis, L.F. Johnson

and E.A. Pier, *Varian High Resolution NMR Spectra Catalog,* two volumes, Varian Associates, Palo Alto, 1962 and 1963. The equivalent publication for mass spectra was F.W. McLafferty, E. Stenhagen and S. Abrahamsson, *Atlas of mass spectral data,* 3 volumes, Interscience Publishers, New York, 1969.

6. Sir Lawrence Bragg (Edited by D.C. Phillips and H. Lipson), *The Development of X-ray Analysis.* (G. Bell & Sons, London, 1975), pp. 176–177.
7. Bragg, *The Development of X-ray Analysis,* p. 189, quoting Hodgkin and Bunn's original paper.
8. W.L. Bragg, "The Growing Power of X-ray Analysis" in P.P. Ewald (Ed.), *Fifty Years of X-Ray Diffraction.* (International Union of Crystallography, Utrecht, 1962), p. 131 and p. 130.
9. Dorothy Crowfoot Hodgkin, "The X-ray Analysis of Complicated Molecules" *Science* (1965), **150**: pp. 979–988; quote on p. 983.
10. Norman Sheppard, "The U.K.'s Contributions to IR Spectroscopic Instrumentation: From Wartime Fuel Research to a Major Technique for Chemical Analysis" *Analytical Chemistry* (1992) **64**: p. 881A.
11. R.B. Woodward, "Structure and the Absorption Spectra of α,β-Unsaturated Ketones" *Journal of the American Chemistry Society* (1941), **63**: p. 1125.
12. C. Djerassi, *Optical Rotatory Dispersion: Applications to Organic Chemistry.* (McGraw-Hill, New York, 1960), p. 19.
13. C. Djerassi, *Steroids Made It Possible.* (American Chemical Society, Washington D.C., 1990), p. 54.
14. Djerassi, *Steroids Made It Possible,* pp. 59.
15. Djerassi, *Steroids Made It Possible,* pp. 59–61.
16. *Life Search.* (Time-Life Books, Alexandria, Virginia, 1989), p. 25.
17. G.P. Arsenault, K. Biemann, Alma W. Barksdale and T.C. McMorris, "The Structure of Antheridiol. A Sex Hormone in *Achlya bisexualis" Journal of the American Chemical Society* (1968) **90**: pp. 5635–6.
18. *Life Search,* pp. 15–28.
19. Carl Djerassi, *The Pill, Pygmy Chimps, and Degas' Horse,* p. 101.
20. Carl Djerassi, *The Pill, Pygmy Chimps, and Degas' Horse,* pp 103–4. Also see Djerassi, *Steroids Made It Possible,* pp. 115–6.
21. Carl Djerassi, *The Pill, Pygmy Chimps, and Degas' Horse,* p. 102.
22. Roberts, *The Right Place at The Right Time,* p. 154.
23. Raymond U. Lemieux, *Explorations with Sugars: How Sweet It Was.* (American Chemical Society, Washington D.C., 1990), pp. 30–1.
24. Lemieux, *Explorations with Sugars: How Sweet It Was,* pp. 105–6.
25. Alex Nickon and Ernst F. Silversmith, *Organic Chemistry: The Name Game. Modern Coined Terms and Their Origins.* (Pergamon Press, New York, 1987), pp. 183–5.
26. Carl Djerassi, *The Pill, Pygmy Chimps, and Degas' Horse,* pp. 97–8.
27. Roberts, *The Right Place at the Right Time,* pp. 173–4.
28. Roberts, *The Right Place at the Right Time,* pp. 227–228.

Further Reading

Theodor Benfey and Peter Morris (Eds.), *Robert Burns Woodward: Architect and Artist in the World of Molecules.* (American Chemical Society, Washington D.C., forthcoming).

Arthur J. Birch, *To See the Obvious.* (1995).

Mary Ellen Bowden and Theodor Benfey, *Robert Burns Woodward and the Art of Organic Synthesis.* (Beckman Center for the History of Chemistry, Philadelphia, 1992).

William H. Brock, *The Fontana History of Chemistry.* (Fontana Press, London, 1992). [Published in the United States as *The Norton History of Chemistry.* (W. Norton, New York, 1993)].

Robert Bud and Deborah Warner (Eds.), *Instruments of Science: An Historical Encyclopedia.* (Garland, New York, forthcoming).

Carl Djerassi, *Steroids Made It Possible.* (1990).

Carl Djerassi, *The Pill, Pygmy Chimps, and Degas' Horse: The Autobiography of Carl Djerassi.* (Basic Books, New York, 1992).

C.B. Faust, *Modern Chemical Techniques.* (Royal Society of Chemistry, London, 1992).

James Feeney, "Development of high resolution NMR spectroscopy as a structural tool" in Robert Bud and Susan E. Cozzens (Eds.), *Invisible Connections: Instruments, Institutions, and Science.* (SPIE Optical Engineering Press, Bellingham, Washington, 1992).

I.L. Finar, *Organic Chemistry,* volume two, *Stereochemistry and the Chemistry of Natural Products,* 4th edition (Longmans, London, 1968).

Jenny P. Glusker, "Brief History of Chemical Crystallography. II: Organic Compounds" in J. Lima-de Faria, *Historical Atlas of Crystallography.* (International Union of Crystallography, Dordrecht, 1990).

Herbert A. Laitinen and Galen W. Ewing (Eds.), *A History of Analytical Chemistry.* (Division of Analytical Chemistry, American Chemical Society, Washington D.C., 1977).

Raymond U. Lemieux, *Explorations with Sugars: How Sweet It Was.* (1990).

John D. Roberts, *The Right Place at the Right Time.* (1990).

J.C.P. Schwarz (Ed.), *Physical Methods in Organic Chemistry.* (Oliver and Boyd, Edinburg, 1964).

"Profiles, Pathways and Dreams" series, edited by Jeffrey I. Seeman and published by the American Chemical Society, Washington D.C.

"The Instrumental Revolution, 1930–1955" in Dean Stanley Tarbell and Ann Tracy Tarbell, *Essays on the History of Organic Chemistry in the United States, 1875–1955.* (Folio Publishers, Nashville, 1986), pp. 335–352.

Dudley H. Williams and Ian Fleming, *Spectroscopic Methods in Organic Chemistry,* 5th edition. (McGraw-Hill, London, 1995).

FIGURE 38.1: THE GIANT BUBBLE CHAMBER AT CERN. © CERN.

CHAPTER 38

Physics Instruments in the Twentieth Century

PAOLO BRENNI

THE TRIUMPH OF BRASS AND GLASS

At the dawn of the twentieth century technical and polytechnical schools, scientific departments of universities, and sometimes industrial laboratories had important collections of scientific instruments. The wealthiest institutions kept in their store rooms up to several thousand different apparatus which usually came from the best British, German, and French workshops. Though most of the instruments were produced by specialized firms, many scientific laboratories had their own workshop where highly skilled technicians could repair and modify the apparatus or, in some cases, build special instruments and prototypes. Large universities and polytechnical schools had well equipped scientific cabinets and lecture rooms. Running water, illuminating gas and in some cases compressed air were generally available. Electricity was supplied both by the local network and by large storage batteries, which could be loaded by dynamos driven by small oil, gas, or steam engines. High voltage electricity was provided by induction electrostatic machines or by powerful induction (Ruhmkorff) coils. Because of their multiple uses (production of sparks, oscillating electrical discharges, electromagnetic waves and X-rays, etc.), the latter were among the most important apparatus of the time. The physicist Emilio Segré wrote: *"The length of the available spark established the rank of importance of the laboratory like perhaps today the energy of particle accelerators."* The lecture rooms, which were normally adjacent to the instrument store, could be darkened by black curtains, because of the current use of arc light lamps and projectors. They were used both for showing didactic glass slides and for performing optical experiments. In their lectures and for their experiments, professors and scientists were helped by one or more *préparateurs*, whose skill in manipulating instruments and preparing the apparatus was often unsurpassed.

Instruments were usually stored in different cupboards following the order of classical physics: metrology, mechanics, hydrostatics, pneumatics, acoustics, thermology, optics, electricity and magnetism, and meteorology. They were mostly made of brass, glass, wood and iron. For centuries brass had been the preferred metal in instrument construction for tubes, plates, tripods, and many other elements. Brass could easily be engraved, worked, turned, machined, soldered and polished, it had a bright and attractive appearance and, when it was lacquered it did not oxidize and it assumed a warm gold color. The apparatus could be divided into four or five broad categories:

1) **Research and measuring instruments**: These instruments were used for scientific research and investigation, which often required accurate measurements. They were the most sophisticated and delicate apparatus to be found in laboratories and they represented the state of the art of the precision making industry.
2) **Teaching and demonstration instruments**: These apparatus (such as an Atwood's machine, or a Pascal hydrostatic apparatus) were normally employed to demonstrate a physical law, to visualize a particular effect and, finally, to teach the fundamentals of physics. They usually added little to scientific knowledge, simply reproducing well-known phenomena. An enormous variety of instruments of this type were available during the nineteenth century though several of them, still in production, originated in the eighteenth century.
3) **Professional instruments**: Several instruments such as theodolites, sextants and planimeters, which were normally used in a specific professional activity (in the fields of engineering, navigation, topography, etc.,) could be found in large physics laboratories where they illustrated the practical applications of scientific apparatus. At the same time they represented special examples of particular assemblies of mechanical, optical and electrical elements.
4) **Machines and models**: In every physics collection of the late nineteenth century it was usual to find telephones, gramophones, telegraphs, as well as models of steam engines, water pumps, turbines, cranes, electric motors, etc. Machines (and for obvious reasons their models in reduced scale) represented in fact 'marvels of technology' and were used to explain and demonstrate how scientific principles could be applied to useful technology. In a period when scientific and technological progress was highly prized, machines and models played an important role in scientific collections.
5) **Instruments of the 'physique amustante'**: Scientific toys and entertaining demonstration apparatus were popular in nineteenth-century physics. In the large physics cabinets it was also possible to find many instruments of the '*physique amusante*' which illustrated, in a spectacular and amusing way, surprising and sometimes unusual phenomena.

Needless to say the boundaries between these categories were fluid; a research instrument such as an induction coil or a microscope could be used both for demonstration and for teaching purposes. Demonstration apparatus were often also sophisticated and very ingenious instruments. It would be impossible to list systematically all the instruments which could be found in a 1900 physical laboratory. Some physics instruments called for appropriate environmental conditions. Large telescopes, which were at the time the most impressive and most expensive apparatus, meridian instruments, astrophotographic and spectroscopic apparatus, precision magnetometers, or recording seismographs needed special buildings which had to be set in special locations. In fact, the glow of artificial lighting, electrical disturbances (power distribution lines and tramways) and tremors (like heavy traffic) in towns had to be carefully avoided.

Chemical and biological laboratory equipment was, during the nineteenth century, simpler than that in physical cabinets. In spite of the fact that balances, microscopes and vacuum pumps were common to both collections, most of the instruments were made of glass and earthenware. Retorts, test tubes and bottles, serpentines and refrigerators, phials and pipettes, flasks, ovens and gas burners were the most important apparatus used both for inorganic and organic chemistry. Only in the last decades of the nineteenth century did a few instruments for physico-chemical analysis, such as spectroscopes, polarimeters, colorimeters and refractometers begin to be regularly used in chemical laboratories.

At the beginning of the twentieth century, then, scientists had at their disposal a huge number of sophisticated and elegant instruments. Some of these were modified and improved versions of apparatus which had been invented during the first half of the nineteenth or, sometimes, even the eighteenth century. Embodying few technical improvements, they were startling examples of technological longevity. Double acting piston vacuum pumps, for example, were not basically different from those commonly used after 1760. Many sophisticated laboratory moving-magnet galvanometers essentially worked on the same principle as the apparatus available early in the nineteenth century.

In the first decades of our century most of the earlier brass and glass instruments rapidly became obsolete. They were useless for research, and they soon appeared old-fashioned for teaching purposes. From World War I onwards copper and its alloys began to be too expensive and they were slowly replaced by other metals such as steel and aluminium. The first plastic materials such as ebonite or guttapercha were utilized in the second half of the nineteenth century generally as electrical insulators. Various types of woods, which were still used in the first decades of the twentieth century, mostly for the base and for the boxes enclosing the apparatus, where slowly abandoned with the adoption of metallic encasing and with the development of new plastic materials. The latter began to be commonly used in the second half of this century. Plastics and ceramics also replaced glass and ebonite in electrical instruments' insulating elements. Finally the design was simplified, the last traces of decorations were eliminated and around 1930 most of the laboratory apparatus had definitely lost their aesthetically attractive, but now outdated '*physical cabinet*' look. Sadly tons of them were relegated to the scrapheap. Indeed it is only since the 1970s that these precious material witnesses of the history of science and technology have aroused the interest of several researchers and scholars. Abandoned collections have been restored and studied and today nineteenth century instruments tend to be more carefully preserved in public museums and in private collections.

The Rise of Electronic Technology in the Twentieth Century

A striking feature of our century, which led to the introduction of a completely new set of instruments, experimental practices and industrial structures, was the rise of electronic technology, beginning with the development of wireless transmission. This field of research, which strikingly amalgamated laboratory practices (measurement and detection of electromagnetic waves) with industrial technology (construction of special alternators, large condensers, transformers, special switches), witnessed the first applications of the thermionic electron valve or tube. Electronic tubes allowed a gigantic step in wireless, radio and cable technology because they were able to solve the three main problems concerning the earlier electromechanical and electrical instruments. The difficulties of amplifying an incoming signal, of converting a weak alternating current to a direct one and of generating high frequency signals, were among the major drawbacks for late nineteenth century apparatus. Around 1920 direct current amplifiers, a very important element, began to emerge. Especially after 1930, the tube found many uses other than communications. Electronic components began to be applied to measurement and control applications. For example in the 1930s the use of vacuum tubes largely improved the sensitivity of pH-meters. Complex circuits and special-purpose valves began to be commonly used since improvements in both circuit design and in tube manufacturing enabled the production of reliable, stable and long life apparatus.

Finally with the development of vacuum tubes, theories of microscopic world began to play a more and more essential role in the construction of instruments. The explosion of this new generation of instruments was, however, only possible when industry could produce a very large quantity of cheap and reliable vacuum tubes, a production that was stimulated by the different uses that tubes found during the Second World War.

A major step in the development of electronic systems and instruments was the realization of the transistor, after a long search for a solid state amplifier, in the Bell Laboratories in 1947. The importance of the development of semiconductors is epitomized in the following statement: "*No single technology and industry underwent such a rapid transformation in the post-war period, and had such a pervasive impact as semiconductor electronics.*"[2] Transistors were a kind of compact solid-state equivalent of the valve, and in the 1950s began to be an alternative to vacuum tubes, which were almost completely abandoned a few years later. At the end of the 1950s the first integrated circuit appeared, and in the following decades the reduction of the dimensions of very complex electronic circuits and the increase of their power led to the explosion of informatic technology. Thanks to miniaturization, the energy required by electronics instruments and apparatuses was dramatically reduced, failures due to the fragility of vacuum tubes were practically eliminated, and the production costs of the hardware decreased sharply. With the reduction of the dimension of active and passive electronic elements, it was possible to introduce new and automated systems to manufacture and assemble circuits. Large open structures with relays and valves were gradually reduced to high-density packed printed boards, integrated circuits and microchips.

Vacuum production and measurement became an essential technology in the laboratory and in industry in the twentieth century and their development was strictly associated with the production of vacuum tubes. At the end of the nineteenth century the most sophisticated vacuum pumps used in light-bulb and X-ray tube production were mercury ones, which had been developed by A. Toepler, H. Geissler and H. Sprengel after 1860. They were fragile, large and delicate apparatus which suffered from a major drawback: their working speed was very low. By 1910 efficient mechanical rotatory mercury and oil pumps had been developed. Greater knowledge of gas kinetics allowed the introduction of a series of high efficiency, fast pumps such as diffusion and molecular pumps, which did not have any mechanical moving parts. Large pumps were designed in the 1940s and in the 1950s sputter ion and cryogenic pumps were available. Ancillary equipment was improved with the discovery of low vapor-pressure oil, with the introduction of much better sealing technologies, and with the use of a series of sophisticated vacuum components like valves, connectors, traps and leak detectors. Vacuum gauges were also considerably improved and refined with the invention of the Pirani thermocouple in 1906, and later with different types of ionization gauges as well as with dedicated mass spectroscopes. The progress of vacuum technology which was strictly related to advances in metallurgy, welding techniques, and new materials science, proved to be essential for the realization of more and more reliable vacuum tubes, for space simulation programs and of course for almost all branches of physics.

For the modern physicist, who became more and more interested in dynamic and transient phenomena, in phase transition, and in short-lived phenomena, continuous recording of parameters appeared to be increasingly essential. Mechanical and electromechanical analogue recording apparatus which were able to register continuously with a pen on a moving chart the temporal modification of a physical parameter, had been in use since the nineteenth century, mainly in meteorology. Electrical recorders of various types began to be introduced at the end of nineteenth century with the spectacular growth of electrical power supply. Recording apparatus for industry and for the laboratory appeared at the end of the century, problems related to the friction of the stylus on the recording paper sometimes being avoided by using a light beam on a photographic sensitive plate. These systems became more common with the introduction of celluloid film one of whose many uses was the recording of oscillations of the moving-mirror electromechanical oscilloscope.

FIGURE 38.2: THE PHYSICS LABORATORY AT LA SORBONNE IN PARIS AT THE END OF NINETEENTH CENTURY. NOTE THE TYPICAL APPARATUS OF THE TIME SUCH AS RUHMKORFF COIL, GEISSLER MERCURY VACUUM PUMP, WIMSHURST ELECTROSTATIC MACHINE AND PELLAT PRECISION ELECTRODYNAMOMETER. (FROM A POSTCARD, COLLECTION OF THE AUTHOR)

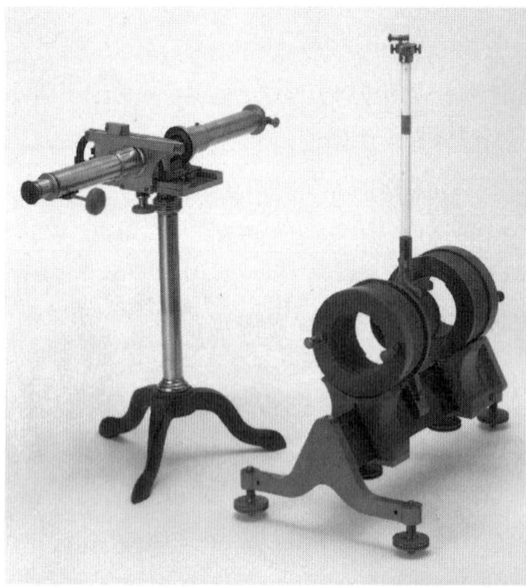

FIGURE 38.3: THE TRIUMPH OF BRASS AND GLASS, TWO TYPICAL LATE NINETEENTH-CENTURY LABORATORY APPARATUS. LEFT: A FÉRY REFRACTOMETER. RIGHT: A WIEDEMANN GALVANOMETER. (FONDAZIONE SCIENZA E TECNICA, FIRENZE)

FIGURE 38.4: ROTATING INTERRUPTOR FOR X-RAY EQUIPMENT AROUND 1920. (FONDAZIONE SCIENZA E TECNICA, FIRENZE)

FIGURE 38.5: LARGE X-RAY APPARATUS OF THE NATIONAL PHYSICAL LABORATORY ABOUT 1930. HEAVY ELECTROTECHNICAL APPARATUS ENTER IN THE LABORATORY. (FROM: THE NATIONAL PHYSICAL LABORATORY, 1933, TEDDINGTON)

FIGURE 38.6: IMPEDANCE BRIDGE, A TYPICAL INSTRUMENT OF THE 1930S. (FONDAZIONE SCIENZA E TECNICA, FIRENZE)

FIGURE 38.7: MID TWENTIETH-CENTURY APPARATUS: AN ABBE REFRACTOMETER AND A GALVANOMETER. NOTE THE DIFFERENCE IN DESIGN AND MATERIALS WITH THE SAME TYPE OF INSTRUMENTS OF FIGURE 38.2. (FONDAZIONE SCIENZA E TECNICA, FIRENZE)

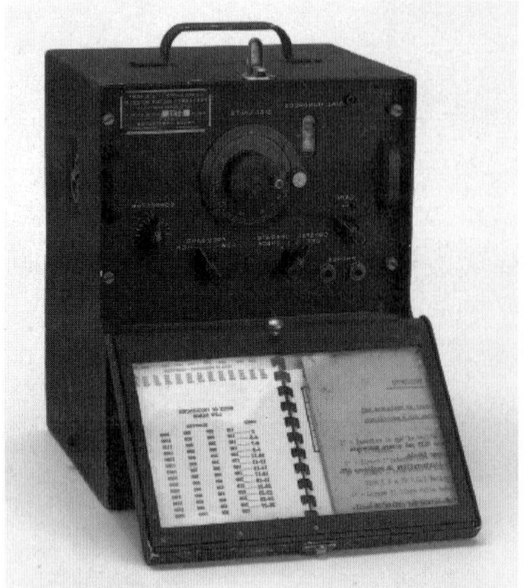

Figure 38.8: Precision frequency meter around 1940. (Fondazione Scienza e Tecnica, Firenze)

Figure 38.9: A series of various types of electronic tubes, which found an incredible number of uses until the invention of transistors and miniaturized components. (Fondazione Scienza e Tecnica, Firenze)

Figure 38.10: Two mid twentieth-century oscillographs. Oscillographs proved to be among the most important and popular instruments in the twentieth century. (Fondazione Scienza e Tecnica, Firenze)

FIGURE 38.11: VARIOUS INSTRUMENTS WITH ELECTRONIC TUBES: COUNTERS, AMPLIFIERS, IMPULSE GENERATORS. (FONDAZIONE SCIENZA E TECNICA, FIRENZE)

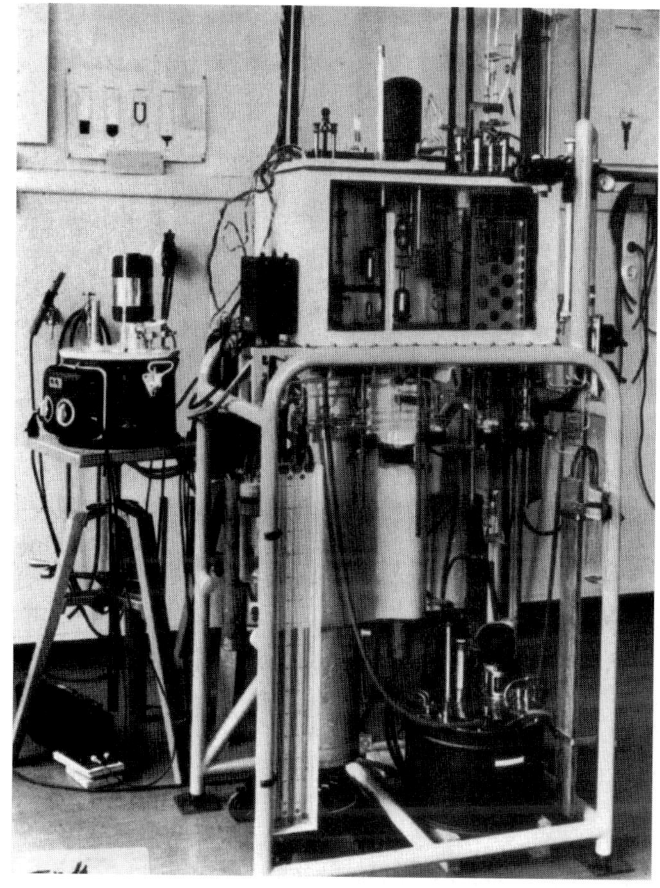

FIGURE 38.12: AN HIGH PRECISION GASTHERMOMETER OF THE PHYSIKALISCH-TECHNISCHE BUNDESANSTALT AROUND 1960. THE 'INSTRUMENT' IS IN FACT A COMPLICATED ARRAY OF VARIOUS APPARATUS. (FROM: FORSCHUNG UND PRUFUNG. 75 JAHRE PHYSKALISCHES-TECHNISCHE BUNDESANSTALT/REICHASNTALT, 1962, BRAUNSCHWEIG)

FIGURE 38.13: TYPICAL MODULAR "BLACK BOX" APPARATUS AROUND 1970. (FONDAZIONE SCIENZA E TECNICA, FIRENZE)

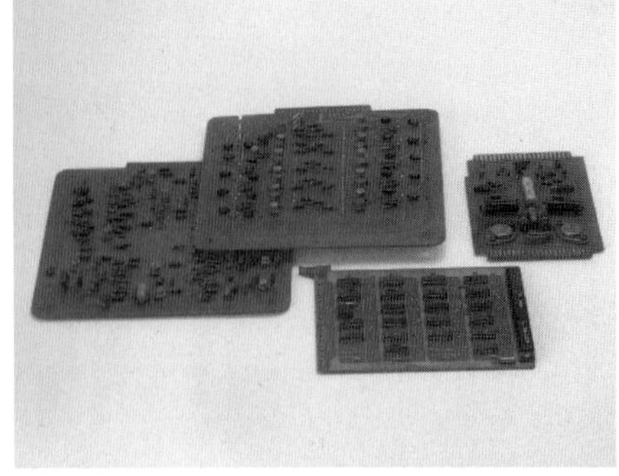

FIGURE 38.14: CARDS WITH MINIATURIZED ELEMENTS AND MICROCHIPS, WHICH ARE NOW THE HEART OF MANY APPARATUS. (FONDAZIONE SCIENZA E TECNICA, FIRENZE)

FIGURE 38.15: MID-CENTURY LABORATORY. GENEVA PHYSICS INSTITUTE. © CERN.

If electronics paved the way for the most spectacular development in twentieth-century instrumentation, it must be remembered that the more 'classical' fields of instrument manufacture such as precision mechanics also made important advances. Indeed fine mechanics has always played a central role in the history of scientific instruments. Accurate mechanisms were needed for high precision metrological instruments, analytical and gravitational balances, gyroscopes, meteorological and electrical recording apparatus, as well as for optical instruments and teletechnique devices. Not only could the different mechanical elements be produced by more efficient machine-tools but their design was also improved and rationalized. Some of the most interesting precision mechanisms have been in use since the late nineteenth century in the mathematical apparatus (calculating machines, tide predictors, harmonic analyzers, planimeters and integrating devices) to which electromechanical components were gradually added (motors, relays, switches). Complex automatic electromechanical control devices were invented during the first decades of the twentieth century and they found many applications both in the laboratory and in industry. In the laboratory, the constant knowledge, monitoring and feed-back regulations of different process parameters such as temperature or pressure allowed experiments of long duration to be carried out without the constant presence of an operator. In the last decades the maintenance of some processes in a steady state (of temperature, pressure, flow, frequency, electric voltage and current etc.) became essential in scientific practices.

Many other techniques developed during the last century could only be profitably developed during the last fifty years thanks to the introducing of new materials and processes. The use of digital devices, whose underlying theory was developed in the nineteenth century, was quite restricted before 1940, even though many of the basic electrical and electronic elements (multivibrators, square-wave generators, flip-flop circuits) were known and had found some applications. Digital systems, which required a completely new series of devices such as counters, displays, and special tubes, began to gain importance with the construction of the first large relays or valve computers. At the end of the 1920s the size and the calculating power of mathematical machines began to grow. The extremely complex problems of the new theoretical physics, as well as the calculations of ballistic tables for artillery led to the realization of larger computing machines. By the mid 1940s electromechanical computing machines were almost completely replaced by electronic ones in which tubes replaced relays. In the following decades progress in computer technology, largely due to the miniaturization of electronic elements, made it possible to undertake complex scientific and engineering projects such as space exploration and, with the introduction of cheap and powerful personal computers, began drastically to change every aspect of human activity. Lately, with the rapid development of microelectronics, computers, computerized machines and instruments, the amount of data that can be recorded, processed and stored has reached an unparallel level. The power of the computer combined with sophisticated software allows the registration and the processing of an enormous quantity of data which an be easily transformed and visualized in the form of charts, graphs, tables, and high-definition images.

To conclude, the increase in quantity and variety of scientific instruments in laboratories after the Second World War has been fast and spectacular. In fact during the last decades, new instruments have not only completely transformed physics but, because of an increasing use of physical techniques they have also greatly modified the practices of chemistry, biology and medicine. A striking example concerning the use of infrared spectroscopy in chemistry has been given by Rabkin: "*The number of infrared instruments, a handful before the war, rose to 700 in 1947, to 3000 in 1958 and to 20,000 in 1969.*"[3] Furthermore spectrophotometers, X-ray spectrographs, mass spectroscopes, NMR apparatus, lasers, and many other items of equipment which originated as scientific instruments for physics laboratory research have quickly found applications in the various technologies of civilian and military industry as well as in activities of everyday life.

Instruments and Industry

The evolution of instruments as well as the profound transformation in the nature of industry during the twentieth century were reflected in a dramatic change in scientific instrument manufacture. Even if at the beginning of the century some firms such as the Zeiss in Jena had the complex and articulated structure of a relatively large and modern firm, until the First World War most scientific instruments were still produced by medium-sized or small firms which were organized in an almost familiar way. Instruments were made by hand by highly skilled craftsmen and, with very few exceptions, there was no mass production. Division of labor appeared only in a few particularly well managed workshops at the end of the century. Until the beginning of twentieth century the owners of such firms were often at the same time scientific leaders, production and financial managers, and technical advisors to the company. They were helped by a few well trained engineers or technicians who supervised the activities of specialized workers. Systematic research and development did not exist in the modern sense of the word and the connections between the scientific community and the workshop were mostly assured by informal personal relationship between instrument makers and scientists on a single instrument basis. There was little or no specialization. In fact until the first years of twentieth century companies such as Max Kohl of Chemnitz or Griffin of Glasgow, listed thousands of very different instruments ranging from simple didactic mechanical models to sophisticated electrical measurement apparatus. Production had to be rationalized however due to the increasing complexity of the instruments. At the same time the old-fashioned trial-and-error, project-by-project realization had to give place to a more systematic instrument design. Most of the old fashioned instrument making firms were absorbed in the first decades of the twentieth century by much larger companies whose special branches manufactured electrical or electronic equipment for laboratories, for industry as well as for the mass market (for example, radios). Large companies could also combine the know-how of special research and development departments with the experience of large mechanical, electromechanical or electronic industries.

It was the First World War which began the transformation in the scientific instrument industry, which for the first time appeared to be strategically important. This was particularly true in the field of optics where, before 1914, optical glass and fine mechanical parts for binoculars, telemeters, range finders and surveying instruments were often supplied by German manufacturers. Indeed the French and English governments only began to take a greater interest in the activities of their precision industry after 1914. Military requirements proved to be essential for the survival of many firms during the war and in the inter-war period, even though mass production of certain instruments which had been introduced for supplying the needs of the army, had to be stopped after 1918. At the same time, new institutions with strong academic links were founded to officially support training and research in optical instrumentation.

If the strategic importance of scientific instruments and of related technologies appeared evident during the First World War (torpedo detection, communications, aircraft guidance, telemetry, etc.) it was only during the Second World War that their role became absolutely essential. But changes were quite slow and, at least in Europe, production techniques were not drastically modified until the 1940s. This trend was accentuated after the war with the increasing importance of Big Science and the rise of the military-industrial complex. The evolution was necessary not only because of the increasing complexity of the apparatus but also because of the demands imposed by a competitive market and, in the case of very large instruments, of minimizing prices and avoiding costly failures. Nowadays the realization of complex instruments requires the cooperation of several different firms as well as the contribution of a network of subcontractors. In fact, the evolution of electronic laboratory instruments has recently witnessed the success of medium-small highly specialized firms which have been able to develop and market a precise typology of instruments or very particular devices.

Instruments and Users

The introduction, during the twentieth century, of completely new series of instruments, which was largely due to the rise of electronic technology, not only deeply changed laboratory practices as well as many manufacturing processes but also radically transformed the relationship between instruments and their users. Until the beginning of twentieth century, most instruments were relatively simple table apparatus, and their function and construction could be quite easily understood and mastered. Apart from the common apparatus such as spectroscopes, microscopes, electrical measurement devices, induction coils, vacuum pumps, and so on, which were routinely on offer by the instrument makers of the time, physicists could easily have their instruments custom-made. The close and often informal relationships between instrument makers and scientists facilitated a constant exchange of information and know-how.

Because of their relative simplicity, 'classical' instruments could be easily modified, disassembled, repaired and overhauled by the researcher himself or by the laboratory technicians who generally looked after the collection in the university and in the scientific institutions. Until a few decades ago, a kind of workshop apprenticeship together with a significant amount of time dedicated to the construction, the modification and the improvement of instruments was an important part of the curriculum of most experimentalists. Today, because of the structure of the instruments and also because of their growing complexity, experimental scientists tend to be more disengaged from what Blume called "*material world of science.*"[4] "Black boxes," compact apparatus whose appearance does not immediately reveal their function and their specific use, became more and more common in laboratories in the second half of our century with the increasing use of compact electronic apparatus. They cannot be dissociated from the notion of mystery. Unlike nineteenth-century scientists, who clearly knew every single element of their experimental set-up, contemporary researchers must very often cope with a series of 'closed instruments' whose internal structure and detailed function are often unknown to them. The material and psychological "impenetrability of black boxes" also contribute to move experimental scientists away from a series of 'hands on' practices, which were once very common.

The distance between experimentalists and the phenomena they are studying has been further increased by the development of electronic modes of detection. Until 1920–30 most optical measurements and observations were made using the human eye as detector and meter. Color, intensity, number and width of spectral lines, scintillations of a fluorescent screen, shadows in the eyepieces of a photometer or a polarimeter etc., were observed, judged and analyzed by the unmediated naked eye. Of course the weight of the subjective factor in these operations was very important. The human eye is nowadays used almost exclusively for diagnosis or pattern recognition. Most signals and phenomena in every type of apparatus are detected, processed and measured electronically. The experimenter examines a long series of numerical tables, graphs or computer images. The latter are very often an 'artistic' display which are created from a series of numbers but which do not reproduce a reality. In this sense, even if the most sophisticated apparatus enormously enhance our possibility of enquiry, they also act as 'filters and transformers' which make it more difficult to interpret their outputs. Furthermore if the observational instruments of classical physics, such as microscopes or telescopes, could be considered passive because they did not interact with the phenomena and entities under observation, the situation is much more complicated in quantum physics. In fact we know that the very act of observing (and thus the presence of an instrument) is not neutral anymore, and that the apparatus itself contributes to modify or create the subject observed.

In the past many experiments could be reproduced by different researchers in different geographical locations because of the relative facility of acquiring the same type of apparatus or having them reproduced. Today with the complexity, the cost and size of many instruments for Big Science, notable high energy physics and much of astronomy, the number of places where a certain kind of experiment or observation can be done have been drastically reduced. What is more the access to such large apparatus has to be negotiated. Big Science instruments cannot be mastered and used by a single scientist or a small group of researchers. They require a large team (sometimes composed of several hundreds of people) which are often comprised of different groups of experts who are responsible for a specific part of the equipment (computers, detectors, beam production, etc.). Many teams of scientists have to wait their turn to be able to 'rent' a certain amount of hours, days or months for the use of the use of instruments which, in certain cases, are unique. The consequent feeling of frustration of some individual scientists has led some to depict Big Science as a "*market-conscious, product-oriented, and capital-intensive activity which, unlike Little Science, has taken on the impersonal and inhumane nature of industrial enterprise.*"[5] Against that, one must not forget that today, even in a small laboratory, scientific instruments are far more complex than they used to be. They may not require multi-million dollar equipment and hundreds of people to run them, but the experimental set-ups are still often complicated agglomerates of instruments including vacuum and cryogenic apparatus, electromechanical and electronic devices, optical components, computer equipment with screens, printers, plotters and so on. In short, the complexity of today's technology of science often makes it more appropriate to speak of an 'instrumental system' than of an 'instrument,' which is perhaps too easily assimilated to the image of the self-contained, relatively simple table apparatus.

The emergence of Big Science has also changed the relationship between the scientific user community and their funders. A hundred years ago the costs of apparatus was relatively secondary compared to the total amount of money needed for running a university or a research institution. Today a large accelerator or a space telescope can require financing in the order of several hundreds of millions dollars and even more. These figures can only be provided with the massive financial support of government agencies and institutions.

The involvement of politicians, industrialists, bankers and philanthropists has thus become essential in these enterprises, and large instruments are now status symbols of power, wealth and scientific and political influence, in the same way as were the precious and highly ornate mathematical instruments, which enriched the cabinets of the princes and noblemen of the Renaissance.

CONCLUDING REMARK

In this chapter, I have pointed out some of the most startling differences between laboratory equipment pre and post 1900 circa. My main argument has been that it was the rise of Big Science in America around 1930, the frantic research during the Second World War, and the explosion of semiconductor technology followed a few years later by the invasion of computers which were the most important factors which led to the instrument revolution whose effects we see around us every day. In spite of the fact that I have mostly considered the field of physics, several of my considerations could be extended to other branches of scientific research such as astronomy or chemistry. As pointed out by Turner,[6] our century shows a universalization of the presence of instruments, not only in the scientific fields where it had already started, but also in every human activity at least, in developed countries. A systematic survey of the evolution of the always increasing number of apparatus which are used by today's scientists and their complexity would certainly require many volumes. During the twentieth century scientific laboratory equipment has been revolutionized. Not only has the number of instruments which are used in physics, chemistry, biology, astronomy, medicine and in all the branches of science grown by an enormous amount but the typology of instruments has completely changed. In fact we can assume that these changes have probably been more significant in the last 50 years than in the previous century. Just like the different branches of physics, that after the late nineteenth century tended to redefine their boundaries and to merge together in a more complex but certainly more realistic panorama of nature, the subdivision of instruments in terms of optical, electrical, or acoustical apparatus became more and more redundant in the course of the twentieth century.

REFERENCES

1. Segré, E. *Dai raggi X ai quark. Personaggi e scoperte delle fisica contemporanea.* (Milano: A. Mondadori, 1976), p. 14.
2. Bothelo, A., Jacq. F., and Pestre, D. "The electronic challenger: on historical perspective", *History and Technology II* (1994), **2**.
3. Rabkin, Y. "Technological Innovation in Science: The Adoption of Infrared Spectroscopy by Chemists", *Isis* (1987), **78**: pp. 31–54.
4. Blume, S. "Whatever happened to the string and sealing wax?" in *Invisible Connections* (op. cit., note 2), pp. 87–101.
5. Capshew, J.H., Rader, K.A. "Big Science: Price to the Present" in Thackray A. (ed.) *Science After '40, Osiris,* II series, (1992), **vol. 7**: pp. 3–25.
6. Turner, G.L.E. *Nineteenth Century Scientific Instruments.* (London: Sotheby Publications, 1983).

FURTHER READING

Aitken, H.G. *Syntony and Spark: The Origins of Radio.* (Princeton: Princeton University Press, 1985).

Aitken H.G. *The Continuous Wave Technology and American Radio, 1900–1932.* (Princeton: Princeton University Press, 1985).

Anderson, R.G.W. and Turner G.L.E. (eds.). *An Apparatus of Instruments: The Role of the Scientific Instrument Commission.* (Oxford: Scientific Instrument Commission, 1993).

Bennett, S. *A History of Control Engineering, 1930–1955.* (London: IEE Peter Perogrinus, 1993).

Eckert, M. and Schubert, H. *Crystals, Electrons, Transistors: From Scholars' Study to Industrial Research.* (New York: American Institute of Physics, 1986).

Frémy, Carnot, Jungfleisch and Terrail. *Les laboratoires des chimies.* (Paris: Dunod, 1881).

Frick, J. and Lehmann, O. *Physikalische Technik.* (Braunschweig 1909).

Galison, P. and Hevly, B. (eds.). *Big Science: The Growth of Large-Scale Research.* (Stanford: Stanford University Press, 1992).

Greenaway, F. "Instruments," in Williams, T.I. (ed.). *A History of Technology Vol VII The Twentieth Century 1900–1950,* Part II. (Oxford: Clarendon Press, 1978).

Hackman, W. "Scientific Instruments: Models of Brass and Aids to Discovery," in Gooding, D., Pinch, T. and Heilbron, J. and Seidel, R.W. *Lawrence and his Laboratory: A History of the Lawrence Berkeley Laboratory,* Vol. 1. (Berkeley: University of California Press, 1989).

Henrivaux, Girard and Pabst. *Les laboratoires de chimie, Supplément.* (Paris: Dunod, 1882).

Medely, T.E. and Brown, W.C. *History of Vacuum Science and Technology: Special Volume Commemorating the 30th Anniversary of the American Vacuum Society, 1853–1883.* (New York: American Vacuum Society, 1984).

Redhead, P.A. *Vacuum Science and Technology Volume 2: Pioneers in the 20th Century.* (New York: AIP Press, 1994).

Schaffer, S. (eds.). *The Uses of Experiment: Studies in the Natural Sciences.* (Cambridge: Cambridge University Press, 1989).

Williams, M.E.W. *The Precision Makers: A History of the Instrument Industry in Britain and France, 1870–1939.* (London and New York: Routledge, 1994).

The Bulletin of the Scientific Instrument Society.

CHAPTER 39

Science in the United Kingdom

A Study in the Nationalization of Science

DAVID EDGERTON

The United Kingdom has been one of the scientific great powers of the twentieth century. Since 1901 it has obtained about the same number of Nobel Prizes as Germany, and about half the number of the United States; all other countries are way behind. The United Kingdom's *comparative* performance has, on some measures, improved during the twentieth century, with a definite *relative* decline setting in only since the 1960s (see Tables 39.1 and 39.2). A qualitative analysis also shows that twentieth-century British science should be

TABLE 39.1: NUMBER OF BRITISH NOBEL PRIZE WINNERS.

	Chemistry	Physics	Medicine	Total
1901–1910	2	2	1	5
1921–1930	3	2	2	7
1931–1940	1	3	3	7
1945–1954	3	5	4	12
1955–1964	6	0	7	13
1965–1974	4	4	2	10
1975–1984	3	1	3	7
1981–1990	0	0	3	3
1901–1990*	22	20	23	65

*Note that these figures are not the sum of the above.

NB: This is a table of the number of prize winners: on a number of occasions two or more British scientists won the same prize in the same year. The allocation of prizes does not correspond to obvious modern disciplinary categories. Thus Rutherford, Soddy, and Aston won Chemistry Prizes; the three British DNA prizes were for Medicine; Sanger (twice), Kendrew and Perutz, and Hodgkin won prizes for Chemistry. In addition Britain won 8 Literature and 9 Peace Prizes, as well as 5 Economics Prizes (since 1969 only). The record in Economics is outstanding, while the Peace and Literature record is meagre. However, contributions to peace and literature have not required the sort of infrastructure science has.

TABLE 39.2: SIGNIFICANT EVENTS IN SCIENCE, MEDICINE AND TECHNOLOGY.

	1901–45			
	UK	Ger	USA	Total
Anthropology and Archeology	2	1	7	16
Astronomy	17	10	51	108
Biology	34	11	49	125
Chemistry	10	21	37	93
Earth Sciences	8	8	11	36
Mathematics	14	14	5	56
Medicine	20	18	44	112
Physics	56	36	43	189
Technology	20	32	66	142
Total	181	151	313	877
	20%	17%	35%	100%
	1945–1965			
Anthropology and Archeology	2	–	4	11
Astronomy	14	1	34	72
Biology	11	2	35	58
Chemistry	5	2	7	19
Earth Sciences	2	–	13	15
Mathematics	–	–	3	4
Medicine	4	1	19	25
Physics	5	4	39	52
Technology	6	–	23	36
Total	49	10	177	294*
*does not add in source				
	17%	3%	60%	100%
	1966–1988			
Anthropology and Archeology	7	1	11	28
Astronomy	5	3	62	89
Biology	10	–	52	77
Chemistry	6	7	16	36
Earth Sciences	2	–	9	16
Mathematics	1	–	10	14
Medicine	8	4	40	61
Physics	5	6	21	48
Technology	9	–	20	32
Total	55	21	241	401
	14%	5%	60%	100%

Source: compilation by Mansel Davies from Alexander Hellemans and Bryan Bunch, *The Timetables of Science* (London: Sidgwick and Jackson, 1988) as reported in 'A Thousand Years of Science and Scientists: 988 to 1988,' *History of Science* Vol. 33 (1995), pp. 239–51.

taken seriously. To take an index given undue weight, the United Kingdom launched the first serious atomic bomb project; and became the third nuclear power. By any reasonable standard the United Kingdom was, and to some extent is, a force to be reckoned with in world science, and worth studying for that reason. However, the strength of British science in the twentieth century does need some stressing. It has been overshadowed by the emphasis on the 'decline' of the United Kingdom's scientific, as well as military and industrial power. It is important to note, however, that Britain's industrial decline is much exaggerated, and that in any case scientific power remains everywhere more concentrated than industrial or military power. The British case should be treated as one of a small handful of scientific great powers of the twentieth century; indeed we should ask the same questions of British science as we do of US science, and indeed of Soviet and German science. There is no typical case from which the others are deviations. It is certainly not to be studied as a case of exceptional resistance to science, or of the failure to turn scientific advantage into economic growth.

It is not at all obvious, however, that we should treat science in relation to particular nation-states. Indeed, it is not obvious that we should treat any aspect of twentieth century history in this way. And yet, national accounts of politics, society, economy and science dominate historical writing. There is nevertheless a strong case for treating science nationally since it has been constituted by national education systems, by national funding (not least for that ultimate purpose of the nation-state — waging war), and by national industrial and economic policies. I will argue that there has been a quite deliberate *nationalization* of science, which compels us to treat the growth of science with reference to the nation-state. This does not mean that the nation-state is the only context in which science should be studied — far from it — but it does mean that it is an important one. But this linkage between science and the nation is itself the product of historical development, and is not to be treated as natural or inevitable.

Nationalization is a complex term. Its most obvious meaning, in a number of languages, is the transfer of industrial production from the private to the public sector. But the very term indicates that these particular 'socializations' of industry involved their transfer to a single *national* authority, one that not only belonged to the nation, but covered the whole nation. Here I will use the term in a broader sense to indicate an increasing proportion of public rather than private *funding*; the shift from private sector to public sector *institutions*; the shift from local and regional funding and institutions to *national* institutions; and finally, the extent to which the *borders* (economic as well as social and political) of the nation-state became a significant factor in determining whether and how science was undertaken by both public and private agents. In the latter case national tariffs and other import restrictions played a particularly important role.

It is interesting that British scientists' own rhetoric was not particularly concerned with the state before the 1880s, but thereafter it was central.[1] It is within

TABLE 39.3: ESTIMATES OF THE NUMBER OF SCIENTISTS IN THE UK DURING THE 1960S. (THOUSANDS)

	1961	1966
Scientists	117.7	147.4
Agriculture and biology	22.3	28.7
Chemistry	36.4	45.1
Mathematics	17.6	21.5
Physics	17.9	23.7
General and other	23.6	28.4
Engineers	142.0	179.9

Adapted from Hilary Rose and Steven Rose, *Science and Society* (Harmondsworth, 1969) Table 3, p. 130.

this nationalistic context that most arguments about British science have developed. Several features stand out. The first is the relentless technocratic critique of existing national elites, for example the "two cultures" argument of C.P. Snow. The second is the insistence on national decline caused by a lack of national investment in science. A strong feature of such arguments is that British science has been insufficiently nationalized: public funding has been inadequate; scientists have been deficient in penetrating *national* political and cultural elites; the national scientific effort was not bent through appropriately powerful central agencies to the purposes of the nation. Such arguments have by no means been confined to British scientists.

In this chapter I will take a very different tack: I will deliberately discount scientists' own accounts of British science.[2] I will show the extent to which science was part of a national culture, and indeed how it became so. I will also emphasize the fact that at the beginning of the twentieth century science was not nationally organized, but that it is not to be ignored for that reason. I will also argue that since the 1960s especially, British science has been both de-nationalized and privatized, though I will note some aspects in which the process of nationalization has been radically extended.

NATIONALISM AND DECLINISM

By the 1950s and 1960s British science was very substantially nationalized in the senses outlined above: public funding of scientific activity was very important; much publicly funded research was carried out in public national laboratories; public (and private) funding was dominated by bodies of national scope; and industrial science was carried out by nationally-organized firms in a context of powerful national economic frontiers. This nationalization was reflected in writings about science of the period, which assume that science should be nationally organized.

C.P. Snow's essay on the "two cultures" is an example. Snow's argument was that as a *nation* Britain was not producing enough scientists and engineers. Its national economy was not growing fast enough: "Why aren't we coping with the scientific revolution?" asked Snow, "Why are other countries doing better?"[3] Snow's response was a social and cultural history setting out to explain why British science did not penetrate the nation's "corridors of power" (a phrase he coined). As Snow wrote, he had to hand figures for the total number of scientists and engineers in the United Kingdom, defined as those with degrees or other qualifications in these subjects: they numbered about 80,000. As a member of the Civil Service Commission, Snow claimed that he and his colleagues had interviewed some 25 percent of them as candidates for service in the central state.[4] Snow found they did not read "books," by which he meant novels, poetry, plays and history. There was a gulf of "culture" — broadly defined — between them and "literary intellectuals." Snow's points of reference in analyzing this gulf were those of a class-divided, and politically-divided nation. Scientists, he argued, tended to be on the political left, and many came from poor families. "Literary intellectuals" were, by implication, more right wing, and richer; crucially, they formed the core of the national elite. One would not guess from Snow that of all university teachers in 1961–2, 26 percent were in all the arts and education, 26 percent in pure science, 20 percent in technology, 20 percent in medicine, with a paltry 8 percent in social sciences. Literary culture — even in this very broad definition — was a distinct minority in the university.

Snow's work is a good examplar of writings on British science in other ways. Despite the breadth of his thesis relating science, the state, the nation, decline, class and politics, the details given are highly specific, without being properly specified. Snow, following the fashion of the 1950s, spoke of a "Scientific Revolution" which took place not earlier than 1920–1930, which was creating an "industrial society of electronics, atomic energy, automation" and which was profoundly different to what had gone before.[5] His analysis is restricted to the industrial applications of *physics*. His specific references are almost entirely to British *academic research physicists*. There are no chemists, much less industrial chemists, or teachers of mathematics (nor does he include doctors and engineers). On the other side of the divide, British *novelists* stand for the whole of the "traditional culture" or the "literary intellectuals" — there are no classicists, philosophers, economists, or historians, not to mention lawyers or clergymen.

The emphasis on the nation, and a small part of 'science' misleadingly standing for the whole is typical of much writing on British science in the twentieth century. The historiography of British science has concentrated on *national* bodies with *science* in their title, and has systematically exaggerated the insignificance. Thus we know a great deal about the Department of *Scientific* and Industrial Research. Often overlapping with this is a literature on national cross-disciplinary organizations of scientists, especially the British *Science* Guild, the National Union of *Scientific* Workers, the Association of *Scientific* Workers, the British Association for the Advancement of *Science*, and on central government committees with science in the title, like the Advisory Council on *Scientific* Policy. Indeed much of the literature concerned with the above is a history and pre-history of 'science policy.' There is an underlying assumption in this literature of the existence of a nation to which a national science corresponds, which is, or should be, regulated by a national science policy. But, as it happens, 'science policy' usually means the policy of that small part of government concerned with *civil* (and not military) *research* (and not development), funded by the state (and not the private sector), largely in universities (and not in firms or government laboratories). The work of the Department of Scientific and Industrial Research cannot stand for the whole of British science.

Ironically, the best general treatments of British science are given in books and articles without science in the title. Thus Michael Sanderson's, *The Universities and British Industry* is by far the best general treatment of British university science there is. Similarly the official history of *The Administration of War Production* of the 1939–45 war gives the best available account of the organzation of military R&D in Britain in the first half of the twentieth century. Margaret Gowing's books on British atomic energy — the best single treatment of an aspect of twentieth-century British science — do not have science in the title or subtitle. This curious phenomenon is little more than a reflection of the striking fact that institutions without 'science' in their titles have been much more important for 'science' than those with it. The main government funders of scientific research have been the likes of the War Office, Admiralty and Air Ministry, the Ministries of Aircraft Production, Supply, Aviation and Technology, and since 1973 (only) the Ministry of Defence. None of the universities, major centers for scientific research, have science in their titles. (Imperial College of Science and Technology (and now Medicine) has always been part of London University; UMIST, UWIST, and the Royal College of Science and Technology (Glasgow), have been part of larger universities). Nor do any of Britain's important research-intensive fims, such as ICI, Metropolitan-Vickers, Courtaulds and so on.

We also need to be very wary of the meanings given to 'science.' The term 'science' is usually taken to exclude 'medicine' and 'technology,' though modern medicine is routinely included in the benefits of science, as are technological achievements. To discuss 'science and society' or 'science and history' without discussing technology and medicine is clearly an absurdity. Indeed the number of scientists only exceeded the number of doctors in the 1940s, while the number of engineers has always remained somewhat higher. But even within 'science' narrowly construed, it is striking the way the literature focuses on that small fraction of scientific activity called 'research,' and within this category on the academic physicist and perhaps biologist. It is worth bearing in mind that industrial chemists, or teachers of mathematics were much more numerous. Indeed industrial research chemists hugely outnumbered academic physicists.

There are, then, systematic biases in the extent of coverage of scientific activity in twentieth century Britain. So deeply ingrained are they that we do not have the resources to begin to write a history of science in twentieth-century Britain that properly reflects the range of scientific activity. This chapter will thus, by default, follow the bias of the literature, especially by concentrating on the history of *research.*

THE SCALE AND SCOPE OF SCIENCE

'Scientists' have no collective existence. There has never been an organization for all scientists which has had more than a fraction of its potential membership. Scientists have been divided into disciplinary groupings, themselves not especially well organized. More important is the fact that scientists have been, with the exception of a few consultants, employees: of educational institutions, companies, and local and central government. A central register of scientists was compiled during the Second World War, and after it the British government routinely estimated the stock of scientists, and counted the number working in particular industries and services. Such statistics have prompted the making of estimates for earlier periods. R.M. Pike's figures, which I believe are substantial underestimates but nevertheless of the right order of magnitude, adequately describe the trends. His first estimate for 1902 or thereabouts, was a direct one. He found that there were some 2,400 scientists in Britain. By 1914 the figure was up to 7,000, and by 1918 to 7,500. In the inter-war years Britain was producing about 2,500 scientists per annum, such that by 1939 the 1918 stock had at least trebled.[6] By the early 1960s output was running at 9,000 per annum. Official figures give a stock of 147,000 for 1966. All the above figures exclude engineers and medical doctors.

When we think of 'science,' especially before the 1960s, we would do well to think of chemistry. Even in 1966 chemists accounted for 30 percent of the national stock of scientists. In the inter-war years the figures were higher still. A clear majority of these chemists were employed in *industry.* The situation for physicists was rather different: about half were in industry in the late 1950s. Mathematicians and biologists were highly concentrated in the educational system. In the mid-1950s half of all scientists were in education, but fully 38 percent were in industry. Table 39.4 shows the numbers of different categories of scientists employed by different organizations in the late 1930s.

The fifty-fold expansion in the stock of scientists between 1902 and 1966 deserves emphasis. The factor of expansion is quite out of line with that of other professional groups such as clergymen, medics, lawyers, or administrative civil servants. This quantitative expansion had consequences for class composition, for professional status, for pay, and for differentiation and stratification with the category 'scientist' which have not been fully appreciated. For example, there have

TABLE 39.4: SCIENTISTS IN THE UNITED KINGDOM, SELECTED INDICATORS, MID-1930S.

I		II		III	
London	304	ICI	464	Army	506
Cambridge	142	GEC	175	Navy	226
Oxford	109	BTH	104	Air	110
Sub Total	555		743		
	(46% of total)		(29% of total)		
Bristol	61	Metro-Vick	87		
Leeds	56	United Steel	86		
Manchester	51	Dunlop	85		
Durham	44	Lucas	84		
Glasgow	43	Lyons	70		
Birmingham	41	Unilever	57		
St Andrew's	41	Courtaulds	51		
Liverpool	39	Anglo-iranian	50		
Nottingham	38	Calico print	48		
Edinburgh	32	Morgan Crucible	46		
Sub Total	446		664		
Total of above	1001		1407		
Total	1217		2566		2300

I – University teaching staff in science (excluding medicine and technology)

II – Qualified staff in research and development in industry

III – Qualified scientists in research in government

Note: R&D staff in industry account for half or less of all scientific and technical staff in industry; university teaching staff are of course only part-time in research. The government figures are for scientists (only) engaged in research. The list of firms probably omits some large employers.

Sources: Universities — J.D. Bernal, *The Social Function of Science* (1939), p. 417; Industry — Edgerton D.E.H. and S.M. Horrocks, 1994. 'British Industrial Research and Development before 1945,' *Economic History Review* Vol. 47 (1994): 213–38; Government — R.M. Pike, *The Growth of Scientific Institutions and Employment of Natural Science Graduates in Britain 1900–1960* (University of London MSc thesis, 1961), p. 46.

been numerous complaints that British scientists' social origins have been more humble than those of other professional groups, due to a supposed lack of elite recruitment into science. But, the lower class origins of scientists may in part be explained by the differential expansion of science. That is, the larger the group, the lower the average class position is likely to be, and the more likely that a wide range of class origins and class positions exists within science. The direct evidence is unfortunately fragmentary and inconclusive. According to Gowing, the atomic scientists and engineers of the 1940s nearly all came from the "lower side of the

TABLE 39.5: EXPENDITURES ON RESEARCH AND DEVELOPMENT, 1937 AND 1961-2, BY SOURCE OF FUNDS, £M.

	1937		1961–2	
	£m	%	£m	%
Govt Civil	1.717	17	110.1	18
Govt military	2.764	27	245.7	39
Industry (1938)	5.703	56	266.2	43
Total	10.184	100	622.0	100

Sources: J.D. Bernal, *The Social Function of Science* (1939), p. 422, 427; Edgerton D.E.H. and S.M. Horrocks, 1994. 'British Industrial Research and Development before 1945,' *Economic History Review* Vol. 47 (1994):213–38; Hilary Rose and Stven Rose, *Science and Society* (Harmondsworth, 1969), p. 128.

middle class, or from the working class" and as such were representative of British science and engineering.[7] On the other hand, Werskey has stressed the elite origins of Cambridge science students of the inter-war years, though his picture of the Cavendish laboratory has been challenged.[8] For the post-war years at least the elite of British science has included a large number of public school boys (that is, they were educated in exclusive private schools) as may be seen in Table 39.6. It is also

TABLE 39.6: THE SCIENTIFIC ADVISORY ELITE: ANALYSIS OF ALL MEMBERS OF THE ADVISORY COUNCIL ON SCIENTIFIC POLICY (1947–1964), THE COUNCIL FOR SCIENTIFIC POLICY (1964–1971) AND THE ADVISORY BOARD FOR THE RESEARCH COUNCILS (1971–8): PERCENTAGES.

	ACSP	CSP	ABRC
Academics	63	65	81
Industrialists	37	30	19
Proportion of all members educated at Public School	29	27	37
Proportion of all members with first or higher degree from			
Cambridge	34	46	31
Oxford	9	5	12
London	20	27	50
Total	63	78	93
Proportion of academic members only by employing University			
Cambridge	23	17	38
Oxford	14	13	15
London	23	33	31
Total	60	63	84
Total Number	35	37	16

Data for 'independent members' only. These bodies were concerned with the research councils only.

Source: P.J. Gummett, *Scientists in Whitehall* (Manchester, 1980), p. 94.

worth noting that the ancient and elite University of Cambridge has educated about half of all British Nobel Prize winners (most of the rest were accounted for by Oxford and London).

The expansion in the number of scientists was to a considerable extent intensive rather than extensive. By this I mean that most of the expansion has taken place within already *existing* institutions. The key institutions of science were creations of the late nineteenth and early twentieth centuries, but the number of new universities, research laboratories and research-performing firms has grown much more slowly than the number of scientists. The upshot of this is that the average size of scientific units increased very dramatically. The growth of institutions has been more significant than their creation. This deepening rather than broadening of science itself suggests the development of stratification and specialization and the extension of scientific hierarchies. Thus, at the beginning of the twentieth century some 30 percent of university staff were professors; by the 1960s it was only 10 percent. Working as part of a team of ten in an Edwardian industrial research laboratory was very different from being one of two hundred or so scientists working in such a laboratory after the Second World War.

Another way of making the same point is to note that scientific activity, despite its huge expansion, has remained highly concentrated in a few institutions. A history of British science could quite easily focus on ten universities, ten firms, and ten government laboratories and still capture a very significant proportion of all activity (see Table 39.4). This may be illustrated by the history of industrial research. Industrial research funding has generally grown much faster than the output of the economy. The result was that R&D represented about 0.3 percent of GDP in the mid-1930s, peaking around 2.5 percent in the mid-1960s. It is currently about 2 percent. Despite this enormous growth of GDP and R&D, a very few firms accounted for a very large proportion of industrial research. In the 1930s seven firms accounted for 30 percent of industrial R&D employment; in the early 1990s five firms accounted for 40 percent of private expenditure.[9]

The Nationalization of Science

Given the concentration of science in a few firms, universities and government laboratories, it might be profitable to follow their individual histories. That, unfortunately is not possible, given the paucity of detailed historical studies. But we know that there was a significant joint development of industry, universities and the state, indeed that the extent of linkage has its own history. It is important to note that this linking was not due to some implicit or explicit 'science policy.' The state did not have one 'science policy,' but many different ones. Even more important is the fact that policies for particular bits of science were appropriately subsidiary to medical policy, energy policy, military policy, industrial policy and so on. Each ministry concerned with science pursued different policies, and in any case, the private sector had a very important role too. But the state, in many different guises, did play an important part.

Scientific activity was hardly nationalized at all around 1900, in that most science was done in non-national organizations and in a non-national context. Universities and some technical colleges were privately endowed, and where they received public money this was local rather than national. Funding for research in industry by government was minimal.

Science was non-national in another sense. It was imperial. Many bodies, public and private, produced scientists and knowledge not merely for the nation but for a highly diversified empire. Thus British universities were in part universities for the Empire. It is important to stress that the UK was not, to the chagrin of many imperialists, connected only to its Empire. Britain was the greatest free-trading economy in the world: only after 1914 did it start closing its economic frontiers. Before 1914 the UK was hardly defined by *economic* borders.

Nevertheless there clearly were some connections between science and the central state around 1900. The military departments had officer-training schools which included science in their curricula. The state maintained some laboratories, and partially financed the National Physical Laboratory. It also owned and ran some higher educational institutions (e.g., the Royal School of Mines); and subsidized universities (other than Oxford and Cambridge). One of the least known connections between science and the state derived from a peculiar relationship between universities and Parliament. University graduates, which most scientists were, had a second vote. They elected, by university, 9 members of Parliament before 1918; 15 for a brief period, and 12 from 1920 to 1950. Scientists represented some 15 percent of the university electorate in the inter-war years.

The war and inter-war years saw some important changes. First, there was a very significant extension in in-house military research by the armed forces. Secondly, civil research laboratories (e.g., the National Physical Laboratory) owned by the central state and agencies grew substantially. There was increased central funding of universities; and the introduction of tariffs and quotas created a national economy for selected products. From the 1930s especially, there were arguments from the scientific left for a more thoroughgoing nationalization of science. The aim was to increase the relative role of the public sector; to plan scientific development from the center; and, to coordinate scientific with social and economic development. The call for the planning of science in J.D. Bernal's *Social Function of Science* was also a call for nationalization, both in the sense of increased public participation in funding, and in that the basis of the plan was the nation. It is significant that the 1947 manifesto of the Association of Scientific Workers was called *Science and the Nation.*

Although the priorities of the scientific left were largely ignored by the central state, and the call for the *centralization* of scientific funding was rejected, there was undoubtedly a further nationalization of science between the 1930s and the 1950s. The purposes of science became increasingly national, were increasingly looked at as a whole from a national perspective, were increasingly state-funded and increasingly took place in public sector institutions. Thus the proportion of public funding of R&D was under 44 percent in the mid-1930s, but increased to 57 percent in the early 1960s, having been larger still in the 1950s. The main reason for this increased nationalization was the defense effort, but that was only part of the story. Nationalization was also evident in the way the central state came to dominate university funding; in the collection of national statistics for stocks of scientists, and for research and development expenditure and employment. To be sure science was never completely nationalized, in any of the senses indicated above, but the trend was clear. The fact that *central*, unified control of science was never achieved nor sought does not mean that science was not increasingly planned and funded by national organizations, whether government ministries, universities, or private firms.

It is worth looking at this process of nationalization in more detail by discussing industry, the universities, and government separately. The dynamics of nationalization were different in each case. In addition looking at these sectors separately allows us to give a more concrete account of the development of British science.

Science in Industry

British private manufacturing industry was of great international importance. Before 1914 it accounted for 27.5 percent of world trade in manufactured goods; 18.5 percent in the 1930s and 21.4 percent in the early 1950s. At least into the 1960s, British firms became more and more scientific. The proportion of employees with scientific qualifications increased as did the proportion of turnover devoted to research and development.

Before 1914 British industry operated from a domestic market into which imports came in freely. This context is vitally important for an understanding of the debates about science in British industry at that time. The failure of the British dyestuffs industry to establish research and development on any scale has since been taken as an indictment of British industry, and indeed of the British state, and British universities. This is despite the fact that the dyestuffs industry was tiny, even in Germany. But more interesting is the point that after the mid-1880s no major research based dye firm was established anywhere, even in Germany. Established German firms dominated the world market and their own domestic German market. Dyestuffs research and production were only properly established behind powerful import restrictions. These arose in the Great War, but continued after it had ended. After the war the dye effort was concentrated in the *British* Dyestuffs Corporation, later part of *Imperial* Chemical Industries (ICI).

The build up of research behind import controls was by no means unique to dyestuffs. ICI's huge oil-from-coal program of the early 1930s relied on differential duties on imported and home-produced oil. Many other classes of 'science-based' goods were protected under the 'Safeguarding of Industries' provisions. It seems likely that the peak period for the formation of research laboratories in industry was the inter-war years. Indeed by the 1930s most large British manufacturing firms had a research and development laboratory. The number of qualified staff in research and development in industry was about twice the number of teachers of science (only) in British universities.

But the nationalization of the economy in the inter-war years should not be overstated, since a general tariff did not come in until the 1930s, and because of the importance of US capital in research-intensive sectors like electricals. After the Second World War, however, the British economy was highly protected such that it was extremely difficult to import manufactures into Britain. This was a key context for the massive development of industrial research by British companies. Indeed, between 1945 and the early 1960s, industry's own expenditures on R&D increased about sevenfold in real terms (a much greater rate of growth than pre-war). Furthermore state expenditure in industry, notably in armaments, but also in some civil sectors, increased massively. The nationalization of many important non-manufacturing industries was intimately linked to the development of all-British technologies for these industries. Much of the R&D for the nationalized industries was state-funded, though not by the nationalized industries themselves. In the case of nuclear power for the nationalized electricity supply industry, the nationalized Atomic Energy Authority had a monopoly over reactor research and design. In the case of the civil airlines, government funded a large proportion of the development of new civil aircraft. These programs would have been senseless if the user organizations had been free to purchase technology on the world market. The nationalization of industrial science was, on this account, the result of the creation of import controls; the nationalization of major purchasers of technology; and direct state funding of research and development activity by the central state. Just how significant each of those factors was, or what the net effect was is difficult to say, but government funding of industrial R&D accounted for around 50 percent of spending in the 1950s.

Industrial science was nationalized in another sense. Already by the 1930s, and certainly by the post-war years, the major British firms were national firms. By this I mean they were typically multi-plant enterprises spread all over the United Kingdom, with headquarters in London. A typical, large firm of 1914 would be headquartered in a particular region and have strong links only to that region. A company like ICI was quite different. It was engaged not with particular localities, but with the national economy, the central state, and the national higher education system. It saw itself, and was seen by others, as a national champion, even if it was in the private sector.

Universities

British universities have been the most important source of scientists. Indeed it was this that made them increasingly subject to state interest. The expansion and nationalization of the university sector, especially in the 1950s, went hand in hand with the increasing importance of science in the university, and the fact that scientists increasingly followed industrial careers. Traditionally, British universities, even if they played a national role, were not funded by the central state. The universities of Oxford and Cambridge, although national universities for England and Wales, did not receive state funds before the 1920s. The English 'civic' universities, created largely in the late nineteenth century, were funded by local government and local philanphropists and served a local clientele. As their collective name indicates, they belonged to large cities, and as the individual names of some of them indicate, they relied on industrial money as in the case of Owens College (Manchester), Mason College (Birmingham) and Firth College (Sheffield). Scotland, Wales and Ireland had their own universities. Already before 1914, a process of nationalization was taking place. The central state funded provincial universities from 1889, and the professoriate was already drawn from a national pool. Particularly interesting is the fact that by the inter-war years the professoriate of the ancient Scottish universities was drawn from a UK-wide pool.[10]

In the inter-war years central state funding was centralized in the (UK-wide) University Grants Committee (UGC), and Oxford and Cambridge started to receive grants. Funding levels were much increased, but there was a dramatic shift in funding to Oxbridge and London. In the mid-1930s London, Oxford and Cambridge received 42 percent of all the Parliamentary grants to Universities in Great Britain. From all sources they had 46 percent of total income.[11] In contrast to the other universities, however, Oxford and Cambridge did not receive local funds; and their student bodies were full-time and drawn from outside the local area.

Following the Second World War, and especially from the late 1950s, there were important changes. All British universities received the vast majority of their funds from central government and their student bodies were overwhelmingly full time. They drew increasingly on a mobile, state-funded national body of students. All universities were now national institutions, subject to national criteria of prestige. New universities were also created nationally. Certain regional technical colleges were designated as national institutions and then turned into universities. Others were created from scratch by the UGC. The national character of the universities may be contrasted with the local and regional character of the technical colleges, and later the polytechnics. This national versus local and regional dichotomy also paralleled a distinction between types of subject. Science, higher engineering, medicine, the humanities and the higher social sciences were national subjects, taught only in universities, while lower engineering and vocational courses were local subjects, allied to local forms of industrial practice.

It is important for us not to confuse the status rankings within the nationalized system (where local or civic would indicate lower status) with the ranking of the pre-nationalized system. Before 1914, perhaps before 1945, the University of Manchester was not national, except in its professoriate, but it was nevertheless an innovative institution of world rank, with connections to a powerful local elite and local industry. There is a strong case for arguing that until 1914 Manchester was the pioneering research university in physics and chemistry. Significantly, its strengths in these areas were transferred in the inter-war years to Cambridge and Oxford respectively. This transfer was connected both to the fact that the ancient English universities now received state funding, and that national industrial firms funded research in Oxford and Cambridge. Furthermore, industrial plants in Manchester, belonging to national firms, had strong links to Oxford and Cambridge science departments. By contrast, in the years before 1914, most philanthropic funding for university research was given by local firms to local universities. It is noteworthy too, that additional government funding for university research, and research students, started during the Great War, was done on a UK-wide basis. The so-called research councils had only a minor impact in the inter-war years. But in the years after the Second World War their impact far outweighed any other support for postgraduate study and academic research. And here too a self-reinforcing system of national ranking was at play, with very high proportions of monies going to the 'golden triangle,' and very high proportions of members of the 'golden triangle' allocating funds and deciding on policy (see Table 39.6).

The Military and the State

A second national cadre of scientific and technical expertise in Britain has been the technical officers of the army, navy and (from 1918) the air force. It is significant that the military schools never achieved the fame or prestige of the Ecole Polytechnique. Indeed one of the important features of the British military industrial complex has been the creation of *civilian*, and thus externally trained, scientific and technical staff in the service and supply ministries, despite tensions between the civilian and uniformed branches. These civilian branches, notably those devoted to research, underwent an enormous growth during the twentieth century. Great state laboratories were created, like the research establishments at Woolwich and Farnborough (both predating the Great War) followed, in the inter-war years, by the Admiralty Research Laboratory and the Porton Down chemical warfare complex. Farnborough had around 200 scientific and technical officers in the late 1930s. After the Second World War the Ministry of Supply, which was at the center of an unprecedently powerful scientific civil service, had easily the largest number of research scientists of any British organization. These laboratories took an increasing proportion of weapons development work, such that by 1945 a good deal of military technological expertise was in the public rather than the private sector. Radar was a good example, but perhaps the clearest was atomic energy. The development of nuclear weapons and nuclear energy was absolutely monopolized by state laboratories. Indeed, the government gave itself powers to ban private nuclear activity.

The state also exercised increased control over weapons development in the private sector. It seems likely that by 1939 most weapons development was financed by the state, even that carried out in the private sector. Aviation is the key case here. Within military research and development we see the state taking a much more direct role leading to an effective monopolization and indeed centralization of research activity. At the same time the proportion of the (entirely notional) national R&D budget going on defense was increasing through the century, peaking in the mid-1950s. Even in the early 1960s, 40 percent of all R&D carried out in industry was funded by the state for military purposes.

De-Nationalization and Privatization of British Science

The high point for the nationalization of science probably came in the mid-1960s. From then on there was a process of de-nationalization. Part of the reason was a loss of faith in national science as a source of national power. It was found that there was no positive correlation between investment in R&D and rate of economic growth. Nor did Nobel Prizes betoken industrial innovations which would transform British industry. Even high levels of production of scientists did not transform the rate of productivity growth. National champions were not winning international contests. British industry had a poor record selling aircraft and nuclear reactors abroad. It was increasingly argued that government laboratories, and government-funded programs were too disconnected from the market, especially the international market. In discussions of national policy for science and technology 'demand' and 'customers' were highlighted, reducing the previous emphasis on the supply-side.

From the 1960s the proportion of private funding increased. As government funding slowed, private funding grew somewhat faster. Foreign-owned enterprises greatly increased their share of all industrial R&D spending in Britain. Funds remitted from abroad for industrial R&D in Britain became by the 1990s, comparable with funds spent in industry by the British government. Similarly British-owned companies increasingly located R&D abroad. In the 1980s, the privatization and liberalization of procurement in formerly nationalized industries has also had a dramatic effect. So earlier did the general removal of intra-European tariff and non-tariff barriers.

At the level of state funding, international collaboration became increasingly important. Most major aircraft projects were now European collaborations, for example. The first such major project was the Anglo-French Concorde, but it was followed by a European space program, and work on major military and civil aircraft and engines. A very large proportion of the state's funding of R&D in industry, which is overwhelmingly for warlike purposes, is now for collaborative programs. Though funding for such programs comes from national governments, the context of decision-making is supra-national.

The great exception to this trend is to be found in the higher education sector. This system has been further nationalized, especially in the 1980s. The local polytechnics are now under the same national control as the universities. The only element of de-nationalization has been the division of university funding bodies into agencies for England, Wales, Scotland and Northern Ireland. Otherwise the tightening of central control has been very clear. The funding of university research has been thoroughly centralized through national rankings of research quality, and much more highly directed and controlled programs of externally funded research. In all these areas the aim of the nationalization has been to bring the actions of higher education institutions in line with national policy, which, ironically, hardly exists for technology, industry or defense.

Conclusions

The increasing privatization and internationalization of British science has had one important consequence for historians of British science. They are less prone to take the nationalistic statist assumptions of the post-war years for granted. Similarly, we can see 'declinism' as itself part of a nationalistic era now passing and can begin to see the history of twentieth-century British science from a more international perspective. It also helps us to see that the history of British science was never just the history of state-funded civil science for national purposes and was hardly so at all around 1900. It is also becoming clearer that the military and the private sector have been systematically neglected in the literature of a previous era. Taking both into account also helps us see that nationalization was a very general process. The identification of science and the nation was a historical process.

References

Many of the ideas in this chapter have come out of discussions over many years with John Pickstone; some are undoubtedly entirely his. Terence Kealey too has forced me to think about the history of British science in new ways. I am also grateful to Jon Agar, Jeff Hughes and Dominique Pestre for comments on an earlier version.

1. Frank Turner, 'Public Science in Britain', *Isis* (1980), **71**.
2. Turner, 'Public Science'; David Edgerton, 'British Scientific Intellectuals and the Relations of Science, Technology and War', in Paul Forman and Jose-Manuel Sanchez-Ron, *National Military Establishments and the Advancement of Science: Studies in Twentieth Century History*. (Dordrecht: Kluwer, 1995).
3. C.P. Snow, *The Two Cultures and the Scientific Revolution*. (Cambridge, 1959), p. 33.
4. Snow, *Two Cultures*, p. 11.
5. Snow, *Two Cultures*, p. 30.
6. R.M. Pike, *The Growth of Scientific Institutions and Employment of Natural Science Graduates in Britain 1900–1960*. (University of London MSc thesis, 1961).
7. Margaret Gowing, assisted by Lorna Arnold, *Independence and Deterrence: Britain and Atomic Energy 1945–1952*, Vol. II, *Policy Execution*. (London: Macmillan, 1974), p. 33.
8. P.G. Werskey, *The Visible College: a collective biography of British scientists and socialists of the 1930s*. (London: Allen Lane, 1978; Free Association, 1988); David Wilson, *Rutherford: Simple Genius*. (London: Hodder and Stoughton, 1983).
9. David Edgerton, 'Research, Development and Competitiveness', in Kirsty Hughes (Ed.) *The Future of UK Competitiveness and the Role of Industrial Policy*. (London: Policy Studies Institute, 1994).
10. R.D. Anderson, 'Scottish University Professors, 1800–1939: Profile of an Elite', *Scottish Economic and Social History*, Vol. 7 (1987), p. 42.
11. J.D. Bernal, *The Social Function of Science* (1939), pp. 420–21.

FURTHER READING

Peter Alter, *The reluctant patron: science and the state in Britain.* (Oxford, 1987).

J.D. Bernal, *The Social Function of Science.* (London, 1939).

Biographical Memoirs of Fellows of the Royal Society.

David Edgerton, *Science, Technology and the British Industrial 'Decline,' 1870–1970.* (Cambridge, 1996).

Margaret Gowing, *Britain and Atomic Energy, 1939–1945.* (London, 1964).

Margaret Gowing, assisted by Lorna Arnold, *Independence and Deterrence: Britain and Atomic Energy 1945–1952,* vol. I: *Policy Making,* vol. II, *Policy Execution.* (London, 1974).

P.J. Gummett, *Scientists in Whitehall.* (Manchester, 1980).

Andrew Hodges, *Alan Turing: The Enigma of Intelligence.* (London, 1983).

Greta Jones, *Science, Politics and the Cold War.* (London, 1988).

Terence Kealey, *The Economic Laws of Scientific Research.* (Basingstoke, 1996).

W. McGucken, *Scientists, Society and the State: The Social Relations of Science Movement in Great Britain, 1931–1947.* (Columbus, Ohio, 1984).

J.B. Morrell, 'The non-medical sciences 1914–1939,' in B. Harrison (Ed), *The History of the University of Oxford, Vol VIII The twentieth century.* (Oxford, 1994).

R.C. Olby, *The Path to the Double Helix.* (London, 1974).

K. Pavitt, (Ed.), *Technical change and British economic performance.* (London, 1980).

Sidney Pollard, *Britain's Prime and Britain's Decline.* (London, 1989).

Michael Sanderson, *The Universities and British Industry, 1850–1970.* (London, 1972).

Keith Vernon, 'Science and Technology' in Stephen Constantine, Maurice Kirby and Mary Rose, (Eds). *The First World War in British History.* (London, 1995), 81–105.

P.G. Werskey, *The Visible College: a collective biography of British scientists and socialists of the 1930s.* (London, 1978, 1988).

David Wilson, *Rutherford: Simple Genius.* (London, 1983).

CHAPTER 40

Russian Science in the Twentieth Century

NIKOLAI KREMENTSOV

The history of Russian science in the twentieth century is full of striking, contradictory, and enigmatic events. Russian science quickly evolved from a scattered network of small laboratories into a gigantic centralized system with thousands of institutions and hundreds of thousands of scientists. Yet, its explosive institutional growth was accompanied by the abolition of entire disciplines, and outstanding achievements routinely co-existed with backward doctrines. The greatest honor Russian science could bestow — membership in an academy — was shared by brilliant scientists and ignorant political functionaries. A scientist could be an adviser to the highest state bodies one day, an 'enemy of the people' the next, and vice versa. Scientists conducted research in the well-equipped institutes of 'Science Cities' and in prison camps. They made impressive showings on the international scene and then vanished behind the Iron Curtain. Furthermore, many of the greatest triumphs of Russian science occurred exactly at the time of the greatest repression: practically all Soviet Nobelists received this highest scientific award for research done when arrests were common and the Gulag camps overflowing.

This contradictory and puzzling history has understandably attracted the attention of Western scientists and historians, many of whom have tended to view Russian science as something unique, alien, and strange. They have often resorted to explanations that relied upon the mysteries of the Russian national character, the 'totalitarian' nature of the Soviet state, the Marxist ideology of the Communist Party, or the personalities and power struggle within the Kremlin. Both the triumphs and tragedies of Russian science, however, can be viewed equally as products of the same thing — the particular science system, Big Science Soviet-style, which emerged in the 1930s and has preserved its definitive features well into the present day.

This system was comprised of two major interdependent and integrated components: science and the state, or, more precisely, the scientific community and the apparatus of state control. In interpreting Soviet science, some scholars have tended to view science and the state as two monolithic and opposing entities locked in an uneven conflict, with the state in the role of dictator and oppressor, and the scientists as victims trying to defend and protect their intellectual autonomy. This description is in certain respects correct, but, at the same time, misleading. In fact, in pursuit of its *interests*, the state established a much more impressive and terrifying system of control over the scientific community than any critic of Soviet 'totalitarianism' could have imagined, and, in pursuit of its interests, the scientific community developed much more elaborate devices to avoid, elude, and exploit this control system than any advocate of 'academic freedom' could have reasonably hoped.

Despite its totalitarian character, the Soviet state was never monolithic. The numerous agents and agencies involved in the state science-policy apparatus pursued their own, often conflicting, objectives. Nor was the Soviet scientific community monolithic. It was fragmented into numerous, often competing subgroups, each of which had its own goals and resources; and each developed its own relations with its special patrons within different divisions of the control apparatus. It is sometimes useful to consider the scientific community and the state apparatus as separate entities, but the actual boundaries between them were often blurred, and their interaction was more symbiotic than antagonistic. Each strove to acquire what it needed from the other, to implement its own policies, and to achieve its own objectives. Each developed various tactics and used various means to deal with its partner.

The control apparatus and the scientific community were parts of the same system. Not only did scientists occupy key positions within various state agencies, not only did their leaders enjoy an elite position within the Soviet bureaucratic hierarchy; some scientific institutions, such as presidiums of Soviet academies, were in fact key elements of the state apparatus itself. The control apparatus and the scientific community became integrated, not only in their internal structures and overlapping networks of individuals and groups, but also by a common and quite peculiar set of images, rituals, and rhetoric. The symbiosis of science and the state resulted not only in institutional integration and individual co-optation, but also in a shared lexicon and culture.

The history of Russian science in the twentieth century, then, was to a large extent defined by this symbiosis — the volatile equilibrium between the scientific community and the state apparatus of control.

Russian Science: 1900–1929

At the end of the nineteenth century Russian science was a variant of European science, and its development from the 1890s and through the 1910s paralleled developments elsewhere in Europe. These similarities were reinforced by a steady flow of people and ideas. Upon graduation, many Russian scientists spent several years in German, French, and British laboratories, and some of the scientific organizations and practices they experienced there were brought back and recreated on Russian soil. The development of Russian science was particularly influenced by the German model. As in Germany, the major institutional base of Russian science was a system of state universities and specialized educational institutions. Again as in Germany, the nobility patronized certain important scientific institutions, and the development of Russian industrial capitalism stimulated corporations and entrepreneurs to subsidize scientific research and education. In short, the expansion and professionalization of science in Russia involved the creation of numerous scientific societies, the founding of specialized scientific periodicals and institutions, the organization of scientific conferences and congresses, and the development of an international network of scholars specialized in newly emerging disciplines and specialties, mirroring developments in Europe and the United States.

During the pre-revolutionary decades many Russian scientists won international fame. Such figures as Dmitrii Mendeleev in chemistry, Vladimir Lobachevskii in mathematics, and Vladimir Dokuchaev in soil science firmly established the reputation of Russian science on the international scene. Russian contributions to medical fields were recognized with the Nobel Prizes awarded to Ivan Pavlov (1904) and Il'ia Mechnikov (1908).

As in other countries, the process of professionalization and institutionalization led to the creation of a professional culture in Russian science. In their discourse, self-image, and professional behavior, pre-revolutionary Russian scientists were almost indistinguishable from their colleagues in other countries. As in the West, specialized scientific periodicals focused on the novelty and objectivity of research. Despite their broad social activity, Russian scientists treated science itself as 'above' politics, ideology, and narrow practical interests. These developments reflected the maturation of the Russian scientific community, and served to insulate and protect it from interference by the ideological authorities, especially by the Church and the tsarist censor.[1]

By 1914 the Russian scientific community comprised about four thousand scientists working in 289 scientific institutions.[2] Russian scientists complained constantly about the slow tempo of institutional development and especially the low level of state support. Like scientists in many other countries, Russian scientists repeatedly criticized their own government for its neglect of science. The scientific community, then, enthusiastically supported the February 1917 revolution, which dethroned Tsar Nicholas II and created a liberal Provisional Government that promised to increase support for education and scientific research. Its Ministry of Enlightenment governed by eminent scientists formed several commissions to develop a plan for the reorganization and expansion of the educational and research system.

These endeavors were interrupted in October 1917, when a radical faction of the Russian Social-Democratic Labor Party, the Bolsheviks, suddenly came to power.

The Bolshevik *coup d'état* led to the Civil War which ravaged the country for almost four years. The Red Army finally triumphed, and at the end of 1922 the Union of Soviet Socialist Republics was established.

The Bolsheviks' primary concern during their first years of power was the restoration and maintenance of the national economy, which they considered crucial for the building of what their political program called 'the first socialist society.' This defined their policy and attitude toward science and scientists: science was to play an important role in the building of socialism in Russia. Like many political parties in Russia and elsewhere (and like most Russian scientists themselves), the Bolsheviks were captivated by a technocratic vision of a future society that would reap the fruits of scientific progress. This technocratic ideal, together with urgent economic needs, defined the dual direction of Bolshevik science policy during the 1920s. On the one hand, the Bolsheviks strove to co-opt the pre-existing 'bourgeois' scientific community and to invite Russian scientists to collaborate with the new regime. On the other hand, they began to create their own 'Communist' science and to prepare their own 'proletarian' scientific cadres.

The initial encounters between Russian academics and the Bolsheviks were colored by considerable mutual suspicion and distrust. Scientists were considered a part of the bourgeoisie inherently alien to the proletariat, whose interests the Bolsheviks claimed to represent. During the first years of Bolshevik power, however, both groups compromised and developed symbiotic relations. The Bolsheviks provided scientists with considerable resources and autonomy; the scientists provided the Bolsheviks with their expert knowledge, helping the new government to revive industry, agriculture, and medicine which had deteriorated over the seven years of almost permanent warfare. During the 1920s the Bolsheviks revived and expanded research institutions, and spent large amounts of precious hard currency to buy scientific equipment, secure foreign publications, and send scientists to study abroad. They financed various conferences and congresses, published scientific periodicals and monographs, and organized numerous expeditions not only within the country but also abroad. They provided scientific institutions with buildings, heat, and electricity, and scientists with housing and salaries, thereby raising the public prestige of scientific work.

The Bolsheviks created a number of special agencies within their ministries (People's Commissariats) that formulated science policies and supervised scientific institutions, while also granting the scientific community considerable autonomy. During the 1920s, such leaders of the scientific community as Vladimir Bekhterev, Abram Ioffe, Vladimir Ipatieff, Aleksei Krylov, Sergei Oldenburg, Ivan Pavlov, Vladimir Steklov, Nikolai Vavilov, and Vladimir Vernadskii established close personal contacts with the heads of state agencies.[3] Lenin himself and his old comrades-in-arms — Anatolii Lunacharskii (head of the educational ministry), Nikolai Gorbunov (Lenin's secretary), Nikolai Semashko (head of the public health ministry), Gleb Krzhizhanovskii (head of the central planning administration) — often intervened personally on behalf of individual scientists and institutions.

Bolshevik science policy generated a large, diversified, and decentralized network of scientific institutions subordinated to various governmental agencies. A prominent place within this network was occupied by the Russian Academy of Sciences. Despite the numerous declarations of militant Bolsheviks who wanted to abolish this odious remnant of the 'bourgeois, imperialistic past,' the academy leadership began to collaborate with the Bolshevik government soon after the revolution. In June 1925 the government issued a special decree declaring the academy "the supreme scientific institution of the USSR."[4] Renamed the USSR Academy of Sciences, it gained considerable influence in governmental circles and expanded quickly.

The science system that emerged in the 1920s closely resembled that of tsarist Russia, with one principal difference: the disappearance of private funding. Science in Soviet Russia had become an exclusively state enterprise. By the late 1920s, the scientific community had been completely co-opted into the system of power relations and occupied a prominent place within the social structure of the Soviet state. The scientific community enjoyed considerable authority and state support, while preserving a high level of professional autonomy. The network of close personal contacts between eminent scientists and heads of governmental agencies (or their trustees) allowed scientists to actively influence state science policy and decision-making. Many Russian scientists no doubt employed Bolshevik science policy to institutionalize their own research interests. A number of disciplines and research fields that had been absent or weakly developed in tsarist Russia were quickly institutionalized, and the number of scientific institutions almost tripled during the first decade of Bolshevik rule.

While co-opting the 'bourgeois' scientific community they had inherited, the Bolsheviks actively prepared their own 'proletarian' scholars and their own 'Communist' science. In contrast to the liberal and accommodating policy toward pre-existing *research* institutions, Bolshevik policy toward *educational* institutions was stern and aggressive. Numerous purges of educational institutions in the 1920s were designed to 'proletarianize' students and to 'Bolshevize' professors. This dual policy, combining relative autonomy in research and severe control in education, created a dichotomy between teaching and research that became a characteristic feature of the Soviet science system.

In addition to the reformed traditional educational institutions, a number of special 'Communist' ones, such as the Institute of Red Professors, were created. The Bolsheviks also created their own Communist research institutions and scientific societies. A special place among these new institutions was occupied by the Communist Academy formed in 1918. First established as a primarily educational institution, the Communist Academy in the early 1920s was transformed into the center of 'Communist' research. It initially conducted research in social and humanitarian disciplines, but by the mid-1920s it also included several institutes devoted to the natural sciences.

These Communist institutions presented a clear alternative to the 'bourgeois' science inherited from tsarist regime. Perhaps some Bolshevik leaders believed it necessary to replace bourgeois science with their own Communist science. Most importantly, however, the Communist Academy presented an alternative model of science organization and professional culture. It did so by replicating the centralized hierarchical organization of the Bolshevik party itself and by involving party rhetoric and rituals in scientific practice. Unlike the democratic Academy of Sciences, which was governed by the general assembly of its members, the Communist Academy was governed by its so-called 'presidium': a self-appointed body of its high-ranking founders. Unlike the Academy of Sciences, which had neither the intention nor the authority to dictate the activities of its members and institutions, the presidium of the Communist Academy actively controlled subordinate institutions and exerted 'party discipline' over its members and workers. Moreover, the Communist Academy played a leading role in the politicization of scientific discourse by introducing a militant style of inner party struggle and a Marxist lexicon into scientific discussions.

Although bourgeois science expanded and was quickly adapting to the new social circumstances, the dependence on its sole patron made it a hostage of the changing domestic and foreign policies of the Bolshevik state. During the 1920s, militant Communists were unable to replace bourgeois science and scientists. Communist science was too weak, too lacking in the expertise required by the state. By the end of the 1920s the seizure and tremendous expansion of the educational system had provided the Bolsheviks with the necessary personnel to implement the new Communist model throughout the entire Russian science system.

Soviet Science: 1929–1939

The year 1929 marked a dramatic change, a "Great Break" (*Velikii Perelom*) as Stalin termed it, in all aspects of the country's life. The Bolsheviks launched a grandiose plan for the collectivization of Russian peasantry and the rapid industrialization of the Russian economy in order to build the "material-economic basis of socialism." Beginning in 1936, the Great Terror ravaged the country for two years, resulting in the arrests of some eight million people and the execution of about one million. The Great Terror completed the establishment of Stalin's regime.

Science was profoundly affected by the radical reorganizations of the 1930s. It was mobilized to serve the new policies of the new state. The grandiose plan of social and economic reconstruction reinforced the instrumental attitude of the Bolsheviks toward science embodied in the infamous slogan — "Science in the service of socialist construction."[5]

The 1930s witnessed the emergence of Big Science Soviet-style. During the early 1930s, the Bolsheviks greatly enlarged their support for science, vastly expanded the network of scientific institutions, and continued to raise the public prestige of science and scientists. They simultaneously began to limit the considerable autonomy enjoyed by the scientific community in the previous decade. The policy of co-optation was replaced by a policy of active command and control.

As in all other spheres of Soviet life, the Bolsheviks created in science a centralized, hierarchical complex of institutions and a bureaucratic apparatus to supervise and control it. Shortly after the Bolsheviks began to plan the Soviet economy, they introduced planning in science, and this became the main way they controlled the direction of research. With the isolationism of Soviet foreign policy came the isolation of Soviet science, and the contacts between Soviet scientists and their foreign colleagues were severed. The Bolsheviks exerted tight control over the personnel of scientific institutions; the scientific community was Bolshevized and its 'commanding heights' were seized by party members and 'non-party Bolsheviks.' The two parallel systems of 'bourgeois' and 'Communist' science were welded into a unified whole.

The establishment of strict party control was accompanied by the politicization of the professional culture of Soviet scientists. They adopted the lexicon, polemical style, and modes of group behavior characteristic of the Communist Party. Mass propaganda campaigns, a series of purges and show trials, and widespread arrests stimulated the development of specific rhetoric and rituals to demonstrate the scientific community's conformity to party policies and to justify its research and institutional agendas in the eyes of the party bureaucrats in charge of science.

As in all other spheres of the country's life, the role of the Communist Party in science greatly increased. Science policy decision-making fell under the authority of the highest party apparatus — its Central Committee, Secretariat, and Politburo headed by Stalin. Special agencies to supervise scientific development were created within the party apparatus. The main direction of party control over science became the 'Bolshevization' of its personnel: the appointment of party members to key posts in scientific institutions. The principal party slogan "Cadres decide everything!" (*Kadry reshaiut vse*) was actively applied to the scientific community. The main instrument of party personnel policy was the system of so-called *nomenklatura*. *Nomenklatura* was literally a list of posts which could not be occupied or vacated without special permission from the appropriate party committee. All party committees, from the Central Committee to the smallest one in the countryside, established special Personnel Departments, whose main function was to approve candidates for appointment to any posts included in their *nomenklatura*. The system was strictly hierarchical; the higher the post, the higher the party committee controlling its personnel.

The policy of active control invented in the 1930s profoundly affected the institutional structure of Soviet science. The model of science organization created and developed during the previous decade within 'Communist' scientific institutions was fully implemented throughout the Soviet science system. The main direction of Bolshevik institutional policy was centralization, concentration, and stratification of the scientific community, which led to the creation of centralized institutions (academies) that governed the development of research in particular subjects and regions. During the First Five-Year Plan (1928–1932), the number of scientific institutions grew by 50 percent, from 1,263 to 1,908.[6] During this time centralized scientific institutions were created under the authority of the Council of People's Commissariats. The Lenin All-Union Academy of Agricultural Sciences (VASKhNIL) was created in 1929, and the All-Union Institute of Experimental Medicine (VIEM) in 1932. Various centralized regional institutions were established, including the Ukrainian Academy of Agricultural Sciences (1931) and the branches of the USSR Academy of Sciences in the Urals (1931), the Caucasus (1931), Kazakhstan (1932), and the Far East (1932). The main goal of all these institutions was to direct scientific development in relevant fields and to serve as intermediaries between the party-state agencies and the scientific community. The policy of centralization led to a significant decrease of the general number of scientific institutions, while increasing the size of separate institutions during the late 1930s. By 1938 the number of institutions had dropped to 1,557.[7]

A special role among the centralized institutions was assigned to the USSR Academy of Sciences. It was the first to be Bolshevized. During the 1930s it grew from an honorific society with a few research facilities into the country's largest and most influential scientific institution, uniting about one hundred institutes, laboratories, observatories, and experimental stations throughout the nation. By 1939 the Academy of Sciences had acquired the leading position among all Soviet scientific institutions, absorbing institutes and laboratories previously subordinate to various ministries. In 1936 the academy even devoured its main competitor, the Communist Academy.

To turn science into a 'productive force,' the party-state agencies strengthened their immediate control over research. Just as planning became a major instrument of control over the Soviet economy, so it was introduced into science. In the early 1930s planning focused upon budgets for equipment and personnel. In the mid-1930s the state agencies also began to plan the direction of research. "Directives and Forms to Complete the 1936 Plan for the People's Economy" issued by the government, for instance, required central scientific institutions to plan the "main problems addressed by scientific works, the system of scientific institutes, the number of scientific workers, and the number of graduate students educated in these institutions."[8] Research plans and annual reports became the control agencies' main source of information on Soviet scientific institutions, and the main basis for decision-making in science policy. In 1938, for example, dissatisfied with the plan proposed by the Academy of Sciences for that year, the government decided to completely reorganize the academy.[9]

Party-state agencies exerted special control over scholarly communications — scientific meetings and publications. The Soviet scientific community's international contacts attracted particular attention. After a short period of relative freedom in the 1920s, all international contacts came under the strict control of the

Central Committee of the Communist Party. As Soviet foreign policy became increasingly isolationist, the control apparatus began to sever Soviet science's international contacts. The isolationist policy developed gradually during the 1930s and accelerated in 1936, when a broad press campaign was launched against "servility to the West."[10] After 1937 Soviet science's international contacts were almost completely broken off.

During the 1930s, then, the party-state apparatus established a complicated system of control over Soviet science and scientists. All aspects of the scientific community's activities (the appointment and certification of personnel, the structure of institutions, the directions of research, and scholarly communications) fell under the tight control of party functionaries. A large number of party-state officials, including Lunacharskii, Gorbunov, Semashko, and Krzhizhanovskii, "migrated" into the new scientific establishment, becoming members of academies and directors of academic institutions. The Bolsheviks created a huge, centralized, hierarchical, and strictly controlled system of scientific institutions; the scientific community was politicized and effectively isolated from its Western counterparts.

The scientific community quickly adapted to the new system of interrelation with the state apparatus. Having enjoyed considerable autonomy during the previous decade, Soviet scientists were compelled to learn the new 'rules of the game,' and did so by incorporating Bolshevik lexicon, polemical style, and modes of group behavior into their professional culture. They greeted every new party line with numerous rituals, signaling their conformity and adjusting their rhetoric to the demands of their sole patron, the Central Committee of the Communist Party. By the late 1930s Russian scientists no longer claimed that their work transcended ideological, political, and practical concerns. On the contrary, they pledged allegiance to a model of science wedded to Bolshevik theory and practice, proclaiming their work as Marxist and utterly practical.

In 1939 the supreme patron of Soviet science established a prize of his own name for scientific research, the Stalin Prize. At the same time, the 'supreme national scientific institution,' the USSR Academy of Sciences, elected him an 'honorary member.' No other event could have symbolized the new reality so perfectly: by 1939 Russian science had been transformed into Soviet science.

Soviet Science and World War II

World War II profoundly changed almost every aspect of Soviet life, including relations between scientists and the party-state apparatus. By the end of the 1930s the Soviet science system had reached maturity. The party apparatus had established strict control over the scientific community, and, concurrently, scientists had developed their skills at influencing the party-state bureaucracy. Isolated from its foreign counterparts, Soviet science seemed to live in accord with exclusively domestic dynamics and policies. In the early morning of June 22, 1941, the Nazis attacked the Soviet Union. Suddenly, everything changed.

With its continued survival threatened, the party-state bureaucracy recognized the vital importance of science and gave its scientific community new responsibility and respect. Scientists again participated, as they had in the 1920s, in decision-making on scientific, economic, and political questions. The war also shattered Soviet isolationism and gave rise to a new internationalism. The anti-fascist coalition formed by the 'Big Three' — the USSR, the USA, and Great Britain — greatly diminished the isolating barriers between Soviet scientists and their Western colleagues. The strict control of party functionaries over the activity of scientific institutions that had been in practice since the mid-1930s was loosened. The war redirected state officials from the evaluation of the ideological loyalty of scientists toward the immediate, practical outcome of their research. The party apparatus delegated considerable authority in science policy to the presidiums of the expanded academies that emerged during World War II, and the administrative apex of the scientific community became a part of the highest state elite. Furthermore, the war created a new, very powerful patron for Soviet scientists: the military. It immersed science into what later would be called the military-industrial complex.

The war greatly accelerated the institutional expansion of Soviet science. Numerous new institutions were established. By the end of the war the total number had increased to 2,060.[11] During the war the government established several new academies: the RSFSR Academy of Pedagogical Sciences, the USSR Academy of Medical Sciences, and academies of sciences (or branches of the USSR Academy of Sciences) in almost all republics of the Soviet Union. Academy presidiums became the *de facto* highest governmental agencies in charge of science.

Personnel policies during the war and the first post-war years reflected the heightened authority of the scientific community. Although the *nomenklatura* system was still in force and, hence, party personnel departments still wielded considerable influence over personnel decisions in scientific institutions, actual appointments generally followed the recommendations of the academy presidiums composed of eminent scientists such as Sergei Vavilov, Petr Kapitsa, Leon Orbeli, and Abram Ioffe. Research policy at that time was also largely defined by scientists, not state officials. The Academy of Sciences' plan for scientific development during the post-war period illustrates this convincingly. This very impressive document resulted from the independent efforts of scientists and identified priorities in almost every field of contemporary research from protein biosynthesis to nuclear physics and from high technology to fertilizers.

The wartime anti-fascist alliance created a new internationalist orientation in Soviet foreign affairs that profoundly affected Soviet science policy. The isolating barriers erected after 1937 between the Soviet and Western scientific communities were dismantled. Cooperation between Soviet and Western scientists became an officially sanctioned part of Soviet science policy. The American detonation of the atomic bomb in August 1945 added a new dimension to Soviet-Western scientific relations. The atomic bomb, which became the embodiment of both the advances of Western science and the superpower status of the United States on the international scene, stimulated Soviet officials to expand their support for science and to spur Soviet science in the competition for superpower status. In his speech before a meeting of voters on February 6, 1946, Stalin declared: "I have no doubt [that] if [we] provide the necessary help to our scientists, they will not only catch up with, but also soon overtake the achievements of science abroad."[12] The last part of the sentence — "catch up with and overtake" (*dognat' i peregnat'*) Western science — would become a well-known slogan of Soviet science in the coming years. The US atomic monopoly forced Soviet officials to recognize the importance of international scientific relations. As minister of foreign affairs Viacheslav Molotov stated in his speech before the General Assembly of the United Nations on 29 October, 1946, the atomic monopoly could not last long because "science and scientists cannot be put in a box and kept under lock and key."[13] The development of science clearly acquired a strategic priority in Soviet policies.

Science made a great contribution to the nation's 'arsenal of victory.' For the scientific community, the fruits of victory over Nazi Germany were plentiful and sweet. During 1943–46 the state rewarded numerous scientists with the highest awards, orders, and prizes; and a number of scientists came to occupy important posts within the state's agencies. Almost every request of the scientific community was granted. When scientists requested that the government increase the size and circulation of a periodical, or establish a new institute, or increase their pensions — these requests were immediately granted. The prestige of science soared, bringing a host of new privileges. In March 1946, the government issued a special decree, establishing high salaries and numerous privileges for scientific workers, especially for those in administrative posts. In short, after the war, the administrative apex of the scientific community joined the highest state elite.

World War II, then, strengthened considerably the symbiosis between the state apparatus and the scientific community, increasing their interdependence and integration. Unlike in the 1930s, when high-ranking party-state bureaucrats had become members of the scientific establishment, during the war scientists became members of the highest party-state bureaucracy. Many of them received the high military rank of general and developed elaborate contacts with military agencies on the level of both institutions and individuals. During and immediately after the war, the Soviet academies were transformed from a stronghold of fundamental research into a domain of military science and military-related scientists. Science acquired strategic significance, which enlarged state support and enhanced the scientists' control over their institutions, personnel, research directions, and foreign and domestic communications and contacts.

Although the wartime Soviet science system preserved certain distinct features it had acquired in the 1930s (the centralization of scientific institutions, the *nomenklatura* system of control over scientific personnel, the stratification of the community, and its politicized professional culture), now it seemed these very features could give Soviet science new vitality. During this period, Soviet scientists were able to take a leading role in their dealings with their 'partners' in the party-state bureaucracy over the general direction of science policy, and to exploit the Soviet science system to their own advantage. Enjoying almost unlimited state support and high public prestige, Soviet scientists conceived grandiose plans for a further expansion of their institutions and new forms of international cooperation. Indeed, after the war it appeared that Soviet scientists might well achieve the independence and respect they had always regarded as their due. What they did not and could not plan for, however, was the Cold War.

Soviet Science and the Dawning of the Cold War: 1946–1953

Beginning in 1946, a new change in the international situation once again had profound effects on the Soviet science system: the growing Cold War. Although it began more slowly than the outbreak of World War II, taking two years to develop fully, its effects on Soviet science were no less profound than that of the World War. First of all, the Cold War resurrected one essential feature of prewar Soviet science: internationalism was again damned and replaced with a militant Soviet nationalism. This shift in foreign and domestic policy re-shaped the cultural terrain of the Soviet science system, once again raising the authority of party ideologists and ideological considerations in science-policy decision-making.

In mid-1947 the Central Committee of the Communist Party waged a militant campaign against the "servility and slavishness of Soviet scientists before the West."[14] Within three months of its inception the 'patriotic' campaign had radically curtailed the international activities of the scientific community. Soviet scientists were forced to resign from foreign scientific societies. Foreign visits were minimized. The publication of Soviet research in foreign languages was forbidden. The growing Cold War enabled party ideologists to seize control of international scientific contacts and to use them for fueling isolationist policies. The campaign not only resulted in the restoration of isolating barriers between the Soviet and Western scientific communities, but also signified the expansion of the party apparatus' influence on and control over the scientific community. Although this once again isolated Soviet scientists from their Western colleagues and proved disastrous for certain individual scientists, during 1947 the community's leadership, empowered by the new structure of the science system, was largely able to sustain its *de facto* control over other aspects of scientific activity — personnel, institutional structures, and research directions.

In spring-summer 1948, the rapid escalation of the Cold War radically reshaped the interactions between the control apparatus and the scientific community. The year 1948 was the crescendo of the Cold War. It marked the final division of the 'spheres of influence' in post-war Europe between the USA and the USSR, and resulted in the establishment of two opposing camps — East and West. In the history of Soviet science, the year 1948 is commonly associated with the culmination of the so-called 'Lysenko controversy': Trofim Lysenko's infamous triumph at the August VASKhNIL meeting and the consequent 'death' of Soviet genetics, which was secured by Stalin's long suspected, and recently proved, personal intervention.[15] The 1948 events, however, far exceeded genetics and even biology and had a profound impact on the Soviet science system as a whole.

The conflict between two groups, Mendelian geneticists led by Nikolai Vavilov and 'agrobiologists' led by Trofim Lysenko, steadily developed during the 1930s and early 1940s. By the late 1930s agrobiologists had seized almost all genetics institutions expelling geneticists from their previous strongholds in the Academy of Sciences and VASKhNIL. Lysenko replaced Vavilov as president of VASKhNIL and director of the Academy of Sciences' Institute of Genetics. At the end of the war, however, geneticists launched a broad attack on Lysenko's ideas with the aim of restoring their institutional base. Using the new internationalism encouraged by the war, they employed their elaborate contacts with the Anglo-American genetics community to undermine Lysenko's authority. This tactic proved successful in the immediate post-war years leading to a significant improvement of geneticists' standing in governmental circles.[16] With the Cold War gaining momentum, however, the international contacts that had served them so well suddenly became a dangerous liability and appeared a major reason for the official abolition of genetics. At a special meeting of VASKhNIL held in August 1948 Lysenko declared genetics "a bourgeois science" and announced that his position "was approved by the Central Committee of the Communist Party."[17] During the following months genetics laboratories were closed, geneticists were fired, genetics courses were abandoned, and books and periodicals were removed from the libraries.

A number of factors contributed to Stalin's decision to intervene on Lysenko's behalf in 1948, but by far the most important was the escalating Cold War. Stalin used the institutional struggle between geneticists and Lysenkoists as a suitable pretext to announce a new party line in domestic and foreign policies. His actions transformed a local institutional conflict between two competing groups within the scientific community into a broad propaganda campaign that spread far beyond genetics, not only over Soviet scientific institutions, but throughout the world.

The party's approval of agrobiology signified much more than approval of Lysenko's doctrine. It signified approval of the particular model of science it embodied. The core of this model was the juxtaposition of 'Soviet' and 'Western' science, which clearly reflected the Cold War confrontation, portraying science as a mere extension of politics. In this model, science had no broader loyalties to anything but the state and had no interests aside from those set by the state. It was merely an instrument for pursuing state objectives. This model implied that science must be completely subordinate to the state, not only institutionally, but also intellectually. This universal model was also applied to Western science, which was depicted as completely subordinate to the political, economic, and ideological goals of Western countries, and especially the United States. It clearly aimed to reassert the image of the USSR as the only force for world progress, as the 'right side' in the Cold War confrontation.

The events of summer 1948 marked a new stage in reshaping the interrelations between two components of the Soviet science system, the scientific community and party agencies, in accordance with the new ideological atmosphere and priorities of the Cold War. The August VASKhNIL meeting and the subsequent propaganda campaign signified the intentions of party agencies to establish complete control over the community, expanding its power from 'external' (political, practical, and ideological) to 'internal' (intellectual and cognitive) aspects of scientific activity. Science policy decision-making once again (as it had in the 1930s) fell under the authority of party functionaries. They seized the right to judge scientific disputes and to dictate to Soviet scientists what theories to follow, what subjects to study, and what lines of research to pursue.

Lysenko's triumph over geneticists provided Soviet scientists with a convincing example and a powerful cultural resource 'approved by the Central Committee.' Various interest groups clearly strove to use the ongoing propaganda campaign for institutional and disciplinary expansion. Furthermore, they themselves initiated analogous campaigns within their own disciplines in order to discredit their competitors and to gain personal and institutional advantage. Numerous individuals and interest groups within and outside the scientific community successfully employed the new reality to advance their own agendas. Intended to establish complete control over scientific activities, this new system instead became an object of manipulation by high-ranking scientific administrators. Built to make science serve the state, it could also be used to make the state serve scientists.

In the last years of Stalin's rule the Soviet science system was completed by a cascade of campaigns in almost all disciplines. A campaign against 'idealist physics' — quantum theory and the theory of relativity, started in early 1949, but, after forty two 'rehearsals' it was not staged as a public show and was finally canceled in the late spring. A campaign in astronomy culminated in a discussion of its 'ideological questions' in mid-1949, while in physiology a campaign for 'I.P. Pavlov's legacy' began in late 1948 and culminated in 1950. In chemistry a discussion of the theory of resonance bonds slowly developed for two years before the theory was 'damned' at a special meeting of 1951. In cytology, a concept of 'non-cellular living matter' was sanctified at two meetings in 1950 and 1952. In linguistics, N. Marr's 'new theory of language' was proclaimed truly 'Soviet' one at a meeting in 1948, only to be dramatically dismantled by Stalin's personal intervention in 1950.

Conducted in a highly ritualistic form and filled with party rhetoric, these campaigns created an impression among Western scientists that Soviet science was completely 'subdued' by political authorities, lost its 'academic freedom,' and, hence, was 'dead.' They published 'obituaries' to Soviet sciences and grieved about the fate of their colleagues behind the Iron Curtain. Very soon, however, they learned that they were mistaken.

Big Science Cold-War-Style: From Khrushchev to Gorbachev

During almost forty years, the Cold War defined the uneasy balance of world affairs and the dynamics of scientific development in both East and West. The dawning of the Cold War engaged Soviet scientists in the continuous 'catching up' with their Western counterparts. Already in the final years of Stalin's rule Soviet science had proved its capability: in 1949 it produced an A-bomb and in 1952 an H-bomb. Two years later the world's first nuclear reactor to produce electricity went to work. Two years later at the Geneva Conference on the Peaceful Uses of Atomic Energy, Soviet scientists shocked the world with their pioneering achievements in controlled thermonuclear synthesis. Within the following decade, Soviet science 'overtook' the advances of its Western counterpart, launching Sputnik and sending the first man into the space. Soviet science convincingly demonstrated to the world that it was very much 'alive' and efficient. Now it was the turn of Western science to 'catch up.'

It was the Cold War that gave Soviet science its final form and enduring character. It was that pattern of interactions, structures, and styles that was 'frozen' by the Cold War, which from 1948 on defined its dynamics. The strategic political and symbolic significance of science during the Cold War dramatically increased the party's interest in and control over science. But, it also provided scientists with the 'armored' shield of Cold War rhetoric to defend their own authority and to protect their own interests in their dealings with the party apparatus.

By the time of Stalin's death in 1953 Soviet Big Science had acquired its ultimate form. A huge, centralized, and hierarchical system of scientific institutions was supplemented by an equally huge, centralized, and hierarchical apparatus, which tightly controlled its structure, personnel, research directions, and communications. Yet the structure of this bureaucratic apparatus, which included both the party-state agencies and the academy presidiums, the overlapping network of high-level academics and party functionaries in charge of science, provided the scientific community with a surprising amount of room for maneuver and with an opportunity to establish a steady equilibrium between the state's interests and those of the community.

Even within a system thoroughly dominated by the party apparatus scientists pursued their own interests by employing the very same system that was established to subdue them. They skillfully played upon the contradictory concerns of the ruling group. Decision-makers at the apex of the party could sometimes be forced to choose between their ideological, political, military or economic agendas. This was facilitated by the instrumentalist attitude of party leaders toward science: since they valued science primarily for its potential service to larger goals, any scientific research could be justified in their eyes through a connection to the party's practical priorities at the moment.

The most important concern of the party was winning the Cold War. Scientists skillfully employed Cold War priorities, especially the real or imagined military applications of their research, to influence the decision-makers. Certain scientists even managed to employ this overriding strategic objective to nullify a policy (the official abolition of genetics) based on lower-priority concerns (that is, propaganda and agriculture), and to preserve their research and institutions. Another element of Cold War politics that scientists frequently exploited was the image of the USSR as the leader of world development. The rhetoric of Soviet priority in all fields — 'The First . . .,' 'The Biggest . . .,' 'The Most Advanced . . .,' 'For the First Time in History . . .,' 'Nowhere in the World But in the USSR . . .,' and the like, was an important component of this image. Scientific administrators employed such rhetoric to present their proposals to the control apparatus on a number of issues, ranging from the building of gigantic new institutes and sophisticated equipment for space and nuclear research to the organizing of wide-scale oceanographic expeditions. Clearly the party monopoly over decision-making in science policy did not reduce Soviet scientists to passivity. They retained the ability to influence decision-makers at the highest level of the party apparatus, and some managed to do so with great ingenuity.

The striking success of Soviet nuclear and space programs in the 1950s greatly empowered the leadership of the scientific community, once again raising its authority in governmental circles. These scientific administrators, including such eminent physicists as I. Kurchatov ('father' of Soviet A-bomb), Nobelists P. Kapitsa and I. Tamm; chemists as A. Nesmeianov and Nobelist N. Semenov; and mathematicians as A.P. Aleksandrov and M. Lavrentev, managed to considerably limit the direct interference of party bureaucracy in their internal policies and to reassert their own control over scientific development. In exchange, however, Soviet scientists pertained to the 'pure' science and surrendered possible political and ideological outcomes of their professional activities to the party leadership.

During following years the community spokesmen successfully employed the Cold War ambitions of political leadership to further expand Soviet science system: to build 'Science Cities' (such as Dubna, Akademgorodok, and Pushchino), to publish numerous books and periodicals, to enhance prestige of science and scientists, and to partially bridge the isolating barriers. In the mid-1960s, they even managed to officially 'revive' genetics in the USSR. During the 1970s and early 1980s, continuous arms race and space race led to tremendous expansion of the Soviet science system, which by the late 1980s included over ten thousand institutions and employed over one million workers.

With all zigs and zags of Soviet domestic and foreign politics under Nikita Khrushchev (1953–1964), Leonid Brezhnev (1964–1982), and Mikhail Gorbachev (1985–1991), then, the Cold War secured the lasting equilibrium between the scientific community and the party-state apparatus, providing the community with almost unlimited resources and the state with the ultimate tokens of Cold War politics — nuclear weapons, missiles, and spacecrafts. It also created large disproportions in disciplinary development, placing a premium on exact and technical disciplines to the detriment of biomedical, agricultural, and ecological ones. It enveloped scientific research with the impenetrable cover of secrecy, under which enormous resources were wasted on ambitious, but useless, dangerous, and sometimes simply phony projects.

In 1991 the Soviet Union ceased to exist, and the Communist Party lost its dominating position in Russia. With its symbiont and patron — the Communist Party — casted out, and with its major justification — the Cold War — ended, Soviet Big Science now searches desperately for new patrons and new justifications, struggling to survive and sustain its previous privileged position within the political, cultural, and institutional landscape of the new Russian state.

Conclusion

The development of science in the USSR has long been viewed as a uniquely Soviet phenomenon. It does indeed possess many peculiarly Soviet features. It also offers, however, a more generally instructive example of the complex, multi-dimensional interrelations that have developed between Big Science and the twentieth-century bureaucratic state. The influence of ideology, economics, systems of financing and organizing research, and foreign and domestic policies upon the internal 'physiology' of science — institutionalization, specialization, debate, criticism, competition for recognition and resources — is surely to be found in countries other than the Soviet Union. However specific its form in the USSR, state control over science policy is a typical component of the Big Science of the second half of the twentieth century.

Of course, if anything, Big Science appeared in Russia first. In the 1930s, because of the nature of the Soviet state, many characteristics developed which would only become prominent in other countries later. Soviet scientists were the first to confront the potentially profound influence that their close collaboration with a state bureaucratic machine could have on science, and their experience has served their Western colleagues well. Indeed, not only did Soviet scientists use the positive and negative examples of the West to gain the attention and support of their government; Western scientists also used positive and negative Soviet examples, both Lysenkoism and Sputnik, for instance, to advance their own agendas with their own Big Science patrons and to develop strategies and tactics for dealings with their own government bureaucracies.

During the last half of the twentieth century, scientists throughout the world employed Cold War ideological, political, and practical arguments in their drive for institutional advantage and their competition for government funding; and throughout the world, politicians employed science and its achievements in pursuit of their domestic and international objectives. Had it not been for the Cold War, it seems unlikely that either the Western or the Soviet systems of Big Science would have survived, developed, and prospered as they did. The Cold War originated two systems of Big Science, two mutually interdependent creatures, mutually isolated, but each almost unthinkable without the other.

References

1. Daniel P. Todes, "Biological Psychology and the Tsarist Censor: The Dilemma of Scientific Development," *Bulletin of the History of Medicine*, 1984, vol. 58, pp. 529–544.
2. *Kul'turnoe Stroitel'stvo SSSR.* (Moscow: Gosstatizdat, 1956), p. 244.
3. Daniel P. Todes', "Pavlov and the Bolsheviks," in *History and Philosophy of The Life Sciences* (1995), vol. 17, pp. 379–418.
4. "Postanovlenie TsIK i SNK SSSR o Priznanii Rossiiskoi Akademii Nauk Vysshim Uchenym Uchrezhdeniem Soiuza SSR," *Izvestiia VTsIK* (July 28, 1925), p. 1.
5. V.D. Esakov, *Sovetskaiia Nauka v Gody Pervoi Piatiletki.* (Moscow: Nauka, 1971).
6. *Kul'turnoe Stroitel'stvo SSSR.* (Moscow: Gosstatizdat, 1956), p. 244.
7. Ibid., p. 244.
8. *Ukazania i Formy k Sostavleniiu Plana na 1936 God.* (Moscow: Gosplan, 1935), pp. 412–413.
9. "V Sovnarkome SSSR," *Pravda* (May 11, 1938), p. 2.
10. A.P. Iushkevich, "'Delo' Akademika N.N. Luzina," in *Repressirovannaia Nauka.* (Leningrad: Nauka, 1991), pp. 377–394; and Alex E. Levin, "Anatomy of a Public Campaign: 'Academician Luzin's Case' in Soviet Political History," *Slavic Review* (1990), vol. 49, no. 1, pp. 99–108.
11. *Kul'turnoe Stroitel'stvo SSSR.* (Moscow: Gosstatizdat, 1956), p. 244.
12. J. Stalin, *Vestnik Akademii Nauk SSSR* (1946), no. 2, p. 11.
13. V. Molotov, "World Arms Cut," *Vital Speeches of the Day*, vol. 13 (October 15, 1946–October 1, 1947), pp. 74–80, on p. 78.
14. Nikolai Krementsov, "The 'KR Affair': Soviet Science on the Threshold of the Cold War," *History and Philosophy of The Life Sciences* (1995), vol. 17, pp. 419–446.
15. Kirill Rossianov, "Editing Nature," *Isis* (1993), vol. 84, pp. 728–745.
16. Nikolai Krementsov, "A 'Second Front' in Soviet Genetics," *Journal of the History of Biology* (1996), vol. 29, pp. 229–250.
17. T. Lysenko, "Concluding Remarks," in *The Situation in Biological Science.* (New York: International Publishers, 1949), p. 605.

Further Reading

Kendall Bailes, *Science and Russian Culture in an Age of Revolutions.* (Bloomington: Indiana Univ. Press, 1990), and *Technology and Society Under Lenin and Stalin: Origins of the Soviet Technical Intelligentsia, 1917–1941.* (Princeton: Princeton Univ. Press, 1978).

Michael David-Fox, *Revolution of the Mind: Politics, Culture, and Institution-Building in Bolshevik Higher Learning, 1918–1929.* (Ithaca: Cornell Univ. Press, 1997, forthcoming).

Sheila Fitzpatrick, *Education and Social Mobility in the Soviet Union, 1921–1934.* (Cambridge: Cambridge Univ. Press, 1979).

Loren R. Graham, *The Soviet Academy of Sciences and the Communist Party, 1927–1932.* (Princeton: Princeton Univ. Press, 1967), *Science and Philosophy in the Soviet Union.* (New York: Vintage Books, 1974), and *Science in Russia and the Soviet Union: A Short History.* (Cambridge: Cambridge Univ. Press, 1993).

Loren R. Graham (Ed.), *Science and the Soviet Social Order.* (Cambridge: Harvard Univ. Press, 1990).

David Holloway, *Stalin and the Bomb.* (New Haven and London: Yale Univ. Press, 1994).

David Joravsky, *Soviet Marxism and Natural Science, 1917–1932.* (New York: Columbia Univ. Press, 1961), and *The Lysenko Affair.* (Chicago: Chicago Univ. Press, 1986).

Paul R. Josephson, *Physics and Politics in Revolutionary Russia.* (Berkeley: Univ. of California Press, 1991).

Nikolai Krementsov, *Stalinist Science.* (Princeton: Princeton Univ. Press, 1997).

Linda L. Lubrano and Susan G. Solomon (Eds.), *The Social Context of Soviet Science.* (Boulder, Colo.: Westview Press, 1980).

Alexander Vucinich, *Science in Russian Culture, 1861–1917.* (Stanford: Stanford Univ. Press, 1970), and *Empire of Knowledge: The Academy of Sciences of the USSR, 1917–1970.* (Berkeley: Univ. of California Press, 1984); and Douglas R. Weiner, *Models of Nature.* (Bloomington: Indiana Univ. Press, 1988).

CHAPTER 41

Twentieth-Century German Science

Institutional Innovation and Adaptation

MARK WALKER

Twentieth-century Germany has been the source for a great deal of institutional innovation in science. This chapter will emphasize how changing political, economic, and ideological forces have (and have not) influenced the relationship between science and the state in Germany during this century. Two themes will contrast the continuity represented by science with the discontinuity with respect to political regime: innovation and adaptation. This is certainly not the whole story of German science during the twentieth century. Individual scientists and scientific discoveries are also important, but because of space limitations can only be touched upon here. Moreover, science policy is arguably best suited for studying the peculiarly 'German' stamp on contemporary science, because scientists and scientific institutions are more malleable with regard to political ideology than science is. Finally, the National Socialist period dominates the historiography of recent German science. A study of scientific institutions before, during, and after the Third Reich illuminates both the peculiar effect of National Socialism on science and the significance of the Nazi period for twentieth-century German science.

This chapter also attempts to revise slightly the conventional picture of German science policy and scientific institutions. Historians often describe the science policy of a given period as decisively influenced, if not dominated, by the economic, political, and ideological environment: imperial science policy under the Empire, democratic in the Weimar Republic, Nazi during the Third Reich, communist in the German Democratic Republic, and federal (and democratic) in the Federal German Republic. In fact, there is much more to this story.

THE GERMAN EMPIRE 1900–1914

Three decades after Otto von Bismarck had used military force and political guile to unify Germany, the Empire was one of the leading industrial nations. This economic power derived to a considerable extent from the fruitful collaboration between the German university system, the first to make research prowess a prime criteria for hiring faculty, and the new industrial research labs in the science-based industries, especially electrical and chemical.

However, by the beginning of the twentieth century, many observers of German science were arguing that the needs of industry and the teaching demands placed on academic scientists were becoming incompatible. Scientists, educators, industrialists, and civil servants began calling for a new type of scientific institution: independent of the universities and thereby of teaching obligations; independent of the support of the individual German states (which financed the various universities); and funded both by private industry and the Empire. It was the 'research imperative' of the German universities that inextricably linked original research and university training and thereby made German science and clinical medicine the model for the rest of the world and, in particular, the United States.

The first such institution in Germany was the Imperial Institute of Physics and Technology (*Physikalisch-Technische Reichsanstalt*), founded in 1887 in order to represent the best both of pure science and industrial technology. The driving force behind the creation of this new type of research institute was the industrialist and scientist Werner von Siemens, who wanted to build an institute devoted to pure scientific research, but which would also address the long- and short-term needs of technology.

Indeed this institute, led by a series of respected physicists beginning with the charismatic and influential Hermann von Helmholtz, succeeded on both counts by performing important experimental work on blackbody radiation, which led to quantum physics, developing electrical standards for science-based industry, and testing and certifying scientific instruments, measuring apparatus, and materials. Perhaps the best demonstration of the Imperial Institute's success is the many imitators it spawned, including the National Physical Laboratory in Great Britain, the National Bureau of Standards in the United States, and in Germany itself the Imperial Institute of Chemistry (*Chemisch-Technische Reichsanstalt*, 1921). The Imperial Physics Institute also inspired two other innovative institutions: the Göttingen Association for the Advancement of Applied Mathematics and Physics (*Göttinger Vereinigung der angewandten Mathematik und Physik*, 1898); and, perhaps most important, the Kaiser Wilhelm Society (*Kaiser-Wilhelm-Gesellschaft*, 1911).

In contrast to the Imperial Institute, the Kaiser Wilhelm Society was designed first and foremost for fundamental research. Separate 'Kaiser Wilhelm Institutes' were founded with specific research programs, indeed often tailored to specific scientists who, together with their assistants and collaborators, could devote themselves full-time to scientific research. But although the goal was fundamental research, it was fundamental research of interest to the German state and industry.

Some of the first institutes, for example the Institute for Physical Chemistry and Electrochemistry (1912), were founded with the help of substantial donations from German industrialists. In general the Society tried to achieve a mixed form of funding, from the Empire, individual German states or cities, and from interested industries, in order not to be beholden to any one benefactor. On the eve of the First World War Kaiser Wilhelm Institutes for biology, chemistry, coal research, experimental therapy, labor physiology, and physical chemistry had been opened or approved.

The Imperial Physics Institute and the Kaiser Wilhelm Society were in a very real sense imperial institutions, funded by and designed for the needs of a powerful empire whose military and economic strength depended to a large extent on the effective exploitation of scientific research. Both the German state and industry were willing to fund research divorced from the university context. Leading scientists were willing to take up research which was 'fundamental,' but which was also of direct relevance for, and of interest to industry and the state. The Kaiser Wilhelm Society was also widely copied as a model for scientific institutions, but before the Society could expand to cover a significant portion of the many scientific disciplines, the first world war intervened and fundamentally changed science policy in Germany.

World War I 1914–1918

Many German scientists and science students enthusiastically welcomed the war in 1914, just like their counterparts in the Allied countries, but they usually went to war as soldiers, not scientists. Universities and research institutes were emptied of their students and younger faculty. However, one of the brand-new Kaiser Wilhelm institutes, Fritz Haber's institute for physical chemistry, was transformed by the conflict and thereby became one of the first examples of a scientific institution and its staff created for fundamental research, but harnessed instead for applied research in the service of the war.

Germany's military leadership initially dismissed the suggestion that scientists or even industrialists could play an important role in the war effort. But once the Schlieffen Plan had failed and the war in the West had bogged down into trench warfare, it became clear that, without the help of science, Germany would quickly lose. Science-based industry provided Germany with synthetic raw materials to help fight the war and synthetic foodstuffs to feed the home front. Of course Germany eventually did lose the war, but without the mobilization of science German defeat would have come much more quickly.

Haber placed his institute at the disposal of the government and turned it into a R&D center for chemical warfare. The institute personnel swelled to 1,500, including 150 scientists, and the budget increased fifty-fold. The institute soon resembled an industrial laboratory and developed new poison gases, gas masks and other defensive devices, gas shells and other delivery methods, and effective strategies for using chemical weapons. After the war, Haber (and many other Germans) considered himself a German patriot because of his wartime work; the Allies labeled him a war criminal.

German scientists and engineers were also mobilized for the research, development, and production of aircraft. Although this work had not borne much fruit by the end of the conflict, a great deal of state money had been invested in interdisciplinary research centers with close ties to German industry. The Treaty of Versailles temporarily halted aeronautic research, or at least drove it underground, but the institutions created during the First World War and the close cooperation between academic scientists and engineers, industry, and the state were recreated in different forms during the Third Reich.

The Weimar Republic 1919–1932

Germany's defeat in World War I was a national humiliation and an economic disaster. German soldiers returned home to political revolution, social unrest, and hunger. The value of the German Mark plummeted immediately after the war, only to be followed a few years later by hyperinflation. Scientists were hardly the hardest hit, but they nevertheless had to struggle to hold on to their jobs and fund their research.

These drastic times created equally drastic reforms in science policy and especially science funding. Before the Second World War, science in many industrialized countries was supported by universities and (especially in the United States) private foundations. Indeed this was considered the wisest and most effective funding system. In the United States, for example, the Carnegie and Rockefeller Foundations were among the most important sources of funding and thereby had considerable influence on the development of science. In Germany, however, money was so tight that German scientists and their patrons were forced to create new scientific institutions which used their money more efficiently and created the system of peer review so common today: the public Emergency Foundation for German Science (*Notgemeinschaft der deutschen Wissenschaft*) and the private Helmholtz Foundation for the Support of Physical-Technical Research (*Helmholtz Gesellschaft zur Förderung der physikalisch-technischen Forschung*).

Although Germany had a few research institutions supported by the central government, most scientific research took place at universities and most funding came from the individual German states. Such funds were usually simply given to the full professors who headed the respective university institutes. These individuals, who dominated their disciplines, dispensed these funds as they saw fit, and often exerted decisive influence over the careers of younger scientists. But the German states and their universities were now strapped for funds. There was so little money available for science that a much more effective allocation system had to be created. Both the newly-established Emergency and Helmholtz Foundations collected funds for scientific research, the former mainly from the national government, the latter from private and especially heavy industry. These foundations also funded different types of research, the latter naturally catering much more to the interests of industry. The two foundations played a crucial role in German science policy between the wars. In physics, for example, they may have doubled the real expenditures available for the direct support of research.

The peer review system meant that research moneys were now provided by different institutions, distributed by different individuals, and given to different scientists. Instead of the automatic support from the state ministry dispensed by the institute director, individual scientists now had to submit requests for their own specific research projects. The foundations determined which applications should be funded by setting up small committees of respected scientists. Although these reforms had not been initiated for political reasons, they effectively made the funding system for science more accountable and democratic than ever before.

These experts examined each application and allocated funds according to merit, although of course the committee members naturally favored certain research subjects over others. Committee member Max Planck ensured that quantum physics and relativity theory, perhaps two of the most famous and important developments in German science between the wars, received relatively generous Emergency Foundation support. Many of the leading scientists of the younger generation, including Werner Heisenberg and Erwin Schrödinger, were the beneficiaries. There were also political reasons for the new funding system. By considering only individual applications for specific projects, the foundations avoided any responsibility for the general support of the universities or for distributing money evenly among the various German states.

The Emergency Foundation and German scientists benefited from an unusual consensus in the national German parliament for support of science. The right and the left agreed, although for very different reasons, that science should be supported where possible. Germany's scientific institutions were a legacy of the Empire and still bore an imperial stamp congenial to the right. The forward-looking Social Democrats distrusted the men who dominated science in Germany, but nevertheless supported science for ideological reasons. Finally, science was seen by many Germans as a replacement for political power (*Wissenschaft als Machtersatz*): what Germany had lost in the political and military arenas would be made up for by science and culture.

Toward the end of the Weimar Republic the Emergency Foundation and other institutions were challenged by the Prussian Ministry of Culture over the fundamental question of who controlled science policy, the government which funded science, or the scientific institutions which did it? The state bureaucrats in the ministry did not want to influence which sort of science would be supported, rather they were concerned with the administrative procedure the Emergency Foundation and in particular its president Friedrich Schmidt-Ott, were following. Ironically, the peer-review system, a much more open and democratic system for funding scientific research than had previously been used, was created and implemented in a most undemocratic manner by the autocratic Schmidt-Ott, who withheld information both from the public and the inner circle of the Emergency Foundation and resisted sharing responsibility for important decisions.

Money was also a problem for the Kaiser Wilhelm Society. When it was founded shortly before the First World War, it relied almost exclusively on private donations, with the exception of the Prussian state's contribution of land, buildings, and the salaries of the institute directors. Among the leading scientists brought into the Society during the first decades of the century was the young (but not yet famous) Albert Einstein, who became the director of an institute for physics (which at this time existed only on paper). By the early twenties the society's endowments had evaporated and industrialists were much more cautious about donating significant sums of money.

The society adopted an effective two-pronged strategy to overcome the economic crisis but not sacrifice too much autonomy to either the public or private sector. First of all, its leaders turned to the Reich and Prussian governments for public support of the science which could no longer be funded by industry. Secondly, the society persuaded industrial leaders to provide funds by complementing the society's fundamental research institutes, built or conceived before the war, with a host of industry-related institutes created primarily in the industrial region of Germany. Various industries saw the Kaiser Wilhelm Society as an effective body to administer and organize their research. In some cases, the society effectively provided a surrogate for an industrial research lab.

Despite the best efforts of left-wing politicians and some ministerial bureaucrats, the society, like the Emergency Foundation, never became democratic during the Weimar Republic. It remained an elitist institution with autocratic leaders. Nevertheless, the society prospered, despite the inflation and depression. Throughout the Weimar period, the society expanded from fifteen institutes in 1920 to almost thirty by the early thirties. Thus it was perhaps no surprise when the Kaiser Wilhelm Society was accused in 1929 of using public support not only to advance research and science, but also to rebuild its former financial independence and promote its power and influence as an institution.

The Third Reich

During the first few years of the Third Reich, the National Socialists purged the entire civil service, including the universities and most state-financed scientific institutions, of racial, political, and ideological enemies. The best-known case is physics: more than fifteen percent of all academic physicists emigrated willingly or unwillingly after 1933, although the actual damage to physical research was much greater than this number implies. Although scientists were thereby targeted during the purge, science was not singled out for special treatment, or seen by the National Socialists as a threat. Albert Einstein is the exception which proves the rule. The National Socialists publicly attacked Einstein, because his great public stature, liberal worldview, and Jewish heritage made him their perfect villain.

The National Socialists handed over two organizations, the Imperial Institute of Physics and Technology and the Emergency Foundation, to Johannes Stark, one of their few loyal supporters among senior scientists. Stark, a Nobel laureate for physics, had openly embraced Hitler's movement early in the 1920s, at a time when the National Socialist leader was very grateful for such recognition and legitimation. Stark enjoyed considerable political influence at the start of the Third Reich.

Despite his rhetoric, Stark did not transform the Imperial Institute or the Emergency Foundation into a new type of National Socialist institution. He did not even have to purge these institutions of Jews, for the outgoing administrations had already complied with the civil service law and fired their Jewish employees. Stark instituted the National Socialist leader principle, whereby everyone had a place in a strict hierarchy, where he was expected to demonstrate unquestioning obedience to his superior but could expect the same from those under him, but this policy also fitted Stark's own autocratic nature.

Stark shut down research he disagreed with at the Imperial Institute, including quantum physics and relativity theory, and stopped all funding from the German Research Foundation (*Deutsche Forschungsgemeinschaft*, the new name for the Emergency Foundation) of the same work. He also brought both institutions into a much closer relationship with the German armed forces, increasing military-oriented applied research. Although Stark definitely brought a National Socialist flavor to these institutions, the substantial changes he made were not uniquely National Socialist, and, perhaps for this very reason, he soon lost his influence in the power struggles within the bureaucracy of the Third Reich.

Stark stopped Research Foundation funding of what he labeled "Jewish science," but at the same time, he balked at supporting research which he considered unscientific at the Ancestral Heritage Foundation (*Ahnenerbe*), the scientific research institute of the SS (see below). This principled stand got Stark fired in 1936 as part of an intrigue launched against him within the National Socialist Party and the Reich Ministry of Education.

Stark was replaced by Rudolf Mentzel, who was an honorary member of the SS and already held posts at the Reich Ministry of Education. Mentzel decided to fund the Ancestral Heritage research, but otherwise did not diverge far from his predecessor's policies. Stark retired from the Imperial Institute in 1939 and was succeeded by the respected technical physicist and influential National Socialist scientist Abraham Esau, who continued Stark's policy of enlisting the Imperial Institute in the war effort.

The Kaiser Wilhelm Society was also hit hard by the purge of the civil service. Max Planck, Nobel laureate for physics and president of the Kaiser Wilhelm Society, did not openly resist the firings, but instead he worked behind the scenes and negotiated for exceptions with the various ministries. However, neither Planck nor the other officials of the society met with much success. Sooner or later, almost all Jewish employees lost their jobs. The National Socialism transformation of the society had only begun with the purge of the civil service. Step-by-step, the entire society was integrated into the National Socialist state. The society's senate, composed of influential industrialists and state officials, broke with tradition and elected in turn two industrialists as successors to Planck, Carl Bosch from the giant chemical company IG Farben and Albert Vögler, president of one of Germany's largest steel manufacturers. The society thereby consciously took a step closer to German industry in order to gain some freedom of movement with regard to the political authorities.

When German rearmament began in earnest with the Four Year Plan under Hermann Göring, the Kaiser Wilhelm Institutes for Physical Chemistry and Leather Research were designated 'Four Year Plan' institutes and were dedicated primarily to applied research relevant for the National Socialist policies of autarchy or rearmament. The Kaiser Wilhelm Society prospered financially during the war by expanding its research into areas compatible with National Socialist ideology and policies like acquiring 'Living Space' (*Lebensraum*) for Germans in the East. A new type of institute with a special status was created, for example the 'Institute for the Science of Agricultural Work in the Kaiser Wilhelm Society.' The names of these institutes illustrate the society's ambivalence about the new institutes and the type of cooperation with the National Socialist state they represented. After the war began, three new institutes for agriculture or biology were founded in Breslau, Bulgaria, and Greece respectively. Another was planned for Hungary. These institutes were designed to facilitate collaboration with local scientists and exploitation of local resources.

National Socialist science policy and the war effort affected the various scientific disciplines and institutions in different ways. Some basic research, including especially biology, continued to flourish. In certain cases, for example work on mutations, research was supported because the National Socialists were interested in it. But while the tradition of basic research was sometimes upheld at the older institutes in Berlin-Dahlem, other institutes outside of Berlin took up more and more applied work. The industry-related institutes accepted research contracts from the military. The Kaiser Wilhelm Institutes for Physics, and of Medical Research, both participated in Army Ordnance's wartime nuclear fission project. In general, the Society's participation in the war effort varied from institute to institute: some institutes contributed little if anything, many contributed something, and a few contributed heavily.

The Prussian Academy of Sciences, which included the humanities and social sciences as well as the natural sciences, presents a different example of accommodation to National Socialism. Whereas the Kaiser Wilhelm Society managed to trade collaboration in some areas for autonomy in others, the Prussian Academy was gradually and completely transformed into a tool of National Socialist cultural propaganda. The Academy had both employees and members. The former were purged as part of the civil service in 1933. Similarly, some members of the Academy lost their university positions in the first years of the Third Reich and left Germany. However, a few individual members, who were designated 'non-Aryan,' but had nowhere else to go and never left Germany, remained in the Academy. Eventually their presence made the Academy a target for National Socialist attacks.

In 1936, shortly after the Reich Ministry of Education began to restructure the various academies of science in Germany, the Prussian Academy chose to do voluntarily what would otherwise be done by force. They changed their statutes in a simple, yet significant way: the word "elected" was replaced by "appointed." Although henceforth the academy superficially appeared to function independently as it always had, in fact it could now only make recommendations to the Ministry of Education.

In the fall of 1938 the Ministry required still more changes in the statutes: the leadership principle would be introduced, together with the office of Academy President; the academy would be expanded, which was an effective way of diluting the influence of existing members; henceforth only Aryans could become members; and finally, as expected, the Jews still remaining in the academy had to go. Most academy members accepted the changes with misgivings, but also without protest. Max Planck and Max von Laue had protested the academy's treatment of Einstein in 1933, and von Laue had successfully fought to keep Stark out of the academy in the early years of the Third Reich, but both Planck and von Laue now had to stand by and watch as Theodor Vahlen, a mathematician and long-standing National Socialist, became president.

The newly-transformed academy went on during the war to become an active participant in both National Socialist propaganda and the cultural rape of occupied Europe. Academies of science traditionally cultivated international scientific relations by exchanging their publications and continued to do so right up to the start of the Second World War. When the German occupation officials, in what had previously been Poland, began their program of 'Germanization' by creating a new Reich University of Posen, they contacted the Prussian Academy in order to acquire German-language scholarly literature. In return, the Posen officials could offer a great deal of Polish literature which had been taken from the various universities and academies of science in Poland, all of which were now closed. Ironically, the Prussian Academy, which had used the exchange of publications for centuries as a gesture of goodwill, now not only accepted this war booty, indeed it ordered specific Polish scientific publications. This may not be the strongest example of German science benefiting from conquest, but it shows that even an institution as devoted to 'pure science' as the Prussian Academy took part in the National Socialist exploitation of Europe.

Historians have long debated what effect National Socialism had on science. Perhaps the closest thing to a purely 'Nazi' scientific institution was the Ancestral Heritage Foundation created by the SS in 1935. This foundation was certainly the most ideologically-driven of all the scientific institutions in the Third Reich. In

1939, an Ancestral Heritage official described its goals as follows: "... to research the space, spirit, deed, and legacy of the racially pure Indogermanic peoples, to mold the research results in a lively way, and to present them to the people." In order to achieve this goal, the SS awarded research and teaching contracts, sponsored research trips, held conferences, and helped finance publications. Although Ancestral Heritage had some of its own researchers, scientists from other state and private research institutions were recruited for work on their projects. Ancestral Heritage itself was funded mainly by the German Research Foundation and as of 1942 by the state directly.

The SS foundation's main emphasis was placed on scholarly disciplines which were of propaganda value. For example, archaeology and anthropology could provide real or apparent scientific support of German claims on territory in the East. Even so, Ancestral Heritage supported a wide range of research. Although some topics would now be considered unscientific or even pseudo-science, such as the 'World Ice Theory,' first-class fundamental biological research, including entomology, and both plant and human genetics were also supported. Finally, Ancestral Heritage was the branch of the SS which planned, financed, and carried out inhuman experiments on prisoners of war and concentration camp inmates at Auschwitz and elsewhere.

It is difficult to judge the 'racial science' supported by Ancestral Heritage because the scientific results cannot be divorced from their often murderous context. For example, there has been an on-going debate about the data gathered from some of the concentration camp experiments done on people without their consent and often leading to their death. Some scientists have argued, and still argue, that data is data, and it should be used where relevant, while others insist that it is unacceptable to use this data like any other scientific result because of how it was gathered.

However, it is not clear that even Ancestral Heritage was a peculiarly Nazi scientific institution. Pseudo-science was also supported by traditional scientific institutions like the universities during the Third Reich. As noted above, the SS funded cutting-edge science as good as that found at any university, and other institutions, including the Kaiser Wilhelm Society, also participated in National Socialist propaganda. Ancestral Heritage was unique in its *deliberate* embrace of the most murderous and racist aspects of National Socialist ideology, as demonstrated by the inhuman experiments of August Hirt, Josef Mengele, and others. But even here it was the goals and the means that were unusual, not the scientific personnel or organization. For this reason, even with an example as extreme as Ancestral Heritage, it is difficult to determine the peculiar effect of National Socialism on science.

There were several attempts to centralize and thereby coordinate and control scientific research during the Third Reich. Such efforts paralleled the general trend towards centralization in the National Socialist state and away from the decentralized character of the Weimar Republic and Empire. These efforts culminated in the Reich Research Council, a new and thereby genuinely National Socialist scientific institution which was founded in 1937 and reached its peak of influence during the Second World War. The Research Council is often dismissed by historians because it was ineffective and fell far short of its goals. Nevertheless it represented a radical change in traditional German science policy and, thanks to the pressures of the war, did achieve a significant centralization and coordination of research. Its failure, or perhaps better said, relative lack of success, had more to do with the general chaotic and *polycratic* structure of the Third Reich (with its many competing and often overlapping centers of power, including the Armed Forces, state bureaucracy, National Socialist Party, SS, etc.,) than with its own efforts.

The Reich Research Council was created in the context of the Four Year Plan, the massive rearmament effort led by Hermann Göring in close collaboration with several top-ranking officials from IG Farben. The aim of this new council was to coordinate more effectively all scientific and technological research in both the public and private spheres and thereby facilitate its incorporation into the rearmament effort. Leading scientists, appointed by the head of the Research Council, would be responsible for their various research disciplines, deciding which research would be carried out and supported. In fact, the Council's experts were rarely able to exert such control and instead had to be content with steering funds, materials, and labor towards particular researchers and projects.

The Council's beginning was auspicious, with Hitler, Göring, and leading dignitaries from the Armed Forces attending the opening ceremonies. However, the Research Council always suffered from a lack of political support because it had to compete with the already entrenched institutions, like the German Research Foundation, all of whom resisted giving up any of their influence or autonomy.

During the Second World War German science was not exclusively geared to the war effort, but researchers who wanted significant funding often had to devote themselves to applied research for the Armed Forces, and young scientists often needed to be involved in such work in order to avoid the draft. In the autumn of 1939 the Reich Ministry of Education took advantage of the war in order to take control of both the Research Foundation and Research Council and thereby cut the latter loose from Göring's huge economic pseudo-ministry and its potentially strong support. The first president of the Research Council, General Karl Becker, had many other responsibilities more pressing than the Council. He taught as a full professor for military technology, physics, and ballistics and was dean of the newly-created military technological faculty of the Technical University of Berlin. In 1938 Becker became head of Army Ordnance as well.

Between its founding and the start of the Second World War the Council saw its funding steadily reduced, perhaps as a result of its relative impotence. After Becker's suicide in 1940, the relatively weak Minster of Education Bernhard Rust took over the Council presidency himself, thereby making it practically irrelevant. In June of 1942 the Research Council was revived by Albert Speer, Hitler's new Minister of Armaments, as part of his efforts to centralize the war economy. Göring received the order from Hitler to create a new Research Council which would subsume its predecessor.

However, the Council still received its funds from the Research Foundation, which in turn was still under Rust's control. Even though Göring was now head of the Council, he nevertheless subsequently exempted the considerable research efforts of the Air Force from Council jurisdiction, again thwarting the attempt to centralize and coordinate all research. In 1943, as the war became more desperate for Germany, Werner Osenberg, an engineer at the Technical University of Hannover and an SS member, was given the task of coordinating all scientific and technological research for the war effort by means of a planning office within the Research Council. His tireless efforts were doomed to failure.

Since the Research Council was unable to take control of research away from influential institutions like the Kaiser Wilhelm Society or the Armed Forces, it could do little more than influence research by offering funding and, more important during the last phase of the war, the priority classifications from Speer's Ministry which made research possible. Since priority classifications, research moneys, and exemptions from military service could only be justified by contributions to the war effort, by this time a great deal of scientific work in Germany had both become inextricably intertwined with the military research efforts of German industry and was often dependent on its support.

Although the Council was a creation of the National Socialist state, and indeed was dissolved with the end of the Third Reich, there was little specifically or peculiarly National Socialist about it. In fact, it resembles the type of science policy institution which was so successful in both the United States and the Soviet Union, the victors in the Second World War, whereby state-appointed scientists were given wide-ranging powers over both academic scientists and industries in order to centralize and coordinate the national scientific effort. This suggests that it had more to do with international developments in science, technology, and warfare, in particular the general acceptance of the idea of comprehensive central planning and coordination, than with National Socialist ideology. This example shows that it was the technocrats within the National Socialist system, and not the ideologues, who most effectively determined science policy.

The most significant new scientific institution created during the war was arguably the 'military-industrial-university' complex created at Peenemünde for the research and development of ballistic missiles, the so-called 'V-2' rockets. Amateur rocketry had flourished during the Weimar Republic, sometimes more as a hobby or sport than serious research, and Army Ordnance had become interested in rockets early in that period. General Becker was convinced that the surprise introduction of a radical new weapon could produce a stunning psychological blow against the enemy. Shortly after the National Socialists came to power, the Army seized control of rocket research, co-opted some of the amateurs, and effectively closed down all potential competitors in Germany.

The rapid growth of the Army rocket project during the 1930s was fostered further by the structural and political conditions inherent in the Nazi regime and the German armed forces. Until the early years of the war the Army had some autonomy and was able to push ahead with some of the armaments projects like rockets it wished to pursue. The massive push towards rearmament during the mid-1930s facilitated the expansion of the project. The competitive nature of the polycratic state ensured that there was no centralized coordination of the nation's armaments research effort. This encouraged the armed forces to pursue their own projects, thereby reinforcing both the Army's ability and determination to develop the ballistic missile. Thus the lack of centralization was both good and bad: it allowed branches of the armed forces to pursue their pet projects, but it also wasted great amounts of materials, manpower, and other resources.

During the first years of the war the leaders of the project, Army officer Walter Dornberger and the young engineer Wernher von Braun, built up a massive, secret in-house development capacity for their project and contracted out for important university and corporate research and development resources. Fortunately for the Peenemünde group, when the war turned against Germany in the winter of 1941–1942 and their generous support threatened to dry up, both Minister of Armaments Albert Speer and subsequently Hitler himself became rocket enthusiasts.

The success of Dornberger and von Braun made the rocket project a tempting prize for the SS, which enjoyed a massive expansion of its influence during the war. SS leader Heinrich Himmler eventually succeeded in grabbing large pieces of the rocket program. After convincing Hitler that rocket production had to be moved underground in order to protect it from Allied bombing, the SS set up the infamous Mittelwerk underground factory, where concentration camp inmates were used to build the last German V-2 rockets. Indeed such slave labor was a common feature of the German industrial complex during the last years of the war.

The weapon itself was a strategic and even psychological failure, diverting massive resources away from other sectors of the war economy. The rocket project cost around half a billion 1940s US dollars, about a quarter of the money spent on the atomic bomb. Since the German war economy was significantly smaller than the American one at its peak, the burden imposed by the Army rocket project on the Third Reich was roughly equivalent to that of the Manhattan Project on the United States. At its height, Peenemünde had roughly 6,000 R&D staff, including around a 100 engineers; and another 6,000 workers and laborers, including forced and concentration camp labor. The German rocket project did shorten the war, but in favor of the Allies. Furthermore, it was the Americans and the Soviets who benefited most from the work of Dornberger, von Braun, and their colleagues.

The most specifically 'Nazi' part of the rocket project was undoubtedly the murderous use of slave labor. However, it is difficult to find much else about the German ballistic missile program that did not happen in large-scale scientific R&D efforts in the Soviet Union or United States during the 1940s and 1950s. Indeed the Soviet Union used forced labor for its nuclear weapons project, even if this labor can hardly be compared to the systematic working of slave laborers to death by the National Socialists. Peenemünde was one of the first examples of what has come to be called Big Science and, just like the Reich Research Council, it had more to do with general and international trends in science, engineering, and industry than with National Socialist ideology.

Perhaps the most controversial war-time project in Germany was the 'uranium project.' The atom bomb was never developed in Germany, rather a research program with 70 to 100 academic scientists began at the start of World War II and investigated all possible economic and military applications of nuclear fission. The administration of the project was under military or National Socialist control for most of the war: from September 1939 to the fall of 1942 Army Ordnance scientist Kurt Diebner oversaw the research; from the fall of 1942 to the end of 1943 Abraham Esau, a respected technical physicist and influential Party member, managed the project as Hermann Göring's plenipotentiary for nuclear physics. An academic scientist, Walther Gerlach, took over as head of the project only in 1944.

The uranium project was in practice broken down into several parts, each of which was delegated to a leading scientist who directed this research at his own institute. The measurement of nuclear constants and other preliminary work was carried out under Walther Bothe in Heidelberg. Isotope separation and heavy water production was pursued by two physical chemists, Klaus Clusius in Munich and Paul Harteck in Hamburg. Harteck was by far the most dynamic and effective member of the project. Finally, nuclear reactor experiments were performed in three places: by Robert Döpel in Leipzig, by Karl Wirtz in Berlin, and under Diebner at Army Ordnance. Werner Heisenberg made an important contribution at the very beginning of the project by working out the theoretical foundation for applied nuclear fission and recognizing that the pure isotope uranium 235 would be a nuclear explosive. In 1940 Carl Friedrich von Weizsäcker discovered that a nuclear reactor would produce transuranic elements (plutonium) with the same explosive properties.

The status of the research project at the end of the war was comparable to what the Americans, British, and émigrés had achieved by the summer of 1942: a nuclear reactor that was close to maintaining a self-sustaining chain reaction and an isotope separation process (the Germans concentrated on ultracentrifuges, the Allies on electromagnetic and gaseous diffusion technologies) that had managed a slight enrichment of the ratio of isotope uranium 235 to uranium 238. This performance was due in large part to the changing fortunes of the war. During the Blitzkrieg phase from 1939 to the end of 1941, it appeared that Germany would win the war soon, so that no 'wonder weapons' were needed. When the war turned sour for

Germany in the winter of 1941–1942 and German Army Ordnance asked its researchers whether nuclear weapons could be produced in time to influence the outcome of the war, the answer was clearly no. Thus these scientists never had to confront the question, should we build nuclear weapons for Hitler?

West Germany 1945–1989

As the Allied armies pushed towards Berlin, special intelligence-gathering squads, often led by scientists, sought 'intellectual reparations' in the form of technology, scientific reports, and the German scientists and engineers themselves. The most famous examples include the US Armed Forces taking von Braun and most of his rocket team to the United States, and the Soviet Armed Forces luring or coercing German scientists to join its nuclear weapons program in the Soviet Union. The Soviets brought around 3,000 researchers and technicians from Germany to the Soviet Union. In comparison, around 500 Germans came to the United States. However, many other scientists or engineers were invited or coerced to work for all four of the Allied powers (Britain, France, Soviet Union, United States) inside or outside of Germany. The victorious Allies were interested in anything with the potential for military or economic applications. In particular, the Allies did not care about the political past of a German scientist or engineer, so long as his research was of use to them.

After the unconditional German surrender in May of 1945, scientists and scientific institutions, like all Germans and German institutions, had to achieve political rehabilitation and distance themselves from the legacy of the Third Reich. As we have already seen, the Kaiser Wilhelm Society was implicated in the crimes of National Socialism: the physics institute exploited war booty in the form of equipment and valuable raw materials for the uranium project, other institutes collaborated with the expropriation of Soviet laboratories and factories, and the infamous Josef Mengele, former assistant at the anthropology institute, sent 'research materials' in the form of body parts, from death camp victims, back to his mentor, Otmar Verschuer. The Society leadership has systematically ignored, downplayed, or misrepresented its consequential collaboration with National Socialism since the end of the war.

The Kaiser Wilhelm Society was a victim of the conflict between the victorious powers. The Soviets tried to take over the society by naming a left-wing scientist in Berlin as head of the organization. This move was rejected by the society's administration which had moved to the British occupation zone. The Americans were willing to recognize individual research institutes, but considered the society dead. The French tried to seize certain Kaiser Wilhelm institutes and move them to France while controlling the rest.

Only the British were willing to recognize the society, but with one proviso: the name Kaiser Wilhelm, reminiscent of German militarism, had to go. General secretary Ernst Telschow and new president Otto Hahn, 1944 Nobel Laureate for chemistry, fought hard to keep the old name. Hahn threatened to resign, then made the empty threat of turning to the Soviet Union for support. In the end the society was renamed after Max Planck.

The society took other, more substantive steps to change its profile in the post-war world. All ties to German industry were cut or hidden, and many of the industry-related research institutes created in the Weimar Republic and Third Reich were closed or not reopened. Of course, the institutes 'in the Kaiser Wilhelm Society' vanished as well. The society was now recast as an institution devoted exclusively to disinterested fundamental research. The society leadership worked very hard to give the misleading impression that it had always been so.

The financing of the society changed fundamentally after 1945. German industry was in no position to donate significant funds, and the post-war Big Science style of research was much more expensive. As a result the federal government took over responsibility for funding the society, making it a private organization completely dependent on public funds.

In many respects, the major problems faced by German science after 1945 were similar to those caused by defeat in the first world war and a severe shortage of funds. It was no surprise that the Emergency Foundation for German Science was revived in its original form (and then given the same name the National Socialists had used, German Research Foundation). However, there was now a competitor. A small group of West German scientists lead by Werner Heisenberg announced the creation of a 'German Research Council' which would coordinate all research and which would be funded directly by the federal government.

The various German states, determined to retain the considerable control over science they had traditionally enjoyed through their control and funding of the German universities, backed the Research Foundation. In turn, Heisenberg and the Research Council was able to count on the personal support of Chancellor Konrad Adenauer. A power struggle followed, with the result that the Research Council was absorbed by the Research Foundation and a decentralized, federal system of science policy, not a centralized approach, was developed.

This conflict mirrored how West German scientists dealt with the legacy of National Socialism. Bringing back the old Emergency Foundation was in many respects an attempt to turn back the clock to the Weimar period and treat the Third Reich as a brief aberration, an attitude widespread in Germany after the Second World War. Heisenberg's Research Council was vulnerable to political attack because it reminded many of the National Socialist Reich Research Council and its attempts to exert tight control over most of German science. Heisenberg and his colleagues fought back against such attacks, but their elitism and ambition only reinforced the doubts of their critics. In fact, as we have seen centralization of science policy was not specifically National Socialist and Heisenberg's plans for a Research Council owed just as much to science policy institutions created in France, Great Britain, and the United States.

FIGURE 41.1: THE PROPOSED IMPERIAL INSTITUTE FOR PHYSICS AND TECHNOLOGY, DRAWN SOMETIME BETWEEN 1884 AND 1887. FROM: DAVID CAHAN, *AN INSTITUTE FOR AN EMPIRE*. (CAMBRIDGE: CAMBRIDGE UNIVERSITY PRESS, 1989), P. 44. ORIGINAL SOURCE: NATIONAL BUREAU OF STANDARDS, GAITHERSBURG, MARYLAND.

FIGURE 41.2: KAISER WILHELM II AFTER THE OPENING OF THE FIRST TWO KAISER WILHELM INSTITUTES FOR CHEMISTRY AND FOR PHYSICAL CHEMISTRY AND ELECTROCHEMISTRY ON 23 OCTOBER 1912 ON HIS WAY TO THE FIRST GENERAL MEETING OF THE KAISER WILHELM SOCIETY. FROM: RUDOLF VIERHAUS AND BERNHARD VOM BROCKE (EDS.), *FORSCHUNG IM SPANNUNGSFELD* VON POLITIKUND GESELLSCHAFT. (STUTTGART: DEUTSCHE VERLAGS-ANSTALT, 1990), P. 60. ORIGINAL SOURCE: BILDARCHIV PREUSSISCHER KULTURBESITZ, BERLIN.

FIGURE 41.3: THE EMPLOYEES OF THE KAISER WILHELM INSTITUTE FOR MEDICAL RESEARCH MARCHING ON 1 MAY 1934 OR 1935. FROM: RUDOLF VIERHAUS AND BERNHARD VOM BROCKE (EDS.), *FORSCHUNG IM SPANNUNGSFELD VON POLITIKUND GESELLSCHAFT*. (STUTTGART: DEUTSCHE VERLAGS-ANSTALT, 1990), P. 362. ORIGINAL SOURCE: ARCHIVES OF THE MAX PLANCK SOCIETY, BERLIN-DAHLEM.

There is no more money for science. The Observatory is falling down.
The astronomy professor unsuccessfully asks the government for funds.
Desperate, he decides to take up astrology.
As an interpreter of the stars, he gives horoscopes.
His predictions soon are highly prized. He makes a lot of money.
Now the professor has the means to renovate both himself and his observatory. The newest and best instruments are acquired.

FIGURE 41.4: THE PREDICAMENT OF GERMAN SCIENCE, CARICATURED IN *SIMPLICISSIMUS* 1921 BY TH. TH. HEINE. FROM: RUDOLF VIERHAUS AND BERNHARD VOM BROCKE (EDS.), *FORSCHUNG IM SPANNUNGSFELD VON POLITIKUND GESELLSCHAFT.* (STUTTGART: DEUTSCHE VERLAGS-ANSTALT, 1990), P. 59. TAKEN FROM: *SIMPLICISSIMUS* 1921, PRECISE REFERENCE UNKNOWN.

FIGURE 41.5: MAX PLANCK, PRESIDENT OF THE KAISER WILHELM SOCIETY, SPEAKING AT THE TWENTY-FIFTH ANNIVERSARY OF THE SOCIETY ON 11 JANUARY 1936 IN THE GOETHE HALL OF THE HARNACK HOUSE. FROM: RUDOLF VIERHAUS AND BERNHARD VOM BROCKE (EDS.), *FORSCHUNG IM SPANNUNGSFELD VON POLITIK UND GESELLSCHAFT.* (STUTTGART: DEUTSCHE VERLAGS-ANSTALT, 1990), P. 383. ORIGINAL SOURCE: ARCHIVES OF THE MAX PLANCK SOCIETY, BERLIN-DAHLEM.

FIGURE 41.6: PRE-WAR GERMAN ROCKET EXPERIMENTS. LIQUID FUEL ROCKET READY FOR TEST-FIRING, 22ND JUNE 1932.

FIGURE 41.7: AN A-4 ROCKET IS LAUNCHED FROM A TEST STAND IN PEENEMÜNDE IN 1942 OR 1943. FROM: MICHAEL J. NEUFELD, *THE ROCKET AND THE REICH*. (NEW YORK: FREE PRESS, 1995), BETWEEN PAGES 210 AND 211. ORIGINAL SOURCE: SMITHSONIAN INSTITUTION, NEG. NO. 83-13847.

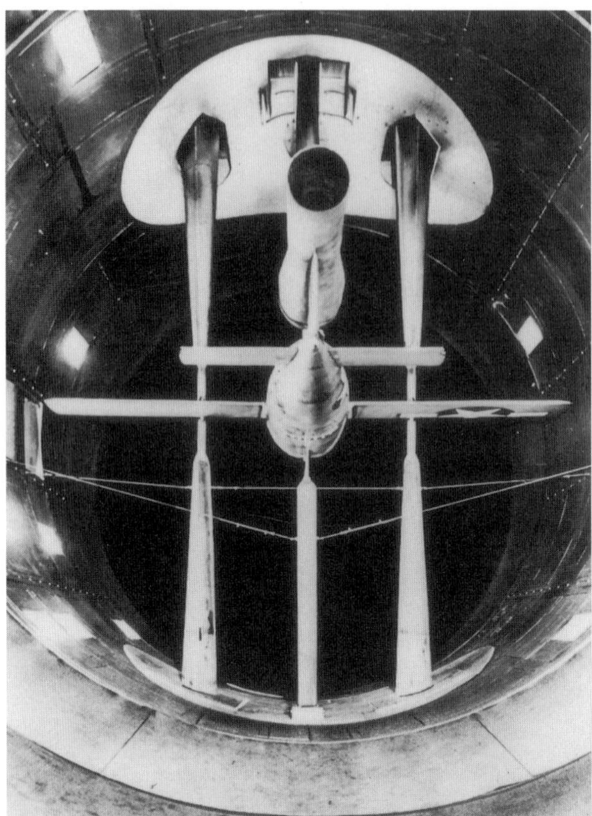

FIGURE 41.8: TESTING OF V WEAPONS, WRIGHT FIELD, DAYTON, OHIO. A USUAL SCENE IN THE AIR TERMINAL SERVICE COMMAND'S MAMMOTH TWENTY FOOT WIND TUNNEL, IN WHICH AN ARMY AIR FORCE V1 REPLICA IS UNDERGOING TESTS.

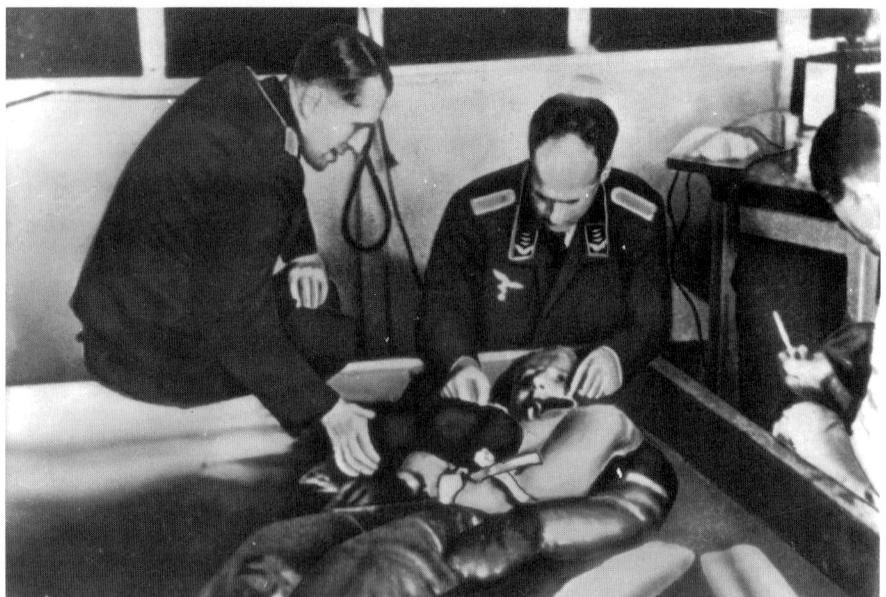

FIGURE 41.9: EXPERIMENTS BY THE GERMAN AIRFORCE WITH PRISONERS AT DACHAU CONCENTRATION CAMP: IN FREEZING EXPERIMENTS TEST PERSONS HAD TO LIE IN A TROUGH OF ICY WATER FOR UP TO THREE HOURS. © ULLSTEIN

FIGURE 41.10: LOCATIONS OF INSTITUTES AND RESEARCH CENTERS OF THE KAISER WILHELM GESSELLSCHAFT IN 1939 (WITHOUT THE INSTITUTES IN ROME, ROVIGNO AND SÀO PAOLO).

Immediately after the fall of the Third Reich, German science had to be de-nazified and de-militarized. Neither program was carried out systematically or effectively, but some German science, including nuclear research, stayed under Allied control for several years. When the last restrictions were lifted in the middle of the 1950s, the West German state began a very ambitious program of national Big Science (*Grossforschung*) research centers.

Most of the first institutions were related to nuclear research and were founded as part of the 'atomic euphoria' that characterized much of the fifties and sixties. Scientists like Hahn, Heisenberg, Weizsäcker, and Karl Wirtz had to walk a fine line between the military and civilian applications of nuclear fission. Along with other colleagues, they published in the late fifties the "Göttingen Manifesto," which was an open letter to their government announcing that they would have nothing to do with the development of 'German' nuclear weapons. On the other hand, they pushed hard for public support for 'peaceful' nuclear energy and nuclear research, even though they knew that the peaceful aspects of this research could not be separated from its warlike applications.

In time West Germany had Big Science centers for aviation and aerospace, various types of nuclear research, applied mathematics, plasma physics, bio-medical research, and high-energy physics. Once again, the West Germans were in large part responding to the establishment of large-scale interdisciplinary research centers in the United States, France, and Great Britain, but more than any other western country, West Germany made this type of Big Science institution into a vital part of its research capacity.

German scientific research today takes place at one of three levels: the traditional university institute, Max Planck Institutes, and Big Science installations. Ironically, the individual German states, which had fought the German Research Council in order to retain control over scientific research, quickly realized that Big Science was beyond their means. Eventually the states abdicated most responsibility for, and support of, science to the federal government. As a result, the centralization of scientific research under the authority of the state, a process begun under National Socialism, has finally overcome the decentralized, federal model of Weimar and the Empire. Centralization was necessary because of the demands now made by Big Science, not because of German traditions in science, and certainly not because of the totalitarian legacy of National Socialism.

East Germany 1945–1989

Science in the Soviet zone of occupation and the German Democratic Republic took a more radical turn away from the science policy of the Empire, Weimar Republic, and Third Reich. The Soviet officials purged the Prussian Academy of Sciences of 'Nazis' just like the National Socialists had purged it of racial and ideological undesirables. Of course these two examples should not be equated, but their similarities should also not be denied. The academy was renamed several times and rebuilt along the lines of the Soviet Academy of Science. In particular, the new Academy of Sciences for the German Democratic Republic now had its own research institutes, including some of the most important in East Germany.

The new structure was pragmatic as well as ideological, for the Academy incorporated the various research institutes that had been orphaned by the Cold War, including several former Kaiser Wilhelm Institutes. The Academy institutes also played an important role in a division of labor within the science policy structure of the East German Communist state. Whereas the university institutes and instruction of students were usually reserved for the politically more loyal scientists, an apolitical or politically unenthusiastic scientist might find a home in the Academy.

One of the few, new, scientific institutions created in East Germany was the Physical Society of the German Democratic Republic, a counterpart to the very old and respected German Physical Society operating in the West. The East German physicists were torn between their desire for strong international relations, and the demands of the East German political leadership. This new society, founded in 1952, was intended to facilitate this compromise. At first the society enjoyed considerable autonomy. After the construction of the Berlin Wall in 1961 it increasingly became a political instrument of the political leadership. However, the society and in particular its many disciplinary subdivisions played an important role in the East German scientific community by providing an institutional framework for scientific activities.

East German science was shaken by some of the same ideological debates which dominated Soviet science, including the Lysenko affair, the attempt to replace modern genetics with an alternative agricultural science, but in the long run it overcame them. Research in East Germany could not compare with its West German counterpart with regard to quantity, and only occasionally was it competitive with regard to quality. However, its scientists did make a respectable, if relatively modest, contribution to the international scientific community.

Epilogue: Reunification, Since 1989

The removal of the Berlin Wall and the subsequent collapse of the East German regime in 1989 also brought an end to its autonomous developments in science policy. With few exceptions, West Germans were placed in charge of reorganizing science policy in the East. They often closed an institute or section, fired its employees, and then reestablished it in an identical or altered form, a process know as *Abwicklung*. The new institute's positions were open to competition between applicants from both parts of Germany, but since the standards were set by West Germans who often had a negative perception of East German science, in practice East Germans often lost their jobs and West Germans took their places.

For political and ideological reasons, East German society, including the scientists and scientific institutions, began to be transformed according to the model of the Federal Republic. The East German universities were reorganized along the lines of universities in the West. The research institutes of the East German Academy of Sciences were detached and either dissolved or reattached to other institutions, leaving behind the traditional form for an academy. The fortunate institutes were attached to an university or industry; the unfortunate were dissolved.

The Max Planck Society did not incorporate any former East German institutes. It sifted through the orphaned researchers and either hired or created new Max Planck institutes for a select few scientists whose work was either very promising or had already earned them an international reputation. The East German Physical Society was merged into its Western counterpart. Within a few years of the East German revolution, there was little trace remaining of the forty years of Communist science policy in Germany.

Conclusion

At the start of the twentieth century German science set the standard for the rest of the world. The Third Reich was on balance parasitic of German science, using the scientists and institutions it had inherited from the Weimar Republic, but wasting the next generation of potential scientists by its racial and ideological purge of the civil service, its politicization of the schools and universities, and the great loss of life in the Second World War. Since 1945 Germany has recovered to the point of being a respected scientific country like Britain, France, or Japan, but it no longer dominates science in general or even any particular scientific discipline.

German science has been successful, in spite of its troubled history, because when forced to adapt to changing circumstances, it has been able to do so. Germany in the twentieth century experienced radical swings in politics and ideology on one hand, and important innovations in science policy on the other. The connection between the two is neither simple nor direct. Institutional innovations were a response to a mixture of professional, economic, and political pressure. Precisely because these three factors were always present, no one was ever dominant.

When existing institutions tried to adapt themselves to new political circumstances, their considerable institutional inertia ensured that they would be altered, not transformed. Each regime had the opportunity to create new institutions, but the latter also had to accommodate themselves to the existing institutions and the requirements of the international scientific community and as a result never became as radical as might have been expected. German scientific institutions, and science policy, were so often innovative and effective precisely because they were constantly being pulled in different directions by different forces and were often placed under considerable pressure. The final result is scientific institutions and policies that have been influenced by all of the regimes, with the possible exception of the German Democratic Republic, but have not been completely determined by any one.

The Weimar Republic did not create lasting democratic institutions, nor the Third Reich fascist, nor the Federal Republic republican, nor the East German communist. The Great Depression did not extinguish scientific research, neither did the West German 'Economic Miracle' of the fifties provide scientists with all the funds they wanted. Pressure from the international community could stop neither the purge of 'non-Aryan' scientists during the first few years of National Socialist rule nor the imprisonment of East German scientists behind the Berlin Wall. Historians and scientists often debate which is most important, 'external' forces like Stalinism, National Socialism and the Great Depression, or 'internal' factors like international competition and standards for science. Germany illustrates that these forces and factors interact in many ways, both productive as well as destructive.

Further Reading

David Cahan, *An Institute for an Empire: The Physikalisch-Technische Reichsanstalt 1871–1918.* (Cambridge: Cambridge University Press, 1989).

John Gimbel, *Science, Technology, and Reparations. Exploitation and Plunder in Post-war Germany.* (Palo Alto: Stanford University Press, 1990).

L.F. Haber, *The Poisonous Cloud: Chemical Warfare in the First World War.* (Oxford: Clarendon Press, 1986).

John L. Heilbron, *The Dilemmas of an Upright Man: Max Planck as Spokesman for German Science.* (Berkeley: U. of California Press, 1986).

Jeffrey Johnson, *The Kaiser's Chemists: Science and Modernization in Imperial Germany.* (Chapel Hill: University of North Carolina Press, 1990).

Kristie Macrakis, *Surviving the Swastika: Scientific Research in Nazi Germany.* (New York: Oxford University Press, 1993).

Michael Neufeld, *The Rocket and the Reich.* (New York: Free Press, 1995).

Monika Renneberg and Mark Walker (Eds.), *Science, Technology, and National Socialism.* (Cambridge: Cambridge University Press, 1993).

Mark Walker, *German National Socialism and the Quest for Nuclear Power, 1939–1949.* (Cambridge: Cambridge University Press, 1989).

Mark Walker, *Nazi Science: Myth, Truth, and the German Atomic Bomb.* (New York: Plenum, 1995).

CHAPTER 42

Science in the United States

LARRY OWENS

US science rose from modest beginnings to become, by the midpoint of the twentieth century, a bulwark of national power. Its foundations were laid in the last quarter of the nineteenth century with the creation of the science-based university and the blossoming of scientific professionalism. The first decades of the new century found corporate America exploring the advantages of industrial R&D and a new breed of scientific expert moving into the political arena, expanding previous footholds in government and embracing the Progressive Era faith that life could be rationally managed. Scientist-entrepreneurs like George Ellery Hale and Robert Millikan exploited opportunities offered by the Great War to articulate expansive, ambitious visions and form a national cadre of scientific leaders based in new institutions like the National Research Council. During the decades following the Great War, this elite assembled a network of academic, industrial, and philanthropic leaders that brought American science up to global standards. Industrial R&D expanded beyond the largest corporations, while business and academia explored cooperative programs. Private philanthropies like the Rockefeller Foundation and the Carnegie Institution of Washington funded disciplinary programs, built scientific instruments, and sponsored exchanges between European and American schools that allowed young PhDs to master the best of European science. By 1933, the MIT physicist John Slater said of foreign visitors at the annual meeting of the American Physical Society that "European physicists were here to learn as much as to instruct."[1]

World War II demonstrated the strength that had been built up since the turn of the century. From science at war came a multitude of new and improved weapons including radar, the proximity fuze, rockets, anti-malarial drugs, blood substitutes, and the quantity production of penicillin. And, of course, there was the bomb. The atomic destruction of Hiroshima and Nagasaki energized the public's appreciation for scientists whose leaders had mobilized in such agencies as Vannevar Bush's Office of Scientific Research and Development (OSRD). OSRD was the leading edge of what the *New York Times* called a vast test-tube army, "more than a hundred thousand trained brains working as one," whose efforts helped win the war. Hiroshima and Nagasaki were "a long way from Washington, yet thought waves sent from Fifteenth and P streets in the capital built themselves into the mighty explosions which rocked the world."[2]

Wartime success, an intensifying Cold War, and the threat of nuclear Armageddon reshaped the landscape of American science. New state agencies, like the Atomic Energy Commission and the National Science Foundation, were established in 1946 and 1950, respectively, and, after the shock of Sputnik in 1957, a Presidential Science Advisory Committee was lodged within the White House. An unprecedented system of national laboratories, presided over by the Atomic Energy Commission, united the interests of science and defense; universities harnessed their missions to federal funding and military research; and the economy pivoted on military R&D. The new climate bolstered fields like oceanography, the chemistry of antibiotics, materials science, space science, nuclear and high-energy physics, quantum electrodynamics, and solid state physics which spawned, for example, transistor radios and computers, and nourished the belief that science would secure both the Good Life and the nation's place in the world. Encouraged as well were thermonuclear weapons and doomsday strategies, spy satellites and electronic battlefields, intercontinental ballistic missiles, laser-guided bombs, and loyalty investigations. The years after the war were bountiful, but also frightened, suspicious, and, at times, paranoid. These qualities deeply influenced the character of post-war science.

Statistics index this story in their own way. Virtually invisible over the first half of the century from 1880, the national budget for R&D began climbing steeply during World War II, soaring to approximately fifteen billion dollars in 1961. As budgets increased, so patterns of patronage changed. Between the wars, most R&D had been done in industrial laboratories, supported by private and philanthropic sources; science financed by the state was done in government labs by government employees. By the early fifties, the vastly increased amount of R&D, still performed predominantly in industrial and university labs, was paid for by the federal government — the bulk (around 83 percent) ostensibly for defense. Scientific workers had always been few. By the era of the Cold War, however, they had multiplied six-fold, increasing both their share of global publication and recognition. Only two Nobel Prizes had gone to US scientists by the outbreak of World War I. After the Second World War, they were winning the lion's share of all Nobel Prizes in science and, by the 1990s, contributing over 30 percent of all articles in mainstream science journals — more than three times the number of their closest rivals, the Japanese.

There are many ways to tell the story of American science. One can narrate the struggles of influential individuals or concentrate on institution-building and professionalization. Alternatively, one can study the economy of science, the development of public policy, or the theory and practice that define particular disciplines. Indeed, all of these approaches can be found in the present volume. In part, the choice depends on the historian's background and purposes. In what follows, the focus is on a set of broad and pervasive factors which, separately and in combination, have helped shape American life and the science adapted to it. In brief, over the course of the last century, both have been organized, industrialized, and militarized.

AN ORGANIZING WORLD

In his 1957 bestseller, *The Organization Man,* William Whyte complained about the bureaucratization rampant in American life. The forces responsible for the man in the gray flannel suit were equally at work, he felt, in the laboratories of science, where the individualistic explorer had been replaced by the blinkered and disciplined technician. This organizational dynamic had been noted earlier, and with more optimism, by James Angell, the chairman of the National Research Council, who had argued in 1920 that "research work is capable of being organized in ways not wholly dissimilar to the organization of our great industries . . .":

> I would accordingly urge that in our conception of research we look beyond the peculiar combination of intellectual traits, which may characterize any one individual, and think of it as the organized technique of science itself for its own propagation.[3]

Angell spoke as if organization could exploit the latent powers of science by overcoming the limitations of individuality and separateness. Indeed, for much of the century, and for many in positions of power, organization has been a dynamic to conjure with. It could hardly be otherwise for a people capable of imagining (in 1908) that the Universe was an "all-including system of perfect organization" whose creation was an act of organization, or who could view man (in 1972) as "a coordinating interface system in the multilevel hierarchy of nature."[4] For many in the US, especially its political and economic elites and a rising middle class, organization became an appealing instrument for understanding and controlling an increasingly complex urban and industrial world.

This dynamic is evident first in the quickening pace of professionalization before 1900. Earlier in the nineteenth century, the American Association for the Advancement of Science was the only scientific society and *Silliman's Journal* the only publication with more than parochial ambitions. In the half century from the Civil War to World War I, almost all of the important societies and journals were established, as well as the modern agencies of licensing and regulation. Furthermore, the theoretical models of this professionalized science sought to lay bare the hidden joints and connections in confused end-of-the-century experience. The rise of sociology in attempts to come to terms with urbanization is only one example. The first Department of Sociology, established as part of the new University of Chicago in 1892, dominated American sociology until World War II. The 'Chicago School' embodied the hope that a complex urban world, and growing working class ghettos, could be as easily explained through natural 'urban ecological' laws as Darwinian Nature was explained by laws of biological competition. Professional scientists like these claimed to articulate the 'connectedness' in seemingly disjointed experience. For many, they displaced the clergy and defined a new type of credentialed intellectual whom Americans could trust to help them make sense of the mysterious forces moving their island communities from familiar self-sufficiency to anonymous interdependence.

But this new science was more than a social role and a mode of thought. It proved, as well, a form of power suited to the organizational changes remaking industry and government. The first modern business structures in the US were created to manage the early railroads. The fully-developed corporation emerged from the struggle to control destructively competitive national markets in the steel, oil, chemical, and electrical industries. For early industry, science had been relatively insignificant. Within the modern corporation, it was crucial for both strategy and structure. General Electric led with the establishment of its Research Laboratory in 1900, followed soon after by Du Pont, AT&T, and Eastman Kodak. Only the largest corporations could afford in-house laboratories before the First World War; but during the twenties and thirties the spirit, if not the fact, of the laboratory spread as more and more companies discovered that science was not only a source of products but a means of market control.

As the corporate world organized, so, in counterpoint, did government. Before the turn of the century, the American state hardly deserved the name, bearing few of the responsibilities we moderns expect, neither social, educational, economic, or scientific, even considering those engineering, exploratory, or agricultural agencies long deemed essential for various federal and state missions. Early in the new century, the creation of a civil service promoted the bureaucratization of government by professional administrators and accelerated the organization of the state. The crisis of the Depression expanded the federal bureaucracy dramatically, as did World War II with the insinuation of national security into wide reaches of American life. And as in industry, structural elaboration multiplied opportunities for experts at municipal, state, and federal levels, from Harvey Wiley's Bureau of Chemistry mandated by the Pure Food and Drug Act of 1906 to Vannevar Bush's OSRD made possible by President Roosevelt's newly expanded Executive Office.

Henry Waite, an engineering graduate of MIT, is a good example of the new breed of intellectual adapted to an organizing age who redefined politics as technical reason and organizational expertise. Inspired both by corporate models and scientific professionalism, Waite was appointed city-manager of Dayton, Ohio, after the disastrous flood of 1913, and proceeded, in that spirit, to rebuild the city and transform its government. He collected garbage with efficiency, cleared overgrown lots, and pasteurized milk. He instructed a new Department of Public Welfare "to study the causes that produce poverty, delinquency, disease and crime."

Waite did more, however, than sweep the streets: out of the shadowy backrooms of City Hall, he swept the rubbish of old-style bossism. That image captured the imagination of a reform-minded public. Different from "those local statesmen . . . gathered in mysterious corners discussing momentous issues in whispers," wrote one journalist, Waite

> does not greet his callers with effervescence; neither does he treat them with disdain. While talking with them he does not glance at the ceiling and nervously finger his mail . . . . He listens attentively, asks questions quickly, smiles pleasantly at the right moment . . . . The fact is that Mr. Waite behaves admirably in character; he is precisely what he has always been — a man with the technical training of an engineer, experienced in problems of public works, accustomed to dealing with figures and facts, and having none of the talents that make the great American politician.[5]

The organizational dynamic, however, implies more than connectedness and control. It touches on an expansiveness and fascination with heroic enterprises that is a distinctive feature of much American science and culture generally. It is evident in the obsession with great feats of engineering like the Panama Canal and with large scientific instruments that appealed to both patrons and public. Emblematic of this spirit are the observatories and great telescopes built by George Ellery Hale, first in Chicago and later in California; the monumental dams built by the federal government as examples of Progressive Era engineering; and the ever larger (and more expensive) cyclotrons which helped E.O. Lawrence build both a new science and a Nobel Prize-winning reputation. Such expansiveness pervades American science, becoming particularly dominant in the decades after World War II. Alvin Weinberg, the first director of the Oak Ridge National Laboratory, observed in 1961 that:

> When history looks at the twentieth century, she will see science and technology as its theme; she will find in the monuments of Big Science — huge rockets, the high-energy accelerators, the high-flux research — symbols of our time just as surely as she finds in Notre Dame a symbol of the Middle Ages. She might even see analogies between our motivations for building these tools of giant science and the motivations of the church builders and the pyramid builders. We build our monuments in the name of scientific truth, they built theirs in the name of religious truth; we use our Big Science to add to our country's prestige, they used their churches for their cities prestige; we build to placate what ex-president Eisenhower suggested could become a dominant scientific caste, they built to please the priests of Isis and Osiris.[6]

Monumental science entails large organizations. Early examples of such projects are the Chemical Warfare Service's development of poison gases in World War I and Du Pont's pursuit of nylon. Post-World War II examples are the system of national laboratories supervised by the Atomic Energy Commission; the Mohole Project of the 1960s (an unsuccessful attempt to drill through the earth's crust); the space missions presided over by the Jet Propulsion Laboratory; and, most recently, the Hubble Space Telescope and the Human Genome Project. However, the archetype of Big Science remains World War II's Manhattan Project, an enterprise of Brobdinagian proportions. Uncertain about the timely production (or even the possibility) of atomic bombs for the current war, the American effort had shifted into high gear only during the summer of 1942. At that point, the Army engineer, Leslie Groves, was brought in to supervise the construction and operation of the facilities needed to build atomic weapons. Over the next three years, on the way to the successful Trinity test in the desert of Alamogordo, New Mexico, Groves organized a nation-wide network of laboratories and production plants that employed a workforce of 600,000 at a cost of two billion dollars. Requiring unprecedented cooperation between academic and industrial scientists, engineers, soldiers, and industrial managers, the Manhattan Project was R&D on a vast scale that turned the country, as Niels Bohr once noted, into a gigantic factory.

Big Science shifts attention from the individuality of scientific achievement to the organization that Angell argued in 1920 would tap the nation's strengths. As Whyte observed, central to such science are those features that mimic the structure of modern business: managerial hierarchies, interdisciplinary team-work, the ramification of project and site-specific skills and specializations, and functional divisions between line and staff, finance, and planning. For the post-war generation, such science has been powerfully symbolic. Arthur Compton, who had worked in the Manhattan Project, argued shortly after the war that the Atomic Crusade was a model of social planning, a timely lesson on how to organize a free people for the Cold War pursuit of peace. Compton and those who planned the post-war scientific regime were merely echoing earlier figures like Angell or Willis Whitney, the head of General Electric's Research Laboratory, who claimed that the power of science was an exponential function of the number of laboratory workers! They would have agreed as well with George Perkins, the organizational rhapsodist for US Steel, who believed such large enterprises an evolutionary leap to a new and better species of knowledge.

But Big Science has had its critics. It is, of course, ferociously expensive, demands the planning of experiments years in advance, and frequently acquires an institutional momentum that makes it difficult to redirect, much less shut down, large projects. Moreover, the deep commitments of time and resources have required scientists to become political operators, adept at identifying, mobilizing, and maintaining the constituencies on which their existence depends. Weinberg himself worried that large organizations presented a three-fold threat: the replacement of investigation by spectacle, thinking by the spending of money, and the experimenter by the administrator, and, "unfortunately, science dominated by administrators is science understood by administrators, and such science quickly becomes attenuated if not meaningless."[7] The Space Telescope illustrates some of Weinberg's concerns. Built with the philosophy that 'big is beautiful,' the telescope may indeed enlarge our picture of the universe, but at what organizational price? With the

Hubble up there and astronomers down here, and government, industry, and scientific agencies in between, NASA (the National Aeronautics and Space Administration) needed to establish an institute just to help astronomers get in line to use the instrument. Moreover, the length and complexity of the Hubble's development raise the issue of instant obsolescence; after all, much changed in the thirteen years it took to get the telescope into space, finally, in 1990. Not least, some argue that Big Science threatens the survival of smaller, more creative science; without doubt, it endangers those who have harnessed their careers to large projects. The recent collapse of the Superconducting Supercollider undermined the community of high-energy physicists as well as the regional economies in which it was sited.

Bacon's Promise

From professionalization, the growth of the state, and the corporate adoption of science, to post-World War II Big Science, the last century is a play on the Weberian theme of bureaucratic reason. Within the calculus of structures that have come to define much of American life, meaning has often been a matter of connectedness and control, and progress dependent on organizational growth. However, science is also an engine of profit and a promise of progress. Francis Bacon, the "Lord Chancellor" of the Scientific Revolution, elaborated best what became a defining doctrine of the industrialized West: knowledge is power. Therein lay the mastery of Nature and the source of endless invention and guaranteed improvement in the material conditions of life.

Over the next centuries, the evolving association between science, expanding markets, and entrepreneurial capitalism shaped the industrial world and helped 'the West grow rich.' By the early decades of the century, corporate America had learned that lesson well and was promoting its industrial laboratories as the fulfillment of Bacon's dream. In *Forty Years with General Electric*, company spokesman John Broderick fantasized that Bacon, in the manner of Rip van Winkle, had reawakened after a centuries-long sleep:

> Visiting Schenectady, he would find the research laboratory there very much to his liking; he would see carried on diligently and systematically in its huge buildings the search, only imagined by him, for "a knowledge of causes and the secret motions of things"; he would have the pleasure of mingling with groups of trained men zealously doing sundry chores like the ones which had been assigned to the "fellows" of Salamon's House[sic]; and presently he would be made agreeably muzzy by a veritable feast of the "fruits" of which he had written with a seer's enthusiasm.[8]

Americans luxuriated in the goods and services that spewed forth from industrial laboratories: from the twenties on, they became "muzzy" indeed as inventions mounted as fast as Henry Ford manufactured model T's, from electrical appliances, plastics, and radio, to anti-knock gasoline for those very same cars. Travel agents trumpeted their scientific approach to vacationing, ladies' bridge clubs turned to science to outwit Nature and defeat the odors of perspiration, and even the five-o'clock shadow met its match. In "Science Turns to Shaving," one author wrote that "for the past five years the writer has engaged in a comprehensive study of shaving technics and devices . . .":

> Facilities for the investigation were provided by a Fellowship at Mellon Institute of Industrial Research in Pittsburgh, maintained by the Magazine Repeater Razor Company. In these studies all phases of shaving were examined, from the physiology of the growth of hair to the metallurgy and sharpening of safety razor blades.[9]

William Meggers, a physicist at the National Bureau of Standards, looked around during the twenties and saw "a fever of commercialized science," while the *Nation* remarked that "a sentence that begins 'Science says' will generally be found to settle any argument . . . or sell any article from tooth-paste to a refrigerator."[10] When the Age of Normalcy was shaken by Depression, mainstream Americans were only mildly disturbed by high-culture, anti-scientific diatribes: for most, science continued to promise both the Good Life and good profit.

The most potent example of commodification in science, however, derives not from the feverish consumerism of the twenties and thirties but from the mobilized science of World War II. Vannevar Bush and an establishment elite were able to mobilize American science in a surprisingly rapid fashion because of key decisions made early: they utilized private, rather than federal, facilities whenever possible; lodged key decision-making power in the hands of civilian volunteers rather than with federal bureaucrats, civil servants, or politicians; and conducted the mobilization itself by means of contracts.

This last point must not be underestimated. The contract was more than "the social and legal machinery appropriate to arranging affairs in any specialized economy which relies on exchange rather than tradition . . . or authority . . . for apportionment of productive energy and of product."[11] During the 1930s when universities began to turn for support to the federal government, Bush had argued that the contract was the ideal means to mediate public/private cooperative R&D in a manner that protected the initiative and prerogatives of the private sector. His decision to organize wartime science in a similar fashion quickly eased the apprehensions of industrial leaders about cooperation with the government and helped get R&D rapidly underway. By the end of 1945, OSRD had let with private agencies almost 2300 contracts worth half a billion dollars and contributed greatly to winning the war. In so doing, OSRD promoted military-civilian partnerships that outlasted the war and insinuated the business of science into the sinews of national power. By helping define national defense as an entrepreneurial market, OSRD's strategies promoted the formation of the military-industrial complex that emerged to dominate Cold War America.

Consumerized science was bolstered in the US by the spread of industrial R&D

after the turn of the century, the rise of a culture of consumption in the 1920s, and by its later association with national defense. Most recently, science has become the matter of financial speculation since capturing the attention of Wall Street in the booming economy of the fifties and sixties. Shortly after the war, Georges Doriot, a Harvard professor of industrial management, went to Wall Street seeking financial support for his new venture capital firm, American Research and Development (ARD). Hoping to attract investors willing to risk support for small and inventive startup companies, Doriot found the Wall Street climate inhospitable, with investors content with mature technologies like television and established firms like RCA. By the end of the fifties, things had changed dramatically. The deepening Cold War and the shock of Sputnik ratcheted upwards the technological conflict between the superpowers and loosened Depression-era restrictions on small businesses designed to protect investors. In 1957, ARD provided $70,000 to two young MIT engineers named Ken Olsen and Harlan Anderson to help them start the Digital Equipment Corporation. For venture capitalists, it was the investment that made the industry. Entrepreneurial scientists soon became the darlings of financiers, leading to a proliferation of small start-up companies whose names were "likely to include such words as data, electronics, computer, instrument, control, space, systems, dynamics, research, microwave."[12] The association between national security, venture capitalists, and entrepreneurial scientists proved potent, helping revolutionize the post-war economy and leading, eventually, to high-tech geographies like Silicon Valley in California and the Rte. 128 circle around Boston.

Taking Up Arms

Until the turn of the century, government had little concern with science beyond the modicum demanded by its regulatory responsibilities and the imperatives of geography and expansion; and, despite precocious involvements during the First World War with radio, gas warfare, sound-ranging, submarine detection, and aeronautics, the military establishment had even less concern. World War II changed that attitude profoundly. After all, science and the bomb had helped win the war in dramatic fashion. As the Cold War intensified, the link between science and national security became an article of faith. Vannevar Bush, the leader of wartime science, expressed it most powerfully in his best-selling manifesto *Modern Arms and Free Men,* published in 1949 just as news spread of the first Soviet A-bomb test: the nation, he wrote, faced a historic struggle between two opposing ideologies, one totalitarian, one free; the future would belong to the people who harnessed most effectively the resources of science.

War-born adulation was a tonic. Karl Compton, the president of MIT and a member of OSRD's inner circle, noticed new respect even before the end of the war when he found himself in Hawaii, the day Hiroshima was bombed, lecturing MacArthur's soldiers on nuclear physics. After the war, with new-found clout and a federal sponsor, scientists reshaped the balance of power in universities and other institutions. Recalling this new confidence, the physicist Robert Wilson remembered a story making the rounds about

> a typical prosaic professor of physics who had lived in academic squalor in an inland state university, bullied in turn by his department chairman, by his dean, and by his president. He had been drafted into a big wartime laboratory; the story concerns his return to the university. For the first time in his academic career he attended a university meeting, but not to play the part of a cowering back bencher. Rather he came in late, walked right up to the presiding president and slapped down his demands. These were that a new laboratory in a new building be set up with him as director, that ten new faculty positions be created to staff it, that adequate funds be provided to match his already promised government monies, that his salary be quadrupled, and that his teaching be cut to zero so he could devote to the laboratory what little time would remain after his tight schedule of consulting in Washington. The president accepted all these demands on the spot — with a speech of gratitude — after which our friend turned on his heel and walked out to a rising ovation of the faculty. Not long after, his wife became pregnant — with triplets, all boys![13]

Federal agencies and committees for science proliferated and budgets, after only a brief lapse from wartime levels, began to climb steeply, spurred by the Korean War, the Communist Revolution in China, and the launching of Sputnik in 1957. At a new generation of defense-inspired institutions — from the system of national laboratories managed by the new Atomic Energy Commission, 'think tanks' like the Air Force's civilian RAND Corporation, numerous defense-oriented contract labs at universities, like MIT's Lincoln Labs and the Applied Physics Laboratory at Johns Hopkins, to the Office of Naval Research, the National Science Foundation in 1950, the President's Science Advisory Committee in 1957, and NASA in 1958 — the urge for security and the new macho spirit of science erected the scaffolding of a Cold War scientific regime.

The new appreciation for science revolutionized military notions of preparedness, with enormous economic, political, and scientific consequences. Realizing that the outbreak of the next war might not leave the nation enough time to mobilize after the fact, in the traditional manner, the military opted for a state of permanent readiness. Out of that change in military thinking, and building on the pervasive civilian-military network that OSRD created for wartime mobilization, came the military-industrial-university complex and arsenals of science in which weapons-systems became the largest of the nation's public works.

The influence of the Department of Defense was wide-spread. First, it encouraged the formation of high-tech client firms, among them companies like Engineering Research Associates, which developed some of the earliest computers, and Thompson-Ramo-Wooldridge, the company that was the primary designer and manufacturer of intercontinental ballistic missiles. Second, it promoted the development of new technologies themselves like the computer, the transistor, and the laser in military and, especially, corporate facilities like AT&T's Bell Laboratories.

Third, it underwrote the manufacturing and assisted the diffusion of these technologies throughout industry, as it did with Bell Lab's transistor, and provided stable, guaranteed markets for their consumption. And fourth, it subsidized scientific disciplines relevant to service missions, as with the navy and oceanography, helping build up, in effect, reserve labor pools of scientists who could be called on for defense work in times of national need. By 1985, the US was spending more per capita on defense than any other Western nation, had become the world's major arms supplier, surpassing even the Soviet Union, and had evolved a defense establishment ready, because of the exigencies of technology, to declare and conduct a war *in advance* of congressional approval.

George Ellery Hale once declared the Great War a great opportunity for American science; his World War II descendants jockeyed even more vigorously to take advantage of new largesse, with schools like MIT, Caltech, and Johns Hopkins shaping programs and molding institutional identities with the new federal sponsor in mind. Traditional career paths swerved dramatically as ambitious scientists moved into large defense-oriented laboratories, acquiring specializations, experimental skills, and criteria of good practice carefully adapted to Cold War institutional niches. In some of these Cold War sites like Los Alamos or the Lawrence Livermore National Laboratory where nuclear weapons are designed, the constraints of secrecy, tightly-compartmentalized experimental practice, and carefully controlled communication between workers inside and outside approved groups have engendered a new form of scientific life with few connections to the wider community and possibly diverging standards and vision.

Yet the Cold War influenced more than the economy of science or even its workplaces and practice. Transformed, as well, were basic commitments and professional loyalties that defined scientific self-identity. Robert Oppenheimer is a case in point. Once the hero of the Manhattan Project, an influential spokesman for science, and a powerful figure within the Atomic Energy Commission, Oppenheimer was taken to task by a farcical court, indicted for early flirtations with communism and opposition to the hydrogen bomb, declared a security risk, and stripped of his role as a government advisor. Oppenheimer was no traitor, the trial board declared, but his "loyalty" did not meet contemporary standards: "The premise of the concrete, contemporary definition of loyalty is the fact of the Communist conspiracy."[14] The Oppenheimer case was no simple spasm of paranoid McCarthyism in an otherwise tolerant time. In the charged atmosphere of the 1950s, the government inexorably narrowed its criteria for "loyalty" from actual evidence of disloyalty, to doubt, to clear consistency with national interests by 1953, and even the prestigious National Academy of Sciences backed away from the defense of embattled scientists. Overt intolerance drove from careers and positions of influence many less famous scientists, men like Edward U. Condon, the director of the National Bureau of Standards, or Dirk Struik, a respected MIT mathematician. But the influence of new loyalties could be much more subtle and insidious.

Loyalty oaths were not, at times, even necessary when contractors and others internalized the "contemporary definition" their federal sponsor expected. As spokesmen for the Office of Naval Research put it:

> ... the arrangements we have found workable are as follows: The contractor is entirely free to publish the results of his work, but ... we expect that scientists who are engaged on projects under Naval sponsorship are as alert and as conscientious as we are to recognize the implications of their achievement, and that they are fully competent to guard the national interest.[15]

Americans have always been fascinated by weapons. An early observer noted that the colonists were "the greatest weapon-using people of that epoch in the world."[16] Indeed manufacture with interchangeable parts — unveiled as the "American System" at the 1851 Crystal Palace Exposition in London — was achieved by federal armories in the pursuit of a dependable musket. But the interest goes beyond mere utility and draws on the nation's hopes for technology, its commitment to the control of Nature, and an uneasiness with the Other that borders, at times, on paranoia. Ronald Reagan's (and Edward Teller's) Star Wars defense is only the latest manifestation of the pursuit of the 'winning weapon' that reaches back through General Westmoreland's Vietnam-era electronic battlefield and the atomic bombs of World War II, to the ironclads of the Civil War. The fascination with weapons has shaped the American imagination and studded its discourse with metaphors of science and war. One thinks of the contemporary wars on drugs, crime, and poverty, and of President Nixon's 1970s war on cancer. The tendency is long-standing. Energized by a booming economy, and inspired by the scientific wizards of the Great War, the bankers of the twenties declared that in "commercial warfare ... research supplies the ammunition."[17] In 1929 there appeared a comic strip that soon became enormously popular, lasting until 1967 and being published by 450 newspapers. It told the story of a young American combat pilot who went to work for a mining company shortly after the Great War. Trapped by a cave-in and preserved by a mysterious gas, the young flier slept for five centuries until he awoke and climbed out, as Buck Rogers, into the precarious twenty-fifth century. It was a future of space flight and antigravity, of radioactive gases, repeller rays, smart weapons, and feminist soldiers, and a garrison state desperately mobilized to thwart the red Mongol hordes that had used advanced science and technology to conquer North America and reduce its cities to glassy slag. All in all, the strip was a pretty fair adumbration of America's future Cold War psyche.

A New Epoch?

American science can be divided into three epochs. The first stretched from the 1870s through the First World War and witnessed the advent of the science-based university and the professionalization of the scientific community generally. The second reached from the end of World War I to the outbreak of World War II and was dominated by the burgeoning of industrial R&D and the commercialization of science. The third originated in World War II and has been marked by the maturation of a regime powerfully driven by the imperatives of national security. Each epoch built synergistically on the achievements of the earlier, producing by the latter half of the century a powerful scientific establishment and culture.

American science may be passing out of this third epoch. Certainly, the collapse of the Soviet Union and the ending of the Cold War have shifted the dynamics of the scientific establishment, affecting budgets and institutional missions and necessitating the redirection of massive sectors of the economy. The result, according to some, has been a dangerous erosion of leadership in key military and civilian technologies as the nation has struggled to remain competitive in the global economy. Moreover, the shock waves generated by the weakening of the Cold War partnership between government, industry, and academia, have aggravated long-standing, if minority, apprehensions about the nation's technoscientific culture that antedate the Soviet collapse, extending back through Vietnam-era worries about the militarization of science and the hidden costs of technology generally, to nineteenth century Romantic opposition to scientific reason itself.

Mainstream science has always had critics. What makes current commentary especially corrosive is the very media by which, in the past, science built the foundations of powerful public support. Well-publicized episodes of alleged misconduct call into question the ethical standards of scientific behavior, and the 'single voice' of scientific authority has devolved, at times, into a babble of conflicting testimony over almost any issue of public importance, from the siting of nuclear power plants to the interpretation of evidence in sensationalist trials. This portrayal of experts at loggerheads, as concupiscent as the patrons for whom they have labored, suggests to many a science no longer transcendent, with authority dependent not so much on the truths of Nature as on the superior mobilization of persuasive resources. How this shifting image will affect the nation's dependence on experts and the public's support for a scientific establishment no longer wedded to Cold War survival is an open question. Some things seem clear, however. Organizational expansion has its limits. The prosperity promised by science has its costs. And the patronage offered by the state demands its peculiar loyalties. Moreover, the innocent association between science and power which inspired, for much of the century, a trust in disinterested expertise (what one historian has called the American "escape from politics"[18]) is no longer sacrosanct. Whatever adjustments shape the new epoch, the transition is sure to be slow, ponderous, and probably uncomfortable, for the momentum generated by cultural investments in science over the last century has been large, and the interplay between its shaping elements of organization, commodification, and militarization, intricate and mutually supportive.

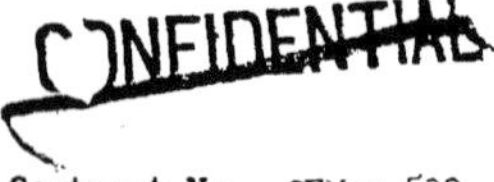

This document contains information affecting the national defense of the United States within the meaning of the Espionage Act, U S C 50; 31 and 32. Its transmission or the revelation of its contents in any manner to an unauthorized person is prohibited by law.

Contract No. OEMsr-530

MEMORANDUM OF AGREEMENT made this 21st day of July 1942, effective as of the **15th** day of **April** 1942, between THE UNITED STATES OF AMERICA (hereinafter called "the Government"), represented by the Executive Secretary (hereinafter called "the Contracting Officer"), Office of Scientific Research and Development in the Office for Emergency Management, Executive Office of the President, and **Independent Engineering Company, O'Fallon, Illinois,** -

- -

(hereinafter called "the Contractor").

WHEREAS, the Contractor conducts and maintains an experimental testing and research laboratory or laboratories and the Government desires that the Contractor conduct studies and experimental investigations in connection with **the development and building of oxygen rectification apparatus and similar oxygen equipment, and report the results thereof;** - - - - - - - - -

- -

- -

- -

NOW, THEREFORE, THIS AGREEMENT WITNESSETH:

1. The Contractor agrees, during the period commencing April 15, 1942, and ending August 31, 1942 (both dates inclusive), to furnish the necessary laboratory facilities and skilled technicians for and to conduct, with the utmost secrecy and dispatch, in accordance with instructions issued and designs furnished by the Contracting Officer or his authorized representative, studies and experimental investigations in connection with the development and building of oxygen rectification equipment and other oxygen equipment of similar character. The Contractor shall report the results of its investigations from time to time as requested by the Contracting Officer or his authorized representative, and upon termination of the period specified above shall deliver to the authorized representative of the Contracting Officer such equipment as may be developed hereunder, together with a final report of its findings and conclusions. The authorized representatives of the Contracting Officer for the purposes of this paragraph include the Chairman of Division B, the Technical Aide to such Chairman, and the Chairman of Section 7 of Division B, of the National Defense Research Committee of the Office of Scientific Research and Development.

FIGURE 42.1: THE FIRST PAGE OF A TYPICAL OSRD R&D CONTRACT. NOTE THE STANDARDIZED LANGUAGE CONTRACTING FOR 'EXPERIMENTAL' INVESTIGATIONS. (SOURCE: NATIONAL ARCHIVES, RG227)

TABLE 42.1: THE TOP CONTRACTORS FOR OSRD-SPONSORED SCIENCE IN THE SECOND WORLD WAR.

Contractor	Total	Obligated
MIT	$116,715,319.00	$116,383,305.67
Caltech	$86,249,330.35	$86,249,306.81
Harvard University	$29,667,829.40	$29,303,243.18
Columbia University	$27,183,486.99	$27,183,486.99
Western Electric Co.	$16,393,127.62	$15,184,627.42
University of California	$14,793,631.39	$14,636,110.03
Research Construction Co.	$13,662,000.00	$12,933,887.52
Johns Hopkins University	$10,750,981.15	$10,750,981.15
General Electric Co.	$8,196,861.08	$7,599,644.75
George Washington University	$6,901,863.00	$6,904,863.00
RCA Manufacturing Co.	$5,874,491.00	$5,706,941.60
E.I. du Pont de Nemours	$5,853,330.00	$5,372,758.32
University of Chicago	$5,828,826.88	$5,709,806.94
Westinghouse Elec. & Mfg.	$5,337,713.51	$4,561,045.12
Remington Rand, Inc.	$4,675,050.00	$3,731,601.02
Monsanto Chemical Co.	$4,536,964.00	$4,508,168.80
Eastman Kodak Co.	$4,498,871.29	$4,259,843.23
Zenith Radio Corporation	$4,175,000.00	$4,175,000.00
Princeton University	$3,703,568.58	$3,561,394.76
National Academy of Sciences	$3,555,587.83	$3,135,010.96
Standard Oil Development Co.	$3,424,056.80	$2,925,683.91
Hygrade Sylvania Corp.	$3,193,497.35	$3,084,331.38
University of Pennsylvania	$3,048,137.94	$2,865,697.54
Northwestern University	$2,737,766.14	$2,640,071.65
Erwood Sound Equipment Co.	$2,690,000.00	$2,690,000.00
Carnegie Institute of Tech.	$2,626,373.23	$2,477,186.42
Douglas Aircraft Co., Inc.	$2,592,500.00	$2,477,500.00
Carnegie Institution of Washington	$2,549,185.00	$2,442,500.00
State University of Iowa	$2,339,175.00	$2,332,458.94
Budd Wheel Co.	$2,330,000.00	$2,313,710.44
University of Illinois	$2,283,557.27	$2,240,321.78
University of Michigan	$2,241,359.00	$2,165,192.71
Woods Hole Oceanographic Inst.	$2,035,000.00	$2,035,000.00
Franklin Institute	$1,997,025.00	$1,980,672.84
University of Rochester	$1,931,024.91	$1,747,399.67
Evans Memorial Hospital	$1,869,058.00	$1,863,278.78
Gulf Research & Development Co.	$1,583,950.00	$1,570,719.90
Delta-Star Electric Co.	$1,451,438.92	$1,451,438.92
Emerson Radio & Phonograph Corp.	$1,333,500.00	$1,321,946.85
New York University	$1,328,454.00	$1,317,788.37
M.W. Kellog Co.	$1,280,000.00	$1,206,380.53

(Source: Larry Owens, "The Counterproductive Management of Science in the Second World War")

REFERENCES

1. Wiener, Charles. "Physics in the Great Depression," *Physics Today* (October 1970), p. 33.
2. *New York Times* (January 3, 1943); *New York Times Magazine* (September 2, 1945), p. 4.
3. Whyte, William A. *The Organization Man.* (New York: Simon and Schuster, 1956); James Angell, "The Organization of Research," *Scientific Monthly* (1920), **11**: 26–27; 37.
4. Perkins, George. *The Modern Corporation* (1908) — Perkins was associated with US Steel and a partner of J.P. Morgan; Ervin Laszlo, *The Systems View of the World: The Natural Philosophy of New Developments in the Sciences* (1972), p. 79.
5. Hendrick, Burton. "Taking the American City Out of Politics," *Harper's Magazine* (June, 1918).
6. Weinberg, Alvin. "Impact of Large-Scale Science in the United States," *Science* (1961), **134**: 161.
7. Weinberg, pp. 161–62.
8. Broderick, John T. *Forty years with General Electric* (1920), pp. 98–99.
9. Casselman, Elbridge. "Science Turns to Shaving," *Scientific American* (November 1937), 261ff.
10. Meggers and Nation quotes come from Spencer Weart. "The Physics Business in America, 1919–1940: A Statistical Reconaissance," in Nathan Reingold (ed.), *The Sciences in the American Context: New Perspectives* (1979).
11. See Roscoe Pound and K.N. Llewellyn. "Contract," in *The Encyclopedia of the Social Sciences* (1931), Edwin R.A. Seligman and Alvin Johnson (eds.).
12. Teitelman, Robert. *Profits of Science.* (New York: HarperCollins, 1994), p. 122 and 127.
13. Wilson, R.R. "US Particle Accelerators at Age 50," *Physics Today* (1981), **34**: 92–93.
14. The quote is from the concurring opinion of the AEC Commissioner Thomas Murray — see *In the Matter of J. Robert Oppenheimer . . .*, p. 1059.
15. Forman, Paul. "Behind Quantum Electrodynamics: National Security . . . ," *Historical Studies in the Physical and Biological Sciences* (1987), **18(1)**: 149–229.
16. The remark is quoted in Alex Roland. "Science and War," *Osiris* (1985), **1**: 247–72; p. 248.
17. The quote comes from Spencer Weart. "The Physics Business in America, 1919–1940: A Statistical Reconnaissance," in *The Sciences in the American Context: New Perspectives.* (Washington, D.C.: Smithsonian Institution Press, 1979), Nathan Reingold (ed.), p. 302.
18. Ezrahi, Yaron. *The Descent of Icarus: Science and the Transformation of Contemporary Democracy.* (Cambridge, MA: Harvard University Press, 1990).

FURTHER READING

Kevles, Daniel. *The Physicists: The History of a Scientific Community in Modern America.* (New York: Alfred A. Knopf, Inc., 1978).

Haskell, Thomas. *The Emergence of Professional Social Science.* (Urbana: The University of Illinois Press, 1977).

Hughes, Thomas. *American Genesis. A Century of Invention and Technological Enthusiasm.* (New York: Viking, 1989).

Mowery, David and Rosenberg, Nathan. *Technology and the Pursuit of Economic Growth.* (New York: Cambridge University Press, 1989).

Noble, David. *America by Design: Science, Technology, and the Rise of Corporate Capitalism.* (New York: Alfred A. Knopf, Inc., 1977).

Owens, Larry. "The Counterproductive Management of Science in the Second World War: Vannevar Bush and the Office of Scientific Research and Development," *Business History Review* (1994), **68**: 515–576 and "MIT and the Federal `Angel': Academic R&D and Federal-Private Cooperation before World War II," *Isis* (1990), **81**: 189–213.

Hooks, Gregory. *Forging the Military-Industrial Complex: World War II's Battle of the Potomac.* (Urbana: University of Illinois Press, 1991).

Sherry, Michael. *In the Shadow of War: the United States Since the 1930s.* (New Haven: Yale University Press, 1995).

Leslie, Stuart W. *The Cold War and American Science: The Military-Industrial-Academic Complex at MIT and Stanford.* (New York: Columbia University Press, 1993).

Mukerji, Chandra. *A Fragile Power: Scientists and the State.* (Princeton: Princeton University Press, 1989).

Galison, Peter and Hevly, Bruce (eds.). *Big Science: The Growth of Large-Scale Research.* (Stanford: Stanford University Press, 1992).

John R. Sutton. "Organizational Autonomy and Professional Norms in Science: A Case Study of the Lawrence Livermore Laboratory," *Social Studies of Science* (1984), **14**: 197–224.

Ian Hacking. "Weapons Research and the Form of Scientific Knowledge," Canadian Journal of Philosophy (1986), **12(Suppl.)**: 237–260,

Thomas Misa. "Military Needs, Commercial Realities, and the Development of the Transistor, 1948–1958," in *Military Enterprise and Technological Change.* (Cambridge: The MIT Press, 1985), M.R. Smith (ed.).

Jessica Wang, "Science, Security, and the Cold War: The Case of E.U. Condon," *Isis* (1992), **83(2)**: 238–269.

CHAPTER 43

Science in Latin America

HEBE M.C. VESSURI

National scientific communities have emerged in Latin America during the twentieth century, under a cloud of tension between the need to join the international scientific community and the desire to achieve an independent voice, i.e., autonomy in the definition of their role and interests. The state of science in the region is highly inconsistent. In some countries there are government-supported laboratories and research teams which command international recognition. In others, the essential infrastructures of education and training are lacking. Often funding is meager and insufficient to maintain even the minimum scientific capability. There are not enough researchers to respond to government programs. Scientists are discouraged by the lack of incentive and poor conditions of work. At the same time, there is widespread criticism that much scientific work is trivial, and the ideology of 'applied science' with which pressure is often exerted on the scientific community, often masks and consolidates mediocre research capacity.

At a time when scientific knowledge means industrial opportunity, most Latin American countries have yet to develop a consensus about the role of science. Should governments support basic science? Should they compete with the research agendas in the United States and Europe? What kind of science, if any, should be funded? Or should budgets be spent on the development of badly needed technologies?

Caught between changing policies and inconsistent budgets, between the desire for economic development and international recognition, science in Latin America is as marginal as ever. In this chapter I try to convey something of the compromises, challenges and restrictions that accompanied the scientific endeavor in this particular region of the world in the twentieth century.

ORDER AND PROGRESS: POSITIVISM BETWEEN TWO CENTURIES

Towards the end of the nineteenth century, many countries initiated a process of economic and political modernization. Education, science, European immigration and foreign capital were the main tools. Exports were expanded and power showed with the civil oligarchies that had emerged in the previous half century. European positivism, with its strong belief in progress, offered politicians and intellectuals a conceptual scheme that combined knowledge of history, science and society. It reinforced the importance of order and stability, by contrast with the civil strife endemic since independence in the early part of the century. To such an order political liberties were sacrificed as unnecessary and perturbing. The unrelenting dynamism of the Northern neighbor and European powers gave a feeling of urgency to the social changes required. In the words of Mexican Justo Sierra (1880):

> We need to become stronger, otherwise, incoherence will increase, the organism will not be integrated, and this society will abort. We would remain defenceless and would be the weakest in the struggle for life Darwin speaks about. While we destroy ourselves, at our side lives a marvellous collective animal, for whose huge intestine there is not enough nutriment, armed to devour us. Confronted with this Colossus we are exposed to be a proof of Darwin's theory, and in the struggle for existence all the odds are against us.[1]

European philosophers like Comte, Stuart Mill, and Spencer had a great influence in Latin America in the late nineteenth century. The ideas of the British positivists would inspire liberal groups, while in Brazil, orhodox Comtians tried to carry through all the ideas of their teacher, including religious positivism, and declared Rio de Janeiro, that had founded the first and only Church in Iberoamerica, the orthodox center of universal positivism.[2] Many republicans, in their turn, supported the early, scientific Comte, particularly in educational reform and scientific ideology. In his major treatises, *Course of Positive Philosophy* (1830–1842) and *System of Positive Polity* (1851–1854) as well as in a number of lesser works, Comte developed his positive philosophy and his sociological theories, and set forth his plan for an ideal society, inspiring the minds longing for order and science in Latin America.

At this time, the number of qualified scientists was minimal and there was an almost complete lack of anything resembling a research position. In the universities, science was subordinated to liberal arts. It was taught merely to discipline the mind and most courses did not go beyond a most elementary level. Students learnt maths and physics, not to become scientists and engineers, but as part of a good education. With a few exceptions, science was taught, not by experiment in the laboratory, but by reading, recitation and repetition. In a continent where education was controlled by the Church, it is not surprising that most teachers were priests with a greater commitment to hierarchical obedience than to free enquiry.

Although positivism promoted a social appreciation of science as a source of progress and practical knowledge, it remained largely rhetorical, only becoming embodied in a true research effort in exceptional cases. Florentino Ameghino and Eduardo Holmberg in Argentina, Luis Razetti in Venezuela, Justo Sierra and

Alfonso Herrera in Mexico are some of the scientists influenced by positivism. In specific disciplines there were original developments in the production of texts, as in the explanation of infinitesimal calculus attempted by Diaz Covarrubias, Gargollo and Ramirez in Mexico; the geological and paleonthological works by Ameghino and Burmeister and the astronomical contributions by Gould in Argentina; and the texts by von Ihering and Goeldi in the natural sciences in Brazil.

Comparison, classification, and generalization became the goals of natural scientists who, funded by museums and other European, American and local institutions, made collections of flora, fauna and human diversity. After having fed the European public and private collections with exotica for many years, a number of Latin American museums of natural history had by the end of the century managed to build remarkable collections, both for their quantity and quality, with many irreplaceable and unique holdings. The La Plata Museum fossils were praised in the 1890s by the British Museum curator, for example, as "extraordinary and of world-wide interest."[3] Quite independently, numerous expeditions to the interior were funded by national and international interests to evaluate natural resources. The possibilities of commercial success supported a growing literature on the flora and fauna of the region.

A real demand for science emerged in the first years of the century as the need for graduates of law, medicine, and engineering increased. Scores of students with scientific ambitions left to study abroad, particularly in France, Germany and the United States. Brazilian microbiologist Oswaldo Cruz, who studied at the Pasteur Institute in Paris, and subsequently built up the Manguinhos Institute in Rio de Janeiro, was typical of the new generation. The original staff of this institute which rapidly acquired international recognition was entirely Brazilian, including future leaders of Brazilian medical science such as Carlos Chagas, Henrique de Rocha Lima and Artur Neiva.

During the first half of the twentieth century, the Southern Cone countries had an advantage over the rest (with the sole exception of Costa Rica in Central America) in having a literate population. Argentina, Uruguay, and to a lesser extent Chile, instituted programs of primary and secondary education for the entire population. Inherited positivist values of secularism, good citizenship, republicanism and scientific veracity were reflected in the school curriculum.

The Foundations of Experimental Science: 1918–1940

Progress remained an illusion as long as stability and order did no more than maintain the *status quo.* The enthusiasm for positivism waned and the allure of an advanced scientific Europe was undermined by the atavistic destructiveness of the First World War.

The inter-war period witnessed a deep transformation of Latin American societies marked by workers' strikes and student revolts. A new stage of political organization of the workers saw the emergence in most countries of communist and socialist parties. A revitalization of Catholic thought was also visible in the reassertion of religious education. Several national armies became professional.

The advancement and prosperity of a new middle class opened a new market for authors, stimulating the expansion of a publishing industry. This growth in the publishing field was a crucial factor in the professionalization and autonomy of intellectual work in Argentina, Brazil and Mexico. Although there were already scientific journals published in these countries before 1899, the first decades of the twentieth century saw a marked growth of periodical publications by learned institutions and scientific societies.

In 1918 Buenos Aires was the second largest Atlantic city after New York. Except for import and distribution trade centers like Holland and Belgium, no other country in the world imported as many goods per capita. The old Argentine universities were ripe for change as demonstrated by the Córdoba Reform Program of 1918, to which most of the Latin American university communities adhered. With the help of a group of German physicists and astronomers the *Universidad Nacional de La Plata* became one of the best centers of Latin American science.

In Mexico, after the first post-revolutionary decade, the Mexican state was restructured unifying the country ideologically within a nationalist model. The *Universidad Nacional* was established on a different base from that of the former Royal Pontifical University. Most of the new research institutes were created within the national university. Although several disciplinary institutes antedated it, the Science Faculty was founded in 1939.

In Brazil, the Sociedade Brasileira de Ciencias, founded in 1916 and transformed into the Brazilian Academy of Sciences in 1922, had as its main aim the development of basic sciences. In the same vein, some engineers, mathematicians, astronomers and natural and physical science teachers in Rio de Janeiro claimed the creation of a Higher Faculty of Sciences devoted exclusively to the education of scientists, without any commitment to technical or professional training. The argument was that in new countries like Brazil, utilitarism and pragmatism, associated with the positivistic tradition in the local context, degenerated and were transformed into a fanaticism of material progress.

The noticeable development of a discipline in a particular country today is often the result of efforts started much earlier, as in the case of geology and geophysics in Peru. Although linked from an early date to government interests in mining, it experienced a qualitative change in 1922 when the Carnegie Institution from Washington installed a Magnetic Observatory in Huancayo, thus beginning the local systematic register of geophysical information. Although the original concern was to know the origin of the Earth's magnetic field, the range of interest gradually broadened to other geophysical parameters, and the Observatory won world renown because of the quality of the data from its unique geographical location.

Agricultural research in most countries had an early start, aiming at enhancing the economic competitiveness of their staples, although the levels attained differed remarkably. The Argentine Rural Society, founded by a group of cattle-breeders in 1866, had already created an Agricultural Institute with experiment fields by 1870 and inaugurated the first agricultural-cattle-breeding exhibition in 1875. In Uruguay a Rural Association came into being in 1871 to bolster the modernization of the agricultural sector. The Campinas Agronomical Institute in Brazil was founded in 1887 in order to study tropical plants, particularly coffee, corn and tobacco. Applied institutes such as these, in most cases led by European or North American researchers, were the roots of the agricultural scientific traditions that would develop later, when National Institutes of Agricultural Technology, in which agricultural basic science was also pursued, were created in several Latin American countries.[4]

Small countries sometimes had a single institution influencing national scientific life, as was the case of the *Universidad de la República* in Uruguay where the introduction of the notion of academic research was due mainly to Clemente Estable, who left a deep imprint in the local evolution of biology. In the late 1920s he was the focal point for the founding core of the current *Instituto de Investigaciones Biológicas Clemente Estable,* devoted to basic research. By contrast, the authoritarian regime of Juan Vicente Gómez (1908–1927) in Venezuela, kept the country in a *sui generis* process of deep repression at the political level, penetration of monopolistic capital in the oil sector, and unification and centralization of the national territory. The small amount of scientific-technical activity carried out was directly linked to concerns of a practical nature, lagging behind other Latin American countries. The universities faced problems with the regime and remained open only sporadically.

In this period the foundations of experimental science were laid in several countries, with a marked influence of foreign professionals and the institutional cooperation of advanced countries.France created the "*Groupement des Universités et Grandes Ecoles de France pour les Relations avec l'Amérique Latine*" in 1907. While not completely indifferent to the 'scientific needs' of Latin America, it had closer contacts with diplomacy than with the universities, and with the humanities and the social sciences rather than the exact and experimental sciences. Without completely abandoning scientific exchanges, the *Groupement* gradually ceased to be a 'scientific-cultural' project to become a 'diplomatic-cultural' one. Two of its activities stood out during the 1920s: the creation of French cultural institutes in Latin America and the journal *Revue d'Amérique Latine.* The *Universidade de São Paulo,* founded in 1934, and especially its Faculty of Sciences and Letters, was probably the most important Latin American knowledge institution in whose creation and early life the French were strongly involved.

The United States, in their turn, were keen to consolidate an empire that extended from Puerto Rico through a large part of Central America to the Philipines. The State Department was supported in its Pan American policies by the largest firms, foundations and educational institutions. Between 1913 and 1940, the Latin

American activities of the Rockefeller Foundation concentrated on public health and the control of epidemics.[5] It also supported physiological research, particularly in Argentina, as a result of the high quality scientific work undertaken there, such as the 1947 Nobel Prize awarded to Bernardo Houssay, director of the *Instituto de Fisiología* of the University of Buenos Aires, for his research on the glandular basis of sugar metabolism.[6] Beginning around 1940 and coinciding with the interruption of scientific relationships between the United States and Europe produced by the Second World War, the Rockefeller Foundation expanded its interest in Latin America. It emphasized scientific education and the support of individual scientists in other physical and natural sciences and in basic and applied agricultural research, contributing, with its Mexican Program, to the Green Revolution.[7]

Germany had made substantial incursions into Latin America at the beginning of the century, particularly Argentina. With the active support of the Imperial Ministry of Foreign Affairs, German science and culture were implanted in Argentina in direct competition with North American interests. The development of physics at La Plata University was entrusted to Emil Bose, one of the first students of Walther Nernst's Physical Chemistry Institute in Göttingen. Bose's premature death in 1911 did not end La Plata's project,[8] that was pursued between 1913 and 1926 by Richard Gans, an assistant to Nobel Prize winner Ferdinand Braun who had already made a brilliant career in Tübingen and Strasbourg. Although a lack of interested students made it difficult to establish a local research base and the school produced many more engineers than physicists or astronomers Gans supervised the first six physics theses at an Argentine university. His most distinguished Argentine pupil, Enrique Gaviola, was awarded a doctoral degree in Berlin in 1926, studying with top figures such as James Franck, Max Born, Max von Laue, Max Planck and Einstein. Afterwards he spent some time in Johns Hopkins University and the Carnegie Institution. His work won recognition in Europe and North America but on his return to Argentina in 1931 he met with difficulties in adjusting to the petty politics of local academic circles. Later he had unsurmountable political disagreements with president Peron and the military over the national atomic endeavor,[9] and his valuable scientific know-how was thus not fully assimilated by the country of his birth.

In the early decades of the century, Spain reinforced its links with Hispanic America. The *Institución Cultural Española*, created in 1914 at the initiative of the Spanish colony in Argentina, aimed to make Spanish science and literature known in Argentina by means of a chair in the University of Buenos Aires to be occupied by Spanish researchers, and the development of other intellectual exchanges. The *Institución* was placed under the scientific auspices of the *Junta para Ampliación de Estudios e Investigaciones Científicas* in Madrid, presided over by neurohistologist Santiago Ramón y Cajal, who had won the 1906 Nobel Prize for medicine and was the scientist with the greatest prestige in Spain at the time. Well-known Spanish intellectuals benefited from invitations from the Spanish Cultural Institution to travel to America.

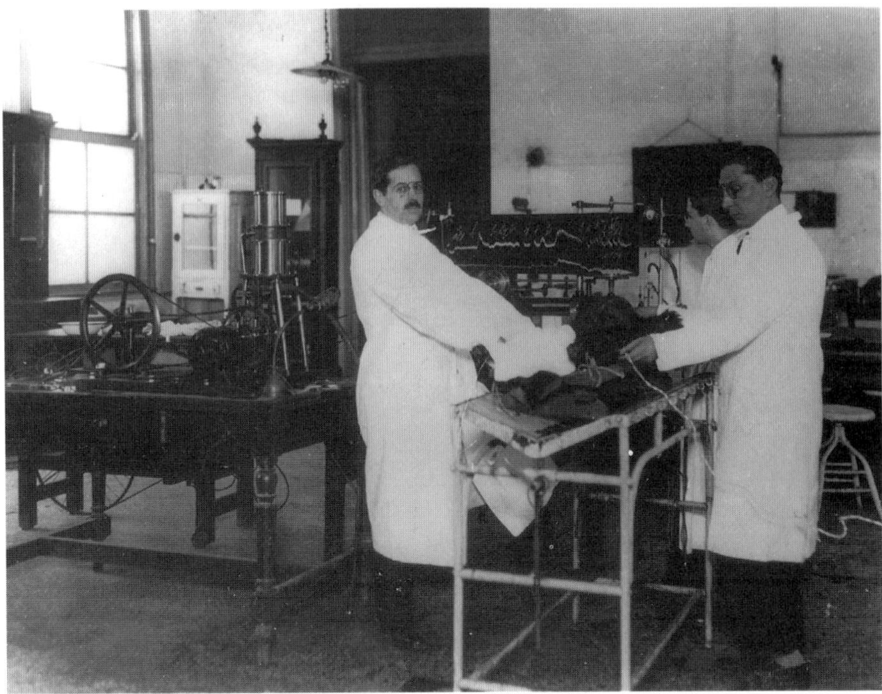

FIGURE 43.1: BERNARDO A. HOUSSAY AND COLLABORATOR DURING EXPERIMENTAL WORK AT THE INSTITUTE OF PHYSIOLOGY, BUENOS AIRES, AROUND 1936.

FIGURE 43.2: MORE READING-ROOM SPACE FOR STUDENTS AT THE FACULTY OF MEDICINE, BUENOS AIRES, IN THE 1920S.

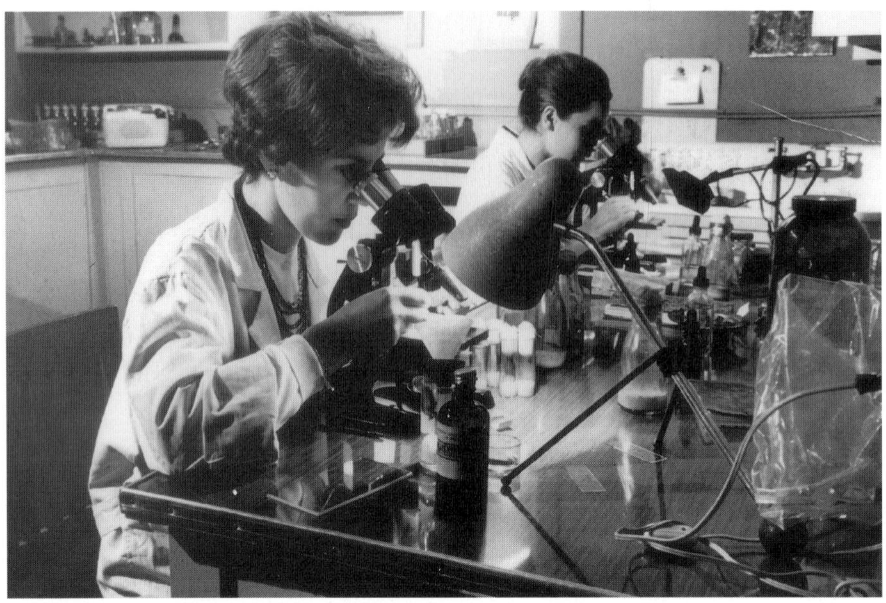

FIGURE 43.3: STUDIES OF EXPERIMENTAL EVOLUTION. UNIVERSITY OF CHILE, 1960.

FIGURE 43.4: VACUUM CREATING UNIT, USP, PHYSICS.

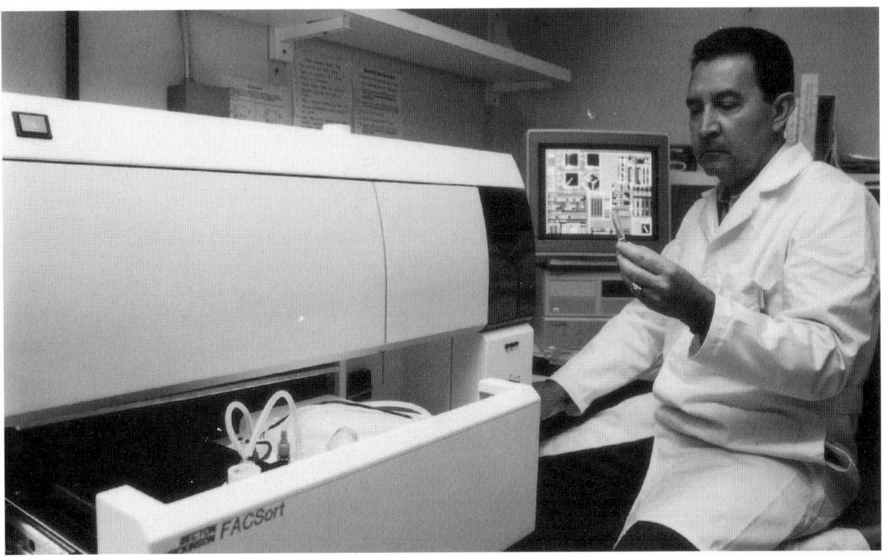

FIGURE 43.5: J.A. O'DALY, RESEARCHER ON *LEISHMOMIESIS* AT HIS IMMUNOBIOLOGY LAB., IUIC, VENEZUELA.

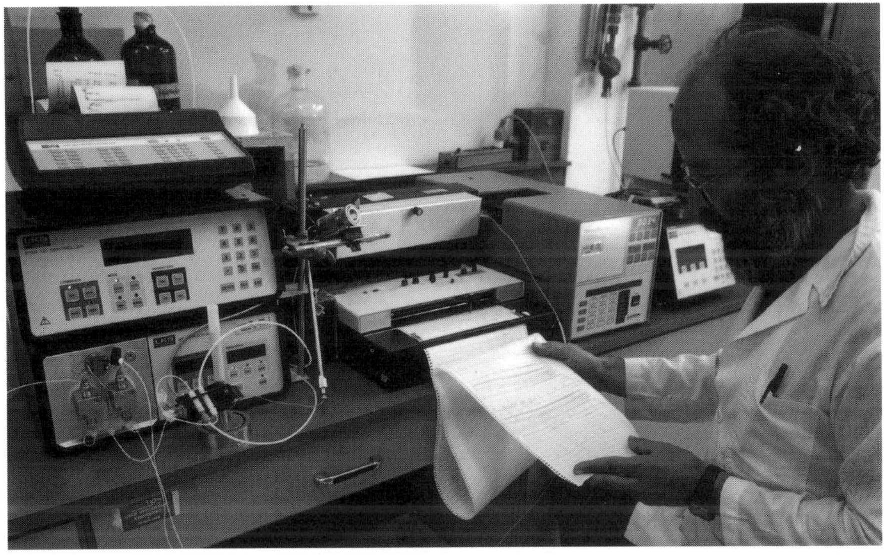

FIGURE 43.6: NEUROLOGY LABORATORY, IUIC, VENEZUELA.

But more important than these official initiatives was the mass collective contribution of the thousands of refugees from the Spanish Civil War who settled in Latin America. Many Spanish scientists and intellectuals, joining other mainly Jewish refugees from Nazism, played a crucial role as catalysts of an institutional transformation whose first phase was completed in the 1950s when substantial numbers of young Latin American scientists, trained mainly in the United States, began to be active in the region.

Italy also contributed considerable quantities of teachers to higher education and research laboratories, sometimes as part of the strong immigrant contingent that moved to the Americas, or as members of official cooperation programs. Great Britain had a smaller though significant role, fundamentally through the British Council and its fellowship program for Latin American students who received training in the famous British universities, as in the cases of Argentine Luis Leloir and Brazilian Mauricio Rocha Silva in biochemistry and pharmacology respectively.

The Development Decades, 1940–1960

During the 1930s and 1940s, scientific leaders claimed government support for basic research, usually on a shared basis with international cooperation, as a means of building scientific communities. World War II inaugurated a period of growth in industrial activity, of rapid expansion of the population in large urban centers and of improvement in the general level of education, in a political context that alternated between the predominance of populism and authoritarism. The notion that science and the universities would play a central role in socioeconomic development was part of the 'developmentalist' ideology emerging from the United Nations Economic Commission for Latin America (ECLA) established in 1949. The works of economist Raúl Prebisch and his associates argued the need to adapt and combine international technological knowledge, to define priorities from the point of view of economic planning and to organize research programs in response to those priorities.[10]

In practice, however, the substitution of imports was favored without systematic concern for technological learning. Most technology transferred to Latin America was embodied in equipment and procedures. The choice, negotiation, acquisition and assimilation of disembodied technology was widely ignored; the same happened with domestic R&D development. Objectively, local investment in technological development became too costly. This explains the late growth of the capital goods sector, the delayed start of graduate education, the marginal structure of experimental R&D and the still very low enterprise participation in the financial support of these activities. All these factors shaped a non competitive industry that continues to exist today.[11]

Despite the fact that the general pattern of industrialization adopted did not foster the growth of dynamic R&D systems, university and governmental research did achieve *momentum* in some fields, particularly since the 1950s. Universities were the centerpiece of the model adopted for national science policies, indeed the only institutions to which it seemed to apply. The purpose was to grow a 'scientific-technical' infrastructure assuming, often in an implicit manner, that when achieving a 'critical mass' in scientific research, an automatic reinforcement of local technology would evolve to exploit raw materials and other resources, thereby increasing production and productivity. The elements of a public policy for science and technology, which came to fruition in the 1960s, were conceived in the 1950s and its most conspicuous spokesmen were leading figures of the scientific community. Physiologist Bernardo Houssay expressed his commitment to science as follows:

> I have devoted my life to three main aims: 1) to cultivate science to know her and make her progress; 2) to train pupils and help those who devoted themselves to science; 3) to work for the development of science in my own country. For me this third obligation is more rigid than the other two. I have been lucky to devote myself to science as I wished and to have received resources for what I did. Science is not a simple mercantile activity and in order to pursue it money is not enough. Discoveries are not made by equipments, buildings or money; they are made by able and idealist men, intensely and devoutly dedicated to science, with enough resources to advance and not stagnate or fall back. On several occasions I have exposed the causes of the slow scientific development of Latin America: 1) lack of scientific tradition; 2) ignorance of the role and importance of science; 3) vanity and other moral defects; 4) technical defects; 5) intellectual defects; 6) character and personality failures.[12]

The Brazilian scientific community began to expand on a par with the nation's industrialization. The new University of Sâo Paulo (1934) was predicated on the assumption that the development of a research capability would help restructure the existing system of higher education. The foreign teachers it hired were crucial in training pupils and establishing research traditions, allowing the University to achieve a scientific density unequalled in the country. Two Europeans opened up fruitful research traditions in theoretical physics. In Sâo Paulo, Gleb Wataghin developed two research lines between 1934 and 1942: one in theoretical physics, with his most distinguished pupil Mario Schenberg, and the other in the experimental side, on cosmic rays, with Marcelo Damy. With all major scientific groups in England and the United States involved in the war effort, Wataghin and his group were for a while considered the only scientists working on cosmic rays in the world.[13] In Rio de Janeiro, Bernhard Gross became interested in the interaction of radiation with matter, publishing the first systematic work on electrets in 1957 in the *Journal of Chemical Physics* and in *Physical Review*. Although his research was basic, a few years later a German and an American researcher, using the same method and theory he had described, made the first practical electretic microphones.[14]

As a symptom of the expansion of scientific activity, the *Sociedade Brasileira para o Progresso da Ciencia* (SBPC) was founded in 1948. The following year the private *Centro Brasileiro de Pesquisas Físicas* was established in Rio de Janeiro, bringing together several high-quality scientists such as Cesare Lattes, José Leite Lopes, Jaime Tiomno, and Roberto Salmerón. In 1951 the *Conselho Nacional de Pesquisas* (CNPq) was created to support science. In both institutions, Admiral Alvaro Alberto, a military man with a strategic view of science and technology, struggled to have a national commission on atomic energy organized within the research council. But in 1954 President Getúlio Vargas committed suicide and Alberto was dismissed from his post as President of CNPq. Having decided to link the Brazilian economy to the international economic system, in 1967–68 Brazil's military rulers attached nuclear institutions to the power industry establishment, excluding local scientists from the decision-making process, killing the domestic program of nuclear development already under way, and buying an American light-water reactor.[15]

Despite the fact that Brazil was endowed with an advanced industrial infrastructure and a more sophisticated physics program than available in Argentina, nuclear policy there became an arena for domestic and international political pressure, failing to become insulated from broad political, economic, and social issues, as was the Argentine nuclear program. In the latter country, by contrast, there was a central institution with the political autonomy and leadership necessary to sell an independent nuclear plan to the ruling elites and ensure its execution. Through careful policies of technology purchase, staff training, R&D backing, the establishment of physics labs and of the nuclear engineering profession, the Argentine *Comisión Nacional de Energía Atómica* generated a critical mass of scientists and a technological infrastructure that enabled this agency to have an excellent track record.[16] Each success generated further political support for the project, which continued well into the 1980s.

The National Institute of Amazonian Research, and the Institute of Pure and Applied Mathematics in Brazil were also established in this period. A small number of elite teaching and research institutions came into being, and served as model and inspiration for the broader reforms that would be attempted in the following phase at the level of the national system of higher education. The first one was the *Instituto Tecnológico de Aeronáutica* (ITA), supported by the Ministry for Aeronautics in close collaboration with the Massachusetts Institute of Technology (MIT), which carried out research and the training of military staff in R&D. ITA's experience was important in the renewal of the university curriculum in that it emphasized the need for a scientific base with an important experimental component aimed at technological applications. Another important institution was the Medical School of Ribeirão Preto, conceived as a model of modern biomedical research. The contribution of the Rockefeller Foundation, was decisive in the early development of the School. Another emblematic institution was the University of Brasilia, an ambitious, imaginative project imbued with developmentalist ideals, which rapidly fell victim to military repression in 1964. Although it continued to be reputed as a good federal university, it never recovered its original mystique and prestige.

Scientific research in Mexico has been closely linked to the Science Faculty at the National Autonomous University of Mexico (UNAM), and to a series of measures geared to legitimate the role of the full time researcher within the University throughout the 1940s. In 1940 the Laboratory of Medical and Biological Studies was founded to accommodate a group of Spanish refugees who were students and followers of Ramón y Cajal. In 1956 it became the Institute of Biomedical Research.[17] When UNAM's University Campus began to be constructed in 1950, the first building was the Science Faculty, followed soon after by the Science Tower, where for the first time in the university specific space was allotted to research Institutes in an organizational pattern that has since separated undergraduate teaching from graduate teaching and research. Some individual researchers began to be hired on a full time basis although research budgets continued to be too small. In 1949 Manuel Sandoval Vallarta returned from the US, where he was a teacher at MIT, keen to develop physics locally. Spain's leading physicist, Blas Cabrera, and Spanish astronomer Pedro Carrasco were also influential in this task. Outside the university structure, Arturo Rosenblueth, a Mexican physiologist who had collaborated with Walter Cannon in Harvard, returned in 1944 to head the physiology laboratory of the new Cardiology Institute, that received strong financial support from the Rockefeller Foundation. In 1960 he became the founding director of the Center for Research and Advanced Studies (CINVESTAV) of the National Polytechnic Institute.

In Venezuela, in the wake of dictator Gómez'death in December 1935, a frantic process of modernization of the State apparatus ensued. The Central University started to build some research capability, basically relying on the contribution of foreign scientists. In the medical faculty alone, the arrival of people of the scientific stature of physiologist Augusto Pi Sunyer, the most distinguished scientific Spanish exile, the Swiss anatomo-pathologist Rudolf Jaffé, the former Hamburg University tropical pathologist Martin Meyer, the Spanish physiologist Rossend Carrasco i Formiguera and others, resulted in the establishment of research in several disciplines and a new mystique with respect to the practice of science. A student of Pi Sunyer, Francisco de Venanzi, would be a devoted heir to this tradition in later years, promoting the growth and social recognition of research in the country.

The Science Policy Era, 1960–1980

Economic and social planning agencies began to operate in the region and science had its share of them. Early reports complained about their shortcomings, such as lack of institutional coordination, incoherence between short, medium and long term plans; scarcity of adequately trained staff, projects and statistics. Sometimes, planning and policy-making emerged as an imposed mechanism for getting funds from international agencies. The dominant modernization ideology was expected to lead to higher levels of autonomy, self confidence and social justice. The post-war period witnessed the ascent of self-reliant, optimistic social movements, aimed at building more equitable societies. The development of local capabilities in science, technology, industry, management, and work-force, introduced significant changes and the emergence of new sets of individuals, better trained and with a better understanding of the art of negotiation. Groups of scientists, engineers, public officials and the armed forces, who tried to put into practice projects such as that of atomic energy in Argentina, electronics in Brazil or oil in Venezuela and Mexico, managed to make an impact on the international competitive game with their unexpected achievements.

But changes were insufficient and a pattern of economic development prevailed based on growth without social equity. Industrialization was geared to the domestic market and biased towards the conspicuous consumption of luxury goods at levels significantly higher than in other countries of comparable income levels. There was a lack of leadership in domestic private firms in the most dynamic industrial sectors (automobiles, chemistry, capital goods), combined with a weak development of small and medium sized industrial firms. The private sector had little participation in R&D even in the most advanced countries of the region, and this was coupled with distorted and underdeveloped entrepreneurial capacities. Growth rates have been unsatisfactory, showing deep regional and sector imbalances, marked income concentration, growing foreign control and substantial increases in the national debts.

In the 1960s, parallel to the abandonment of the post-war euphoria with science in the industrial countries, public criticism of science emerged in Latin America with regard to a double challenge: redirecting the aims of the scientific endeavor, and its ability to provide solutions to local and regional problems. An influential voice was that of Oscar Varsavsky, Argentine mathematician and physicist, who became impatient with the so called 'scientificist' researcher

> who has become adapted to this scientific market, that renounces to worry about the social meaning of his activity, dissociating it from political problems, and devotes himself entirely to his 'career', accepting the norms and values of the large international centres, embodied in an academic scale ... The mission of the scientific rebel <by contrast>, is to study in all seriousness and using all the arms of science, the problems of social change, in all their stages and theoretical and practical aspects. This is to make *politicized science.*[18]

As a result of the growing number of unresolved problems authoritarian regimes emerged in Brazil, 1964; Peru, 1968; Ecuador, 1969; Bolivia and Uruguay, 1970; Chile, 1973; and Argentina, 1974. When authoritarian governments in the Southern Cone tried to suppress the social sciences, the large North American Foundations came to their rescue, contributing to the emergence of private research centers. The Ford Foundation was particularly generous. In Argentina, it helped Gino Germani's Institute of Sociology in the University of Buenos Aires. In Brazil, support for the social sciences, which was already significant at the beginning of the 1960s before a politically conservative military government took power in 1964, practically quadrupled in the wake of the deep reorganization of higher education in 1968. Chile was host to the most important Ford Foundation social science program in the region in the early 1970s.

In several countries, scientific and technological research moved out of the universities. The new groups that benefited from growing R&D resources tended to be young and politically indifferent, or at least had few personal links with the recent past. Working in isolated and protected places, with salaries unrelated to the university budget and without having to teach undergraduates, they often came to think of themselves as long-term reformers waiting for the political storm to wane so they could set the foundations of their country's future scientific and technological self-reliance.

In universities, the North American model of centralized institutes and departmental organization was adopted, but mass education came to be the crucial problem of universities, whose budgets were chronically insufficient to attend to the growing demand. Several traditional universities, which historically hosted research groups, suffered a progressive deterioration and lost their attractiveness as privileged *loci* for research. The small research communities had to compete with an increasing number of university teachers who did not do research but who nonetheless had access to the full-time regime and the stability of employment that had historically been reserved for researchers.[19] Scientists and engineers tried, when they could, to organize their work outside of the universities or around isolated graduate programs. However, the high mortality rate and the diversity of aims and objectives of the courses, as well as noticeable quality differences, insured that the old mechanism of sending students on scholarships to the developed countries continued to be operational whenever funds and opportunities were available. The National Science and Technology Councils began to fund the research that could not be supported by academic institutions and tried timidly to define priorities and guide scientific activity.

Since the end of the 1960s Brazil embarked on the broadest government attempt to direct the scientific endeavor as a function of economic development. Resources for scientific and technological research came from government sectors responsible for economic planning and long term investments. The government improved its ability to carry out policies through the Funding Agency for Studies and Projects (FINEP) under the Ministry of Planning, CNPq, CAPES (the agency within the Ministry of Education providing fellowships for graduate studies) and the Sâo Paulo Foundation for Research Support (FAPESP). Large-scale centers for R&D also emerged during this period, such as the Coordination for Graduate Programs in Engineering of the Federal University in Rio de Janeiro (COPPE) and the University of Campinas, geared to technological research and graduate education in engineering and science. Although after 1980 the sector entered a period of great instability and uncertainty, Brazil managed to build a system which currently has some 35.5 thousand scientists and engineers active in R&D, and about one thousand graduate programs in most fields of knowledge. Several thousand students are on fellowships paid by the Brazilian government in North America and Europe with about 1.5 million students enrolled in undergraduate university programs, 30 thousand in masters and 10 thousand in doctoral programs.

Venezuela's modernization, initiated in previous decades, assumed a fast pace in the 1960s and 1970s, supported by the dramatic expansion of the oil income. Although there had been a small crop of science institutions since 1936, and even though the Venezuelan Association for the Advancement of Science (AsoVAC) came into being in 1950 at the initiative of a small group of scientists in the country, it was necessary to wait until the fall of the dictatorial regime of Pérez Jiménez in 1958 for science to begin to institutionalize on a more continuous basis. Education received a great boost at all levels. A new university law was approved that emphasized research as one of the basic functions of the universities. The Science Faculty at the Central University of Venezuela was founded in 1958 with the aim of training scientists. It was followed in 1959 by the Venezuelan Institute of Scientific Research (IVIC), a center of excellence with the exclusive purpose of doing world-class research and teaching, and in 1967 by the National Council of Scientific and Technological Research (CONICIT) for the promotion and funding of scientific research. With the nationalization of the oil industry in 1976, there was a need for an oil research institute. Thus, together with the creation of PDVSA (*Petróleos de Venezuela, S.A.*), the Venezuelan Institute of Oil Technology (INTEVEP) came into being as an R&D Center, but with the status of an operating oil company integrated to the PDVSA holding. The importance of this late development for a country that depends on oil exports for 70 percent of its foreign exchange is obvious. Scientific, technological and managerial capabilities grew considerably, as exemplified by INTEVEP's development of ORIMULSION®, thereby introducing a new fuel into the world energy market; a rare event, particularly when it has the potential to impact in a significant way such an important fuel consumer as the power industry.

The current phase of turbulence and macroeconomic unbalance in the transition towards more open and disregulated economies under new domestic conditions in the Latin American countries and a changed international economy, has not found them well prepared to face the challenges ahead. Particularly since the 1980s the Latin American industrial sector has presented relatively high margins of idle capacity in several branches and shown an increased technical obsolescence. In the public sector, the combined effects of restricted sources of investment, concentration on short-term problems with the consequent neglect of strategic thinking and drastic salary reductions weakened the support given to critical areas like R&D and also became manifest in the crisis of educational systems at all levels. What is at stake today is the entire productive and social system.

The possibilities of change and modernization in the universities on a global scale seem remote. But some public universities busy themselves with scientific research and the education of future researchers. Given the diminished levels of State patronage, many scientists who decide to remain in the academic context seek funding for basic research beyond the traditional governmental support. Others prefer to approach customers who might appreciate them for their ability to educate, innovate and give expert advice, rather than for their publications and scientific recognition, whilst profiting from their prior prestige as academic researchers. A new alliance between university science and utility begins to develop. The rhetoric of industrially useful science, however, faces two difficulties. On the one hand, the opportunities for an industrial science and for a highly qualified work-force are not many. The very low yield in industrial innovation is not explained by the existence of a 'useless' science but rather by the very low amount that industries invest in R&D. On the other hand, a pernicious gap widens between what is supposedly 'useful' or at least 'saleable' and what is purely cognitive.

The main R&D customers of the universities have been the large public enterprises. This situation did not come about as a result of political or ideological decision making; it was a consequence of the fact that in countries like Argentina, Brazil and Mexico State companies have represented the most important and advanced segment of the productive sector. However, in view of the ongoing privatization process of a significant number of these public enterprises this may not continue to be significant.

Conclusion

This century has seen remarkable changes in Latin American science, although it continues to occupy a marginal position in both national society and international landscape. The scientific enterprise at the international level has grown and changed dramatically during this century and continues to move with a dynamism that reduces the space of maneuver available to Latin American societies, putting at risk not only the markets for raw materials and industrial exports which they had slowly and painfully built up on the basis of low wage production but also the institutions, industrial firms and social learning processes that accompanied modernization in the last half a century.

The world economy appears clearly favorable to the most industrialized countries. Latin American science today needs to dramatically redefine the social contract with its host societies in order to help them face the new conditions, confront the threats of science-based technological development in many sectors of economic life and, at the same time, devise creative responses aimed at exploiting new opportunities. The risk of not doing it, is to be relegated to a subordinate position as appendixes of researchers from North America and Europe who head south in search of unique resources, cheap scientific talents and new topics for research.

References

1. Sierra, J. *La libertad*, vol. IV. (México, 1880).
2. Zea, L. *Filosofía y Cultura Latinoamericanas.* Consejo Nacional de la Cultura. Centro de Estudios Latinoamericanos 'Rómulo Gallegos'. (Caracas, 1976), p. 204.
3. Lydekker, R. The La Plata Museum, *Natural Science* (1894), **IV**: pp. 27–35, 117–28, quoted by S. Sheets-Pyenson, *Cathedrals of Science. The Development of Colonial Natural History Museums during the Late Nineteenth Century.* (Kingston and Montreal: McGill-Queens University Press, 1988), pp. 291–92.
4. Dean, W. The Green Wave of Coffee: Beginnings of Tropical Agricultural Research in Brazil (1885–1900),*Hispanic American Historical Review* (1989), **69, 1**: pp. 91–116.
5. Cueto, M. *Excelencia Científica en la Periferia. Actividades Científicas e Investigación Biomédica en el Perú 1890–1950.* (Lima: GRADE/CONCYTEC, 1989).
6. Foglia, V.G. The History of Bernardo A. Houssay's Research Laboratory. Instituto de Biologia y Medicina Experimental: the first twenty years, 1944–1963. *Journal of the History of Medicine and Allied Sciences* (1980), **35**: pp. 380–96.
7. Fitzgerald, D. Exporting American Agriculture: the Rockefeller Foundation in Mexico, 1943–1953. *Social Studies of Science* (1986), **16**: pp. 457–483; Cotter, J. The Rockefeller Foundations Mexican Agricultural Project: A Cross-Cultural Encounter, 1943–1949. In Cueto, M. (Ed.), *Missionaries of Science. The Rockefeller Foundation and Latin America.* (Bloomington and Indianapolis: Indiana University Press, 1994), pp. 97–125.
8. Mariscotti, M. *El Secreto Atómico de Huemul. Crónica del Origen de la Energía Atómica en la Argentina.* (Buenos Aires: Sudamericana-Planeta, 1985).
9. Pyenson, L. The Incomplete Transmission of a European Image: Physics at Greater Buenos Aires and Montreal, 1890–1920, *Proceedings of the American Philosophical Society* (1978) **122**: N°2, pp. 92–114.
10. Economic Commission for Latin America and the Caribbean, *Raúl Prebisch: un Aporte al Estudio de su Pensamiento.* (Santiago de Chile: CEPAL, 1987).
11. Fajnzylber, F. *La Industrialización Trunca de América Latina.* (México: Nueva Imagen, 1983).
12. Houssay, Bernardo A. Closing Conference, Second Joint Meeting of the Regional Commissions of the Science and Technology Council of Argentina, Buenos Aires, April 4th 1960. Archives of the 'Bernardo A. Houssay' Museum, Buenos Aires.
13. Bernhard Gross. Interview in *Ciência Hoje* (Rio de Janeiro, 1986), **4, 22**: pp. 74–80.
14. Schwartzman, S. *A Space for Science. The Development of the Scientific Community in Brazil.* (Pennsylvania: The Pennsylvania State University Press, University Park, 1991).
15. Leite Lopes, J. Interviewed by E. Candotti, *Ciência Hoje* (Rio de Janeiro, 1985), **4, 20**: pp. 18–24.
16. Adler, E. *The Power of Ideology. The Quest for Technological Autonomy in Argentina and Brazil.* (Berkeley, Los Angeles, London: University of California Press, 1987).
17. Fortes, J. and L. Adler Lomnitz, *Becoming a Scientist in Mexico.* (Pennsylvania: Pennsylvania State University, University Park, 1994).
18. Varsavsky, O. *Ciencia, Política y Cientificismo.* (Buenos Aires: CEAL, 1969).
19. Brunner, J.J. *Educación Superior en América Latina, cambios y desafíos,* (Santiago de Chile: Fondo de Cultura Económica, 1990); Kent, R. *Modernización conservadora y crisis académica en la UNAM.* (México: Nueva Imagen, 1990).

Further Reading

Babini, J. *La evolución del pensamiento científico en la Argentina,* La Fragua, Buenos Aires. There is a revised 1986 edition under the title *Historia de la ciencia en la Argentina* with a historiographical introduction by M. Montserrat. (Buenos: Solar, 1954).

Brunner, J.J. and A. Barrios *Inquisición, mercado y filantropía. Ciencias sociales y autoritarismo en Argentina, Brasil, Chile y Uruguay.* (Santiago de Chile: FLACSO, 1987).

Cueto, M. (Ed.) *Missionaries of Science. The Rockefeller Foundation and Latin America.* (Bloomington and Indianapolis: Indiana University Press, 1994).

De Gortari, E. *La ciencia en la historia de México.* (México: Fondo de Cultura Económica, 1963).

Díaz, E., Y. Texera and H. Vessuri (Eds.) *La ciencia periférica. Ciencia y sociedad en Venezuela.* (Caracas: Monte Avila, 1984).

Glick, T.F. (1989), *Darwin y el darwinismo en el Uruguay y en América Latina,* Universidad de la República, Facultad de Humanidades y Ciencias, Montevideo.

Glick, T.F. Science and Society in Twentieth Century Latin America. *The Cambridge History of Latin America, vol. VI, Latin America since 1930. Economy, Society and Politics. Part I Economy and Society,* edited by L. Bethell. (Cambridge: Cambridge University Press, 1992), pp.463–535.

Herrera, A. *Ciencia y política en América Latina.* (Mexico: Siglo XXI, 1985), 9th edition.

Piñeiro, M. and E. Trigo *Procesos sociales e innovación tecnológica en la agricultura de América Latina,* Instituto Interamericano de Cooperación para la Agricultura -IICA-. (Costa Rica: San José, 1983).

Pyenson, L. *Cultural Imperialism and Exact Sciences: German Expansion Overseas, 1900–1930.* (New York: Peter Lang, 1985).

Stepan, N. *Beginnings of Brazilian Science.* (New York: Science History Publications, 1981).

Vessuri, H. (Ed.) *Ciencia académica en la Venezuela moderna.* (Caracas: Fondo Editorial Acta Científica Venezolana, 1984).

CHAPTER 44

Big Science and the University in India

DHRUV RAINA AND ASHOK JAIN

For the historian of science, as much as for the historian of education, it would be a matter of wonderment that towards the end of the twentieth century the community of scientists in India are as perplexed by the status of the university as an examining body, as were their forbears a century ago. In the last decades of the nineteenth century, a burgeoning community of scientists in colonial India strived for the inclusion of research in the charter of university education. The source of inspiration, as elsewhere, was the German university. Almost a century after the processes that resulted in the formation of an elaborate academic research system had been inaugurated, the role of the university continues to be a subject of heated discussion. In fact, a currently raging polemic relates to a proposal for the establishment of a National Science University in India. The present chapter chronicles the evolution of the academic and the scientific research systems. The dualism, it is suggested, currently characterizing the institutions of the scientific and technological research system was structured by the requirements of a rapidly evolving knowledge form as much as by the imperatives of the modern post-colonial state.

This chapter discusses, in part, the unfolding of the history of Big Science in India. Further, it suggests that the emergence of Big Science required the emergence of new institutions and the concomitant supersession of the university considered as 'the age old site for the production of knowledge.' The phenomenon is not specific to India, though there are elements that are distinctive of nation-states where the scientific and technological research system acquired concrete form in the first decades of the twentieth century. The supersession of the university as the primary center of scientific research is the outcome of a number of processes concerning the production of knowledge. Of these, some relate to the commoditization of scientific knowledge, and others may be visualized as a consequence of the pact signed between institutions of scientific research and the state, and embodied in the entity called defense research. There would certainly be elements in the Indian experience that others share. The dissimilarities arise from the time lag between the commencement and legitimization of the tradition of modern science in India and Europe. Further, this history is subsequently complicated by India's colonial past. In the pages that follow, we shall document the emergence of the academic research system within the university, and then outside it. In doing so, we shall show how the university system has shown tremendous growth as an examining body, but its status as a site for the production of knowledge has declined.

The Beginnings of the University Research System in India

One account of the emergence of the institutions of science suggests a sequence of four stages. In the first we have the founding of societies of a varied and comprehensive character. This is also referred to as the age of 'The Great Surveys,' spanning the period 1761 to 1903. In a manner of speaking these societies were founded by the British to extend their dominion over the country. Studies on the history of modern science in India detailing this stage in the evolution of the sciences adopt a position that is, to put it mildly, problematic. The first has to do with the fact that these institutes were seen as the primary conduits for the dissemination of the traditions of scientific research among the indigenes. Second, these institutes were not merely the symbols of imperial power, but the 'tools of empire' championing the imperial program to expropriate and control.

The second stage that followed three quarters of a century after the founding of the Asiatic Society is marked by the founding of the universities, first in the Presidency towns of Calcutta, Bombay and Madras in 1857, followed by Panjab University, Lahore, now in Pakistan, in 1882 and Allahabad University in 1887. By 1900 there were 170 colleges, 4 colleges of medicine, 4 colleges of engineering, 28 medical schools, and 12 engineering schools. The subsequent stages have to do with the institution of an annual meeting of scientists initiated in 1914: the Indian Science Congress, and with the founding of an all-India coordinating body, embracing all modern scientific research in the country: the National Institute of Science (now the Indian National Science Academy). But this account is inadequate.

The changing cast of actors/scientists participating in the emerging nationalist struggle brings the specificity of the Indian experience in to the foreground. In 1876, an Indian doctor Mahendra Lal Sircar, founded the Indian Association for the Cultivation of Science (IACS). The Association was founded in response to a pressing demand from educated Indians that the university must cease to be merely an examining body. Throughout the second half of the nineteenth century, Western educated Bengalis were demanding a fully fledged scientific research system, a demand that was not accepted by the Imperial administration until 1904, when the Indian Universities Act was passed (almost half a century after the founding of Calcutta University) permitting postgraduate teaching and research in the humanities and the sciences. Meanwhile, the IACS had already become the fount of inspiration for a whole generation of Indian scientists entering the profession of science.

The Association sought to combine the character, scope and objectives of the Royal Institution of London, and the British Association for the Advancement of Science. Nevertheless, its objective, Sircar pointed out was to "carry on the work with our own efforts, unaided by Government . . . I want freedom for the institution. I want it to be entirely under our own management and control. I want it to be solely native and purely national".[1] IACS was to provide an umbrella for the first generation of Indian research scientists, and even those academics who were associated with Calcutta University, such as J.C. Bose and the chemist P.C. Ray, either taught at IACS or were associated with it. The Association blossomed in the wake of the emerging nationalist struggle in the country, and the greater the reluctance displayed by the imperial administration, the greater the resolve on the part of the incipient scientific community to ground their independent efforts.

A cursory discussion of the genesis of two institutions during this period would illustrate the point. The first had to do with the founding of the Bengal National College and the Bengal Technical Institute, Calcutta, through the efforts of the National Council of Education (NCE), and the second the Indian Institute of Science, (IIS), Bangalore through the efforts of the industrialist J.N. Tata. These attempts were themselves stimulated by a utilitarian vision of modern science and technology, a vision that had acquired currency amongst the Bengali Bhadralok class as well as other sections of the Western educated Indian community.

The NCE was founded by Satish Chandra Mukherjee in the last decade of the nineteenth century, and included amongst its active members and associates the scientific and cultural constellation of modern Bengal. It was attempting to extend the research charter of the IACS into the domain of pedagogy: namely, that of founding an educational system on 'national lines' and under 'national control.'[11] A noteworthy feature of this effort was its conception of what national education meant. While eschewing the framework of Westernization propounded by the Macaulayans, it proposed instead critical assimilation from both the West and East. Through their cultural organ, *The Dawn* the members of the Council sought to institute a critical examination of tradition and modernity.

At stake were two conceptions of science, the first, to use a phrase from Thackray, as a radical ratifier of a new world order, and the second as the harbinger of economic prosperity and well being. This was to find expression through the NCE's protracted deliberations on what ought to be the nature and content of a scientific and technical education. Following the partition of Bengal in 1905, and growing unemployment among the new intellectual proletariat, the NCE decided to found its own college whose educational agenda would be at variance with that of Calcutta University which was considered tainted now that it was under British control. Scientists such as Ray and Bose were clandestinely involved in the enterprise.

The Council split in 1906 into the National Council of Education and the Society for the Promotion of Technical Education. The cause of dissension appears to have related to the manner of imparting a scientific and technical education along 'national lines,' with the scientists and the engineers going one way — drop the cultural and moral component they said — and the members with a liberal arts background going the other way. However, the immediacy of the nationalist struggle, and the growing demand for professional engineers and scientists to man India's mushrooming industry resulted in a patch up amongst the two camps. In any case, Ray had given credence to the idea that wealth flows out of the portals of the laboratory.[2] Ray was to write: "The history of the modern supremacy of Germans in the industrial world is the history of triumphs achieved by succeeding generations of silent and patient workers in the laboratory."As a leading chemist employed by Calcutta University, Ray had established an industrial enterprise through public contributions, the Bengal Chemicals and Pharmaceuticals whose products would substitute the exorbitant imports from Europe. The new knowledge form and activity thus came to be coupled with aspirations for freedom from British rule. A prerequisite for self-rule (*swarajya*) was economic self-reliance (*swadeshi*).

The important feature of this episode in the history of the institutions of modern science in India, is that in characteristic late nineteenth century fashion, two ideas had acquired palpable currency: first, in contributing to science prestige accrues to the nation; second, the path to political independence must be paved by economic self-sufficiency that in turn requires an adequate scientific and technological base. By 1915, a substantial school of research in the areas of physics and chemistry had been established within Calcutta University through the research programs initiated by J.C. Bose and P.C. Ray. C.V. Raman gradually moved from IACS to a Professorship of physics at Calcutta University. The research agenda for the university stood legitimized.

The Indian Institute of Science (IIS), Bangalore, was not created through the efforts of a highly educated professional middle class, seeking funding for their efforts from the rural gentry (landowning class) of Bengal, but emerged as an idea of a leading Indian industrialist, Jamsetji N. Tata. Tata like his compatriots in Bengal was dissatisfied with the state of India's teaching universities (an euphemism for examining bodies), and proposed setting up a 'real university' that he referred to as the Research Institute of Science (when finally established it was christened the Indian Institute of Science), through an endowment of his own, with matching support from the Government of India. Responsible for setting up India's first modern textile and steel mills, Tata was inspired by the German model of the university that had motivated so many in other parts of the world. This real university would give a "fresh impulse to learning, to research, to criticism, which will inspire reverence and impart strength and *self-reliance* to future generations of our and your countrymen."[3] [emphasis added].

For the members of the NCE and for Jamsetji Tata the source of inspiration for the university was the German one, but the model of emulation was Johns Hopkins University, Baltimore, USA. The historian of science, Subbarayappa proposes two possible reasons as to why this was so for Tata. The first had to do with the fact that it was the first university in the world founded as a postgraduate institution: this is what Tata's institute turned out to be, and the Indian Institute of Science continues in this way to the present day. Secondly, when founded in 1875, Johns Hopkins was in the center of a district in which vast industrial development was in progress. The latter reason, we conjecture, might have appealed to the members of the NCE. As far as the IIS was meant to be a post-graduate institution, P.C. Ray disagreed with Tata's proposal on the ground that an institute where only scientists of acknowledged position carried out their research would be out of place in India at the end of the nineteenth century. The Indian student still required to be apprenticed to a researcher at an existing university. These universities needed to be well endowed and widened in scope. Ray's proposal was to build up capabilities from below. Ray responded to the founding of the IIS: "What is . . . needed is encouragement in the shape of handsome postgraduate fellowships . . . giving the holders thereof full option to carry on their research at any well recognized place or institution . . . In this way a kind of healthy inter-provincial emulation would also be set-up". Whereas Tata was suggesting leap-frogging and proceeding with the task of building the new nation.

This is not to say, that Tata's conception was purely instrumental. On the contrary it was a "liberal" one. However, as in the case of the NCE, the general consensus was that scientific and technical subjects would receive priority, medical investigations next, and philosophical and educational subjects were last on the list. In any case, the industrialization of scientific research programs was underway in the minds of those shaping the future of India's scientific institutions. The mood of the nationalist uprising was such that these efforts in university pedagogy were driven by the urgency of acquiring economic sovereignty.

The Promise of the Millenium

By 1920, a school of physics had been instituted in Calcutta. The members of this school were situated at the Presidency College, Calcutta, and Raman while a Professor of physics at Calcutta University continued performing experiments at the Indian Association for the Cultivation of Science. By 1917 Raman had transformed the Proceedings of the Association into the *Indian Journal of Physics*. The next decade was to prove very fruitful in terms of the contributions that came out of this school. Raman had initiated work on molecular scattering by 1925, M.N. Saha had written his most influential papers that were instrumental in the formation of the discipline of theoretical astrophysics, and S.N. Bose was to author with Einstein one of the papers that closed the quantum theory phase of quantum physics.

The close intellectual influences and ideological predispositions that subsequently conditioned the destiny of Indian scientific institutions and programs can now be clearly mapped. On the one hand chemists and applied chemists who had congregated around, or were under the influence of Ray, were now moving into the phase of industrial research, and the trope of legitimation of their efforts was economic and scientific self-reliance. 'Berthelot's millennium' would prove an effective rallying score for the organization of the industrial research system, a task that Shanti Swaroop Bhatnagar was to successfully stage manage, through the founding of the Council of Scientific and Industrial Research (CSIR). On the other hand, the physicists emerging from or associated with the Calcutta school either embarked substantially into the most rapidly advancing areas of physics of the 1920s — the trope of national sovereignty and prestige proved effective — or were themselves to take recourse to the trope of the energy millennium to establish nuclear research facilities in India.

The moving figures behind these initiatives, as celebrated in the hagiology of science in modern India, are Bhatnagar, Saha and Homi Bhabha. Between them, it is suggested, two research imperatives, the *industrial research imperative* and the *nuclear research imperative*, provided the frame for the emergence of the scientific and technological research system. Essential to the legitimation of these imperatives, was the promise of economic and social progress, national sovereignty and prestige. In the pre-independence era this agenda would have run contrary to the imperial program. This essential argument was quickly grasped by Bernal, for in the first edition of *The Social Function of Science* he wrote:

> It is inevitable that in science, as in other aspects of life, the Indian should feel the need for national self-assertation, but his attitude is always an uneasy one . . . In order to release the enormous potentialities for scientific development in the Indian people, it would be necessary to transform them into a free and self-reliant community. *Probably the best workers for Indian science today are not the scientists but the political agitators who are struggling towards this end*[4] [emphasis added].

As far as the basic sciences were concerned, it is clear that once scientists like Ray, Raman and Saha registered their presence in the international arena of science, they consciously switched their publishing strategies in favor of Indian journals. The intent was to intervene in the process of building up Indian journals and instituting the importance of publications to the professionalisation of science, to the creation of a scientific community in India. Saha on this count is a particularly fascinating figure, for he embodies the tensions between the basic sciences, and what is alluded to in the policy discourse as relevant research. For him the relationship between science and society went beyond the creation of a scientific community. The question of 'relevance' was to crop up within policy circles and academic research system in the 1970s in India, when the latter was going through a crisis of dysfunction. Part of this dysfunction was imputed to the feeling that the principal agenda of research in India was being set elsewhere, and hence the research system was 'alienated' from the culture and needs of the Indian people.

This is not the place to examine Saha's central role in the National Planning Committee, under the Chairmanship of Nehru. But it was Saha who pressed for Nehru's candidature as Chairman of the Committee, since that would give the efforts of the committee legitimacy, and would in addition ensure the salience of science and its infrastructure in the planning process. Though the concern of science and planning is not central to this chapter, what must be noted is that Saha, like his mentor P.C. Ray espoused the development of infrastructure and research capabilities within the university system. Between the years 1922 and 1938 he was at Allahabad University, and the difficulties he encountered in establishing a laboratory have been chronicled by his biographers. However, he did leave behind a substantial school of ionospheric physics at Allahabad. By the time he returned to University College of Science, Calcutta, his career had taken a turn and he was immersed in the central political concerns of the time as well as in establishing the infrastructure absolutely essential for scientific research. His left wing politics put obstacles in his path, for Saha had difficulty in mobilizing resources or carrying the wave of political opinion with him despite the fact that international recognition had come his way, and that he had become a Fellow of the Royal Society by the time he was 34.

But in 1946, a year before India became independent, Saha decided to have a separate institute for research in nuclear physics, an institute that was subsequently carved out of his department at the University College. The institute was formally inaugurated in 1950, and was named the Saha Institute of Nuclear Physics after his death: it is currently supported by the Department of Atomic Energy. In the 1950s when the budget for R&D was probably smaller than it is now, even a modest investment in a nuclear research facility would have overshot the expenditure of several university research departments.

The above is crucial, for the theory of the nation, the imperatives of the cold-war, and political realignments had their place in structuring the R&D system the world over. The late 1940s were very decisive for the emergence of the institutions of Big Science related research in India. In the 1940s, another important actor entered the Indian scene, and this was Homi Bhabha, still fresh from the metropolises where the startling discoveries of nuclear physics were made in the 1930s and 1940s. He was still a student when Saha had already acquired eminence. On his return to India, he worked initially at the IIS, and later went on to found the renowned Tata Institute of Fundamental Research (TIFR), Bombay, in 1946 as an institute for fundamental research at the cutting edge of the scientific research system.

Studies in the history of science and scientific institutions, as well as policy oriented studies, have highlighted the significant differences that characterized the vision of Saha and Bhabha. However, it could be shown that differences in style and *savoir-faire* apart, there was much that they shared in common, and that Bhabha, like Saha, gave concrete form to a vision of Big Science, embedded in a culture of advanced technology: a vision, that Saha possibly espoused in his disagreement with those oriented towards the 'cottage industries' and Gandhian model of development. As has been pointed out the forces of political economy defined the arena within which Saha and Bhabha operated. In their proposals the two problems central to India's future development were related to the sources of power generation and that of self-reliance. The latter was a particularly effective and sensitive weapon for a society coming out of colonial rule, and the former would serve as the motor of development.

Further, both Saha and Bhabha could combine their commitment to theoretical physics and the development of its infrastructure with their commitment to the development of a high technology industry, and the public sector. In some way, the TIFR provided not merely the specialized manpower needed for the "tasks of national development", but became the cradle of the country's atomic energy program.[5] The idea to create a Department of Atomic Energy (DAE) that would overlook the nations atomic energy program was proposed to Nehru in 1954, and a new laboratory was suggested where the required technology would be developed since this task could not be undertaken at the TIFR. In any case, the TIFR served as a platform for the arrival of the Atomic Energy Establishment Trombay (AEET), renamed the Bhabha Atomic Research Centre (BARC) after Bhabha's death. As Bhabha was to suggest, the success of BARC : "... is due to the assisted take-off which was given to it by the Institute (TIFR) in its early stages of development".[6]

The dualism characterizing scientific institutions in India begins to become apparent. One set of institutes, such as the TIFR, (In a recent popular exposition on Bhabha, the Tata Institute of Fundamental Research, is referred to as the 'National Centre for Nuclear Science and Mathematics') the IIS and the five Indian Institutes of Technology (IITs — located at Bombay, New Delhi, Kanpur, Madras, Kharagpur and Varanasi) were research institutes where: "young men of the highest intellectual calibre in a society" are trained to think and analyze problems "with a freshness of outlook and originality which is not generally found."[7] And on the other, there was the large body of universities. The universities for Bhabha were to be the centers for "pure or long range research," whereas the institutes of the first tier would be moulded in the mission-oriented research paradigm. The realization of this institutional innovation, that came with its own organizational apparatus, required that Bhabha, like Saha, be accredited with intellectual leadership — or what Gibbons calls "paradigmatic leadership"[8] — credentials he had acquired through his own contributions to the discipline of the physics of elementary particles, and his close association with other paradigmatic leaders such as P.M.S. Blackett and P.A.M. Dirac. A number of factors aided the realization of Bhabha's vision. His success and thereby the limitation of Saha, lay in the manner in which he was to muster support from the industrial lobby, in the person of

Sir Dorabji Tata, and his proximity to a charismatic figure such as Nehru's in the post-independence era.

The pressure for the creation of facilities for nuclear physics research was significant, and this pressure came from within the scientific community and from outside. Raman, Saha, D.M. Bose, Bhabha, R.S. Krishnan and K.S. Krishnan were among the principal actors at the time. Raman had deputed R.S. Krishnan to Cavendish Laboratory in the 1930s to undertake research in nuclear physics. In 1946, after returning to the Indian Institute of Science, he presented a proposal to the Atomic Energy Board founded by the Government of India, to establish a nuclear physics laboratory at the school of physics at the IIS. Krishnan's proposal was rejected by Bhabha and Taylor on three counts. The first had to do with budgetary considerations, for as they mentioned in their report: "we would observe that nuclear physics cannot be regarded simply as another branch of physics . . . The immense progress in this field in recent years, and its fundamental character places it in a category by itself."[45] They then went on to suggest, alternatively, that the department must go on to develop its ongoing programs in the areas of the physics of solids, and low temperature physics. The justification for a large center for nuclear research stifled the claims of the other contenders: "the sums that are likely to be available in India in the near future for nuclear research are small compared with those that are spent in the United States and Great Britain, it is necessary that all large scale research in nuclear physics in the near future should be concentrated in one centre in the country . . . ."[46] Further, the entity called 'science,' that Bhabha and the community of nuclear physicists were schooled in, was already a transformed one, requiring the sort of funding that the starved Indian universities could ill afford.

In the subsequent years BARC emerged as the primary center for research into nuclear physics, and the academic research system that would provide the foundation consisted of the TIFR and IIS, and much later the five IITs. These five IITs were established in the late 1950s and early 1960s, as elite institutes. To date over a hundred thousand students take the entrance examination at the undergraduate level, and about two hundred and fifty are admitted at each of the IITs for degrees leading up to a Bachelor's in Technology. Similarly, half the number are admitted for the master's and doctoral degrees. The IIS at Bangalore is primarily a research institute, awarding doctoral degrees, and students are granted entry after they have graduated and passed an entrance examination. The five IITs and the IIS together will not have more than twelve thousand students in any given year, which would be less than the strength of an average Indian university. The impoverishment of the university has resulted in a perceived decline of students entering the elite institutes for research. Thus institutes like IIS have sought to institute integrated Ph.D programs, where students are directly registered for Ph.D after having obtained a Bachelor's degree. This is an interventionist strategy that hopes to rejuvenate the innovative potential of students before they are set in their ways. In fact,

TABLE 44.1: NUMBER OF MSc/MAs AND PH.DS IN THE NATURAL SCIENCES BETWEEN 1984 AND 1989.

Institutes	1984	1985	1986	1987	1988	1989
5-IITs & IIS	548	563	506	530	559	514
Cumul Fig for all Univs.	27990	29874	31401	32732	32108	47126
IITs and IIS as percentage of Univs.	1.95%	1.88	1.61	1.61	1.74	1.09

Source: *Outturn of Scientific and Technical Manpower in India 1984–89, Vol.1, Natural Sciences.* Human Resources Development Group, CSIR, 1993.

Table 44.1 gives an idea of the number of students graduating in the sciences (with a Master's degree or a Ph.d) out of the IIS, Bangalore; or the IITs in Bombay; Kanpur; Madras; and Delhi. These figures are then tabulated against the cumulative figure for all the Indian universities. The institutional dualism having come to stay, a number of conjectures may be proposed. Immediately following independence, a poorly endowed university research system in India could not have countenanced the novel institutional paradigm of scientific research. Students of international politics have documented the factors that influenced India's commitment to non-alignment while simultaneously according its nuclear program a particular urgency. As a result, while strategic factors, amongst others, favored the formation of the mission oriented research paradigm within new institutional and organizational arrangements, the explosive growth in the university system was oriented towards modifying the knowledge base of the newly independent republic along modernist lines.

The pivotal innovation that Bhabha gave institutional form had to do with his ability to convince the new political leadership that while the state would intervene in the creation of these institutions of science, through fiscal support, this did not necessarily entail governmental control. In fact, Bhabha possibly took his lead from the founding of the IIS; that was organizationally set up through a triangular arrangement between the Tata Trust, the Government of India and the Government of Mysore (presently the state of Karnataka). Thus the TIFR was founded through an arrangement between the Tata Trust, the Government of India (that currently operates through the Department of Atomic Energy) and the Government of Maharashtra: this has been alluded to as "Bhabha's formula." The impact of this two tier structure on the allocation of resources for research and infrastructure shall be seen in the last section. A similar arrangement was later worked out between the Government of India, through the Department of Atomic Energy and various state governments, first to extend support to the Saha Institute of Nuclear Physics, and later for subsequent institutes supported by the Department of Atomic

Energy, for example, the Institute of Mathematical Sciences, Madras. The Department of Science and Technology likewise offers grant-in-aid to a large number of research institutes.

The Industrial Society

The emergence of such high profile institutions naturally raises questions concerning science policy. Issues relating to Industrial Policy appear in the forefront of development discourse in India towards the beginning of this century. During the years of the nationalist struggle the swadeshi movement epitomized this deliberation on India's industrial future. The debates within the NCE related to the scientific and technical institutions required for a development scenario that would be commensurate with India's technical skills and resource endowment. The Indian Industrial Commission headed by Thomas Holland in 1916, included Dorabji Tata and Madan Mohan Malaviya. Viswanathan writes that the seven volumes of the report of the IIC and the journal founded and run by Saha, *Science and Culture*: "constitute the two most important texts in the archives of Science Policy in India since 1900. Strangely, the reports of the Commission have never been systematically considered in the histories of science and technology. It is true that its recommendations were never accepted; yet defeated documents often contribute to later victories. *Viewed as a discourse, the Reports remain one of the most systematic attempts to outline the making of an industrial society.*"[emphasis added]

The history of the industrial research system, that was projected as one of the founding pillars of the industrial society, is by and large the history of the Council of Scientific and Industrial Research (CSIR), the idea for which grew out of the deliberations of the National Planning Committee, the journal *Science and Culture,* and took final shape following the visit of Prof. A.V. Hill, Biological Secretary of the Royal Society, to India in 1943. In fact, the compulsions of the 1930s appertained to the organization of research and further, applying science to the larger goals of national development. For example, *Science and Culture* carried extensive and elaborate communications on the planning of resources, and the taming of India's vast riverine system. So much so, since it was the physicist Saha who initiated this discussion, this discourse has been labeled "river physics."[11] However, it is uncontestable that Domodar Valley Corporation was conceived in the pages of *Science and Culture.* The commencement of the Second World War, and the necessities of a war economy resulted in the establishment of the Board of Scientific and Industrial Research in 1940. Among the omnibus objectives of the Council, the principle ones included the promotion, guidance and coordination of industrial research in India, and the utilization of the results of these researches under the auspices of the Council towards the development of industries in the country.

There is a substantial corpus of literature discussing the emergence of the CSIR and the institutions of industrial research founded by it.[12] The important feature to note is that S.S. Bhatnagar became the first head of the CSIR in independent

India. Bhatnagar, since the 1920s, when he was at Panjab University, Lahore, was concerned with the applications of chemistry. He had made a name for himself as a colloidal chemist. It was through his proximity to Nehru that he was able to work towards the establishment of a network of twenty two of the 'national laboratories' under the CSIR, between 1948 and 1958 (as of now the number stands at forty one). In the first phase of the CSIR, natural resource utilization appears to have been a priority area. As Secretary of the Ministry of Natural and Scientific Resources, Bhatnagar was instrumental in orienting the direction of industrial research and development along lines that aligned with the objectives laid out in the plan documents. Thus industries such as the India Rare Earths Limited, for example, were initiated to examine the processing of monazite sands; this effort would provide the fuel required for atomic power generation.

The period between 1948 and 1964, within the discourse on planning, policy, or even for purposes of political history, is alluded to as the Nehruvian era. The important issues of this era may be listed as follows: the path of socialist development was to be pursued, in the realm of international politics India had committed itself to non-alignment, the cultural goal of the age may be referred to as a brand of scientistic imperialism: (the core of the new state would be organized around "the temples of science," further the Scientific Temper was enshrined in the Indian Constitution); in the domain of economics, the model of the planned economy was the exemplar. And the relationship of scientific and technological knowledge to the developmental process was seen in technological determinist terms.

In the annals of science in twentieth-century India, the year 1950 could well be referred to as the year of the CSIR laboratories. During that year six very important CSIR laboratories were inaugurated: the National Physical Laboratory, the National Chemical Laboratory, the Central Fuel Research Institute, the Central Glass and Ceramics Research Institute, the Central Fuel and Technological Research Institute, and the National Metallurgical Laboratory. This proliferation of scientific and technological institutions in a nation which until a few decades before had suffered from a shortage of them, could be realized only by enlisting the political leadership into the endeavors of scientists. For science and scientists this was the best decade in post-independence India. The first jolt was in the 1970s, when it was felt that the scientific and technological research system had failed to live up to its promise: the crisis was however precipitated by a larger one in development itself — but that is a separate issue.

While Bhatnagar had a seminal role to play in the founding of the industrial research system in India, in independent India he was for sometime the Chairman of the University Grants Commission. But even in this capacity he was committed in his view that mission oriented research, with short term objectives clearly defined, was best undertaken outside the academic research system. The universities, according to him, were suited for open ended research of a foundational nature. This was to become a major source of disagreement between Saha and

Bhatnagar. Bhatnagar was to point out that the "activities of national laboratories lie essentially in the domain of applied research, though these laboratories are not precluded from taking up research of a fundamental character." However, as has been indicated above, even Saha's institute after his death was drawn under the umbrella of BARC, an institute that had been virtually built from the department of physics of the University College, Calcutta.

The feature of the first half of the century is that the political leadership amongst the progressive nationalists drew upon science as a narrative of freedom, and the scientists responsible for the post-independence organization of science offered it as an activity that would herald the economic millennium. In this manner the institutional frame for Big Science was established in the country. For all the rhetoric on building the university research system, the university in real terms merely figured as a supplier of trained manpower for Big Science. The period referred to as the Nehruvian era in India, within policy discourse Solomon and Lebeau call the 'golden age of scientism,' for it was a period when scientific elites legitimated their existence within local cultures.

Saha and Bhatnagar are certainly to be credited for both the organization of scientific research and the establishment of scientific societies and institutions of science in modern India. Alongside Bhabha they constitute the trinity that has left an indelible mark on subsequent scientific research in the country. More than half a century later it is possible to infer that they recognized the crucial transformation which science was undergoing globally in the 1930s and the convoluted years of the Second World War. Furthermore, they were able to reflect these transformations in the institutions established in post-independence India ensuring that India leapfrogged into the era of Big Science. But in order to sustain the advantage gained from the historical experience of others, a complementary increase in support was required in infrastructure and logistics at the university level. The next section discusses the consequences of what happened when this support was not forthcoming.

The Dualism of Scientific Institutions and the Aftermath

When India attained independence from British rule in 1947, there were about twenty universities in India, and the figure rose to one hundred and sixty by 1987. In addition there has been a proliferation of national laboratories, institutes of national importance, centers of excellence, deemed universities, central universities, defense laboratories. While the growth in the number of universities reflects the urgency the government of India ascribed to higher learning, the proliferation of the varieties of scientfic institutions discloses the undercurrents that have a bearing on the demands of the state on science, as well as the transformation of science itself. The university system had to carry the burden of serving as an 'examining body' and undertaking open-ended long range research. The research institutes or national laboratories were compelled by either the urgency of the market or by the demands of the nation-state.[13] The warrant for this phenomenon is reflected in the discourse on science policy in India over the last fifty years. National science policies emerged against the backdrop of strategic considerations and competition for prestige. In this environment scientific institutions find it indispensable to respond to either the stimulus or pressure exerted by the state. As a result scientific institutions in India responded to the clamor of national goals through the industrial research option and the 1970s and 1980s were marked by debates that were plagued by the epidemic of relevance.

The autonomy for scientific institutions and scientific research that Bhabha, Saha and the progenitors of India's scientific institutions had negotiated went uncontested until the 1970s. During this period policies relating to science were articulated by a strong scientific elite with whom decision making rested, and they were backed by the political leadership. The crisis in development characterizing the 1970s is a landmark in policy discourse, and as far as India's scientific and technological research institutions are concerned it carried the intimations of a new notion of accountability. Since the state was the major supporter of scientific and industrial research (as of now about 85–90 percent of the R&D expenditure is met by the government of India, the remaining coming from the private sector), the sentiment that Science and Technology (S&T) policy must articulate national goals resulted in efforts undertaken by the states as well as the central government to dovetail the science and technology plan with the broader trajectory of the socio-economic plan.

Thus the core research infrastructure that consisted of the chain of laboratories coming under CSIR, the DAE, the Indian Council of Agricultural Research (ICAR), and the Defense Research and Development Organization (DRDO), had to be supplemented by agencies addressing themselves to environmental concerns, ocean development, and as the frontier of knowledge advanced, micro-electronics. Planners were compelled to apportion the R&D budget differently, thereby accommodating the novel requirements of the S&T research system. Thus there was block funding for R&D that was meant for the S&T agencies (such as the Department of Science and Technology) and a second block that was to be administered through the socio-economic ministries. Of the total budgetary allocation in the plan about 52 percent was meant for the S&T agencies.

The expenditure on research and development in terms of the source and user agency is given in Table 44.2. Of a total R&D expenditure of approximately $753 million, close to 55 percent was disbursed by user agencies that included the sectors of defense, space, and atomic energy, another 18–20 percent by the CSIR and the Department of Science and Technology (through its Science and Engineering Research Council, (SERC) which funds basic researches). This study has not paid heed to the agricultural sciences, which constitute a separate empire that has evolved through its own institutional innovation — the State Agricultural University. The Defense Research and Development Organization, the Department of

TABLE 44.2: R&D EXPENDITURE IN INDIA IN 1990–91 BY USER AGENCY

User Agency	Amount (in million $)
Defense	200.00
Space	111.76
Agriculture	94.12
Atomic Energy	82.35
CSIR	73.52
Environment	58.82
Department of Science and Technology	52.94

Source: E.C. Subbarao, 'Industry-Defined Research and Development: An Indian Case Study', in S. Chandrasekhar, *Physics and Industrial Development: Bridging the Gap*, Wiley Eastern Limited, 1995.

Space and the Department of Atomic Energy do in turn fund research institutes and centers and though these figures are difficult to obtain, we provide an example to illustrate the functioning of this two-tier system ahead.

The SERC of the Department of Science and Technology is the major supporter of research in basic science and engineering. One would expect to see a greater proportion of university research reflected in the funding profile of the SERC, rather than that of the national laboratories, deemed universities, and the centers of excellence. But the problem of the concentration of skills and facilities distorts the picture. This problem was sought to be redressed in the 1973 Approach document of the National Council for Science and Technology, which recognized that the absorptive capacities of scientific institutions was in direct proportion to their existing infrastructure and capabilities — a vicious circle was the inescapable consequence.

Let us take a look at the SERC's analysis of its projects for the period 1985–1990 and 1990–1995. Figures 44.1 and 44.2 list the top twelve institutions receiving SERC grants. In Figure 44.1, of the twelve institutes cited, five belong to the university system: Roorkee (Roor), Hyderabad (Hyd), Benaras Hindu University (BHU), Jawaharlal Nehru University (JNU), and Bombay University (Bomb). In Figure 44.2, we see how the IIS, and the IITs at Kanpur, Madras, Delhi and Bombay have more or less dominated the space for research in the basic sciences. Other than Osmania University and Pune University there are no new entrants. The reports highlight that the only difference between the two periods is that the number of institutions supported increased from 189 in 1985–1990 to 258 in 1990–1995. But what is common to both is that "5 institutes get the maximum share of SERC funding." However, both in terms of number of projects and share of project funding, the share of the IIS and IITs as well as the national laboratories increased (in terms of project funding the figure went up from 48.7 percent to 55 percent, in terms of the number of projects from 42.3 percent to 47.5 percent). Despite

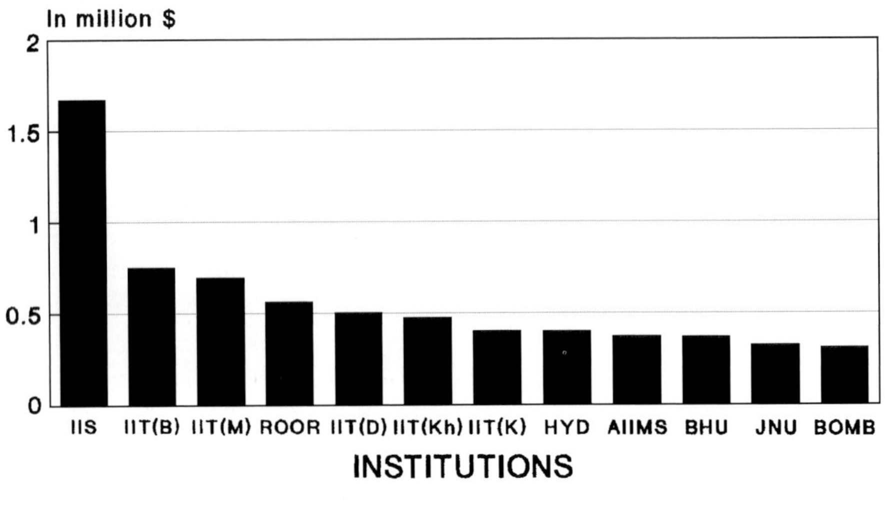

FIGURE 44.1: SERC SUPPORT: 1985–1990.

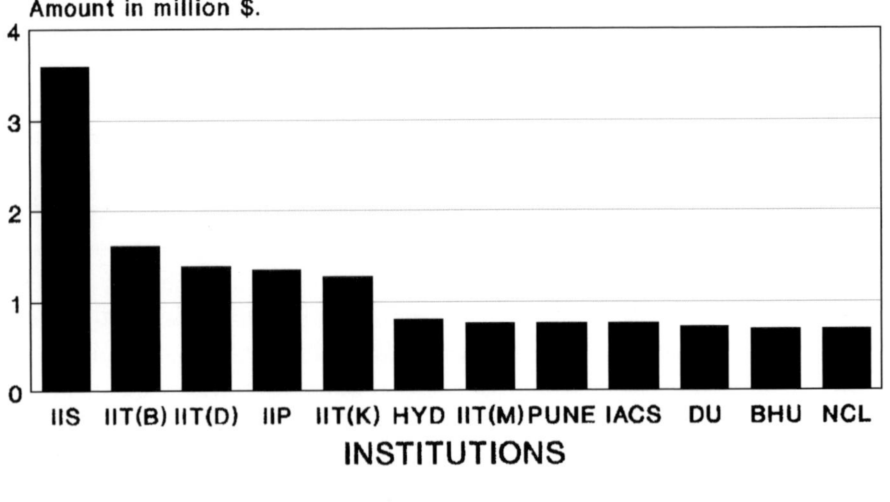

FIGURE 44.2: SERC SUPPORT: 1990–1995.

the fact that the SERC had extended its support to a larger number of universities, a transformation in the topography of the institutions of science required more radical measures. If we take into consideration, the total expenditure on research and development, and not just the SERC expenditure, it is found that only 7–8 percent is incurred within the university system. This figure has remained more or less constant over two five year plan periods.

The question of inter-organizational linkages has been a subject of extended debate within Indian scientocracy, and everytime the research system is subject to cutbacks, the issue is raked up afresh. The CSIR has through each of its reviews of the laboratories coming under its control sought mechanisms of deepening the linkages with industry. Contacts between research institutes and industry have been relatively few, and the cause for this has been traced back to the reward system of research institutes being out of tune with the imperatives of industrial research. This has largely been the state of affairs until recently, except for laboratories such as the National Chemical Laboratory, the Central Leather Research Institute to mention a few. Within research institutes scientists who have links with industry sustain these links as consultancies. On the other hand, linkages between research institutes and academic institutes have been weak. It has been observed that meaningful linkages have been forged between the two when research institutes have undertaken large projects that have required the development of generic capabilities. Some of these include programs undertaken by the Wadia Institute of Himalayan Geology, the large scale embryo transfer experiments of the National Institute of Immunology, or TIFRs cosmic ray program.

A recent survey of the state of astronomy and astrophysics in India neatly bears the point. The work simultaneously reviews the state of astronomy and astrophysics in India, and seeks to attract students to research. The book mentions eleven observational facilities that are actively networked with the international community at the frontier of research in the discipline: only two of these are within the university system. This is despite the fact that the pioneers of astrophysics in India created the discipline in the university departments: M.N. Saha, N.R. Sen, V.V. Narlikar, and D.S. Kothari. Thus the Inter-University Centre for Astronomy and Astrophysics, Pune, was created in order to overcome the deficiencies in infrastructure and to widen the research horizons of drained universities.

The debate on the proposal for a National Science University could, against this backdrop, be construed by some as an occassion for thrashing an extinct horse, when what really appears to be warranted is a fresh look at the dualism in scientific institutions. P. Balaram, a leading scientist from the Indian Institute of Science set the terms for the debate within academia. We shall briefly offer a precis of his editorial. Despite India's successes in atomic energy and space programs, Indian presence in the basic sciences in the international sphere was poor. Science teaching and research at the universities was at a low ebb: the universities were undergoing a financial and moral crisis, the national laboratories were afflicted by a paucity of ideas and a dearth of resources. Two major threats to Indian science were governmental indifference and the absence of collective dedication to common goals. The challenges were those of free markets, the growing crisis of accountability and the pace of growth of modern science.

The proposal for the National Science University was put forward by an Indian plasma physicist at the University of Texas at Austin, Swadesh Mahajan. For him the dismal state of Indian science reflected the dismal state of the universities. The situation was produced by scientific elites who exercised control through financial channels, the patronage system and by setting research priorities. This elite had abandoned India's university research system and established a large number of research institutes that were disengaged from the university — this was a strategic blunder that had damaged Indian science. The national science university is meant to be an exemplar for turning the scales around. In fact an avid perusal of the issues of *Current Science* over the last three years would reveal a certain commonality of perspective divulged through shared metaphors: the metaphors of impoverishment, desolation and vacuity are often used to describe the situation in the universities. Naturally, this rhetoric must be read in the context of the better facilities available at the research institutes. Further, there is a proposition that binds the entire community together, and that relates to a vision that the university must continue to be the age-old site for the production of knowledge, though it need not remain the only site for this activity. This is despite the fact that most of what passes for internationally acclaimed science is undertaken outside the university system which means that the crisis of the university is visualized not so much as a result of the modification of the nature and relationships of science with society and the state, but as a purely logistic one.

Big Science made its entry into India at a particular historical juncture, and came in a cultural and institutional mould. In nations where the university system was highly developed, resources were more easily available, among other things, mission oriented research could not secure the entire space available for the scientific research system. In India, at the time of independence, the university system was still weak: the number of university departments boasting of an international reputation were few and far between. In such a post-colonial society, the push factors — the need for catching up with the West, the neurosis of development, questions of national sovereignty and prestige, the scarcity of resources — left the scientific community with few options. The nuclear and industrial research imperatives provided the frame for the emergence of the new research institutes. The mission oriented research paradigm was rapidly instituted, whereas the larger agenda of academic research appeared lower down in the list of priorities. The crisis today is whether science policy still remains a national concern, whether the support given by the state for research in the basic sciences can be justified by the scientific community and finally, can the scientific community again mobilize political support.

References

The authors wish to thank S. Irfan Habib, V.V. Krishna, and Rajeswari S. Raina at NISTADS for clarifying some of the points raised in this chapter.

1. *A Century: Indian Association for the Cultivation of Science.* (Calcutta, 1976,) p. 5–8.
2. P.C. Ray, *Essays and Discourses.* (Madras: G.A. Natesan and Co. 1918), p. 4.
3. Tata quoted in B.V. Subbarayappa, *In Pursuit of Excellence: A History of the Indian Institute of Science.* (Tata MacGraw Hill, 1992), p. 20.
4. J.D. Bernal, *The Social Function of Science.* (London: George Routledge & Sons, 1939), p. 207–8.
5. Bhabha as quoted in B.M. Udgaonkar, 'Homi Bhabha on Growing Science', in B.V. Sreekantan, Virendra Singh, B.M. Udgaonkar (Eds.), *Homi Jehangir Bhabha: Collected Scientific Papers,* TIFR (1985), p. iv.
6. Bhabha quoted in *ibid.*
7. Bhabha quoted in *ibid.*, p. IV.
8. Michael Gibbons, 'The Changing Role of the Academic Research System, in Gibbons and Wittrock (1985), *op. cit.*, pp. 2–20, p. 10–11.
9. Bhabha and Taylor quoted in Subbarayappa (1992), *op. cit.*, p. 171.
10. Bhabha and Taylor, quoted in Subbarayappa (1992), *op. cit.*, p. 172.
11. G. Venkatraman, *Saha and his Formula.* (Hyderabad: Universities Press Limited, 1995).
12. Jean-Jacques Solomon and Andre Lebeau, *Mirages of Development: Science and Technology for the Third World.* (Boulder & London: Lynne Rienner Publishers, 1993).
13. Source: (1) *Analysis of SERC R&D Projects 1985–1990,* (2) *Analysis of SERC R&D Projects 1990–1995.* These reports have been prepared by the SERC Secretariat, Department of Science and Technology, New Delhi.

Further Readings

Claude Alvares, *Homo Faber: Technology and Culture in India, China and the West: 1500 to the Present Day.* (The Hague: Martin Nijhof, 1979).

Zaheer Babar, *The Science of Empire: Scientific Knowledge, Civilization and Colonial Rule.* (State University of New York Press, 1996).

Bipan Chandra, *The Rise and Growth of Economic Nationalism in India.* (New Delhi, 1977).

Debiprasad Chattopadhyaya, *Studies in the History of Science in India,* 2 vols. (New Delhi, 1982).

Dharampal, *Indian Science and Technology in the Eighteenth Century: From Contemporary European Accounts.* (New Delhi, 1971).

Irfan Habib, 'Potentialities of Capitalist Development in the Economy of Mughal India', *Enquiry,* New Series III, no. 3. (1971), 1–56.

A.V. Hill, *Scientific Research in India.* (London: William Chowers and Sons, 1944).

Ashis Nandy (Ed.), *Science, Hegemony and Violence: A Requiem for Modernity.* (New Delhi, 1988).

Ahsan Jan Qaiser, *The Indian Response to European Technology.* (Delhi, 1982).

A. Rahman, *Science and Technology in Indian Culture: A Historical Perspective.* (NISTADS, New Delhi, 1984).

Vandana Shiva, *Staying Alive: Women, Ecology and Survival in India.* (New Delhi and Zed Books, London, 1988).

G. Venkatraman, *Journey into Light: Life and Science of C.V.Raman.* (Indian Academy of Sciences, Bangalore, 1988).

Mathew Zachariah and R. Sooryamoorthy, *Science for Social Revolution: Achievements and Dilemmas of a Development Movement: The Kerala Sastra Sahitya Parishad.* (New Delhi, 1994).

CHAPTER 45

Science in Twentieth-Century Japan

JAMES R. BARTHOLOMEW

Japanese science has undergone enormous change in the twentieth century, but these changes have often followed some well-defined parameters. From a small, elitist project based in two universities and a few state-supported laboratories at the beginning of the century, it has evolved into the massive enterprise of academic, government, and corporate laboratories of the present day. In 1900 its mission was limited to either a disinterested form of inquiry modeled on that of the German university or to narrowly applied work designed to improve rice strains or find a variety of sheep better suited to Japan's humid climate. By the mid 1990s one influential form of science still had an obvious practical thrust, oriented as it was to crowding ever more circuitry onto a silicon chip or enhancing the efficiency of small batch production. But this science was now being done in corporate laboratories which scarcely existed at the turn of the century, and it benefitted from the preferential treatment of a sympathetic government. As in other countries which contribute to science, the twentieth century experience in Japan has featured the growing involvement with, and querolous support of, scientific and technical research by ever more social elements with diverse motivations. In 1900 the Japanese military establishment had little to do with scientific research, as is true in Japan today. But for several decades, the Japanese Army and Navy were significant sponsors of science, with occasionally disastrous results. Similarly, the universities were the dominant research institutions in the century's early years but lost their preeminence after World War II. It will be a central task here to explain these shifts of direction and emphasis.

FROM THE BEGINNING OF THE CENTURY TO WORLD WAR I

In 1900 medicine was by far the strongest field among Japanese technical research specialties. Though rarely the preferred undergraduate or graduate major, it enjoyed a number of assets not possessed by any other discipline: leadership, money, access to the powerful, and major discoveries. Some of this was owing to Shibasaburo Kitasato (1852–1931), co-discoverer of natural immunity, the plague bacillus, and tetanus antitoxin, who in 1892 returned to Japan after seven years of research in Germany to found the Institute of Infectious Diseases. Both a distinguished researcher and a political infighter of unusual skill, Kitasato joined forces with sympathetic officials and physician members of parliament to create and sustain a laboratory which, over the next twenty years, evolved into one of the world's three leading centers of non-academic medical research.[1] Some of medicine's prestige also reflected its relatively strong presence at the Institute's major rival, Tokyo University. Deriving in part from an eighteenth century medical academy, Tokyo's modern Faculty of Medicine had by 1902 entirely replaced German guest professors with Japanese academics and had begun contributing to medical knowledge. It had no achievements quite so grand as those of Kitasato, but important contributions would come during World War I.

In contrast to the physical sciences or engineering, medicine became the first technical field in which Japan gained a significant reputation, partly because it began with a substantially stronger base. Tokugawa rule (1600–1867), which immediately preceded the reformist Meiji era (1868–1912), is often associated with political isolation and the status quo in cultural life. And while this description is not completely inaccurate, it obfuscates the complex reality of differential treatment of various disciplines by a regime which generally allowed considerable freedom to medical practitioners while keeping any activity — translation, observation, experimentation, teaching — related to the physical sciences on a very tight leash. The reasons for this policy cannot be dealt with here, but its longterm consequences were substantial. By the 1850s Japanese scholars had translated important European medical texts, performed anatomical dissections and published the results, begun vaccinating against smallpox, independently developed anesthesia, and were teaching European medical specialties in more than two dozen schools and academies.[2] Nothing like this foundation had been laid for any other technical specialty.

Medicine and other technical fields in Meiji Japan developed in a particular institutional context, mostly state dominated. There were private institutions of higher learning — Keio, Waseda, Doshisha, Aoyama Gakuin and others — but few managed to establish successful programs in medicine, engineering, or the sciences prior to World War I. Alongside a network of experiment stations and state laboratories like the Institute of Infectious Diseases was the slowly expanding system of imperial universities, and these were the institutions in which academic science was based. Tokyo, founded in 1877, was the first; but in 1897 a second foundation was made at Kyoto, and by 1914 two additional universities had been created at Sendai (Tohoku Imperial University) and Fukuoka (Kyushu Imperial University). All were organized after 1893 on the basis of the so-called chair system (*koza seido*). Loosely derived from an original German model, a full professor who occupied a chair supervised the teaching and research activities of one or more assistant professors, lecturers, and various graduate students while keeping control of the budget. There were no Japanese *privatdozenten* (private lecturers) to compete with the full professors as in Germany. However, a considerable measure of competition was assured as a result of Japan's explicit rejection of the German one-chair rule and very self-conscious adoption of the French multiple chair system in which any particular discipline at a single institution could be represented by more than one full professor, more or less in accord with student enrollments.

Every major feature of the university system proved controversial over the years; but perhaps the liveliest debate focused on the role of researcher, or its absence. A curious feature of Japan in this period was that the country's political *and* academic establishments created institutions for scientific research and teaching while laying down an elaborate rhetoric about the importance of research, yet did not quite create the role of researcher which remained inchoate, vaguely defined and surrounded by hostile forces. Some of this was due to the emphasis placed on study abroad by Japanese graduate students and professors which on some occasions could take on as much ritualistic as substantive importance. Overseas study in this period was inevitable and desirable on one level, but the relative sums of money expended on it as compared to those expended on research at home were very large for a nation of modest resources and attracted criticism from those urging greater scientific independence of Europe. Much the same was true of the practice of *naishoku* or part time employment by professors who were drawing regular salaries from their academic positions. Because of the high demand from private business or government ministries for consulting services and the scarcity of expertise in all technical disciplines, university scientists were much in demand and often gave more attention to external forms of employment than they did to their regular academic duties.[3]

A fundamental question to ask is what the political and commercial establishments expected of Japanese science and of scientists in this period. A cynical answer provided years later by a physicist historian of science is that Japanese scientists were seen by the business community and government officials as little more than "tools for the extraction of Western forms of knowledge."[4] Certainly there is truth in this characterization. Only a small number of firms developed in-house research facilities; most preferred buying patents from foreign companies to developing their own technology. University chairs had no budget for research as such but only for instruction, and this was in spite of the relentless exhortations to research directed at professors by members of parliament and government officials. Kitasato's Institute of Infectious Diseases was exceptional in having significant funds for research. Most research projects which received funds from government sources did so almost entirely on an ad hoc basis. Dr. Kanehiro Takagi was able to carry out important studies on the treatment of beriberi by alterations in diet because a network of influential friends gave him access to high officials, and through them to special funding. Private sources of support for science were also extremely limited. For example, Jokichi Takamine, an industrial chemist who needed private support for his work in industry had to leave Japan for opportunities in the United States where in 1900 he isolated adrenaline, the first known hormone. Medicine again was a partial exception, but only for those who worked with Kitasato.

World War I and its Impact

World War I was probably the most important turning point in the history of modern Japanese science. The war ultimately forced Japanese elites to provide support for virtually all major fields of scientific study on a systematic, rather than the previously ad hoc, basis and led the military to pay attention to Japanese scientific research. It created the institutional conditions which would support the rise to world prominence of Japanese nuclear physics in the inter-war period. It induced a certain shift in Japanese attention away from German science and toward that of the United States. It stimulated greater research independence. And it drew Japanese researchers and research endeavors into a more intimate relationship with scientific communities in the world at large. Despite the disruptions of World War II discussed below, these changes proved to be long term in nature.

While Japan, though formally allied with Britain and France, was not a major combatant, it may well have been the biggest single casualty of its European allies' decision to blockade Imperial Germany. Before the war, Japan had depended heavily on Germany for industrial chemicals, pharmaceuticals, dyestuffs, and precision instruments to say nothing of Germany's importance to the ever increasing flow of Japanese scientists, engineers, and physicians who visited its universities and industrial laboratories. Thus, the earliest effects of the blockade were to create a powerful sense of crisis, leading to finger pointing and accusations against academic scientists for supposedly letting down the country by neglecting research. In actual fact, Japanese elites had received from the scientists almost exactly what they had paid for, no more and no less. In the years immediately preceding the war, indifference from the business community had caused the Tokyo Industrial Experiment Laboratory, a well conceived chemical research institute founded in 1900, to languish. And between 1908 and 1913, several strong proposals for expanded facilities in physics and chemistry had been rejected as unnecessary, unrealistic, or simply too expensive.[5]

But in the radically changed climate created by the war, many important innovations became not only possible but almost inevitable. One was the establishment of the Research Institute for Physics and Chemistry in 1917, an event that largely fulfilled the frustrated aspirations of the pre-war years. Another, curiously, was the disruption of Kitasato's Institute of Infectious Diseases in 1914 and 1915. A third was the creation in 1918 of the Aeronautics Research Institute at Tokyo University, and a fourth was the founding of an entirely new university (Hokkaido Imperial University) at Sapporo in 1918. Nor did these developments exhaust the list. A project for fixation of nitrogen received support from parliament. Private universities, legally equivalent to the state-supported institutions, were finally authorized; and several of them managed to create programs in technical fields. Private foundations dedicated to the support of scientific research appeared for the first time. A system of peer reviewed grants for researchers was set up under the auspices of the Ministry of Education. And Japan became a founding member in 1918 of the International Council of Scientific Unions.

Establishment of the Research Institute for Physics and Chemistry and the disruption of Kitasato's Institute of Infectious Diseases were strangely linked together by the political initiatives of wartime prime minister, Shigenobu Okuma (1838–1922). Long interested in issues relating to higher education, medicine, and science, Okuma took the lead in arranging the coalition of elites which after many difficulties saw to the creation of Japan's first major facility for research in the physical sciences. These forces included members of parliament, a number of prominent academic scientists, the nation's wealthiest business leaders, and various government officials. The funding obtained featured an impressive combination of public and private resources, and the institution that resulted became Japan's most important single research facility in the inter-war period, providing important assistance to Japan's first two Nobel laureates in science.[6]

The same prime minister who led the campaign to create the Research Institute for Physics and Chemistry was also primarily responsible for disrupting the Institute of Infectious Diseases. This was ironic because Okuma had long admired Kitasato and his achievements. However, he was persuaded by his minister of education, Kitokuro Ichiki, that Kitasato's political independence and ability to achieve almost any political or funding goals he sought were a threat to the political integrity of the state and even more to the nationally coordinated effort in scientific research which the prime minister deemed essential in the face of the war. By moving the Institute from the Ministry of Home Affairs where Kitasato had many admirers to the Ministry of Education and Tokyo University where his detractors were numerous, Okuma had hoped to exert some control over Kitasato but instead brought about the eminent bacteriologist's resignation from government service and that of his entire research staff. The Tokyo University medical scientists who thereby gained control were unable for several years to manage effectively the technically difficult tasks of serum and vaccine production which had been Kitasato's exclusive responsibility for 20 years. This outcome resulted in accelerated death rates for the victims of several infectious diseases and contributed to the Okuma Cabinet's fall from power later in the war. More fundamentally, this combination of events — the establishment of the Research Institute for Physics and Chemistry and the disruption of the Institute of Infectious Diseases — marked a significant loss of political influence by the medical research establishment and substantially greater influence for certain branches of chemistry and physics.

Prior to the war, the Imperial Army and Navy had, of course, established medical and engineering programs of a practical nature; and Tokyo University had taught such subjects as naval architecture and ordnance engineering for more than twenty years. But the Japanese armed forces had never become involved in anything which could be described as scientific research. World War I changed this situation permanently, though not immediately to the military's satisfaction. Their research interests were stimulated by the new technology of military aviation. As far back as the Russo-Japanese War of 1904–1905 the armed forces had been interested in the military possibilities of balloon reconnaisance. But the short duration of that conflict and the armed services' own lack of scientifically trained personnel precluded most development, and only reports on the use of military aircraft along the Western Front in 1915 caused interest to revive. The Navy went its own way, but the Army elected — reluctantly — to join forces with the physicists and engineers at Tokyo University in 1918, creating the Aeronautics Research Institute. This research facility would later play a crucial role in training the noted aviation engineer, Jiro Horikoshi, developer of the famed Zero fighter plane which burst onto the world military scene in 1940.[7]

In the meantime, Japanese researchers were unable to study in, or travel to, Germany and thus began to search for alternatives. Britain was generally the most attractive venue while the war was underway; but a growing number, including Japan's first woman scientist (Kono Yasui), travelled to the United States for advanced training. (In Yasui's case, the destination was Harvard). After the conclusion of hostilities in 1918, foreign study in Germany resumed; but it never again attained the overwhelming predominance for Japanese scientists that had typified the years before 1914.

The war, as we have seen, had other important effects, including the expansion of academic and research infrastructure such as the newly founded Hokkaido Imperial University. But the most radical departure from earlier patterns attributable to the war was undoubtedly the appearance of formalized systems for supporting research. Prior to 1914 the only formal mechanism for supporting research (that is, excluding ad hoc arrangements) was a very small program created in 1912 and administered by the Imperial Academy of Sciences. This arrangement was soon shown to be completely inadequate, and in 1918 the Ministry of Education folded the Academy program into a more ambitious scheme, the Ministry of Education Science Research Grants Program. This program was never free from controversy: the funding was inadequate (the equivalent of $73,000 annually in 1918 dollars); too many grants supposedly went to researchers at the more prominent institutions; and peer recommendations were not always followed. Nonetheless, the program set Japan on the same general funding path for science as was being adopted contemporaneously in Germany, Italy, the United States and elsewhere.

An equally controversial initiative was Japan's participation in founding the International Council of Scientific Unions (ICSU) in the period 1918–1920, because of its implications for relations with German science. On the one hand, Japanese scientists were honored to participate as citizens of a victorious Allied power with European and American colleagues on a basis of formal equality in launching an important international scientific venture. On the other hand, many were troubled by the ICSU's explicit intention to ostracize scientific colleagues in Germany and Austria more or less indefinitely. For some Japanese scientists, the issue turned on their relative standing in particular fields — chemistry, physics, geology, biology, medicine or mathematics — compared to that of colleagues in the West and for those scientists the lure of formal, equal standing was too strong; and the decision was made to join the ICSU and accept its restrictions on German and Austrian science.[8]

For many in medicine, however, equality was not an issue. In fact, a reminder of the excellence which Japanese medical research could achieve was the remarkable work of Katsusaburo Yamagiwa of Tokyo University. A professor of pathology in the Faculty of Medicine, Yamagiwa attracted considerable attention for artificially inducing tumors in laboratory animals by painting their skins with coal tar. This achievement, executed and published in stages between 1915 and 1918, confirmed Rudolph Virchow's chronic irritation theory of cancer and made Yamagiwa a leading candidate for the Nobel prize in medicine and physiology in the mid 1920s.[9]

The Inter-war and Wartime Eras, 1920–1945

The inter-war and wartime years in most respects marked the achievement of maturity by modern Japanese science. For the first time the Japanese produced work which would eventually win Nobel Prizes. More diverse research activities were being carried out than ever before. Science's clientele base underwent enormous expansion, now encompassing civilian government agencies and laboratories, private corporations, the Army and Navy, private universities and foundations, and of course the leading state (so-called imperial) universities. The universities in particular were further expanded with foundations at Nagoya and Osaka, and outside Japan proper in Korea and Taiwan (at Seoul and Taipei). Nonetheless, serious problems relating to funding remained. Whereas the war-time era, 1914–1918, had witnessed a massive industrial boom in Japan, the 1920s was a period of sluggish growth or even contraction made worse by the massive Kanto earthquake and fire which virtually leveled Tokyo, the capital, and Yokohama in September 1923. In addition, civilian politicians in an ostensibly liberal era, were more influential in making public policy than ever before; and they generally preferred to reduce taxes and public expenditures than to spend money on something so speculative as scientific research. Thus, funding for the Ministry of Education's Science Research Grants Program was essentially frozen at about $75,000 annually for the entire decade. There were new funding opportunities for scientists from private foundations; but their awards were dependent on endowments, which did not expand in an unpropitious economic climate. And since the number of researchers seeking grants was far larger than before, per capita funding remained flat or declined.

Despite some financial difficulties, university-based science prospered to an unprecedented degree in the inter-war period, assisted by greater direct contacts with eminent European scientists and by Japan's expanded participation in international scientific conferences. Albert Einstein, Werner Heisenberg, Paul Dirac, and Nils Bohr all visited Japan in these years — Einstein in 1922, Heisenberg and Dirac in 1929, and Bohr in 1937; while in 1924 Japan hosted its first major international scientific meeting, the Pan-Pacific Science Congress.[10] At the same time, Japanese scientific education was now capable of honing and nurturing scientific talent of the highest caliber. Hideki Yukawa and Shin'ichiro Tomonaga were both the sons of Kyoto University professors and Kyoto graduates themselves who had the good fortune to study with Kajuro Tamaki at an advanced level. A specialist in classical mechanics and relativity theory, Tamaki allowed these talented young physicists to make their own way in nuclear physics.[11] Yukawa's work first came to fruition when he announced his meson particle theory in 1934 while teaching at Osaka University. Tomonaga's studies were somewhat delayed by ill health, but in 1943 he published his "super-many-time" covariant theory of quantum electrodynamics which brought him world fame after its translation into English in 1946. They received Nobel Prizes in 1949 and 1965 respectively.

The dominant influence of this period for science and everything else in Japan was, of course, World War II and the preceding tide of militarism. In September 1931, the Japanese field army which was stationed in Manchuria to protect local Japanese investments, illegally and without warning, seized control of the area and set off a political explosion. The pattern of political control by civilian officials indifferent to science was destroyed, and Japan began moving to a quasi-wartime economic system. International opposition, led by the United States and China, intensified local feelings in Japan against the rest of the world and in so doing strengthened the case of those who insisted that greater self-reliance in science was necessary. In this radically altered climate a militarily influenced government in 1932 created the Japan Society for the Promotion of Science (JSPS), a funding agency for scientific research with far greater resources than any previously available.[12] Having no laboratories or other facilties of its own, the hallmark of the JSPS was its dedication to large-scale, expensive projects of a kind not generally undertaken in Japan prior to that time.

The new climate of militarism which increased funding for science in the 1930s also allowed the Imperial Army to launch a new program dedicated to producing biological and chemical weapons. While the origins of this initiative can be traced back to Japanese observations of German chemical weapons in World War I, this new program was launched by a medically trained Army officer, Shiro Ishii, who began lobbying for it in 1930. It concluded only with Japan's defeat by the Allied Powers in August 1945. For ten years at elaborate, specially constructed facilities

FIGURE 45.1: A CYCLOTRON BUILT AT THE PHYSICAL AND CHEMICAL RESEARCH INSTITUTE IN 1936. © NISHINA MEMORIAL FOUNDATION.

FIGURE 45.2: A CYCLOTRON BUILT IN 1937 AT THE INSTITUTE FOR PHYSICS AND CHEMISTRY. TAKEN FROM TAMAKI AND EZAWA (EDS.) *NISHINA YOSHIO: THE DAWN OF ATOMIC SCIENCE IN JAPAN* (1991).

FIGURE 45.3: IMAGES FROM THE AERONAUTICS RESEARCH INSTITUTE AT TOKYO IMPERIAL UNIVERSITY BEFORE WORLD WAR TWO. COURTESY OF CENTER FOR ADVANCED SCIENCE AND TECHNOLOGY, UNIVERSITY OF TOKYO.

in Manchuria, Ishii supervised a large staff of chemists, physicians, microbiologists, and other medical specialists who sought to develop chemical and biological weapons through direct experimentation, not on rats or other laboratory animals, but involuntarily on healthy human subjects apprehended by Japanese military police. Most of the victims were Chinese, but members of other ethnic groups were also targeted by the Ishii researchers.[13]

Ishii came from a prominent family and was the son-in-law of the noted pathologist, Torasaburo Araki, President of Kyoto University. These personal connections and the ruthless militarism of the time allowed him to mobilize not medical fringe elements but the very best young researchers trained by leading research universities. Deliberately concealed from the scrutiny of most US Occupation personnel after the war, it has become known that American military authorities at the highest level authorized lenient treatment — usually no punishment at all — for the perpetrators in return for their turning all scientific data over to the US biological and chemical weapons research facility at Ft. Detrick, Maryland.[14] Ishii's activities bequeathed a legacy to post-war Japanese science which has yet to be calculated in full.

As the bio-chemical weapons initiative shows, Japan was much like other belligerent nations in mobilizing scientists for military projects during World War II. Under the loose direction of Yoshio Nishina at the Research Institute for Physics and Chemistry, and Bunsaku Arakatsu at Kyoto University, Japan pursued two projects aimed at developing an atomic bomb. Hidetsugu Yagi worked on military field communications systems. Masatoshi Okochi and several coworkers at the Research Institute developed piston rings for military aircraft.[15] Ken'ichi Fukui and other chemists at the Army Fuels Research Institute in Fuchu worked to convert ordinary gasoline into high octane fuel suitable for use in military aircraft.[16] But Japanese scientists, unlike their colleagues in Britain or the United States, did not become involved in operations research; in fact, the nature of Japanese militarism all but precluded such a role. Scientists in Japan did not propose and the military dispose, as was the case with the Manhattan Project in the United States. Instead, Japan's military decided what it wanted; and Japanese scientists were expected to deliver.

This does not mean that the wartime experience was either unimportant for modern Japanese science or completely negative in its implications. Funding for research increased by several orders of magnitude compared to the conditions of the 1920s. Many scientists received training in research specialties which would find their audience after 1945. Science's institutional base grew far more rapidly than would ever have been possible under peacetime conditions. And under JSPS auspices there began the practice of organizing the large scale group initiatives called "Gakushin Tokutei Kenkyu" or JSPS Priority Research Projects. These programs continued into the post-war era, receiving their sponsorship from an influential government agency known as the Ministry of International Trade and Industry (MITI).

The Post-war Era, 1945–1973

Military defeat in August 1945 marked the inception of a quite different era for science. Not only Japanese military forces themselves but all kinds of military research were expressly forbidden by the US Occupation forces. The Research Institute for Physics and Chemistry was accordingly disbanded with its cyclotron being thrown into Tokyo Bay. Ishii's researchers were interrogated but not punished. However, the academic perpetrators of vivisection experiments on captured American pilots at Kyushu University were hanged. The new climate for science in Japan encouraged a sweeping, sometimes painful mood of self-reflection and criticism among Japanese scientists. Occupation authorities actively encouraged self-criticism; but some of its manifestations went well beyond anything considered desirable by the Americans. Occupation science advisors, especially the physicist, H. C. Kelly, expounded a very particular vision for Japanese science. Taking the position that basic science was a 'luxury' for Japan under the current conditions of economic chaos and physical destruction, Kelly argued that scientists should devote themselves entirely to applied research beneficial to economic reconstruction.[17] Though not all Japanese scientists agreed with him, many did. Equally important, this message resonated well with the contemporary thinking of many influential people in politics and in the business community.

This renewed emphasis on applied research was not discontinuous with the dominant attitudes of Meiji and the inter-war period, but there was one important difference. In Meiji Japan especially, universities were few in number and elitist in outlook; and dominant political elites almost always considered them to be essential to the nation's modernization prospects. Post–1945 attitudes were almost universally anti-elitist, dedicated to the expansion of higher educational opportunities for Japanese youth, and above all, regarding universities as only one among several institutions important to the cause of economic rebuilding. Corporate laboratories were clearly to be another. A particularly clear illustration of this newly emergent pattern was the early career of Reona (Leo) Esaki in semiconductor research and the venue in which it occurred. Though an early post-war graduate of Tokyo University, Esaki went to work not in the academy but in industry, first at the Kobe Kogyo Corporation and then, between 1956 and 1960, at the Sony Corporation.

While no simple generalizations are possible regarding scientists' relationship to industry, Sony proved to be a particularly favorable environment for Esaki. Sony's founders, Akio Morita and Masaru Ibuka, were formally trained in physics and engineering respectively. They believed that their competitive position vis-a-vis other corporations required them to attract the highest possible caliber of talent since their status as a post-war start-up gave them little else to fall back on in the event of commercial failure. Thus, they regularly hired the employees they needed from other firms when raiding was considered unorthodox. They also had somewhat iconoclastic views, compared to those of many other Japanese corporations.

In 1964 Morita published a book which attacked the importance of formalistic credentials, much esteemed in Japanese big business, and insisted that it was the particular firm's responsibility to determine what an employee's actual talents were, whatever the diploma said. Sony hired Esaki, who was still a graduate student, when Ibuka heard him give a paper in 1955. Their gamble paid off scientifically when their new employee discovered the tunneling phenomenon in semiconductors, a physical reality deemed impossible in classical physics but long predicted in the light of quantum mechanics. Ultimately, this discovery, which Esaki developed as the Esaki tunnel diode, became part of modern computer architecture and won its discoverer a share of the 1973 Nobel Prize in physics.[18]

Esaki's departure for the United States in 1960 (where he went to work for IBM in Westchester County, New York State) shows how far post-war Japan was from developing anything like the contemporary American pattern of the military-academic-industrial complex. Government sponsored work, as on large scale integration in semiconductors, flourished under the Ministry of International Trade and Industry (reorganized in 1949) with additional patronage from private business. But substantive achievements in science, as opposed to engineering, were few in this sector; basic science remained the province of the research universities, which for years were starved for funds. This was owing both to the outlook epitomized by H.C. Kelly and to the political estrangement which developed between political and business elites on the one hand and academic elites on the other. Conservative politicians and business leaders acknowledged the destructive effects of the war; but academic scientists, often leftist in orientation, repudiated their earlier involvement in wartime research and failure to oppose militarism. After 1946 they banded together in the League of Democratic Scientists to assure that nothing similar would occur again; business and political leaders considered the League little more than a Communist front.[19]

These incompatible views and fundamental tensions, while diminishing somewhat with the oil crises of the early 1970s, were not completely dissipated until the 1980s, partly because of Japan's close political alignment with the United States during the Cold War. And they were not conducive to productive relations between academic scientists and other elites. Japan's difficulties with nuclear energy clearly illustrate the pattern. Following President Dwight Eisenhower's "Atoms for Peace" address to the UN General Assembly in December 1953, certain influential politicians and business leaders came to believe that nuclear power offered *the* solution to Japan's long recognized energy shortages, and decided to import Calder Hall reactors from Britain. Prominent members of the scientific community — especially Hideki Yukawa — cautioned against this excessive optimism and warned that nuclear energy was not at the time a proven technology. In Europe, enthusiasts of nuclear power accepted scientists' arguments that additional research to adapt the technology to local conditions was essential. But Japanese leaders rejected this advice and dismissed the scientists' arguments as self-interested and alarmist. The results of this communications gap, intensified by politics, were an inefficient industry and power plant breakdowns into the late 1970s![20]

Anti-elitist policies regarding universities and scientific research in general were espoused by many in the academy, in government, and in business; and they had important results for the evolution of Japanese science in this period. In a somewhat unusual but fortunate development, Ken'ichi Fukui was able to ascend to a full professorship in industrial chemistry at Kyoto University in 1951 at the youthful age of 33. This allowed him to set the research agenda for his university chair (i.e., laboratory) in a way that maximized the formidable research talents he was able to deploy. His unit was originally dedicated to research in fuel mechanics, reflecting long time Japanese concerns with energy supplies. Under Fukui's leadership the research program moved gradually away from the physical chemistry of hydrocarbons to the theory of chemical reactions. In a series of papers published in English during the 1950s, Fukui developed his "frontier orbitals" theory of chemical reactions, ultimately sharing the 1981 Nobel Prize in chemistry with Roald Hoffmann of Cornell University.[21]

These post-war trends of anti-elitism and political estrangement among elites had less propitious results under other circumstances. Because of early post-war allegations that Japanese science had been overly hierarchical in the inter-war period, the Ministry of Education's Science Research Grants Program was considerably revamped in the late 1940s to make at least some research funds available to nearly all investigators, regardless of institutional affiliation or political point of view. In practice this often meant that senior professors received general lump sums for research and then distributed them to younger people, whose political independence was thus supposedly protected. Critics argued, with some justification, that these procedures ignored merit and rewarded mediocrity. When Susumu Tonegawa completed his chemistry program at Kyoto University in 1963, he decided to pursue graduate work in molecular biology at the same institution. His adviser recommended that he do graduate work in the United States instead, partly to circumvent the constraints of the Research Grants Program. Tonegawa remained abroad, gaining his Ph.D at the University of California, San Diego and later doing research at the Basel Institute of Immunology in Switzerland.[22] From 1981 he was professor of biology at MIT and in 1987 became the first Japanese to receive the Nobel Prize in medicine and physiology, for clarifying the genetic origins of antibody diversity.

Especially during the Occupation years, 1945–52, anti-elitism had also expressed itself in the rapid proliferation of new universities and hasty expansion of existing schools. This was to have significant, unforeseen consequences. While expansion offered educational opportunities to some who might not otherwise have received them, it also led to inadequate facilities, overcrowded classes, unhappy professors, and dissatisfied students. From the mid 1950s onward, Japan had a vigorous, highly politicized student movement which regularly opposed the generally conservative national government. But at the end of the 1960s this student activism reached a level of antagonism for which certain government officials sought unprecedented solutions. Their preferred strategy was a dispersion of educational institutions (and, it was anticipated, government agencies) from Tokyo to other parts of the country. The creation of the new Tsukuba University and the accompanying Tsukuba Science City in the early 1970s was the first fruit of this initiative. Despite Tsukuba's excellent facilities, most professors and well established institutions refused to consider the prospect of a move from Tokyo to the seemingly remote region of Ibaragi Prefecture where the University and Science City were built. But the relatively small Tokyo University of Education, headed by Shin'ichiro Tomonaga, did agree to move; and it became the nucleus of the important new venture.[23] In subsequent years the Tsukuba model had some attraction for other, less developed parts of Japan.

From the Oil Shock to the Present, 1973–1996

An important turning point for post-war Japanese science arrived late in 1973 when the Organization of Petroleum Exporting Countries (OPEC) dramatically, and in a short period of time, increased the price of oil to consuming nations by about 400 percent. Like the naval blockade against Germany imposed in 1914, the effect in Japan was immediately perceived as potentially devastating; yet once certain adjustments were made, the crisis had the effect of creating valuable opportunities. During the two decades immediately preceding the oil crisis, spending on science in Japan had regularly lagged behind that in other affluent industrial democracies, even during the high growth years of the 1960s. Whereas nations like Germany and France had spent over 2.5 percent of GNP on research, the Japanese commitment had been more like 1.5 percent. The oil crisis of 1973–75 changed this pattern fundamentally. Despite high inflation (about 20 percent in 1974) and harsh readjustments, research expenditure levels rose dramatically; at many corporations spending rose by nearly 400 percent over two years, and overall research spending as a percent of GNP gradually increased to nearly 3 percent. Universities did not immediately share in the greater largesse for science, but even they were shortly to see some improvements.

The oil crisis and the reaction to it was similar to the naval blockade crisis of 1914 in another way. Because of post-war rebuilding, diminished self-confidence among elites, and scientists' own need to catch up with developments abroad during the wartime years (which was a period of particular isolation for Japan), there had been a pronounced tendency to rely almost entirely on the United States (and to a lesser extent Europe) for basic research. Yukawa's 1949 Nobel Prize and to some extent the Tomonaga Nobel Prize in 1965 did have some positive benefits for physics; but the more typical situation was that of Susumu Tonegawa who was advised in 1963 to travel to the United States to study the new field of molecular biology. The events of 1973–74 broke this twenty-eight year pattern; but it took several years for a new pattern to emerge. This new pattern exists at present; and it includes a reestablishment of intimate ties between universities and corporations, higher levels of spending, enhanced interest in basic science, what many refer to as the internationalization of Japanese science, and (after 1990) more interest among political and business elites in universities and university-based research.[24]

Toward the end of the previous era, especially from about 1972, the Japanese government, acting through the Ministry of International Trade and Industry (MITI), had begun to revive the pre-war JSPS practice of defining certain broad research themes and initiatives for which special resources were available. One important initiative in the 1970s was the VLSI (Very Large Scale Integration) project relating to computers. Another targeted the development of flexible manufacturing systems using lasers. Still another focused on optical measurement and control systems. And as always, several such initiatives were directed to improving energy supplies. These programs were almost entirely based in government or corporate laboratories. But in the 1980s MITI sponsored major research efforts involving, and even headed by, university professors. Most notable were the relatively unsuccessful 5th Generation Computing Project and the Superconductivity Initiative whose directors were both professors at Tokyo University.[25] During the same decade, major corporations were increasingly inclined to allow academics free access to their (almost invariably) superior research facilities and in certain indirect ways, to support university research.

Major enhancements of university research began to appear in the late 1980s. For most of the twentieth century, it had been impossible for foreign nationals to hold tenured professorships at the leading government universities. In 1987 this restriction was thrown out; and in the name of internationalization, efforts were being made to attract foreign professors. (Similar efforts to attract foreign graduate students were launched in 1985 and have met with some success). Combined with the initiative regarding professors was a growing recognition among at least some leading corporations that they should offer greater direct financial support to universities than had been done after 1945. Thus, in 1987 an American scientist from Bell Laboratories was hired to fill a new chair in computer science at Tokyo University established with a major gift from NEC Corporation. And in April 1992 Reona Esaki left the United States after more than 30 years abroad and returned to Japan as President of Tsukuba University. As though to set the stage for this decision, the Keidanren organization of large corporations in 1990 had issued a special report which criticized the delapitated state of many university facilities — widely reported in the press — and advocated spending an unprecedented $100 billion in new money on universities and university research. For the moment, unsettled economic conditions in Japan have prohibited the full realization of this proposal; but indications are that the plan will eventually be fulfilled.

Finally, it seems significant that there has been a much heightened interest in Japan since 1987 in the Nobel Prizes, especially those for science. The *Yomiuri*

Shimbun, a leading newspaper, started its annual Yomiuri Forum of Nobel Prize Recipients that year, partly to publicize the prizes and partly to stress the need for more fundamental research in Japan. The first Japanese language biographical dictionary of Nobel Prize recipients was published in 1990; a second such directory appeared in 1994. Many books and articles in the popular press, together with extended commentary in science-oriented journals, about the Nobel prizes and Japan's historical involvement with them have appeared in the last decade or so. In 1981 a prominent Swedish official closely connected with the Nobel Prizes predicted that Japanese scientists would win an increasing number over the next twenty years, and in April 1994 Dr. Esaki told the Emperor and Empress at their spring garden party that he and others were working hard to create universities whose professors *would* win more Nobel Prizes.[26] There is some distance to go if this goal is to be reached. To date, only five Japanese scientists have received Nobel Prizes in science, and the only one awarded for research done in the past forty years was to Dr. Tonegawa, whose work was done in Europe.

Nonetheless, there are favorable auguries — internationalization; the end of the early post-war estrangment between government, business and the academy; greater commitments to basic research; enhanced funding; and so on. And if they come to fruition, Japanese science may enter a golden age.

References

1. John B. Blake, "Scientific Institutions since the Renaissance: Their Role in Medical Research," *Proceedings of the American Philosophical Society*, Vol. 101, No. 1 (February, 1957, passim).
2. Shigeru Nakayama, "Japanese Scientific Thought," in Charles C. Gillespie (Ed.), *Dictionary of Scientific Biography*, Vol. XV, Supplement 1. (New York: Charles Scribners' Sons, 1978), passim.
3. James R. Bartholomew, *The Formation of Science in Japan: Building a Research Tradition.* (New Haven and London: Yale University Press, 1989), Chapter 3.
4. Yuasa Mitsutomo, *Kagaku goju nen.* (Tokyo: Jiji Tsushin Sha, 1950), p. 73.
5. Bartholomew (n. 3 above), Chapter 7.
6. Kiyonobu Itakura and Eri Yagi, "The Japanese Research System and the Establishment of the Institute of Physical and Chemical Research," in Shigeru Nakayama, David L. Swain and Eri Yagi (Eds.), *Science and Society in Modern Japan.* (Tokyo: University of Tokyo Press, 1974), passim. See also Bartholomew (n. 3 above), Chapter 7.
7. Takashi Nishiyama, "Legacy of Japan's Wartime Technology in Postwar Japan and the United States: The Case of the Mitsubishi Zero Fighter," unpublished seminar research paper, Department of History, Ohio State University, June 1995.
8. Bartholomew (n. 3 above), Chapter 8.
9. See *Medic. Nob. Kom. 1926 P. M. Forsandelser och Betankanden.* (Stockholm: Karolinska Institutet, 1926), passim.
10. Ippei Okamoto, "Albert Einstein in Japan: 1922," *American Journal of Physics*, Vol. 49, No. 10 (October, 1981), pp. 930–940, trans. Kenkichiro Koizumi.
11. Hideki Yukawa, *Tabibito: The Traveler.* (Singapore: World Scientific Publishing Company, 1982) Trans. L. Brown and R. Yoshida, pp. 163–164.
12. Sugiyama Shigeo, *Nihon no kindai kagaku shi.* (Tokyo: Asakura Shoten, 1994), p. 123.
13. Peter Williams and David Wallace, *Unit 731: Japan's Secret Biological Warfare in World War II.* (New York: The Free Press, 1989).
14. Sheldon H. Harris, *Factories of Death: Japanese Biological Warfare 1932–45 and the American Cover up.* (London: Routledge, 1994).

15. Michael Cusumano, '"Scientific Industry': Strategy, Technology, and Management in the Riken Industrial Group, 1917–1945," in William D. Wray (Ed.), *Managing Industrial Enterprise: Cases from Japan's Prewar Experience.* (Cambridge: Harvard University Press/Council on East Asian Studies, 1992).
16. Fukui Ken'ichi and Esaki Reona, *Kagaku to ningen o kataru.* (Tokyo: Kyodo Tsushin Sha, 1982), pp. 166–167.
17. Shigeru Nakayama, "The American Occupation and the Science Council of Japan," in Everett Mendelsohn (Ed.), *Transformation and Tradition in the Sciences.* (Cambridge: Cambridge University Press, 1984), pp. 357–358.
18. Shakuntala Jayaswal, "Leo Esaki," in Frank N. Magill (Ed.), *The Nobel Prize Winners: Physics,* Vol. 3. (Pasadena, California and Englewood Cliffs, New Jersey: Salem Press, 1989), pp. 1019–1025.
19. Shigeru Nakayama, *Science. Technology and Society in Post-war Japan.* (London and New York: Kegan Paul International, 1991), pp. 18–26.
20. Hideo Sato, "The Politics of Technology Importation in Japan: the Case of Atomic Power Reactors," pp. 7, 16, 22, 25, 48–49. Prepared for the Conference on Technological Innovation and Diffusion (sponsored by the Social Sciences Research Council), Kona, Hawaii, February 7–11, 1978.
21. Scott A. Davis, "Kenichi Fukui," in Frank N. Magill (Ed.), *The Nobel Prizes: Chemistry,* Vol. 3. (Pasadena, California and Englewood Cliffs, New Jersey: The Salem Press, 1989), pp. 1061–1066.
22. Stephen Kreider Yoder, "Native Son's Nobel Award is Japan's Loss," *The Wall Street Journal,* Vol. CCX, No. 75. (October 14, 1987), p. 26.
23. Sharon Traweek, *Beamtimes and Lifetimes: The World of High Energy Phvsicists.* (Cambridge: Harvard University Press, 1988), pp. 134–138, passim, and Nakayama (n. 19 above), pp. 137–138.
24. "Reforming Japan's science for the next century," *Nature: International Weekly Journal of Science,* Vol. 359, No. 6396 (15 October 1992), pp. 573–582.
25. "A Fifth Generation: Computers That Think," in *Business Week* (December 14, 1981), p. 31 and "Finishing First with the Fifth," *Time,* Vol. 122, No. 5 (August 1, 1983), p. 57.
26. Professor Sune Bergstrom, President of the Nobel Foundation, quoted in "Opening Address," The Nobel Foundation (Ed.), *Les Prix Nobel 1981.* (Stockholm: Almquist and Wiksell International, 1982), p. 20 and Esaki Reona, quoted in "Yorokobi no hi ni enyukai," *Yomiuri Shimbun,* No. 42384 (May 12, 1994), p. 26.

CHAPTER 46

The Politics of European Scientific Collaboration

JOHN KRIGE

A new form of scientific cooperation has been put in place in Europe in the second half of this century. It is a form in which the governments of several countries — often as many as a dozen — agree to finance an organization dedicated to collaborative research. Such agreements are both formal and binding in that they are enshrined in a 'convention,' a kind of international treaty which has to be ratified by national parliaments and wherein the 'member states' of the organization, in consultation with their national scientific communities lay down its basic aims, structure and functioning. What we have here is both more and other than simply the post-war alliance of science and the state. It is the emergence of a new structure and a potent source of funding and of legitimation for expensive fields of scientific research and technical development. It is the development of a lucrative strategy in which scientists and their allies have mobilized political support in peacetime for essentially nonmilitary research, bypassing and supplementing their national research programs and budgets, and the related decision-making and priority-setting mechanisms.

Two main considerations, one scientific the other politico-economic, have underpinned this new collaborative mode. Scientifically, it has been a response to the transformation in the nature of experimental work which began to take hold in the United States before the war, and which became ever more widespread after it. The availability for some fields of science of the resources required to build and use increasingly heavy equipment containing state-of-the art technology has redrawn the contours of disciplines (like nuclear physics), created entire new fields (as in high-energy physics or space research using rockets) and put otherwise unthinkable projects on the research agenda (like the human genome project). Experimental work now brings together scientists and engineers from a variety of disciplines who demand massive financial support from the state to build and to buy the instruments and machines required for their research, and to put in place and to manage the teams of researchers to exploit them.

European governments could not remain indifferent to these developments. The pace and nature of change was set by the world leader and they either had to follow suit or leave the United States free to dominate the research frontier. Opportunities were limited for individual countries, however, by the relative scarcity of human and material resources at their disposal, particularly after World War II. Battered at home by first one and then another brutal war, forced to relinquish their colonies abroad, they came to recognize, often with great difficulty, that as individual nations they could never hope to be more than medium-sized world powers with scientific programs commensurate with their diminished status. Hence the interest in exploiting an alternative path, a path opened up by the restructuring of the European politico-economic space which got under way after the war, and which has expanded and deepened over the last half-century.

Beginning in the 1940s a number of influential and charismatic political figures again promoted the idea that the time had come to build a 'united' Europe, to create new institutions which would make another bloody conflict in Europe 'impossible,' to put in place mechanisms which would hold in check the destructive forces of nationalism and economic autarky which, they felt, had been the root causes of two world wars. This movement towards some form of European economic and political union gathered momentum in the immediate post-war period, propelled by the worsening economic situation and the need to expand markets, by the felt need to contain and control German revival, by the onset of the Cold War which called for a collective response in the West to what was perceived as a Soviet military threat, and by US pressure for collaborative European action. Its success was not, and indeed never has been, guaranteed. The dilution of national sovereignty which integration entailed meant that any formal intergovernmental collaborative project necessarily raised doubts and opposition, and limited its sphere of application. Within these limits however, a new 'European' economico-political space has been gradually put in place over the last half-century, a space which has been populated by a huge variety of European scientific, technical and industrial programs covering diverse fields, sometimes under the umbrella of the European Union.

Consistent with the aims of this volume I shall concentrate on scientific research, as opposed to technological and related industrial development. I shall also limit myself to that form of scientific collaboration in which European governments agree to pool their resources and to build together multinational organizations. Broadly speaking, dedicated European research structures of this type have been set up in three main fields of science: the nuclear and solid-state physics, astronomy and space, and molecular biology. It is meaningless to date their births precisely: their establishment has often involved long and complex negotiations between the participating actors. Table 46.1 identifies some of the most important of these organizations and their facilities or functions, and uses the year in which the decision was taken on where to site them as a convenient temporal marker.

This form of European scientific collaboration, I will argue, is to be situated at the heart of the process of European economic and political integration. One

TABLE 46.1: SOME OF THE MORE IMPORTANT EUROPEAN ORGANIZATIONS FOR SCIENTIFIC COLLABORATION

Acronym	Date[a]	Site	Name	Facility/Function
CERN	1952	Geneva (CH)	E. Organization for Nuclear Research[b]	High-energy physics with accelerators
JRC[c]	1958/ 1959	Main lab in Ispra (I)	Joint Research Centre, Euratom	Nuclear, and then diverse
ESRO	1962	HQ in Paris (F)[d]	E. Space Research Organization	Space science with satellites and sounding rockets
ELDO	1962	HQ in Paris (F)[e]	E. Launcher Development Organization	Development of a 3-stage heavy satellite launcher
ESO	1964	La Silla (Chile)[f]	E. Southern Observatory	Ground-based telescopes
EMBL[g]	1973	Heidelberg (G)	E. Molecular Biology Laboratory	Diverse instruments
ESA[h]	1975[i]	HQ in Paris (F)	E. Space Agency	Scientific & application satellites and heavy launcher development
JET[j]	1977	Culham (UK)	Joint European Torus	High current tokamak
ESRF[k]	1984	Grenoble (F)	E. Synchrotron Radiation Facility	Electron synchrotron radiation source for solid state studies

Notes

a. The date is that on which the site for the organization was chosen.

b. CERN is today called the European Laboratory for Particle Physics.

c. The JRC is funded by the European Union.

d. Other centers were established in Noordwijk (NL) for developing space technology and integrating payloads, in Darmstadt (G) for satellites operations, in Frascati (I) for scientific research, in Kiruna (S) for launching sounding rockets, as well as a world wide tracking network.

e. ELDO used the Woomera launch range in Australia.

f. In 1975 it was agreed that ESO's headquarters would be at Garching, near Munich (G).

g. The establishment of EMBL was one of the main objectives of EMBO, an organization set up under Swiss law in 1964, and supported by an intergovernmental council, EMBC, established formally in 1970.

h. ESA merged the functions of ESRO and ELDO. It handed the launching base in Kiruna back to Sweden, transferred all activities at Woomera to the launching base in Kourou, French Guyana, and re-oriented the work at Frascati.

i. Date of the signature of the convention.

j. Jet is funded by the UK (20 percent) and the European Union (80 percent).

k. The ESRF shares a site with the Franco-German Institut Laue-Langevin, set up in 1966, and originally equipped with a neutron high flux reactor.

might sometimes imagine that scientists and governments alike collaborate in pursuit of 'noble' ideals, that in working together they renounce self-interest, subjugating it to the values of 'universalism' and 'internationalism' or 'Europeanism.' Doubtless such ideals do partly shape their actions. At the same time in an age where competition between scientists and between nation states is a dominant feature of their behavior, scientists and governments essentially collaborate with others because they believe that it is to their competitive advantage to do so. This is as true of, say, particle physicists setting up an international team to build and use a huge detector as it is of governments agreeing to join with one another to establish a European research organization. Collaboration, then, and European scientific collaboration in particular, is not undertaken at the expense of self-interest; it is rather, the pursuit of one's interests by other means.

Collaboration in a European scientific organization always involves a loss of, or at least a dilution of, national sovereignty. This loss is accepted but not taken for granted. Its scope is limited, carefully monitored and constantly re-evaluated. In what follows I shall illustrate some of the ways in which the interface between the national and the European dimensions of collaboration is defined and controlled by governments. The fusion of scientific need with national policy considerations in European collaboration occurs in selected areas. That of foreign policy is intrinsic to the integrative process itself, and it is this that I shall discuss first. I shall then look at how governments protect national interests in various contexts — the choice of the site for the research facility, the application of industrial policy, the definition of the scientific program. These case studies will be made essentially from the national perspective. Reversing the arrow of analysis I shall then explore how the organizations themselves view the national state apparatuses, how they have kept the centrifugal pull of national interest in check by creating a space which has been deliberately 'denationalized,' how they live at a level which transcends the vicissitudes of the national political and science systems. To conclude I shall quickly step outside the sphere of European scientific collaboration narrowly conceived in order to place it in a global, more specifically US context, indicating the long-term fragility of the structures which have been put in place over the past forty years.

Science and Foreign Policy

The visibility of the link between the scientific and foreign policies of the member states of European organizations is highly conjunctural. It is particularly evident when the bodies are being set up since the highest levels of the national political apparatus need to be mobilized and the ministers concerned need to reach agreement with their homologues in other countries. The connection also emerges whenever a major re-orientation in the policy of such organizations is on the agenda, a redirection which requires a debate at ministerial and cabinet level. For much of the time though foreign policy concerns are implicit, a taken-for-granted structural presence which frames the day-to-day operations of the institutions and which de-escalates the majority of the crises and difficulties through which it passes. Disgruntled governments have threatened to withdraw from European scientific organizations on more than one occasion. The foreign policy implications have, at least to date, forced them to draw back from what is, or has been, the 'unthinkable.'

The birth of CERN, the European Organization for Nuclear Research (as it was originally called), and the first intergovernmental organization of the type that concerns us here, can be used to illustrate some of these points. In the immediate post-war period a number of young nuclear physicists, notably in France and Italy, felt that some kind of collaborative European nuclear program was essential if the US was not to dominate the field, and attract the brightest and the best of a gifted and strategically important elite across the Atlantic. Capitalizing on their newly acquired status and influence in high political circles, which were now particularly receptive to all things nuclear, they played with a number of possible schemes, but to little avail. Many of their colleagues felt that priority should be given to building national atomic programs. There were also fears that the US would not tolerate a collaborative European effort in the nuclear field, particularly if that included countries with strong Communist parties.

The fears that the US would impede such a venture were lifted in June 1950. At a UNESCO conference in Florence, the US Nobel Prize winner for physics, and scientific statesman, Isidor I. Rabi, officially suggested that continental European states might like to establish what he called "regional research centres and laboratories" in fields like high-energy physics and biology where any one country lacked the requisite resources. Seizing the opportunity, a handful of scientists and science administrators, with the assistance of Pierre Auger, who was responsible for the exact and natural sciences inside UNESCO, sketched a collaborative European nuclear project. At its core lay a high-energy accelerator with a design energy at least as great as the most powerful machine then under construction in the US. This project was deemed big enough to attract politicians into a collaborative scheme without obviously touching national susceptibilities. Two intergovernmental conferences, at the end of 1951 and early in 1952, defined the technical and political structure of the project. In June 1953 the convention establishing CERN was signed. It entered into force after ratification by national parliaments in September 1954.

Foreign policy considerations in both France and the US were particularly important in the creation of CERN. In May 1950, the month before the UNESCO meeting, the French Foreign Minister Robert Schuman proposed that Franco-German coal and steel production be placed under a common authority. This was a turning point in his country's policy towards her former enemy; an effort to protect France and Europe from German revanchism, not by confrontation as had previously been the case, but by a new kind of *supranational* organ through which

France hoped to control two key sectors of a resurgent (West) German economy. This resonated perfectly with the idea of building a collaborative nuclear center which was, like the coal and steel plan, regional, sectorial, and in a field which was associated with energy production.

Rabi's initiative is, similarly, to be situated in the framework of the policy of the US State Department at the time. The announcement of a Soviet atomic explosion towards the end of 1949 demolished the US nuclear monopoly and Washington began to think of ways to involve allied scientists more closely in the defense of the 'Free World.' In April 1950 the so-called Berkner report on "Science and Foreign Policy," with which Rabi had been involved, was published. It recommended that the US play a more active role in the reconstruction of continental Europe's shattered scientific infrastructure, science and technology being "the foundation of security and welfare." Indeed Rabi himself saw his initiative as an expression of the US's new willingness in the Cold War context to encourage and extend European collaboration beyond the economic and military fields. It was a kind of Marshall Plan for European science.

The weights attributed to scientific and foreign policy considerations differ according to the actors and their changing perceptions of the scientific and political issues at stake. When Britain finally decided to join CERN, for example, it was against the trend of her foreign policy objectives at the time. In the immediate postwar period the UK had three overlapping and mutually exclusive spheres of interest, the USA, Europe and the Commonwealth. She was also, at least for the first decade or two after the war, the leading scientific and technological nation in Europe. Initially her physics community, led by James Chadwick, and her government, were not interested in joining CERN. They had embarked on a major accelerator construction program after the war and, as a matter of policy, they were against setting up new international organizations. Chadwick was assured by Lord Cherwell, Churchill's adviser on scientific matters, that if they needed a new accelerator the money could be found to build it at home. It was only when it was decided, in summer 1952, that CERN should try to build a technically novel machine, far more powerful than any yet planned in the world, that a group of engineers at the atomic energy research establishment at Harwell managed to persuade a reluctant Chadwick, the Foreign Office and the Treasury that the United Kingdom should take the plunge.

It would be wrong to assume that scientific interests necessarily predominate over foreign policy considerations in bringing a government into a European project. The contrary can also be the case. This happened with British participation in the European space effort through ESRO and ELDO (see Table 46.1). The scientific and technical (as well as industrial and military) advantages of participating in a European program to build satellites and a launcher were evident. But the main force driving Harold Macmillan's Conservative government in the early 1960s was the conviction that the scheme could satisfy both domestic political exigencies and foreign policy requirements in all three of her spheres of interest. In April 1960, the government announced that it had decided to cancel development of its nearly completed, but already outdated, intermediate range ballistic missile, Blue Streak. Rather than abandon the project, however, it proposed to recycle Blue Streak as the first stage of a multistage civilian satellite launcher built in partnership with other countries on the continent. This gesture was domestically significant. It was intended both to 'save' the money already spent on the rocket, and to keep the engineering teams together in the event that the UK decided to develop an independent launch capability (as the French were doing) later in the decade.

Foreign policy considerations were also crucial. The idea was congenial to the US, and to the Kennedy administration in particular, who favored the UK's presence in Europe as a counterweight to the growing Paris-Bonn axis. The Commonwealth was protected by London insisting that the launch pad for the European rocket should be its missile testing range in Woomera. And indeed Australia was one of the seven founding member states of ELDO. As for Europe, and de Gaulle in particular, it was a move by the Macmillan government towards the newly founded European Economic Community, a 'proof' of their goodwill towards their potential partners across the Channel. Such a proof was all that more imperative since the same government had already tried to wreck the establishment of the EEC by setting up the rival EFTA (European Free Trade Association). In January 1961 Macmillan persuaded the French President that they should collaborate in the joint development of a European rocket, with Germany as a third major partner. That summer, and against the advice of many of his closest colleagues, the British Prime Minister submitted his country's (first) application to join the Common Market.

The domination of foreign (and indirectly domestic) policy considerations over scientific ones is also visible in the entry of Germany into CERN. Nuclear physicists, like Heisenberg, who had remained in Nazi Germany during the war were not readily reaccepted into the international scientific community immediately after it. A European collaborative venture which included Germany (imperative in the light of the USA's insistence that she be readmitted rapidly into the Western fold) was a perfect way for her scientists to regain international respectability and to take the first steps along the path of a civilian nuclear energy program. The interest of membership, Heisenberg remarked to his government, was 20 percent scientific and 80 percent political. The same could be said of Germany's membership of ESRO and ELDO. Acceptance into the European space effort signaled the removal of the stigma surrounding those scientists and engineers who had worked under the Nazi regime, and the relaxation of the limitations on missile/rocket development imposed by the 1954 Paris treaty. It legitimated the eventual start up of German national space programs in parallel to the European effort. And, in the case of rocketry, it dovetailed neatly with Minister of Defence Franz Josef Strauss' firm conviction that the strength of the Western alliance, and of the Federal

Republic's place in it, rested on the development of high technologies and the underlying industrial structure that called for.

Scientific collaboration can thus be an important tool of foreign policy. Indeed scientific cooperation, just because it often does not directly touch a strategic (i.e., military or commercial) nerve, can be one of the first areas in which new and bold foreign policy initiatives are taken between governments. Paradoxically, just because it is seen as being a 'non-political' activity, scientific collaboration can be a particularly useful first and tentative step in a politically delicate context of alliance building. There is a striking example of this at CERN in the mid-1960s (a case which also shows that it is not only at key turning points in the lives of these organizations that the link between the political and the scientific comes to the surface).

In summer 1967 CERN signed an official agreement with the Soviet authorities to design, build and test the equipment required to feed a proton beam from their new 76 GeV synchrotron to a bubble chamber called Mirabelle. This move enabled European scientists to collaborate with their Soviet colleagues on the most powerful machine in the world at that time. It also helped the French government reinforce its bridge building with the Soviets. It was not just that Mirabelle had been built at Saclay, just outside Paris. The agreements were signed shortly after de Gaulle, disenchanted with German Chancellor Ludwig Erhard's 'Atlanticism,' had persuaded the Soviets to adopt the French television system SECAM rather than the German system PAL, and had withdrawn his forces from NATO command. Collaboration through the provision of equipment and the sharing of the experimental workplace was an instrument at the service of European science and French foreign policy. It fostered the fledgling Franco-Soviet alliance without further damaging a tattered NATO and a strained Western solidarity. This use of scientific collaboration (in a field of no strategic interest) as a bridgehead into deeper, more complex forms of alliance building continued after the collapse of Communism and the fall of the Berlin wall in 1989. Many Eastern European states eager to be accepted into the European family took one of their first steps in that direction by joining . . . CERN.

Furthering Domestic Policy Objectives through European Collaboration

European collaboration and integration is not antithetical to the protection and the strengthening of the nation state; rather it can be seen as one strategy among others for furthering national interests. Extending my analysis beyond that of foreign policy, I shall now illustrate some of the ways in which governments use European collaborative projects to pursue domestic policy objectives.

The choice of the site for a major laboratory is a typical occasion on which domestic, not to say local political demands carry enormous weight in the decision-making process (be that in the USA or in Europe). The esteem in which science, and scientists and engineers are held, and the benefits which flow from hosting a major international research center transform the site question from one of simple geography into an issue of local and national pride and prestige. This prestige is fabricated through news-making and image-building visits of kings and queens, presidents and popes, ministers and secretaries of state. It is cashed in terms of contracts for local industry, notably the construction industry, the generation of high-technology satellite firms and even science parks in the vicinity of the laboratory, and the advantages to the local community of counting among their number a relatively cultured and well-educated, well-paid and well-disciplined professional elite. Member states vie to be host states.

The choice of a site for Europe's 300 GeV super proton synchrotron accelerator illustrates the point well. The first tentative attempts to settle the question got under way in 1964. At the time, it was assumed that the location would be somewhere other than Geneva, both to avoid concentrating all Europe's front-line high-energy physics capacity at CERN, and to allow other partners to capitalize on the presumed advantages of a major research center. The delicacy of the issue quickly emerged. Nine of CERN's member states submitted (sometimes multiple) bids for the laboratory. Fearful of losing control over the decision to their national state bureaucracies, the members of CERN's supreme governing body, the Council, selected 'Three Wise Men' from among their number to make an 'objective' assessment of the merits of the offers received. Their impartiality was 'assured' by choosing them from countries which had not offered sites, by their eminence — and by the shared understanding that the Council wanted above all to make a 'rational' choice and so denationalize the issue. The Site Evaluation Panel duly assessed the merits of the proposals along three axes pertaining respectively to the construction, operation and conviviality of the site concerned. No site got top marks in all three categories, though locations in France and in Italy came out best in two of them.

The panel laid its report before the Council towards the end of 1967. The document was warmly praised, and promptly buried. Its sought-for objectivity was predictably impugned by the disagreements between the member states over the appropriateness of the evaluation criteria used. More to the point, a Germany increasingly selfconfident about its right to protect its interests on the European scene, argued that the project was technically obsolete and financially extravagant. It demanded that if the machine was built at all it had to be located at the site it had offered, no matter what the position of that site was on the panel's league table. The German delegation was so recalcitrant on this point that even after they had agreed to join the project they forced the last minute cancellation of an interministerial meeting called by the Swiss to settle the site question in January 1970, fearing that the decision would not be in their favor. The problem was only resolved by a fundamental revision of the terms of the debate. In February 1970, John Adams, the 300GeV Project Director suggested that the accelerator be built on the existing CERN site, so reducing sharply the costs of construction and operation of the machine. Germany could not but yield.

The choice of a site for a European ground-based astronomical observatory is the exception that proves the rule. It took ten years partly because there was no question of building it in Europe. Discussions got under way in 1953 between astronomers from six countries (Belgium, France, the Federal Republic of Germany, The Netherlands, Sweden and the United Kingdom) who proposed to equip the European community with two new powerful telescopes. Astronomy, however, lacked the aura of the nuclear. The research was remote from any practical use and the technology required for the two planned instruments was relatively straightforward. What is more no member state was a candidate for the host state: the site was to be in South Africa. Scientific interest apart, the decision to proceed or not was thus a purely financial one, bereft of any broader industrial or local, national or global political considerations. That financial burden was made all the more difficult by the UK's decision in summer 1960 to withdraw from the project in favor of participation in a Commonwealth telescope to be built in Australia.

As the politicians hesitated, successive expeditions of astronomers and their assistants left Europe between 1955 and 1963 to search for the most suitable site for their instruments on the bleak and isolated South African scrubland. During these trips they made in-depth investigations of the factors affecting the quality of observations, and also conducted small research programs. Then, in the space of 18 months, the whole site question was reoriented, and in May 1964 the choice was finally made. The observatory would not be built in South Africa after all but at La Silla in the Andes Mountains in Chile. This decision was taken without any comparable study of the suitability of the location. It was also a decision to refuse an offer to erect the telescopes on US-owned land somewhat further south where an association of a few American universities was building its own observatory.

There is no doubt that observing conditions were better in Chile than in South Africa. It was obvious that if American astronomers were going to take advantage of them Europe had to follow suit. The deteriorating political climate in South Africa in the early 1960s did not help either. But fundamental too was the role played by ESO's (see Table 46.1) provisional director, Otto Heckmann, who had been appointed in November 1962. Heckmann took advantage of the presence of a large number of Germans in powerful positions in Chile to win government support for the project, to make government land available for it, and to enter into an intergovernmental agreement with ESO. The advantages of scientific collaboration (and infrastructural cost-sharing) with the US on a geographically proximate site were sacrificed in the interests of capitalizing on these links into Chilean high politics, of retaining European autonomy of action, and of having the status of an international organization (the US agreement was with the University of Chile) — which meant a number of diplomatic privileges and immunities for the establishment and its staff. The lack of national rivalry over the site of the observatory created an opportunity for a determined and ambitious person like Heckmann to impose his definition of the project on national bureaucracies which had been authorized to proceed and did not much care where in the southern hemisphere, or in Chile in particular, the telescopes were installed.

Member states also use European projects to foster their domestic high-technology industries, ensuring whenever possible that their firms win the more 'noble' contracts. In the early 1950s, when CERN was set up, governments were not that sensitive to this issue, and the only specific measure they took to protect national industry was to require that as few contracts as possible were placed outside Europe. It also has to be said that repeated enquiries have revealed that, on average, no more than about 10 percent of CERN contracts by value were 'technically interesting.' Things were quite different when ESRO (and ELDO) were set up a decade later. Governments were fearful of a widening 'technological gap' with the USA, and recognized that scientific space technologies (sounding rockets, satellite platforms, information retrieval and relay systems, etc.,) could be transferred from the 'scientific' to the commercial and military space sectors. They formalized procedures to ensure that their industries were protected. In particular they adopted the policy of 'fair return' i.e., the policy that the percentage by value of 'noble' contracts placed in any member state should be as close as possible to that member states' percentage share of the budget. (An even tighter form of 'return' was imposed for ELDO, where each of the three major participating countries, Britain, France and Germany insisted on building a stage of the rocket on its home soil using national firms.) The European space effort was thus, from its inception, an integral component of the national industrial policy of the member states.

This point can be generalized. Research fields of strategic interest, i.e., with relatively direct industrial and, above all, military relevance sharpen the conflict between national autonomy and European cooperation. Here the need in the major powers to protect national autonomy is particularly strong, an everpresent centrifugal force which threatens to rent the fabric of the collaborative venture. The founders of CERN, confronted by the mingling in the minds of politicians and public alike of all things nuclear with the bomb, deliberately fabricated a distinction between basic and applied research so as to defuse their project of immediate strategic importance, and to clothe it in the garb of an 'academic,' civilian project. This strategy was not available to the proponents of a European atomic energy community (Euratom), foreseen in the Treaty of Rome of March 1957. Their hope was to exploit the atom, seen as an essential source of energy which Europe would fail to exploit at her peril. Unfortunately for them, by the end of the decade all the main countries concerned had important national programs, often involving privileged bilateral arrangements with non- 'EEC' countries, which they wanted to protect. The resonances between national power and atomic power were so strong at the time that Euratom lacked a clear program and so a meaningful budget, and was condemned to situate itself in those rather dreary parts of the nuclear field which were deemed less central to national programs.

European collaboration does not necessarily involve abandoning a national program. On the contrary they can be complementary. Indeed a two-track approach, national and European, allows both scientists and governments to benefit selectively from the advantages of both. For scientists, it increases research opportunities and the funds available for their field. For governments, the European path saves resources and reinforces political and industrial alliances. A parallel national program is a protective net and a bargaining counter which can be used in negotiations with partners to limit the loss of sovereignty that must be tolerated in a European joint effort.

Space research is a case in point. Getting an experiment into space is an irregular and perilous affair. Individual satellites are mission-oriented, i.e., dedicated to specific kinds of research and research communities. Payload weights are tightly restricted. Rockets frequently fail to perform as planned. Spreading risks helps to reduce conflict inside the community and to ensure that the field can advance on a broad front without systematic prejudice to any one part of it.

When serious discussions started among scientists in 1960 about building a European organization for space research they already had two paths open to them. Countries like Britain, France, Italy and also Sweden and Norway, had their own national sounding rocket programs, and the larger nations among them were also actively considering national satellites. Even more important was the American option. In 1959 NASA offered to help build and to satellize space experiments proposed by scientists in friendly nations using US rockets. In this context ESRO was a third alternative, desirable only if it guaranteed additional resources under the control of national communities. There were two symptoms of this in the early negotiations that led to the establishment of the organization. Firstly, the scientists were emphatic that ESRO should have no inhouse scientific program, or inhouse scientific staff of its own. ESRO's resources were to be used to pay for satellite platforms and her engineers were to be employed to interface with industry and to integrate payloads — payloads containing individual experiments built and paid for by national groups using additional national funds. Secondly, and contrary to all that we usually see in the contemporary scientific community, the scientists who founded ESRO accepted without complaint a relatively low eight-year budget envelope. Their attitude to the European program was essentially instrumental, one among several means to an end, a supplement to other, existing opportunities, not a replacement for them, and they were happy to get what they could. This relative detachment by the scientific community from ESRO's program facilitated the reorientation of the organization's mission towards applications which was consolidated when ESA was established in 1975. And it partly accounts for the fact that only that program was made mandatory in ESA's portfolio of activities. At the time it was feared that if it was optional, as were the others, it would wither on the branch for lack of support.

FIGURE 46.1: THE GIANT BUBBLE CHAMBER AT CERN KNOWN AS BEBC. © CERN (FOR ENLARGEMENT SEE FIGURE 38.1).

FIGURE 46.2: ASSEMBLY OF ONE OF THE SUPERCONDUCTING MAGNET COILS OF THE BIG EUROPEAN BUBBLE CHAMBER (BEBC). © CERN.

FIGURE 46.3: THE CAST-IRON SHIELD ENCLOSING THE MAGNET COILS SURROUNDING THE BUBBLE CHAMBER. © CERN.

FIGURE 46.4: INAUGURATION OF THE GARGAMELLE BUBBLE CHAMBER IN 1971. © CERN.

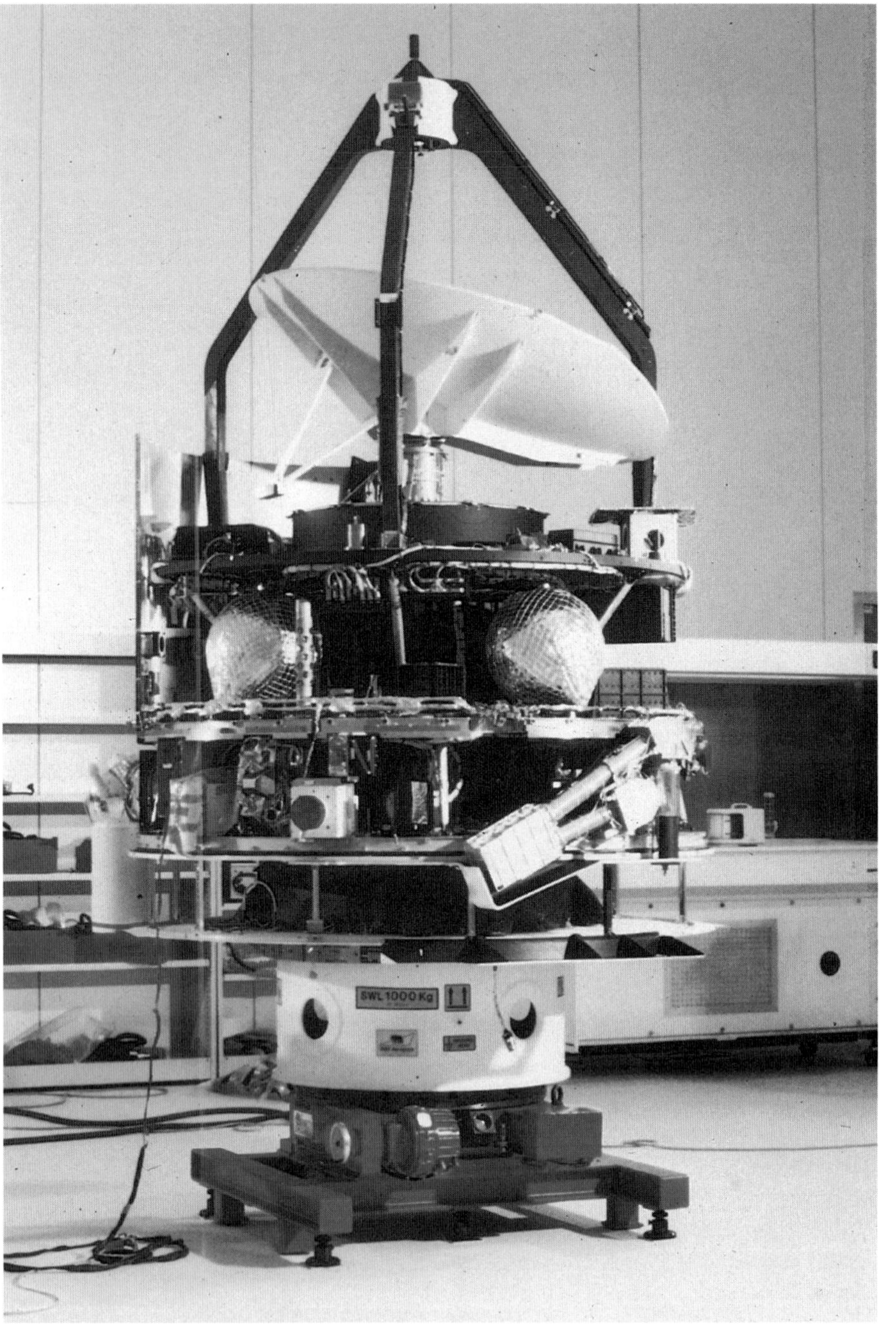

FIGURE 46.5: GIOTTO SPACECRAFT DURING FINAL INTEGRATION AT THE EUROPEAN LAUNCH FACILITY AT KOUROU, FRENCH GUYANA. © ESA.

FIGURE 46.6: GIOTTO SPACECRAFT COMPLETELY ENCLOSED. © ESA

FIGURE 46.7: GIOTTO DURING INTEGRATION ACTIVITIES. © ESA

FIGURE 46.8: LIFT OFF OF ARIANE FLIGHT 14 CARRYING GIOTTO AT 11HRS 23 MINS UTC ON 2 JULY 1985. © ESA

FIGURE 46.9: THE ISO SATELLITE WILL BE VISIBLE FROM ESA'S SATELLITE TRACKING STATION AT VILLAFRANCA NEAR MADRID FOR ABOUT 14 H PER DAY. © ESA

FIGURE 46.10: A VIEW OF THE MAIN CONTROL ROOM OF THE EUROPEAN SPACE AGENCY OPERATIONS CENTRE, DARMSTADT, GERMANY. © ESA

The maintenance of a parallel national program, deemed imperative scientifically and politically in one context, can be judged an intolerable luxury in another. When the British physics community agreed that the country should join CERN, this was not to be at the expense of a national accelerator program. National expansion continued with the decision, taken in 1957, to build a high intensity proton synchrotron at the Rutherford laboratory near Oxford, followed by the decision in the early 1960s to build an electron synchrotron at Daresbury, further north. This twin-track approach eventually collapsed in the late 1970s when CERN commissioned its new big acccelerator, the 300 GeV Super Proton Synchrotron. The nuclear had lost its erstwhile political prestige: indeed protests against nuclear power made it a political liability. Fears over the 'technological gap' led many to think that state money should be spent on R&D that had some prospect of improving industrial competitiveness — i.e., virtually any field of research apart from high-energy physics. The UK thus chose to concentrate its high-energy physics effort in Geneva and to reorientate its national program towards more practical ends, like nuclear spallation. Similar conclusions had already been reached in France and in Italy. Indeed only Germany, with its electron synchrotron and other machines in Hamburg has managed to maintain a major national high-energy physics facility alongside its participation in CERN.

European programs can also provide an avenue for maintaining a strong, if more curtailed, national presence. They do so by 'Europeanizing' all or part of the national effort. The idea for JET (the Joint European Torus) was inspired by the fears of a 50 percent budget cut over five years to the British fusion effort at Culham near Oxford, coupled with the improvement of political relations between Britain and the EEC on the abdication of de Gaulle in April 1969. Saving a national program by Europeanizing it has also occurred repeatedly in the space sector. Britain's enthusiasm for ESRO was partly inspired by the wish of her space science community to build a large astronomical satellite which, if only funded nationally, would have swamped the other projects they wanted. By sharing costs with European partners, and trying to maintain overall management of the project, they hoped to increase the funding for their program with little loss of control over it. The same thing happened in ELDO, where Britain successfully Europeanized Blue Streak, keeping tight control, as did all her partners, over her industry's share of the work on the European rocket.

France did the same thing a decade later when it embarked on the Ariane rocket program. The French were determined to maintain an autonomous European launcher capability despite the scepticism of their major allies, and US insistence that the future lay in reusable launchers like the Shuttle. They invited their partners to contribute through ESA to a program which was maturing nationally and for which they were prepared to pay about 60 percent of the development costs of a 'conventional' launcher, provided overall project control remained in their hands. At the same time the French insisted that their equatorial launching base in Kourou, French Guiana, become a collective asset and responsibility. They also decided to Europeanize their meteorological satellite. There is no conflict here between the national and the European, no question of subjugating national interest to European collaboration. On the contrary these are 'perfect' illustrations of the claim that the European option is simply the pursuit of national self-interest by other means.

The Denationalization of European Scientific Space

European scientific collaboration aims both to further the interests of the nation state and to contribute to the European integrative process. A delicate balance has to be struck between these potentially conflicting objectives. The defense of national interest must not be perceived by one's European partners as unrestrained nationalism. The willingness to subjugate national interest to the integrative process must not be seen by one's national state apparatus as supine betrayal. It is up to the members of the supreme governing body of these organizations, the Council, to walk this tightrope. Composed of senior science administrators and scientific statesmen, their task is to protect 'their' institutions from the vagaries of the national political process, to create a space in which national political concerns are respected but never allowed to override the more fundamental aim of getting the organization to function successfully as a collaborative European body.

Council members are the lynchpins of the European collaborative scientific process. Benefiting from the prestige and the legitimation that international collaboration brings them, jealous of their right to decide its broad lines of development, they strongly resist any move which might give national authorities greater leverage over the organizations which are their responsibility. To this end, they strive to localize any dispute or disagreement, to settle it between themselves, and never to allow it to spill over into the national arena, where their power and control will be reduced. To give just one indication of this: the dread in the Council of having to revise the intergovernmental convention that underpins the organization, the members' determination to exploit every possible legal trick to keep changes in policies or in orientation out of the national political limelight and above all out of national parliaments. For who knows what revisions will be imposed, who knows what changes will be made to the structure of the world that they have put in place so carefully over so many years . . .

The shared perception which they have of their role in the European scientific space is reinforced among the core of the Council members by the bonds of trust and collegiality which have been built up between them over decades. As one ESA Council delegate recently put it, "Experience has led to the conclusion (and its implementation) that a body meeting regularly in a basically unchanged composition develops mutual knowledge of personalities, their habits and their ways of thinking, as well as human relationships, which in turn facilitate the shaping of compromises making everybody equally happy, or unhappy, but being necessary for the well-being and even survival of a multi-member organization."[1] This survival is ensured not simply within each organization but at the European level: perhaps no more than a hundred people circulate between the Councils of the bodies we are discussing, and the trust and understanding between them is a glue which holds the collaborative projects together.

The autonomy of the Council vis-à-vis the national state bureaucracies is underpinned by three considerations. First, insofar as the organizations are seen to be doing 'basic' science, member state governments are prepared to take a back seat, counting on the scientists to propose research programs which are affordable and which keep Europe at the research frontier. This serves to train an elite, reinforce an image of scientific and political leadership, enhance prestige and strengthen European and indeed 'Western' alliances. Correlatively the closer the organizations move towards strategic areas of research the more the Council finds itself hemmed in by national policy makers. In ESA, for example, major policy choices are decided at Ministerial level. The regular Council delegates are not perceived as having the authority to take fundamental programmatic and policy decisions.

The second related, if somewhat different, factor is the denationalization of scientific work itself and, by extension, of the internal life of the organization. Since the eighteenth century scientists have constructed their professional identity around a number of axes one of which was nationhood. This identification of oneself as scientist with one's country was facilitated by the gradual association of national vigor with scientific achievement, a link which cemented the ties between science and the state. In European organizations this identification of science with nation has all but disappeared, surfacing only in jokes and anecdotes which, for the moment anyway, are mostly harmless. This denationalization of science has certainly been facilitated by the overall post World War II political climate. But it rests on deeper changes, changes in the very nature of scientific work and the meaning of what it is to be a scientist.

Experimental practice during the last two or three decades has been dominated by the long term aim of designing, procuring, building and using complicated and heavy equipment, and then of processing the masses of data that flow from it. This has forced the scientist to subordinate all other concerns to the pragmatic demands of getting the equipment working on time, and extracting the maximum possible credit from it. He or she has goals which make no sense individually but only collectively, is no longer primarily a member of a national community but of an international team, is identified not by country but by project (a member of the 'Aleph' or 'Giotto' collaboration), is no longer tied to a national laboratory but to a supranational organization. And acts only as a national when called upon to procure support for the European research laboratory at home.

Finally, denationalization at the everyday level has only been possible because there has been the political will to build a united Europe, a determination to keep the European family together and, correlatively, the fear of the 'domino effect'

— the fear that failure in one collaborative venture will have serious repercussions on others. The willingness to search for compromises, even between points of view which initially seem irreconcilable, has only been possible because governments have seen the building of European scientific and technological organizations as of a piece with that of building a European economic and, eventually, political structure. If the latter project crumbles the former will surely risk crumbling with it.

THE GLOBAL CONTEXT

For much of the postwar period European scientific collaboration has been reactive. The pace and nature of scientific development has largely been set by the United States which, competition oblige, has imposed its priorities and practices on all who wish to remain in touch with the research frontier, want to avoid a 'braindrain' of the most talented scientists, and who have some pretensions to being a leading scientific nation. The physicists who defined CERN's first big accelerator were emphatic that it had to be at least as powerful as the biggest then being planned across the Atlantic. The astronomers who planned ESO originally envisaged building two telescopes comparable in size to those in leading US observatories, and shifted their site from South Africa to Chile as soon as it was clear that the Americans favored the Andes. The pioneers of Euratom were convinced that the development of the continent itself depended on exploiting the 'industrial revolution' heralded by atomic energy and that they would fall irreversibly behind the USA if this new source of power were not harnessed. The fear that a vast industrial, technological and managerial gap had opened up between the two sides of the Atlantic was a major argument for persisting with the European space program in the face of growing difficulties in the late 1960s. In short, for at least 20–30 years after the war, developments in the US have largely set the agenda for the European collaborative scientific effort.

Generally speaking, US scientists welcomed these developments, even encouraged and stimulated them — often using the European 'threat' as a bargaining chip with their own government. An entirely novel idea for accelerating particles to high energies was shared immediately with CERN engineers on a visit to Brookhaven National Laboratory in 1952. The idea of building a European observatory in the southern hemisphere, and its research program, was shaped in discussions with a distinguished astronomer from Mount Wilson and Palomar observatories. Many European scientific satellites were launched from US bases using US rockets and with the help of NASA. Scientific statesmen too, in line with the conviction of successive US administrations, have been favorable to European scientific collaboration, seeing it as essential to the integration and political stability of the West, an essential component of US foreign policy.

It must also be said that whenever influential administrators in the State Department or, of course the military establishment, have feared that scientific (and above all technological) cooperation would jeopardise US interests or, more precisely, US hegemony in strategic fields, they have been quick to withdraw their support. Believing that accelerators might be useful for particle beam weapons and for producing fissionable material, the US Atomic Energy Commission repeatedly wrapped Brookhaven over the knuckles for being so cooperative with CERN, and stopped information flowing out of the US on more than one occasion between 1952 and 1954. The plans to make a uranium enrichment plant the backbone of Euratom's joint program were not only scuttled by rivalry and distrust between European states; they were sabotaged by the US, in 1956, deliberately putting enriched uranium on the market at a price with which no one could compete. European discussions in the late 1960s on whether or not to build a new, expendable heavy launcher (precisely to break the US monopoly in this sector) were overshadowed by US claims that the technology was obsolete and that they were better advised to join with the NASA in its so-called post-Apollo program. In short the US has encouraged European cooperation but only so far as it did not threaten its dominant position. And until recently it has had the power to decide, more or less unilaterally, when that domination was in jeopardy, and where the limits of scientific and technological cooperation with Europe would be set.

It has taken European science one generation, perhaps two, to reach parity, or better, with the US in some parts of basic research (and technological development). Organizations like CERN, EMBL, ESA, ESO and JET have played an important part in establishing that equilibrium. On the other hand, as we have stressed, their creation and expansion has taken place in a specific global political epoch, an epoch marked by superpower rivalry and the Cold War, and the building of a Western alliance in which a united 'Europe' was one key element. This chapter is a child of this rapidly receding past; a past in which the boundaries and the priorities of the European politico-scientific space were more or less fixed. The histories it recounts are histories of achievements, often against adversity: histories which come uncomfortably close to hagiography. Several decades hence an entirely different history of these organizations might be needed. One which begins with the deconstruction of the Berlin wall in November 1989, and the stop to the construction of the Superconducting Supercollider (SSC) in Ellis County, Texas, four years later.

Reference

1. The quote is by R. Loosch, "Decision-Making and Voting," in *The Implementation of the ESA Convention. Lessons from the Past*, Proceedings of the ESA/EUI International Colloquium, Florence, 25 and 26 October, 1993. (Dordrecht: Martinus Nijhoff, 1994), pp. 59–71, at p. 63.

Further Reading

A. Blaauw, *ESO's Early History*. (Garching: ESO, 1991).

R.M. Bonnet and V. Manno, *International Cooperation in Space: The Example of the European Space Agency*. (Cambridge: Harvard University Press, 1994).

M. Dumoulin, P. Guillen and M. Vaïsse (Eds.) *L'énergie nucléaire en Europe, des orgines à Euratom.* (Berne, 1994).

D.W. Ellwood, *Rebuilding Europe. Western Europe, America and Postwar Reconstruction.* (London: Longman, 1992).

P. Galison and B. Hevly (Eds.) *Big Science: The Growth of Large-Scale Research.* (Stanford University Press, 1992).

A. Hermann, J. Krige, U. Mersits and D. Pestre, *History of CERN, Vols. I and II.* (Amsterdam: North Holland, 1987, 1990).

J. Krige (Ed.) *History of CERN, Vol. III.* (Amsterdam: North Holland, 1996).

J. Krige and L. Guzzetti (Eds.) *European Scientific and Technological Cooperation.* Proceedings of a Conference held at the EUI, Florence in November 1995. (Brussels: EEC, 1997).

J. Krige and A. Russo, *Europe in Space 1960–1973.* (Noordwijk: ESA, 1995).

A.S. Milward, *The European Rescue of the Nation-State.* (Berkeley and Los Angeles, 1992).

L. Sebesta, *The Availability of American Launchers and Europe's Decision 'To Go it Alone,'* Report HSR-18. (Noordwijk: ESA, 1996).

E.N. Shaw, *Europe's Experiment in Fusion, the JET Joint Undertaking.* (Amsterdam: North Holland, 1990).

Index

B

D

E

N

O

S